# THE PAPERS OF THOMAS A. EDISON

## FINANCIAL CONTRIBUTORS

*Public Foundations*
National Science Foundation
National Endowment for the Humanities
National Historical Publications and Records Commission

*Private Foundations*
The Alfred P. Sloan Foundation
Charles Edison Fund
The Hyde and Watson Foundation
Geraldine R. Dodge Foundation

*Private Corporations and Individuals*
Alabama Power Company
Anonymous
AT&T
Atlantic Electric
Association of Edison Illuminating Companies
Battelle Memorial Institute
The Boston Edison Foundation
Cabot Corporation Foundation, Inc.
Carolina Power & Light Company
Consolidated Edison Company of New York, Inc.
Consumers Power Company
Cooper Industries
Corning Incorporated
Duke Power Company
Entergy Corporation (Middle South Electric System)
Exxon Corporation
Florida Power & Light Company
General Electric Foundation
Gould Inc. Foundation
Gulf States Utilities Company
Hess Foundation, Inc.
Idaho Power Company
IMO Industries
International Brotherhood of Electrical Workers
Mr. and Mrs. Stanley H. Katz
Matsushita Electric Industrial Co., Ltd.
Midwest Resources, Inc.
Minnesota Power
New Jersey Bell
New York State Electric & Gas Corporation
North American Philips Corporation
Philadelphia Electric Company
Philips Lighting B.V.
Public Service Electric and Gas Company
RCA Corporation
Robert Bosch GmbH
Rochester Gas and Electric Corporation
San Diego Gas & Electric
Savannah Electric and Power Company
Schering-Plough Foundation
Texas Utilities Company
Thomas & Betts Corporation
Thomson Grand Public
Westinghouse Foundation
Wisconsin Public Service Corporation

**THE PAPERS OF THOMAS A. EDISON**

Volume 3

*Thomas A. Edison at thirty, in a cut from the* Phrenological Journal.

Volume 3

# The Papers of Thomas A. Edison

## MENLO PARK: THE EARLY YEARS

*April 1876–December 1877*

VOLUME EDITORS
Robert A. Rosenberg
Paul B. Israel
Keith A. Nier
Martha J. King

EDITORIAL STAFF
Grace Kurkowski
Helen Endick

DIRECTOR AND EDITOR
Reese V. Jenkins

SPONSORS
Rutgers, The State University of New Jersey
National Park Service, Edison National Historic Site
New Jersey Historical Commission
Smithsonian Institution

THE JOHNS HOPKINS UNIVERSITY PRESS
BALTIMORE AND LONDON

Volume 3 of *The Papers of Thomas A. Edison* appears through the generous publication subsidy of the National Historical Publications and Records Commission, the National Endowment for the Humanities, and Rutgers, The State University.



The Johns Hopkins University Press
2715 North Charles Street
Baltimore, Maryland 21218-4319
The Johns Hopkins Press Ltd., London



Printed in the United States of America

The paper used in this book meets the minimum requirements of the American National Standard for Information Sciences—Permanence of Paper for Printed Library Materials, ANSI Z39.48-1984.

Library of Congress Cataloging-in-Publication Data
(Revised for volume 3)

Edison, Thomas A. (Thomas Alva), 1847–1931.
The papers of Thomas A. Edison

Includes bibliographical references and index.
Contents: v. 1. The making of an inventor, February 1847–June 1873—v. 2. From workshop to laboratory, June 1873–March 1876—v. 3. Menlo Park. The early years, April 1876–December 1877.
1. Edison, Thomas A. (Thomas Alva), 1847–1931. 2. Edison, Thomas A. (Thomas Alva), 1847–1931—Archives. 3. Inventors—United States—Biography. I. Jenkins, Reese.
TK140.E3A2 1989 600 88-9017
ISBN 0-8018-3100-8 (v. 1): alk. paper)
ISBN 0-8018-3101-6 (v. 2. : alk. paper)
ISBN 0-8018-3102-4 (v. 3. : alk. paper)

TO THE MEMORY OF THEODORE AND ANN EDISON

# Contents

# Calendar of Documents

# List of Editorial Headnotes

# Preface

When Thomas Edison moved his family, a few co-workers, and his laboratory equipment from Newark to Menlo Park at the end of March 1876, he traded a city bustling with manufacturing activity and skilled workers for a railroad stop in pastoral New Jersey. Yet for all his apparent isolation at Menlo Park, Edison commanded resources—financial, human, and material—probably unmatched by any other inventor anywhere. In the twenty-one months following the move he would prove to himself, the Western Union Telegraph Company, and the world at large that his faith in his own inventive ability, organizational capacity, and leadership was well founded. He explored several telegraph technologies and a number of minor projects before turning his attention in the spring of 1877 largely to the development and improvement of Alexander Graham Bell's new speaking telegraph, or telephone. Before the end of the year Edison had designed and produced an entirely new receiver, a circuit that amplified the signal from the transmitter, and a transmitter embodying a principle that would be used for a century. In the midst of this intensive activity, to his own surprise, he invented the device that would bring him world fame—the phonograph.

In Volume Two the editors noted that selection pressure was changing the nature of the editorial annotation from an extensive description of Edison's environment to an epitome of his full archival collection. That shift is conclusive in Volume Three. The documents presented in this volume are representative of the historical record Edison left, and we hope that our selection and discussion will satisfactorily illuminate for most researchers the many facets of Edison's work and world. In the case of his business dealings the documents, although incomplete, are likely with their annotation to prove sufficient. In order to follow the tortuous, ramified legal trails his work engendered, researchers will have to read court documents and Patent Office files too voluminous—and often too peripherally relevant to Edison's experience—for these volumes; most are included in the microfilm edition, and we have noted those that closely involved or concerned Edison. Most importantly and unfortunately, however, we can no longer illustrate many of Edison's creative blind alleys, half-successful attempts, or even all the steps in the development of the ideas and techniques that were important to his career. We have tried to highlight his changing knowledge and understanding of problems, and the many ways he found to approach and solve those problems, but researchers must understand that they cannot fully appreciate Edison's genius without poring over the entire glorious variety of his ideas as they unfold in his drawings and writings. Readers desiring to understand Edison and his work more deeply should consider these documents and the accompanying annotation a guide into the much fuller record presented on microfilm—a starting point from which they can depart knowledgeably into the richness of the archive.

The Thomas A. Edison Papers project has benefited from many supporters, including the Sponsors, other financial contributors, academic scholars, Edison specialists, librarians, archivists, curators, and students. All of the representatives of the four Sponsors have assisted the project with this volume and the editors thank them for their loyal concern and attention. The strong support of public and private foundations and of their program officers has sustained the project and helped it maintain editorial productivity.

The preparation of this volume was made possible in part by grants from the Division of Research Programs (Editions/Texts) of the National Endowment for the Humanities, an independent federal agency; by grants from the Program in History and Philosophy of Science and the Engineering Division of the National Science Foundation (Grant No. SBR-9112303); by grants from the Alfred P. Sloan Foundation and the Charles Edison Fund; and with the assistance of Rutgers, The State University of New Jersey, the National Park Service (Edison National Historic Site), and the National Historical Publications and Records Commission. The editors appreciate the interest and support of many program officers and trustees, especially Douglas M. Arnold, Margot Backas, Ronald Overmann, Nancy Sahli, Richard N. Sheldon, Kathryn A. Jacob, Mary Giunta, Arthur L. Singer, Ralph E. Gomory, Paul Christiansen, David Shantz, John P. Keegan, and Thomas L. Morrissey. Any opinions, findings, conclusions or recommendations expressed in this publication are solely those of the editors and do not necessarily reflect the views of any of the above federal foundations or agencies, the United States Government, or any other financial contributor.

The Edison Papers project is indebted to the National Park Service for its multifaceted support. The editors express particular thanks to Marie Rust, John Maounis, and Edie Shean-Hammond of the Boston Regional Office and Fahy Whitaker, Maryanne Gerbauckas, Nancy Waters, George Tselos, Douglas Tarr, and Edward Wirth of the Edison National Historic Site in West Orange, New Jersey.

Many associates at Rutgers University have contributed significantly to the Edison Papers. Paul L. Leath and Joseph A. Potenza, Provosts at Rutgers in New Brunswick, and Bruce Newman and Richard Lloyd of the Rutgers University Foundation have made the project a genuine priority. The editors value the efforts of colleagues in the History Department, especially James W. Reed, Rudy Bell, Susan Schrepfer, Philip J. Pauly, Michael Adas, William Aspray, and Richard L. McCormick; and members of the Rutgers University Libraries, notably Ruth Simmons, Ronald L. Becker, Jeanne Boyle, Peggy Weniger, and Patricia Ann Piermatti. The support of the Rutgers University administration has encouraged the editors and facilitated their work. The staff of the project thank Joseph J. Seneca, Richard M. Norman, Nancy Winterbauer, Donald B. Edwards, William W. Owens, Jr., Peter Klein, Robert F. Pack, John S. Salapatas, Barbara J. Callaway, James B. Coe, Dorothea B. Hoekzema, Lisa Hendricks Richardson, Richard F. Foley, Andrew B. Rudczynski, David A. Rumbo, Barbara A. Reis, Albert Hanna, Ronald Thompson, Felicia T. F. Hu, Donna J. Estler, Mildred A. Timmons, Beth A. Palaia, Patricia M. Anderson, Jacquelyn Halvsa, Olga Horodecky, Cynthia Lydia Jucks, Joseph M. Harrigan, Constance J. Bornheimer, Judith A. Garelick, Irma Lucy Cardinale, and Sameerah Diaab Allen.

Many scholars have shared their insights and assisted the editors in a variety of ways. For this volume, notable help came from Nathan Reingold, Thomas P. Hughes, Edward C. Carter II, Melvin Kranzberg, Brian Bowers, David Kohn, Charles Hummel, Allen Koenigsberg, Carolyn Miller, Bernard S. Finn, Robert Friedel, Edward Pershey, Bernard S. Carlson, David Gooding, and Jeffrey Sturchio.

Institutions and their staffs have provided documents, photographs, photocopies, and research assistance. The staff of the Henry Ford Museum & Greenfield Village, especially William S. Pretzer, John Bowditch, and Jeanine Head, have provided many invaluable services.The City of Ft. Myers, Florida, and the trustees and staff of its Edison-Ford Winter Homes have assisted the Edison Papers. Notable for their encouragement were Thomas Smoot, Jim Newton, Les Marietta, Robert Halgrim, and Mary Fitzpatrick. Likewise, Sheldon Hochheiser of the Corporate Research Archives at AT&T has advised and assisted the project. The editors appreciate the professional courtesies of Florence Bartechevski of the Baker Library at Harvard University, Margaret Jasko of Western Union Telegraph Company, Lenore Symons of the Institute of Electrical Engineers in London, James J. Holmberg and James R. Bentley of the Filson Club in Louisville, Kentucky, Mary Bowling at New York Public Library, Ward J. Childe and Allen Weinberg at the Department of Records of the City of Philadelpia, and the staff at the Historical Society of Pennsylvania.

Staff members, interns, and students not mentioned on the title page but who have contributed to this volume include Thomas E. Jeffrey, David W. Hutchings, Gregory B. Field, Theresa Collins, Lisa Gitelman, Karen Detig, Jacquelyn Miller, Gregory Jankunis, Matthew Mehalik, Jennifer Bannister, Eric Boyles, and Karen Palumbo.

The Johns Hopkins University Press has continued its splendid production of the challenging Edison documentary resources. For the devoted efforts of the staff of the Hopkins Press, the editors thank Jack Goellner, Henry Y. K. Tom, Robert Brugger, Barbara Lamb, Lee Sioles, Penny Moudrianakis, Jim Johnston, Martha Farlow, Douglas Armato, and Mary Yates.

# Chronology of Thomas A. Edison

*April 1876–December 1877*

| | |
|---|---|
| 1876 | |
| April | Sets up new laboratory and household at Menlo Park, N.J. |
| 3 April | Executes first patent applications (on acoustic telegraphy) since moving to Menlo Park (his second and third applications after a yearlong hiatus). |
| 5 April | Sells his stock in Domestic Telegraph Co. |
| 11 April | Atlantic and Pacific Telegraph Co. files suit in New York state court against Edison, Lemuel Serrell, Western Union, and George Prescott over rights to Edison's quadruplex telegraph patents. |
| 1 May | Signs agreement with Marshall Lefferts regarding foreign rights to the electric pen. |
| | Begins experimenting in new laboratory, working on a system of acoustic transfer multiple telegraphy. |
| 10 May | Centennial Exhibition, at which Edison's inventions are displayed, opens in Philadelphia. |
| 11 May | With George Harrington and Josiah Reiff files suit against Western Union, George Prescott, Interior Secretary Zachariah Chandler, and Patent Commissioner Rudolphus Duell in Washington, D.C., to prevent any of Edison's quadruplex patents from issuing jointly with Prescott and being assigned to Western Union. |
| 17 May | With George Harrington and Josiah Reiff sues Atlantic and Pacific and Jay Gould in federal court in New York to force payment of money owed to old Automatic Telegraph Co. investors for Edison's automatic telegraph inventions. |
| 26 May? | Renews five-year agreement with Gold and Stock Telegraph Co. regarding his patent rights and salary. |
| 31 May | Executes an extensive caveat covering his acoustic transfer telegraph system. |

| | |
|---|---|
| 7 June | Learns of impending patent interferences with Elisha Gray and Alexander Graham Bell over acoustic telegraph inventions. |
| 25 June | Alexander Graham Bell demonstrates telephone at Centennial Exhibition. |
| June | Given awards at Centennial Exhibition for his automatic telegraph system and his electric pen and autographic press. |
| 3 July | Marshall Lefferts, Edison's mentor and president of Gold and Stock, dies. |
| c. 10 July | Sells British rights to the electric pen to John Breckon and Thomas Clare. |
| 18 July? | Visited at Menlo Park by British scientist/engineer William Thomson. |
| 24 July | Conducts etheric force experiments in response to published criticisms by Elihu Thomson and Edwin Houston. |
| 25 July | Begins tests of acoustic transfer telegraph over line to Philadelphia. |
| July | Conducts first sustained series of telephone experiments. |
| 3 August | Begins two months of experiments with his electromotograph, including its use as an automatic telegraph repeater and as a galvanometer. |
| c. 15 August | John Breckon and Thomas Clare establish the Electric Writing Co. to market the electric pen in Great Britain. |
| c. 28 August | Receives offer for his Port Huron and Gratiot Street Railway Co. stock. |
| August | Files statements in patent interference conflicts with Elisha Gray over acoustic multiple telegraph designs. |
| 7 September | Conceives Morse telegraph recorder/repeater. |
| 13 September | Begins experimenting with octruplex acoustic transfer over line to Philadelphia. |
| Summer | Experiments with carbonized paper in an effort to make electrical resistances, and also considers using carbonized paper for battery carbons, chemical crucibles, and other purposes. |
| 18 October? | Designs and soon experiments with sewing machine motor that consists of tuning forks set in motion by electromagnets. |
| 30 October | Executes two patent applications, one for a tuning fork motor and the other for a telegraph recorder/repeater that uses a punching apparatus to record messages. |
| 3 November | Discovers that the laboratory's chemical stocks have been largely damaged by sunlight and begins extensive series of observations of and experiments with the chemicals, publishing some of his observations in the *American Chemist.* |

| | |
|---|---|
| 28 November | Agrees to have Western Electric Manufacturing Co. become manufacturer and domestic sales agent for the electric pen and press copying system. |
| | With Edward Johnson incorporates the American Novelty Co. to sell miscellaneous inventions. |
| c. 1 December | Invents "duplicating ink" for making multiple copies of documents, which is subsequently marketed by American Novelty. |
| 10 December | Laboratory building damaged by storm winds, and chemical stocks further damaged by subzero temperatures. |
| 1877 | |
| 6 January | Resolves two patent interference cases over acoustic multiple telegraphy designs with Elisha Gray by making and receiving formal concessions of priority on various points. |
| 8 January | Sketches several ideas for the application of small electric motors that could be marketed by American Novelty. |
| Mid-January | Begins extensive series of chemical and etheric force experiments. |
| 17 January | Begins two months of development work on a rotary, high-speed press for electric pen stencils, much of which is done by Charles Batchelor, who begins to resume role as Edison's chief experimenter. |
| | With Charles Batchelor proposes a plan to George Bliss for establishing a foreign electric pen company. |
| 18 January | Suit against Edison by former Newark landlords of American Telegraph Works, Ezra and Roscoe Gould, commences. |
| 19 January | Agrees with Charles Batchelor and Edward Johnson to assign his duplicating ink, Batchelor's door indicator, and Johnson's ribbon mucilage to American Novelty. |
| 20 January | First experiments with a telephone transmitter that varies electrical resistance of a circuit by changing the pressure on carbon rather than the amount of carbon. |
| 29 January | Proposes new agreement with Western Union for support of his laboratory. |
| | Stockholders of the two competing Port Huron street railways agree to consolidation after receiving a proposal from Edison. |
| 2 February | Complains to Jay Gould about his relations with Thomas Eckert and Atlantic and Pacific. |
| 3 February | Executes a patent application for an embossing recorder/repeater. |
| 8 February | Conducts first experiments with a two-plate embossing recorder/repeater. |

| | |
|---|---|
| 9 February | Western Union electrician George Prescott visits the Menlo Park laboratory to examine the recorder/repeater and Edison's telephones. |
| 15 February | Edward Johnson proposes name change from American Novelty Co. to Electro Chemical Manufacturing Co. |
| Mid-February | Begins sustained work on telephone. |
| 19 February | Urges Jay Gould to settle dispute with Western Union over rights to his quadruplex inventions. |
| 26 February | Anson Stager and George Bliss visit Menlo Park to discuss foreign rights to electric pen. |
| 28 February | Leaves for Port Huron to effect the merger of the Port Huron and Gratiot Street Railway Co. with the opposition line and to discuss his interest in the Sarnia Street Railway Co. with other stockholders. |
| | Charles Batchelor accompanies Edison to Port Huron and then goes to Chicago to settle electric pen accounts with Western Electric, returning to Menlo Park on 8 March. |
| 12 March | Returns to Menlo Park from Port Huron. |
| 14 March | Displays two-plate embossing recorder/repeater at Western Union headquarters in New York. |
| 18 March | Devises first electromotograph telephone receiver. |
| 19 March | Demonstrates his telephone instruments to Western Union officials in New York over a line to Menlo Park. |
| 21 March | Edison's father, Samuel, arrives in Menlo Park for a visit. |
| Winter–Spring | Experiments on waterproof varnishes for paper barrels for the New York Paper Barrel Co. |
| 22 March | Signs agreement with Western Union for regular financial support of the Menlo Park laboratory in exchange for all of Edison's future telegraph and telephone patents. |
| 23 March | Begins to prepare applications for a telephone patent and for a patent on his new sextuplex (six-message) telegraph system. |
| 30 March | Receives permission from William Orton to have Joseph Murray build six two-plate embossing recorder/repeaters. |
| 1 April | Devises telephone design that becomes the basis for his later claims to invention of the microphone. |
| 6 April | Laboratory staff's "pet" bear gets loose and they kill it. |
| 10 April | Begins experiments on telephone transmitter designs relying upon variations of pressure on semiconductors (primarily carbon) and devises pressure relay. |
| 18 April | Executes the first of his 1877 telephone patent applications. |
| 24 April | Signs agreement with Batchelor, George Bliss, and Charles Holland for marketing the electric pen in Europe. |
| 26 April | Begins testimony in New York in the "Quadruplex Case," *Atlantic & Pacific v. Prescott & others.* |

| | |
|---|---|
| 28 April | Begins weeklong exhibition of his "musical" (electromotograph) telephone at the Newark Opera House. |
| 3 May | Finishes primary testimony in Quadruplex Case. |
| 16 May | Signs agreement with his nephew, Charles Edison, and former electric pen agent George Caldwell to exhibit his musical telephone. |
| 18 May | Visited at Menlo Park laboratory by British telegraph engineer William Preece. |
| 31 May | Begins monthlong investigation of plumbago mixtures for telephone transmitter. |
| | Signs an agreement with George Prescott and Gerritt Smith to pool their British quadruplex patents. |
| 4 June | Electro Chemical Manufacturing transfers all of its duplicating ink and ribbon mucilage business to George Bliss and Charles Holland, who had acquired foreign rights a month earlier. |
| 5 June | Loses a patent interference case on acoustic multiple telegraphy to Elisha Gray. |
| c. 6 June | Designs combination telephone transmitter–electromotograph receiver. |
| c. 8 June | Tests sextuplex system on Western Union line from New York to Boston after extensive experiments in laboratory and then abandons its development in early July. |
| 28 June | Quadruplex Case final arguments conclude. |
| 2 July | Samuel Edison leaves Menlo Park for Port Huron. |
| 3 July | Demonstrates embossing recorder/repeater to Western Union officials and British telegraph officials William Preece and Henry Fischer. |
| 17 July | Conceives telephone message recorder/repeater. |
| 18 July | Conceives phonograph. |
| 19 July | Edward Johnson and George Barker exhibit Edison's musical telephone before large crowd at the Permanent Exhibition Hall in Philadelphia. |
| 19 July? | Finishes preparing preliminary specification for British telephone patent, which includes description of phonograph. |
| 21 July | Visited at Menlo Park laboratory by scientists George Barker and Henry Draper, and electrical manufacturer William Wallace. |
| 27 July | Visited at Menlo Park laboratory by Alexander Siemens, nephew of William Siemens. |
| 30 July | Begins to develop carbon "fluff" for telephone transmitters. |
| 1 August | Edward Johnson gives second telephone exhibition concert in Philadelphia and then begins a series of concerts along the eastern seaboard that continues through the end of the year. |

| | |
|---|---|
| 3 August | Begins monthlong correspondence with Henry Draper regarding spectroscopy. |
| 4–10 August | Conducts telephone experiments with Thomas David at Menlo Park laboratory and over a line in lower Manhattan. |
| c. 7 August | Sends Charles Edison to Port Huron to represent his interest in street railways. |
| 10 August | Exhibits combination transmitter–electromotograph receivers built by Joseph Murray to Western Union officials and begins telephone line tests in New York City. |
| Mid-August | Tests various telephone designs over lines in New York City. |
| 20 August | Demonstrates a new telephone design to William Orton and other Western Union officials. |
| 21 August | Western Union announces agreement to purchase control of Atlantic and Pacific, thus ending commercial and legal conflicts between the corporations. |
| 24 August | Begins designing hand-held telephones. |
| | Begins using newly installed gas lighting machine at laboratory. |
| 25 August | Visited at Menlo Park by Robert Watson of Montreal, who wants to introduce Edison's telephone in Canada. |
| 9 September | Resigns his position as Atlantic and Pacific electrician. |
| | Conducts experiments with electric arc lights. |
| 10 September | Files depositions by himself, Batchelor, and Adams in electric pen patent interference case against Henry Trueman and gives evidence in hand-stamp patent interference with A. E. Hix. |
| c. 10 September | Conducts first experiments with incandescent electric lighting. |
| 11–13 September | With laboratory staff goes fishing in Raritan Bay. |
| 15 September | Signs an agreement with Franklin Badger, associate of Robert Watson, regarding the marketing of his telephone in Canada. |
| 17 September | Demonstrates a new transmitter to William Orton, who orders 150 sets of transmitters and receivers for Western Union subsidiary Gold and Stock. |
| 18 September | Proposes to George Bliss that Bliss and Charles Holland sell rights to European telephone patents. |
| | With George Prescott signs agreement with Stephen Field and Cornelius Herz regarding European quadruplex patents. |
| 28 September | Learns of successful quadruplex telegraph tests conducted on British Post Office lines by Gerritt Smith and George Hamilton. |
| 29 September | Learns first telephone patent application will be suspended pending resolution of patent interference conflicts. |

| | |
|---|---|
| 5–8 October | Redesigns telephone transmitter being manufactured by Joseph Murray. |
| 6 October | Wins electric pen patent interference against Henry Trueman. |
| 18 October | Participates in Edward Johnson's telephone exhibition at Jersey City. |
| 22 October | Begins using disks of a plumbago-rubber mixture in place of fluff in telephone transmitter. |
| 26 October | Begins using disks of lampblack and rubber in telephone transmitter. |
| 17 November | Western Union establishes the American Speaking Telephone Co. to combine its interest in Edison's telephone patents with the Harmonic Telegraph Co.'s ownership of Elisha Gray's patents. |
| Mid-November | Devises induction-coil circuit for telephone. |
| November | Explores alternative, noncarbon transmitter designs. |
| 1–3 December | Tests induction-coil circuit for telephone. |
| 1–6 December | John Kruesi makes first tin-foil, cylinder phonograph. |
| 7 December | Demonstrates his cylinder phonograph at *Scientific American* office in New York. |
| 12 December | Wins electric pen patent interference against Edward Stewart. |
| 15 December | Executes first phonograph patent application. |
| 16 December | Receives old "Telegraphy" entry for *Appleton's Cyclopedia of Applied Mechanics,* which he has been commissioned to revise. |
| 17 December | Signs agreements with George Bliss and Hungarian promoter Theodore Puskas regarding the sale and exploitation of his European telephone and phonograph patents. |
| 27 December | Confers at Menlo Park with Edward Johnson, Uriah Painter, and Gardiner Hubbard about their proposals that he break his contract with Western Union and let them form a company to market his phonograph. |
| c. 29 December | Proposed as a scientific member of the American commission to the 1878 Paris Universal Exposition. |
| 31 December | Demonstrates phonograph to William Orton. |

# Editorial Policy

The editorial policy for the book edition of Thomas Edison's papers remains essentially as stated in Volumes One and Two. The additions that follow stem from new editorial situations presented by documents in Volume Three.

### Organization

Edison and his staff often worked into the early morning hours. Sometimes they noted quitting time, but usually they did not. In order to maintain the chronology of events, such documents are generally assigned to the day they began. If such a record bears only the finishing date, it appears as the first technical document on that date.

### Form

*Notebook entries and technical notes.* Many documents from 1876 onward are signed repeatedly by one or more staff members. To avoid cluttering the text and inflating the textnotes, this practice is simply noted in the general textnote by "Document multiply signed [and/or] dated." Edison's name appears first in all such documents. Although members of the staff often signed one another's records as witnesses, the degree of interpretation required to identify witnesses as separate from authors precludes the categorization of signatures in almost all cases. It is important to understand that a signature does not necessarily imply that the signer was an eyewitness or participant; the writer often had the others sign to attest to the date on which the idea or experiment was recorded.

*Authorship.* Notebook entries and technical notes (documents with "X" as a physical description) are, as before, written by Edison unless otherwise noted. If another writer is indicated in the general textnote (for example, "Written by Adams"), this should be understood to include drawings as well.

# Editorial Symbols

~~Newark~~ Overstruck letters
Legible manuscript cancellations; crossed-out or overwritten letters are placed before corrections

[Newark] Text in brackets
Material supplied by editors

[Newark?] Text with a question mark in brackets
Conjecture

[Newark?][a] Text with a question mark in brackets followed by a textnote
Conjecture of illegible text

⟨Newark⟩ Text in angle brackets
Marginalia; in Edison's hand unless otherwise noted

[ ] Empty brackets
Text missing from damaged manuscript

[---] Hyphens in brackets
Conjecture of number of characters in illegible material

Superscript numbers in editors' headnotes and in the documents refer to endnotes, which are grouped at the end of each headnote and after the textnote of each document.

Superscript lowercase letters in the documents refer to textnotes, which appear at the end of each document.

# List of Abbreviations

## ABBREVIATIONS USED TO DESCRIBE DOCUMENTS

The following abbreviations describe the basic nature of the documents included in the third volume of *The Papers of Thomas A. Edison:*

| | |
|---|---|
| AD | Autograph Document |
| ADf | Autograph Draft |
| ADfS | Autograph Draft Signed |
| ADS | Autograph Document Signed |
| AL | Autograph Letter |
| ALS | Autograph Letter Signed |
| D | Document |
| Df | Draft |
| DS | Document Signed |
| L | Letter |
| LS | Letter Signed |
| M | Model |
| PD | Printed Document |
| PDS | Printed Document Signed |
| PL | Printed Letter |
| TD | Typed Document |
| TL | Typed Letter |
| X | Experimental Note |

In these descriptions the following meanings are assumed:

*Document* Accounts, agreements and contracts, bills and receipts, legal documents, memoranda, patent applications, and published material, but excluding letters, models, and experimental notes

*Draft* A preliminary or unfinished version of a document or letter

*Experimental Note* Technical notes or drawings not included in letters, legal documents, and the like

*Letter* Correspondence

*Model* An artifact, whether a patent model, production model, or other

The symbols may be followed in parentheses by one of these descriptive terms:

*abstract* A condensation of a document

*copy* A version of a document made at the time of the creation of the document by the author or other associated party

*duplicating-ink copy* A copy made with Edison's duplicating ink

*fragment* Incomplete document, the missing part not found by editors

*historic drawing* A drawing of an artifact no longer extant or no longer in its original form

*letterpress copy* A transfer copy made by pressing the original under a sheet of damp tissue paper

*photographic transcript* A transcript of a document made photographically

*telegram* A received telegraph message

*transcript* A version of a document made at a substantially later date than that of the original, by someone not directly associated with the creation of the document

## STANDARD REFERENCES AND JOURNALS

### Standard References

| | |
|---|---|
| *ADR* | *Allgemeine Deutsche Biographie*, 1967 reprint ed. (Berlin: Duncker and Humblot) |
| *BDAS* | *Biographical Dictionary of American Science* |
| *DAB* | *Dictionary of American Biography* |
| *DNB* | *Dictionary of National Biography* |
| *DSB* | *Dictionary of Scientific Biography* |
| *MEB* | *Modern English Biography* |
| *NCAB* | *National Cyclopedia of American Biography* |

| | |
|---|---|
| *OED* | *Oxford English Dictionary* |
| Royal Society *Catalog* | Royal Society of London, ed., *Catalog of Scientific Papers,* 1968 (Metuchen, N.J.: Scarecrow Reprint Corp.) |
| *TAEB* | *The Papers of Thomas A. Edison* (book edition) |
| *TAEM* | *Thomas A. Edison Papers: A Selective Microfilm Edition* |
| *TAEM-G#* | *A Guide to Thomas A. Edison Papers: A Selective Microfilm Edition, Part #* |
| *WWW-HV* | *Who Was Who in America, Historical Volume* |

**Journals**

| | |
|---|---|
| *Am. Chem.* | *American Chemist* |
| *Elec. Engr.* | *Electrical Engineering* |
| *Elec. W.* | *Electrical World* |
| *J. Frank. Inst.* | *Journal of the Franklin Institute* |
| *J. Soc. Teleg. Eng.* | *Journal of the Society of Telegraph Engineers* |
| *J. Teleg.* | *Journal of the Telegraph* |
| *Phil. Mag.* | *Philosophical Magazine* |
| *Proc. Am. Acad. Arts Sci.* | *Proceedings of the American Academy of Arts and Sciences* |
| *Sci. Am.* | *Scientific American* |
| *Teleg. and Tel.* | *Telegraph and Telephone Age;* formerly *Telegraph Age* |
| *Teleg. J. and Elec. Rev.* | *Telegraphic Journal and Electrical Review;* formerly *Telegraphic Journal* |
| *Telegr.* | *The Telegrapher* |

## ARCHIVES AND REPOSITORIES

In general, repositories are identified according to the Library of Congress system of abbreviations. Parenthetical letters added to Library of Congress abbreviations have been supplied by the editors.

| | |
|---|---|
| DNA | National Archives, Washington, D.C. |
| DSI-NMAH | Archives, National Museum of American History, Smithsonian Institution, Washington, D.C. |
| Filson | Filson Club, Louisville, Ky. |

| | |
|---|---|
| MdSuFR | Washington National Records Center, Suitland, Md. |
| MiDbEI | Library and Archives, Henry Ford Museum & Greenfield Village, Dearborn, Mich. |
| MiDbEI(H) | Henry Ford Museum & Greenfield Village, Dearborn, Mich. |
| NjBaFAR | National Archives, Northeast Region, New York, N.Y. (formerly Federal Archives and Records Center, Bayonne, N.J.) |
| NjWAT | AT&T Archives, Warren, N.J. |
| NjWOE | Edison National Historic Site, West Orange, N.J. |
| NN | New York Public Library, New York, N.Y. |
| NNHi | New-York Historical Society, New York, N.Y. |
| PHi | Historical Society of Pennsylvania, Philadelphia, Pa. |
| PPCiA | City Archives of Philadelphia, Pa. |
| PU | Archives, University of Pennsylvania, Philadelphia, Pa. |
| UkCU | Cambridge University Library, Cambridge, UK |
| UkLIEE | Institution of Electrical Engineers, London, UK |
| UkLS | Science Museum, London, UK |

## MANUSCRIPT COLLECTIONS AND COURT CASES

References to documents included in *Thomas A. Edison Papers: A Selective Microfilm Edition* (Frederick, Md.: University Publications of America, 1985–) are followed by parenthetical citations of reel and frame of that work; for example, Cat. 1185:34, Accts. (*TAEM* 22:562). Documents found at NjWOE after the microfilming of contemporaneous material will be filmed as a supplement of the next part of the microfilm edition.

A&P Executive    Atlantic and Pacific Executive Commit-

| | |
|---|---|
| | tee Minutes, Western Union Collection, DSI-NMAH |
| A&P Stockholders | Atlantic and Pacific Stockholders Minutes, Western Union Collection, DSI-NMAH |
| Accts. | Accounts, NjWOE |
| *American Graphophone v. U.S. Phonograph* | *American Graphophone Company v. The United States Phonograph Company, Victor H. Emerson and George E. Tewksbury,* Lit., NjWOE |
| American Speaking Telegraph Minutes | American Speaking Telegraph Co. Minutebook, Western Union Collection, DSI-NMAH |
| *Anders v. Warner* | *Anders v. Warner,* Patent Interference File 5603, RG-241, MdSuFR |
| Batchelor | Charles Batchelor Collection, NjWOE |
| BC | Biographical Collection, NjWOE |
| Centennial | Reports of Awards, 1876, Bureau of Awards, United States Centennial Commission, Records Series 230.19, PPCiA |
| CRD | Charles Darwin Archive, UkCU |
| DF | Document File, NjWOE |
| EBC | Edison Biographical Collection, NjWOE |
| Edison Caveat | Edison Caveats, RG-241, DNA |
| *Edison v. Dickerson* | *Edison v. Dickerson*, Patent Interference File 10,132, RG-241, MdSuFR. |
| EP&RI | Edison Papers and Related Items, MiDbEI |
| *Field v. Pope* | *Field v. Pope,* Patent Interference File 7976, RG-241, MdSuFR |
| *Gamewell v. Chester* | *Gamewell v. Chester,* Patent Interference File 5406, RG-241, MdSuFR |
| *Gamewell et al. v. Chester and Chester* | *Gamewell et al. v. Chester and Chester,* Equity Case File 183, NjBaFAR |
| GF | General File, NjWOE |
| GFB | George F. Barker Papers, PU |
| *Harrington v. A&P* | *Harrington, Edison, and Reiff v. Atlantic & Pacific, and George Gould et al.,* Equity Case File 3980, RG-276, and Equity Case File 4940, RG-21, NjBaFAR |
| HD | Henry Draper and Anna Palmer Draper Papers, NN |

| | |
|---|---|
| Kellow | Richard W. Kellow File, Legal Series, NjWOE |
| Lab. | Laboratory notebooks and scrapbooks, NjWOE |
| *Ladd v. Seiler* | *Ladd v. Seiler,* Patent Interference File 8765, RG-241, MdSuFR |
| Lbk. | Letterbooks, NjWOE |
| LBO | William Orton Letterbooks, Western Union Collection, DSI-NMAH |
| Libers Pat. | Libers of Patent Assignments, U.S. Patent Office Records, MdSuFR |
| Lit. | Litigation Series, NjWOE |
| Meadowcroft | William H. Meadowcroft Collection, NjWOE |
| Miller | Harry F. Miller File, Legal Series, NjWOE |
| ML | Marshall Lefferts Papers, NNHi |
| Pat. App. | Patent Application Files, RG-241, MdSuFR |
| Pioneers Bio. | Edison Pioneers Biographical File, NjWOE |
| PS | Patent Series, NjWOE |
| Quad. | Quadruplex Case (Vols. 70–73 and Telegraph Law Cases [TLC]), NjWOE |
| Royce | Royce Miscellaneous Papers, Manuscript Department, Filson |
| *Sawyer & Man v. Edison* | *Sawyer & Man v. Edison,* Lit., NjWOE |
| *Sawyer & Man v. Edison (U.S.)* | *Sawyer & Man v. Edison,* contained in *Edison Electric Light Co. v. U.S. Electric Lighting Co.*, Lit., NjWOE |
| Scraps. | Scrapbooks, NjWOE |
| Supp. # | Supplement in *TAEM,* Part # |
| Telephone Interferences | *Edison v. Gray v. Dolbear et al.*, Patent Interference Files 6627, 6628, and 6630, RG-241, MdSuFR |
| TI | Telephone Interferences (Vols. 1–5), NjWOE; a printed, bound subset of the full Telephone Interferences |
| UHP | Uriah Hunt Painter Papers, PHi |
| WHP | William H. Preece Papers, UkLIEE |
| WU Coll. | Western Union Collection, DSI-NMAH |
| WU Executive | Western Union Executive Committee |

| | |
|---|---|
| | Minutes, Western Union Collection, DSI-NMAH |
| WUTAE | Envelope of Edison letters, Letterbox 8, Western Union Collection, DSI-NMAH |

# MENLO PARK: THE EARLY YEARS
# APRIL 1876–DECEMBER 1877

*A map showing Menlo Park, N.J., and the nearby cities and towns where Edison worked and tested his inventions in 1876–77.*

# –1– April–June 1876

As a result of a lawsuit brought by a former Newark landlord, Edison had decided in late 1875 to purchase land and build his own laboratory.[1] He bought land and a house in Menlo Park, a whistlestop on the New York–Philadelphia line, and built a laboratory during the winter of 1876.[2] Edison and a small group of co-workers spent April settling themselves and their families and equipping the new building.[3] Western Union soon provided Edison with a connection to its main East Coast line, which allowed him to experiment on loops to Philadelphia and Washington, D.C.[4] It is not clear exactly who moved with Edison at the start, but at least Charles Batchelor, James Adams, John Kruesi, and Charles Wurth accompanied him.[5] According to Batchelor, Menlo Park was "a beautiful country place where we live and we all feel considerable benefit from the change," although, to calm his wife, he kept "one big Newfoundland dog and two smaller ones and a seven shooter under my pillow nights."[6] Edison equipped another building as a factory for Ezra Gilliland's manufacturing operation, which moved from Newark in late April.[7] The staffs of the laboratory and the factory, under the leadership of the railroad station agent Marcus Hussey, organized the Menlo Park Brass Band. At some point the laboratory staff also acquired a "pet"—a young black bear.[8]

The laboratory staff resumed experimental work during the second week of May, although by the end of spring Batchelor was again primarily occupied overseeing the electric pen business and commuted daily to New York City. The first work in Menlo Park continued the explorations, begun at Newark, of multiple telegraphy using acoustic instruments.[9] They briefly

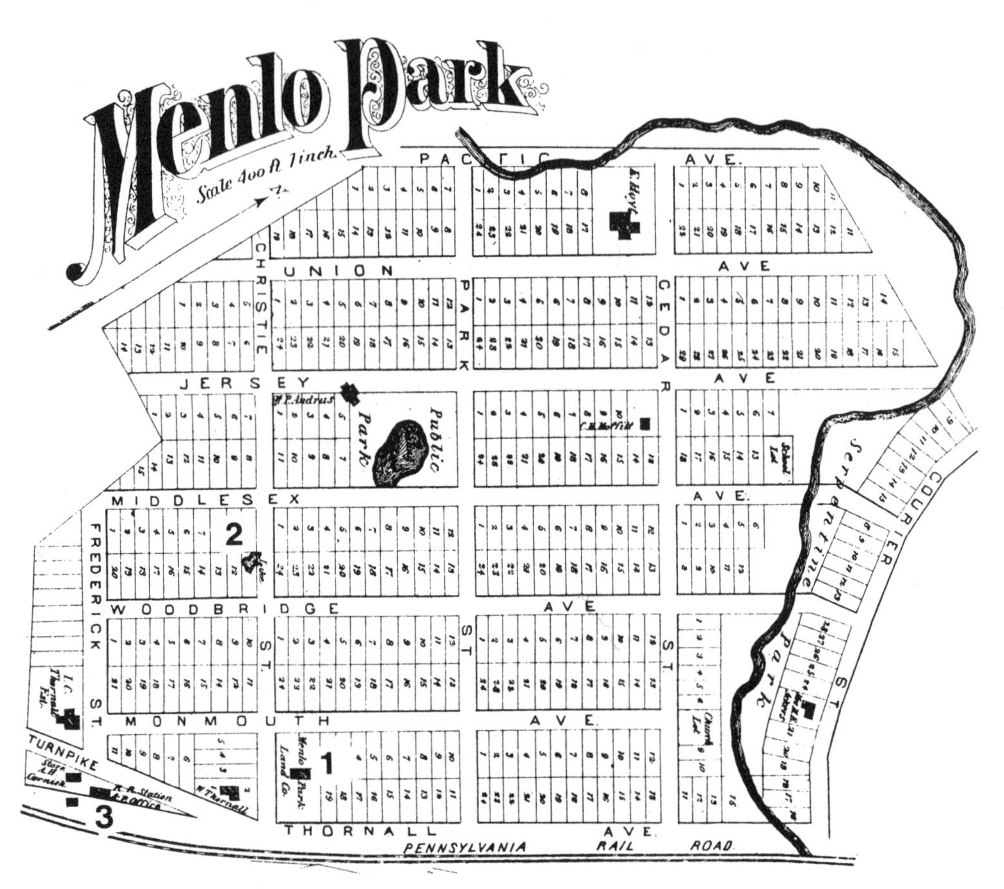


*An 1876 map of Menlo Park, N.J. (1) Edison bought the office of the Menlo Park Land Co. for a home. (2) He built the laboratory up a hill two blocks away, between Middlesex and Woodbridge Aves. The Pennsylvania Railroad line runs along the bottom of the map; (3) marks the railroad station/post office.*

resumed their experiments on circuits that used paired tuning forks or reeds vibrating at particular frequencies, but they soon turned to the development of a new system—"acoustic transfer" telegraphy—that used tuning forks to switch the circuit synchronously between sets of instruments. By the end of May, Edison had executed five patent applications for acoustic telegraphy and had drawn up an extensive caveat describing several acoustic transfer circuits.

The U.S. Centennial Exhibition opened in Philadelphia on 10 May. Although Edison prepared a large display in conjunction with Western Union, that exhibit was not ready when the fairgrounds opened on 10 May.[10] The Atlantic and Pacific Telegraph Company, however, which ran the fair's public telegraph office, was using Edison's automatic telegraph system (as well as standard Morse instruments) in an office visible to the public.[11] A committee headed by the British physicist Sir William Thomson awarded Edison's automatic system a prize medal, calling it "a very important step in land-telegraphy."[12] Thomson also praised Edison's electric pen as an invention of "exquisite ingenuity and . . . usefulness."[13]

Edison's business relations with Western Union and Atlantic and Pacific remained complex and turbulent. The March reactivation of the review process for Edison's quadruplex patent applications[14] precipitated a flurry of legal activity. In April, Atlantic and Pacific filed a suit against Western Union, asserting ownership of Edison's quadruplex patents.[15] Josiah Reiff, one of the principal investors in Edison's automatic telegraph, initiated two suits in the names of Edison and George Harrington. The first, against George Prescott and Western Union, sought to restrict the assignment of Edison's patents to anyone but Edison.[16] The second, against Jay Gould and Atlantic and Pacific, sought to void the latter's purchase of Edison's automatic system and their subsequent claim to his quadruplex.[17] Reiff had earlier organized a new venture, the American Automatic Telegraph Company, as the legitimate owner of Edison's automatic telegraph patents.[18]

Although Edison had moved to Menlo Park with the expectation of a reliable income from his contracts with Gold and Stock and Western Union,[19] the expenses of the new laboratory kept him looking for additional sources of money. He received income from electric pen sales and was trying to sell foreign rights for the invention. Edison also sold stock he held in the Domestic Telegraph Company, which brought him several thousand dollars in April and May, although he did not receive the final payment until early in 1877.

1. Edison remembered the landlord to be Thomas Slaight, a padlock manufacturer at 115 New Jersey Railroad Ave., where Edison and Joseph Murray had established Murray and Co. in February 1872 (see *TAEB* 2:583 n. 8). However, the landlord may have been Ezra and Roscoe Gould from whom Edison and Harrington leased space for the American Telegraph Works in October 1870, and whose suit against Edison commenced in January 1877 (Doc. 108 n. 3; Cat. 1233:13, Batchelor [*TAEM* 90:58]; also see *TAEM-G1*, s.v. "Manners, A.").

2. On 7 April, Edison insured the household furnishings for $3,000: $1,000 for Mary's "Bronzes and Articles of Virtu [antiques and curios]," $1,500 for "Household furniture useful and ornamental beds bedding linen pictures printed books and family wearing apparel," and $500 for a "Piano-Forte." On 13 April Edison insured the laboratory for $2,666.67: $1,000 for the building, $267.67 for the engine and boiler, and $1,400 "On his Telegraphic Instruments, Electrical and Other Machinery and Apparatus, Chemical Apparatus and Materials used in experimenting. Printed Books Shop & Office furniture and fixtures, Shafting, Belting, Pulleys, Hangers, Piping. Tools and Implements of

*The Edison family home in Menlo Park.*

trade all contained in the above described building." DF (*TAEM* 13:918, 921); for the purchase itself see *TAEB* 2:583.

3. Bills and receipts related to the building and equipping of the new laboratory are in 76-002, DF (*TAEM* 13:799).

4. Philadelphia is approximately 60 miles (95 km) from Menlo Park; Washington is approximately 140 miles (225 km) past Philadelphia.

5. Edison and Batchelor bought homes in Menlo Park; the living arrangements of the other first workers are not known. Batchelor and Adams signed the first laboratory entries from Menlo Park, and Charles Wurth remembered moving there in March (Doc. 749; Vol. 10:26, Lab. [*TAEM* 3:838]; "Charles Nicholaus Wurth," Pioneers Bio.). Kruesi had been with Edison in Newark up until the move. Although there is a dearth of payroll records for the middle of 1876, his name appears in the first extant records (from October 1876; Cat. 1213:7, Accts. [*TAEM* 20:8]). Finn (1989) places William Carman and Charles Stilwell at Menlo Park in 1876 as well. Stilwell, Edison's 15-year-old brother-in-law, later remembered moving to Menlo Park to work, but it is very unclear when ("Charles F. Stilwell," Pioneers Bio.). William Carman's name does not appear in any records until he signed two letters for Edison on 5 February 1878 (TAE to Condit and Hanson, TAE to Rohrbeck and Groebler, Lbk. 1:335–36 [*TAEM* 28:208–9]); his handwriting then appears in an account book in late February or early March (Cat. 1185, Accts. [*TAEM* 22:549]).

6. Batchelor to Thomas Batchelor, 9 May 1876, Cat. 1238:51, Batchelor (*TAEM* 93:63). On Batchelor's family, see Doc. 922 n. 1.

7. Insurance policy of 24 April 1876, DF (*TAEM* 13:924). It is not clear whether Edison bought or built this building.

8. Charles Wurth remembered it as having been brought by Samuel

Edison from Port Huron. "Charles Nicholaus Wurth," Pioneers Bio.

9. Elisha Gray and Alexander Graham Bell were the two other prominent researchers in this technology, and on 7 June Edison received a letter from his patent attorney, Lemuel Serrell, alerting him to an interference involving both other men. DF (*TAEM* 13:1031).

10. See Docs. 601 and 657; and "Opening of the Centennial Exhibition.—The Electrical and Telegraphic Department," *Telegr.* 12 (1876): 123.

11. "A Visit to the Centennial Exhibition," *Telegr.* 12 (1876): 160.

12. Doc. 757.

13. See Doc. 803 n. 10. Batchelor told his brother that he had "four men in the Centennial." The electric pen exhibit was in the main building. Batchelor to Thomas Batchelor, 9 May 1876, Cat. 1238:51, Batchelor (*TAEM* 93:63); electric pen pamphlet of George Caldwell, Agent, Centennial Exhibition, Supp. III (*TAEM* 162:990); see electric pen copies, Cat. 593, Scraps. (*TAEM* 27:623, 625, 628, 634, 641).

14. See *TAEB* 2:770 n. 6.

15. *Atlantic & Pacific v. Prescott & others,* Quad. 70, 71, 73 (*TAEM* 9:288–10:797). Edison, although still the electrician for Atlantic and Pacific, was at first a defendant in this case and participated in the preparation of an answer to the complaint. Quad. 70.3 (*TAEM* 9:317); on the Quadruplex Case, see *TAEB* 2, app. 3.

16. *Edison and Harrington v. Western Union & others,* Quad. 72, TLC (*TAEM* 9:8–287, 10:798–997). This suit also named the Commissioner of Patents and the Secretary of the Interior as defendants.

17. *Harrington v. A&P;* see *TAEB* 2:469 n. 2.

18. See Docs. 676 and 729.

19. Docs. 164 and 695.

**–738–**

*To Norman Miller*

Menlo Park Apl 1=76[a]

Friend Miller[1]

I have a chance to sell mine yours & Batchelor's[2] Domestic Stock to the A&.P people for $5 per share ½ Cash & ½ approved notes=[3] I think I can let you have more than ½ if not all cash for your 100 shares

I am to be ready to deliver them Wednesday next if the sale goes on so please put your certificate in a letter and leave it ~~on the~~ with[b] General Lefferts,[4] or in his care, so that I can make the delivery if the trade goes on[5]

Answer qk Yours

Edison

ALS, NjWOE, DF (*TAEM* 13:1209). [a]Electric Pen Co. letterhead illustration appears above. [b]Interlined above.

1. Norman Miller, long associated with Edison through the Gold and Stock Telegraph Co., had overseen the operations of the electric pen business from early October to mid-January. See Doc. 639.

2. Charles Batchelor was Edison's chief experimental assistant. See *TAEB* 1:495 n. 9, 2:72.

3. The trustees of the Atlantic and Pacific Telegraph Co. had decided earlier in the year to gain a controlling interest in the Domestic Telegraph Co. through an exchange of stock, principally with "various persons interested in the success of the Atlantic and Pacific Telegraph Company" who had purchased Domestic stock "for the purpose of preventing its falling into the possession of rival interests to the injury of the Atlantic and Pacific Telegraph Company." The purchase was delayed until February 1877, at which point the company paid cash for the Domestic stock. A&P Stockholders (1875–88): 73–74; A&P Executive (1873–78): 159–63.

4. Marshall Lefferts, Edison's friend and promoter since shortly after Edison's 1869 arrival in New York, was head of the Gold and Stock Telegraph Co. See *TAEB* 1:170 n. 6.

5. According to signed receipts on the back of this document, Miller delivered 100 shares of stock to Lefferts on the next Wednesday (5 April) for which he received $462.50 on 13 April. *TAEM* 13:1210.

**–739–**

*Receipt from Alfred Nelson*

New York April 5— 1876[a]

Rec'd from Mr T. A. Edison Certificates as per account for Thirty three hundred shares of the Capital Stock of the Domestic Telegraph Company, for which I agree to pay him Sixteen thousand, five hundred dollars one half Cash and one half notes due Six months from date without Interest— Owing to the absence of some of the parties who have decided to purchase this stock I am unable to make the settlement in full today, but now pay him Three thousand dollars in cash on account of the purchase, and am to have a few days allowance in which to make the further payments the notes and cash to be placed in Mr Edison's hands as soon as I receive them from the several parties, and meantime all the Stock Certificates to be left in my custody— The entire settlement to be completed by April 10th 1876.[1]

| No. 47. | T. A. Edison | 200. | shs |
|---|---|---|---|
| 48 | " | 200. | " |
| 49 | " | 200. | " |
| 50 | " | 200 | " |
| 51 | " | 200 | " |
| 52 | " | 200 | " |
| 53 | " | 200 | " |
| 54 | " | 200 | " |
| | For'd[b] | 1600 | |

| | | | |
|---|---|---|---|
| 55 | Thos. A. Edison | 100. | " |
| 56 | " | 100 | " |
| 60 | " | 100 | " |
| 62 | " | 20 | " |
| 61 | " | 20 | " |
| 79 | Alex Morten[2] | 10 | " |
| 99 | T. A. Edison | 1250 | " |
| 21 | Chas Batchelor | 100 | " |
| Thirty three hundred | | 3,300 | — |

Alfred Nelson[3]

ADS, NjWOE, DF (*TAEM* 13:1211). Letterhead of Atlantic and Pacific and Franklin Telegraph Cos. Executive Office. [a]"New York" and "187" preprinted. [b]This line begins new page.

1. The final sale of stock took place on 9 December 1876, although a small amount of cash remained due in January 1877 (statement of 24 Jan. 1877, DF 77-018, [*TAEM* 14:754]). Also see an undated statement showing who owned the 3,300 shares in DF 75-013 (*TAEM* 13:532); and Alfred Nelson to TAE, 15 Sept. 1876, and Albert Chandler to TAE (with TAE's figures on the back indicating that he was still owed $2,000), 27 Nov. 1876; both DF (*TAEM* 13:1230, 1240–41).

2. Alex Morten, who had served as acting secretary and treasurer of the Domestic Telegraph Co. in 1875, also invested in the Automatic Telegraph Co. in 1871. See Doc. 159 n. 7.

3. According to company letterheads, Alfred Nelson was vice president of the Domestic Telegraph Co. and treasurer of the Atlantic and Pacific Telegraph Co.

**-740-**

*From Josiah Reiff*

N.Y. April, 8/76

My dear Edison

Eckert[1] from some source knows I am going—[2] Whilst I am absent he & Gould[3] will renew their attempts to get you away from me—

With Hs[4] signature which is <u>finally</u> necessary, although we can move without him now, we can of course secure something—[5] now they only desire & hope to get you to yield & my ruin they think will be complete—

Your Domestic money will serve your present needs. You will hear from me in 14 days.[6] Your pen[7] is in good shape, so that prospectively you have ample resources, & for present your needs are meet—

My Leondon trip[8] of course is for our mutual good, hence I simply rely on your promise to me <u>in which I have implicit faith</u>, that you will do nothing in Automatic or Quadruplex to complicate until I return— You cannot finally lose because they assure you they are willing to care for you, hence you have a sure thing anyhow—

I further rely on you when I return to carry out your promise about my continued interest for the future as a means for indemnity[a] for the past & our mutual interest hereafter.[9] You told me once—Good faith needed no written contracts— Yrs tr

J C Reiff[10]

ALS, NjWOE, DF (*TAEM* 13:1125). [a]Obscured overwritten letters.

1. Thomas Eckert was at this point president of both the Atlantic and Pacific Telegraph Co. and the Domestic Telegraph Co. See *TAEB* 2:120 n. 6.

2. Reiff soon left for Europe and did not return before mid-May; see note 8.

3. The financier Jay Gould, who controlled the Atlantic and Pacific Telegraph Co. at this time; see *TAEB* 2:369 n. 12.

4. George Harrington had been president of the Automatic Telegraph Co. until it was absorbed by Atlantic and Pacific early in 1875; he was in Europe at this time. See *TAEB* 1:190 n. 1.

5. Harrington had sold his own interest in Edison's work to Gould a year before (*TAEB* 2:462), but his sworn signature was needed to complete the case against Gould being prepared by Reiff, Edison, and other partners in the former Automatic Telegraph Co. who had never been paid. That suit (*Harrington v. A&P*), filed on 17 May, asserted that Harrington and Automatic Telegraph still held the patent rights to Edison's automatic and quadruplex telegraph systems since Gould and his company had failed to pay for them. To strengthen the case, the plaintiffs wanted documents from Harrington detailing the original arrangements and transferring Automatic Telegraph's assets to the new American Automatic Telegraph Co. After Harrington signed those papers the plaintiffs initiated legal action (Doc. 750). Another case (*Edison & Harrington v. Western Union & others,* Quad. 72, TLC [*TAEM* 9:8–287, 10:798–997]), asserting Edison's and Harrington's (and indirectly Automatic Telegraph's) rights to patents on quadruplex telegraphy against the claims of Western Union, also was filed only after Reiff had met Harrington (see Doc. 763 n. 5).

These two suits constitute portions of the congeries of litigation surrounding the main legal battle over quadruplex telegraphy. In the central case (*Atlantic & Pacific v. Prescott & others,* Quad. 70, 71, 73 [*TAEM* 9:288–10:797]) filed on 11 April 1876 in Superior Court in New York, Atlantic and Pacific sought to establish title to Edison's quadruplex work and to deny Western Union the use of that technology. See "Telegraph Litigation (Quadruplex Cases)," *TAEM* 9:2–6; and *TAEB* 2, App. 3.

6. No such cable or letter has been found.

7. The electric pen, developed the previous year. See *TAEB* 2, chap. 9 introduction; and Doc. 595.

8. Harrington had left America the previous April. Although Reiff

apparently thought Harrington was in England, Harrington signed the necessary documents in Paris, France, on 2 May. See *TAEB* 2:462 and Doc. 750 n. 5.

9. See Doc. 876.

10. Josiah Reiff, president of the American Automatic Telegraph Co., had been one of the principal investors in Edison's automatic telegraph work and remained a lifelong associate. See *TAEB* 1:243 n. 7; and Docs. 676, 729.

**–741–**

*To Marshall Lefferts*

Menlo Park April 11= 76

Dear Genl

Did you get the list of prices for foreign pen patents. Put your negotiations in trim as Yeaton[1] is so busy with the American Co that he will undoubtedly allow his foreign option to expire as he has only 19 days more.[2] Yours

Edison.

ALS, NNHi, ML.

1. Charles and Lily Yeaton had set up a company to promote Edison's electric pen. See Doc. 731.

2. On 4 March 1876 Edison had signed an agreement with Lily Yeaton for foreign sales of pen patents, to be negotiated by Charles Yeaton acting as agent. The agreement was to expire after two months if the American, English, French, and Belgian patent rights were not sold (Miller [*TAEM* 28:1014]). On 8 April, Amasa Mason, a London-based business acquaintance of Lefferts's, approached him with a proposal to purchase foreign rights to the electric pen. Lefferts subsequently sent Mason a pen outfit for use in attracting foreign investors in the pen. Lefferts's death and Edison's existing contractual arrangements precluded Mason's involvement in foreign pen rights (Mason to Lefferts, 4 and 13 Apr.; 5, 25, 27, and 30 May; and 3 and 30 June 1876; ML). Mason later had other business dealings with Edison (Doc. 859).

**–742–**

*From Edward Johnson*

~~New York,~~Philada Apl 24 1876.[a]

My Dear Edison

I've been thinking a good deal about the Segar[1]—and in order to get a general judgment on it—have mentioned it to a good many people— Everybody says "Capital"—not found one dissenting ~~p~~voice= and this all—accords exactly with my conclusion everytime I think about it=

At all events whether ~~or not~~ it shall immediately become popular or not— If I had it in time to place in the Centennial[2] & on sale in this City—by Opening day—enough could be made out of it—simply viewing it as a novelty, to nett you &

I a good round Sum this summer Its a thing which could be nicely handled at such a show as this= I am all warmed up for it & am therefore impatient of even a moments delay in getting it in shape— I am of course just in my usual luck in having to go away on the Eve. of having something of a personal nature to give my attention to— But I will be back within 10 Days—perhaps ere this week is out— then I'm coming to Menlo—to help you through with it—If indeed you dont have it complete by that time.

I feel as if at last I was in a fair way to make something— —& I never needed it more— —and am in a great hurry to get at it. I have thought of a number of points in which this will excell the Pipe—& of none in which it is inferior to it—& only very few in which the Segar could be Inferior—(all of which however[b] are offset by the one element of cost—)[c]

In this—the smoker has during the entire "smoke"—what in the Pipe he only has on[d] the 1st lighting—that is the cool unmoistened whif which carries with it the flavor of the tobacco he smokes—making the smoking of 1st class tobacco a reality in everyway— in a pipe after a few whifs you may as well have 3 ct tobacco as 50—[3]

In this—the fire is always equal in body—never gets any nearer the mouth—nor down into the nicotine—nor increases in Bulk by igniting a larger Body of Tobacco no in intensity by heating the ~~Body of~~ material[b] which composes the ~~material of~~ Body[b]—of the Pipe—

The mouth Piece—alone can become strong— then make it very cheap— unscrew it, & throw it away retaining the Barrel & Spring &C—

The barrel—being open at bothe ends may be swabbed out like a Rifle—& Dundrearys[4] may even have theirs made with a Belgium twist & mirror polish inside—

Make the spring so when pushed back it will lock—so as to permit of the handy & easy filling of the barrel—then lock the barrel by the end cap—& set the spring free—

Get this up—& let me have an ample supply— Come & pay me a visit—& I'll guarantee to show the money for one days sale on the Eve of the 10th of May—of 1000 of—"Edisons Perpetual Segar,"—

Dont lose any time in order simply to imitate the appearance of a Segar— Let that come in good time— my idea is that a nicely gotten up & elaborated arrangement will be even more popular than an imitation of a Segar— it certainly will immediately sell as a novelty—

I propose to sell Tobacco—Tobacco Pouches, &c—in connection with it—

If I was in the Country today I'd whittle one out of a Piece of Alder Bush ƟAs it is, I'm not at all sure I wont improvise one from a piece of Oil Driving Pipe as soon as I reach that Country Yours once more Enthuzed

E.H.[5]

ALS, NjWOE, DF (*TAEM* 13:753). Letterhead of Automatic Telegraph Co., Edward Johnson, General Manager. [a]"New York," and "187" preprinted. [b]Interlined above. [c]Parentheses cancel dashes. [d]Obscured overwritten letter.

1. Nothing is known of this proposed invention beyond the information in this letter.

2. Edison, in combination with Western Union, staged a large exhibit at the the United States Centennial Exhibition, held in Philadelphia (see Docs. 601, 657, and 730). The Exhibition opened on 10 May.

3. Here "ct" probably means "cent."

4. A Dundreary was an affected fop, after a character in Tom Taylor's 1858 play, *Our American Cousin.* Farmer and Henley 1970, s.v. "Dundreary."

5. Edward H. Johnson had known Edison since Johnson's hiring by the Automatic Telegraph Co. in 1871. See *TAEB* 1:505 n. 13.

**-743-**

*From Marshall Lefferts*

New York 25. April 1876.[a]

T. A Edison, Esq Menlo Park. Grand Laboratory—and general investigating Department, of Spiritual and Material Matters, Electro Motive force I × U + G ÷ A.&P. = W.U. Consolidated—[1b]

You will have to come over and see me. so as to close up several matters. Am prepared to talk about your Salary[2] for next five years—

Close Pen matters.

Dont notify Yeatons, until I see you. Yours truly

M. Lefferts.

ALS, NjWOE, DF (*TAEM* 13:758). Letterhead of Gold and Stock Telegraph Co. [a]"New York" and "187" preprinted. [b]Followed by centered horizontal line.

1. That is, Lefferts ("I") and Edison ("U") and Gould ("G") could effect the consolidation of Atlantic and Pacific with Western Union.

2. Under an 1871 agreement (Doc. 164), the Gold and Stock Telegraph Co. paid Edison a $2,000 annual salary for his work related to printing telegraphy. That agreement was due to expire on 26 May 1876.

**–744–**

*From Alfred Nelson*

New York April 26—1876[a]

Dear Sir

Yours of inst is at hand[1] covering receipt for $500. cash, and for $2500 in notes, all in payment for Domestic Tel Co Stock.

I am obliged for your graceful compliment as to ingenuity, but really it is not deserved, as I simply carried out the agreement between us. However "alls well" &c I now enclose Note of F. W. Roebling[2] to his own order and endorsed by him for $1500. due Oct 27/76. being payment for one half of Six hundred shares for which please return me receipt— I will send you more cash tomorrow or next day— As suggested in yours of 19 inst[3] I called on Mr Morten to secure the information as to Fire Alarm system, and enclose a circular which he had prepared[4] It seems to me this should have much additional matter to render it profitable for circulation? No doubt you can furnish it? Please do so that we may have a neat circular or pamphlet prepared— Will Mr Batchelor send his drawings to Mr Murray?[5] Mr M will set to work as soon as he is posted— Time is an important element for a successful issue so please do what you can to aid us in this as in other particulars— I hope your new home is agreeable, and that the boy thrives—[6] Our Annual election today was entirely Goulden in its character[7] I enclose the ticket[8] Respectfully

Alfred Nelson

ALS, NjWOE, DF (*TAEM* 13:1219). Letterhead of Atlantic and Pacific and Franklin Telegraph Cos. Executive Office. [a]"New York" and "187" preprinted.

1. Not found.

2. Ferdinand Roebling was secretary-treasurer of John A. Roebling Sons Co. *NCAB* 35:386–87.

3. Not found.

4. Not found. Edison had been working on a fire alarm system since the winter of 1874. He executed a patent application for a fire alarm system two weeks later, on 9 May 1876 (U.S. Pat. 186,548). See Doc. 654.

5. Joseph Murray, Edison's erstwhile manufacturing partner. See *TAEB* 1:282 n. 1.

6. Thomas Alva, Jr., nicknamed "Dash," was born 10 January 1876. Pioneers Bio.

7. This refers to the election of officers for the Domestic Telegraph Co., which had been formally taken over by the Jay Gould–controlled Atlantic and Pacific Telegraph Co. two months earlier. A&P Stockholders (1875–88): 73–74.

8. Not found.

-745-

*Draft to William Orton*

[Menlo Park, April, 1876?][1]

Want Merrihew[2] ordered to loop two good[a] Washington wires into Laboratory[3b]

Order on Phelps[4] ~~for~~ allowing me to select from disused instruments what I require; making a list of same for your approval. Also for 100 cells Callaud battery, latter to be charged to me on acoustic account=

Also order on Division Supt whose district covers Boston & NY to allow me and assistant the use of the w~~e~~ires at night when the same are not used for business, that is to say the same kind of order that you gave me when I was experimenting nights on the Quadruplex[5]

Then I shall be happy—

| | |
|---|---|
| Gartland.[6] | 85 |
| Wiley.[7] | 120. |
| Wurth[8] | 200 |
| Kruzi.[9] | 160 |
| Adams.[10] | 260 |
| Man | 40 |
| Forks. | 40 |
| Platina | 15 |
| Blue Vitrol | 38 |
| Bichromate & $SO_3$ | 20. |
| Office wire | 16. |
| Power | 50 |
| Freight | 6 |
| Brass Casting | 28 |
| Iron Casting | 15. |
| Gas & Kerosene. | 36. |
| Alcohol Files & Mercury | 40 |
| | $1169. |
| Spice.[11c] | 28 |
| | $1297. |

Spent on Acoustic to date, not paid for—

If the experiment & progress so far is satisfactory I should like for you to resume specie payments, but will not require but ½ as much as stated in Contract.[12]

ADf, NjWOE, DF (*TAEM* 13:1266). [a]Interlined above. [b]Followed by centered horizontal line. [c]Followed by horizontal arrow.

1. Since Edison spent most of April preparing his new laboratory for experimental work, the references to a Washington loop (note 3) and to resumption of payments for acoustic work (note 12) make this date a reasonable conjecture. Furthermore, neither Wiley nor Gartland ap-

pear in the acoustic labor accounts after 19 April. 75-020, DF (*TAEM* 13:722–32).

2. James Merrihew, who had been connected with telegraphy since 1849, was district superintendent in Philadelphia for Western Union's Southern Division. Reid 1886, 657; Western Union 1876.

3. Edison acquired the use of a Washington loop sometime before early July. NS-76-002, Lab. (*TAEM* 7:354).

4. George Phelps supervised Western Union's factory. See *TAEB* 1:135 n. 2.

5. See Docs. 288 and 454.

6. Thomas Gartland first worked in Edison's Newark laboratory in December 1873; see *TAEB* 2:119 n. 6.

7. Osgood Wiley, the son of publisher John Wiley, was employed by Edison in winter 1875–76; see Doc. 682.

8. Charles Wurth was a Swiss machinist first employed by Edison in 1870; see *TAEB* 2:519 n. 2.

9. John Kruesi was a Swiss machinist first employed by Edison in 1871 or 1872; see *TAEB* 2:633 n. 6.

10. James Adams had been an agent for Edison's inductorium before becoming a laboratory assistant in summer 1874; see *TAEB* 2:250 n. 4.

11. Robert Spice, professor of chemistry and natural philosophy at Brooklyn High School, had been connected with Edison since 1874. See *TAEB* 2:301 n. 1.

12. The contract, signed 14 December 1875 (Doc. 695), called for Western Union to cover weekly expenses up to $200. A receipt of 28 August 1876 for payments covering Edison's acoustic work from November 1875 to February 1876 suggests that payments ended in February. DF (*TAEM* 13:774).

**-746-**

*Agreement with Marshall Lefferts*

[New York?,] May 1, 1876

This Agreement made this 1st day of May 1876, between the undersigned, Thos A. Edison of Menlo Park, state of New Jersey, and[a] Marshall Lefferts, of the City and State of New York.

The said Edison having invented and patented in the United States, an instrument and device known as the "Edison's Electrical Pen and Duplicating Press" and has taken out a patent for the same in England,[1] and now agrees to patent the same in France, Belgium, Austria, Prussia, and various other countries, and hereby further promises and agrees that the said Lefferts is to have the option of purchasing from the said Edison, either or all of the Patents aforesaid, and for the countries for which patents may be taken out, such option to continue for the period of four months from and after the fourth day of May 1876, and at the following prices.

| | |
|---|---|
| For England— | Ten thousand Dollars |
| " France & Belgium | Ten thousand Dollars |
| " Austria | Five thousand Dollars |
| " Prussia | Four thousand Dollars |
| " Russia | Four thousand Dollars |
| " Italy & Sardinia | Three thousand Dollars |
| " Australia | Five thousand Dollars |
| " Spain | Four thousand Dollars |
| " India | Ten thousand Dollars |

to be paid by the said Lefferts upon the assignment in due form of law, of any or all of the above mentioned Patents, and the said Edison agrees that he will not, during the said term of four months, sell, assign, or otherwise dispose of any of the said Patents, but will hold the same, subject to the sale of said Lefferts.— and the said Lefferts agrees that he will use his best endeavors to introduce the said "Edison's Electrical Pen and Duplicating Press" to the notice of parties in Europe, and to make arrangements for their introduction, and that upon his exercising the option of purchase herein granted by the said Edison, the said Lefferts, will, upon the completion of the papers by a proper and usual deed of Assignment as prescribed for the Country for which the said Patent may be purchased, pay to the said Edison in Cash, the sum of money herein before stated for such Country which may then be required by the said Lefferts to be assigned.[b]

Said payments to be made in Gold.

Thos A. Edison Marshall Lefferts.

Witness L Eugene Lefferts[2]

DS, NNHi, ML. Date from text, form altered. [a]Followed by mark to fill space at end of line. [b]To this point written by Eugene Lefferts; following sentence and "Witness" written by Edison.

1. Brit. Pat. 3,762 (1875); see Doc. 748 n. 5.

2. Louis Eugene Lefferts was Marshall's youngest son. Bergen 1878, 139–40.

**-747-**

*From Thomas Eckert*

New York May 2d, 1876[a]

My dear Edison:—

I would like to have the 200 signal or call boxes[1] finished at the earliest moment possible, cash on delivery.[2]

I want to contract for 500 to 1000 more to be made at once. What will be the lowest figure for a complete improved box?

I have arranged with Mr Nelson to have Applebaugh[3] take charge of the "Domestic" in addition to the Manhatten[4] lines, and by the end of this month hope to be able to say we are in a fair way to compete for business.

When can the 200 boxes be ready? Yours truly

Thos T. Eckert

ALS, NjWOE, DF (*TAEM* 13:1221). Letterhead of Atlantic and Pacific and Franklin Telegraph Cos. Executive Office. [a]"New York" and "187" preprinted.

1. That is, transmitters for the district telegraph system. See *TAEB* 1:411 n. 2.

2. Ezra Gilliland was manufacturing these instruments in Menlo Park. Insurance policy of 24 April 1876, DF (*TAEM* 13:924).

3. W. K. Applebaugh (see *TAEB* 1:622 n. 16) was associated with the Manhattan Quotation Telegraph Co. (see note 4) and did in fact become involved with the Domestic Telegraph Co.

4. The Manhattan Quotation Telegraph Co. was organized in 1872 as a competitor of the Gold and Stock Telegraph Co. for the financial reporting market. *TAEB* 2:436 n. 1.

**-748-**

*To Marshall Lefferts*

[Menlo Park, May 5, 1876?][1]

My Dear Genl

I suppose you received my telegram asking you to delay negotiations on the English Patent for a fews days.[2] It appears that I gave an English Gentleman named Gloynn[3] an option in the first part of Febry Running 60 days after the filing of the Complete Specifctns at that time I had paid Serrell[4] $280. Gold to send forward the papers and supposed he had done so and that their Option had expired[5] but they[6] came down here on saturday and said the patent was sold to T. D. Clare of Manchester the person who exhibited it before the Royal Society.[7] I told them that was no way to do to get an option and then never say a word for months & suddenly jump on me besides I said your option expired long ago. They replied no—we have been to Serrells and he says the patent has only just been filed. =if this is true then I am fast. I have sent Batchelor over to ascertain the date of[a] filing of the patent from Serrell. Should it prove that my option with them is run out what do you advise that I do; sell it to them or not,— ~~pls send~~ This option is only on English never had any arrangement with any one on other countries except Yeaton= Pls send G & Stock Contract[8] by Batchelor.

I would come in today but I am working on the acoustic Telegh & I would lose entire day Yours

Edison

ALS, NNHi, ML. Letterhead of T. A. Edison, Ward St., Newark. [a]"date of" interlined above.

1. See note 2.

2. This telegram, dated 5 May 1876, is in ML.

3. John Fox Gloyn also held the electric pen agency for Ontario (see Doc. 639 n. 11); see Doc. 752.

4. Lemuel Serrell was Edison's primary patent attorney from May 1870 until the early 1880s. See *TAEB* 1:196.

5. In a 10 May letter to Edison (DF [*TAEM* 13:1028]), Serrell indicated that he had given Lefferts the history of the patent (Brit. Pat. 3,762 [1875]). Edison had signed the provisional specification on 15 October 1875 and it was filed 29 October, giving Edison until 29 April 1876 to file the full specification. Serrell had sent the full specification on 3 April, and it was sealed (filed) on 26 April 1876.

6. Gloyn and George Walter.

7. Thomas Clare, a Birmingham merchant, is Gloyn's "friend" mentioned in Doc. 723. He also presented the pen at the 26 April meeting of the Society of Telegraph Engineers. *J. Soc. Teleg. Eng.* 5 (1876): 180–82; see also Doc. 925.

8. This is the contract Lefferts referred to in Doc. 743.

## UNBOUND NOTEBOOKS (VOLUMES 8–18) AND TELEPHONE INTERFERENCE EXHIBITS

The laboratory records designated Volumes 8–18, covering the period 1875–79, are collections of notebook entries relating to etheric force, acoustic and multiple telegraphy, the telephone, phonograph, electromotograph, and electric light, as well as a few other subjects. Most of the drawings and notes were originally made in soft-cover tablet notebooks that Edison distributed around the laboratory. Volume 8 is thirteen such notebooks bound together. To create the other volumes, Edison's staff removed the pages from various tablet notebooks and arranged the material topically, making tracings of the drawings.[1]

The page/volume numbers were probably assigned when the volumes were assembled in 1880, after the Patent Office declared Edison's telephone patent applications to be in interference with those of Alexander Graham Bell, Elisha Gray, and several other inventors. In letters of 16 February and 1

April 1880, Edison's patent attorney, Lemuel Serrell, suggested that one of Edison's assistants "sort out the drawings and evidence according to date" and "lay out all matters relating to Telephones" so that Serrell could go over the material with Edison and "number and list the drawings."[2]

The drawings and notes that were taken from Volumes 9–15 for use as exhibits in the telephone interferences are not at the Edison National Historic Site.[3] However, photolithographed facsimiles do appear in Volume 2 of the printed court record.[4] In assembling these materials into documents, pages excised for inclusion in the court record and pages still in the volume have been treated as equivalent—that is, as technical notes—and, because they were arbitrarily ordered when the volumes were assembled, the individual pages have been reordered if a logical progression of ideas demanded it.[5]

1. See Edison's and Carman's testimony (TI 1:59–60, 6:62–63 [*TAEM* 11:50–51, 971]). Other notebooks at the Edison National Historic Site may at one time have been considered part of this series. There are bound notebooks among the Miscellaneous Shop and Laboratory Notebooks which are labeled "Vol. 2" and "Vol. 5" (Cat. 30,094, Cat. 30,095). Another bound notebook (Cat. 1175) may have been Edison's Volume 7. Its pages contain penciled page/volume numbers similar to those in this series.

2. *TAEM* 55:36, 77.

3. According to the court record, the original documents and artifacts entered as evidence remained in the possession of Edison or his attorney. TI 1:7 (*TAEM* 11:24).

4. A substantial number of the documents entered as evidence are tracings of the originals. On these pages drawings are less clear and words more often illegible.

5. Defining a document—that is, delimiting it from other materials of the same date—involves consideration of continuities or resemblances of content, wording, and style.

**–749–** [Menlo Park,] May 8 1876

*Technical Note: Multiple and Acoustic Telegraphy*

Keeping 1 line on two sets of instruments.[1]

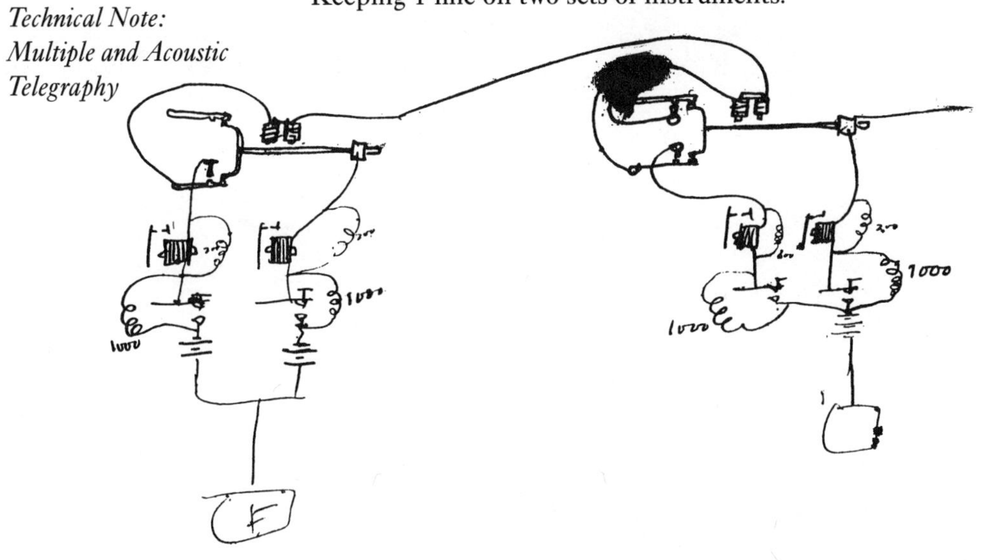


2 wires ac[ous]t[ic] t[ransfer][2]

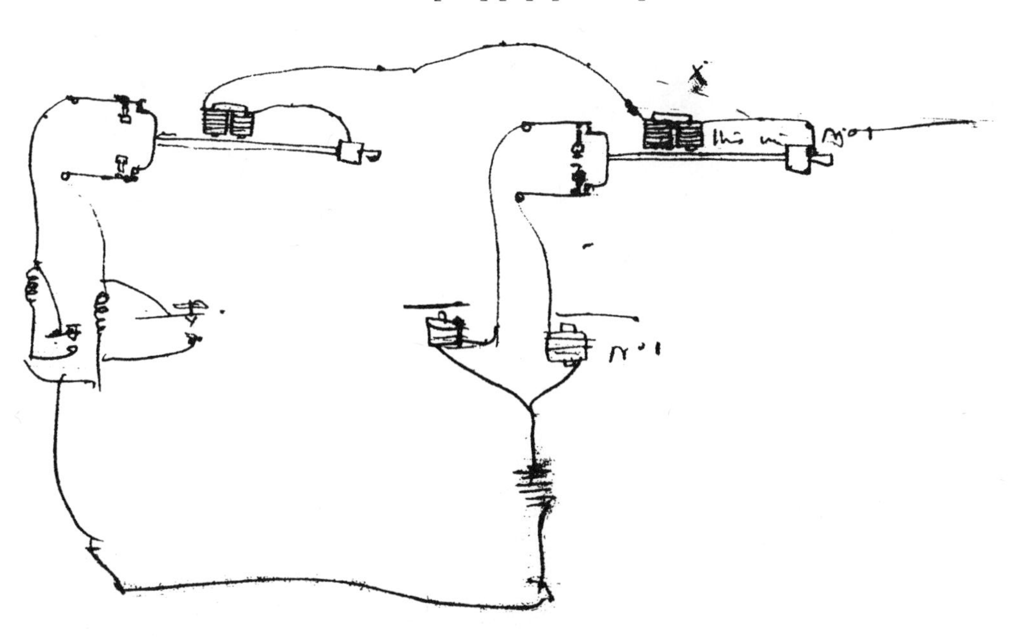


main line direct to Reed= X is 2 ohm spool the variation in resistance makes it hard to adjust hence failure

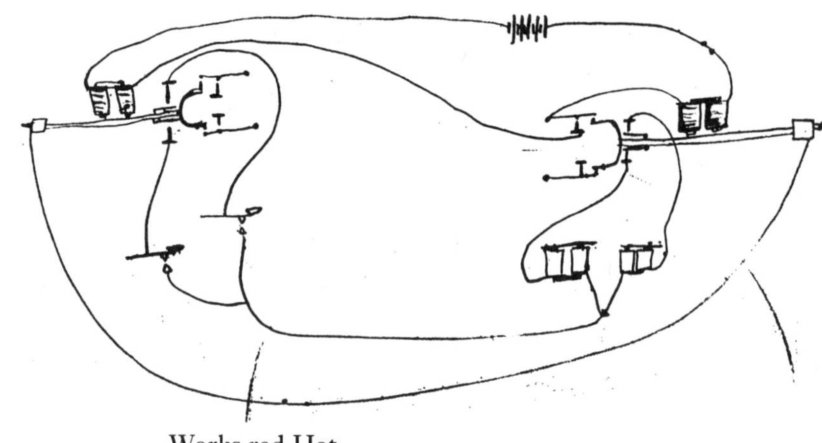

Works red Hot

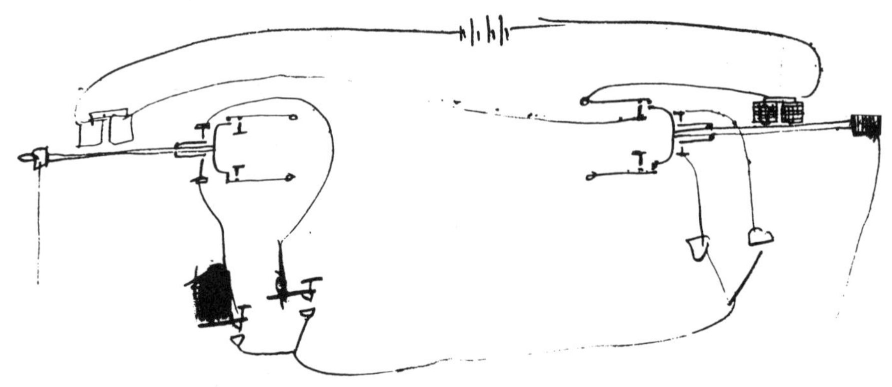

Apparatus whereby 1 wire with acoustic vibraters is made to thrown another wire on two sets of insts at once all mkg 2 clomplete wires of a single one= for instance bet ny & Washn 1 wire can be the directer with acoustic Reed this reed can mdo[a] make contact so as to double 10 or 15 more wires wi both through & with way station ⟨Works beautifully⟩

Same with a way station

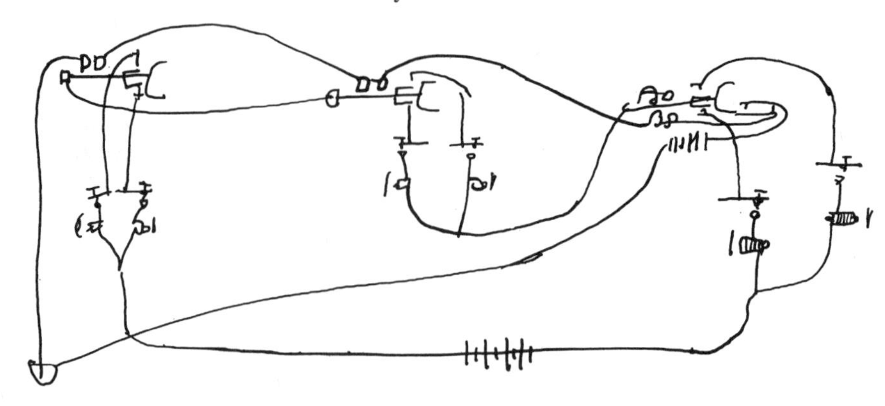

Acoustic Tel[3]

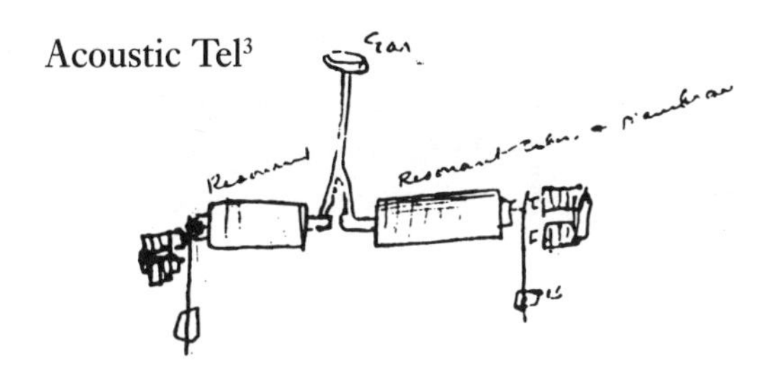

One reed one note higher than other. 1st part of dot sent on one reed & other part on other one reed thus giving up & down stroke distinctive causing the op[erato]r to read without trouble

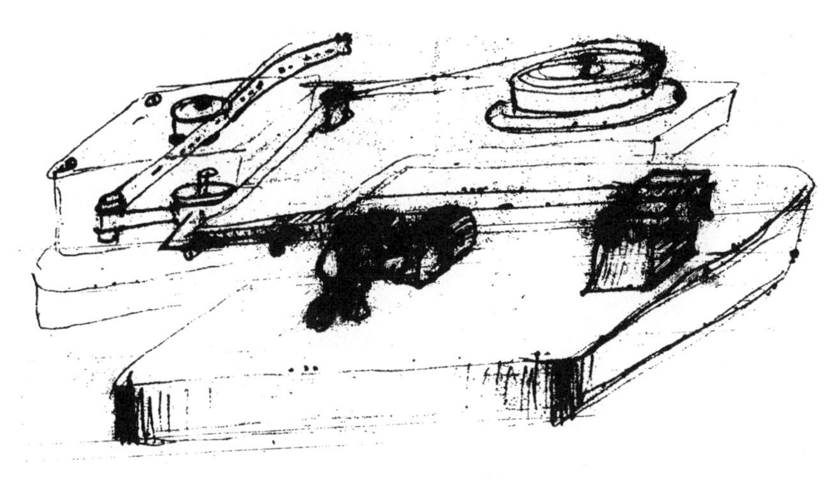

Acoustic Chemical Recorder

Edison Chas Batchelor

X, NjWOE, Lab., Vol. 10:21, 24, 22; Lab., Vol. 15:107; Lab., NS-76-002; TI 2, Edison's Exhibit 25-10; Lab., Vol. 10:23 (*TAEM* 3:836, 837, 836; 4:423–24; 7:384; 11:214; 3:837). Document multiply signed and dated; continuity of text links the loose note to the other material. [a]"d" overwrites "m".

1. Figure labels are "200" at each resistor shunting a relay and "1000" at each resistor shunting a key. These notes are the first extensive indication of Edison's work on a kind of synchronous multiplex telegraphy that he called "acoustic transfer telegraphy" (see headnote, p. 27). Its origins appear to be a transformation of one of his versions of acoustic telegraphy (see Vol. 15:133, NS-76-002, both Lab. [*TAEM* 4:440, 7:353]; and Doc. 709, esp. fig. 5). Edison started designing and testing such systems about a week earlier and continued for many months (Vol. 10:19–20 [*TAEM* 3:834–35]).

2. Figure labels are "this is No 1," and "No 1."

3. This is an acoustic, rather than an acoustic transfer, telegraph design. Figure labels are "Resonant," "Ear," and "Resonant tubes & membrane."

–750–

*Robert Russell to Atlantic and Pacific Telegraph Co. and Jay Gould*

Dated, New York, 9th May, 1876.

*To the Atlantic and Pacific Telegraph Company and Jay Gould:*

I am instructed on behalf of Mr. George Harrington and Mr. Thomas A. Edison and their associates, to demand of you the return of the deeds dated respectively 1st January, 9th March and 9th April, 1875,[1] purporting to be assignments from the said George Harrington to the said Jay Gould, and which were placed in the hands of said Jay Gould for a specific purpose, which has not been accomplished, and which cannot now be accomplished.[2] And also to demand assignments or releases to the said George Harrington, trustee of the interests assigned by the said deeds, and the surrender of any powers of attorney therein contained. And also the delivery to the said George Harrington, trustee, of all original deeds and contracts delivered to the said Gould by the said George Harrington.[3] And I am also instructed on behalf of the said George Harrington, as President of the Automatic Telegraph Company, to demand of the Atlantic and Pacific Telegraph Company, the surrender and relinquishment of possession of the telegraph lines and offices and telegraph apparatus formerly held and used by the Automatic Telegraph Company, and now held and used by the Atlantic and Pacific Telegraph Company.

And I am also directed, on behalf of the said George Harrington and T. A. Edison, to require the Atlantic and Pacific Telegraph Company to account for and pay over to the Automatic Telegraph Company the value of the use and occupation of the said lines and offices and the apparatus therein by the Atlantic and Pacific Telegraph Company, up to the time of the surrender thereof as aforesaid.

And also to require the Atlantic and Pacific Telegraph Company to desist from the use of the patented inventions of the said T. A. Edison, and to account to the said George Harrington, as trustee, for the use thereof since the said, The Atlantic and Pacific Telegraph Company, was notified in the month of August last to discontinue the use of the same.[4]

Mr. Harrington and Mr. Edison recognize and ratify the notices heretofore given to you, in August last, on behalf of the parties holding the major part of the beneficial interest in the said patents held in trust as aforesaid.[5]

R. W. Russell.[6]

PL, NjBaFAR, *Harrington v. A&P,* Box 17B, Exhibit C, 1:143.

1. Quad. 70.1, pp. 18, 22; Quad. 71.2, p. 31 (*TAEM* 9:303, 305; 10:242). The associates referred to, who had interests in the old Auto-

matic Telegraph Co., are listed in Doc. 561. The interests of most of them had been transferred to the new American Automatic Telegraph Co. Docs. 676 and 729; see also note 5.

2. The purpose was the transfer of all of Automatic Telegraph's assets to the Atlantic and Pacific Telegraph Co., in exchange for which the owners of the former company were to receive stock in the latter. Gould was to transfer the properties each way, but no stock was distributed. Docs. 522 and 561.

3. The several items are specified in Quad. TLC.2, p. 156; see also Quad. 71.2, p. 44 (*TAEM* 10:898, 249).

4. This notice is Exhibit D in *Harrington v. A&P* (Box 17B, 1:145); see also Doc. 676.

5. The parties were the trustees of American Automatic, including Edison, Josiah Reiff, and Russell. Reiff had obtained Harrington's signature supporting their claim on 2 May 1876 (not 6 May as was incorrectly stated in n. 2 to Doc. 676; Quad. TLC.2, p. 165 [*TAEM* 10:902]). Gould and Atlantic and Pacific did not meet the demands of this letter and, on 17 May in the U.S. Court for the Southern District of New York, Reiff sued them in the name of Harrington and Edison (*Harrington v. A&P;* see Quad. 71.2, p. 44 [*TAEM* 10:249]). Thirty-seven years later, in 1913, the U.S. Supreme Court ruled that the District Court had not had proper initial jurisdiction and therefore voided all prior action in the case; the surviving parties did not renew it. See Docs. 522, 561, and 676.

6. Robert Russell was a trustee of American Automatic and one of Edison's lawyers in the quadruplex litigation, although not in *Harrington v. A&P,* for which the filing solicitors were William Butler, Thomas Stillman, and Thomas Hubbard of New York. See Quad. 71.2, p. 86, TLC.1, TLC.2 (*TAEM* 10:270, 800, 810); and *TAEB* 2:491 n. 1.

**-751-**

*To Frederick Royce*

Menlo Park May 16 76.

Friend Royce[1]

Yours recd.[2] am much obliged for the information. am anxious to get the patent for the pen etc issued as it wish to raise the wind. besides its been dragging along for a great length of time.

Can you ascertain what formality it has to go through yet, if it has passed through "printing" "stationary" & "perforating Machine" Classes safely or have they dug out of the Assyrian stone tablets of Babylon an antiduluvian reference;

Regarding that despicable puppy Sawyer[3] I never believed a word he ever said. Hes nothing but a bag of miasma under pressure. He's got his two Autographic machines ~~connected~~ placed on a table and connected together by a rod and bevel gear under the table, to obtain synchronism. So I am told. Yours

Edison

P.S. in wkg from NY to .W. he will have to employ a revolving shaft 280 miles long. & I think he is equal to the occassion —on paper

ALS, Filson, Royce.

1. According to the records of the Filson Club, in Louisville, Ky., Frederick Royce had been a telegrapher during the Civil War. Edison probably met him while working in Louisville.

2. Not found.

3. William Sawyer developed and patented a facsimile telegraph system. The United States Postal Telegraph Co. was incorporated to exploit the system. "A New Idea in Telegraphy," Clippings 1876, NjWOE; Reid 1879, 225.

*William Sawyer's facsimile telegraph.*

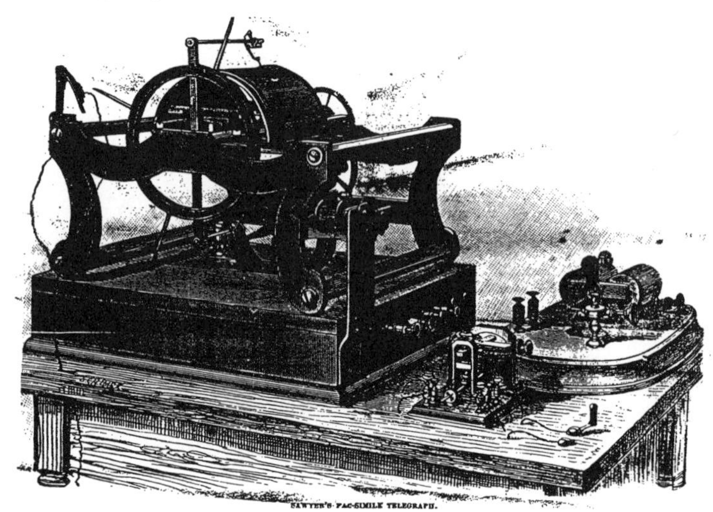

**–752–**

*From George Walter*

New York May 23d 1876

Dear Sir

You will please to take notice that you must fulfil in all respects your contract with me[1] bearing date the 28th day of January 1876 as under said contract or refusal or option I did on the 29th day of January last negotiate for you the sale of the Patents for Edisons Electrical Pen and Duplicating Press in & for the Kingdom of Great Britain (according to the terms you authorised me to make) to one Jno F Gloyn who will be ready to pay for the same within the time stated in your contract with me

You will please have the transfer or assignment ready for delivery upon payment to you of the price agreed upon.

If you neglect or fail to fulfil the contract on your part you will be held liable for all damages resulting from such failure[2] Truly Yours

Geo B Walter[3]

ALS, NNHi, ML.

1. Copies of the Edison-Walter agreement and Walter's 31 January letter notifying Edison of the sale are in ML; the original letter is in DF (*TAEM* 13:965).

2. See Doc. 925.

3. George Walter was secretary and a principal promoter of the National Telegraph Co.

**–753–**

*To Frederick Royce*

Menlo Park, N.J. May 29= 76

Friend Royce

The next time you drop into the Patent office please ascertain how long the Electric Pen will probably be detained in the printing stencil or stationary class if it has to pass through many classes. It makes me uneasy to see fellows running around with a wagon load of Legal tenders asking when that patent will be out=

Are the WU working the click click clicks click click clicks, from N York to Chicago regularly now.[1]

Im going to send something within next six weeks to patent ofs that will make the Teleghers eyes stick out a little.[2] Yours;

Edison

ALS, Filson, Royce.

1. Edison is probably referring to the quadruplex, which had begun commercial operation between New York and Chicago in early December 1874. *TAEB* 2:361 n. 1.

2. Probably the acoustic transfer telegraph.

## ACOUSTIC TRANSFER TELEGRAPHY

Docs. 749, 754, 761, 765, 766, 768, 769, 773, 793, 795, 797, 800, 801, and 808

Edison's acoustic transfer telegraph was a novel type of synchronous multiplex system. These systems used a kind of time-sharing of the transmission line, switching rapidly between several sets of telegraph instruments. Each set was connected for only a fraction of the total time, but the switching was so fast that each set responded as if it were continuously and exclusively connected to the line. A system like this could only work if the high-speed switching devices at both ends of the line remained precisely synchronized, so that each transmitter and its designated receiver were on the line at the same time.

Although numerous inventors had sought to design synchronous multiplex systems, Edison had never explored the technology in his extensive previous work on multiple telegraphy.[1] Most attempts to realize this operating principle employed rotating disks with multiple contacts for the switching.[2] Edison instead tried using the electrically driven harmonic oscillators—tuning forks and metal reeds—with which he was familiar from acoustic telegraph experiments.[3]

1. The earliest proposal for multiple telegraphy (by Moses Farmer in Boston in 1852) had described a synchronous multiplex, and contemporary French inventors were developing successful systems of this sort. Lines 1876, 22–29; Prescott 1877, 862–66; Butrica 1986, 88–108; King 1962b, 308–15.

2. Maver 1892, 336–43; Fleming 1921, 61–70.

3. For acoustic telegraph devices with characteristics and functions similar to those used in acoustic transfer circuits, see Doc. 708, esp. fig. 7 and discussion and notes, and Doc. 709, esp. fig. 5 and discussion.

**-754-**

*Caveat: Multiple Telegraphy*[1]

New York, May 31, 1876[a]

To all whom it may concern,

Be it known that I, Thomas A. Edison, of Menlo Park, in the State of New Jersey, have invented an Improvement in Acoustic Transfer Telegraphs, of which the following is a specification.[2]

The object of this invention is to regulate all kinds of telegraphic mechanism and causing the synchronious movement of two different machines placed at a distance from each other and connected by a telegraphic circuit.

The invention consists of two acoustic reeds, tuning forks or other bodies following the law of the pendulum[3] operated directly by electromagnetism or indirectly by the intervention of sonorous bodies or columns of air between such reeds, tuning forks or other bodies following the law of the pendulum and the electromagnets, such electromagnetic acoustical reeds are placed at the two ends of a telegraphic circuit and being provided with suitable contact breaker points, switches connections and other proper devices vibrate when properly adjusted in perfect or practically perfect unison with each other.

The invention further consists in combining with such acoustical instrument of other contact devices than is necessary to separate the reeds themselves, which contact devices serve to throw the wire upon which the electromagnets operating such reeds are placed, upon a new set of electromagnets entirely disconnected from the reed magnets, at both ends of the wire simultaneously at theat particular period of time when the wire is not transmitting an acoustical wave.

The second set of electromagnets contained in a branch circuit at both ends may be common Morse relays operating sounders and controlled by a key; the two receiving the same number of waves as the reed but not at the same time and occuring with great rapidity act upon the relay in the same manner as a constant current.

The invention further consists, in the various contact points to ensure a perfect transfer of the wire from the signalling instruments to the reed instruments and vice versa.

The invention further consists in various forms of reeds and electromagnets operating the same to suit the various conditions which arise upon long, short, and inferior telegraphic circuits.

The invention further consists, in the various methods of compensating for the mutilation of the acoustical and signalling waves by the static current and by derived currents from contiguous wires.[4]

The invention further consists, in the method of transferring the main wire from the acoustic magnets and circuit at terminals and at way stations to the signalling magnets at the same stations.

The invention further consists, in the method of transmitting from one circuit into another.

The invention further consists, in the method of translating into branch wires,

The invention further consists in the method of duplexing and quadruplexing the signalling circuit obtained by acoustical transfer.

The invention further consists in the method of transmitting two messages in the same direction upon the circuit obtained by acoustical transfer.

The invention further consists in the method of signalling (upon the circuit obtained by acoustical transfer) by the use of polarized relays and reversed currents.

The invention further consists, in the method of transferring several separate circuits from into two branches each by means of local reeds operated by contact points upon the main line reed.

The invention further consists in transferring a single circuit on several branches simultaneously at both ends of the wire so as to allow the transmission of several distinct messages in either direction at the same time.

The invention further consists in the method for regulating machinery so as to run synchronously at both ends of the line, and the transmission of telegrams over the same wire at the same time.

The invention further consists in the method of transmitting messages upon the acoustic transfer branch as well as the other branch by a reversal of the current.

It is well known that if two bars of steel of equal length breadth and thickness be secured to a pillar that they will when acurately adjusted give out the same note when struck and make the same number of vibrations or movements in the same time. It is also well known that the reeds can be automatically kept vibrating by placing electro magnets upon one side of the bars in such a manner that they will attract the bar when an electric current passes through them, and when the current ceases the natural elasticity of the bar will cause it to spring away from the cores of the electromagnet. If a contact spring connected to a battery thence to the electromagnet thence through its wire to the bar, be properly adjusted the bar will separate from the contact spring at the moment when the bar nearly reaches the cores of the magnet, this severing of contact between the spring and the bar will open the local circuit, the magnet will lose its magnetism and the reed will recede from the cores but will again touch the spring and close the local circuit and be again attracted by the magnet and so on adinfinitum.[5]

When both reeds are allowed to break the circuit, one contracts[6] the other so that it is not absolutely necessary to obtain the same vibration in each that they should be acurately timed, one may even be a note higher or lower than the other the result will be that the vibrations will be the mea~~r~~n of the two, but the amplitude of vibration will not be so great as it would were they in perfect accord.

These two reeds are shown in Figure 1. one at one end of a telegraphic circuit and its fellow at the other end. When arranged in this manner they run accurately together, both making the same number of vibrations in each second and independent of the battery power, which only serves to increase or decrease the amplitude or sweep of the bars, which do not effect the proper adjustment of the various contact points which I will show in the subsequent drawings. Only a constant current circulating through the main circuit will destroy in a measure the adjustment and that is easily compensated for by simply adjusting the magnet to and from the reed.

Fig. 1.

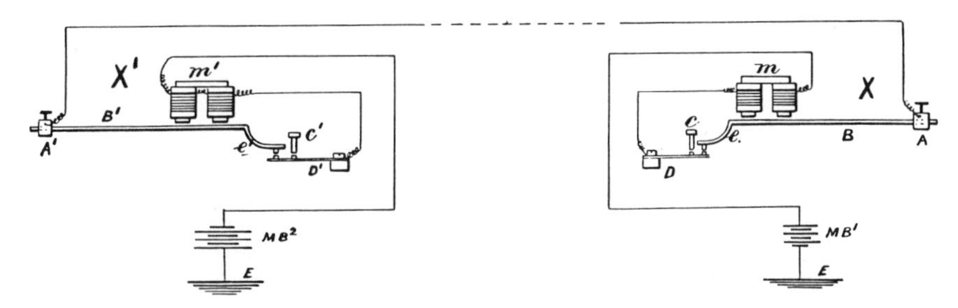


In fig. 1. X and X′ are the two reeds, B. B′ are the bar vibrators kept in motion by the power of electromagnets m. and m′.

The main line containing batteries MB$^1$ and MB.$^2$ at each terminal is connected to the vibratory bars of both instruments. A and A′ being the rigid pillars or blocks that serve to secure the reeds, D. and D′ are the contact springs connected to the magnets and to batteries and earth. C. and C′ are limiting screws which serve to check the spring at that point where the reed stands when in a state of rest, and allows the main circuit to be broken at both ends at the same time, the momentum acquired carries the bar nearly up to the face of the magnets both then fly back and touching the springs D and D′ simultaneously close the circuit through the~~r~~ir respective magnets which again attract them.

It will be noticed that no current passes over the line at the particular period of time when the bars have left their contact springs D. and D′. hence the line is only used one half of the time.

It is these intervals or time when the line is not used for the transmission of acoustical vibrations that I transfer it over upon morse signalling instruments connected to batteries and earth at both terminals simultaneously and as the vibrations are much more rapid than it is possible to make the relay lever respond to the~~n~~m it follows that if the batteries are connected to the signalling branches at both ends a practically continuous current will pass through the relays at both ends which may be interrupted for the purpose of signalling by opening one of the branch circuits.

I am thus enabled to use the wire in the ordinary manner for sending and receiving in either direction and at the same time obtain two perfectly synchroneously moving instruments for controlling and regulating the special applications of which I shall describe in another portion of this communication.

The above described arrangement is shown in figure 2.[7]

Fig. 2.

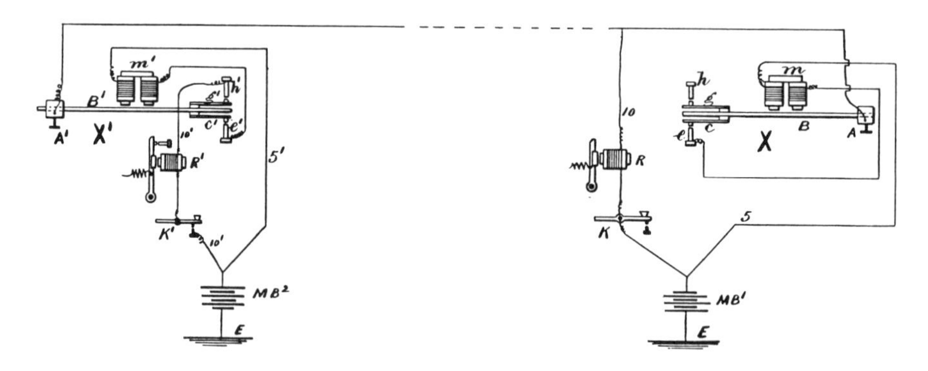


X and X′ are the acoustic instruments placed at the terminals of the line. A. and A′. are the two pillars which hold the reed bars B and B′. and to which the main line is connected at both ends of the line. Upon the extreme end of each reed are two contact springs c. g. and c′. g′. on either side and facing them are contact points e. and e′. h. and h′.

To the contact points h. and h′. are connected branch wires[8] to Morse apparatus consisting of relays, keys and a battery MB.$^1$ and MB.$^2$, one at each end of the wire.

The contact points e. and e′. ~~is~~are connected to other branches to the battery in which is included the electromagnets M. and M′. which give the reed motion. The points e. h. and e′. h′. are so adjusted that when the reeds are in a state of rest the springs upon it do not come in contact with either points, but when the reed is vibrating it alternately comes in contact with e. e′. and h. h′. thus transferring the line at both ends simultaneously from the acoustic instrument to the Morse apparatus and vice versa with great rapidity and if the keys of both Morse instruments are closed and the reed makes two hundred movements each second there are 200 waves of electricity sent over the wire while the line is in contact only with the Morse instruments and these waves occurring so closely together and ~~i~~at so rapid a rate do not allow the core of the magnet to discharge, hence a weak but practically through not a really continuous current passes through both instruments attracting their levers and closing the local circuit containing the sounder (not shown) and none of these waves ever pass through the electromagnets of the reed as a different and distinct series of waves are sent to actuate the reed magnets at a time when the line is entirely disconnected from the Morse apparatus by a break on the circuit between the springs g and g′. and the contact points h. and h′. The wires 10. and 10′. form the Morse or signalling branch while the wires 5. 5′. form the second branch containing the acoustic magnets.

K and K′ are the signalling keys.

Having now shown a method of transmitting the acoustic waves to obtain a perfect syncronous movement of two machines and at the same time the transmission by Morse apparatus in the ordinary manner and practically obtaining a complete wire, I will now show the method by which I transmit two messages over this wire so obtained; by means of the duplex principle;

This method is known as the differential method and is shown in Fig. 3.[9]

Fig. 3.

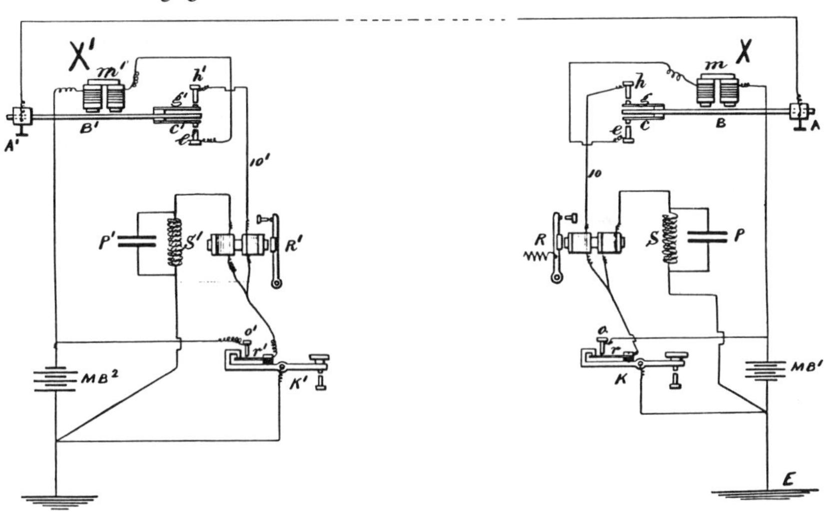


X and X′ are the acoustic regulating instruments, operated as shown in Fig. 2. R. and R′. are the two Morse instruments placed in the branch circuits.[10] These instruments are provided with double coils, through which the currents are made to pass in opposite directions, one coil is connected in the branch wire, 10. and 10′. at both stations, while the other coils are connected within an artificial circuit formed by resistance coils S and S′. shunted with condenser when by the adjustment of the resistances S and S′. the current circulating in the artificial line is equal to the current in the main line or its effect upon the cores of the relay a perfect balance is obtained and the battery M.B.[2] may be included or excluded from the circuit by the signalling key or sounders K or K′ without producing any effect upon the relay at the same station, but the distant relays will respond thus allowing the transmission of two different messages over the wire obtained by the acoustical transfer.

The condensers P and P.′ serve to compensate for the static charges of the line.

It might be though,[11] that whereas a series of vibrations pass through one set of spools upon the differential relays that it would be impracticable to obtain a balance by passing a continuous current through the other spools in an opposite direction, but in practice no trouble arises from this cause, should it arise on working very long wires it can easily be obviated by providing the reed with another contact point which shall break the equating circuit at the same time and in the same manner.

In figure 4. is shown the method of duplexing the transfer wire by the use of a Wheatstone breidge and it will only be necessary to describe the connections at one end of the line.[12]

Fig. 4.

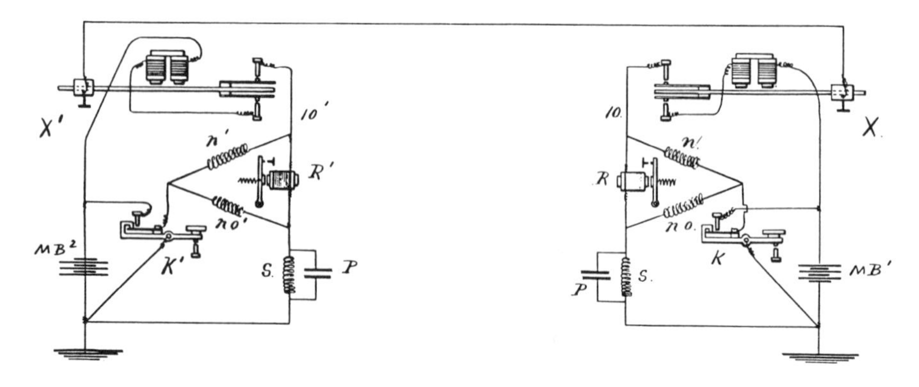


X. is one of the acoustic contracters.[13] 10. is the branch circuit containing a part of a wheatstone bridge, n. and n.o. are two[b] resistances forming two sides of the balance while the relay R. is placed in the bridge wire.

S. is the resistance forming the equating or artificial line.

K. is the transmitting key whereby the current from the battery MB′ may be made to pass through the bridge and its connections over the line, or not, as it is depressed or raised.

When the current from MB′ which passes through the artificial line is equal in strength to the current which passes through wire 10. to the distant station, its taking off or putting on will not produce any movement of the lever of the relay R′ but will act upon the relay at the distant station while the current from the distant station will not act upon its relay at that station will upon the relay R thus allowing two operatives to transmit the~~r~~ir respective messages in opposite directions.

I will here mention that the ordinary relays. R. R′ may be replaced by polarized relays and worked by single currents or by double currents, in which case the transmitting keys (or sounders which I prefer to use) are made double and connected with the main battery as shown in figure 5.[14]

Fig. 5.

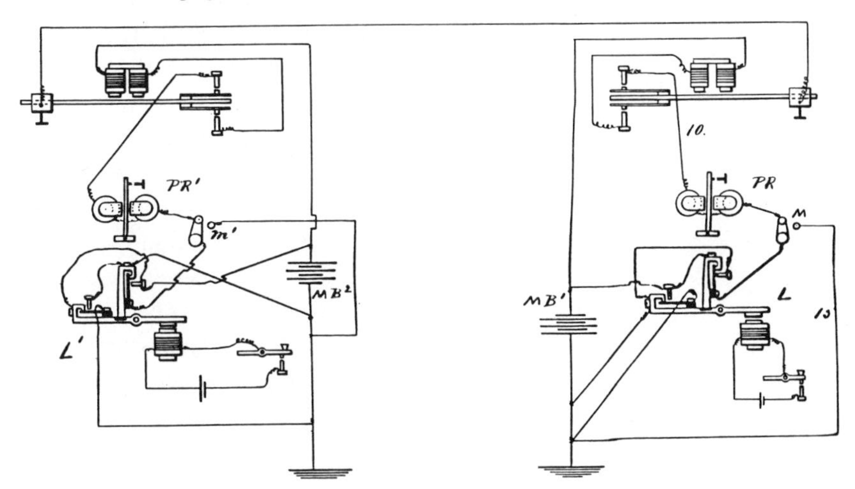


PR and PR′ are the two polarized relays which are shown connected in the ordinary manner for working Morse.

m and m′ being switches to throw the main circuit to earth when not transmitting.

L and L′ are the reversing transmitters which serve to send positive and negative currents over the wire by altering the direction of the flow of current from the main batteries.

It is obvious that if the polarized relays PR′ and PR are provided with extra current[15] one of which is included in an equating circuit as in fig. 3. or is placed in the bridge wire of a Wheatstone balance as in fig. 4. the transmission of two messages at the same time is obtained.

In this case the use of the switches m. and m′ are unnecessary.

Fig. 6.

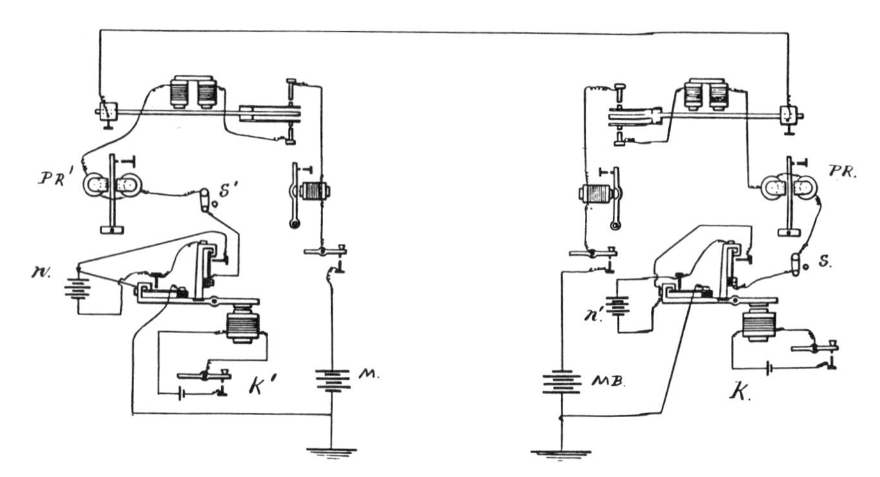


In Fig. 6. is shown connections f~~r~~or transmitting one message on the transfer Morse circuit in the manner already described and another message upon the transfer circuit containing the acoustic magnets,[16] the transmission of a second telegram by the use of the acoustic circuit is obtained by taking advantage of the fact that if the reeds are made of Bell metal with an armature near their ends secured to them and facing electromagnets, the reversal of the current will not in the least effect the transmission of vibrations or decrease their amplitude, as the electromagnet not being[c] polarized responds equally to both positive and negative currents.

Now the only way that an interference is possible is that in time the cores of the electromagnet which serves to give motion to the reed may ~~be~~ have its cores permanently magnetized.

In this case a permanently magnetized steel bar placed across its yoke can be made to balance or neutralize it all together, as the waves which give motion to the acoustic reed are always passing over the line.[17]

I include polarized relays PR. and PR′ in the branch circuit containing the electromagnets of the acoustic.

The operation is as follows:—

If the switch S′. is turned so as to connect the branch line direct to earth the battery n′ will send a current into the line through both electromagnets of the reeds and both polarized relays, and the tongues of both will be thrown on one side, if now the key K be closed the current will be reversed and the tongues of the polarized relays thrown to the opposite sides thus giving the signal, but the reeds will continue to vibrate, as their electromagnets not being provided with a directi~~on~~ve force are indifferent to the reversal of the direction of the flow of the current; at the same time that messages are being transmitted with the polarized relays, messages can be sent with the common relays in the manner previously described.

In figure 7. is shown the connections for duplexing both of the transfer circuits (ie) both branches.[18]

It will only be necessary to describe the connection at one end to explain the principle.[19]

Fig. 7

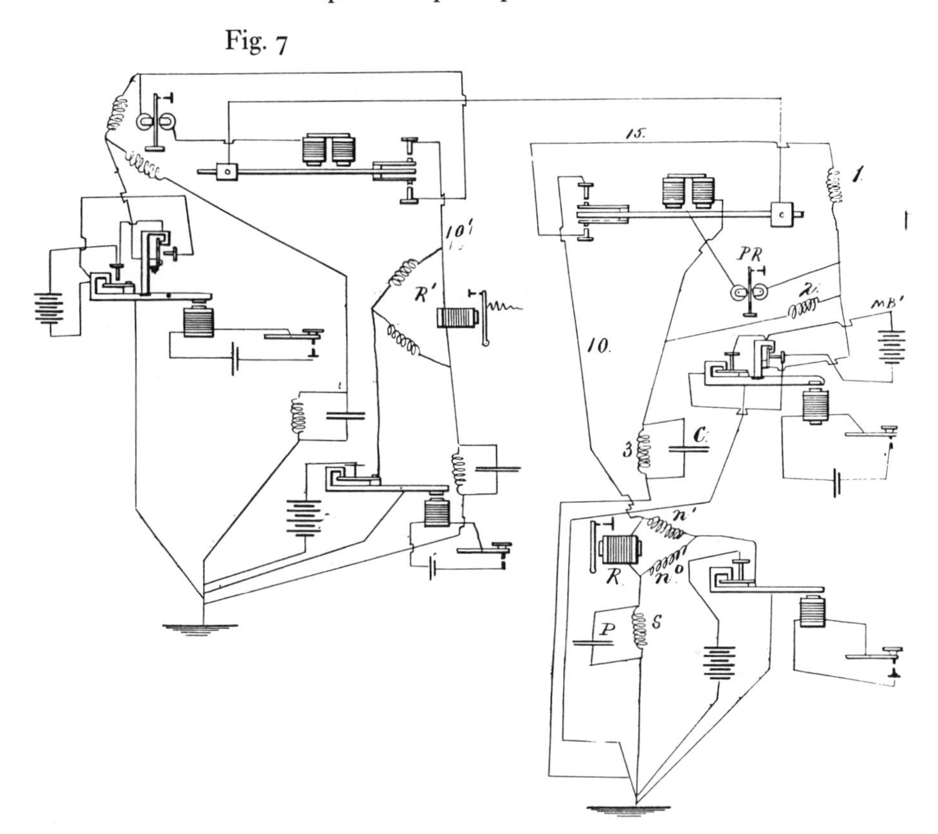


10. is one of the branches which only contains a Wheatstone balance in the bridge of which is a common relay. All the connections in this circuit are the same as in the wire 10. fig. 4. at X. or the connection may be as in wire 10 Fig 3. or in wire 10.[d] fig 5. Two communications being sent in opposite directions at the same time.

The connection of wire 15. containing the reed magnets is similar to that shown in fig. 4 with the exception that the polarized relay PR and the electromagnet of the reed are placed in the bridge wire of the Wheatstone balance formed by the resistances 1 and 2.

3. is a resistance forming the artificial line for balanceing[b] the currents from the battery MB′ so that they pass into the line but do not effect PR or the reed magnets.

The same operation takes place at the distant end; by this means I am enabled to transmit two messages by means of the instruments in wires 10. and 10′ and two messages by reversed currents polarized relays through the wire containing the reed magnets without at all interfering with the proper vibration of the two reeds, in all 5 distinct series of signals.

In figure 8. is shown the two reeds worked alone upon a single wire,[20] the bars serving to transfer another and distinct wire from one set of instruments to another.[21]

Fig. 8.

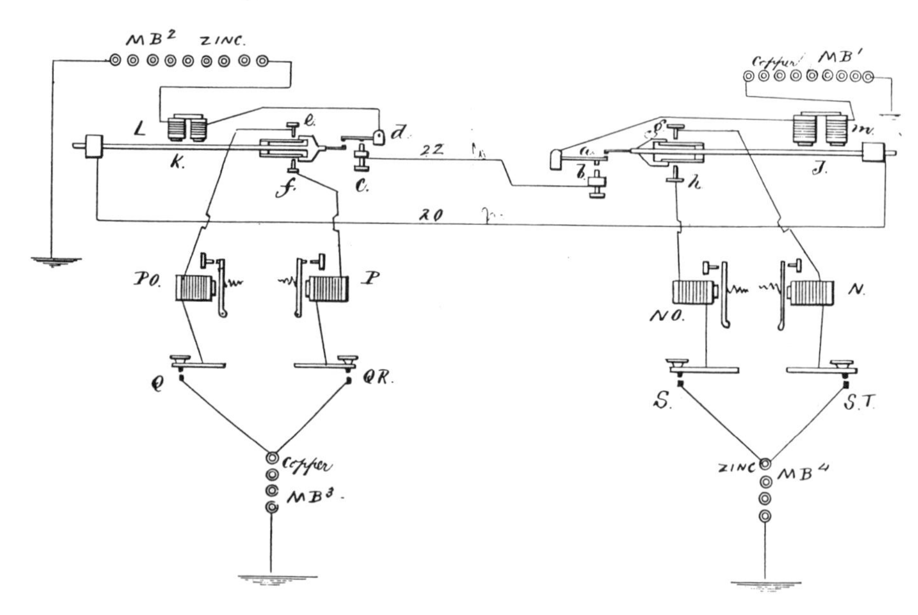


22. is the controlling or directing wire containing the reeds at both terminals in circuit.

The contacts c. d. and a. b. serve to automatically open and close the main circuit and keep the reeds vibrating in unison. MB.[1] and MB.[2] are the two main reed batteries.

Upon the extreme ends of the bars are contact springs facing contact points e. f. and g. h. to which the two sets of instruments are connected. The other main line 20 is connected to the bars themselves so that at each vibration the line 20. is alternated at the rate of 200 times per second from N and PO. to NO and P. thus giving the operator two distinct circuits to transmit upon, each of which may be duplexed or quadruplexed in the manner already shown.[22]

Fig. 9.

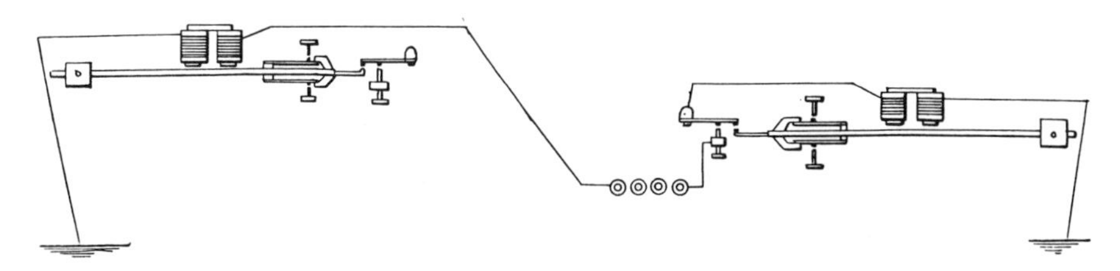

In figure 9. is shown a method of connecting where one reed only breaks the circuit thus controlling the other, but it is only useful where the wire upon which they are placed is used only to transmit these waves alone and not for signalling purposes.

Fig. 10.

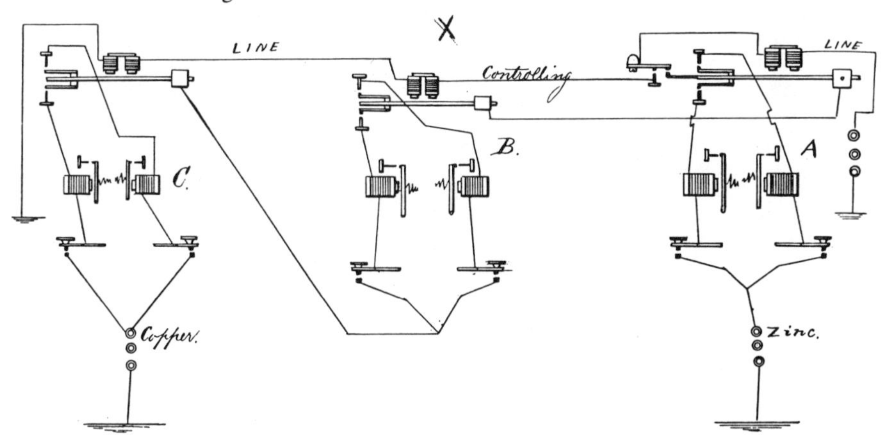


In figure .10. is shown the connections for a way station, X being the way station.[23] In this case the wire is transferred to A. B. and C. simultaneously and then to the other set of relays the keys in any one of the branches serving to interrupt the current and produce signalling. It does not matter how many way stations are connected in this manner to the main line, all can work as in the usual manner.

Fig. 11.

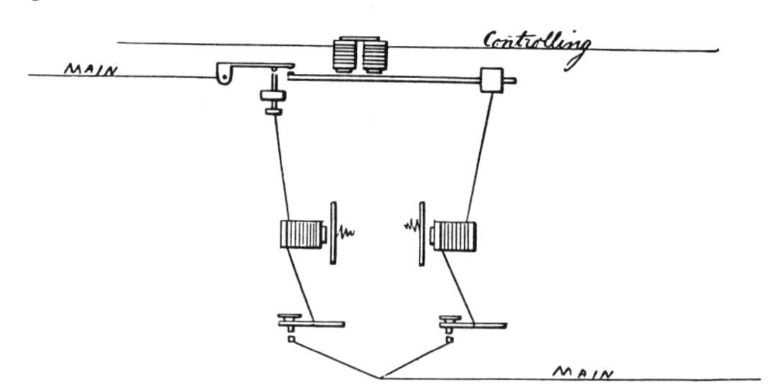


In figure 11. is shown a better form of contact for transferring on second circuits as the line is only disconnected from one set of instruments at the moment it comes in connection with the other set of instruments.

I will now show the method of synchronously vibrating a number of reeds placed in one circuit at both ends of the line; each reed serving to transfer upon two sets of instruments separate circuits. The connections are shown in fig 12.[24]

Fig. 12.

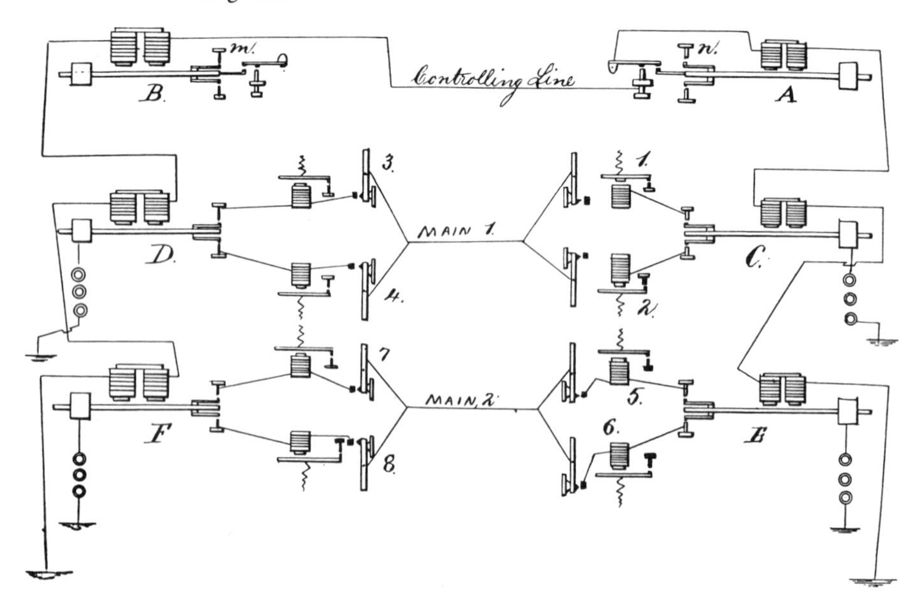


A and B. are transfer reeds serving to transfer if so connected the controlling line from its own magnets to branch signalling instruments connected, n. and m.[25]

C and D are included in the same main or controlling circuit and serve to transfer the main line number 1. from the signalling instruments 1. and 3. simultaneously and twhen to 2 and 4. creating two distinct signalling circuits.[26]

E and F. are two more reeds which serve to transfer the main line No. 2. on 5 and 7. and 6 and 8. in the same manner as with line No. 1.

It is not essential that the reeds C. D. E. F. should be placed in the main controlling circuit as they may be reeds operated by magnets placed in local circuits which receive breaks from an extra contact point and spring in connection with the main reed. The connections for the local reeds are shown in fig. 13.

Fig. 13.

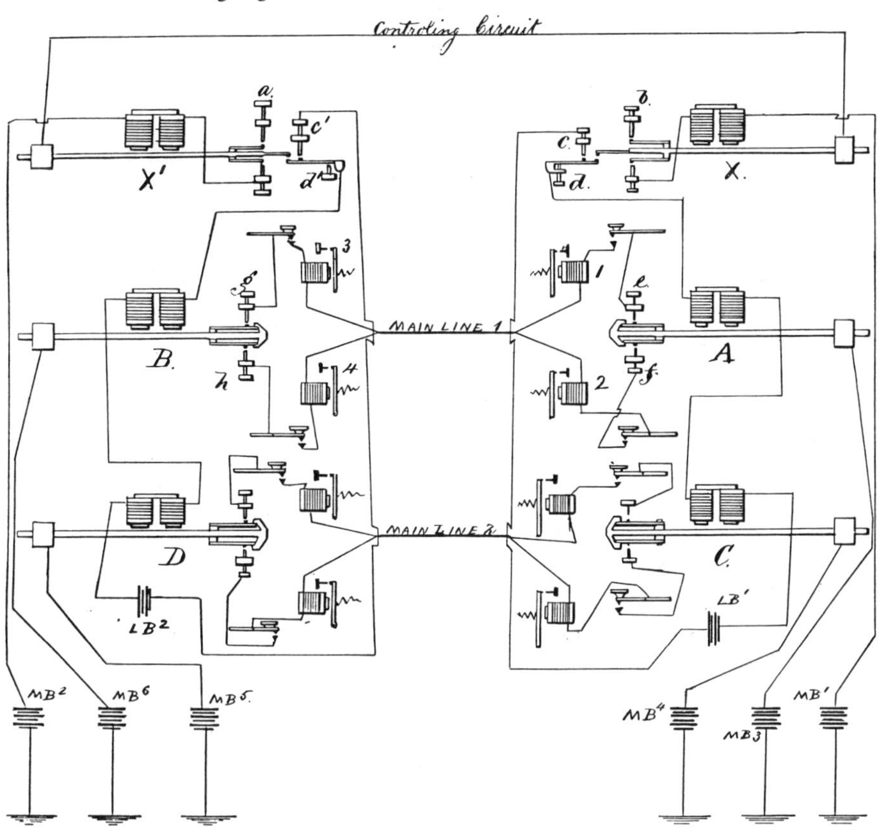


X and X′ are the primary controlling reeds, placed at the two ends of the controlling circuit and the reeds vibrate continuously in the manner already described. To the points a and b. of X and X′ may be attached the branch lines to the batteries MB$^1$ and MB.$^2$ in which branches there may be inserted a key and relay with sounder attached and the same may be worked with reversed currents or the branch or transfer circuit thus obtained may be duplexed in the manner already described. The reeds are both provided with extension upon their extreme end, opposite these extensions are springs d. and d′. whose platina tips are opposite the platina tipped contact screws c and c′. and at every vibration of the reeds the springs d and d′. are separated from the points c′ and c. thereby breaking the local circuit simultaneously at both ends of the line. In this local circuit at both terminals are several reeds, (two at each end are only shown) and local batteries $LB^1$ and $LB^2$. It follows that these reeds at both ends will vibrate simultaneously together and the bars of each being connected together by a line wire the wire will be transferred from one set of instruments to another and independent set of instruments as many times per second as the reeds vibrate at always at the same moment.

A and B are one set of local reeds to which the main line No 1. is connected. e. and g. are the contact points for one branch, f and h. are the contact points for the other branch; in these branches are the keys and relays 1. 2. 3. and 4.

The arrangement of C and D are precisely the same, thus each wire is made by the acoustic transfer into two[e] wires each of which is sas perfect for signalling purposes as the single wire in every respect. It is obvious that the primary vibrator may be made to close and open a local circuit in which any number of local vibrations might be included. For instance between New[b] York and Washington where one company has 15 wires one of them would be set apart to work the primary reed and by the points a. and b. signalling could go on just the same as if the vibrators were not worked upon the same wire. At the same time the remaining 14 wires could each be provided with a local vibrator all placed in one local circuit which was opened and closed by the reed of the primary vibrator, and these local vibrators would split, so to speak, by transfer each of the 14. wires into two wires, each making 29 complete wires out of 15. If it was desired to work Philadelphia and Baltimore, extra primary vibrators could be inserted upon the controlling wire at those places and these made to actuate the 14. local vibrators and the wires transferred simultaneously at all points from one set of instruments to the other in a similar manner to that shown in figure 10.

I have shown two main batteries $M.B.^3$ and $MB.^4$ at X. but in practice the line No. 2. could be attached to the battery $MB.^3$, and $MB^4$— dispensed with

It is obvious that the branch or transfer circuits 1. 2. 3. 4. of A. and B. as well as those of C and D could be worked by reversed currents or duplexed either by the use of Wheatstone balance or differential relays. It is not even necessary to employ a separate local vibrator for each line as one very powerful local reed or vibrator might be used and all the contact points attached to it by being properly insulated.

The local vibrators A. B. C[b] and D may be replaced with the vibrator shown in figure 14. whereby the wire is split by transfer into 4 circuits instead of two.[27]

Fig. 14.

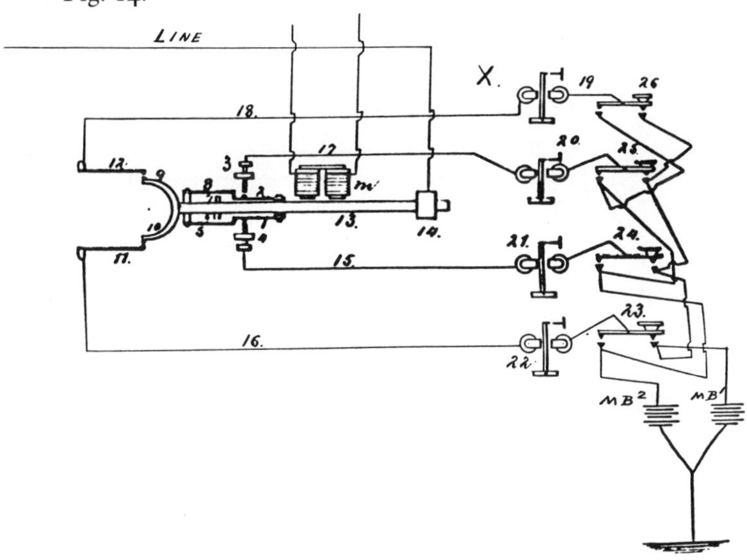


X is the local vibrating reed its magnet m. being included in a local circuit opened and closed by the contact points of the primary vibrator. 13 is the vibrating rod. 1 and 2. are springs with platina tips in contact with the platina points of the springs 5 and 8. The springs 1. and 2. are insulated from the reed 13. while 5 and 8. are in metallic contact with it. 6 and 7 are two limiting pins, 4 and 3. are two contact points to which are connected the branches 15 and 17. containing the polarized relays and keys 21 and 24. 20 and 25. When the reed is in a state of rest the springs 1 and 5 and 2 and 8. are in contact with each other and the tension on 1 and 2. being slightly greater than on 2[28] and 8. separate them from the limiting pins 6 and 7.

The points 4 and 3. do not touch, but when the reed moves to the right the spring 2. comes in contact with 3. thus throwing the line to the branch 17 until the spring 8 touches the limiting pin 7. when connection is broken, just as this takes place the extension 9 comes in contact with the spring 12. throwing the line upon the branch 18. on its return contact is made again between 3 and 2 and 8. when the reed goes to the left and that side being provided with exactly the same devices the main circuit is thrown alternately into 15 and 16 but at no time is the main circuit thrown onto two branches at the same time; of course the polarized relays in the branches receive very short waves of current but as[b] the reed makes about 300 vibrations in each second there is abundant current to give good signalling. $MB^1$ and $MB^2$ are two batteries one with its carbon pole towards the line while the other has its zinc pole towards the line.

23. 24. 25 and[c] 26 are the double point reversing keys the lever of each being in connection with the branches or main line and the front points with the zinc pole of the battery $MB^1$. while the back points are connected to the carbon pole of the battery M.B.[2] the manipulation of the key by the operator transmits positive and negative currents into the circuit.

It is not necessary that polarized relays should be used as ordinary relays and straight currents could be used and each branch could be duplexed but I prefer to use the polarized relays as they work upon a weaker current than the unpolarized relays. I do not wish to confine[b] myself to the method shown of reversing the current as a single battery could be employed and each key could reverse the same as regards its particular branch. I will also mention that switches could be inserted between the polarized relays and the keys so that the branch could be put to earth in the act of receiving. The arrangement with common relays and primary vibrator is shown in figure 15.

Fig. 15.

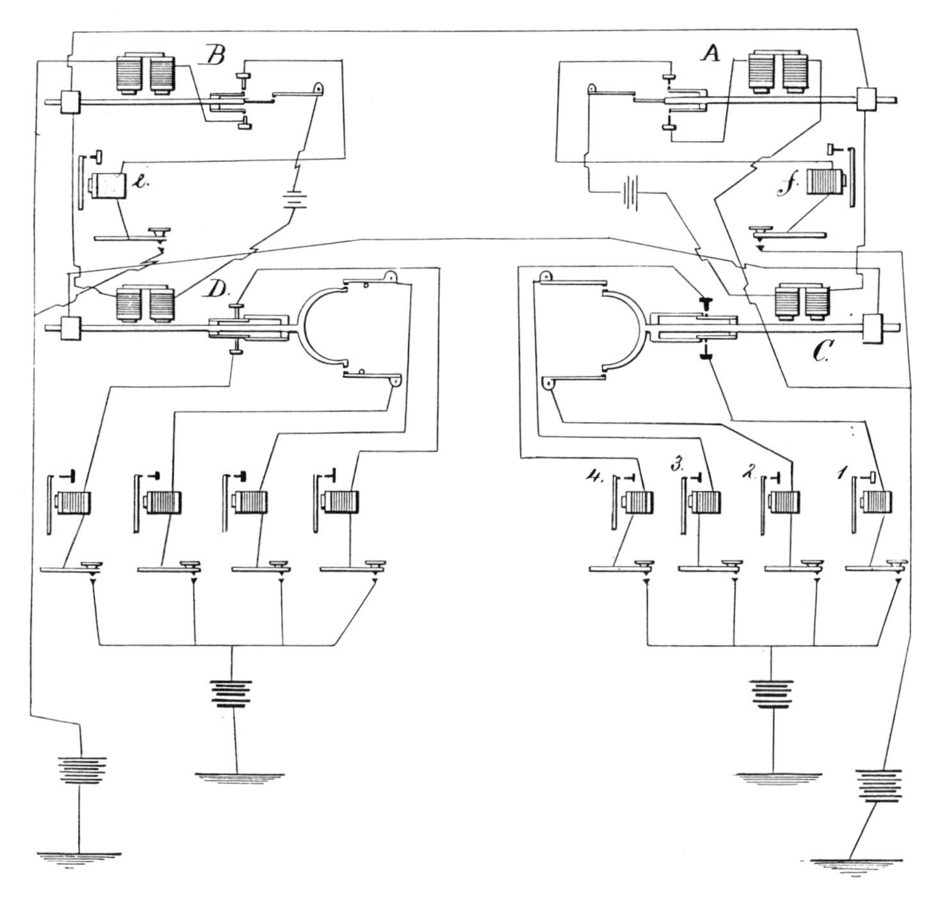


A and B are the primary vibrators connected and operated in the manner already described. f and e. are the two signalling relays the circuit of which is obtained by the transfer of the primary controlling circuit.

The magnets of C and D are in a local circuit opened and closed by the primary reeds.

1. 2. 3 and 4. at C are ordinary relays provided with keys the front anvils of which are all connected to a common battery, the operation is about the same as in figure 14.

In figure 16. is shown a modification of fig 15, whereby the contact points are dispensed with and an ivory surface inlaid with contact teeth is used,[29] it is also different as regards the number of transfer circuits, 6 being easily obtained by the use of this device.

Fig. 16

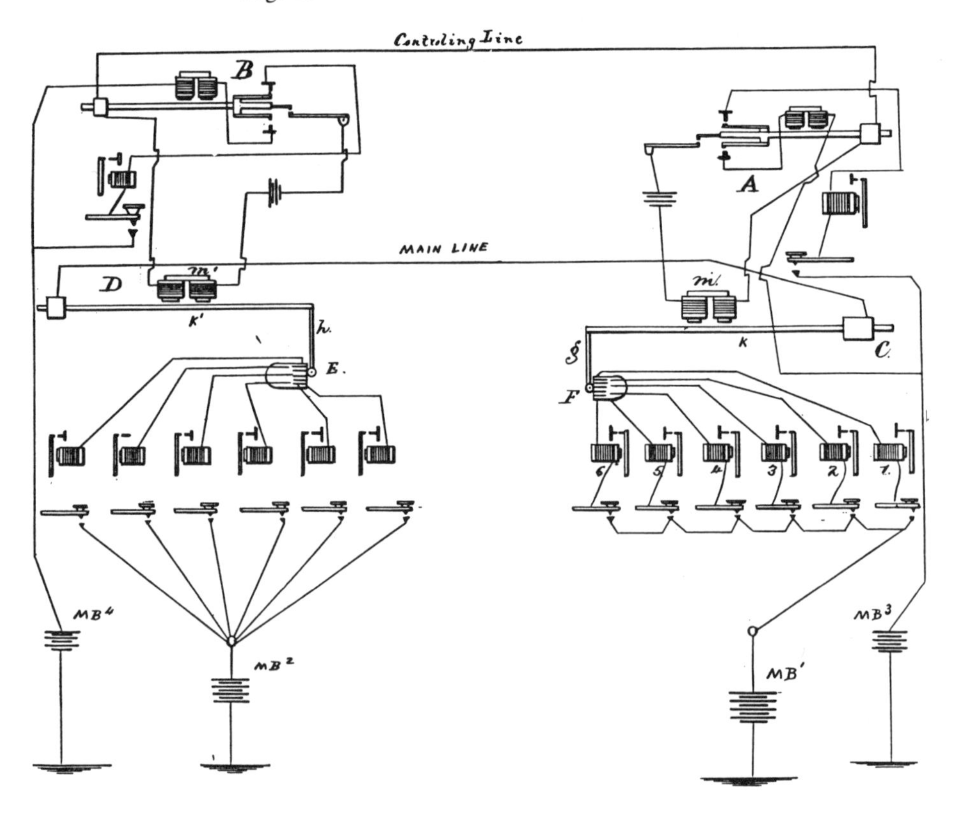


A and B. are the primary vibrators, the reeds of which open and close the local circuit containing the local reeds C and D.

K and K′ are the reed bars; upon the extreme end are two springs g and h. tipped with a platina roller running on a flat ivory[e] surface F and E inlaid with platina faced teeth to each of which is connected a wire to the several relays and keys at both ends of the line.

$MB^1$ and $MB^2$ are the main batteries which work the several derived circuits.

When the reeds C and D. are vibrating the rollers g. and h. pass back and forward over the teeth simultaneously at both ends, thus connecting the line alternately in several instruments at the same time and several hundred times in each second the number of circuits obtainable being only limited to the amplitude of vibration which it is practicable to give the reeds and the delicacy of the receiving instruments to weak circuits. I have shown ordinary relays in the branch circuits for the sake of simplicity in explaining the operation but in practice I prefer to use very high resistance and sensitive polarized relays worked by reversed currents.

Fig. 17.

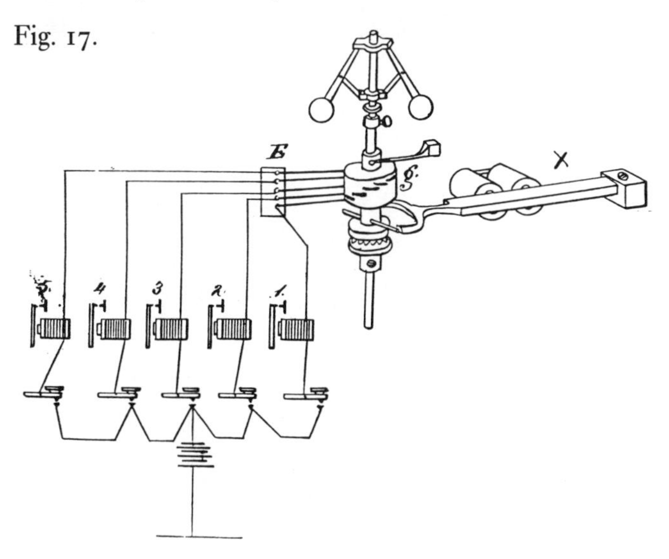


In figure 17. is shown a modification of 16. whereby the ivory is in the form of a break wheel rotated with great rapidity by an electric engine regulatable by a governor ~~or~~ and the ivory wheel is regulatable by the reed X.[30] E is a block to which is secured several contact wheels and springs in line with the inlaid~~e~~ metallic teeth of the break wheel g. each of which is connected to a relay key and sounder. The main line being connected to the break wheel; at each vibration of the reed X the line is alternately thrown through the relays 1. 2. 3. 4 & 5.

In practice I prefer to use a break wheel g. divided in several sections upon each of which there are several teeth one above the other and leaving off just as the next one commences, thus reducing the number of revolutions the engine must make.

This wheel would also be provided with an escape[b] wheel released by the prongs of the acoustic reed X. The break wheel and escape wheel is carried around by friction. I do not confine myself to the use of an electric engine as any other regulatable source of power may be used as a clockwork.

In fig. 18. is shown a method of using tuning forks which are superior for long circuits.[31]

The connection for one terminal station is only shown.

Fig. 18

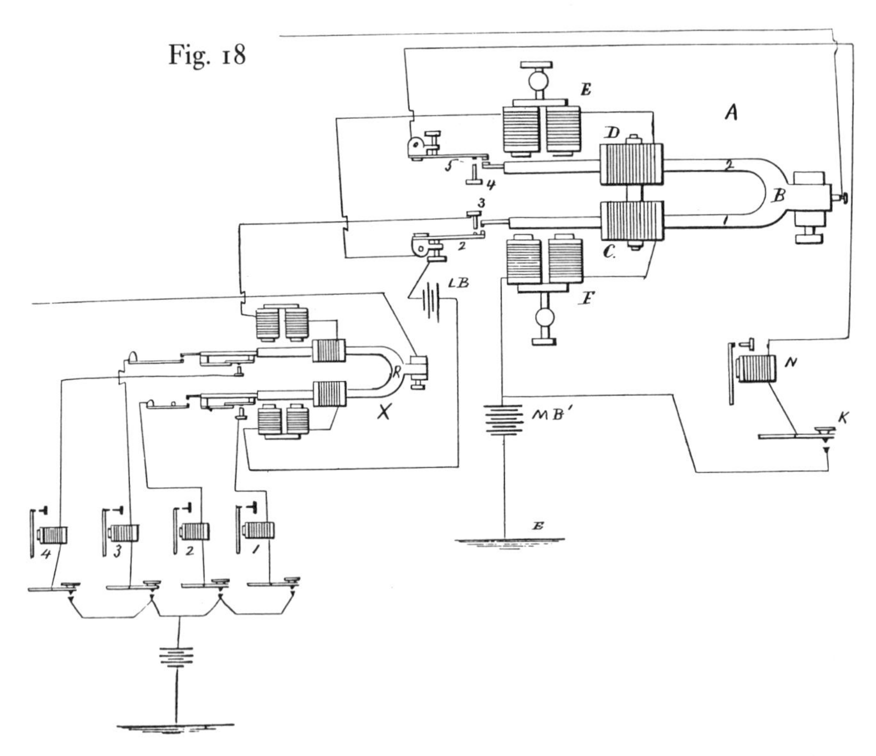


A. is the primary vibrator which consists of a tuning fork B. The two prongs of which pass through the helices C and D;[32] arranged on either side are the magnets E and F. and all are connected together in one branch running to the battery MB′. the other end being in contact with the spring 2. and obtains contact with the line by the contact of the prong of the fork with 2. the former being connected to the line. In this manner the forks at both ends are kept powerfully and continously vibrating. The spring 2 also serves to open and close the local circuit by contact with the point 3; in this local circuit is included a second tuning fork provided with 2 magnets and 1 pair of helices the same as A. which are energized by the local battery LB. This local set X vibrating in the same time as A. at the moment when the prong of $B^1$ leaves the spring .2. the point of $B^2$. comes in contact with the spring 5 thereby throwing or transferring the main circuit from E. C. D. F to the relay N key K to the battery MB′. The connections being similar to those already described.

X. is the local vibrator which is substantially the same as A. with the exception that the contact points upon either side of the reed shown in figure 14. are placed on the left hand side of each prong of the fork R. 1.2.3 and 4. are the branch circuits containing relay and keys.

I will now describe a method in which all the connections for the creation of signalling circuits by rapid transfer are substantially retained and at the same time each one of the circuits so obtained are again divided and retransferred, thus producing a great multiplicity of circuits. This arrangement is shown in figure 19.

I employ reeds in this case so as to make the description more easily but in practice prefer to use tuning forks similar to those shown in figure 18.

Fig. 19.

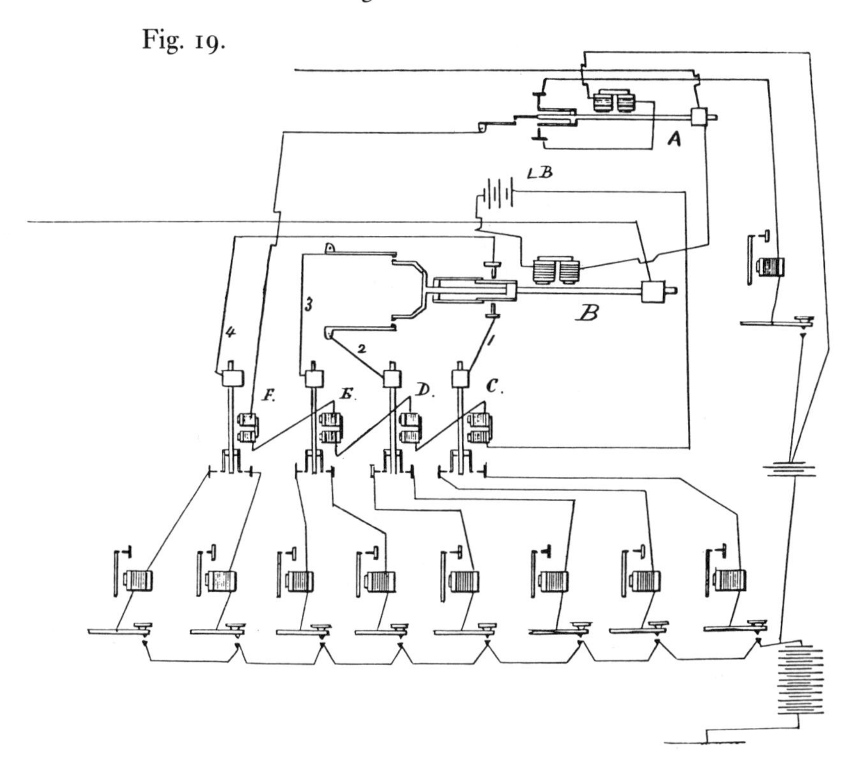


A is the primary vibrator connected in the usual manner and provided with a local point which serves to open and close the local circuit containing the local battery LB and reeds B. C. D. E and F.

The circuits 1. 2. 3. 4. obtained by transfer of the main circuit by the vibration of the reed of B, do not in this case run to the relays and keys to pass as have been previously described but are connected each to the reed of the local vibrators, which play between two contact points on either side of them, each of which is connected to a relay and key ~~to~~ and runs thence to ~~a~~battery, thus by the vibrations of B. we obtain 4 branches or signalling circuits, and these by the vibrations of C ~~and~~ D. E and F are again subdivided into two[b] each and they passing through the signalling instruments allow eight different operators to transmit as many different messages in either direction without interference with each other.

The subdivisions might be carried still further by the employment of another set of vibrators in the 8 branch circuits making 16 circuits but I do not think this would be practicable except with very powerful batteries and delicate signalling instruments.

Fig. 20.

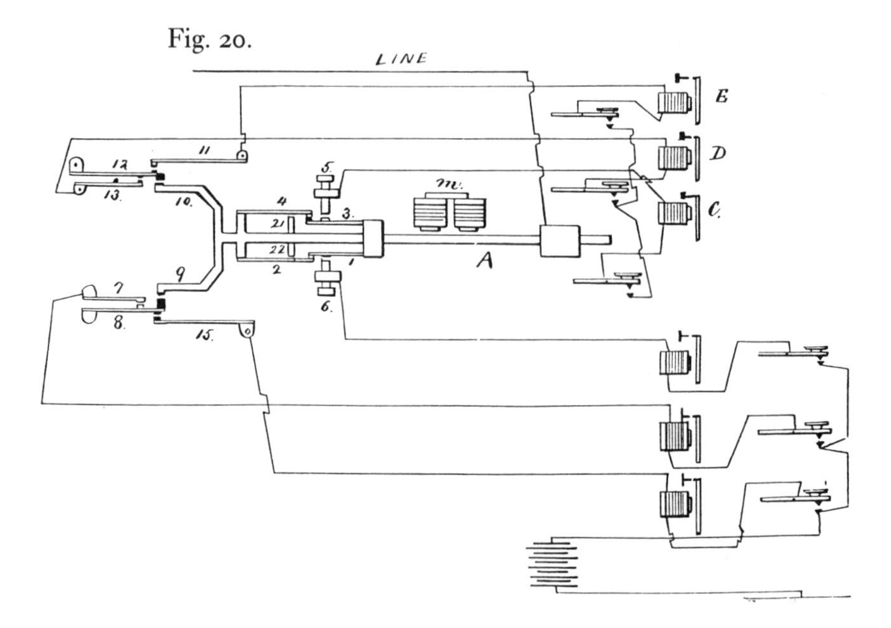


Inventor Thomas A. Edison
per Grosvenor P Lowrey[33] Atty.[f]

Witnesses[g]

I will now close by describing a local reed whereby 6 momentary contacts are obtained without any two occuring at the same moment of time. This is shown in figure 20. A. is the vibrating reed, provided with contact points and springs on either side which operates in a similar manner to those shown in figure 14.

7. 8. and 15 as well as 11. 12. and 13. are the same in principle as 1. 2. 3. 4. 5 and 6. When the reed is in a state of rest the line which is in contact with the reed, A. does not touch either 5. 6. 8. 15. 13. 12. or 11. but when the reed moves to the right the spring 3. comes in contact with 5 and continues to make contact until the spring 4. comes in contact with its limiting pin 21. when contact is broken, as 3 is insulated from

everything, just at the point where the circuit is broken the prong 10 comes in contact with the spring 12. and continues to connect until the spring 13. is stopped by its limiting pin when contact is broken and 12. is brought in contact with 11. thus the main line circuit is alternately thrown through the relays C. D and E and at no time is it in contact with two, at the same time; on the passage of the reed to the left the same action takes place with the duplicated points on that side.

I shall probably claim.

The use for telegraphic purposes of two or more continuously vibrating acoustic reeds, tuning forks or other bodies following the law of the pendulum on a telegraphic circuit for the purposes set forth.

2nd The various methods shown of obtaining a number of signalling circuits by means of continuously vibrating acoustic instruments.

3rd The various methods shown of duplexing[b] the signalling circuits so obtained.

~~Signed by me this~~ ~~day of~~ ~~AD 1876~~. ~~Witnesses.~~

Signed by me this Thirty first day of May 1876.[34]

Thos. A. Edison

Witnesses[35] Geo. T. Pinckney[36] Geo. D Walker.[37]

DS, DNA, RG-241, Edison Caveat 79. Written by Pinckney; petition and oath omitted. [a]Place taken from oath; date taken from text, form altered. [b]Obscured overwritten letters. [c]"ing" interlined below. [d]"Fig 3. . . . fig 5." interlined above. [e]Interlined above. [f]Drawings appear on nine separate pages, the first signed at the bottom. [g]"Inventor . . . Witnesses" in an unknown hand except for Lowrey's signature.

1. See headnote above.

2. This caveat was in part a product of experimental work done over the previous month (Doc. 749; Vol. 10:19–33, Vol. 15:125–32, NS-76-002, all Lab. [*TAEM* 3:838–44, 4:436–39, 7:373–99]) and in part a theoretical exercise, as Edison incorporated devices and principles used in other work, from his simplest duplexes to his most recent acoustic telegraph designs. A copy of an undated draft of this caveat is in Cat. 30,103:47, PS (*TAEM* 8:805).

3. By "acoustic reeds" Edison meant bars of steel, clamped at or near one end, that would vibrate regularly; he also called them "acoustic vibrators" (e.g. Doc. 749). The phrase "law of the pendulum" had no single clear meaning in scientific or technical literature (Atkinson 1890, 69; Avery 1885, 71–73; Peck 1866, 60; Prescott 1879, 235–39). Edison primarily meant that small-scale, simple vibrations of any kind were practically isochronous (see Docs. 664, 708, and 715; cf. Doc. 132). He used the term "law" at least as often to refer to properties and phenomena as to mathematical relationships among variables (e.g., Docs. 40, 46, 258, 409, 410 n. 2, 666, 677, and 678).

4. Edison had been dealing with problems of "static" induction in various modes of telegraphy for years; see, for example, Docs. 283, 317, and 426.

5. On "vibrators" see Doc. 658. Edison had known of them since at least 1873 (see Doc. 327) and possibly since 1868 (see Doc. 36).

6. Should be "controls"; see draft, Cat. 30,103:52, PS (*TAEM* 8:810).

7. Cf. Vol. 10:27, Lab. (*TAEM* 3:838).

8. The diagram does not match the description; wire **10** is misdrawn and should go from **R** to **h**, and the line from **MB**[1] should go not to the central pivot of **K** but to the contact point under its right end, matching wire **10′** at the other end of the line.

9. Cf. Vol. 10:32, Lab. (*TAEM* 3:843). On the duplex principle in general and the differential method in particular see *TAEB* 1:31–32.

10. In this and following diagrams the armature levers of relays like **R** and **R′** act as switches, opening and closing local circuits (not shown) that contain sounders or other receiving instruments.

11. Should be "thought"; see draft, Cat. 30,103:61, PS (*TAEM* 8:819).

12. Cf. Vol. 10:32, Lab. (*TAEM* 3:843). On the bridge method of duplexing a telegraph line see Doc. 285 n. 17.

13. Should be "controllers"; see draft, Cat. 30,103:62, Pat. (*TAEM* 8:820).

14. Cf. Vol. 10:26, 28, Lab. (*TAEM* 3:838–39). "Double current" here refers to a signaling current that reverses polarity, sent by a key that switches the battery connections. On polarized relays, which respond to such reversals, see Doc. 12 n. 1.

15. This should read "coils" to match the drawing.

16. Cf. Vol. 10:20, Lab. (*TAEM* 3:835).

17. This is explained at length around fig. 2 in Doc. 534.

18. Although figure 7 includes instruments using both current reversals and changes in current strength, as did Edison's major quadruplex system (Docs. 348, 472, 494 n. 2, and 515 n. 2), this circuit has two duplexes trading the line back and forth without ever using the line simultaneously for three or four signals.

19. In this diagram a wire is missing on the left side of the circuit. It should run from the right side of the relay magnet coils above the acoustic reed (top center) straight down to join the angle above the resistance coil and condenser that correspond to **3** and **C** on the right side of the circuit. Also, a battery has been left out of the local circuit for the signaling key at bottom right.

20. Unlike the previous circuits, the reeds here do not carry any part of the circuit that regulates their own motion.

21. Cf. Vol. 10:30, Lab. (*TAEM* 3:841).

22. The diagram only shows two signaling circuits and requires two line wires. If each of the signaling circuits were quadruplexed, line **20** would serve as an octruplex and the whole would offer eight signals handled by two wires, matching Western Union's regular quadruplex. However, there is no way "already shown" in this caveat to quadruplex a single signaling circuit.

23. Cf. Vol. 10:29, Lab. (*TAEM* 3:840).

24. Cf. Vol. 10:30, Lab. (*TAEM* 3:841).

25. That is, connected as in figures 2 or 8, assuming also that the main reeds were connected as in those figures instead of as shown.

26. The drawing lacks wires connecting the coils of relays 1 and 2 to the contacts of the appropriate keys.

27. Cf. Vol. 10:31, Lab. (*TAEM* 3:842).

28. Should be "5."

29. Cf. Docs. 469 n. 3, 708 (fig. 3), and 715 (fig. 8).

30. Cf. Vol. 10:33, Lab. (*TAEM* 3:844); see also Docs. 689, 708 (fig. 7), and 709 (fig. 5).

31. Edison's basis for this assertion is unknown.

32. These helices might serve to magnetize the prongs, thus strengthening the vibrations imparted by the magnets E and F. See Doc. 715 (figs. 10, 11).

33. Grosvenor Lowrey had become Edison's attorney for Western Union–related patent matters in December 1875 (see Doc. 695). He was the company's general counsel, and his law firm—Porter, Lowrey, Soren, and Stone—served as its legal department.

34. This caveat was not filed until 8 July. Edison executed patent applications based in whole or part on these designs on 16 and 26 August and 30 October 1876 (U.S. Pats. 185,507, 200,993, 235,142, and 200,032).

35. The identities of the witnesses indicate that Lemuel Serrell's staff prepared this document and sent it to Porter, Lowrey, Soren, and Stone; see *TAEB* 2:710 n. 1.

36. George Pinckney witnessed and transcribed documents in Serrell's office.

37. George Walker witnessed and transcribed documents in Serrell's office.

## -755-

*To Elisha Andrews*

Menlo Park June 21 1876

Dr Sir,

I have received from General Lefferts a letter with additions ~~a~~by you,[1] refusing to pay me anything for~~e~~ experimenting on Autographic I admit that the Contract was for ~~six~~ 6 months & that it was abandoned at the expiration of that time & I agreed to do my own experimenting thereafter But the money I ask for was expended (by my books) before that Contract terminated.[2]

You do not say that you agree with General Lefferts Letter I therefore write you direct.[3] Yours

T. A. Edison

ALS (copy?), NjWOE, DF (*TAEM* 13:1136).

1. Not found.

2. See Doc. 92.

3. It is not known whether Edison sent this letter. No reply to it exists. A rough draft is in 76-014, DF (*TAEM* 13:1135).

–756–

*From William Wastell*

Port Huron, Mich. June 29 1876[a]

T. A. Edison Esqr

I am sorry to inform you of the condition of St RR[1] it is worse now than ever. I have done my best have submitted to every thing imposed upon me, have sacrificed my stock paid my money but at last the whole thing is a miserable failure— the consolidation is as far off as ever— Mr Stewt[2] stands firm against all. I am sure we are largely increasing our debt— I paid one claim to the Sheriff 129.00 cost 49.00 and today he has another and will make a levy this afternoon and I see no way of escape— if you Sir had confidence in me and WPE[3] would have listened to me we could have saved Thousands of dollars— what do now is the question if I can in any way assist you & W.PE I am at your service—and only economy honesty and thoughtful business consideration can save us from loosing all we have in this Gratiot St RR—our expenses are simply ridiculous—it cost two dollars to earn one

I did hope to see a change but all the change is for the worse

I think you will have to come out here to do any thing with the road and something must be[b] done to economize expense— Yours truthfully

W. Wastell[4]

ALS, NjWOE, DF (*TAEM* 13:1063). Letterhead of William Wastell. [a]"Port Huron, Mich." and "187" preprinted. [b]Covered by attached tab.

1. The Port Huron and Gratiot Street Railway Co. See Docs. 175, 176, 530, and 651.

2. William Stewart, a major hardware dealer in Port Huron, owned 35 shares in the Port Huron and Gratiot Street Railway Co. See Doc. 175.

3. Edison's brother, William Pitt Edison, managed the street railway.

4. William Wastell, a Port Huron druggist, was one of the original organizers of the Port Huron and Gratiot Street Railway Co. Jenks n.d., 2, 4, 7; *History of St. Clair County* 1883, 600.

–757–

*Centennial Exhibition Award Report: Edison's American Automatic Telegraph*

Philadelphia, June 30, 1876[a]

REPORT ON AWARDS.[1b]

Group No.[c] XXV[2]

Catalogue No.[c] In Telegraph Department (Public Comfort)[3]

Product,[c] Edison's[d] American Automatic Telegraph

Name and Address of Exhibitor,[c] Atlantic and Pacific Telegraph Company

The undersigned, having examined the product herein described, respectfully recommends the same to the United States Centennial Commission for Award, for the following reasons, viz.:[c]

It gives on [----][e] land telegraph lines of all lengths insulated on poles in the air, speeds of practical working which are from two-fold to ten-fold the speeds attained by the best of the other[f] systems hitherto in use in America or any other part of the world.[4] I have myself, in the General Telegraph Office of the Centennial Exhibition, Philadelphia[g] witnessed the receiving ~~of 1015 words in~~ in 57 seconds of 1015 words from New York, and I retain for verification[h] the slip on which the signals were received, and its transcription in ordinary ~~char~~ writing by the receiving clerk. I am informed that the speeds actually obtained for satisfactory ~~practical~~ practical working through different lengths of telegraph line are as follows:—

| Length of line | Practical working speed |
|---|---|
| 200 miles [----][e] | 1000 words per minute |
| 300 ~~miles~~ " | 500 " " " |
| 400 " | 250 " " " |
| 600 " | 150 " " " |
| 1000 " | 100 " " " |

These ~~splendid~~ important results are obtained by the simplest and surest of apparatus, and with remarkable economy of personal skill and labour. The system is Bain's original automatic system in all its beautiful simplicity.[5] Where something of mechanical complication was needed for practical convenience, Mr Edison has not shrunk from it and he has given a perforating-machine[i] (all the details of which I have examined, and admired exceedingly) with a key for each letter, and one or two more keys for stops &c, by which any ~~one without skill~~ unskilled person can punch his message on the sending slip[j] with perfect sureness and accuracy at a slow speed, ~~and~~ moderately skilled young operators at from 25 to 30 words per minute, and well skilled first class operators at 60 words per minute There is no other mechanism in the whole system except the simplest of appliances, worked by hand, for pulling ~~the sending and receiving slips~~ the sending slip and the receiving slip,[k] through the single instruments used at the two ends whether for sending or receiving. ~~the sending slip and the receiving slip~~ Mr Edison's double spring with nickel rollers seems a perfectly satisfactory solution of the problem of making the sending contacts for Bain's system in a trustworthy manner which had been found very troublesome by many other inventors. In Mr Little's resuscitation of Bain's system,[6] from which Mr Edison took his departure, various chemical solutions had been tried for moistening the receiving slip, and iodide of potassium had alone been found capable of marking the signals at the high speeds aimed at but it ~~did not answer for practical~~ was not found convenient for practical use, as the marks faded away too rapidly, sometimes before ~~a single~~ the[d] message could be transcribed from the slip. On this important point I am favoured ~~by Mr E. H. Johnson with~~ the following statement by Mr E. H. Johnson[7] [~~who assisted——~~?][e] who assisted Mr Edison throughout all his experiments and trials for the practical development of the Bain Automatic telegraph and to whom I feel much indebted for very intelligent explanation of ~~details~~ all the peculiarities of the system ~~now~~ "which has resulted from these labours and is now reported on for award.

> "All the iron-solutions recorded in the electrical books were tried repeatedly but proved to be unequal to more than 100 to 150 words per minute. It became necessary then that a solution should be discovered having a sensitiveness near that of the iodide, without its fleeting character. Mr Edison gave his attention thenceforth to chemistry and chemical experiments, and during six weeks of study and labour, day and night, made many thousands of different formulas, resulting at the end of that time in the discovery that Ferrid-cyanide of Potassium[8] was almost as sensitive as the Iodide and that the record made upon it by an iron stilus was permanent and had other good properties." ~~Th~~

The electromagnetic shunt with soft iron core invented by Mr Edison, utilising Prof Henry's discovery of electromagnetic induction in a single circuit to produce a momentary reversal of the line current at the instant when the battery is thrown off and so cut off the chemical marks sharply at the proper instant is the electric secret of the great speed he has achieved.

The ~~main features~~ main[d] peculiarities of Mr Edison's automatic telegraph shortly stated in conclusion are ~~(1) The perforator, (2) the contact maker, (3) the electromagnetic shunt and (4) the Ferrid-cyanide of iron solution. It deserves award as a great~~ (1) The perforator, (2) The contact-maker ~~of the sending~~[f], ~~(3) the Ferrid-cyanide of iron solution, and (4)~~ (3) The electromagnetic shunt, and (4) The Ferrid-cyanide-of-iron solution. It deserves award as a ~~valuable addition to~~ very important step in land-telegraphy.

conclusion of report W.T.

APPROVAL OF GROUP JUDGES.[c]

| | |
|---|---|
| James C. Watson[9] | H. K. Oliver[10] |
| J. E. Hilgard[11] | F. A. P. Barnard.[12] |
| Ed. Favre Perret[13] | Joseph Henry[14] |
| J. Schiedmayer[15] | E. Levasseur[16] |
| | Chas. E. Emery[17] |

SIGNATURE OF THE JUDGE.[c]

William Thomson[18]

ADS, PPCiA, Centennial. Report form of United States Centennial Commission Bureau of Awards. [a]"Philadelphia," and "187" preprinted. [b]Paragraph preprinted and followed by centered horizontal rule; "Catalog No. ~~5~~ 4 [~~As Received~~?]" written in left margin in unknown hand. [c]Paragraph preprinted to this point. [d]Interlined above. [e]Canceled. [f]"of the other" interlined below. [g]"in . . . Philadelphia" interlined below. [h]"for verifiction" interlined above. [i]"perforating-" interlined above. [j]"his . . . slip" interlined above. [k]"the . . . slip," interlined above. [l]"~~of the sending~~" interlined above.

1. Items placed on exhibit could be entered into competition for medals. A small group of qualified individuals, half from the United States and half from elsewhere, judged the exhibits. Instead of having several grades of awards, winning entries received uniform medals accompanied by a report and diploma that described the individual award. Wilson 1876–78, 3:cxi.

2. Group XXV included "instruments of precision, research, experiment, and illustration, including telegraphy and music." U.S. Centennial Commission 1880, 319.

3. Centennial telegraph services were provided under exclusive contract with Atlantic and Pacific Telegraph Co. from the building of the Department of Comfort. Wilson 1876–78, 3:clii; Nuf Ced 1876, 4.

4. For transmission speeds see *TAEB* 1:150 n. 8.

5. On Alexander Bain's automatic telegraph see *TAEB* 1:65 n. 3.

6. On George Little and his automatic telegraph system see *TAEB* 1, especially pp. 186 n. 2, 302–4.

7. Johnson to Thomson, 27 June 1876, DF (*TAEM* 13:1137).

8. This was the principal ingredient in Edison's standard automatic-telegraph solution; see Doc. 410.

9. James Watson, astronomer, directed the observatory at the University of Michigan and taught physics there until 1879, when he moved to the University of Wisconsin. *DAB*, s.v. "Watson, James Craig."

10. Henry Oliver, a teacher and musician, was prominent in Massachusetts state government. *DAB*, s.v. "Oliver, Henry Kemble."

11. Julius Hilgard, a geodesist, had been connected with the U.S. Coast Survey since 1844 and later became its superintendent. In 1875 he served as president of the American Association for the Advancement of Science. *BDAS*, s.v. "Hilgard, Julius Erasmus."

12. Frederick Barnard was president of Columbia University. *DAB*, s.v. "Barnard, Frederick Augustus Porter."

13. Edouard Favre-Perret represented the Swiss federal council on the clock and watch industry. Favre-Perret 1877; Landes 1983, 319–20.

14. Joseph Henry, director of the Smithsonian Institution, was one of the elder statesmen of American science. His major research concerned electromagnetism. *DSB*, s.v. "Henry, Joseph."

15. Julius Schiedmayer, a son of the co-founder of the famous Stuttgart piano makers, Schiedmayer und Söhne, headed with his brother Paul the internationally successful company J. & P. Schiedmayer, making harmoniums and also pianos. He served as a judge for the international exhibitions at Paris in 1867 and Vienna in 1873, among others. *ADB*, s.v. "Schiedmayer, Julius."

16. The geographer Emile Levasseur, who later wrote *L'instruction primaire et professionnelle en France soux la Troisième République* (Paris: 1906), was a member of the Institut and professor of the Collège de France. Royal Society *Catalog*, 16:746–47; Cassino 1883, s.v. "Levasseur."

17. Charles Emery, engineer, was consultant to the U.S. Navy on steam engines. *DAB*, s.v. "Emery, Charles Edward."

18. Sir William Thomson (later Lord Kelvin), professor of natural philosophy at Glasgow University, was among the world's premier contemporary physicists and electrical engineers. *DSB*, s.v. "Thomson, William."

# -2- July–September 1876

Edison spent the summer of 1876 working hard in his new laboratory. His principal concerns were his acoustic transfer telegraph system and experiments with his electromotograph. Spurred by Alexander Graham Bell's epochal demonstration at the Centennial Exhibition, Edison also undertook new telephone experiments. In addition to these main lines of work, he continued his investigation of etheric force, designed several precision instruments, and experimented with the electrical and chemical properties of carbonized paper. Charles Batchelor worked in New York City for most of the summer, overseeing the electric pen business, but Edison and James Adams were sometimes joined in the laboratory by Ezra Gilliland and Edward Johnson.

Acoustic transfer telegraphy occupied the laboratory staff for the first part of the summer. In July they conducted tests on telegraph lines looped to Philadelphia and back. As August began Edison designed an "octruplex" (eight-message) acoustic transfer system, executing three covering patent applications later in the month.[1] When octruplex instruments were ready in mid-September he tested them on the Philadelphia loop.[2]

At the end of June, Edward Johnson and Josiah Reiff witnessed Bell's telephone demonstration at the Centennial.[3] Stimulated by their reports, Edison, assisted by Gilliland and Adams, began more intensive experiments in this avenue of acoustic research.[4] Using different materials and configurations of diaphragms, different sizes and numbers of resonant tubes, and alternative arrangements of contact points, the laboratory staff transmitted vocal and musical sounds with varying success. According to Edison, James Adams conducted most of these experiments, using "any person who happened to be about to do talking upon instruments that [Adams] had constructed."[5]

At the end of July, Edison received a shipment of electrical instruments from England. He began using them immediately and soon was comparing them to instruments of his own design. Among Edison's designs were mirror galvanometers using his electromotograph. This marked the beginning of renewed experiments with the electromotograph, some of which were soon directed at the old problem of finding a repeater that was sufficiently fast and sensitive for the automatic telegraph system.[6] By the second week of August they were testing an electromotograph repeater on the Philadelphia loop. Edison also considered using the electromotograph in a system for high-speed retransmission of Morse messages that were to be recorded on a newly conceived perforating instrument.

The most expensive item in the English shipment was a set of resistance coils.[7] In an effort to make their own electrical resistances, the laboratory staff baked many different kinds of paper to produce strips of carbon. Edison also thought of using carbonized paper for such things as battery carbons and chemical crucibles and carried out extensive experiments toward that end.[8]

The electric pen became an international business when British businessmen John Breckon and Thomas Clare bought the rights to Edison's British patent and established the Electric Writing Company. Their agent, Frederic Ireland, was in the United States in July to consummate the deal with Edison, and on his return they established a London office to sell the pen.[9] By early August Batchelor began shipping pens. Ireland, who became general manager of the Electric Writing Company, also expressed interest in acting on Edison's behalf for the sale of electric pen patent rights on the Continent.

During the summer the laboratory had its first prominent visitors. In early July, Ernst Fleischl, a Viennese scientist interested in electricity, visited Menlo Park. Later that month Sir William Thomson, fresh from judging exhibits at the Centennial Exhibition, spent a day at the laboratory.

Edison had a steady income from his May contract with Gold and Stock and Western Union's support for his acoustic experiments. More important, he received over $6,500 from the July sale of the rights to his British electric pen patent.[10]

By contrast, Edison's involvement with his brother's street railway in Port Huron continued to be an annoyance, as other principals tried to buy him out or to get him to take sides in management disputes.

Perhaps the most important personal event of the summer for Edison was the death of his long-time friend and mentor, Marshall Lefferts, who died en route to the Philadelphia Exhibition. Edison never recorded his reaction to Lefferts's passing, but he undoubtedly felt his absence.[11]

1. U.S. Pats. 185,507, 200,993, and 235,142.

2. About this time Elisha Gray was in New York demonstrating his own octruplex system between New York and Philadelphia. Doc. 786 n. 4.

3. "Gray's Electro-Harmonic System," *Operator,* 1 July 1876, 7; Reiff's testimony, TI 1:279 (*TAEM* 11:116).

4. Edison later claimed that he had Gilliland and Adams begin making telephone experiments about May 1876. Doc. 772 n. 2.

5. The notebook in which Adams recorded these experiments was lost on his death in England. Edison's testimony, TI 1:37–38 (*TAEM* 11:39–40).

6. On the problem of automatic telegraph repeaters see *TAEB* 2:284 n. 4, 463 n. 1. For Edison's earlier experiments using an electromotograph repeater for automatic telegraphy see, for example, Docs. 573, 604, and 605.

7. The importance of resistance measurement to electrical technology and science is discussed in Hunt 1992.

8. See Doc. 829 n. 11.

9. See Doc. 925; and Serrell to TAE, 23 June 1876; TAE to Breckon, 29 June 1876; Ireland to TAE, 8 July and 15 Aug. 1876; all DF (*TAEM* 13:1032, 1035, 968, 969).

10. An undated memo lists the costs of the British, French, and Belgian electric pen patents (for which Serrell billed Edison on 28 June) and subtracts them (and other miscellaneous expenses) from $10,000. Edison, Batchelor, and Adams split the remaining $9,036 as per their agreement (Doc. 637), Edison getting $6,578, Batchelor $2,214, and Adams $244. DF (*TAEM* 13:952, 1034).

11. Edison purchased a portrait of Lefferts in October 1876. Cat. 1213:6, Accts. (*TAEM* 20:8).

**-758-**

*From Josiah Reiff*

New York, July 5th 1876[a]

Confidential[b]

Dear Edison.

The death of Lefferts was fearfully sudden.[1]

I heard it in Phila. on evening of 3rd.

I have written to Treasry. Dept to try & get Elliott instruments in free.[2]

Will have reply I hope Saturday.

Months & months ago you assured me contracts were not needed between parties who intended to keep faith— Hence you will bear me witness I have never troubled you much. You simply know that as long[c] as I had money, Newark[c] Shop[3] profited by it. When people invest the Capital we have spent, some return is expected.

You know I have not profited personally to extent of one penny in 5 Years. I have spent over $170,000 cash, without interest. You have often told me, you looked forward to the time (regardless of outcome of auto) when I should be your partner & sort of general agent with ⅓ interest—

I am now ready to devote myself[c] execlusively to it, on some fair understanding

You have been annoyed during last five years, but your brain has been tremendously developed. You have had wonderful experiences— You have lived & accumulated considerable property, besides larger interest in Patent yet unsold. Where am I?

Thos. A Edison Knows what is right He has promised to stand by me & by the parties in London.[4] Yrs J C Reiff

ALS, NjWOE, DF (*TAEM* 13:1160). Letterhead of J. C. Reiff. [a]"New York," and "187" preprinted. [b]"34" in unknown hand in top margin. [c]Obscured overwritten letters.

1. Marshall Lefferts died suddenly on 3 July while traveling with his Civil War army company on a train to the Centennial Exhibition in Philadelphia. "Obituary. Marshall Lefferts," *Telegr.* 12 (1876): 168.

2. Edison had ordered over $450 worth of instruments from Elliott Bros. in London, including three kinds of galvanometers, a set of resistance coils, and a condenser. Shipped to Edison at Reiff's address on 16 June, they arrived in New York on 28 June with $180 import duty due. The import duty was paid and the instruments picked up 27 July (Elliott Bros. "Catalogue of Electrical Test Instruments," *Catalogues, Electrical and Telegraph Instruments,* Library, NjWOE; bills and receipts, 76-002, DF [*TAEM* 13:863, 865, 870, 881]). At this time the import duty on "philosophical and scientific apparatus" was 40% ad valorem unless "imported in good faith, for the use of any society or institution incorporated or established for philosophical, educational, scientific, or literary purposes. . . ." (The wording is from the Tariff of 1870 [U.S., *Statutes at Large,* vol. 16, "An Act to reduce Internal Taxes, and for other Purposes," sec. 650]; the rate of 40% was established in the 1874 revision to that act [ibid., vol. 18, "Duties on Imports," sec. 1383].)

3. That is, the American Telegraph Works, Edison and Unger, and Edison and Murray (see, e.g., Doc. 436).

4. Smith, Fleming & Co. See Doc. 350.

## TELEPHONY

Edison had been investigating acoustic telegraphy intermittently since mid-1875, occasionally using the voice as a source of sound, but his serious work on the telephone began in early July 1876 after Edward Johnson reported seeing Alexander Graham Bell's invention at the Centennial Exhibition. During the next seven months the laboratory staff pursued telephone research as a sideline, with James Adams the principal experimenter. The research picked up in early 1877, as Edison was negotiating a new contract with Western Union and Charles Batchelor began easing out of the electric pen business and returned to laboratory work. By the middle of the year the telephone was the principal focus of work at Menlo Park, and it remained so—excepting sporadic attention to the phonograph—until the next spring. Edison filed nine telephone patent applications in Washington during 1877, eight of which would eventually issue.[1]

Edison's telephone work centered on the transmitter. In Bell's system, the speaker's voice produced vibrations in a metal diaphragm that was almost touching the core of a magnet. The diaphragm's motion created induced currents in the line wire, which was coiled around the magnet, and those currents recreated the vibrations of the voice at the telephone receiver. Edison believed such currents were too weak for practical work and approached the problem differently. His system had a battery-produced current on the line, and he sought arrangements where sounds would cause fluctuations in that current's strength. He was primarily looking for a dependable, stable way to cause maximum variation in the signal. In the course of his research he re-examined Bell's transmitter and other schemes—including make-and-break circuits—several times, but his main avenue of pursuit was variable-resistance (or variable-current) transmitters.

Edison's early transmitters changed the line current by having a vibrating diaphragm move a contact in and out of a liquid or shunt resistance coils or batteries in and out of the circuit. In October 1876 he first used carbon as a conductor in a transmitter, coating a hard rubber surface with plumbago and running the line current through the coating; as the diaphragm vibrated it moved a metallic surface that short-circuited more or less of the plumbago and so varied the current. By mid-February, when Batchelor rejoined the laboratory staff full-time, carbon was at the center of the transmitter research, and for the next year Edison and his crew struggled to find the best way to take advantage of its electrical properties, particularly its sensitivity to vibration. They also had to explore all the other aspects of the transmitter, such as the diaphragm's shape, size, fastening, thickness, and material; whether to use more than one diaphragm; the shape of the mouthpiece; the instrument's overall design; its electrical connections; and the means of manufacture. They also studied various receivers, in the process inventing one that employed the electromotograph and which they used to receive musical transmissions in public exhibitions. And finally the line circuit itself became an important part of the telephone, as Edison used induction coils to boost the transmitter's signal.

1. Edison wrote an account of his telephone research for George Prescott's *Speaking Telephone, Electric Light, and Other Recent Electrical Inventions* which focused on the development of the carbon transmitter. A partial manuscript is in 78-005, DF (*TAEM* 17:180–196). Prescott went on to describe a number of Edison's other telephone designs and later research. Prescott 1879, 218–34, 526–50.

**–759–**

*Technical Note: Telephony*

[Menlo Park,] July 6th 1876

Experiments with talking telegraph[1]

Tried a piece of felt saturated with water  salt & water and other chemicals[2]

tried on a long Brass tube Brass Diaphram with magnet in ~~front~~ and Iron armeture Pasted on to the Parchment[3]  get a good many words Plain such as How do you do

X (photographic transcript), NjWOE, TI 2, Edison's Exhibit 3-10 (*TAEM* 11:213). Written by James Adams.

1. Testifying in interference proceedings, Edison produced this sketch as evidence of a series of experiments begun in "February or March, 1876" that tackled the problem of "transmitting articulate

speech, using transmitters which varied . . . the resistance by various means in which water was employed or other chemical solutions" (TI 1:31–32 [*TAEM* 11:36–37]; see Doc. 736). Bell's lawyers used this sketch to argue that Edison had reached only a relatively rudimentary stage of telephonic technology by mid-1876 (TI 4:196–97, 200 [*TAEM* 11:802, 804]).

2. This is the transmitter. The felt is represented by the small dark rectangle at the right, held against a diaphragm at the end of the tube. The circuit (not shown) passes through the saturated felt and then through the receiver's electromagnet (cf. the fifth sketch in Doc. 765). Sound entering the tube from the left vibrates the diaphragm, intermittently compressing the felt and thus varying the current. A similar drawing exists from the previous day. Edison's testimony, TI 1:33; Edison's Exhibit 90-15, TI 2 (*TAEM* 11:37, 622).

3. This is the receiver (Edison's testimony, TI 1:32 [*TAEM* 11:37]). Edison had been using receivers much like this in acoustic telegraphy experiments since the fall and winter of 1875–76. Docs. 674, 675, 699; see also illustration, Jehl 1937–41, 1:75.

**–760–**

*From Josiah Reiff*

New York, July 10th 1876[a]

My dear Edison.

Treasry Deptmt advises me the law will not allow them to enter the Instruments free.

Be careful how you proceed with O[1] in any matter.

When I see you I can tell you of a remark he has made about you within past 4 days. I dont think you owe him anything.

A&P evidently intend to be in position if compelled to pay by law to say auto is worth very little to them. They seem to be depending for future on Duplex.[2]

When you next come in, I want you to sign & swear to Custom House paper whole cost will be about $725 currency. Yrs tr

J C Reiff

ALS, NjWOE, DF (*TAEM* 13:1163). Letterhead of J. C. Reiff. [a]"New York," and "187" preprinted.

1. Western Union president William Orton; see *TAEB* 1:237 n. 1.

2. Reiff may have been responding to the following editorial in the *Operator* of 1 June 1876 (p. 7):

> It seems that *The Automatic* is not a success. We have been informed that Mr. A. B. Chandler, secretary of A.&P. T. Co., was heard to remark that he could purchase eight sets of De Imfreville's quadruplex at the cost of four setts of the automatic, and that one set of the quadruplex could be repaired for ten or twelve dollars, while it would cost from one hundred and fifty to two hundred dollars per set to repair the automatic. If this be true, and the

> opinion is concurred in by the other officers of the company, it will virtually do away with the automatic system, and telegraphers themselves will not be sorry, as the automatic is not looked upon with much favor by the fraternity.

As part of Atlantic and Pacific's defense in the patent infringement suit brought by Edison, Reiff, and George Harrington, President Thomas Eckert later claimed that the company began removing automatic instruments from its lines in the spring of 1876, although according to General Superintendent David Bates the system continued to be used at some offices until the August 1877 pooling agreement with Western Union (Eckert's deposition, for Final Hearing on Part of Respondents, *Harrington v. A&P,* Legal Department Records, NjWOE; Bates's testimony before the Master, 1:25–26, Box 17A, *Harrington v. A&P*). On Atlantic and Pacific's use of the D'Infreville duplex see Reid 1879, 590.

## TALKING TELEGRAPH Doc. 761

In testimony, Edison described the drawing at the top of the following document as

> a transmitter consisting of a tube with a diaphragm to the center, of which is fastened a number of strings. These radiate out some distance from the face of the diaphragm and are connected to contact springs in front of contact points.
>
> The springs and points are all included in a closed circuit containing a battery and a telephone receiver; upon speaking into this tube the contact points were vibrated, and resistance was thrown in and out of the closed circuit at one or more points by the vibration of the springs produced by the voice acting upon the diaphragms and springs.
>
> This instrument was tried; the results were not satisfactory. [TI 1:34–35 (*TAEM* 11:38)]

Edison had drawn a similar sketch a week earlier.[1]

The three lower sketches apparently involve acoustic transfer telegraphy. There is another, more detailed version of the second one (the reed), labeled "Acoustic Transfer," in another notebook.[2] Edison made several other acoustic transfer drawings on this date.[3]

1. Vol. 10:41, Lab. (*TAEM* 3:850).
2. Vol. 10:42, Lab. (*TAEM* 3:850).
3. Vol. 15:70 and NS-76-002, both Lab. (*TAEM* 4:392, 7:355–59).

**–761–**

*Technical Note: Telephony and Multiple Telegraphy*[1]

[Menlo Park,] July 12 1876[a]

Talking Telgh

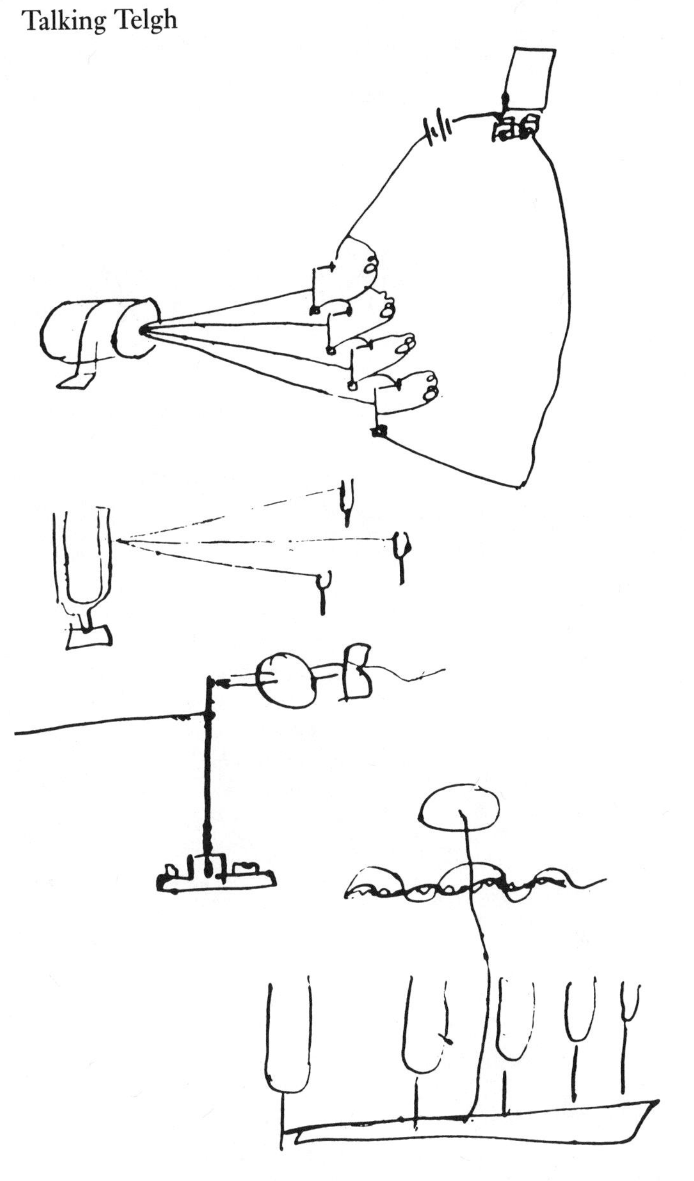

Edison

X (photographic transcript), NjWOE, TI 2, Edison's Exhibit "Talking Telegraph July 12, 1876."(*TAEM* 11:203). [a]Text, date, and signature inverted on original.

1. See headnote above.

**–762–**

*From Lemuel Serrell*

New York, July 1[5, 1876?][1a]

Dear Sir

I have a notice this morning on the Autographic Printing[2] stating that Edward Stewart of Fort Madison Iowa filed an application May 23, 1876 that is considered to interfere with it.

Your preliminary statement, giving in full the dates of the conception, drawing, experiments, and actual operative apparatus must be filed before 29th ins.[3]

Do not lose any time in sending me these particulars fully, so that we shall be in time—

Your failure to give particulars on ~~auto~~ acoustic telegraph case has perhaps proved fatal. I am trying to get time for it. On this case you had better also get out your statement, so that I can have it for filing if I succeed in re-opening the case[4] Yours truly,

Lemuel W. Serrell

ALS, NjWOE, DF (*TAEM* 13:1040). [a]Document damaged.

1. The notice of interference is dated 14 July. Pat. App. 189,857.

2. Edison's application for autographic printing (U.S. Pat. 180,857), filed on 13 March 1876.

3. On 26 July Serrell notified Edison that "priority of invention has been decided in your favor on concession of Stewart" (*TAEM* 13:1037). The interference was dissolved on 3 August (Pat. App. 180,857).

4. On 7 June, Serrell had notified Edison that an acoustic telegraph patent application (Case 118, executed 9 May 1876 and filed 16 May) was in interference with applications of Elisha Gray and Alexander Graham Bell (DF [*TAEM* 13:1031]). Edison's preliminary statement in that interference had been due on 1 July. Serrell succeeded in reopening the case, and after several rounds of amendments Edison received U.S. Patent 198,087 on 11 December 1877 (Pat. App. 198,087).

**–763–**

*From Josiah Reiff*

New York, July 17th 1876[a]

Dear Edison.

Sir Wm Thomson is in town— I wrote him this am asking him to visit Menlo if at all possible. & if absolutely impossible you & I would call on him here. If I get word from him that he will go to Menlo, I will telegraph you hour of our arrival etc.[1]

He sails Wednesday for Europe, & I expect to make an arrangement with him as an associate.[2] I find Varley has become a confirmed spiritualist & thereby seriously injured his influence.[3] I go to the country with O[4] this evening—

I did not receive a check from you today, but hope to do so

not later than tomorrow— We had occasion to retain a lawyer in Washington Saturday about Quadruplex[5] & [Robert] Russell loaned me the money &[b] if he gets more money he is now expecting thro an appropriation before Congress,[6] our affairs may take on a different shape at once— The old gent will put it up if he has it— That's proof of his faith in our success. Y

JCR

ALS, NjWOE, DF (*TAEM* 13:1164). Letterhead of J. C. Reiff. [a]"New York," and "187" preprinted. [b]Remainder written in left margin.

1. Reiff wrote to Thomson in early July to arrange for this visit. Although Thomson had planned an excursion on the Hudson River with his wife for the 17th or 18th, he must have visited Menlo Park on the 18th, as he departed for England on the 19th. In Edison's twentieth century reminiscences—in which he misremembered Thomson's visit as being in 1879—he recalled showing Thomson experimental arrangements for eight-circuit acoustic transfer telegraphy. Reiff to TAE, 7 July 1876, DF (*TAEM* 13:1162); Thomson to Henry Draper, 14 July 1876, HD; Doc. 803; *Telegr.* 12 (1876): 180; App. 1.B59, D340.

2. Nothing is known of such an arrangement.

3. British telegraph engineer Cromwell Varley was an expert in cable telegraphy and Thomson's business partner (along with another British electrical engineer, Fleeming Jenkin) in a firm controlling key patents for submarine telegraphy (*DNB,* s.v. "Varley, Cromwell Fleetwood"; Smith and Wise 1989, 701f). Spiritualism, a religious movement founded on a belief in communication with the deceased (usually via a medium), claimed several adherents in British technical and scientific circles, among them Varley, whose wife had become a medium (Oppenheim 1985, 9, 474 n. 64; Podmore 1897, 12, 37).

4. Unidentified; not likely to be William Orton.

5. A suit against Western Union, George Prescott, the Commissioner of Patents, and the Secretary of the Interior, had been filed in the names of Edison and George Harrington on 11 May 1876 in the U.S. Supreme Court. It sought to stop Western Union from using the quadruplex technology and the government from granting the related patent applications jointly to Edison and Prescott (Quad. TLC.3, 5 [*TAEM* 10:909, 948]). On Wednesday, 12 July, a motion had been filed by Western Union to dismiss the case against them. Although John Latrobe of Baltimore drew up the initial bill of complaint in the case (see *TAEB* 2:491 n. 3), the lawyer referred to here was probably R. D. Mussey, who apparently did most of the later work in the case for the plaintiffs.

6. Unidentified.

## –764–

*Technical Note: Etheric Force*

Menlo Park July 24 1876

Some further experements on the Etheric Force.[1]

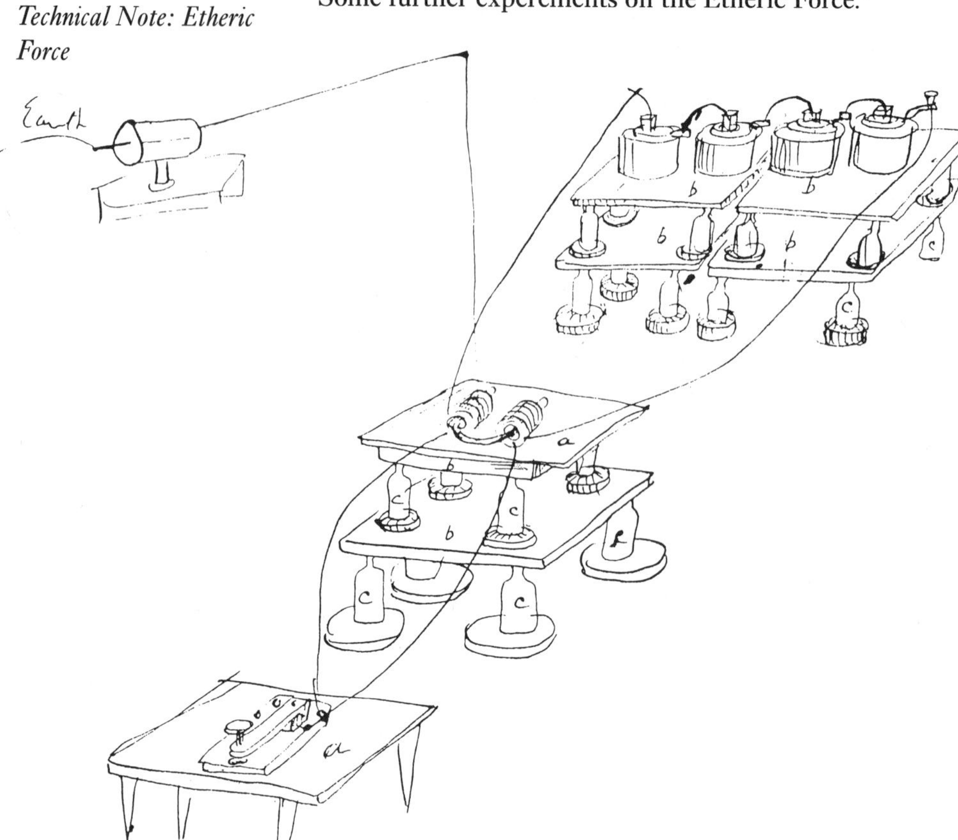


a a are hard rubber sheets= b b b are wooden trays= black walnut)—very dry=[a] c c c etc are white glass bottles wiped dry ½ inch diamter 6 inches high with round stopped leaving but little bearing for the wood to rest against= under each bottle is a peice of hard rubber polished on both sides 2½ inches diameter.=[2]

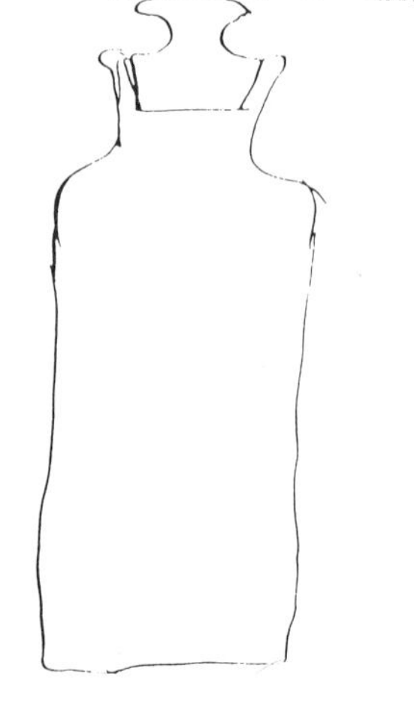

4 cells were used. 1½ size with new zincs & just amalgamted= ~~then~~ 2 cells on one ~~set~~ of the trays & 2 on the other=

The last row or bottom bottles of the battery stand stand on hard rubber slabs 3 by .6. polished & ½ inch thick=

The magnet was composed of 2 bradley spools[3] of naked wires, each spool[b] having [--][c] exactly 6 ohms resistances & were of the same diameter &[b] length=[4] Inserted in the coils was a horseshoe of brass— diameter of brass wire 5+/16 length 3½ inches width between prongs=

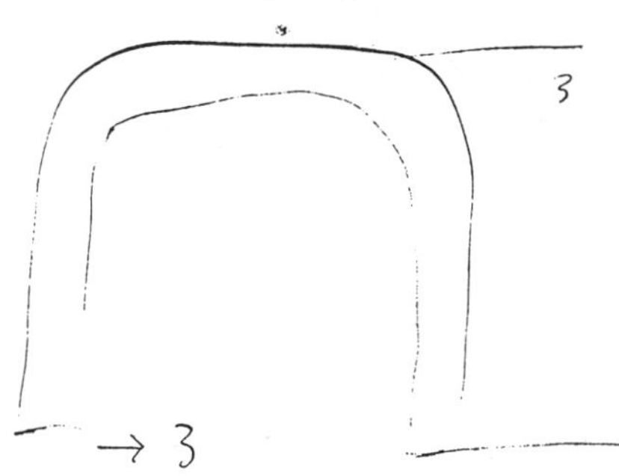


The prongs as they entered spools were covered with thick she~~e~~llac~~k~~ over which was tissue paper= the magnet ends were conducted to the key & battery by office wire= parafined cover= attached to the centre of the horseshoe was a wire also of office w~~e~~ire leading to the dark bo~~o~~x=[5] this wire run to a staple driven in the ceiling thence to another staple down to the box= This bo~~o~~x was the one used in previous expermnts & consists entirely of hard rubber polished into the ends of which are two screws provided with ~~b~~hard rubber heads, these screws run in brass[a] sleeves fixed in the head of the round rubber box on their end are the ~~lead~~ graphite pencil points pointed at the End=

Could get no spark at first ascertained that the graphite points did not meet properly= had them fixed, could not get a spark ~~for~~ attached a ~~one~~[b] condenser to one pole of the battery thus—[6]

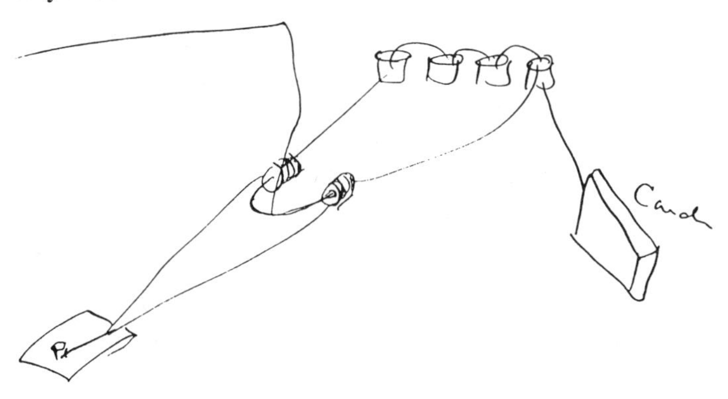

We then after few minutes trial got several sparks— detached condenser set it down on floor 5 feet from the apparatus on one side= got several sparks= took it away altogether still got several sparks= ~~all~~ then for ½ hour could ~~n~~get no~~t~~ne= put condenser within 5 feet got after long trial one spark= took condenser away but after ¾ of an hour failed to get a spark either with condenser on one side or away altogether shewing perhaps bad adjustment of the graphite points= in fact the Screws went too tight to get a fine adjustment= are now having screw eased= afterwards couldnt get anything either with condenser or without put condenser ~~at~~on~~e~~ one side at pole of batty at other end of spool & at key—
inserted key as follows.

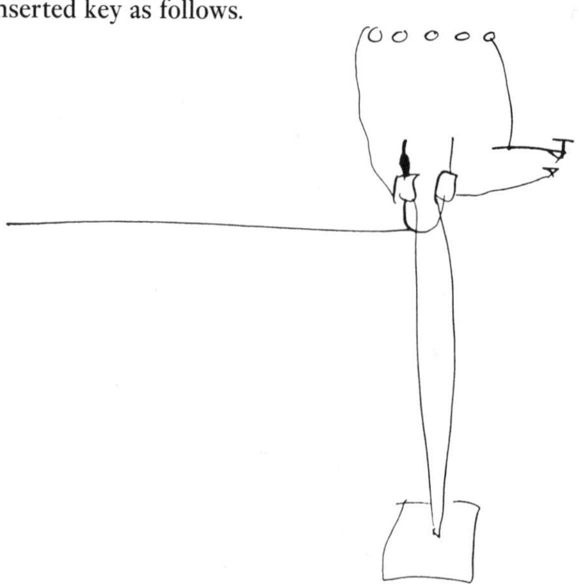

after 20 minutes careful adjustment got very brilliant spark but couldnt get it afterwards= after trying for over an ~~a~~hour added 4 more carbon 1½ cells to battery making 8 alltogether=

Replaced the spools with a G & Stock printer[7] spools with iron cores, & placing key by side of it thus got ~~s~~plenty sparks b

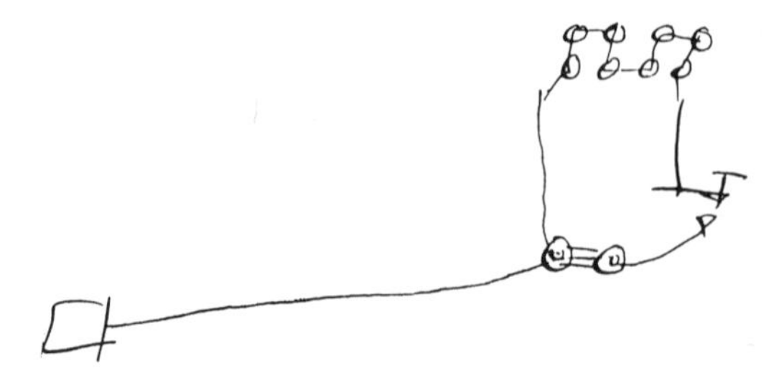

Zinc cylinder laying across cores; when key was connected thus got nothing after 5 minutes trial

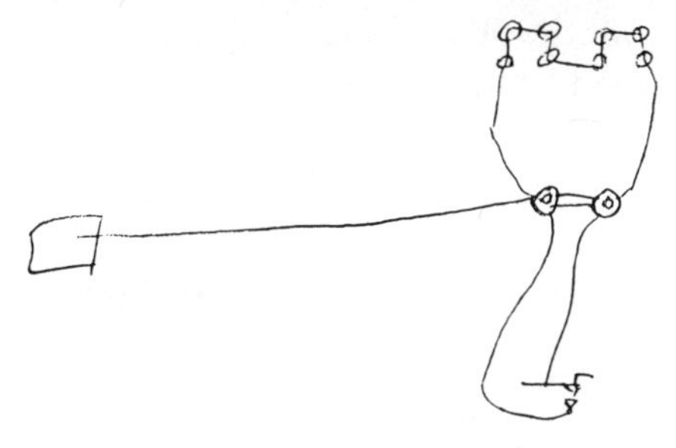

used the zinc cylinder laying on cores of electromagnet= ~~if~~We had the zinc thus—

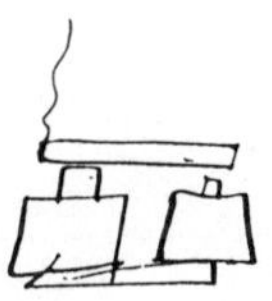

For fear that this threw it out of balance fixed the zinc thus

So the whole arrangement was thus=

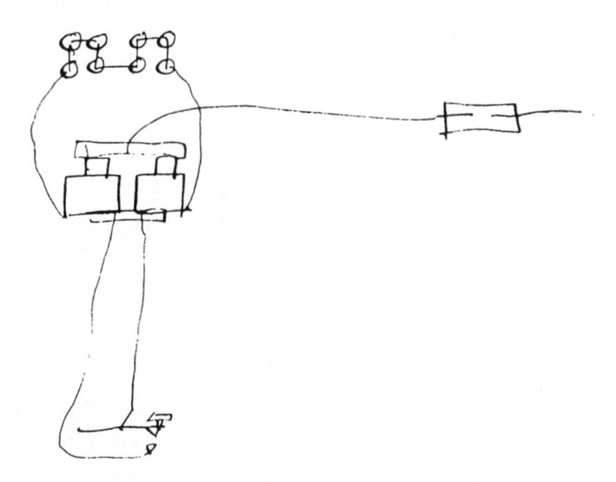

Replaced key, got nothing= put key back on side of magnet got it a few seconds, then replaced key in centre=& got it beautiful every time[d] appears to want a different adjustment= had my body divided so that equal portions should be

on a side  removed every object from vacinity of magnet that would give a greater amount of surface or metal on one side or the other=then I ~~e~~closed key by a glass rod 3 feet long still got brilliant spark=

So this expermnt would seem to indicate that the so called polarity expermnts of Huston & Thompson[8] w~~a~~ere incorrectly made—

In addition to this we replaced the 6 ohm spools thus

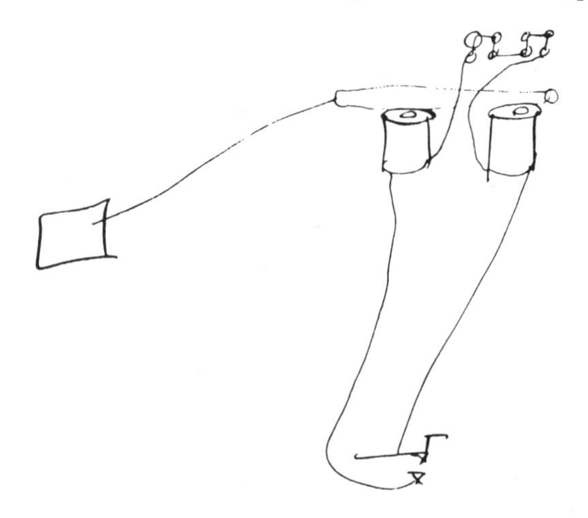

and got it first time. this apparently shews that ~~it is in~~. Thompson & Huston ~~w~~error was in the adjustment of the points= placing the key on one side allowing of a greater length of wire in one side at the moment of break & giving a greater spark, which was only reduced in strength by putting the key in the middle=

We then put the zinc cylinder ~~ins[id]e~~[c] ~~of~~ inside of one of the spools got spark good= then put a cylinder in each spool the zinc cylinders connected together with Copper wire  got it ok= we then replaced the first horseshoe brass pieces== 2 am have just tried another expmnt that apparently confirms Huston & .T & Nulifies our previous ones tonight  by putting one prong of brass Horseshoe[e] in one spool get spark put it in the other get spark but when both prongs were inserted in both spools it took a long time to obtain a spark shewing either a reduction in intensity or else that something must have thrown one[a] side slightly out of bal if H & Ts theory is correct  Tried it with zinc cylinders connected together with copper wire & spools placed wide apart, by careful adjustment sometimes got a spark but when one cylinder only was used got it without any trouble adjusting

So far I think that H & T are confirmed as perhaps one spool of ~~of~~ the G.&.S was stronger than the other or the zinc cylinder did not touch  this would destroy the balance if there is one & allow the sparks but yet ~~it is possible~~ according to T & H possition of the key etc preponderance of mass of matter is everything ~~this~~ is the cause this I do not find to be the case— but further experiment will determine this; perhaps better insulation of the cylinders from the coils is desirable=

One experiment[a] remains to be explained

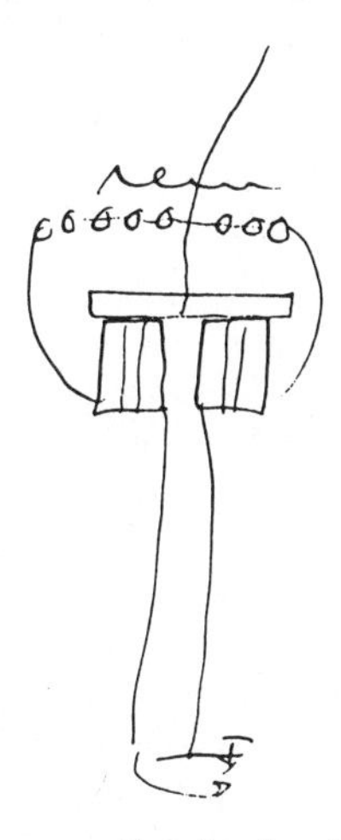

The spark was just as easily obtained by laying the cylinder on top of both spools as when put in the hollow of one

Note.

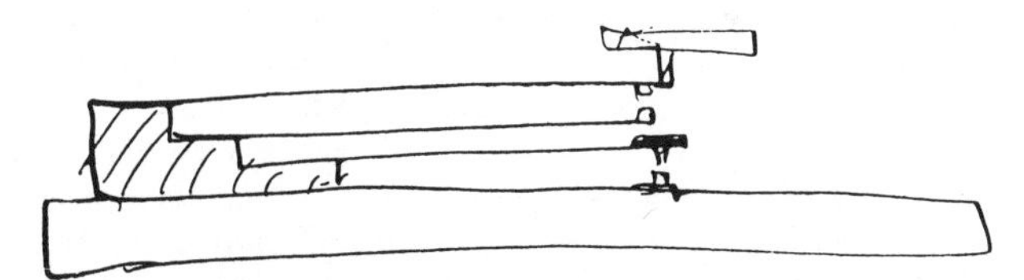

Its almost i[mpossib?]le[f] to close this key with[9] giving a spark in closing shewing there is a rebound[a] & break on closing=[g]

put hard rubber bushings in magnet where upon inserting zinc cylinders in each we got the spark every time with everything Central ~~o~~but it was at least ~~2~~ twice as weak as when only one cylinder in one spool= if the spools were laid flat on rubber sheet didnt get spark with both cylinders in perhaps on account bad adjustment of points.

put magnet [u]pright[f] again  both cylinders in  got the weak spark every time  added a condenser between key &

spool ie[h] tied the earth sheets to wire to increase surface of metal it did not ~~augqument~~[i] increase spark as it should according to H. &. T.[10]

X, NjWOE, Lab., NS-76-001, Supp. III (*TAEM* 162:510). Document multiply dated. [a]Obscured overwritten letters. [b]Interlined above. [c]Canceled. [d]Underlined twice. [e]"of Horseshoe" interlined above; "brass" interlined above that. [f]Paper damaged. [g]Followed by centered horizontal line. [h]Circled. [i]"ug" overwrote "q" before entire word overstruck.

1. The experiments Edison recorded here involve a detailed, critical examination of experiments and conclusions of the Philadelphia science teachers Edwin Houston and Elihu Thomson, who claimed to have proven that all of Edison's etheric force effects were due to instantaneously reversed induced electric currents. They had disparaged his work in print, leading Edison in February 1876 to challenge his critics in *Scientific American* to "back up their assertions by experiment, and give me an equal chance as a critic" (Doc. 726). Thomson and Houston's subsequent experimental report to the Franklin Institute was reprinted in May in the *Scientific American Supplement* (1 [1876]: 326). In June that journal also briefly reported the critical presentation by Silvanus Thompson to the Physical Society of London ("Etheric Force," p. 405). For Edison's original experiments on etheric force, see *TAEB* 2, chaps. 10–11.

2. Thomson and Houston had emphasized the need for extremely thorough electrical insulation and physical isolation of the experimental apparatus.

3. Leverett Bradley was known for his patented magnetic helices and electrical measuring devices. See *TAEB* 1:141.

4. The circuit and physical layout, particularly the electrical and material bilateral symmetry down to such details as the orientation of the key, were designed to correspond to the experiments of Thomson and Houston; the two sides of their arrangement had been designed to produce mutually canceling effects, thereby demonstrating electrical polarity in supposedly non-polar etheric force phenomena and confirming their claim that only induction was involved.

5. Edison called this box the etheroscope. See Doc. 670.

6. Thomson and Houston contended that any appreciable disturbance of the symmetry of this arrangement, from the touch of a finger on one side to the mere approach of something metal, would immediately result in the production of characteristic sparks in the etheroscope.

7. Probably Edison's universal stock printer (Doc. 195).

8. Although Thomson was apparently the primary experimenter, Houston was the senior figure and his name came first in the published report. Carlson 1991, 60–62.

9. Edison probably meant "without."

10. Beyond the etheroscope designs in Doc. 766, there are further etheric force experiments in a notebook entry of 5 November 1876 using apparatus that is a modification of that shown here (Cat. 1172:70, Lab. [*TAEM* 3:314]); see also Doc. 840.

**–765–**

*Technical Note: Multiple Telegraphy and Telephony*

[Menlo Park,] July 26 1876

It is probable that when we get the transfer telgh with several splits working that we can use one of the splits to work the local thus dispensing with the Controlling line=

Tried this last night and it worked well on wire between Menlo[a] Pk and Albany & return with only 40 cells small carbon to transmit on

Same at other end[1]

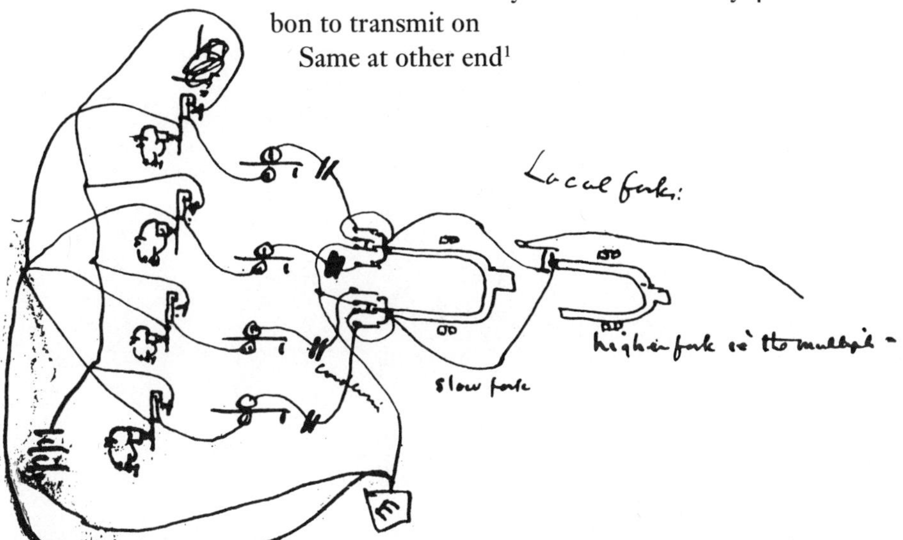


On closing the key the ground is disconnected and the battery connected this rushes over line to the distant station (which in practice will if receiving have its condenser cut out ie[b] in the particular split receiving) and this waves continue only until the Condenser is charged which takes place with tremendous rapidity this short wave which is perhaps only ~~a~~ the[c] length of a single vibration of the local forks proceeds over the wire & scarcely charges it but is suffcient to throw the tongue of the polarzed relay to one side. When the battery is disconnected & the split put to earth the condenser which is now fully charged—discharges[a] through the line in an opposite direction to the wave which passed into it & over t[he?][d] line & ths serves to throw the tongue of the polarzed relay over to the other side. The great advantage of using these short waves & a polazed relay consists in using a very weak current which said polarzed relays readily respond to and also a very short wave of current of the 2 polarties which are all that is ~~supli~~ required to make dots & dashes; hence the wire is not charged anywise equal to what it is when permanent batties are used at both ends and the effect of the escape is entirely prevented in this case by receiving on a ground. When the wire is thrown on one set of instruments the wave is transmitted and then the wire before it is thrown on another set is put to earth simultaneously at both ends so as to discharge ~~it of~~ as far as possible[e] the[a] static charge of the line caused by the passage of this signalling wave & thus prevent it from rushing to earth through the next set of instruments on which the line is thrown   instead of the condenser we tried using the induced current from an electromagnet   Thus[2]

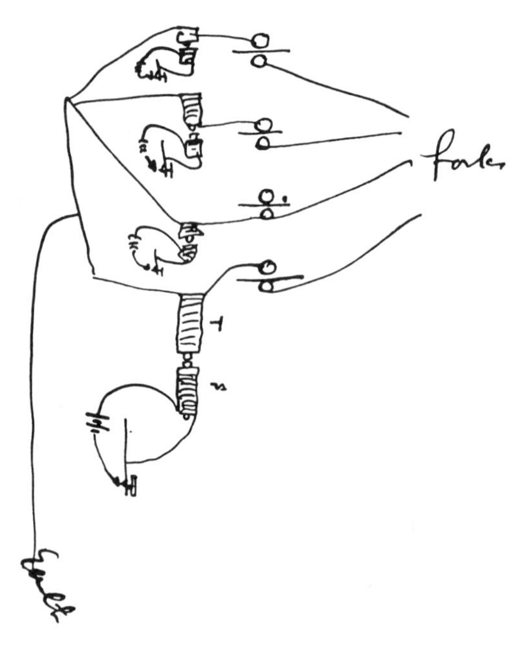

found that S must be of considerable resistance & that T should be quite long so as to lengthen the wave as it is exceedingly short & hardly sufficient to throw the polazied relay over to the other end. by nursing etc perhaps this would work satisfactory   it has one advantage that if way stations are used local battrs will be all that is required to work the induction magnet thus dispensing with main battery= We also tried in place of S. & .T a diffrential[a] magnet one set of wires in line the other in local with key   this was not so powerful but could be made so—

There is a different plan for transfer which we tried and which appears to work very well which is the cutting in and out of the main circuit two receiving instruments simultaneously at both ends of the circuit at the same time and then doing the same one after the other of several instruments. We found that it worked perhaps more satisfactory with common instruments[3] that by the transfer of the wire from one set of instruments to the other probably on account of the short circuit which closes on each instrument immediately after being thrown in the line acting as a shunt to allow of the circulation of the self induced current of the magnet itself which has a tendency to make the magnetism of the cores more continuous. The drawing is shewn on the next page; there is a difficulty with this which is the difficulty of discharging the wire after or between each transfer  we tried innumerable devices such as condensers magnets secondary batteries arranged in various ways with the instruments & contact devices to compensate for the static charge of the line but with small success= Drawing shews one end. (Conceived about month ago)[4]

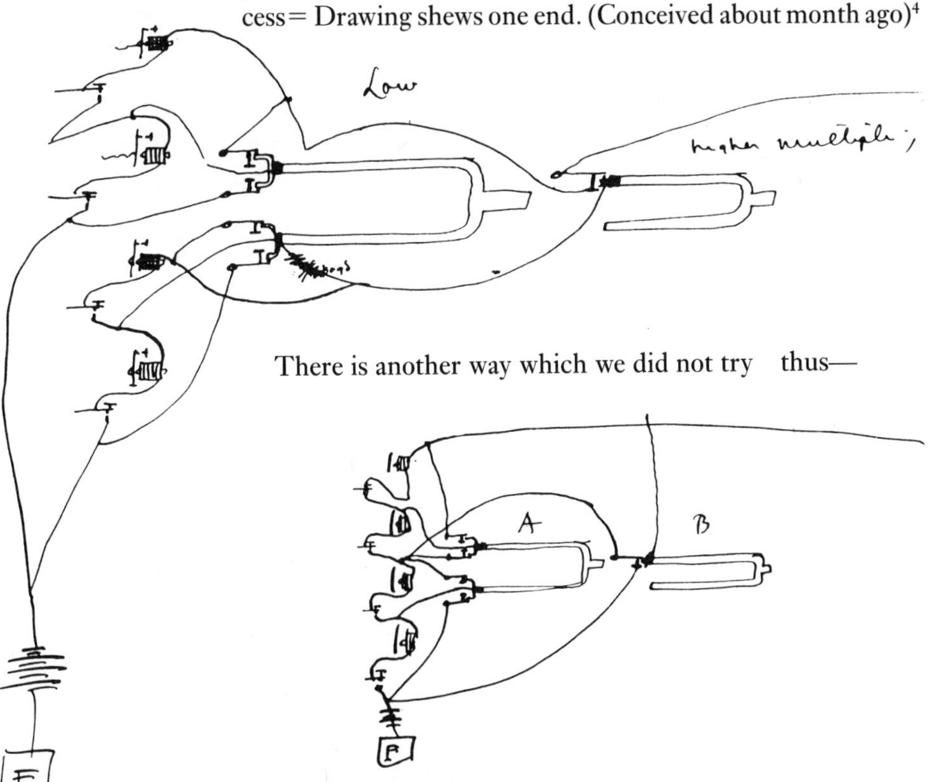


There is another way which we did not try  thus—

B short ckts 2 when on one side & 2 when[c] on the other side  A short ckts 1 on one side 1 on other

In place of reverse induction[c] currents of short duration, inverted battery currents could be used and special mechanical device used so that upon closing the key, it or a sounder operated by local & key could be made to connect the battery p to line for an instant & then ground  Then upon opening the key the ground could be removed and ~~a~~Z pole of the battery connected to the line for an instant followed by dead earth  the idea is illustated as below; of course there are many ways of ~~th~~doing this which any skilled[a] person in the business could accomplish=[5]

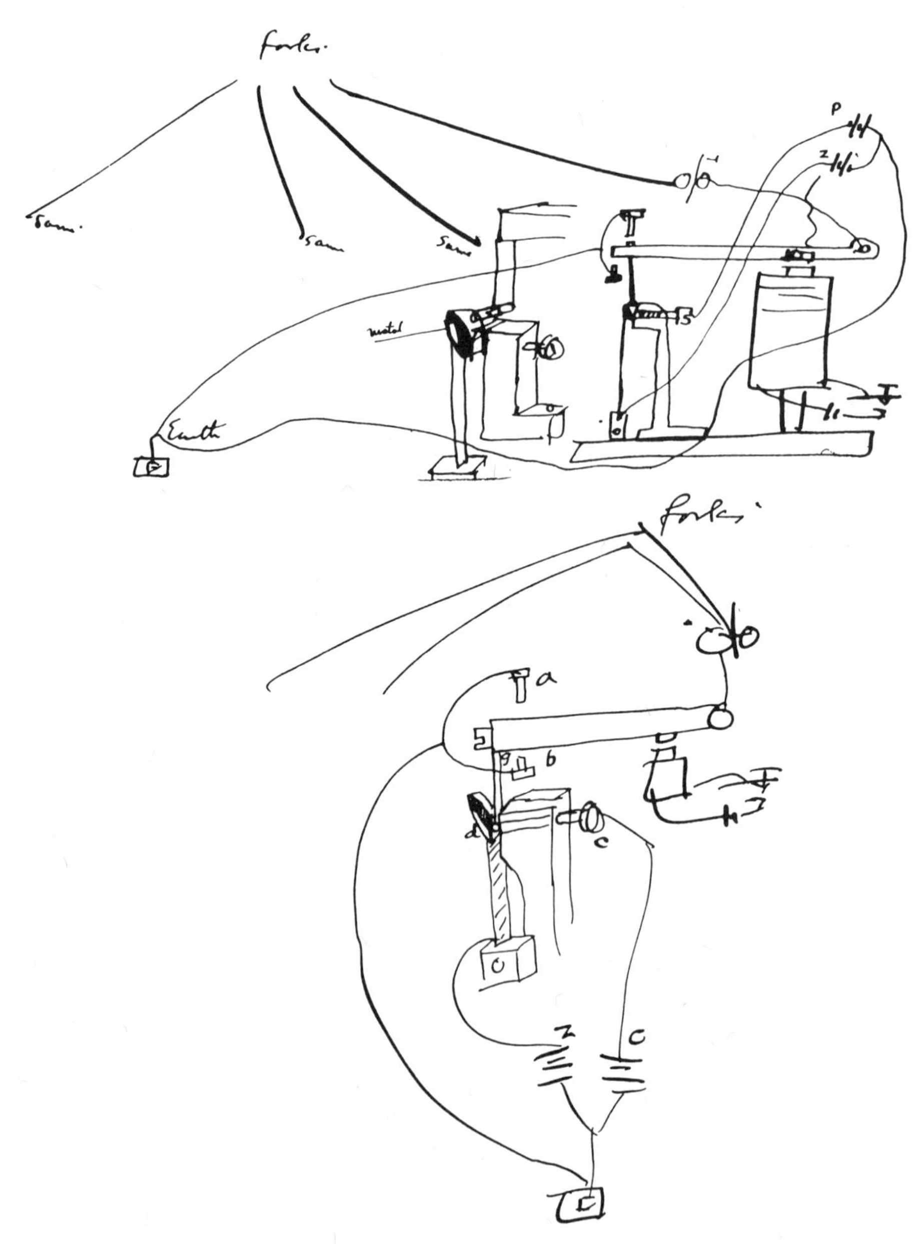


When key opens lever of sounder rests on a  puts split to Earth  in passing downward rubs on face of c inserted in ivory or other insulating substance= That sends copper current of short duration over wire. When lever of sounder touches b & grounds split again= When lever rises platina cylinder on end of spring g[6] passes on the metal d & connects Z pole of battery to line for instant when it rises over top of d when lever touches a & grounds again It may be necessary to shunt the polarized relays[7] with either a condenser or a secondary battery so as to hold the lever or tongue to either side while no current is being received over wire thus preventing Extraneous currents from Effecting the relays= or perhaps by the use of reverse batteries a strong wave could be sent to throw the lever over immy followed by a weak continuous wave to hold it & so on always keeping weak current on line to hold lever on one or the other sides.

Last week we tried numerous experiments with a speaking acoustic telegraph whereby the human voice could be transmitted over a wire & which I have neglected to record=

~~firs~~[8]

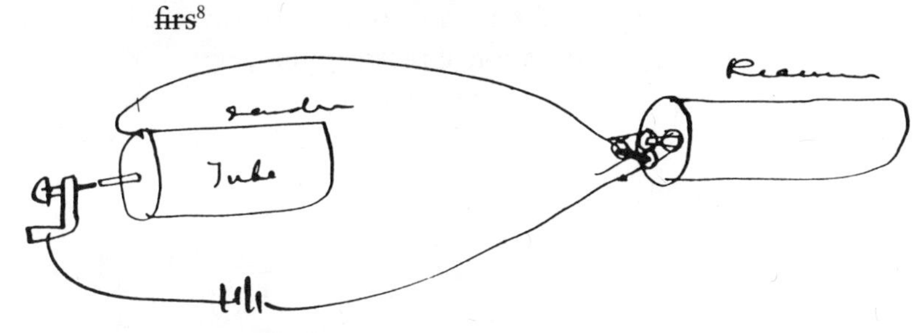


brass diaphram also used parchment. sending ~~p~~Had a projection from middle of diaphram to a contact screw by speaking in tube diaphram set in motion & breaks transmitted to receiver ~~set~~ the magnet acting on armature secured to diaphram set it in motion. Speaking was inferior but the singing was very clear & nice.

Tried tubes of various diameters & lengths diaphrams of various thicknesses also[9]

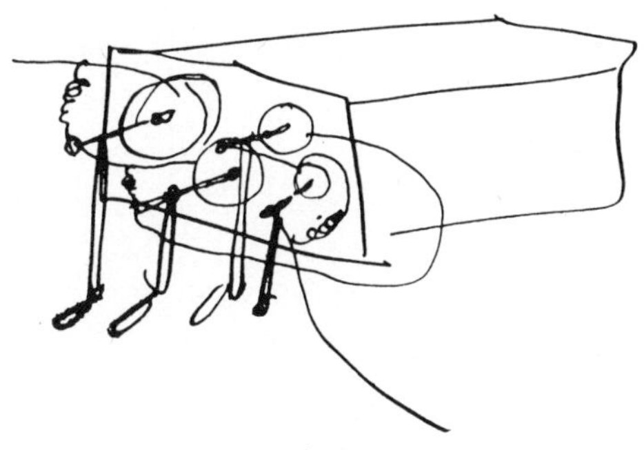

same Rieceiver In end of box placed several diaphrams of different diameters each of which had a contact point the points serving to cut in & out of the circut resistors but none breaking the continuity of the circuit. This owing to bad construction of apparatus appeared to be a failure yet I think if properly made it would be a success[10] also tried a single diaphram[11]

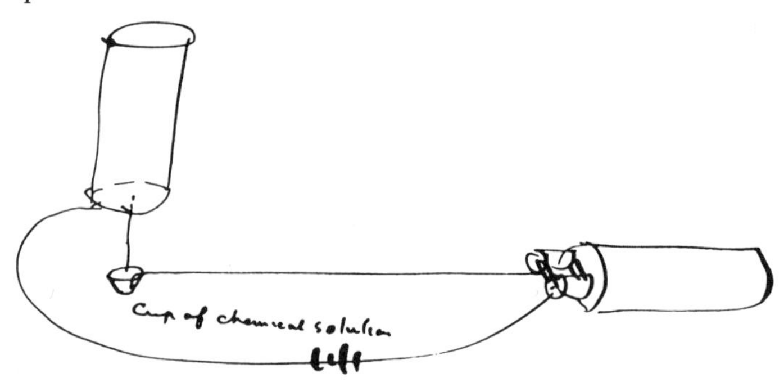


With platina point dipping in a ~~s~~electrolytic Solution the vibration of the diamphram serving to ~~incr~~ ~~s~~transmit waves to the receiver by increasing and decreasing the resistance of the circuit.[12] This appeared to give best results.

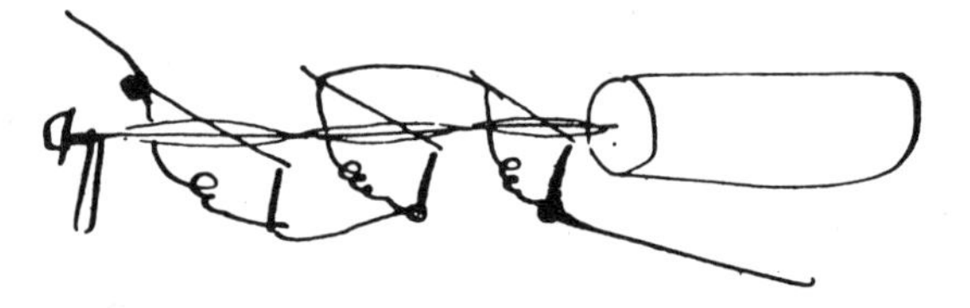

Tried vibrating thread over which were contact levers cutting in & out resistance Experiment was tried in very bad way & got no results but think there is something in it.

This may work[13]

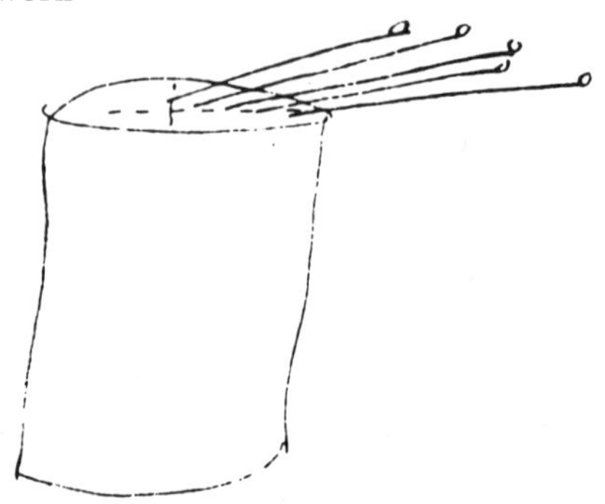

I think the high notes are made by the middle of diaphram only vibrating while as the notes are lower more of the diaphram vibrates. hence by placing contac~~in~~t points from centre to rim may get different rates & have levers cut in & out resistance

Good night  I am—[f]

T A Edison  James Adams
Chas Batchelor

X and X (photographic transcript), NjWOE, Lab., Vol. 10:49–52, 59–60, 53–57; TI 2, Edison's Exhibit 58-10 (*TAEM* 3:855–58, 864–65, 859–63; 11:215). Document multiply signed and dated. [a]Obscured overwritten letters. [b]Circled. [c]Interlined above. [d]Obscured by ink blot. [e]"as . . . possible" interlined above. [f]"Good . . . am—" written under Edison's signature.

1. Figure labels are "condenser," "slow fork," "Local forks," and "higher fork ie the multiple="Another sketch of this arrangement appears in NS-76-002 (Lab. [*TAEM* 7:360]). Edison also drew a variant acoustic transfer transmitter on this date (NS-76-002, Lab. [*TAEM* 7:353]).
2. Figure labels are "forks," "T," "S," and "Earth."
3. That is, Morse instruments.
4. Figure labels are "Low," "higher multiple," and "no g[oo]d."
5. Figure labels on the first sketch are "forks," "same," "same," "same," "P," "Z," "metal," and "Earth," and on the second are "forks," "a," "b," "d," "c," "z," and "c."
6. Not labeled.
7. The polarized relay appears at upper right in the drawing.
8. Figure labels are "Tube," "sender," and "Receiver."
9. See Doc. 873 n. 1.
10. Cf. Doc. 767.
11. Figure label is "cup of chemical solution."
12. See *TAEB* 2:772–73.
13. Figure label is "speak here."

**–766–**

*Technical Note: Miscellaneous*

[Menlo Park,] July 27, 1876

Acoustic annunciator for stable private house etc.[1]

Use tin tube covered with a diaphram & use silk thread as conductor of the vibrations  turn corners by use of wheels and also diaphram= Have tried large number of experiments with this system=

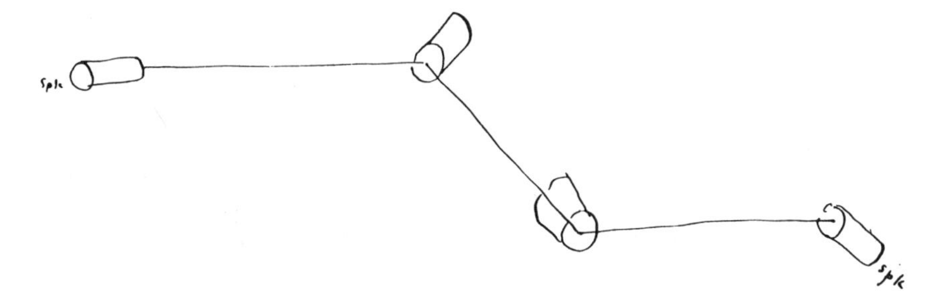

Also[2]

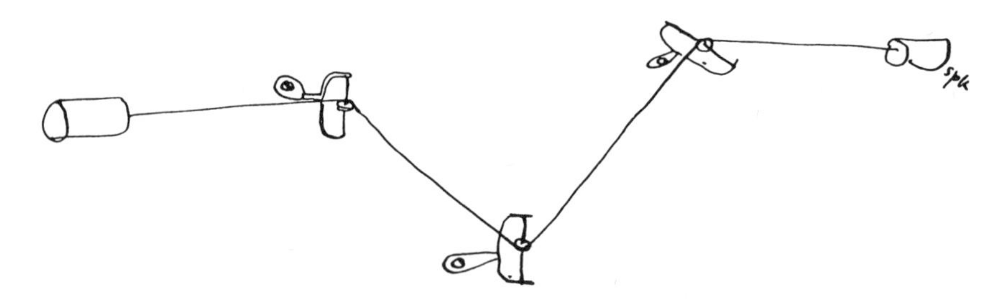

Tried this 2 weeks ago    works fairly

A Repeater

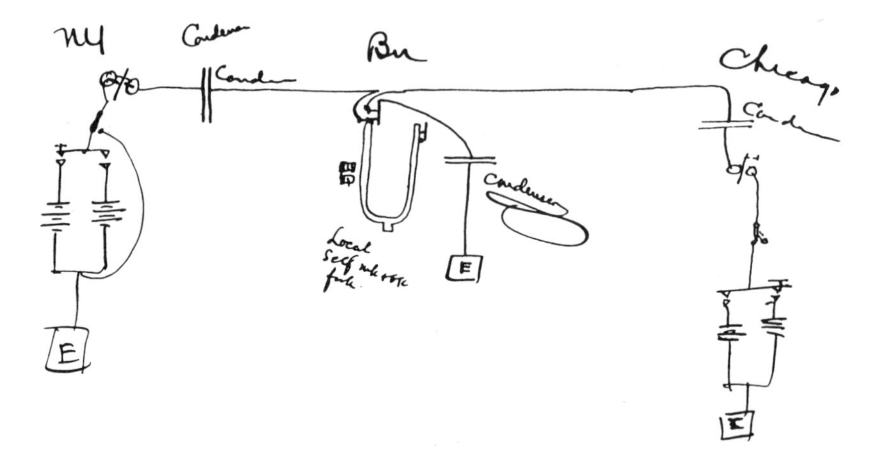


Can be several transfer forks put in along wire; this circumvents the law of the Squares=[3]

Conceived about 3 weeks ago—

Perhaps the following would work upon a line with great static capacity as the reeds would only respond to their particular rate & independent of all extraneous currents out of time with it no matter if very strong

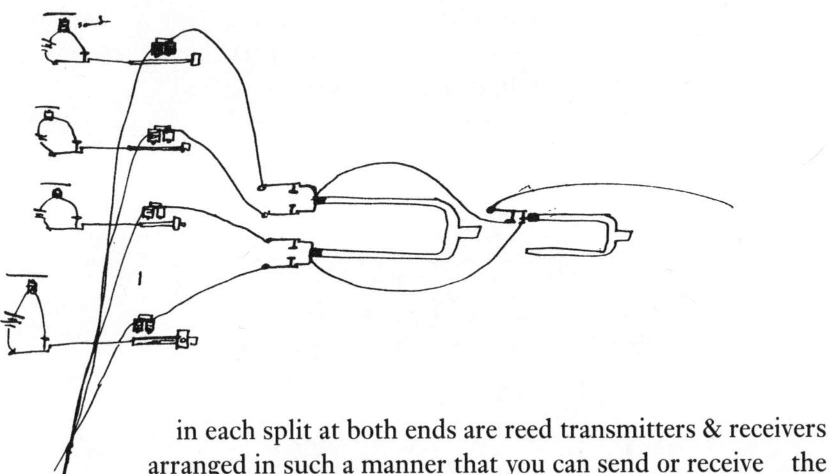

in each split at both ends are reed transmitters & receivers arranged in such a manner that you can send or receive the rates of vibration being very much greater on lowest Reed than the transfer or Switch forks=

found that it[a] was no go=[4] With keys open any particular one would work OK but when you closed say 3 & worked 2 & adjusted right 2 would work OK but when 3 was opened it would change adjustment of 2 materially=

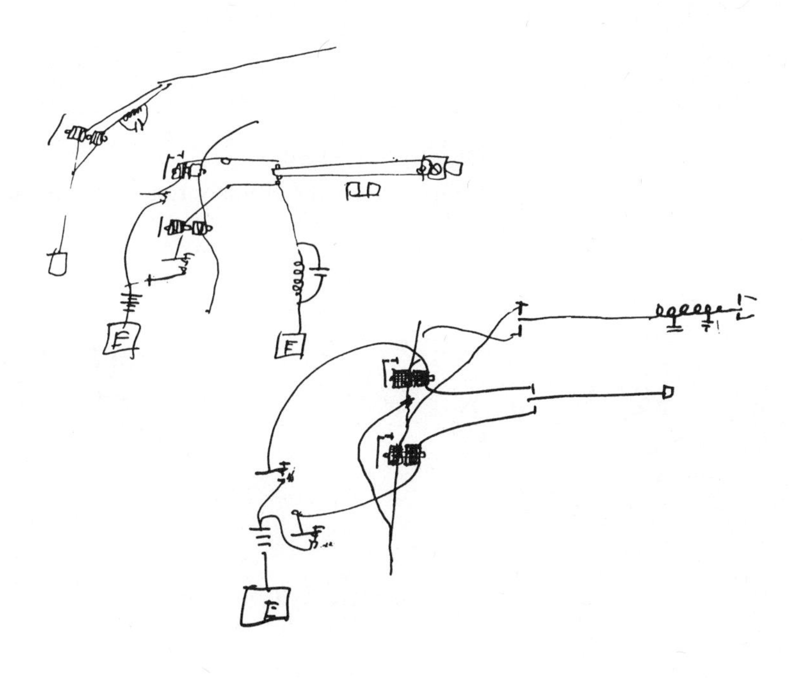

It seems probably that a system of compensation could be obtained by using artificial line[5] & extra splits using also diffrential relays[6] thus[7]

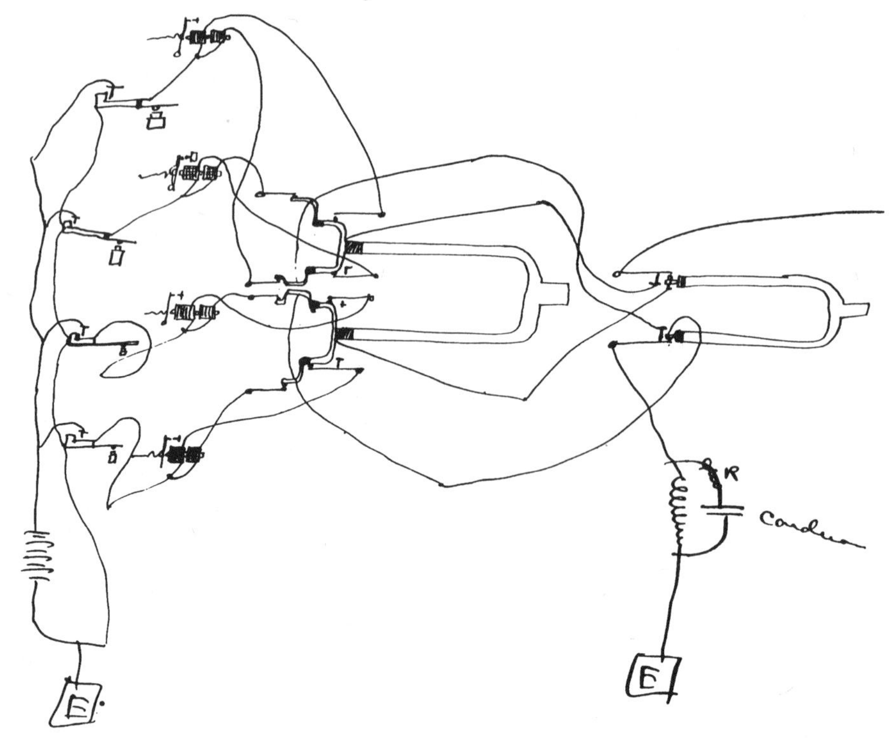

Same at other End

Acoustic transfer Roman Letter chemical printing telegraph=[8]

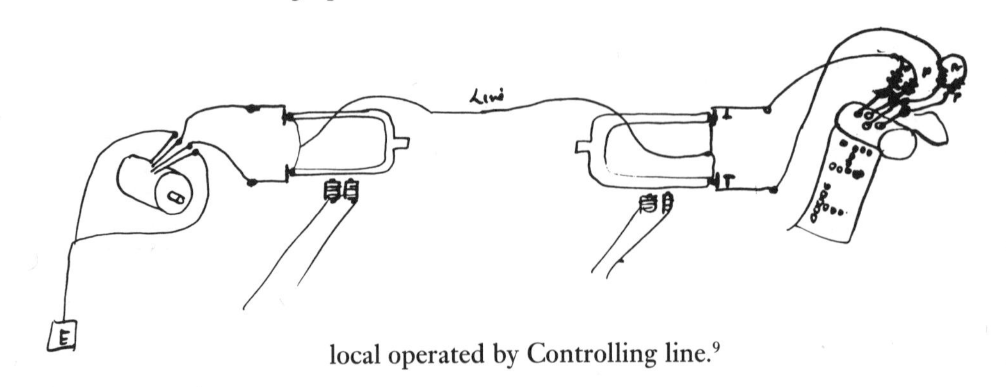

local operated by Controlling line.[9]

I propose to make a Ethericscope thus[10]

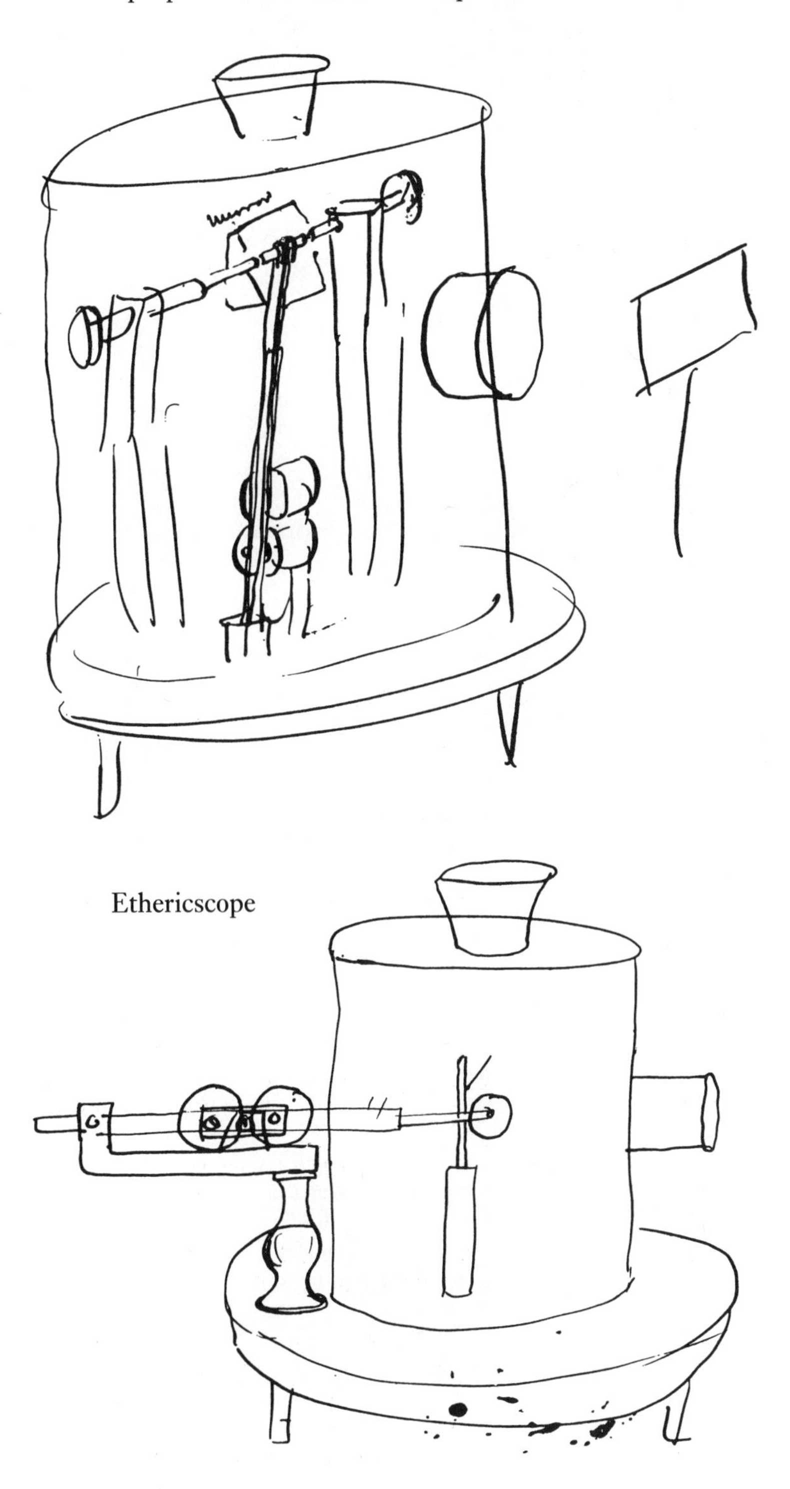

Ethericscope

Ethericscope[11]

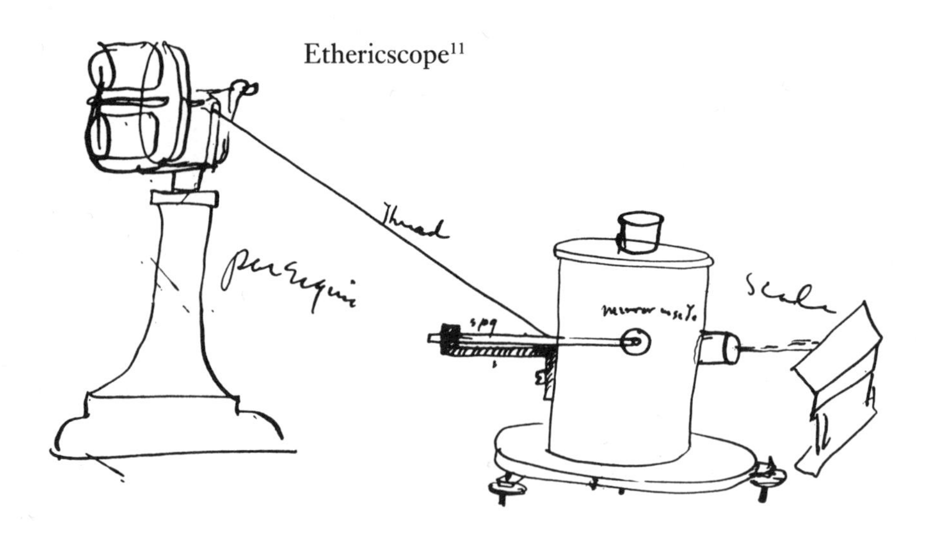


T A Edison | James Adams
Chas Batchelor

X, NjWOE, Lab., Vol. 10:64, 63, 61; Vol. 15:135; Vol. 10:62, 70, 65–67 (*TAEM* 3:869, 868, 866; 4:442; 3:867, 874, 870–72). Document multiply signed and dated. [a]Obscured overwritten letters.

1. Doorbells, call bells, and the like were essentially short-distance telegraphs (although many long-distance telegraph systems required such signals to gain an operator's attention), and were used widely in the second half of the century. It was also common to consider electrical systems together with mechanical means of communication, such as pneumatic tubes. Shaffner 1859, 346–53; Prescott 1879, 375–99; idem 1877, 883–92; Tunzelmann 1900, 282–84.

2. Figure labels are "spk," twice in the first drawing, once in the second.

3. A repeater used a receiving device to operate an independent transmitting circuit that had its own battery, thus forwarding a strengthened signal to the next station (see *TAEB* 1:30–31). This New York–Chicago telegraph circuit shows a repeater at Buffalo ("Bu") using a "Local Self mk & bk fork," with condensers at both ends and at the repeater electrically isolating the line, as was common in cable telegraph practice. The "law of the Squares" Edison hoped to circumvent was William Thomson's 1854 analysis of signal speed on telegraph cables, showing that transmission time increased as the square of the length of the line (Smith and Wise 1989, 446–58, 660–67). Edison hoped that dividing a long route into segments would increase signal speed if the repeaters between segments worked fast enough. See also Doc. 769.

4. This paragraph and the following drawing are on a page archivally separated from the rest of this document. They appear to fit into these notes either here or below after "Same at other End." See headnote, p. 19.

5. An artificial line used resistors and condensers to balance the electrical properties of a telegraph line. See *TAEB* 1:550 n. 2.

6. A differential relay split an outgoing signal into two circuits. See *TAEB* 1:48 n. 1.

7. Figure labels are "R" and "Condenser."

8. Figure labels are "Line," "N," "P," "N," and "P."

9. Adams's version of this drawing is in Vol. 10:72 (*TAEM* 3:874).

10. The figure label is "mirror." It is not clear what is being measured by these "ethericscopes."

11. In this design the armature's vibration is provided by the motor of an electric pen (in its stand at left). The figure labels are "pen engine," "Thread," "sp[rin]g," "mirror inside," and "Scale."

**–767–**

*Technical Note: Telephony*

[Menlo Park,] July 27— 1876

Acoustic[a] Speaking Telgh[1]

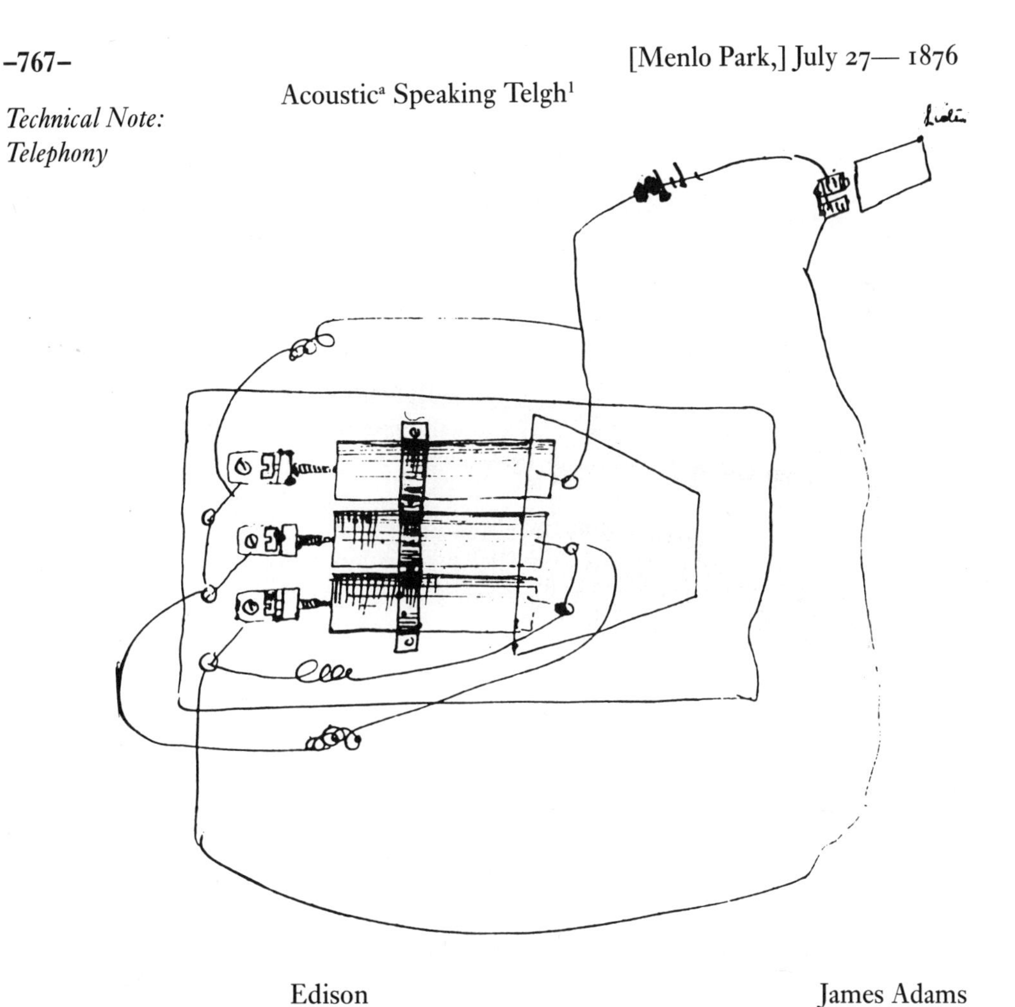

Edison James Adams

X (photographic transcript), NjWOE, TI 2, Edison's Exhibit 73-10 (*TAEM* 11:217). [a]Obscured overwritten letters.

1. Figure label is "Listen." The operator spoke into the trapezoidal chamber from the right. Diaphragms at the left end of the tubes made intermittent contact with platinum points, shorting out resistance coils connected in parallel with each tube. The resistor/tube combinations were connected in series (cf. the multidiaphragm transmitter in Doc. 765). In testimony, Edison stated that this instrument did not work well but "you could tell that someone was talking and if you knew what they were saying it sounded awful like what they were saying" (TI 1:35–36 [*TAEM* 11:38–39]). Another drawing of this instrument is in Vol. 10:69, Lab. (*TAEM* 3:873).

## -768-

[Menlo Park,] July [27–]28[1] 1876

*Technical Note: Multiple Telegraphy*

Position of lines[2a]

Acoustic Transfer

Experiments with a chemical recording instrument at the receiving end with the 4 splits connected to ~~the~~ four iron recording points drum to ground at ~~g~~other end 60 cell small battery Relays in splits, Phila wire, weather moist Kennedy[3] reports wires little soggy= Samples[4] marked —A. result of wkg 1 key at time making long dashes;= ~~Noticed that No 3 reed~~

Exhibits .B. are the record with the ground wire inserted in the low forks the supposition being that immediately after the line has been taken from 1 set instruments & before it is put on another the idea being that putting wire to dead earth would free it of the return charge

Exhibit .C. is the wire on short ckt the ground wire having been inserted on the multiple forks at both ends thus[5]

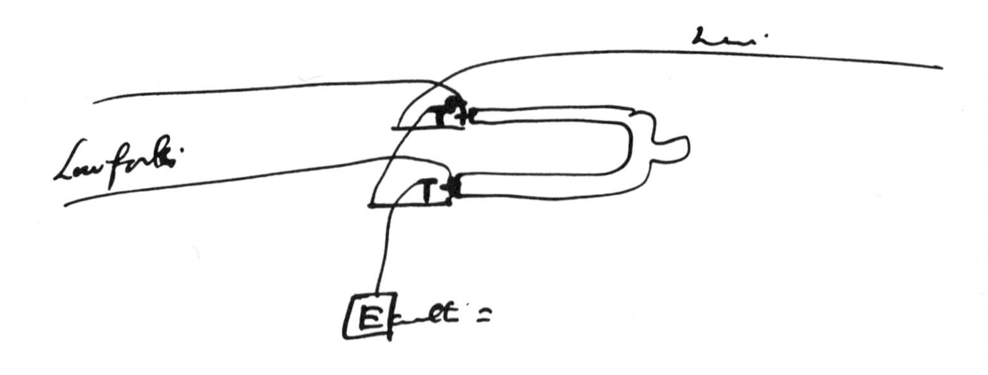

Exhibit D shews same with Phila on= These shew where the mysterious bug was: the device for "Earthing" the wire on the low forks was a failure while that on the multiple is a success= Now this explains why all my rinkles failed

Exhibit E will shew ~~that~~ the effect of the return charge ~~so~~ ascertain if the ground wire is sufficient to discharge it & prevent its entry into the signalling splits= The end sending Exhibits A. B. C. & .D is now to Earth with keys closed Nos. 4 1 & 2 splits are to pens at recg ~~station~~ end while 3 is connected to key & batty 36 cells Callaud.[6] our next exhibit will shew the return chg from 60 small carbon=[7]

Result: No return charge from 3 on [altern?][b] 36 cells Callaud Now for 60 carbon small & 36 Callaud combined= This will be exhibit F

F. as carbon was to line return charge was zinc to pens hence Exhibits F are delusive as marks came underneath paper

We reversed battery to make it correct and get exhibit G when the loop to Phila got foul of another wire with battery on & stopped us.

going to get RR station on our short to get N.Y. so we can have it fixed= 12 Midnight

Phila wire having busted we now make an artificial line composed of 5 resistance Boxes[c] Phelps' Duplex patent.[8] with .5. Duplex Phelps' Condensers Thus[9]

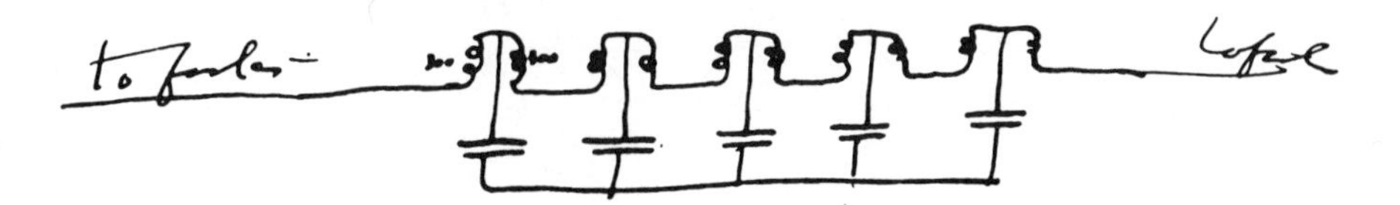

Total .R. 3000 ohms 4 sheets[10] in each condenser.

We now resume the test of the return charge from 60 cells small carbon & 36 Callaud together.—

Exhibit .G.[a]

Exhibit H shews same condensing power but with—7,000 ohms R.R.[11]

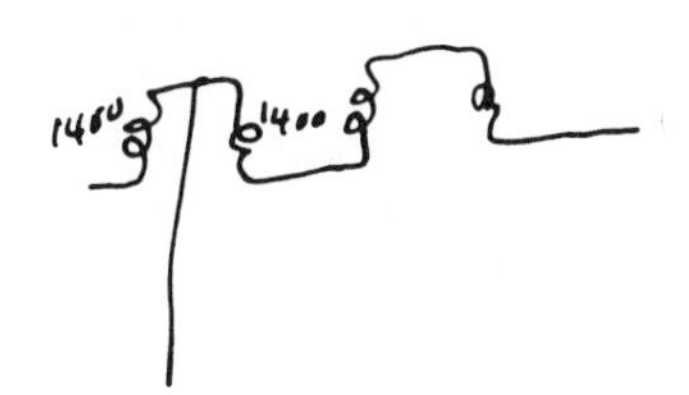

Exhibit I shews 12 sheets in each condenser—~~nothing~~ think[d] can just detect a little return charge

Exhibit J shews with same resist but ~~2~~with 28 sheets on each condenser= detection of return charge myth—

Exhibit K shews same resistance but 72 sheets condenser in each. It see~~e~~ms evident something is wrong— We reverse battery ~~Exhibit~~ No— the ground wire being taken off we get Exhibit .L. which shews return charge= we have copper to line— L shews on .2 best 4 next best then .1.[a]

We have in Exhibit M. the return charge with fork ground wires disconnected at both ends= 72 sheets ~~cond~~ in ~~cod~~each condenser & 7000 ohms resistance.

Exhibit N same with ground on fork at distant end=

O. shews with both grounds on

We now reverse battery, zinc to line ~~P Shews~~ get nothing—

I do not understand why we send out a carbon current & get back [-][e] a carbon current.

Pens are arranged thus[12]

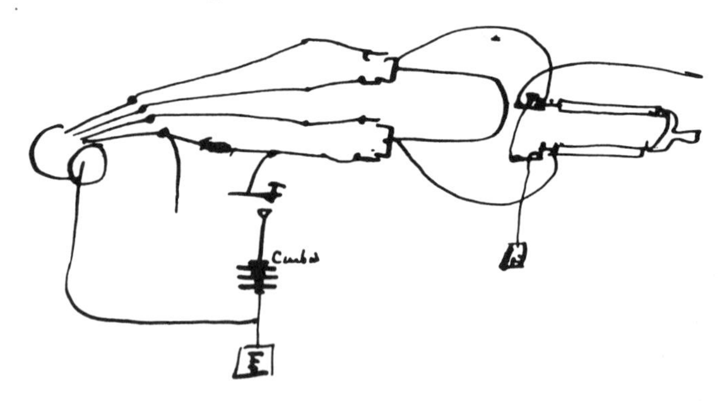

Perhaps the forks are little out unison and it is leakage back through contact points.

Return Chg—

Exhibit .P. shews ground wire off at recg station on at ~~dif~~ distant station with splits to Earth, also with both fork grounds off also with all splits & fork ground off at distant= 7000 ohms in 5 resistance Coils= 74 sheet condenser on each of 5 condensers it also shews with distant entirely open with ground here=

Exhibit Q. shews distant end open splits ground & all= at this end our main carbon to line, in our ground we have a reverse battery, zinc to line with 15 cells— kills return completely.

$R^1$ shews 5 cells on neutralizing battery
$R^2$ " 10 " "
$R^3$ " 15

added 84 sheets to the 5 condensers $S^1$ shews 5 cells on neutralzing battery

Now, 15 000 ohms 91 sheets C on each of 5 condensers

T.$^1$ 5 cells on neutraling battery
$T^2$ 10 "
T.$^{3.}$ 15 wiped out=

New experiment.

7000 ohms, 91 sheets each of the 5 condrs Key closed putting 60 carb & 36 Cal to line 15 cells opposite in ~~g~~fork ground

Exhibit ~~V.~~$U^1$—~~o~~key closed. other end all open.—[f] no exhibit nothing= working key—no exhibit, ~~n~~Nothing[a]

Other end closed, our battery to line nothing.— making dashes, nothing.=

=Found that by putting back instruments for signalling that with 91 sheets condensers the rush in charge ~~when~~ acting like an escape and although we could only just get dashes the correctness of the infererance drawn from the above experiments cannot be doubted that is that the insertion of a wire in the multiple forks at both ends to discharge the wire after each wave is a big success & gives me the required compensation & which can be made as powerful as desired by the insertion of ~~a~~neutralzing & opposte pole batteries in these ground wires. If we receive on a ground undoubtedly the signals will be perfect on very long ~~esp~~ lines, & I am certain they will be if way statins are used whose multiple forks have also a ground= = with batters at both ends we find that we havnt near the margin but I think there will be no difficulty in wkg from ~~W~~NY to Washington direct with the z ground in the multiple fork— We close our laborous night at 5.30. AM

T. A. Edison

A & Gil[13] on—

X, NjWOE, Lab., Vol. 10:74 (*TAEM* 3:876). Document multiply signed and dated. [a]Followed by centered horizontal line. [b]Illegible. [c]Obscured overwritten letters. [d]Interlined above. [e]Canceled. [f]"other . . . open—" interlined above.

1. As indicated in the document, these experiments began on 27 July and continued until 5:30 the next morning.

2. The lines are numbered top to bottom: 3, 4, 1, 2.

3. Unidentified.

4. The sample tapes made during these experiments have not been found.

5. Figure labels are "Low forks," "Line," and "Earth=."

6. Developed by A. Callaud of France, the Callaud battery was a simplified version of a Daniell gravity battery (see *TAEB* 1:615 n. 3). King 1962a, 243, 245; Pope 1872, 106–7.

7. On the carbon (Bunsen) battery, so called because its anode was a carbon rod, see *TAEB* 1:599 n. 3.

8. Phelps had no patents for duplex telegraphy or for resistance boxes.

9. Figure labels are "to forks," "300," "300," and "to forks."

10. Condensers were constructed of interleaved sheets of conducting and insulating materials. By attaching circuit wires to different terminals on a condenser's casing, more or fewer of the sheets could be charged.

11. Figure labels are "1450" twice.

12. Figure label is "carbon."

13. James Adams and Ezra Gilliland. Gilliland (1848–1903), an electrical manufacturer and inventor, had worked as a telegraph operator with Edison at Adrian, Mich., and Cincinnati, Ohio. In 1875 he and Edison had formed Gilliland & Co. *TAEB* 1:16; Doc. 622.

## –769–

*Technical Note: Multiple Telegraphy*

[Menlo Park,] July 28 1876

Acoustic Transfer[1]

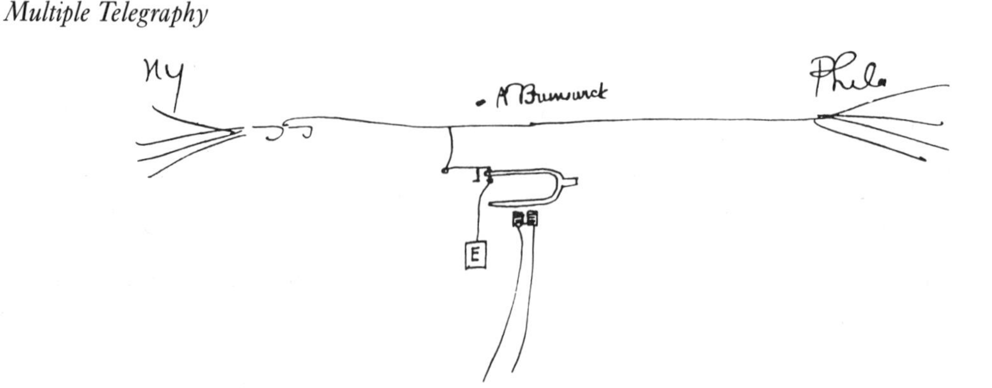


to control[2] this is a multiple of the highest transfer fork

The way station fork serves to discharge wire in midle and thus circumnavigate the law of the squares;[3] I think it probabl~~ye~~ that the line could be put to earth after each wave simultaneous at several stations.[4a]

T A Edison

X, NjWOE, Lab., Vol. 10:86 (*TAEM* 3:888). [a]Open box drawn at bottom of page.

1. Figure labels are "NY," "N Brunswick," and "Phila."
2. This text refers to the wires descending at the center of the sketch.
3. See Doc. 766.
4. Edison drew such an arrangement the following day. Vol. 10:87, Lab. (*TAEM* 3:889).

## -770-

*Technical Note: Telephony and Galvanometer*

[Menlo Park,] July 29 1876

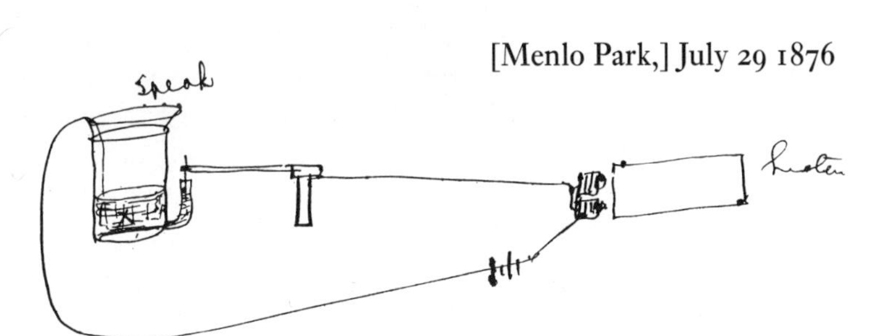


X water[1a] Idea being to speak in beaker & set the water vibrating this giving large amplitude in small side tube & thus increase & decrease the resistance giving the proper waves to magnet so as to carry on a conversation=[2]

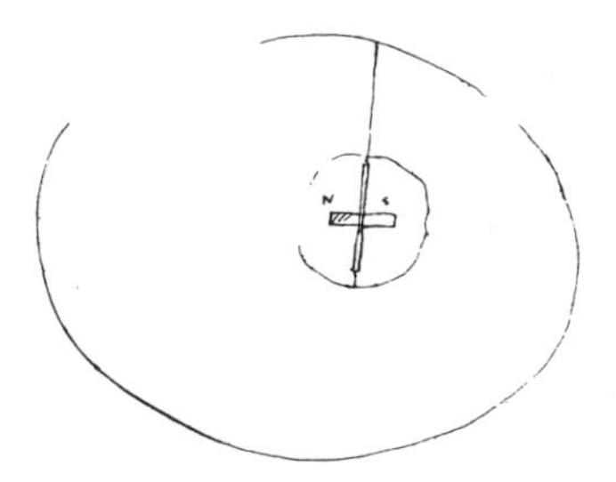

Instead of employing a ~~leens~~ Lens mirror on the needle of a Thomson. mirror galvanometer I propose to employ a think mirror about ½ inch[a] long & 1/32 inch wide or even Smaller=

T A Edison

X (photographic transcript), NjWOE, TI 2, Edison's Exhibit 71-10 (*TAEM* 11:216). [a]Obscured overwritten letters.

1. Figure labels are "Speak," "X," and "Listen."

2. Both Edison and Kruesi testified that this device was constructed a day or two after the exhibit date, and Edison added that it did not work. (TI 1:33–34, 265 [*TAEM* 11:37–38, 109]).

## -771-

*Notebook Entry: Batteries and Cable Telegraphy*

[Menlo Park,] July 30 187~~6~~—

I have an idea that each particular combination of elements & exciting fluids in a galvanic battery give out electricity with different rates of vibration[1] in the same manner as the va~~b~~pors of metals in a flame are proved by the prism to have different rates of vibration[2] hence I have conceived the idea that with a cell filled with certain chemical solutions & platina electrodes that 100 units of current from one source will pass through that solution in greater quantity that 100 units from a different source; in the Same manner that ~~a~~the heat from Hydrogen will pass through a solution of ~~i~~iodine in Carbon Bisulph than from a candle or even from another source of light & heat of greater intensity= If the rate of vibration of the electricity is the same as the vibrating molecular time of the liquid through which it passes ~~w~~very little current will pass but if different it will pass readily. Hence I conceive that the Static Current[3] passing out of a Submarine Cable has a different rate of vibration than that due to the primary battery current hence I propose to filter the primary ~~& pre~~ from the Static or secondary current by a Liquid ~~which~~ whose time of vibration is the same or a multiple thereof of the static charge's Should I be unsuccessful in obtaining this I can employ another liquid and two signalling both of which produce the same charge in the cable but whose vibrating time is different and I insert a certain liquid that will allow the current from one battery to pass through ~~regulativly~~ easily while the other does not pass so readily—hence the Cable[a] will be constantly charged & Rapid signalling will be obtained.= I try a preliminary experiment this evening[4]

T. A. Edison

X, NjWOE, Lab., Vol. 8:285 (*TAEM* 3:713). [a]Obscured overwritten letters.

1. There existed at this time no generally accepted theory of battery action. Compare Edison's earlier idea on the wavelike nature of galvanic electricity (*TAEB* 2:159 n. 2).

2. Spectroscopy—the study (usually with a prism) of the characteristic frequencies of light emitted or absorbed by particular sub-

stances—was a significant part of late nineteenth-century physical science. Edison had acquired a spectroscope in early 1875, experimented with spectroscopy as part of a telegraph system, and had considered manufacturing a toy model. See Docs. 428, 544, and 989.

3. That is, the discharge following each dot or dash, caused by self-induction and capacitance. It lengthened each signal on the cable, slowing down communication. It was a problem Edison had also encountered in automatic and long-distance telegraphy. See, e.g., Docs. 299 and 317.

4. After testing this idea experimentally for several days, Edison concluded that "It is evident that the phenomena so far recorded is nothing more than a destruction of balance & electromotive force of batteries  nothing apparently in it." Vol. 8:288–304, Lab. (*TAEM* 3:716–32).

**-772-**

*Technical Note: Telephony*[1]

[Menlo Park,] August 2nd 1876

Experiments in Electric Talking Telegraphs.[2]

Up to the present date we have tried a great many plans, a few of which are described below:

In our first experiments we found the parchment diaphragm in the transmitter could not be kept in adjustment owing to the expansion and contraction from moisture of the breath— a thin brass diaphragm was substituted which entirely overcame this difficulty but otherwise gave no better results= a brass diaphragm in the receiving instrument is inferior to parchment owing to a metallic ringing sound it gave.

Tubes of various sizes ranging from 3½ inches to 1¼ have been tried but our best results were obtained with a two inch diaphragm brass in transmitter parchment receiver with 3 ohm spool 2 cups of carbon short circuit shunting the sender with about 40 ohms—singing in any key can be transmitted accurately. Solid and spring contacts have been tried small platinum points and large brass discs, also discs & knife blade contacts with no gain.[3]

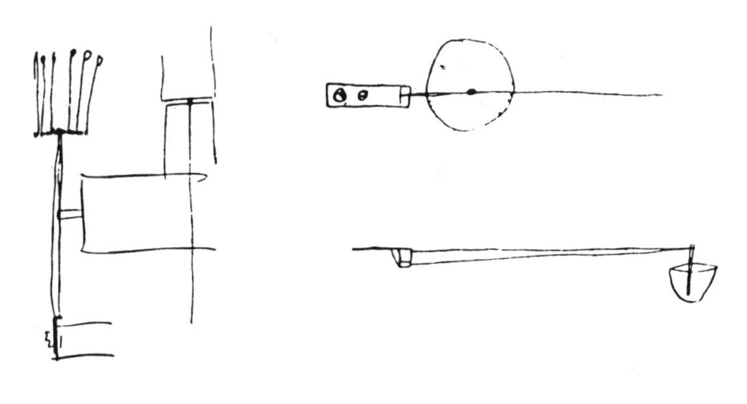

An arrangement of one box with 5 diaphragms of different size ranging from 4 in to 1 inch, no advantage.[4]

A box with 3– 2 inch diaphragms was tried each working independent & connected so as to throw in and out resistance various other changes were made in the connections.

Induction, static discharge, primary & secondary currents, on all the above & on nearly all of the connections singing could be transmitted perfectly—also playing on the cornet can be transmitted perfect— a plan which for want of perfect apparatus was not thoroughly tested is constructed on a very satisfactory plan & will undoubtedly be a great improvement. A good model will be constructed a thorough test made at once.

A platinum wire is attached to the diaphragm & immersed in a solution or liquid resistance  the strength of the wave is by this means varied according to the amplitude of the diaphragm  Various improvements & modifications of this plan are proposed such as shown for increasing amplitude diamond shaped immersing point of aluminum.[5]

Another plan is to be place chemical paper between contact points & the variation in pressure of diaphragm gives corresponding variations.[6] strength of wave. The sketch on the back of this sheet[7] is a plan proposed by Edison & will be tried in due time= The introduction of saturated felting in place of chemical paper is also proposed.[8]

Fig. 1

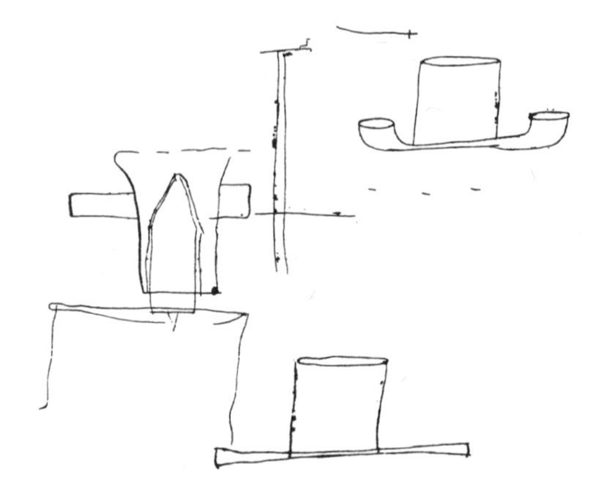

The following plan will be tried at once represented in Fig. 1. on opposite page[9]—diaphragms arranged to force out water against a counterbalancing diaphragm the talking diaphragm is placed in close proximity to metal plate leaving only small amount of solution to be vibrated the two counter-balancing diaphragms furnish elasticity.[10]

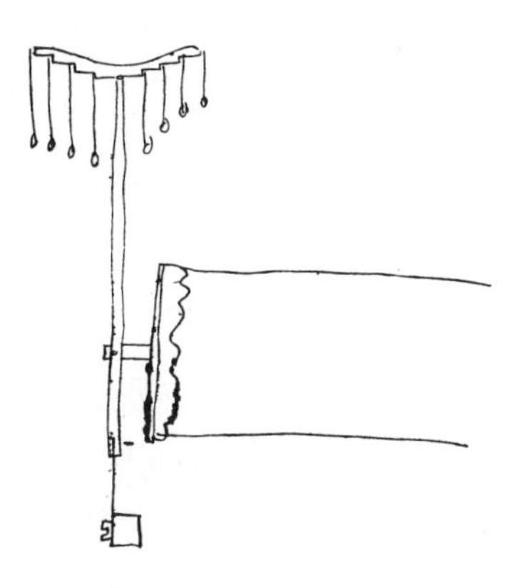

X (photographic transcript of handwritten transcript), NjWOE, TI 2, Edison's Exhibit 101-10 (*TAEM* 11:218).

1. This transcription of Ezra Gilliland's memorandum (see note 2) was entered as an exhibit in the Telephone Interferences. It contains several discontinuities and awkward phrasings that appear to be artifacts of the transcription process.

2. In discussing this set of notes Edison testified that he set Ezra Gilliland to work "about May, 1876," with James Adams to assist him, and that they continued experiments to "about August" (TI 1:34 [*TAEM* 11:38]). However, both the general pattern and details of this research seem related to experiments undertaken during July. (cf. Docs. 759, 761, 765, 767, and 770). Alexander Graham Bell's attorneys later argued that the internal evidence of these notes suggested that they did not represent three months' experiments (Brief for Bell, TI 4 [*TAEM* 11:197]).Undated drawings that may be related to these notes are in Vol. 15 and NS-Undated-006; for example, see the item from Vol. 15 entered in TI 2 as Edison's Exhibit 79-15 (*TAEM* 11:617).

3. In his testimony Edison described the instrument shown in the drawing below as consisting of "a multiplicity of contacts, in connection with a vibrating spring, set in motion by a diaphragm," and said that it was "made and worked to a sufficient degree to allow us to hope." (TI 1:34 [*TAEM* 11:39]).

4. What may be a related set of experiments is in Doc. 765.

5. See Doc. 765 for related experiments. A possible example of the "diamond shaped immersing point" is in TI 2, Edison's Exhibit 79-15 (*TAEM* 11:617).

6. This is the end of a page; it is not certain that the remaining text immediately followed in the original.

7. Because this is a transcription, it is not clear whether this refers to a drawing copied onto these pages. The original is no longer extant.

8. See Doc. 759.

9. In his testimony Edison described this design as

> a tube, with a diaphragm resting upon a flexible pipe joining two reservoirs of water. In each reservoir was an electrode, and the same were included in a circuit containing a battery and a telephonic receiver.
>
> The resistance of the circuit was altered by a compression of the flexible tube altering the size of the column of water between the reservoirs.

Edison claimed that this instrument was made in August or September 1876 and that "it was said to work very good." TI 1:36 (*TAEM* 11:39).

10. According to Edison's testimony, the following design

> consists of a resonant tube, or a chamber provided with a diaphragm immediately opposite is a spring, secured at one end by a screw, and near the center to the diaphragm, and upon the end of the spring is a T, having notches, against which a number of contact springs faced or rested.
>
> These springs were arranged with resistances, so that by the vibration of the springs connected to the diaphragm a complicated system of contacts were made, which put in and took from a circuit resistance, the circuit containing a battery and a telephonic receiver.

Edison noted that this instrument was constructed in September or October 1876 and that "it did not work satisfactorily." TI 1:36 (*TAEM* 11:39).

## –773–

*Technical Note: Multiple Telegraphy*

[Menlo Park,] Aug 3 1875[1876][1]

Acoustic[2] T[ransfer ][a]
Duplex[ed ][a] on one w[ire][3a]

[*Drawing on facing page*]

Adams[4]

X, NjWOE, Lab., Vol. 9:3 (*TAEM* 3:795). Drawing by Edison; date and heading by Adams. Drawing turned 90° here to fit printed page. [a]Document damaged; amount of missing material indeterminate. See note 2.

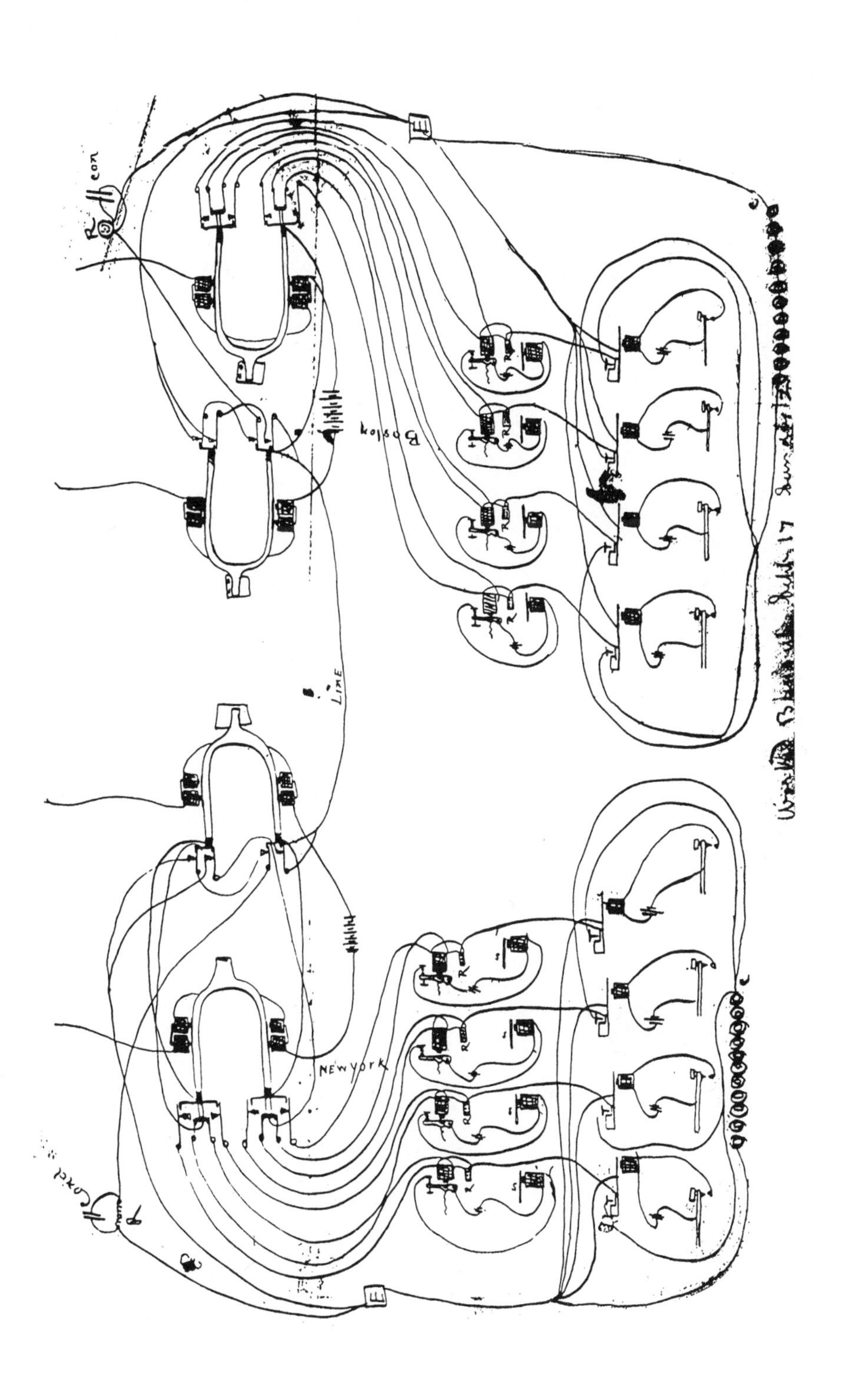
Boston
LINE
NEW YORK

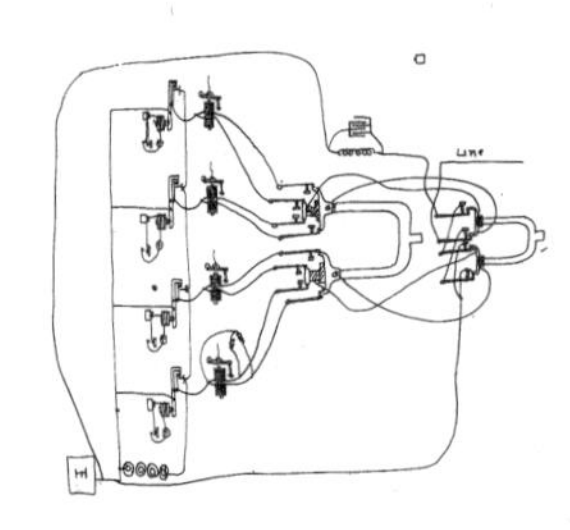

*A drawing of the octruplex acoustic transfer circuit that was tested on the Philadelphia loop circuit with "fair success."*

1. "Worked Bang up Sept 17 Sunday" was added below the drawing by Edison; that date was a Sunday in 1876, not 1875. Moreover, none of Edison's dated work on circuits of this type occurred in 1875. See Doc. 795.

2. On this date Edison also made a closely related drawing labeled "Acoustic Transfer Duplexed making 8 messages on one wire" and noted "Tried last night on Phila loop fair Success—" (Vol. 8:334, Lab. [*TAEM* 3:764]). For other work on acoustic transfer from early August see Vol. 10:88–108, Lab. (*TAEM* 3:890–908).

3. Figure labels are: "R," "Cond," "New York," "Line," "Boston," "R," "con," eight more "R's," and battery pole labels "z" and "c" twice.

4. Edison's signature was probably on the portion of the page now missing. He and Adams signed (and Edward Johnson witnessed) the other sketch of this type from this date (see note 2).

## –774–

[Menlo Park,] Aug 3rd 1876

*Notebook Entry: Electromotograph*

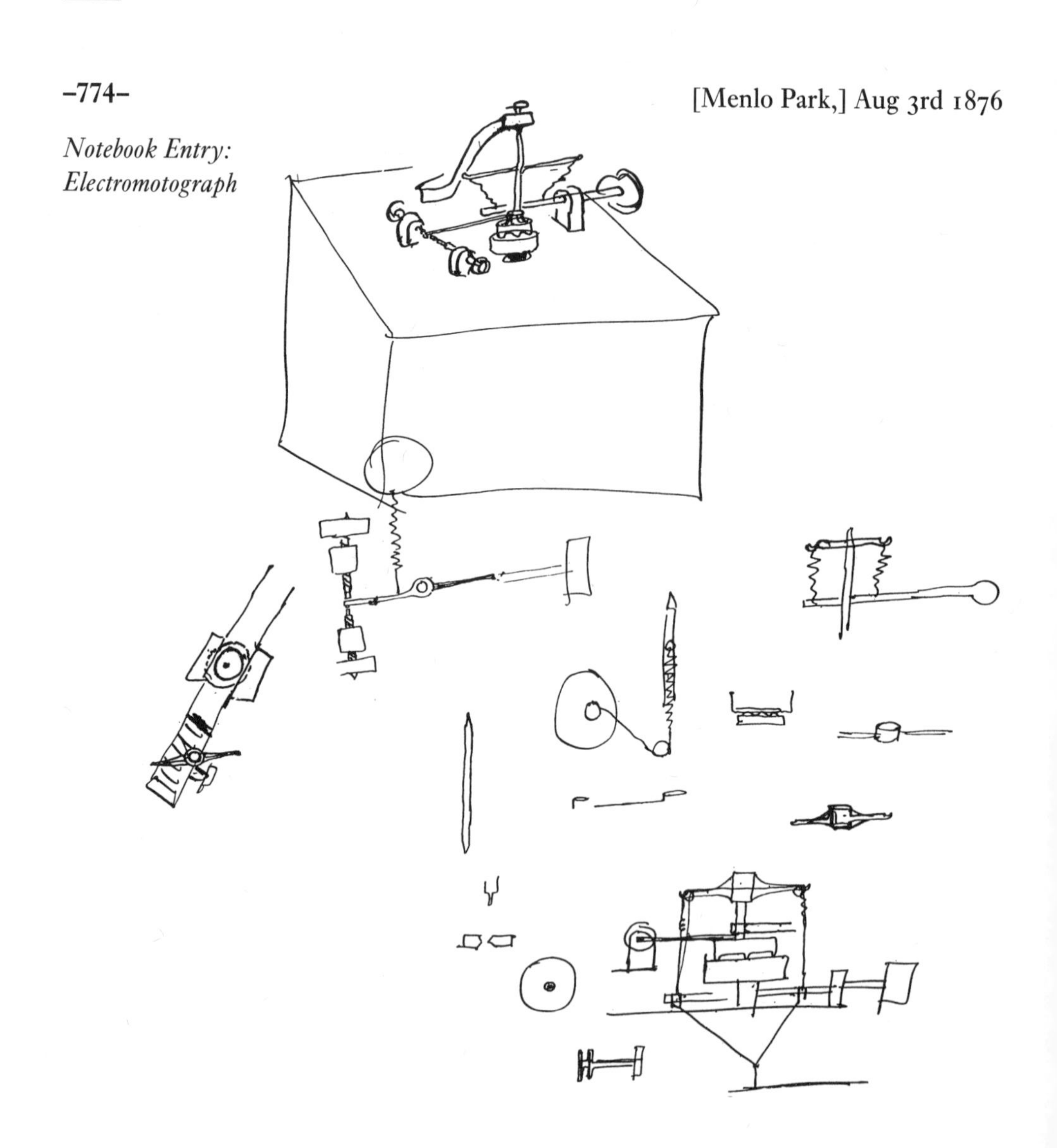

Electromotograph

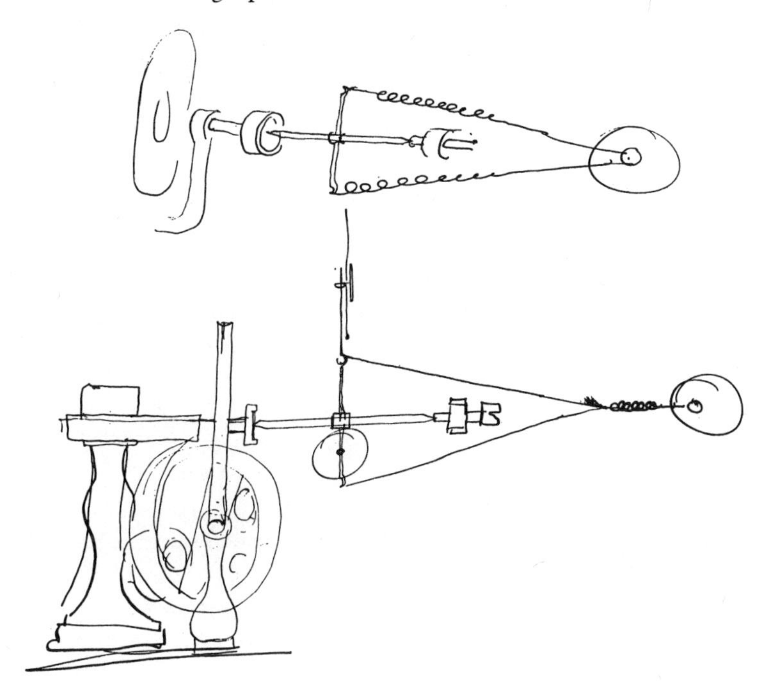

Electromotogh[1]

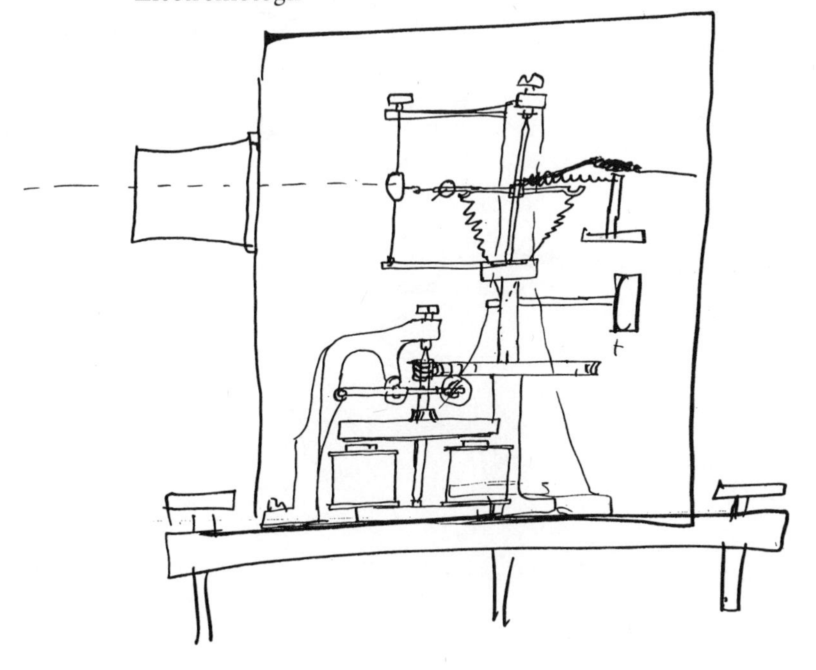

Delicate Electromotograph[2]

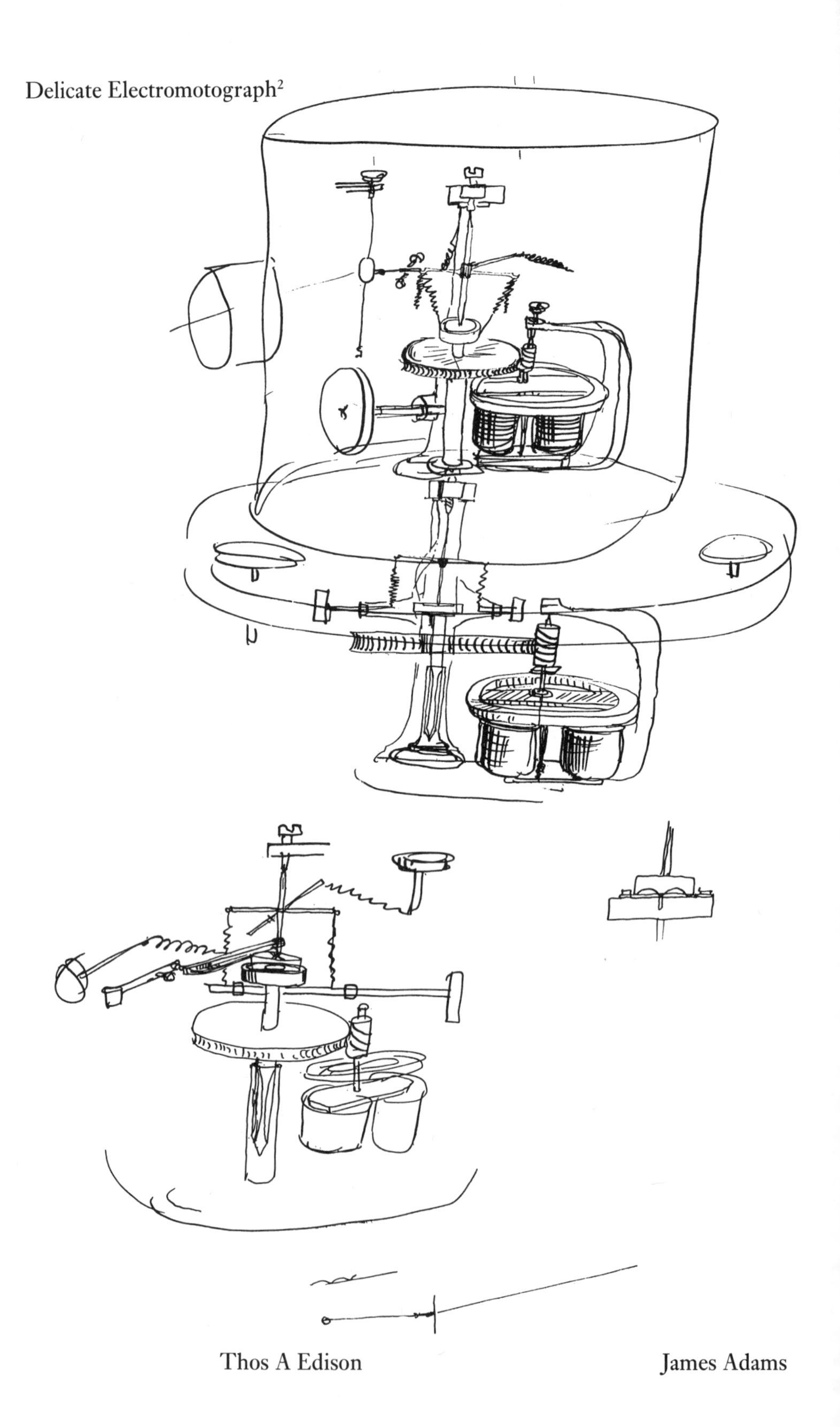

Thos A Edison James Adams

X, NjWOE, Lab., Vol. 8:329 (*TAEM* 3:759). Document multiply signed and dated.

1. These electromotographs appear to use a small mirror to indicate deflection caused by increased or decreased friction. The design may have been inspired by William Thomson's visit to Menlo Park in July, as Thomson's mirror galvanometer used a similar reflecting device (see *TAEB* 2:754 n. 1).

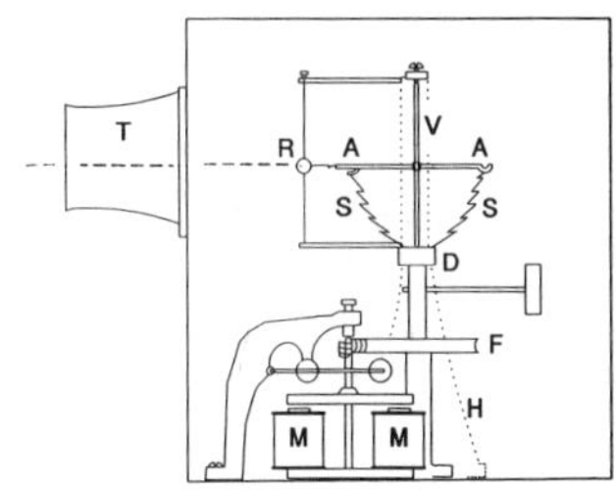


A small motor **MM**, comparable to the one Edison used in his earlier work (e.g., Docs. 165, 211, and 262), drives flywheel **F** through a worm gear. The flywheel in turn spins a disk inside casing **D** (visible in the next drawing). The electromotograph action (friction change) occurs where the end of shaft **V** rests on the spinning disk. The two horizontal arms **A A** mounted on the vertical shaft are held by springs **S S** (secured to **D**) to prevent the shaft's rotation. The presence or absence of a current through the disk and **V** alters the friction at the disk, which changes the torque on the vertical shaft and thus the displacement of arms **A A** from their resting position. The small mirror **R** is attached to the left horizontal arm and reflects a beam of light (dashed line) through tube **T** onto a scale (not shown). **H** is the frame of the instrument.

The previous drawings show other mechanical embodiments of this idea.

2. The next drawing shows two overlapping designs, not one device.

**-775-**

*Notebook Entry: Electromotograph*

[Menlo Park,] Aug 4 1876

Electromotograph Experiments[1]

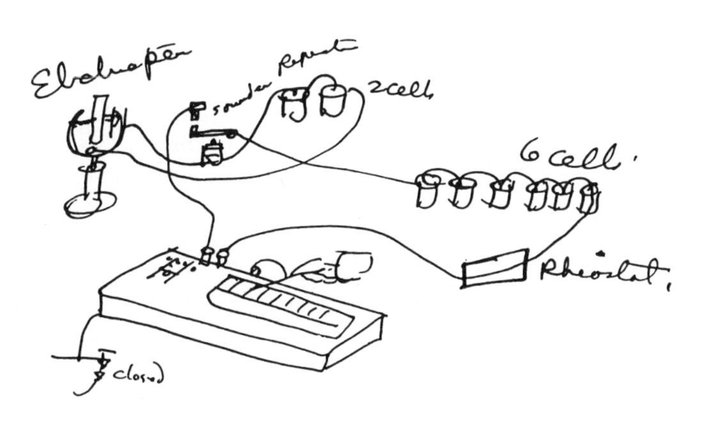


Standard 3000 ohm Pyrogallic paper same as made for Domestic platina pen=

Hydrogen[2]—vibration just ~~vi~~ felt[a]

O. & .H—BiSulphide Carbon Acetic Acid—nothing on 3000.

H. Alcohol & Caustic Soda, will work equal to pyrogallic through 30,000, ohms the paper owing to evaporation of alcohol is quite dry in few minutes gets perfectly dry. it works quite strikingly through 70,000, ohms all we had— by opening & closing the key & stopping & starting the vibrations the ~~effect is enormously~~ detection is greatly convenienced as the hand can readily detect the stopping & starting of the vibrations= ~~Very plain through 70~~

We now take Caustic ~~Potash~~ Soda[b] & Alcohol as a standard, vibrations when key is worked being clearly felt through 70 000 ohms

O & H Alcohol & Oxalic .A. nothing felt at 70 000.

Silver &. Cop on Ag—Alcohol & Tartaric acid vibrations just perceivible not so good as standard

O & H. ~~Noth~~ Oil Sweet Almonds & Iodine nothing surely felt

O & H Oil Turpentine & Creosote—nothing felt

O &. H Oil Rosemary Conc—nothing surely felt—

O & H. Spirits Turpentine & Glacial Acetic Acid nothing felt yet when key open apparently felt some vibs it may be that this sol is of extreme delcy & requires perfect ~~insl~~ insulation to stop vibs=

O & H Bisulphide Carbon Glacial Acetic. A. nothing felt sure=

BiSulph Carbon—Creosote nothing dries almost instantly=

Spirits Turpentine & Iodine nothing felt=

Bilsulph Carbon—Spirits Turp= nothing

Spirits Turpentine Hyposulphite Soda nothing

Spiritts Turpentine Carbozotic[3] Acid—. nothing

By putting vibrating apparatus away from hand EMG so as couldnt feel mechanical vibs—could tell better;

Alcohol & Fused Chl Zinc. just feel on H=[4b]

T A Edison

X, NjWOE, Lab., Vol. 8:266 (*TAEM* 3:693). [a]Followed by horizontal line across page. [b]Interlined above. [c]Followed by centered horizontal line.

1. Figure labels are (clockwise from top) "Electric pen," "sounder Repeater," "2 cells," "6 cells," "Rheostat," and "closed." See *TAEB* 2:540 for an illustration of the hand electromotograph.

2. That is, with the electromotograph contact attached to the cath-

ode (the battery pole that will evolve hydrogen gas from an electrolyte). Similarly, "O" in this document refers to the anode.

3. Carbazotic (picric) acid.

4. At this point there follows the notation "820 PM we start again=" followed by several more pages of similar experiments (*TAEM* 3:695–703).

**–776–**

*Notebook Entry: Galvanometer*

[Menlo Park,] Aug 6th 1876.

Tonight tested our own mirror gal found that it was hard to keep it at a zero.= with 1 Daniell or rather Callaud found that it was extremely delicate unexpectedly so, if there are no sources of error[1]

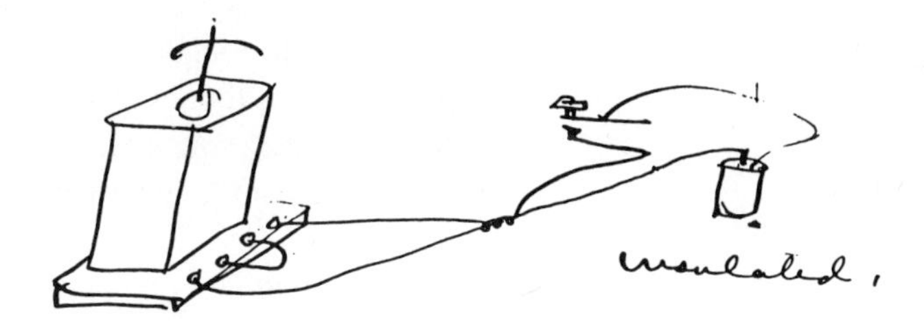


Got deflection of 12 degrees with wires twisted as shewnn. Then brought up The Elliot Thomson[2] & had lot of trouble to get it to zero & at last broke the fibre in the endevor hence couldnt compare=

I think the Elliot mirror is too clumsy=

X, NjWOE, Lab., Vol. 8:278 (*TAEM* 3:705).

1. Figure label is "insulated."

2. The Thomson mirror galvanometer from Elliott Bros.

**–777–**

*Notebook Entry: Electromotograph and Automatic Telegraphy*

[Menlo Park,] Aug 7 1876—

Electromotograph

We have just got our new paper wetter finished it works tolerably well.

Have wet some pyrogallic paper Ferridcyanide of Potassium ditto & Some of the Electromotograph paper of Acetate of mercury & Caustic Potash=

We are going to attempt repeating through the Phila Loop:[1] connecting in the manner shewn in the diagram[2a]

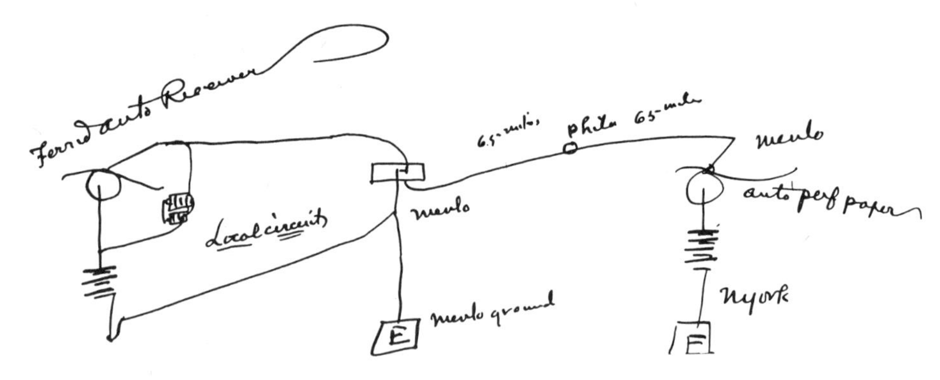


Perhaps we shall have to shunt the E.M.G with a magnet, also to use a plain resistance shunt at Transmitting end even perhaps insert a magnet in the said shunt=

We have made a slight alteration in the method of putting on the springs on the lever of the E.M.G. in all previous Experiments we have put springs near to fulcrum thereby increasing the friction in its pivots. we have now put the springs ~~nearly~~ nearer the extremity of the lever reducing the friction in pivots & making a much better adjustment=[3]

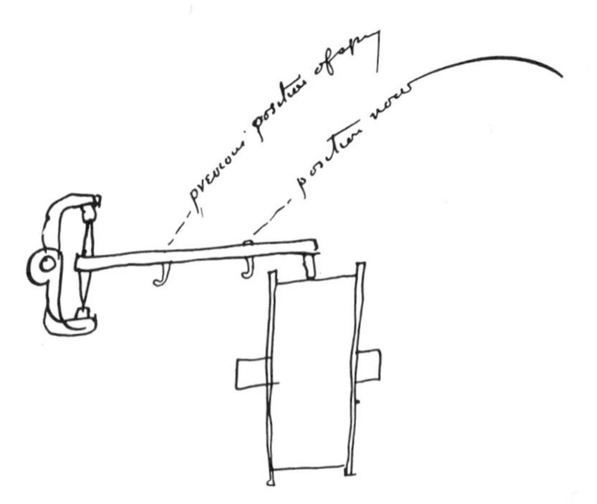


I think that the lever should be arranged thus

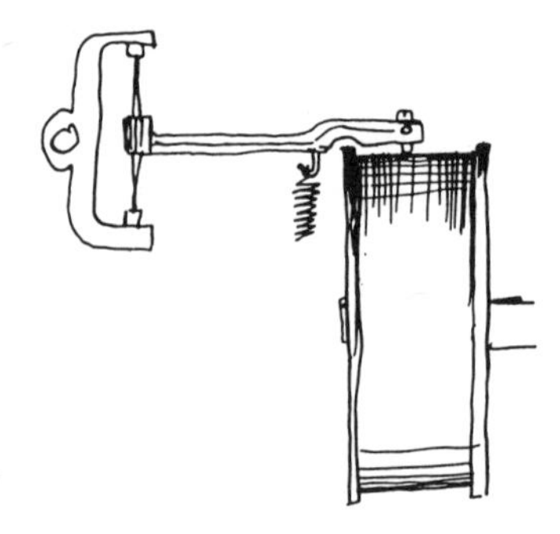

The point wants to be near edge of lever so there shall be now twist= The lever should also be made with ribs= Thus[4]

T A Edison

X, NjWOE, Lab., Vol. 8:304 (*TAEM* 3:732). Document multiply signed and dated. [a]Followed by "over" to indicate page turn.

1. Edison had tried for years to devise a repeater that was fast enough and sensitive enough for the automatic telegraph system (see, e.g., Doc. 105 and *TAEB* 2:521 n. 3, 662 n. 8). He had hoped for some time that an electromotograph instrument would solve the problem (see, e.g., *TAEB* 2:284 n. 4). Later that night and the next day they repeated at 200 words per minute through the Philadelphia loop using a 600 ohm shunt magnet at the transmitting end and 90 cells Callaud battery. (A sample tape from those experiments, supposedly placed in a scrapbook, has not been found.) Vol. 8:309–10, Lab. (*TAEM* 3:737–38).

2. Figure labels are "Ferrid auto Receiver," "Local circuit," "Menlo," "65 miles Phila 65 miles," "Menlo," "auto perf paper," "Menlo ground," and "N York."

3. Figure labels are "previous position of spng" and "position now."

4. Before placing the electromotograph repeater in the Philadelphia loop, they discovered that the point was too sharp to allow the lever to move freely. They flattened the point to solve this problem. Vol. 8:309, Lab. (*TAEM* 3:737).

**-778-**

*Notebook Entry: Electromotograph and Automatic Telegraphy*

[Menlo Park,] August .9. 1876[a]

Electromotograph

Last night we had the most successful of all experiments with the E.M.G= worked it through Phila Loop with 90 cells— first class without a false dot & perfect at 300 words per minute[1] & it did not change its adjustment whether we went at 20 or 300 per minute also run to 800 per minute & got it good but occasionally there would be no breaks between a dash & dot= Shunted[b] E.M.G with 1 spool of a WU 120 ohm relay & sending shunt 600 plain found that whole WU relay was not so good as ½ thus shunted the lever moved with great force at 300 so you could ~~re~~ hear every dot but of course couldnt read=[2]

These results were attained by the use of a very[c] strong solution of acetate of mercury, reduced to metallic state by Caustic Potash= The solution made night before was all together too weak. We worked point up pretty close= I remark here that a successful E.M.G paper must have an initial Lubrication to prevent false vibrations & that is the reason why pyro —caustic soda—silicate soda & acetate Hg or metallic Hg gave best results, because these are themselves Lubricating materials. We also found that with the acetate solution & at 300 per minute we had considerable range of adjustment we used considerable pressure= afterward tried it on a loop to Albany & back 380 m~~u~~iles ~~it s~~ worked very fairly at 100. per minute, though we had too small battery added 30 cells carbon making bat[c] equal 150 cells Callaud= This didnt increase speed[c] or improve it= ~~we qu~~ put WU relay in sending shunt didnt improve it much—varied shunt= ~~it wa~~ came 6 am when we quit= Are having lever stiffened for tonight—[3] & are in meantime trying to wet paper with a Hg Sol & letting roll dry & rewetting with a reducing agent so as to reduce the Hg to metallic state on paper itself & not mechanically as previously—

T A Edison

X, NjWOE, Lab., Vol. 8:313 (*TAEM* 3:741). Document multiply dated. [a]Underlined twice. [b]Obscured overwritten letter. [c]Interlined above.

1. James Adams noted this result in a separate entry. Vol. 10:108, Lab. (*TAEM* 3:908).

2. A fast Morse operator could send or receive upwards of forty words per minute. Three hundred words per minute (roughly seventy-five signals per second) was far too fast to understand.

3. The next entry, dated "evening of the 9th," begins, "Stiffened lever didnt appear cure any of the evils." That entry reports some success receiving at 1,000 words per minute. Edison concluded that "E.M.G is infinitely more sensitive than very Sensitive chemical paper." Vol. 8:316–18, Lab. (*TAEM* 3:744–46).

On 10 August, Edison, Adams, and Edward Johnson experimented further with the electromotograph repeater, receiving 200 words per minute on a Menlo Park–Boston loop of 600 miles (950 km) and receiving 125 words per minute on a Menlo-Philadelphia-Boston-Menlo circuit of 700 miles (1100 km). Vol. 8:341–43, Lab. (*TAEM* 3:771–73).

## -779-

*From Frederic Ireland*

London E.C. August 15, 1876

My dear Sir

The first consignment of Pens ha~~ve~~s—I believe—arrived in Liverpool but I have not yet received them here.— probably I shall have them tomorrow.—

After a discussion with Mr. Breckon[1] we he decided to take offices at the above address and to carry on all business connected with the Pen from there—[2] please therefore address all future Communications to me to New Broad Street, and send Pens there at the rate of twenty per week unless I ask for consignment elsewhere either by letter or wire.[3]

You have not yet sent the particulars of price of the various parts of the pen.—[4] please let me have these by return— I think it may be useful to both of us.

The offer you made to Genl Lefferts has now expired—[5] Will you be good enough to write Confirming the offer you made to me for France Belgium &c—

As soon as I have organized the sale in England[6] &c I will take the Continent in hand—

Pray push the Pens forward as fast as possible— if you can send an extra number please do so I am Yours faithfully

Frederic Ireland[7]

ALS, NjWOE, DF (*TAEM* 13:970).

1. Nothing is known of John Breckon beyond his involvement with Edison's electric pen in Great Britain. See Doc. 925.

2. On 21 August, John Breckon wrote to Edison, assuring Edison of his good will and best efforts regarding the success of the pen in England. By 1 September he and Ireland were calling their business the Electric Writing Co. Breckon to TAE, 21 Aug. 1876; Ireland to TAE, 15 Aug. and 1 Sept. 1876; all DF (*TAEM* 13:972, 969, 974).

3. At this time the pen business was being run by Charles Batchelor out of the office at 41 Dey St., New York, and the pens and presses were being manufactured by Gilliland & Co. in Menlo Park. Charles Batchelor's testimony, TI 1:229 (*TAEM* 11:91); advertising circular "Centennial Exhibition," 76-007, DF (*TAEM* 13:962); insurance policy of 24 Apr. 1876, DF (*TAEM* 13:924).

4. A price list dated 3 October 1876 is in Lbk. 3:6 (*TAEM* 28:498).

5. The agreement (Doc. 746) had presumably expired with Lefferts.

6. In another, apparently earlier, letter of the same day, Ireland asked Edison to confirm Ireland's authority to sell the German patent rights. DF (*TAEM* 13:969).

7. Nothing is known of Ireland beyond his acting as an agent for John Breckon and Thomas Clare in the purchase of rights to Edison's British electric pen patent and his subsequent role as manager of the Electric Writing Co. By October 1877 he had left Electric Writing for the City Circular Co. Ireland to TAE, 29 Oct. 1877, DF (*TAEM* 14:420).

**-780-**

*Notebook Entry: Electromotograph*

Menlo Aug 17th 1876—

Electro Motograph

Experimenting on different porus substances to replace Paper in the E.M.G.—

Found Black Lead[1]—very absorbant —Lamp Blk Lead—Pyro Gallic—solution—Good results—but difficult to keep moist—admits too free passage of current—dont give time for Decom. of Solution—Gets soft on surface—Powdered—& rubbed on metal gives better results.[a]

Chalk & Pyro Gallic— As good as Paper—good conductor at same time admits of motgraph action Holding solution well—chalk[b] not being of itself a conductor—allows the solution to perform that function alone—

Licorice and Pyro Gallic— Licorice soft & rubbed on the Iron—when, nearly dry is quite good[a]

Note It is evident the Electro-Moto action—is due entirely to the solution—It only requires a solid to hold it.[a]

Cork—Pyro Gallic. Pyro-Gallic on surface gives conductivity—but no E.M.G. action[a]

Gum Tragacanth—Pyro Gallic No result—except conductivity[a]

Grape Sugar—Pyro Gallic—No good[a]

Sal. Amoniac—Pyro Gallic No good[a]

Soap Stone—Pyro Gallic Made into Paste—No good

Carbonate of Amonia—Pyro Gallic Works well with good Pressure—but dissolves quickly in the solution[a]

Borax—Pyro Gallic—"Afraid its a Failure"![a]

Camphor—Pyro Gallic—No icks[a]

Note It continues to be apparent that surface conduction dont give E.M.G. action—The action would seem to be direct through the substance[a]

Plaster of Paris Pyro Gallic Thin Layer—mixed with Pyro Gallic Excellent Good Conductor—& Good. E.M.G. action

Pine Wood—Pyro Gallic very good—Will try again when more thoroughly wet.[a]

Tried it again—about same[a]

Chalk is Best so far[a]

Edison E. H. Johnson

X, NjWOE, Lab., Vol. 8:349 (*TAEM* 3:779). Written by Johnson. [a]Followed by centered horizontal line. [b]Interlined above.

1. Black lead (also called plumbago or graphite), lampblack, and gas retort carbon were all forms of carbon. Although the names were sometimes interchanged casually, each designated a material with specific properties. Graphite, the lustrous, slippery form of carbon commonly used in pencils, was employed as well for stove polish and rustproofing. Lampblack was a very finely powdered, flat black form of carbon often used in inks. Both were in turn very different from the shaped carbon

used in batteries and arc lights, formed of gas retort carbon (which will scratch glass) or manufactured from repeatedly baked mixtures of coke, coal tar, or other crude carbon materials. Manufacture of the various types was not standardized, however, and any sample might have properties different from another. For example, plumbago sold for household use was often a much poorer electrical conductor than plumbago from scientific instrument suppliers, and commercial lampblack (also called spirit black) left a white ash if burned, whereas pure lampblack left no residue. Bloxam 1869, 97–98; Sprague 1875, 102–4, 278–79; Gideon Moore to TAE, 25 July 1877, DF (*TAEM* 14:103).

**-781-**

*Notebook Entry: Electromotograph*

[Menlo Park,] August 18 1876

Electromotograph=

Have been making a large number of Compounds of plaster of Paris HO[1] & other things such as Phosphate of Iron, Chalk, Litherage[2] etc to ascertain which is the best substances to make disks of for the ~~mot~~ E.M.G= We having ascertained that the action is entirely due to the decomposition of the fluid. These will not be dry ~~f~~ until tomorrow= Had Wurth fix the EMG from a three point. tripod (which was a failure owing to want of a proper substance to hold the solution) to a disk, thus

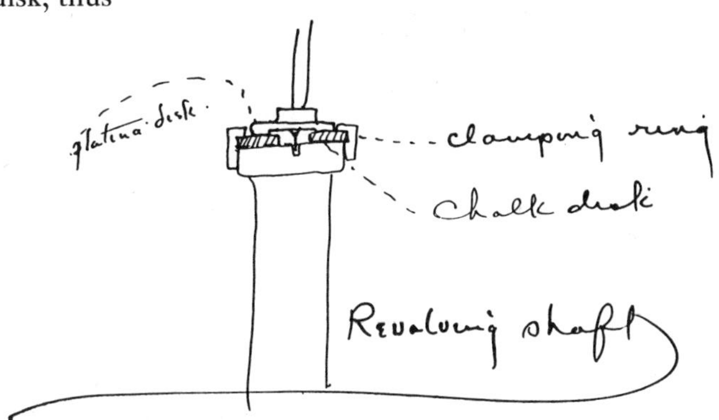


We put on a washer turned from natural chalk, which had hardly been wet with pyrogallic[a] when we found that the Mechanical Vibrating action had ceased[b] and that apparently one of the main obstacles had been removed= by Remoistening and increasing friction spring[c] found that it Repeated at about 100 per min 5000 ohms 20 cells battery perfect[d]

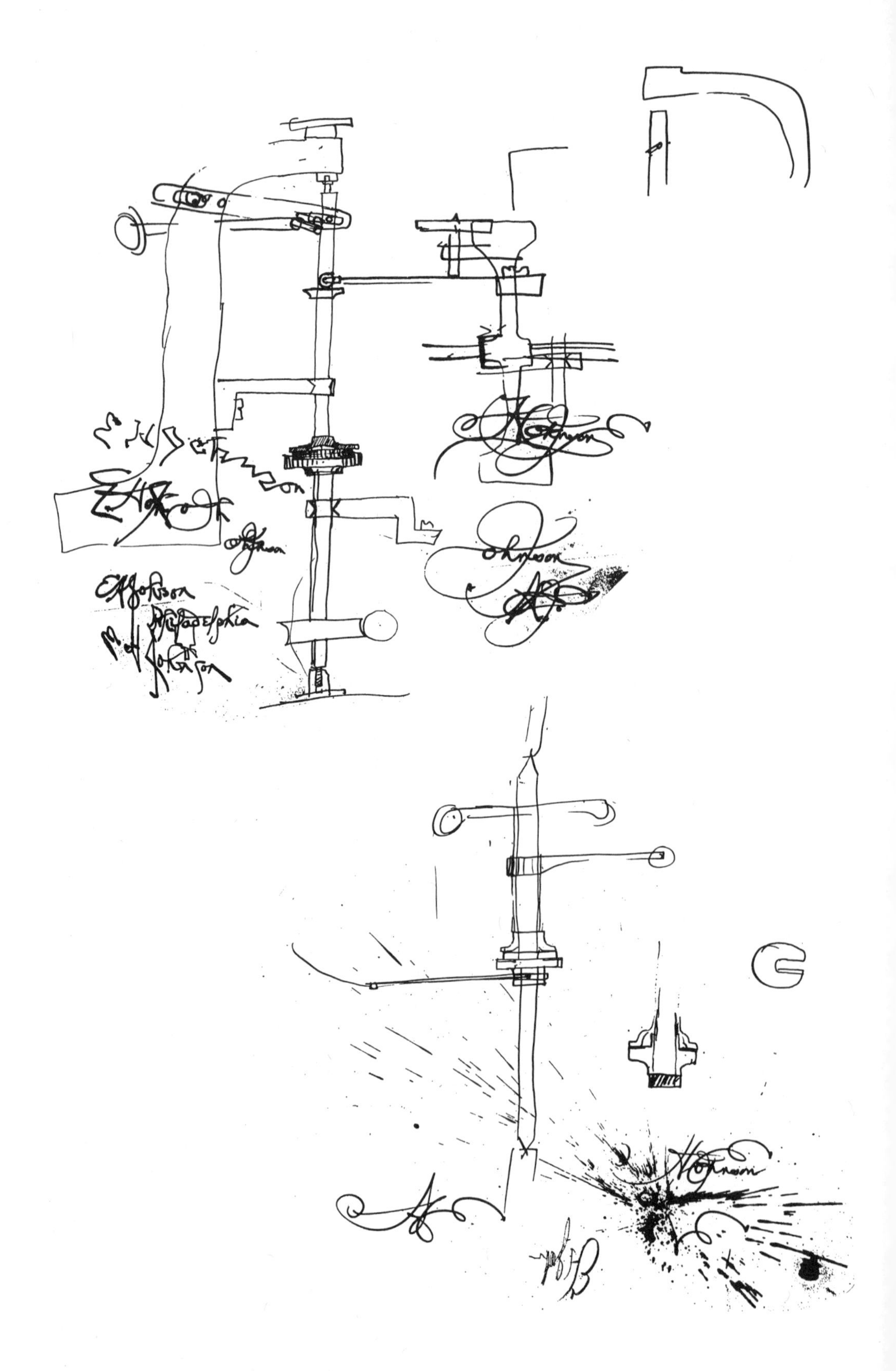
E H Johnson
Philadelphia
Johnson

E.M.G.

With the new E.M.G. & chalk pyrogallic we receive hundreds of wds over loop to Phila= at rate of 200 per minute every dot[e] perfect not withstanding that the shafts run out. We are going to have the shafts all put in clockwork centre to get rid of shake.[f]

X, NjWOE, Lab., Vol. 8:352 (*TAEM* 3:782). Document multiply dated. [a]"with pyrogallic" interlined above. [b]Multiply underlined. [c]Interlined above. [d]Edison marked this sentence with an octothorp in the left margin. [e]Underlined twice. [f]Followed by centered horizontal line.

1. Water. Edison sometimes followed a common chemical notation and designated water in this way; other times he used $H_2O$.

2. Litharge, a fused lead monoxide.

## –782–

[Menlo Park,] Aug 20th 1876

*Technical Note: Printing Telegraphy*

Also with friction[1a]

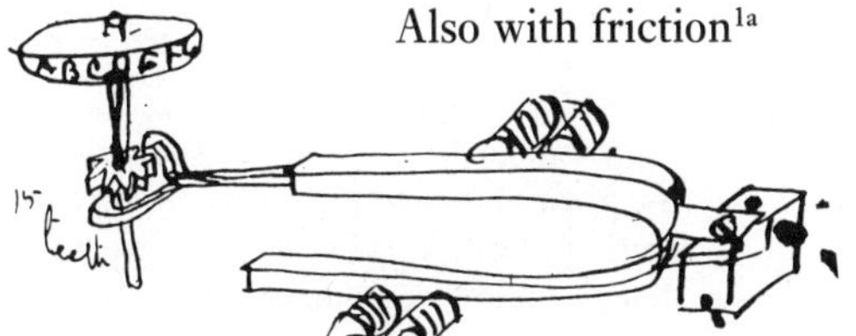

Acoustic Printing Telegraph[2]
Caveat[3]

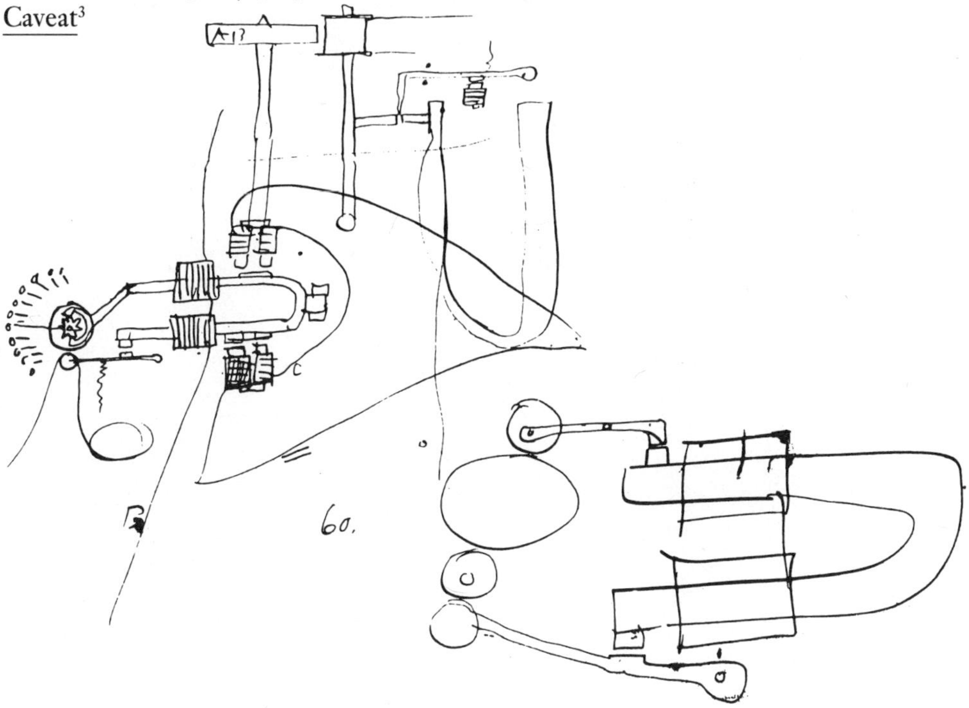

Invented by
T A Edison Chas Batchelor
James Adams John Kruesi

X, NjWOE, Lab., Vol. 10:110 (*TAEM* 3:909). Text written by Batchelor; drawings by Edison. [a]“120” and “60” written below drawing.

1. Figure label is “15 teeth”; the letters “A . . . G” are on the wheel.

2. Additional sketches of this design are in NS-Undated-005, Lab. (*TAEM* 8:295–96, 298–99) and in Cat. 997:26, Lab. (*TAEM* 3:365).

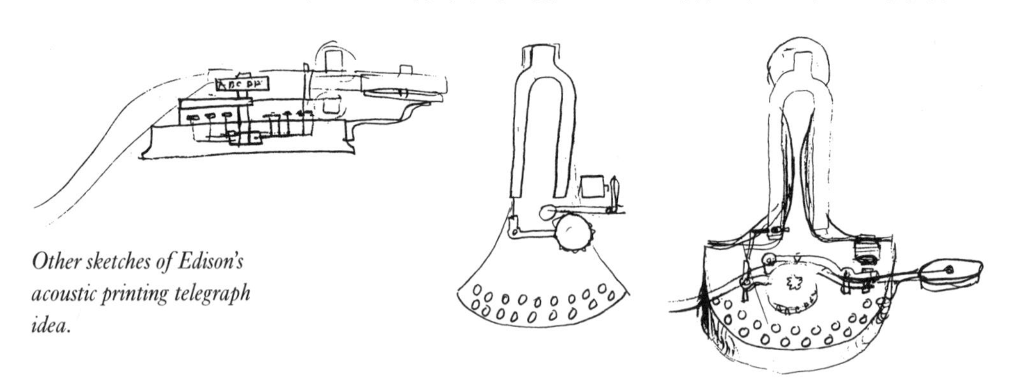

*Other sketches of Edison's acoustic printing telegraph idea.*

3. Edison filed neither a caveat nor a patent application for an acoustic printing telegraph. However, on 30 October 1876 he executed a patent application (U.S. Pat. 200,032) for a synchronous movement for telegraph instruments based on this design and indicated that it could be used for “controlling the movements of type-wheels in printing-telegraphs at distant stations.”

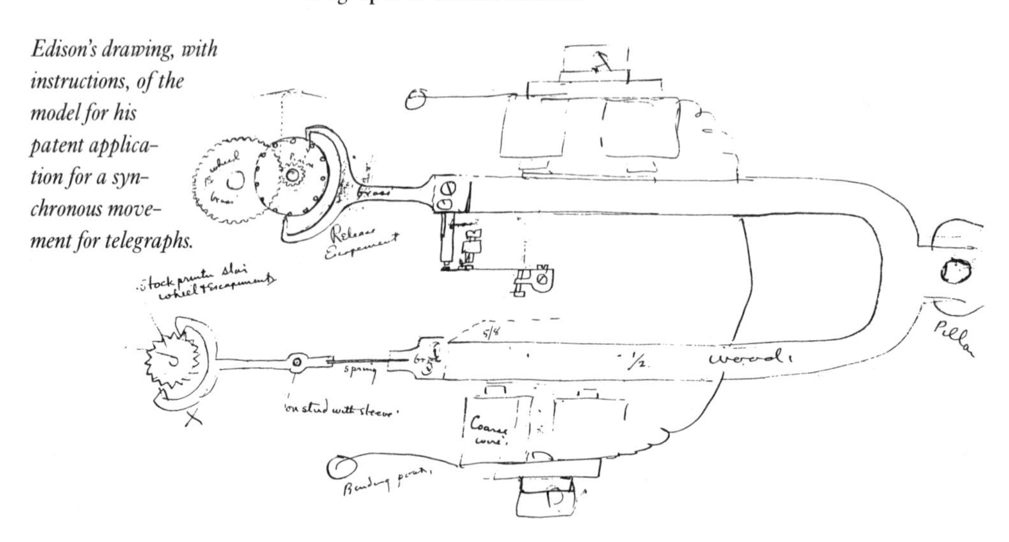


*Edison's drawing, with instructions, of the model for his patent application for a synchronous movement for telegraphs.*

-783-

*Notebook Entry: Introduction to Electromotograph Experiment Series*

[Menlo Park,] Aug 24/76.[a]

Electro Motograph.

Experiments today— To obtain a solution which will show greatest difference in friction with the minimum of Polarization[b]

Note We have found (Aug 23/) that passages of current through any substance containing strong solution of Caustic Soda—that great polarization is had. If the Decomp. points are constantly moved this Polarization is not noticeable & no inverted Current is discharged from the Electrodes.[1] But if the current is allowed to remain closed—the point resting upon one spot—Enormous Polarization takes place—so great that a current from 25 Carbon Batty. is insufficient after 2 or 3 mins to work promptly an Ordinary Relay—but if the Point be moved to a new surface on[c] the chalk the current passes with full strength, & the ~~re~~Relay works strong. Or if the point remains on same spot & the current be reversed—the same effect is had—but again rapidly weakens We have also found that this Polarization is not so great in some solutions as in others for Instance if we wet 4 strips with 4 different solutions respectively Ferrid Cyanide Pottassi. Pyro Gallic Acd—Caustic Soda—& Acetate Mercury & Caustic Potash—they Polarizes very[d] rapidly, except the last, which requires a long closure of the current—to ~~e~~show any Polarization This explains ~~the re~~why the best results have heretofore been obtained by the Mercury Salt. And also why a shunt ~~ca-n~~ould not be used with solutions which Polarize rapidly—the electrodes sending an Inverted Current thus disturbing their own adjustment, by circulating on the Shunt

The above we consider a very important "Bug" discovery— although noticed many times previously—we were not forcibly impressed by it—because of the feebleness of the shewing—The caustic Soda however shewed the Bug so well defined as to compel his recognition—whereupon a great many heretofore unexplained Phenomena—becomes easy of explanation

It is noticeable that Caustic Soda—is an element of Maximum Effects. In whatever direction used it appears to produce results more marked than other solutions.

With this New Element in mind—we now proceed to test for a non-Polarizing agent either to use Independently or to combine with Caustic Soda—[2e]

Ckt. 600 ohms. 25 cells Cold Carbon

Hand Motograph—WU Relay 140 ohm in ckt.

Hand Points—Round 1/16 In Diam (oval end) Plat 1/4 [In][f] Sqr Surface

In each case we use our 20 Gramme Sol. or next below in case No 20 on hand—

X, NjWOE, Lab., Cat. 996:1 (*TAEM* 3:279). Written by Edward Johnson. [a]Undated unrelated sketches appear on facing inside notebook cover. [b]Followed by centered horizontal line. [c]Obscured overwritten letters. [d]Interlined above. [e]Followed by centered flourish. [f]Ditto marks in original.

1. As in a battery, polarization caused by the migration of ions from the solution (acting as an electrolyte) to the points of the electromotograph (acting as electrodes) set up a counter electromotive force.

2. Over the next several days (24–26 and 30 August), the laboratory staff conducted 248 experiments using various solutions with metal points of aluminum, copper, lead, nickel, platinum, silver, tin, and zinc. They compared the resistance when the pens were stationary or moving, the polarization with current flowing in either direction, and the friction with current flowing in either direction. Experiment 248 for a "Caustic soda strip quite Dry" showed the best results and they concluded that "This Fills the Bill." Cat. 996:3–35, Lab. (*TAEM* 3:280–96).

## -784-

*Notebook Entry: Miscellaneous*

[Menlo Park,] Aug 28 1876.

Delicate Electromotograph[1]

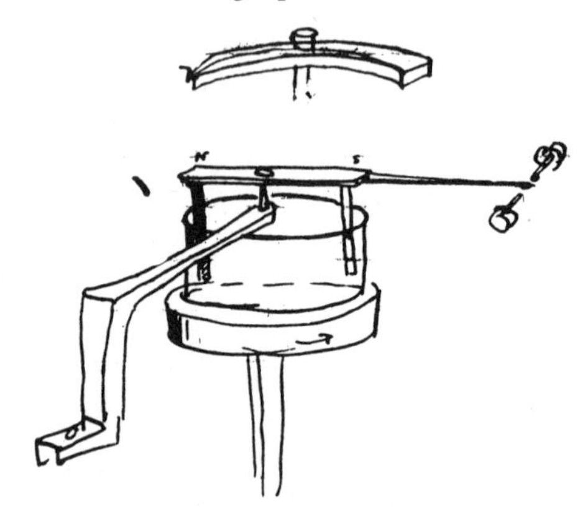

rotated slowly by an Electric Engine.

Reverse Current EMG[2]

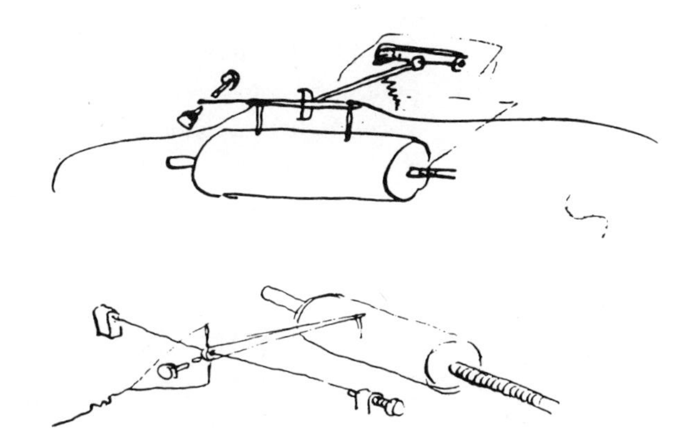

EMG principle  S is a scraper which cuts the chalk off even as it rises & thus gives the recording point & lever a[a] clean fresh surface[3]—

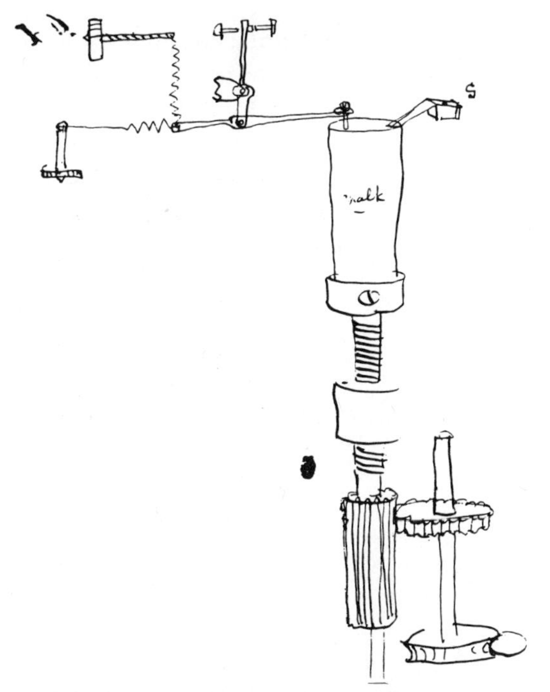


Scheme for Autographic Printing.

Write on a sheet of paper with strong solution of Bichromate of potash, then have dry gelatine covered sheets, slightly moisten these, then lay on Bichromatised writing sheet press. This will transfer writing to Gelatine  then expose to light  this will render parts where Bichromate writing is insoluable, then either wash & print by porousity of paper or by photolithographic style.

Ink Spirter Motograph[4]

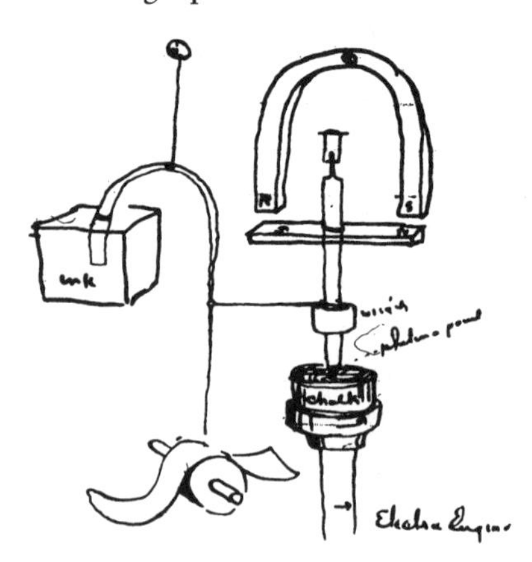


I find that if a glass tube be inserted in a bottle containing HS. gas and the point of the tube drawn out very fine that HS passes all right but by holding paper wet with a Solution of Plumbic Acetate front of the orifice that the mark so obtained is to broad it ~~spreak~~ spreads 1⁄16 even with very narrow orifice. Tried holding little piece of Sulphuret of potash near lead paper but it dont give off enough gas; the peice was rather old.

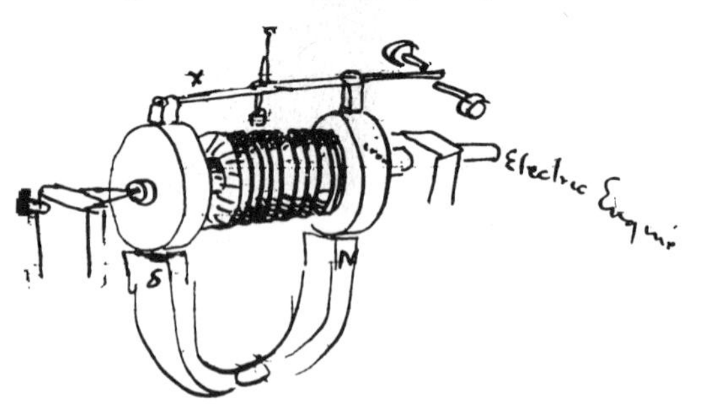


Magnet[5] Constantly rotated=

placing the large permanent magnet where it is, is a mistake the working tongue X should be polarized so that both ends should have the same polarity. This arrangement is a partial application of the EMG principle to a polarized relay.

I have an idea that an artificial rose could be made of s~~o~~uch an nature that it would remain stiff and at the same time absorb by capilliary attraction atto of Rose[6] perfume from a bulb which is filled with it & which acts as the holder in the button hole.

I have an idea that a cheap 25¢ pocket Spectroscope would sell on the street the tube being cast in lead or zinc,[a] Canada Balsam used as a magnifier & common chandalier prisms broken in short peices used for prisms.[a]

I have an idea that the electric pen could be attached to a stiff brass pantagraph and very nice drawings made with it=[7]

I have an idea that perforated stencil sheets similar to those produced by the electric pen only done by a stamping machine & with larger holes representing outline statuary, Landscapes etc could be sold in large quantities to Druggest etc to ornament mirrors windows etc by marking through the stencils with French Chalk; etc.

in addition books accompanied with Colors etc could be made whereby with a number of stencils any kind of a pic~~a~~ture, could be produced, & these books sold to the trade to be used for educational purposes & for amusement. Spoke of this over 8 mos ago=

Battery.

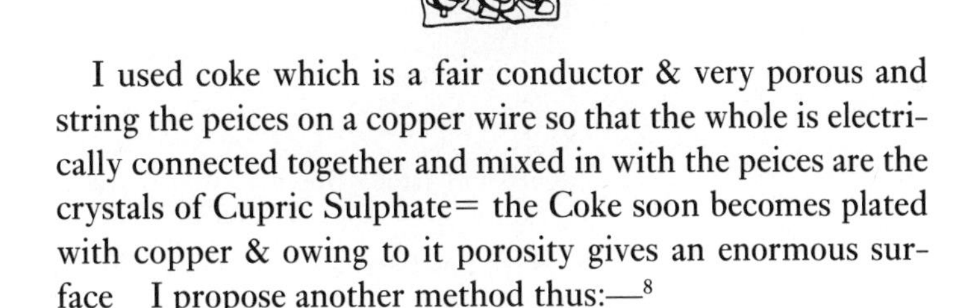

I used coke which is a fair conductor & very porous and string the peices on a copper wire so that the whole is electrically connected together and mixed in with the peices are the crystals of Cupric Sulphate= the Coke soon becomes plated with copper & owing to it porosity gives an enormous surface I propose another method thus:—[8]

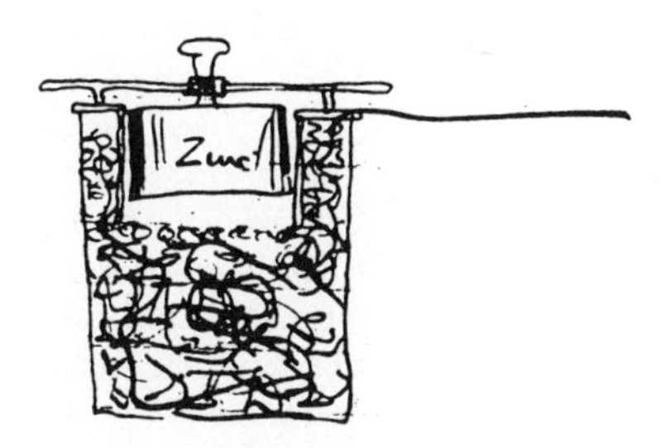


I use a porous tube open at both ends in which I place the zinc hence I am enabled to place the coke from bottom to top and near the zinc.

Another method is to place a disk of Copper at the bottom and fill in granulated tin copper or[b] lead.

Powerful inexpensive condensers.

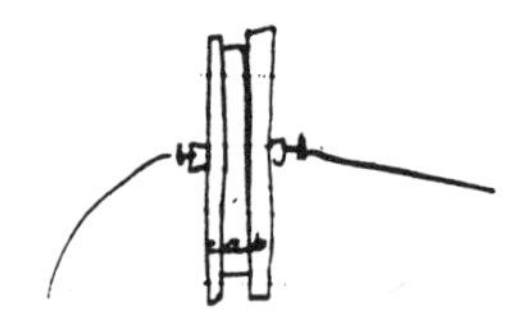

c & b are platinized disks, a is a block of chalk or other porous substance[c] saturated with polarizable chemical such as Caustic Soda. (Strong.)

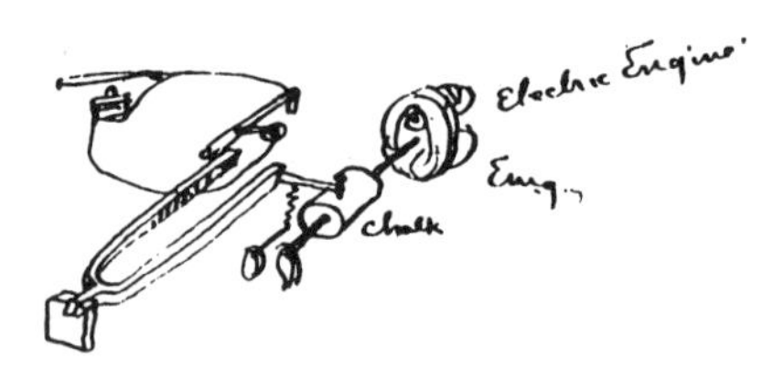


1st Acoustic Telegraph[9] I propose to use a number of tuning forks upon the ends of which are lever resting up revolving chalk cylinders moistened with a chemical solution, and so arranged that the Em.g. principle sets the fork in motion, one single cylinder may be used to set several forks in motion.

T. A. Edison

Chas Batchelor
James Adams

X, NjWOE, Lab., Cat. 997:5 (*TAEM* 3:354). Document multiply signed and dated. [a]Obscured overwritten letters. [b]Interlined above. [c]"or . . . substance" interlined above.

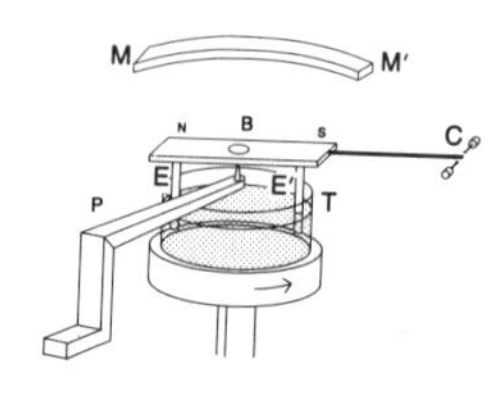


1. In this electromotograph design, rotating tub **T** contains a liquid into which electrodes **E** and **E′** project. Magnet **MM′**, with its poles opposite to those of bar magnet **B** (with poles **N** and **S**), holds **B** in place against the rotation of the liquid. Apparently the electromotograph action—a change in friction—is to occur at the surface of the electrodes as the liquid flows around them, changing the torque on **B** and causing point **C** to move between the contact points. The line current would pass through the liquid and into the electrodes, but it is not clear whether the current would then pass into bar magnet **B** and through the pivot arm **P**, or be taken off the electrodes directly by wires. The local circuit, which would pass through contact point **C**, is not shown either.

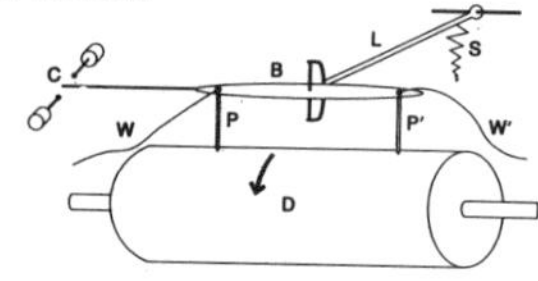


2. This is the electromotograph equivalent of a polarized relay, in that it is quite sensitive and responds to reversing currents. Here the current flows through wire **W** and point **P**,

drum **D**, and point **P′** and wire **W′**. Edison had found that some combinations of metal and solution change friction oppositely depending on the direction of the current. Because the current passes through points **P** and **P′** in contrary directions, using such a combination causes the bracket **B** (at the end of lever **L**, held down by spring **S**) to pivot in reaction to changing currents, causing point **C** to move to one or the other contact point.

3. Figure labels are "chalk" and "Electric Eng."

4. Figure labels are magnet pole designations and "ink," "weight," "platina point," "chalk," and "Electric Engine." A drawing of 12 September 1876 shows this device in a circuit. Cat. 997:22, Lab. (*TAEM* 3:363).

5. Figure labels are "X," "S," "N," "iron," and "Electric Engine."

6. Attar of rose, a fragrant essential oil made from rose petals.

7. On 10 September, Edison proposed using a pantograph in combination with an electric pen for perforating messages for an autographic telegraph (Cat. 997:21, Lab. [*TAEM* 3:362]; see also Doc. 934). On 29 August he also proposed two alternative methods for operating the pen; one used a flexible spiral spring and the other incorporated an air engine (Cat. 997:9–10, Lab. [*TAEM* 3:356–57]).

*An electric pen connected to its power source by a spiral spring.*

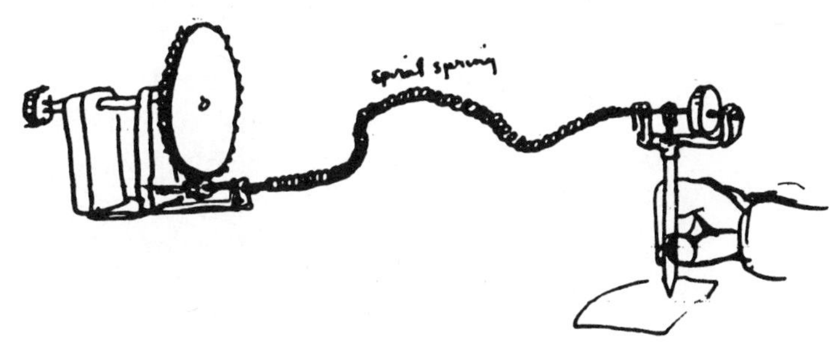


8. Figure label is "Zinc."

9. Figure labels above are "chalk," "Emg," and "Electric Engine."

## -785-

*Notebook Entry: Relays*

[Menlo Park, August 1876][1]

An Idea[2a]

The idea has occurred to me that the EMG principle of wkg by a difference in friction[b] might be applied to an electro magnet. I took a magnet[c] extended one core by placing an armature upon it & by opening & closing the key & passing along a flattened iron wire, the thumb[3] produced by the friction caused by magnetization and the normal friction ~~the~~ was apparant. I propose to embody this principle in a machine thus[4]

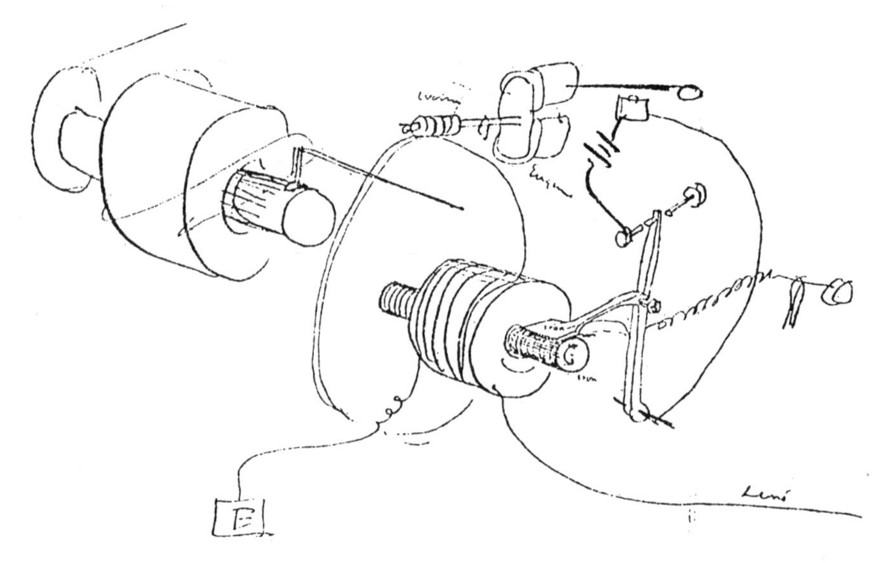

Of course th~~ere~~is arrangement ~~doesn't work by~~ isnt exactly like the principle of the EMG working by difference of friction, its ~~the~~a [p]rinciple[d] of increasing delicacy of magnetic relays by working in actual contact, when the end of the iron [co]re[d] has a ring of ivory of considerable thickness to prevent the effect of [re]sidual[d] magnetism  it will be [n]oticed that the power with which [a][d] relay lever moves is always the [sa?]me;[d] no addition over a certain amount can change the adjustment  [In?][d] fact this principle opens a new [fi]eld[d] in magnetic telegraphy [and?][d] in electromechanics

Actual Contact Magnet Relay  EM.G. [iron?][e] Magnet[5]

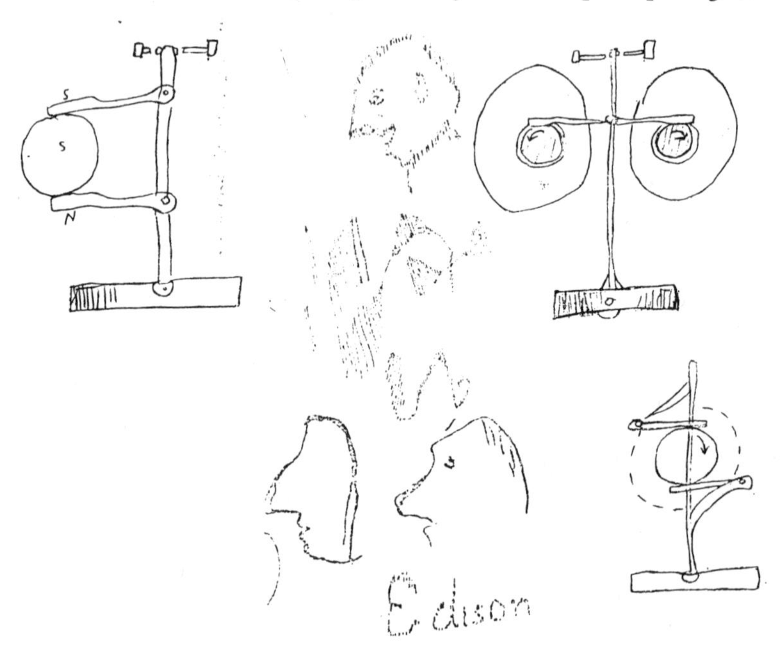

X (photographic transcript), NjWOE, Lab., Vol. 8:320 (*TAEM* 3:750). [a]Multiply underlined. [b]"of . . . friction" interlined above. [c]Obscured overwritten letter. [d]Left margin not copied on photographic transcript. [e]Illegible; interlined above.

1. Date based on Edison's testimony in TI 1:83 (*TAEM* 11:62).

2. Edison later testified that this notebook entry illustrated a principle that he did not embody in an instrument until February 1879, when his nephew Charley Edison constructed an electromotograph telephone along these lines. Edison's testimony, TI 1:77–83 (*TAEM* 11:59–62).

3. Edison meant "thump."

4. Figure labels are "worm," "Engine," "iron" (at lower right center on contact and below shaft), and "Line."

5. "Actual . . . Magnet" may have been added later.

**-786-**

*From Norman Miller*

New York, Sept. 4 187[6][a]

Friend Edison,

You dont seem to get up here "much" lately.

Mr. O.[1] says that he has set so many times in his own mind to go down to see you and failed to be able when the time came that he will not try to say "when" again but start if any time he is able—

Meantime he thinks for your good some one should make you a visit. He understands why you do not want certain persons to do so, and I know fully appreciates your reasons.[2]

But he suggests that as you are on Supt Van Horn's[3] wire that you give him (Van) a written invitation to come down.

He will take it to Mr O. of course, and will get his instructions, which[b] will be to "go and see and say nothing to any one but the man he goes for." You will like Vans quiet observing way and you can trust him surely. He knows but one man in New York.

Batch has got a good office here and every thing in it that the heart of man can desire except a decent pen. Yours

N. C. Miller.

P.S. V. was away all last week, just home now send up to him early as O. says Gray is crowding things, and is now here.[4] M

ALS, NjWOE, Supp. II, 1876 (*TAEM* 97:593). Letterhead of Edison's Electrical Pen & Duplicating Press Co. [a]"New York," and "187" preprinted. [b]Interlined above.

1. William Orton.

2. See Doc. 732.

3. John Van Horne (b. 1827) was general superintendent of Western Union's southern division. He entered telegraphy in 1850 as an opera-

tor and during the Civil War was president of the Southwestern Telegraph Co. In 1878 he was elected Western Union vice president in charge of the electrical department, statistics, and contracts. Reid 1886, 668–69.

4. A 23 September *Telegrapher* article indicated that Elisha Gray had been at the Western Union office in New York for several weeks demonstrating his electro-harmonic telegraphic system for sending four messages each way on a single wire. The article described a successful test between New York and Philadelphia on 21 September. "The Electro-Harmonic Telegraph System," *Telegr.* 12 (1876): 232.

**-787-**

*Notebook Entry: Electromotograph*

[Menlo Park,] Sept 4th 1876.

Electro Motograph Continued—[1]

Pending the construction of an Instrument designed to present a new surface to the friction point so as to avoid the effects of Polarization[2]—the tests found on the preceding Pages were made having in view the attainment of the same object in a different way—ie—by finding a solution having great frictional Difference with a minimum of Polarization[3]

Of the ~~L~~Results of the latter, record will follow—[4]

Of the result of the constantly changing surface—that is to say—the Instrument with chalk cylinder & travelling Pen we have only to record as yet that it fails of its purpose—

We now go back to the small chalk Disks & Platina Point—& find certain results say 200 words per min. On[a] 30 000 ohms nearly[b] always obtainable but not reliably[c] we find that it is not Pressure that is required—but friction created by the ~~p~~chemical solution in which the chalk is steeped—so that when the current passes & this friction is neutralized by it the normal friction of the chalk alone is acti~~on~~ve— this latter by reason of the use of but slight tensions on the springs being slight—the retractile spring is left free to operate with great promp~~t~~tness.

We now proceed to Experiment with this apparatus steeping the small chalks in the solutions recommending themselves as efficient in the properties we require Viz.

~~Strong Increase of~~

Strong Normal Friction upon Open current—& Complete Neutralization of same upon passage of current through it—

Our previous notion has been that our normal friction was to be obtained by great Pressure on the Point. We now find this is wrong, & that we must have a normal friction generated by the chemicals alone to the end that it may be neutralized by the current—& the retractile spring have only to overcome the light pressure of the spring we are by this means enabled to use—

1 Caustic Soda. 2. Gr. Balsam Canada 5 Gr. Caustic Potash 2 Gr. 100 cc HO No good—

2 5 Gr. Balsam Copaiba 1 Gr. Caustic Potash— 100 cc HO No good—

3 5 Gr. Gum Guiac[5] 5 Gr Caustic Potash 100 cc HO no good—

4 Caustic Potash—Gamboge. Sesqui chloride of Iron 100 cc HO no good

5 8. Gr. Bi Carbonate Soda. 100 cc HO. no good

6 Got led off into general Experiments with the point of the pen Tried it shovel shaped thus  but found Fluff would collect underneath it & destroy action— Turned the same point perfectly upright thus — also failed — then twisted it to a position at right angles thus so that it travelled edgewise— This apparently works very well— but may be found less effective than a point.

Tried on these various points[a] various solutions of Gamboge Caustic Soda & Caustic Potash—as follows

Saturated solution of Caustic ~~Potash~~ Soda[b] & Gam Strong Sol. Caustic Soda—Alcohol & Gamboge Caustic Soda & Pepsin[a] &c &c Getting very fair result. But not at all satisfactory—

We tried placing a piece of paper upon the chalk—& found good results— Immediately[a]—following them up & steeping the paper in a solution of Caustic Soda Gamboge 100 cc HO by nice adjustments we got 250 words per min Perfect on 36,000 ohms—good—[6d]

X, NjWOE, Lab., Cat. 996:37 (*TAEM* 3:297). Written by Edward Johnson. [a]Obscured overwritten letters. [b]Interlined above. [c]"but not reliably" interlined below. [d]Followed by wavy horizontal line across page.

1. See Doc. 783.

2. See Doc. 784.

3. See Doc. 783.

4. Actually, "the results of the latter" statement in the above paragraph are in Doc. 783 and preceded this entry.

5. Guaiac (or guaiacum), a resin.

6. The following morning Edward Johnson noted that "We find great difference in Paper—some Papers giving good results others failing to operate at all—." The staff then tested a number of papers using a solution of 5 g caustic soda, 5 g gum gamboge, and 100 cc water. At 11:30 A.M. they concluded that the best paper tested to that point was J. F. Luhm & Co. chemical labels. Cat. 996:40, Lab. (*TAEM* 3:299).

**–788–**

*To Norman Miller*

[Menlo Park,] Sept 6 [1876]

N C Miller

Yours rec'd. just as soon as I get me new Acoustic Insts[1] finished will invite Mr Van Horn down. I had previously been experimenting with a rather rough kind of apparatus & started some time ago to make a complete set. as I have only 2 men, and not much money it goes slow=[2] Yours

Edison

ALS, DSI-NMAH, WUTAE.

1. Docs. 773 and 795.
2. Edison's labor records of work on the acoustic for the week ending 8 September indicate that there were two full-time machinists and a third machinist working half-time. 75-020, DF (*TAEM* 13:731).

**–789–**

*Notebook Entry: Electromotograph and Telegraph Recorder/Repeater*

[Menlo Park,] Sept 7 1876—

New system based on the Electromotograph. first I propose to use Morse registers or other means such as a magnetic engine to move the strip of paper along and I arrange a single punch & mechanism on the lever of the register which is moved by a magnet in a local circuit operated by a key. This little punch is vibrated with immense rapidity in the same manner as the needle in the electric pen & with similar mechanism= & the whole is so arranged that when a message is sent on the key the lever will bring the punching devices in play & punch hole in the paper instead of embossing the characters as when the register is used in the ordinary manner=. This ~~e~~punched strip is then taken to the transmitter which consists of a platina faced drum rotated by suitable mechanism & the paper passed rapidly through it & the contact roller serves to open & close the circuit & transmit the characters to the distant station. It ~~is~~ there sets[a] in motion the lever of the Electromotograph which ~~m~~responds to several hundred words per minute this lever opens & closes the circuit of a local battery & local Electromotograph whose lever moves with great force this second lever serves to bring into play a perforating apparatus similar to that employed in preparing the transmitting slip. this reproduces the messages on a strip perforated. this strip is then passed between a point & drum; rotated slowly and connected to a sounder & the operator copies the message by sound[1]

Puncher

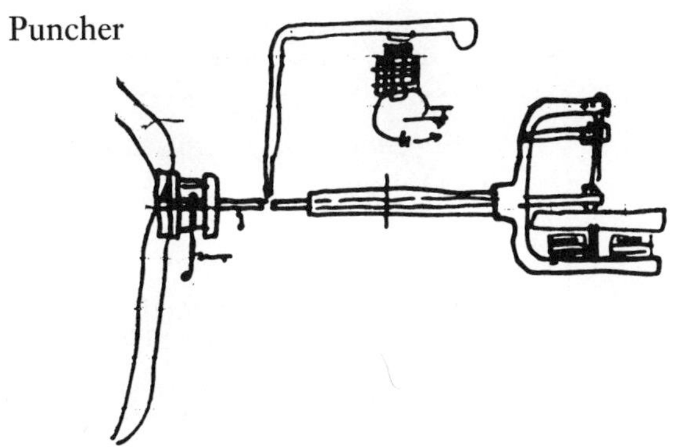

T A Edison

ADDENDUM[b]

[Menlo Park,] Sept ~~7~~9 1876

Another form for transmittin[g?][c] puncher[2]

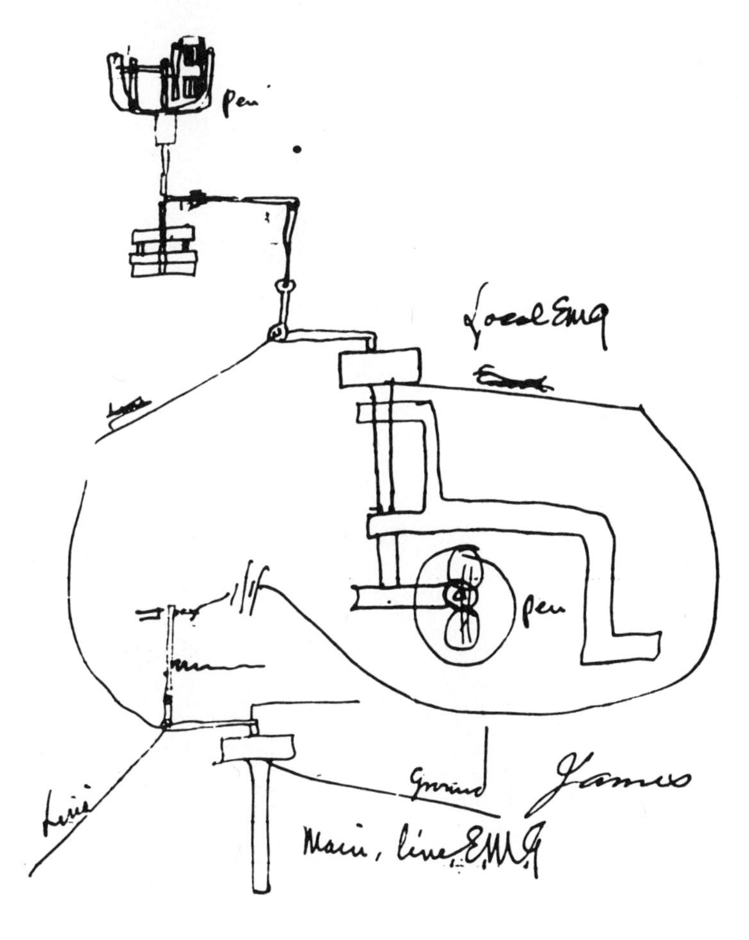

Specimen of transmitted & reproduced perforations.[3]

T A Edison Chas Batchelor

James Adams

X, NjWOE, Lab., Cat. 997:16 (*TAEM* 3:360). [a]Obscured overwritten letters. [b]Addendum is an X. [c]Illegible.

1. Five years earlier Edison had proposed using vibrating punchers in automatic telegraphy. He had also conceived a system in which receiving punchers were used to record messages that then operated a printer, rather than sounders as proposed here. See Docs. 151A and 194.

2. Figure labels are "pen," "~~line~~," "Local Emg," "~~Earth~~," "pen," "Line," "ground," and "main, line, E.M.G."

3. Over the next several days Edison designed new levers for the electromotograph (see Doc. 792) and devised alternative puncher plans, including one using a tuning fork like his proposed acoustic printing telegraph (see Doc. 782).

*A paper-tape puncher powered by a tuning fork.*

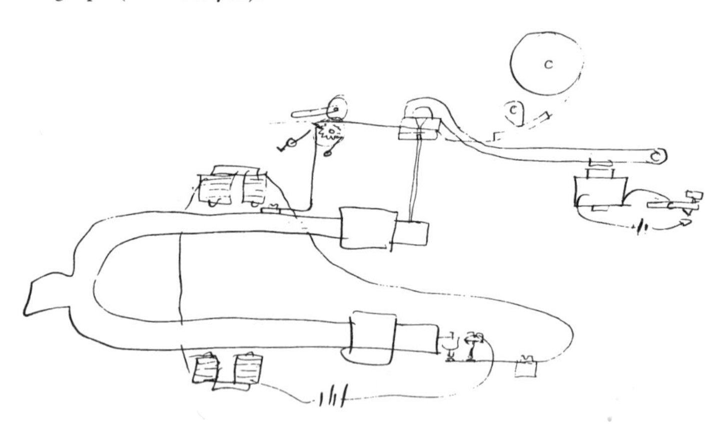

A note of 12 September indicates that Charles Wurth was beginning to make a "complete magnetic puncher" (Cat. 997:19–25, Lab. [*TAEM* 3:361–64]). This instrument probably resembled the mechanism shown in Edison's U.S. Patent 200,994 (a perforator and transmitter similar to these designs, but not employing the electromotograph), which application he executed on 30 October 1876.

*In this perforator/transmitter, the paper tape leaves the motor-driven puncher (figure 1 and at left in figure 2) and feeds directly into the transmitter (at right in figure 2).*

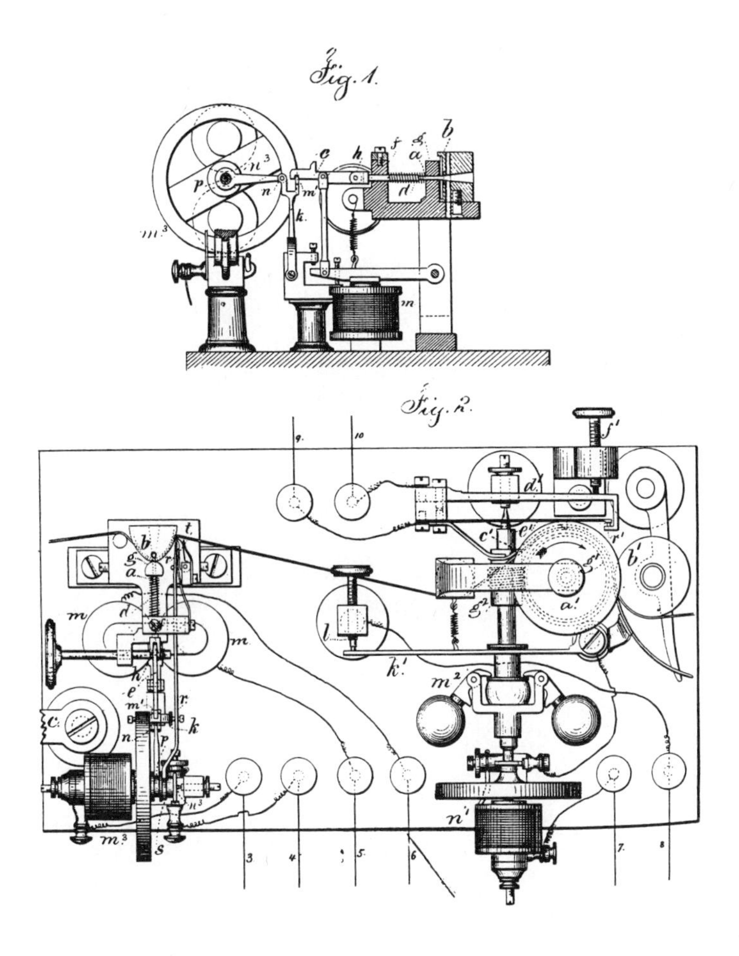


**-790-**

*From Pitt Edison*

Port Huron, Mich, Sept 11 1876[a]

Dear Bro

Your received Wastell informs you that the GTR has comenced suit[1] thay did the amt thay suid for was $1400.00 and at the same time the GTR owed the StRR Co about $1000.00 on Cartage and $208.00 on mail and I did not think best to anoy you as the diffences could and has ben all arranged to the sattisfaction of both partys what Wastell wants is to destroy any confidence you have in me in the managemt of the StRR he boasted the other day that he started in to bust the old road and by G—— he would do it thare will be nothing verry arlarming happen the Road with out you knowing all about it in time so let Wastell write and send me his letters so I can keep posted on his movemets he is as much of a enemy[b] to the road as he ever was Abot Stewart buying the $250.00 in the Sarnia road[2] he bot it fare and square as he ever bot a hat but he supposed before the note became due he could sell at par  he did not sell and he wanted to forse me to take it and thare is whare his mad comes in

I see that Wastell wants to make you believe that the road is all agong to the devil and make you sick and[b] get you to take less  Gouldens[3] offer was $13 000.00 or the old agreemnt[4] carrried out and so I Telegraph you but when I told him all right he said thay would give you the amt with out intrest and wanted me to advise you to take it  I told him I would not it would only be a waste of time and I thought you would sinck your stock first  did you get the statemts[5]

WPE

Atkinson[6] has gon east thay said he would see you[c]

ALS, NjWOE, DF (*TAEM* 13:1066). Letterhead of Port Huron and Gratiot Railway Co. [a]"Port Huron, Mich," and "187" preprinted. [b]Obscured overwritten letters. [c]Written in left margin.

1. This may be related to the Grand Trunk Railway's subsequent effort to remove the Port Huron and Gratiot Street Railway Co. from its yards. See Doc. 832.

2. Pitt was secretary of the Sarnia Street Railway Co. across the river from Port Huron. *TAEB* 2:612 n. 2.

3. James Goulden, a liquor and cigar wholesaler, was a shareholder in the competing City Railroad Co. Endlich 1981, 98; Jenks n.d., 8.

4. In a letter of 28 August 1876, Pitt told Edison that Goulden offered $3,000 for William Stewart's stock, but Stewart refused anything less than $3,500, and that Goulden would also "see that the old agreement that is the bonds & mortgage on the road would be carried out in one year if we would pay the difference on $500.00." DF (*TAEM* 13:1065).

5. In his 28 August letter Pitt stated that he would begin sending Edison weekly financial statements of the Sarnia Street Railway. These have not been found. DF (*TAEM* 13:1065).

6. Probably attorney O'Brien Atkinson, who was connected with the City Railroad. *History of St. Clair County* 1883, 554; Jenks n.d., 15.

## –791–

*Notebook Entry: Chemistry and Telegraphy*

[Menlo Park,] Sept 11 1876

Discovery[1]

When Aniline oil is treated with Nitric Acid (fuming) and then thrown into a large quantity of H.O. a precipitate takes place ~~b~~and a yellow coloring matter disolves in the water; If this be filtered off it may be enormously diluted without destroying its delicacy to electrolysis. If a strip of paper be wet in[a] the solution it gives on the Oxygen platina point a reddish mark not very brilliant rather dull, ~~by increasing~~ 3 cells carbon batty[b] this mark turns greenish, if you increase the resistance to 1000 ohms the mark is green when formed  20 000 ohms may be inserted without any material weakening of the mark=[c] But the most remarkable thing is that if to the solution Nitrate of Lime & Chloride of Calcium be added no mark is obtained at all on short circuit[d] & it only commences to appear after 2000 ohms have been inserted and gradually increases in sensitiveness until 10 or 12,000 ohms have been put in. the mark is seen at 20 000. ~~Thus~~ This is probably due to the fact that with a certain battery power the Nitrate of Lime or Chl ~~C~~Sodium is decomposible & the constutants prevent the appearance of the mark[e] and it is not until the current is so weakened that these substances are not decomposed that that free oxygen is given off at points & oxidizes the Aniline. This opens a new field in telegraphy, as we may by balancing certain solutions, transmit ½ of a Roman letter by a weak current & the other ½ by a strong current & in many other ways use it even in the dot and dash for we can send a heavy[a] current to make the spaces which will not record & use the weaker current of the static discharge to record=[2]

Good EMG Solution

Nitro[f] Aniline as described on last page mixed with Chloride Calc~~o~~ium HO Solution & neutralized so as to leave excess Caustic Soda= this decreases on 20 000 ohms 3 cells Carbon quite plain & increases on .O. Mistake can't ~~fell~~ feel through 20 000 ohms.

Cellulose Solvent=[3]

I placed some tissue paper in a bottle containing crystals of chloral hydrate three weeks ago, having noticed that the cork stoppers (which are cellulose) were dissolved by the vapors. I now find that a portion of the paper is dissolved, at least it has every appearance of it, having[a] aglutinized & is somewhat transparent, while the cork also thrown in at the same time among the crystals is unmistakably. dissolved. I took the paper out and put in a test tube with alcohol. this didnt appear to dissolve it more; heat it, also put camphor in, also chloraform  no increase in solvent power noticed. I have c~~ro~~orked the tube & laid it away & put fresh paper in the bottle. I took some crystals & mixed them with paper & put in a test tube & melted the chloral by heat. it appeared to ½ semi dissolve the paper  put chloraform in this didnt help it— corked & set away= put some paper in tube with crystals chloral & Bisulfide carbon. have set it away

T A Edison

X, NjWOE, Lab., Cat. 996:47 (*TAEM* 3:302). Document multiply signed and dated. [a]Obscured overwritten letters. [b]"3 . . . batty" interlined above. [c]"material . . . mark" underlined twice. [d]Multiply underlined. [e]"& . . . mark" interlined above. [f]Interlined above.

1. This entry follows a series of experimental observations begun on 9 September with chemicals that Edison had used in early August (see Doc. 775).

2. Edison used strong and weak currents in his diplex circuit designs. *TAEB* 2:299 n. 6.

3. For Edison's earlier work on cellulose solvents see Docs. 583, 586, 645, and 655.

**-792-**

*Notebook Entry: Electromotograph*

[Menlo Park,] Sept 14 1876

Electromotograph Experiments for last 5 days.

After trying a great variety of disks of paper upon the chalk, both hard glazed, biblus and high lead glaze; we thought of silk. Cut a washer of very thin silk wet it with a solution of Caustic Soda and Gambooge and laid it on the chalk when the results were splendid and exceeded ~~anything~~ everything tried (papers & chalkly subs) afterwards we tried a number of silk disks & different solution but we did not afterwards obtain such good results as with the first disk of silk which may be attributed to some trifling change in the conditions,= We notice one thing about silk which is different from anything yet tried and that is its beautiful eveness as seen under the microscope; and the great pressure you can place upon the EMG lever without abrasing the surface. In fact the point run in a circle on the silk disk for ½ an hour & with considerable did not shew the circle or an abrasure even when placed under the microscope 1¼ power and even when it has been used for 2 or 3 hours, a thorough soaking of the disk caused this circular abrasion or track to disappear. My opinion is that a disk of silk would last for a[a] week at least, which cannot be said of any other substance known to me;

Another curious thing about silk as the rotation is that it only holds the chemical solution in its interstices, the solution not penetrating the fibre itself as is the case with paper ~~and~~ it holds water only by capilliary attraction in the same manner as a little pile of powdered glass would attract & hold water by capilliary attraction

After trying a very great number of experiments we found that the mechanical vibration which on silk laid on a smetal disk & dry is small became great when moistened and caused the mutulation of the signals; We also found that although the

Caustic Soda & Gamboge Solution had great delicacy it could not be shunted to any extent like Acetate of Mercury & Caustic Soda or potash which is attibuted solely to the fact that it slips on Oxygen whereas Acetate of Hg & Caustic Soda or potash increases friction on Oxygen thus with the Gamboge solution a dash is sent with one current lubrication takes place & when the Counter charge of the magnet comes it also lubricates so in this case compensation by a reverse current does more harm than good in fact prevents the EMG from working while if we use Acetate of Hg & Caustic K or Na ~~the~~we may shunt it with a magnet extraordinarily low as in this case the counter current acts to increase the friction & we thus get double the working difference besides allowing of the use of a Condenser & thus cutting out constant currents. However after finding the important part played in mutilating signals and redering ~~re~~even results pecarious by the uneveness in friction of ~~s~~moistened substances it occured that th~~er~~is might be compensated for by securing to a shaft several independant leves provided with platina[a] points & resting upon the disk of material each ~~leave~~ lever being held down by a spring and working entirely independant of ~~o~~the others yet all serving to give motion to the lever which closes the circuit. A drawing of this device is shewn on page 23, Vol. 5[1] we had it made and although made in a very crude manner it gave great deal better results ~~tha~~ with the inferior solution we were using than any arrangement yet devices. The theory of using several independant levers to obtain an even friction was this that any bad spot on the disk would entirely destroy a dot or dash if one point were used or one half of the disk might be wetter than the other half ie[b] from edge to centre; ~~an~~causing say ½ doz words to come heavy & ½ doz light wherase with several leves a bad spot ~~w~~acting on only one lever would be insufficient to give a false signal it being necessary that the bad spot should pass at least under 2 if not 3 levers; and if one half of the disk had more friction than the other ~~o~~the other half will have more henc 2 leves may be on a less & 2 on a more frictional surface, hence eveness; but we found that owing probably to the inferior manner in which the lever were made & their loseness upon their fulcrum the run about on the disk.

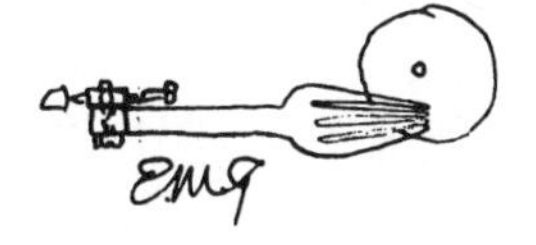


*A four-pronged electromotograph lever.*

We then changed the style slightly and made a 4 ~~spring~~ pronged spring as shewn on page .20. Vol 5.[2] This worked beautifully smooth on dry silk= We had previously been out of acetatc Hg & having recd 2 oz we made a solution with

Na. & .K and henceforth use it= ie[b] Na Acet-Hg= found that on[c] wetting this dry silk mechanical vibrations came in quite strong= and this proved that the four points did not give enough compensation which we attribute to the fact that they are all rigid together and if one lifts by a piece ~~p~~of acetate Hg or grit it tends to lift its neighbors thus decreasing the total friction which was not the case with the four lever= ~~and~~ another thing which militates against this form is that the material passes under each point at different rates of speed and a bad spot in the inner spring would not be so effective in mutilating as on the outer as the lat[t]er moves faster;

I believe the relay lever style is the best, but the springs should be seperate and run independant all the way to the lever and have their pressure put on their ends & at the same time be arranged to run all or nearly so in the same channel & nearly equidistant from the centre;

While experimenting with this form ~~w~~there was considersable muttilation of Signals with silk laid on platina faced disk & moistened with thick black curdy AcHgNa-Solution— and we could shunt on 140 ohm WU relay & 500 ohms through 25 000, 26 cells old Carbon; no sendg shunt[d] & get occasionally good words at 100 wds per minute speed. Having[c] tried an experiment previously of allowing a point to run on a silk di~~p~~aphram with only air underneath and noticing the the absence of mechanical Vibrations it occured that perhaps the silk laid on too hard a substance  we then put a piece of ordinary chem paper moistened with the Reg solution under the silk so as to allow of a give, the <u>results we striking</u>[e]  Mechanical Vibrations nearly ceased and with[a] a single spring pressing on silk, we could receive nicely at about 100 per minute through 25 000, 26 cells carbon old= shunted with 140 ohm WU Relay only= and with 800 plain & relay we could receive at 200 wds per minute 100s of feet[3] without a bug or bad signal  thus giving best results ever obtained on the E.M.G  Hence it is proper to infer that 1st[a] silk is the best for a <u>surface of all substanes yet tried</u>

2nd  <u>that Acetate Hg & NaO. is the best solution both on account of it delicacy & increase of friction on .O. by reversing the current</u>

3rd  that a giving biblus substance must be placed under the Silk. ~~th~~

4th  that several levers or spring all independent of each other with pressure on their ends and each sufficing to move the contact lever and all arranged so that the are ~~in a~~ same[f] distan~~t~~ce or nearly so from center of disk—

5 that the points on these levers or spgs should be as small as possible and be very smooth.

6th that the relay lever movement is the best

~~th~~7th that the slower the engine runs the less the effect of mechanical vibration;

8 That great care should be taken to eradicate all friction & shake in the mechanical moving parts.

9th To put the solution even on the silk & to use it quite dry=

10 to obtain your normal friction by the use of a chemical & not by excessive pressure;

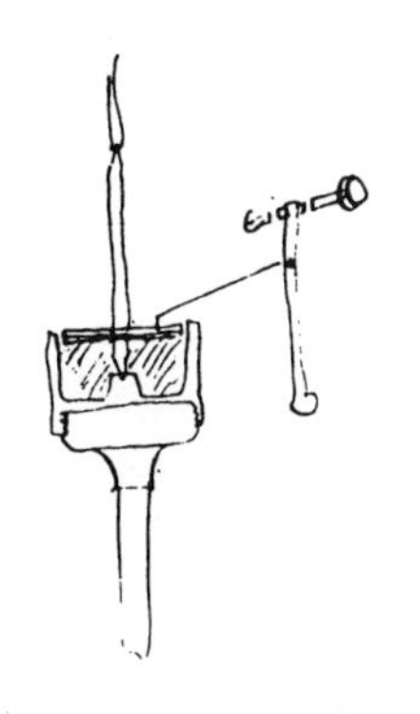

*Edison described this electromotograph as a "disk running in a thick muddy solution of Kaolin & acetate Hg & Caustic Soda."*

W~~e~~ith the spring 4 points silk & paper underneath we staticd the line with 60 sheet of condensr at sending end and recorded direct on auto paper at speed that couldnt read owing to weakness & tailing & which could not be shunted= with EM.G. under same condensters we could get it fair at same speed and sometimes 100s of wds without a mutilation shunted with a WU Relay & 600 ohms=

We are now trying a little side experiment placing a cup on disk with a hard rubber centre (.page 24 Vol 5)[4] & shalft with disk running in cup in which is a muddy solution of Kaoline & the Reg Solution. This will give a beautiful Smooth friction & if there is any working difference it will be OK'= I will mention that the disk is of brass whereas it should be platina to give it a fair trial=

ADDENDUM[g]

[Menlo Park,] Sept 14 1876

We[h] have just finished trying the dish, with a ~~S~~muddy (thick) more like butter;) of Kaolin & Reg HgNa Solution in which the brass disk runs we get a very strong movement of the disk with six cells on short circuit but scarcely ~~th~~[a] any with 150 ohm ie[b] through it= The surface was probably too great besides the surface was brass. ~~O~~To obtain proper friction had to run engine rather fast= Noticed one thing & that is that lever is sluggish it closes rather quick but normal friction dont "catch" quick enough to draw it away rapidly it being mushy "gives" whereas if it was ri~~d~~gid stuff such as paper it would get its normal friction instantly= it proves one thing and that is that a galvanometer could be made with an immersed disk[c] in a Liquid of Considerable Specific Gravity Afterwards we disconnected disk and let wire connecting disk with point lever rest in mushy stuff the action was much more delicate & rapid, & was sufficient to work moderately well through the 150 ohm relay thus

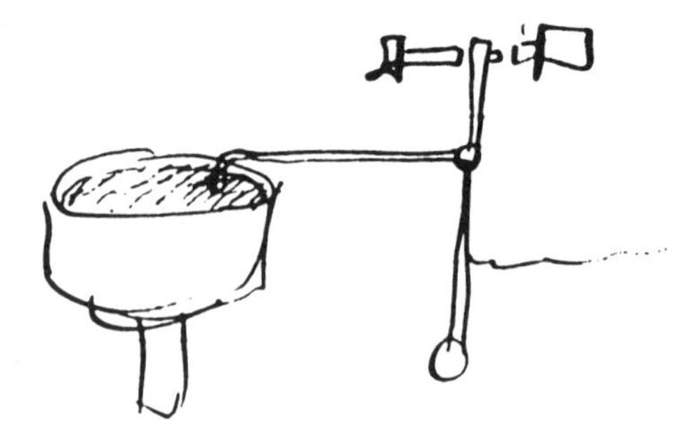

This also proves that you must use small surfaces to concentrate your action on the smallest point that it is possible to use  We are now trying what I think is going to be a success; thus

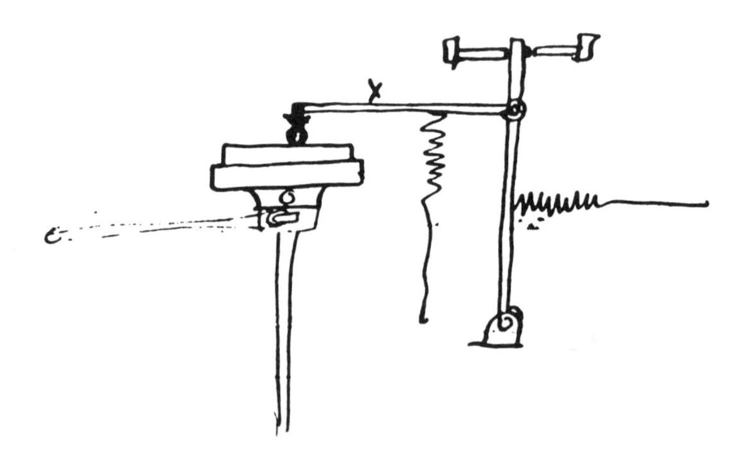


on the end of X is a long platina point with a small ball of cotton or other giving substance covered with silk tightly and wet with the regular Solution it[e] resting on a highly polished platina disk  with this giving point and a highly polished disk perfect even friction ought to be obtained & with great difference= EHJ is now making it=

We have just tried it and it works very well. The bag flattens out owing to pressure and this causes after few moments an uneven friction but the first few moments gives, absolute even friction[e]  Could this bag be made to stand longer or could be easily changed it would be Valuable but E.H.J has gone back and made the last arrangement which gave such good results thus

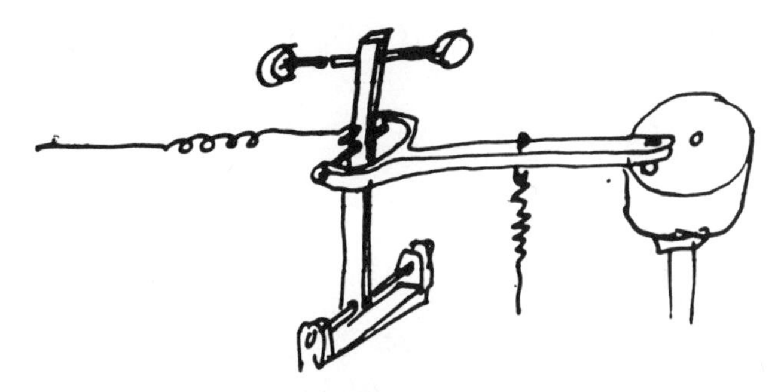

This works very nice has some bugs but current wipes them out= A disk of paper moist but rather dry is laid on platina over this is laid the silk tightly drawn  the silk is then covered with a thick paste of the Acetate of Hg & NaO, which has just sufficient moisture to ~~m~~connect the paste through the silk with the paper underneath. This paste surface ought to be polished, but after a few moments the point polishes a groove which if it shines all the way round gives even friction, but gives a false mark if a portion is seen to be unpolished  still this false jerk is weak & current wipes it out

We notice one thing very prominently with 6 cells through 150 ohms shunted with a 12 ohm sounder giving reversals the margin is considerable & when shunt is disconnected and all the Current passes through paper the working difference is greatly reduced[i] Whereas if the reverse current performed no function it should be stronger as it gets a full current= This is what makes the Acetate of Hg & NaO Solution so invaluable.

T A Edison

X, NjWOE, Lab., Cat. 996:51 (*TAEM* 3:304). Document multiply signed and dated. [a]Interlined above. [b]Circled. [c]Obscured overwritten letters. [d]"through . . . shunt" interlined above. [e]Underlined twice. [f]Added in margin. [g]Addendum is an X. [h]Preceded by "Page 56. Top=", indicating from where text continues; this text begins well down page 57. [i]Multiply underlined.

*An electromotograph with three independent levers.*

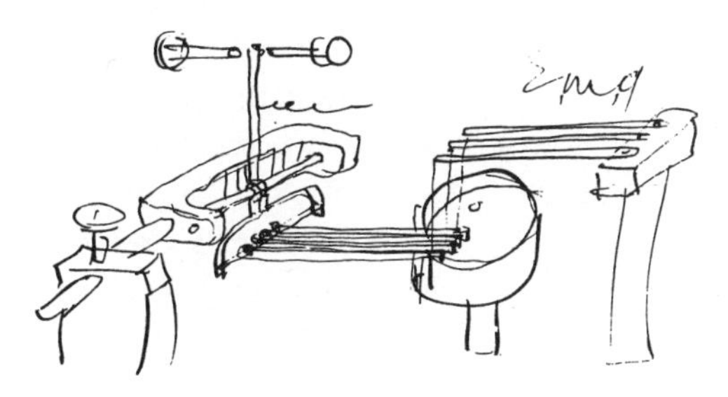

1. Cat. 997:23, Lab. (*TAEM* 3:363).

2. They made this on 11 September. Edison noted it was a "new style lever we are putting on today." Cat. 997:20, Lab. (*TAEM* 3:362).

3. That is, paper tape ("chem paper").
4. Cat. 997:24, Lab. (*TAEM* 3:364).

**–793–**

*Notebook Entry: Multiple Telegraphy*[1]

[Menlo Park,] Sept 14 1876

ACOUSTIC <u>Transfer System Octruplex</u>

Last night got the new set of local forks set up for working 8 messages. the forks especially the high on work beautifully, but the others do not work so even one making an excessive amplitude and striking the magnet which if it is not allowed to do gives scarcely sufficient amplitude but the other one does not give suffcent for perfection Even when striking the magnet ⟨☞ caused by shortness of wand⟩[a] & one weight has to be set ¾ inch ahead of the other to balance; this is not so on the one that gives great amplitude, but perhaps it is due to the working of the low resistance magnet of the high forks with the high res magnets of the low forks & diffrence in battery on one fork & the other. the connections had to be made as follows on account of the difference in resistance of the two magnet.

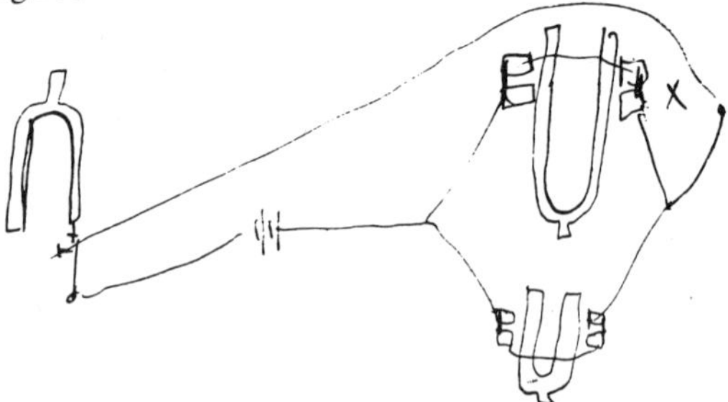


X magnets have together 24 ohms while th~~e~~ose on the high forks have only 6 ohms  this gives them an undue advantage especially if there is much internal resistance in batteries. if both are in one circket X works with great power but high fork magnets do not work.

We adjust up and found had connections wrong. Changed it= then found that spiral connecting wires kept breaking fixed them, then found 3~~0~~8 cells carbon at one end would scarcely move the common WU 140 ohm relays. increased its effect by shunting relays with ~~3~~200 ohms thus causing discharge from magnets themselves to bridge over space between waves, still too weak, put on NY battery which was very weak this made it about same; added our  this made it stronger got it on one; found bug in transmitter at ~~sen~~recieving end fixed. got it on .3 but couldnt get it on other wires kept breaking on forks. We are now engaged in obtaining & testing continuity by a sounder, closing proper points with a metallic wedge;

T A Edison

X, NjWOE, Lab., Cat. 996:56 (*TAEM* 3:307). Document multiply signed and dated. [a]Written in margin; index points to "striking".

1. This entry is continued in Doc. 795.

**-794-**

*From M. Fleming*

Sarnia 15 Sept 1876[a]

Dr Sir

Adams[1] Prest. is confined to bed so cannot at present transfer Stock. I have made enquiries about your Port Huron matter Do not think those parties will come to <u>any</u> arrangement They are afraid of each other and each man has a poor opinion of the others honesty. Stewart is so peculiarly constituted that is not easy to get him to talk business He has gone completely back on your brother & says openly he will not touch anything Pitt is connected with to such an extent does he carry this idea that he yesterday sent ~~my~~ me Five shares of Sarnia St. Ry. Stock to sell for his account at any price no matter what that it would sell for. I told him that I could not sell at all the amount being so small and the credit of the Road bad however he insisted on leaving it with me Pitt does not somehow seem to manage well. There are so many dunning or suing the Road makes things worse. W. B. Clark[2] has put a Six Hundred Dollar claim in suit Wiley[3] of the Grand Trunk tells me they are suing for some Nine Hundred Dollars Ticket and Freight money. Chalmers[4] holds note past due for about $200. C MacKenzie[5] note $125. R & H MacKenzie 230.33 Jas. Lowrie note $50. Stewart Pt Huron $123. and about $200. for rent. So you see there are pressing debts amounting to $2500. besides monthly notes falling due to G.T.R for Iron $150 and interest but I believe they are paying those latter notes as they mature out of the earnings. When Executions issues against the Road they will demolish its credit completely. In its present condition I would not care to undertake the management of your stock. If the liabilities were paid then I would consent to accept the proxies for the purpose of seeing that the Road was managed in your interest that is honestly and economically worked. Cowan told me that some man was trying to buy the Road If that man would offer anything near what stock cost, I would feel inclined to let him have Major Clarks $2000 but I do not think there is much chance and that Cowan sees thro Pitts spectacles. You had better come on here and stay long enough to satisfy yourself how the Road ~~want~~ really stands and what money is necessary to end its troubles If Pitts stock was paid for there would be plain sailing. I shall see those Pt. Huron people before long and let you know result. Yours Truly

M. Fleming[6]

ALS, NjWOE, DF (*TAEM* 13:1068). Letterhead of American Express Co. [a]"187" preprinted.

1. Joshua Adams was president of the Sarnia Street Railway Co. Agreement between Sarnia Street Railway Co. and Grand Trunk Railroad Co., 29 Dec. 1874, DF (*TAEM* 13:29).

2. W. B. Clark, R. and H. MacKenzie, James Lowrie, Cowan, and Major Clark are unidentified.

3. Unidentified; see Doc. 835.

4. Probably Robert Chalmers, an incorporator of the Sarnia Street Railway Co. Sarnia Street Railway Incorporation Act of 1874, DF (*TAEM* 13:26).

5. Probably Charles MacKenzie, an incorporator of the Sarnia Street Railway Co. Ibid.

6. Unidentified.

**-795-**

*Notebook Entry: Multiple Telegraphy*[1]

[Menlo Park,] Sept 17 1876=

Acoustic. Continued from page 57.

We found that the continuity test was OK but the trouble was all caused by Gilliland giving[a] us a wrong diagram of the connections. We straightened this and then found the side that ~~was t~~ it was better and we succeeded on short circuit in balancing and sending in opposite directions even when all the other keys besides the ones sending were closed or open= We then tried it on Phila line and it worked about as well, 4 sheets condenser= We then closed for the night and Adams then in PM run the lose connections on table nicely we then started again worked several hours and I then discovered that he had misplaced one of the wires, rendering it impossible to balance one end while the other was OK having got this OK we worked fairly although troubled with want[b] of battery power= I then changed the method of operating the controlling forks & line

I placed in a local circuit one of the old low forks, used one prong & spring to open & close its ~~main~~ own circuit, and the other prong with two ~~p~~springs & points one to close the line containing high Resistance magnet forks and the other to close a local containing the other high resistance magnet forks. after making this connection found that the other local forks worked even & nicely at both ends which is not the case with the oth old way, the reason of which has already been explained, but could not get anything through the line at either end= closed= next evening This evening started forks with 60 cells carbon (Little battery) each end found circuit ok, but couldnt balance well yet could work on all four without mixing, found that the timeing was very much out readjusted controlling forks and got nearly even—current increased in strength and we succeeded for the first time in Sending Eight messages over the one circuit,[c] a feat never before accomplished twith maorse apparatus[2d] 800 ohms was in the line. Owing to the fearful storm we are unable to get our Philadelphia Loop= I predict that this apparatus will require about 400 cells of Callaud at each end in a 300 mile circuit. The Quadruplex requires 200, but with this do battery several lines may[e] be worked out of the same battery so the excess over that used in the Quad does not amount to much in fact it is economical. besides 8 messages is the product= aI also predict that this apparatus will work better & have more margin[3] than the Quad when it shall be improved as much as that apparatus.

After Running for four hours we put main battery in and find circuits OK; Put 1600 ohms in artificial line with 2300 ohms and as there is 700 in each side bridge making 350, and Relay Resistn is 240, this makes bal correct according to resistance. We send on all 8 keys & all comes clear on 8 Relays= We now shunted line (not to Earth) with 75 sheets condr takes 48 to balance on artificial. We now put condenser in centre of line. 800 ohms each side & ground other end and find that we get everything clear requires no change in resistance but only requires 224 sheets condr at each end to balance;

X, NjWOE, Lab., Cat. 996:60 (*TAEM* 3:309). [a]Obscured canceled letters interlined above. [b]Obscured overwritten letters. [c]Multiply underlined. [d]Last two words underlined twice. [e]Repeated at end of one page and beginning of next.

1. This entry is a continuation of Doc. 793 and is continued in Doc. 797.

2. Edison may be referring to the circuit shown in Doc. 773, which "Worked Bang up Sept 17," but also see the circuits shown in Cat. 997:13–15, Lab. (*TAEM* 3:358–59). At this time Elisha Gray was in New York demonstrating his eight-message electro-harmonic system on Western Union wires (see Doc. 786 n. 4).

3. Difference in current strengths.

–796–

*From Frederic Ireland*

[London,][a] Septr 20 1876[b]

My dear Sir

Thanks for your letter of 29th ulto—[1] I think it must have been delayed in transmission— I am also still without the details of price which Mr Batchelor promised me.—

I have had a great deal of trouble with the first lot of pens (5) owing

1st To the rollers pins all working out—
2nd To two of the Pens working very badly.—
3rd To the ink being very defective & light in color.

Of course I did not at first know what was the matter and dare not send out pens till all was in order— The Second lot are all right but badly packed—i.e. Three pen boxes are entirely without fittings—7 others have no files or screwdrivers.—none have extra needles or cords— Three presses have the clips—ie. the folding part—smashed—one before make four— Please send me spare ones to fix on

The Second lot of Ink is blacker but I could wish it blacker still— I send you specimens of writing— No 1. is your first ink— 2. your Second 3 is ink used by Clare— he says ½ printers and ½ Castor.— I cannot get this to work—

Can you suggest anything that will answer & give deeper color.— It seems almost essential here— Your screw drivers are too weak. Cannot you make them stronger.—

The cords wear out very fast— Please send an extra supply—say 100 and say where they can be had hee[2]—if you know— We have already competition by a Pneumatic Pen— which works a needle— I think it is an infringement—if not it will be a serious competitor.— We have given notice.—[3]

I wired you stopping supplies until I could call your attention to the above matters—especially ink Yours very faithfully (In haste)

Frederic Ireland

ALS, NjWOE, DF (*TAEM* 13:976). Letterhead of Electric Writing Co. [a]"9 New Broad Street, E.C." preprinted above. [b]"187" preprinted.

1. Not found.

2. Ireland probably meant "here."

3. On 11 October, Ireland again wrote Edison about the pneumatic pen and informed him that the inventor's name was Pumphrey (DF [*TAEM* 13:980]), possibly a Josiah Pumphrey of Birmingham. Although no Pumphrey patent has been found for a pneumatic pen, he did take out American and British patents for a copying process (U.S. Pat. 200,759; Brit. Pat. 895 [1877]).

**–797–**

*Notebook Entry: Multiple Telegraphy*[1]

[Menlo Park,] Sept 26 [1876]

between Sept 17 and this date we have been trying Various experiments and find that owing to the magnet discharge our plan of overcoming the excessive resistance of the magnet of the ~~hi~~low fork and the low R of the high fork by connecting thus

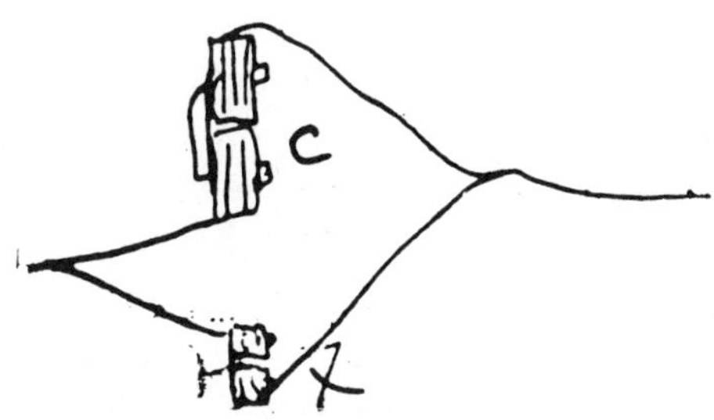


is a failure as the discharge from the magnets are unequal that from X being the greatest  hence C discharges instantly the extra current from X passing through it in a contrary direction while X is prolonged by its own extra current  This necessitated a great change in the controlling fork springs as more or less battery on the local forks generated extra current of different strengths. We obviated this by using only one spool on each magnet of C and connected both it and X in the same circuit thus destroying the closed circuit in which the extra current could circulate and do mischief= We also tuned the forks so that they would start of their own accord when the magnets were some distance away  I find that the tension on the contact springs and the portion of time when the fork is touching and not touching is an important factor in keeping all the forks in unison; to obtain a perfect unison the springs should be in contact with the forks all the time, and the tensions should be very nearly alike. I also find that tuning forks may be adjusted to any rate by a spring pressing against their end, instead of weighting them. We now have the two controlling forks in the same circuit which ckt is operated ie[a] opened & closed by a spring & point on a low fork local, and I have the switching fork local points adjusted alike on both forks so they close simultansly and I find this to be co~~nt~~rrect and the ckts didnt change last night at all after we made the change of C. & .X

Having got the unison practically and theoretically Correct and reliable, we find can do first class work on Resistance line with 80 cells Callaud each end with 2 3 4 5 6 or 8 thousand ohms ~~ion~~= But when we attempted to work the Phila loop

120 miles we found that the static charge invariably changed the circuits throwing No 1 mesg on No. 2 Relay & No 2. mesg on No 1 Relay  I can only account for th~~u~~is by supposing there is a retardation of one pulsation or the 1/120th of a second, in the passage of the current. this would do it= Allowing th~~u~~is I could get 4 message at either end ~~fail~~ fairly but couldnt balance fine enough to get it clear although you could read it, but taking in consideration that there was very considerable escape and there was only required 1000 ohms to balance & we having only ~~7~~80 cells Callaud this was a good showing. We ought to have had 3 times as strong a received current and as you can balance a strong outgoing current as finely as a weak one the interference would not be~~ar so~~ so near the strength of the recd current= We use back point relays now and close the sounder by stopping of the current  We think it works better= Repeating Sounders appear to improve signals but they are hard to manage;

X, NjWOE, Lab., Cat. 996:62 (*TAEM* 3:310). [a]Circled.

1. This entry is a continuation of Doc. 795.

# -3- October–December 1876

During the early fall of 1876 Edison and his staff concentrated their efforts on practical tests of acoustic transfer technology. They demonstrated the system to Western Union and at the end of December Edison began negotiating a new financial arrangement with Western Union on the strength of his expectations of success. At the same time, Edison was working on his new repeating recorder for Morse telegraphy. At the end of October he executed a patent application for a perforating repeater and began work on an embossing repeater, which he would continue to develop into the next year.

Edison also experimented briefly with sewing machines driven by large, electrically powered tuning forks; with various copying processes, including a duplicating ink and a device based on the electromotograph principle; and occasionally on the telephone. He twice made a virtue of necessity when his stock of chemicals suffered damage—first from sunlight and then from freezing weather. Both times Edison recorded extensive observations of the condition of the chemicals, even using some of the observations as the basis for a short publication in the *American Chemist.*

A few changes in the Menlo Park crew occurred during the fall of 1876 as Tony Bronk, Edison's boyhood friend, and Jim Gilliland, Ezra's brother, temporarily joined the regular staff.[1] Charles Batchelor also began to spend more time in the laboratory, primarily experimenting with the sewing machine motor.

The electric pen business continued to grow and prosper. Batchelor hired assistants in the New York office of Edison's Electric Pen and Duplicating Press Company. By the end of the year the company had sold 205 pens and presses and showed a profit of over $1,100. At the end of November, Western Electric Manufacturing Company contracted to manufacture the copying system. The Electric Writing Company of London faced new competition from a pneumatically powered pen system, and the manager of the company pushed Edison to develop an improved battery and a high-speed press. The Continental rights to Edison's electric pen patent remained unsold.

The experiments on carbonized paper conducted during the summer and the development of duplicating ink led Edison and Edward Johnson to set up the American Novelty Company in late November. The company was devoted to the manufacture and sale of these and other miscellaneous inventions, including Johnson's ribbon mucilage.

Edison grew increasingly exasperated by the Port Huron street railway business. He continued to support his brother's involvement, sending $600 in early December but stopping short of the $4,000 Pitt requested. At the end of that month, individuals connected with the opposition line extended an offer to Edison to settle the competition. Although Edison did not accept the offer, the negotiations thus begun led to the eventual merger of the two street railroads into a new company.

On 10 December, a man named Horace Day visited Menlo Park and offered Edison and Batchelor each a one-sixth interest in copper and gold mining rights to a piece of land in Vermont rich enough "to make 50 companies and last a century." Although they apparently let the opportunity pass by, Day did return later to test some ores in the laboratory.[2]

1. Cat. 1213:6–7, Accts. (*TAEM* 20:8). Alfred Swanson, whom Jehl 1937–41 (1:128) identifies as a night watchman, also appears in these accounts.

2. Cat. 1240, item 15, Batchelor (*TAEM* 94:10–11). Day returned in mid-February and spent "all day testing some ores." Cat. 1233:49, Batchelor (*TAEM* 90:77).

**–798–**

*From Alfred Nelson*

New York, Oct 3 1876[a]

Dear Sir

We are notified that the suit of Mr. Morten against this Co.[1] is to be called on 6th inst, and as you are a very important witness and, I am very sorry to hear, are now confined to your house, we should be much obliged if on receipt of this, you would mail us a physicians certificate of the fact to be presented as a reason for delay[2]

In ordinary cases, it would be sufficient to notify the opposing counsel[3] but as he has already acted with discourtesy, we do not desire to run any risk. Trusting you will soon recover
Respectfully

Alfred Nelson V.P

ALS, NjWOE, DF (*TAEM* 13:1233). Letterhead of Domestic Telegraph Co. [a]"New York," and "187" preprinted.

1. The nature of the case has not been determined.

2. On 8 November, Nelson notified Edison that *Morten v. Domestic Telegraph Co.* was to be heard in Part III of the Supreme Court at New Court House in New York on 10 November. Nelson requested that Edison join his former manufacturing partner James Murray in attending the trial (DF [*TAEM* 13:1235]).

3. Unidentified.

**-799-**

*Technical Note: Telephony*

[Menlo Park,] Oct 12th 1876

Speaking Telegraph[1]

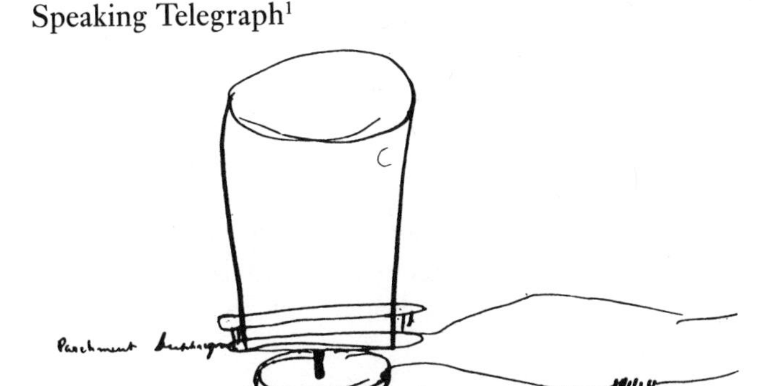


This works but not so good as Plumbago on hard rubber.[2] Tried mercury instead of water but it is a dead make & Break.
T A Edison James Adams

X (photographic transcript), NjWOE, TI 2, Edison's Exhibit 105-10 (*TAEM* 11:222). Written and multiply signed by Adams.

1. Figure labels are "Parchment Diaphragm" and "tin cup filled with water." In his testimony Edison noted that one electrode was formed with a stick of Arkansas oilstone attached to the center of the diaphragm and coated with plumbago. This was dipped into a cup containing water and an immersed plate that formed the other electrode (TI 1:43–44 [*TAEM* 11:42–43]). Edison had experimented in fall 1873 with "resistance coils of Kansas stone & plumbago graphite also on glass hard rubber etc" in connection with his artificial Atlantic Cable experiments (NS-74-002, Lab. [*TAEM* 7:102]).

2. Edison's testimony indicated that this meant the "short circuiting of a film of plumbago on hard rubber, by a spring connected to a diaphragm, and resting on the film, which served in its motion to cut in and out resistance due to the film of plumbago" such as was shown in

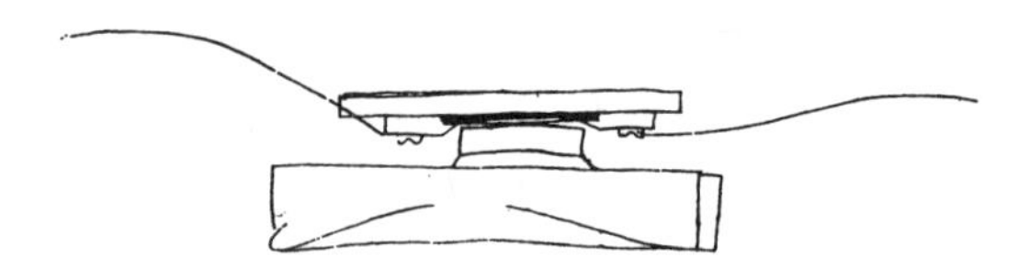

*A telephone transmitter in which metallized felt, resting on a diaphragm, short-circuits the current across a thin layer of carbon.*

an undated drawing that Edison placed between April 1876 and March 1877 (Edison's Exhibit 117-15, TI 2 [*TAEM* 11:624]). There "a piece of either hard rubber or stone, coated with carbon, is used with a circuit passing through it, a piece of felt resting on the diaphragm and provided with a metallic surface, being in contact with the surface of carbon. The vibration of the diaphragm, cutting in and out, or rather short circuiting a portion of the film" (Edison's testimony, TI 1:44 [*TAEM* 11:43]).

## -800-

*Notebook Entry: Miscellaneous*

[Menlo Park,] Oct 18 1876

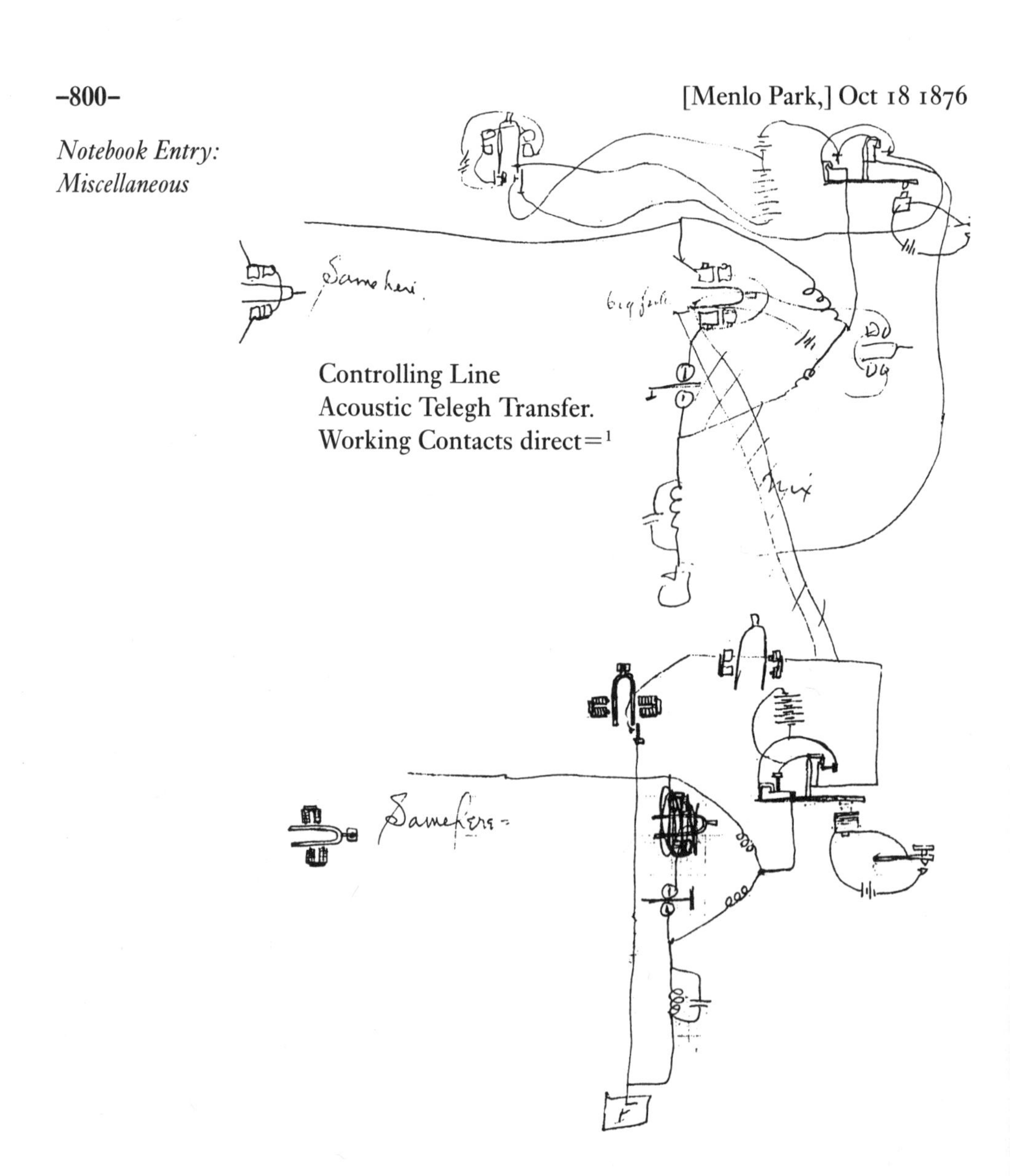

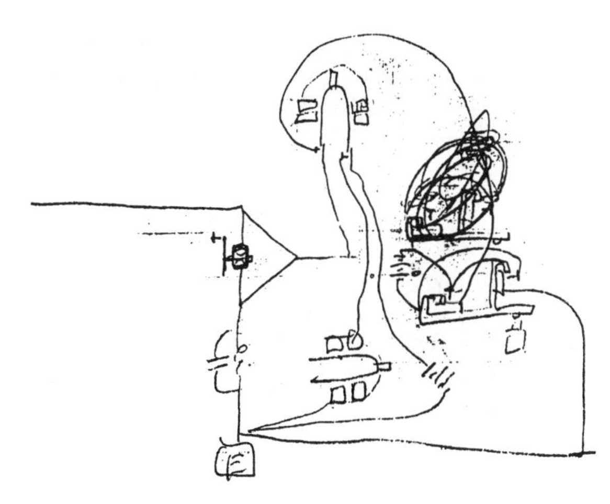

if the switching forks are worked direct on the controlling line, duplex, it will probably be necessary to add a Condenser shunting a Rheostat in the bridge wire to weaken the effect of the Static charge—otherwise one fork will lag on the other & change the circuits=

Idea has occurred to me to arranged two needles with electric pen one having an eye at the end with a thread the other provided with a hook so that by passing pen over cloth streatched it will make a loop stitch thus allowing beautiful embroidery etc to be made with speed=

Stencil copying press felt Roller saturated with ink over which stencil is placed previously having ink worked in= The rollers are geared or otherwise fastened together & paper is passed between them[2]

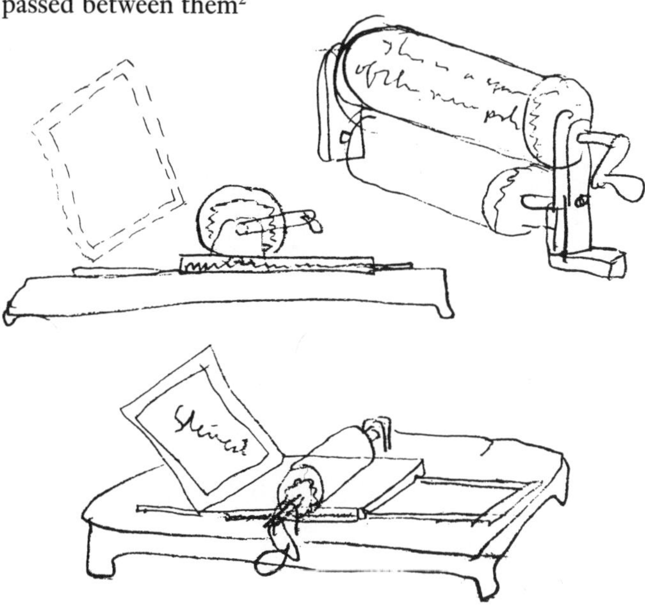

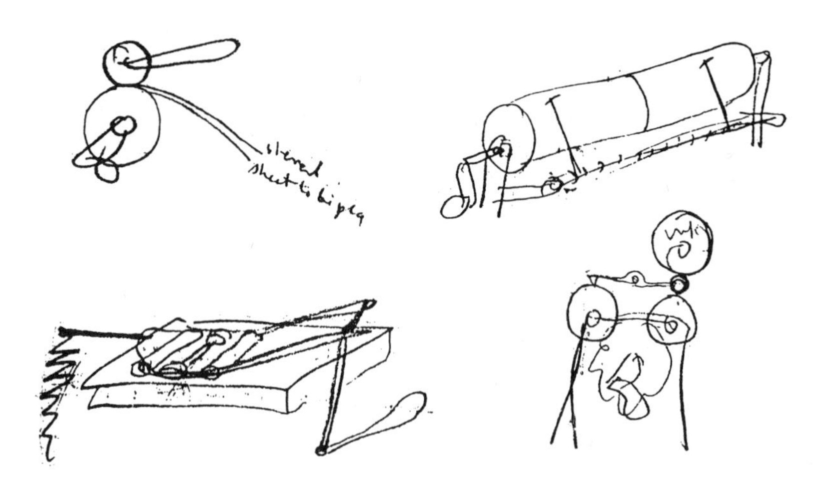

Electric Sewing Machine  Tuning fork applied to drive machine which fork set in motion by Electromagnets[3]

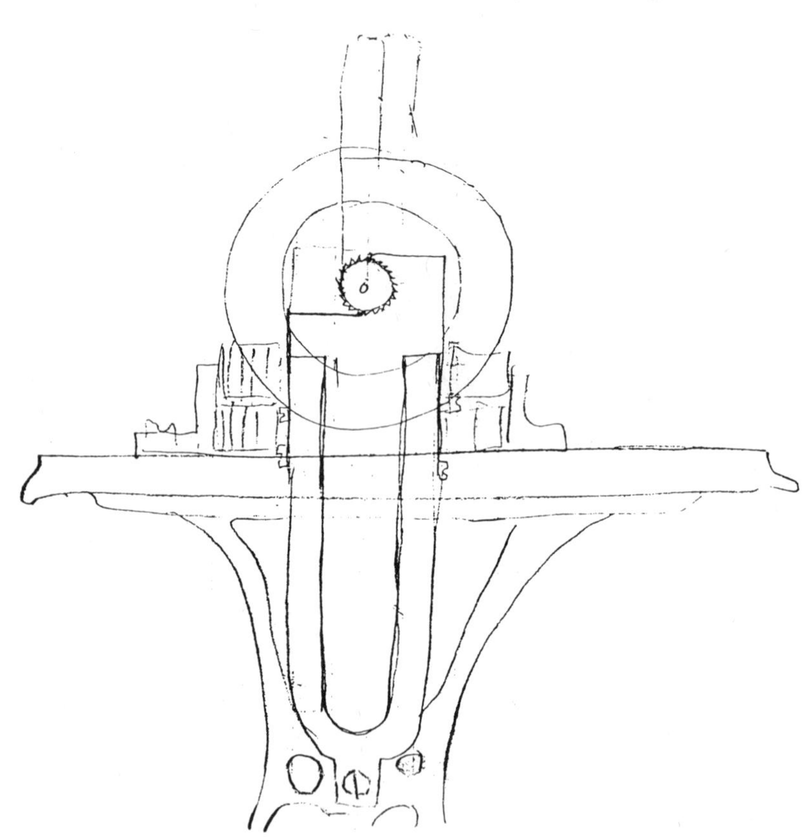

May want use Hinkly & Lyon SM[4] wheel feed=

Forged steel would perhaps be cheaper than bell metal[5] secured thus[6] ☞

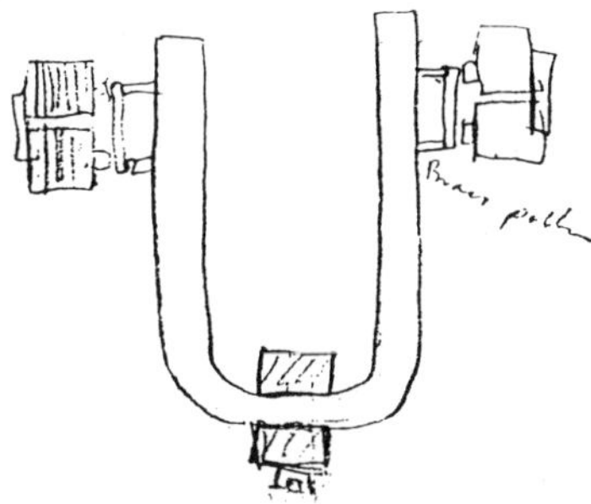

Rubber rim    disk to vary speed[7]

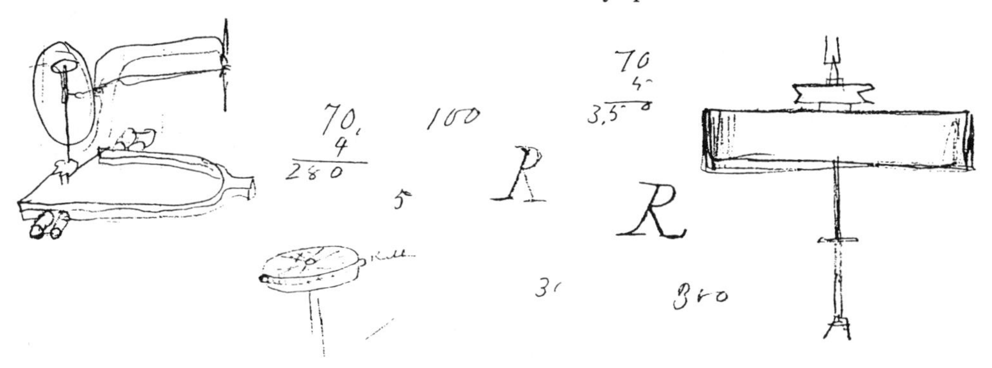

T A Edison

Chas Batchelor
James Adams

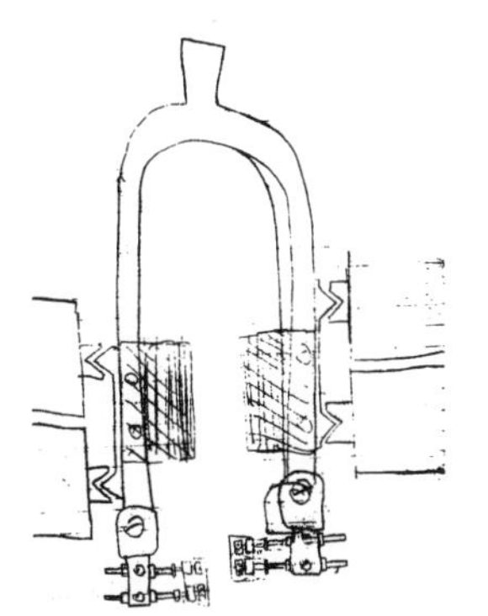

*A 13 October 1876 design for a big tuning fork for the acoustic transfer system.*

X, NjWOE, Lab., Cat. 997:28, 30–31, 29 (*TAEM* 3:366–67). Document multiply signed and dated.

1. Figure labels are "Same here," "big fork," "nix," and "Same here." The big fork may refer to the design drawn in the same notebook on 13 October (Cat. 997:27, Lab. [*TAEM* 3:365]) or to a fork like the ones at the Edison Institute, Dearborn, Mich. (Acc. 00.1382.537, 00.3.11616). See also the multiple contact arrangements for acoustic transfer forks drawn on 6 October (Vol. 10:112–14, Lab. [*TAEM* 3:911–13]).

2. Figure labels are "This is a copy of the new prtr"; "Stencil"; (p. 152) "stencil" and "sheet to be p[rin]t[in]g," and "ink."

3. Figure label is "teeth very fine."

4. Edison may be referring to the Finkle & Lyon Sewing Machine Co. of Boston, which began production about 1859 and continued until 1867. Cooper 1976, 69.

5. A bronze of three or four parts copper and one part tin.

6. Figure label is "brass pillars."

7. Figure label at bottom left is "Rubber."

*One of the large tuning forks used in Edison's acoustic transfer system.*

**–801–** [Menlo Park, October 18, 1876?][1]

*Technical Note: Multiple Telegraphy*

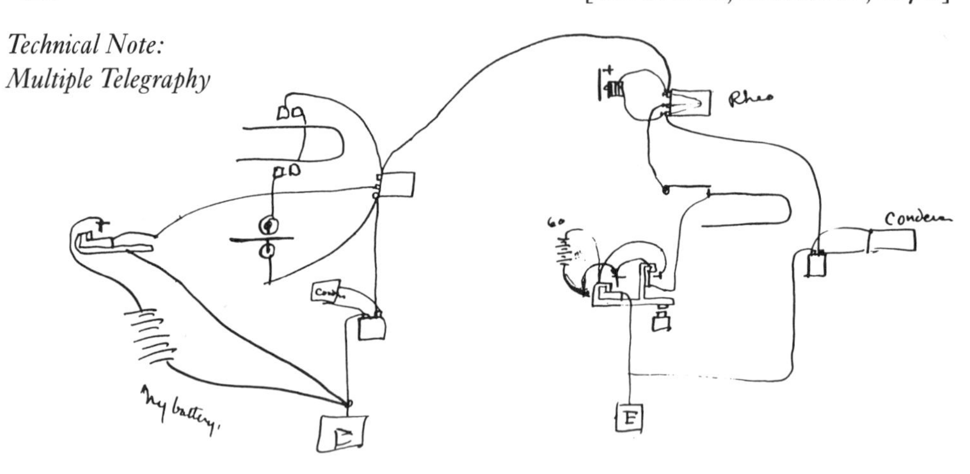


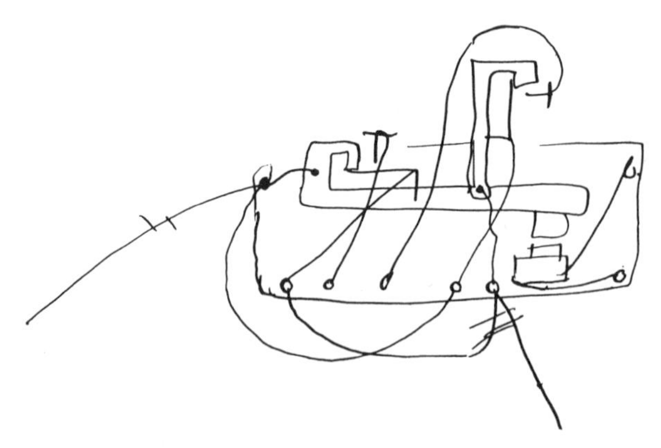

We now duplex the controlling wire as above shewn[2] and find that Duplex itself works with the vibrator going but the receiving vibrator does not work strong enough with the 60 cells Callaud we have on circuit. the loop to phila & return is used transmitting with NY battery at one end & our 60 cells at the other[3]

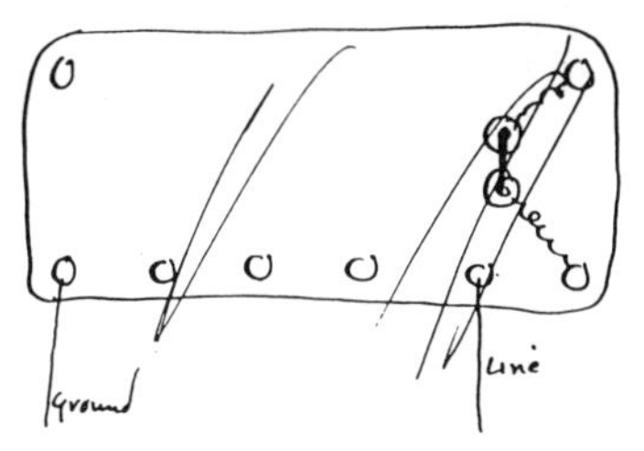


I think that the addition of 100 cells will make it work perfectly satisfactory  we are now going to add 25 cells;

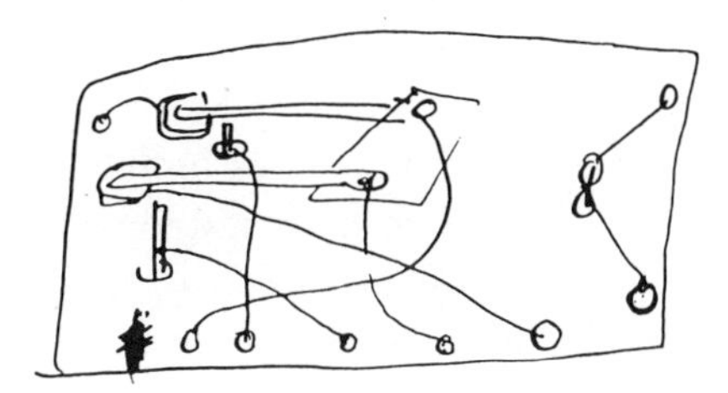

We tried it on direct loop with NY battery at each end consisting of 100 & .90. Callaud & the strength of vibration was ⅔ if not ½[a] more than we should ever want

⟨Wrks OK with 90 cells Callaud one end, ie[b] its 90 cells that wks fork= but had to have 4000 ohms each side bridge at the reed which is in main⟩

X, NjWOE, Lab., Vol. 15:167 (*TAEM* 4:171). [a]"if not ½" interlined above. [b]Circled.

1. Edison apparently wrote this note soon after Doc. 800. Although he seems to have continued these experiments through the fall (see Doc. 833), few notes remain. However, for work during November see Vol. 10:115–16 and Cat. 1169, both Lab. (*TAEM* 3:914–15; 6:574–75).

2. Figure labels are "NY battery," "60," "Rheo," and "Condenser."

3. Figure labels are "ground" and "line."

**–802–**

*Notebook Entry: Electric Motor*

[Menlo Park,] Oct 22nd 1876[1]

Electric Sewing Machine

Edison's idea that the greatest amount of power can be got out of a magnet by the aid of a tuning fork seems to stand good so far.[2] We applied the principle to the driving of a Wilcox & Gibbs[3] sewing machine and succeeded in stiching through 6 thicknesses of cloth at a speed of 82 stiches per minute with the following device:—[4]

Fig 1[5]

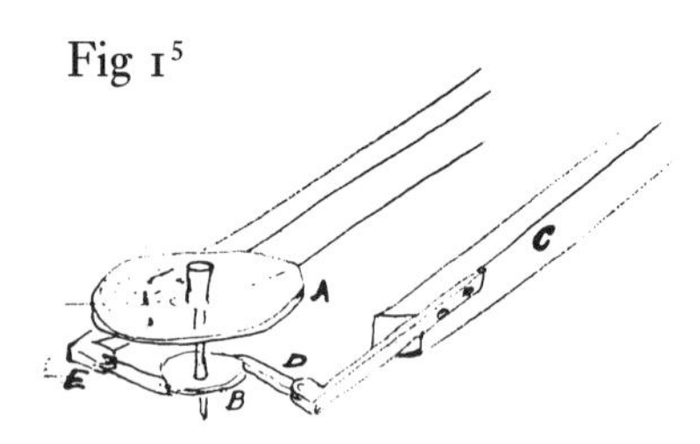


A is a fly wheel on same shaft as 300 tooth ratchet B. This ratchet is driven by 2 pawls D & E working by prongs of 12 in fork giving about 60 vib. per minute.[6] This wheel A was connected by belt to sewing machine This device was made from stock in hand and was only a preliminary experiment and as such far exceeded our expectations[7] It seems to want a much larger fork, each prong having large weights on & each prong having both push & pull ~~pro~~awls on so as to utilize both backwards and forwards movements of prongs. I now altered the ratchet to 30 teeth and put it directly on the sewing machine shaft thus:—

Fig 2

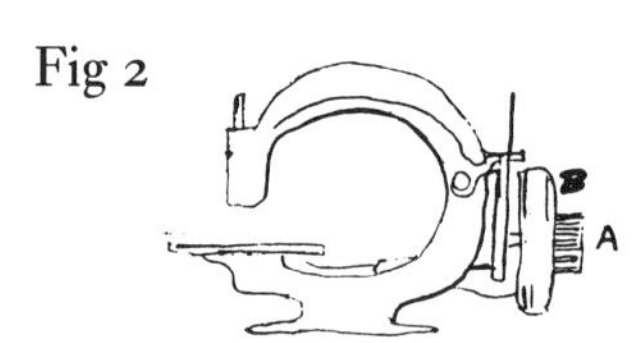


A is ratchet fastened directly on to shaft carrying flywheel B. this was run in the same manner as Fig 2 only clicks were altered to suit the upright ratchet wheel. This did not work so well it seemed to require more power as ratchet was smaller.

It seems to be necessary to get some motion on there that will not allow the fork to throw out.

Chas Batchelor.

X, NjWOE, Lab., Cat. 996:65 (*TAEM* 3:311). Written by Batchelor.

1. Redated at end: "Monday Oct 23 3 a.m. 1876."

2. In his private notebook Charles Batchelor indicated that this idea arose out of experiments they were doing with large tuning forks. Cat. 1317:28, Batchelor (*TAEM* 90:671).

3. Inventor James Gibbs and machinist Charles Willcox formed Willcox & Gibbs Sewing Machine Co. in 1857. They produced one of the first low-cost home sewing machines. Cooper 1976, chap. 4.

4. Batchelor noted that they stitched through "six thicknesses of shirting with 4 ordinary cells of Bunsen battery. This of course is nothing very great but it convinced us that the apparent great strength of a tuning fork when vibrated by a magnet can be utilized if you only strike the right way of applying it." Cat. 1317:29, Batchelor (*TAEM* 90:671).

5. On 16 October, Batchelor had drawn another form of driving ap-

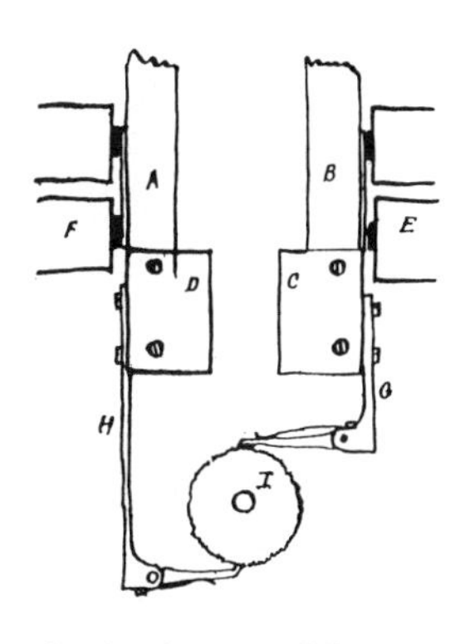


*Another drawing of the sewing machine motor, showing the magnets and weighted prongs.*

paratus for an electric sewing machine in which the tuning fork drove the flywheel by means of a smooth disk rather than a toothed ratchet. Cat. 1307:57, Lab. (*TAEM* 90:642).

6. In the other version of this entry, Batchelor corrected this to "60 vibrations per second." Cat. 1317:29, Batchelor (*TAEM* 90:671).

7. In his notebook Batchelor indicated that they used a bell metal fork with ¾-inch prongs, each prong carrying a two-pound moveable weight, and he drew a diagram (at left) to show the arrangement.

**A B** are the prongs, **C D** the moveable weights, and **G H** are arms extending from the prongs and carrying driving clicks for ratchet wheel and shaft **I**, to which was attached the twenty-pound flywheel (not shown here; **A** in figure 1 of the document).

Batchelor went on to note that "click H is much further from the fork than G and consequently the prong A has much more work to do than B; this throws the fork out of tune (when it is very weak and will do no work at all) but by moving the weights on prongs they can be brought into tune and the maximum strength of the fork is gained when both prongs are exactly in tune. I have now commenced on a large fork which I think will give us a surplus of power." Cat. 1317:29, Batchelor (*TAEM* 90:671).

**-803-**

*To Frederic Ireland*

Menlo Park N.J. Oct 26 1876.

Dear Sir,

Yours of Oct 9 recd.[1] I saw the article in the ~~m~~Mechanics Magazine [d]escribing[a] the pnuematic Pen; you do not often see a clearer [ca]se[a] of infringement than that.[2] I do not know what your laws are [in][a] England but in this country the parties could be stopped [im]mediately.[a] You will find in the patent the very same thing, [ie?][a] turbine or leaf'd wheel within a circular box propelled by [w]ater[a] falling from a high to a low level. now our courts would [sa]y[a] that substituting air for water is no invention, as it does not [req]uire[a] the exercise of the inventive faculties. It dont matter what [the othe?]r[a] power is that drives the machine, if the patent for the machine [ ][a] If those people have a peculiar machine for getting air [pres]sure[a] applicable to the pen that does not give them any right [to][a] use our pen.

[The tr?]ade[a] here is increasing rapidly caused 1st by a revival [of][a] business & 2nd by our establishing agencies. Some of the [telgh?][a] supply houses keep it on sale, notably Tillotson & Co[3] who [ ][a] last week. They also sell our pen batteries for other purposes. I think you should put them on sale [ ][a] your opitician & Philosophical apparatus houses; they are peculiarly adapted for Medical coils, etc. in ~~t~~fact it is the only "Table" battery yet devised=

We have an embroidering pen ie[b] for perforating lace work designs in paper through the holes of which colored dust is passed upon cloth= its nothing more than a very large & peculiar shaped needle.[4] Shall we send you one; We are going to roughen the ends of the pen Tubes, and if satisfactory will hereafter make them rough= The pattern for the large press 11 × 9 is now ready shall we send you one as sample;=

Batchelor promises to attend to the points you speak of= What[c] are the improvements on the press. I have several but I fear to make changes as you never can tell how it will turn out, besides it cost considerable to get up special tools= I will have assignments of Patents for France & Belgium sent to Serrells Agent in London immediately. will give you address as soon as I get it.=[5] Telegraph= Had a test last week over line to Phila. it worked fairly, am waiting now for a set of more delicate relays[6] Preece,[7] one of your PO Engineers has left Eng[land][a] for this Country to inspect our telegraph. I shall get him down to Menlo & fill him up with Telgh and will try & get a good impression for the new machine. Sir Wm Thomson was here at Menlo & had a very interesting time. You see he speaks of my Automatic Telegraph before the B.A.[8]

=Notes= We are selling a number of machines to Colleges =Catholic priests, the latter issuing notices to the faithful with them= We are printing a 300 page book with the Pen; (foolscap size) being the lectures of Prof Thurston[9] of the Stevens Institute of Technl'gy= heretofore each student had to make his own copy. they clubbed together and pay $5. each which gives us 25¢ per stencil and ½¢ for each impression= besides three profs have bought them=

We obtained a medal at the Centennial, but have not yet recd the judges report=[10] Yours

T. A. Edison

ALS (letterpress copy), NjWOE, Lbk. 2:103, 101–2 (*TAEM* 28:437, 435–36). [a]Missing text did not reproduce properly. [b]Circled. [c]Obscured overwritten letters.

1. In this letter Ireland discussed a number of issues related to the electric pen. He indicated that problems remained with the manner of packing pens shipped to Great Britain and suggested changing the screwdrivers used to adjust the pen, improving the grip of the pen barrel by roughening it, and creating a pamphlet specially suited to the British market. He also indicated that they had acquired a greatly improved press. *TAEM* 13:978.

2. See Doc. 796. In December, Charles Batchelor's brother Tom sent him an illustrated clipping from *Cassell's Family Magazine* about the pneumatic pen (Cat. 1240, item 11, Batchelor [*TAEM* 94:8]).

3. A major telegraph and electrical manufacturer. See *TAEB* 2:340 n. 6.

4. See Doc. 800.

5. Ireland wrote to Edison on 11 October asking him to separate the French and Belgian assignments and to send signed transfers for these countries (DF [*TAEM* 13:980]). On 26 October Edison's patent attorney Lemuel Serrell wrote Edison to indicate that he would take care of it (DF [*TAEM* 13:1047]). The agents were probably Brewer and Jensen (see Doc. 1033).

6. This probably refers to Edison's acoustic transfer telegraph system.

7. William Preece (1834–1912) was at this time a divisional superintendent of the British Post Office telegraph system. Later appointed engineer-in-chief, he was one of Britain's most prominent and influential electrical engineers. He and Henry Fischer, controller of the central telegraph office in London, crossed the Atlantic in April 1877 to study the American and Canadian telegraph systems. Baker 1976, chap. 16; Fischer and Preece 1877; see Doc. 904.

8. Edison is apparently referring to Thomson's opening address as president of the mathematical and physical section of the British Association for the Advancement of Science. In his remarks Thomson spoke briefly of what he had seen at the Centennial, focusing primarily on Alexander Graham Bell's telephone. Of Edison's automatic he stated that he had seen it "delivering 1,015 words in 57 seconds; this done by the long-neglected electro-chemical method of Bain, long ago condemned in England to the helot work of recording from a relay, and then turned adrift as needlessly delicate for that." "British Association," *Nature* 14 (1876): 427.

9. Robert Thurston was a pioneer mechanical engineering educator. See *TAEB* 2:509 n. 2.

10. Sir William Thomson, who authored the award report for Edison's automatic telegraph system (see Doc. 757), also drafted the award report for Edison's electric pen and duplicating press, calling it an invention of "exquisite ingenuity and . . . usefulness" (draft of 20 June 1876, Centennial). The actual report, printed in U.S. Centennial Commission 1880 (194–95), was apparently authored by Joseph Henry (draft of 1 Aug. 1876, Centennial). Both Thomson and Henry extracted material from one of Edison's electric pen circulars in their reports. For Batchelor's recollections concerning the medal, see App. 2.

**-804-**

*From Frederic Ireland*

[London,] Octr. 26 1876[a]

My dear Sir

I have just received your telegram at 3.20.—[1] I had waited till midday before wiring you as per Enclosed copy.—[2]

I will now try and act promptly with regard both to the French & Belgium Patents—[3] the latter I think can get taken up at once  I am also negotiating about Spain.

Our sales here have not been so large as I could wish— we

have much prejudice and stupidity to overcome.— We have found it impossibbe to get people to recharge their batteries once a week and their general neglect of the pen is something difficult to comprehend— "Out of Evil Cometh good."— We have succeeded in obtaining a battery quite as strong and contentent[4] as our own and it will work for three months without attention— we are assured it will work for Six[b]— of this we have not had time to obtain proof but our Experience leads us to believe in it.—[5]

Shall we we send you battery & press.— They are not patented in the States,[6] but if you are willing to give the inventors something for their trouble, they would, no[c] dobubt, appreciate it.— Yours very truly

Frederic Ireland

ALS, NjWOE, DF (*TAEM* 13:983). Letterhead of Electric Writing Co. [a]"187" preprinted. [b]Underlined twice. [c]Followed by overstruck comma.

1. Not found, but see Doc. 807 n. 3.

2. No enclosure has been found, but the following copy was written in an unknown hand at the end of the letter: "Telegram Must sell samples Continent to sell patent. Wire authority. Belgium three thousand dollars Wire." The original telegram is in DF (*TAEM* 13:982).

3. Edison had filed the French patent (No. 112,719) on 6 May; it issued on 10 July. He filed the Belgian patent (No. 39,502) on 8 May; it issued on 30 May. Undated memo, 76-012, DF (*TAEM* 13:1026).

4. Ireland probably meant "constant."

5. This was the Fuller battery, a particularly powerful bichromate of potassium cell invented by John and George Fuller (Brit. Pat. 3,339 [1877]). Relatively inexpensive to maintain, untended Fuller cells could stay in good condition for a few months if little used. Steady usage required attention to the cells at intervals of four to six weeks. Ireland to TAE, 27 Nov. 1876, DF (*TAEM* 13:992); Niaudet 1884, 211–20; Maver 1892, 17–18.

6. Although Ireland soon reported that the battery was patented in the U.S., no such patent has been identified. Ireland to TAE, 28 Oct. 1876, DF (*TAEM* 13:986).

**-805-**

*From William Wastell*

Port Huron Oct 26. 76

T A Edison

Tel received this evening[1] did not clearly understand, called on J P Sanborn[2] and asked him; he informed me he had a talk with you in N.J. and had written you at length since he arrived home=[3] if Sanborn can be induced to take hold of this matter he can do much towards a positive arrangement— You may rely on me acting with you as a square man, it is my wish to do any thing to benefit my stock in GStRR and if I am benefited you must be if you are I must be— I have not seen WPE for weeks—he and family are away from home I supposed he was with you or on his way there I have heard no more about Stewart defalcation—it is all smothered for the present—

I am still receiving the earnings of the road and only pay out as ordered by W.P.E. our earnings are not improving will keep you posted if anything turns up—

Write Sanborn your ideas and let him try and see what can be done Yours Respectfully

William Wastell

ALS, NjWOE, DF (*TAEM* 13:1074).

1. Not found.

2. John P. Sanborn, U.S. Collector of Customs in Port Huron, was an incorporator of the City Railroad Company and brother-in-law of William Wastell. *History of St. Clair County* 1883, 593; Jenks n.d., 7–8.

3. No Sanborn letter to Edison dating from this time has been found.

## –806–

*Notebook Entry: Telegraph Recorder/Repeater*

[Menlo Park,] Oct 28 1876

Electric Engine or other source of power clockwork etc giving continuous feed or an intermittant feed can be used either with a fork Reed electromagnet or other source of power[a]

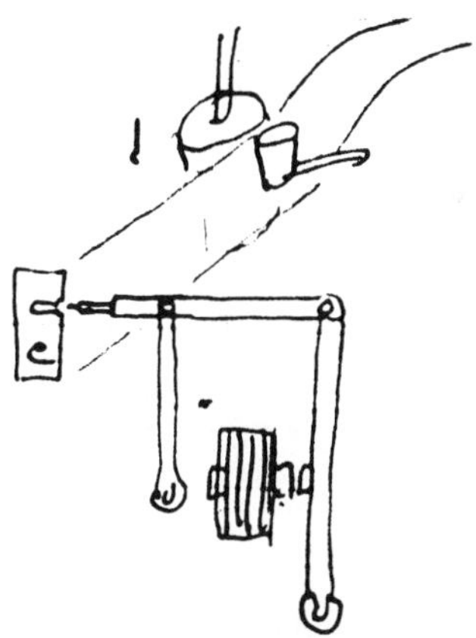

by continuously moving the paper and having a groove in a plate C the deepest part being slotted & having an embossing point have a knife on its e~~a~~xtreme end it embosses & at the same time cuts a slot in the paper in transmitting the contact lever is provided with a knife edge point and the emboss or rather indentation as I prefer to use the ~~inden~~ latter guides the point & it by presses drops into the indentation & through the slot allowing its ~~e~~end to touch a contact point thus closing the circuit.[1]

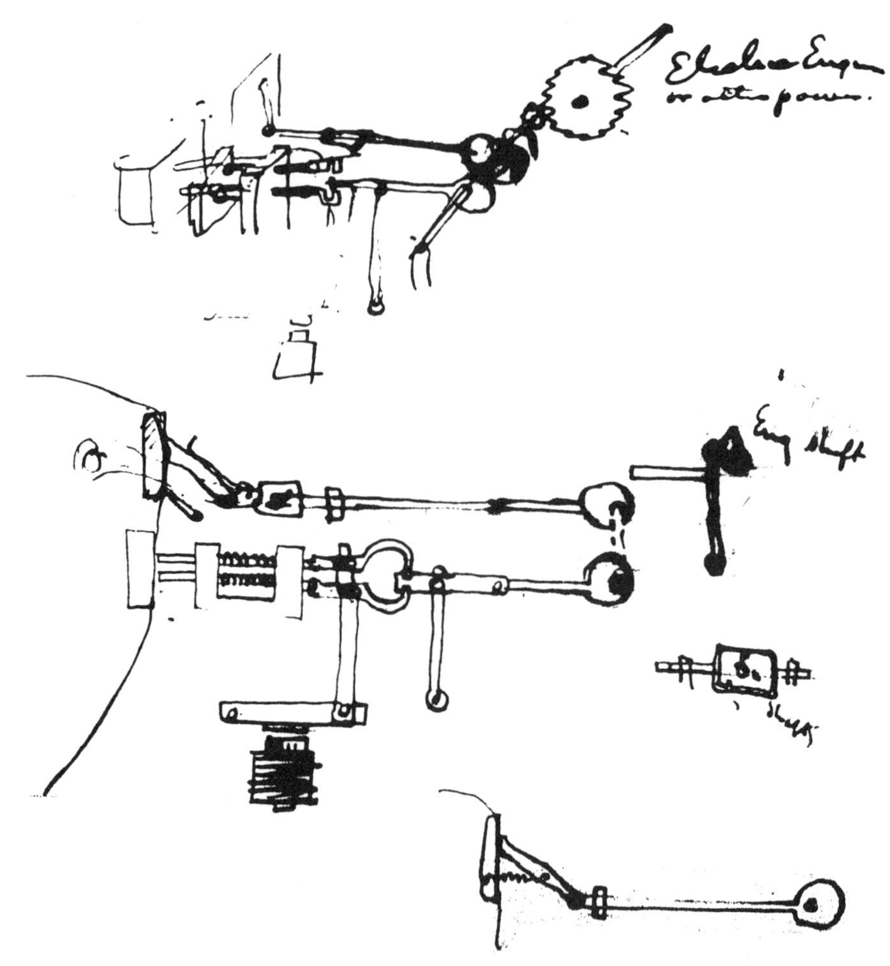

I propose to arrange the punching and feeding devices in such a manner that I can punch the messages ~~o~~in a continuous circle like the record on a Bain chemical disk,[2] Starting the message 2 inches from the centre & thence outward  I can use a square blank & this will be very convenient for handling. The disk may be revolved by one of the ~~e~~eccentrics in the motor feeding it around step by step, and the shaft carrying the disk secured to a moveable slide is moved continuously by another motor.[3]

Idea is to have rolls of tin foil, strips[b] 1/8 wide, and form characters by ~~p~~bringing it down on pasted paper by magnet which keeping closed allows foil to run out from roll & ~~p~~be pasted to roll but when lever raises breaks foil & again allows it to run out when closed again=

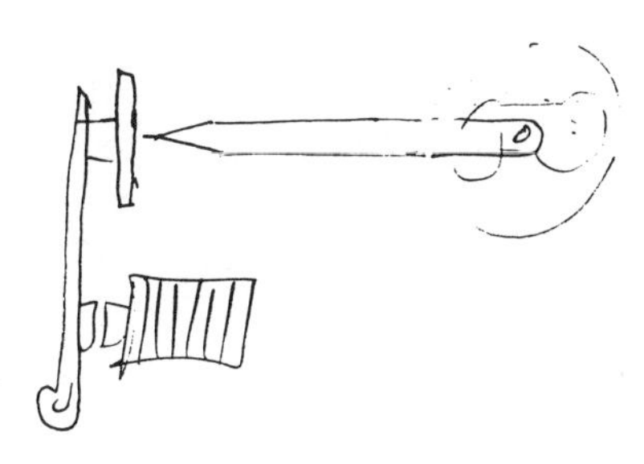

Electric[4] pen large needle. dots & dashes formed transmit by having full thickness paper close sounder & character formed of the dots[c] vibrate lever & open circuit & Reverse points=[5]

Embosser[6]

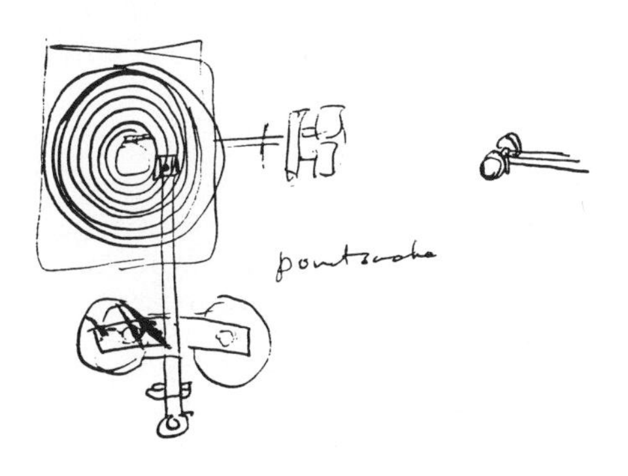

deep plate for feeding the jointed lever out

T A Edison

Chas Batchelor
James Adams

X, NjWOE, Lab., Cat. 997:32 (*TAEM* 3:368). Document multiply signed and dated. [a]Paragraph inserted above drawing. [b]Interlined above. [c]Obscured overwritten letter.

1. Figure labels are "electric engine or other power," "eng shaft," and "shaft."

2. Bain's receiver is illustrated in *TAEB* 1:65.

3. Figure label is "worm driven by another engine."

4. The Morse code preceding this word spells "this."

5. Edison's first automatic transmitter patent (Doc. 101; U.S. Pat. 114,656) used the thickness of the paper tape to open and close a reversing circuit.

6. Figure label is "point [-----]."

**-807-**

*Charles Batchelor to Thomas Batchelor*

[New York,][1] Oct 29th [187]6

Dear Tom,

Your report received with thanks.[2] I have just wired Ireland of London authority to sell Belgium as he has a customer for it.[3] Do you know any one going out to India or the other colonies that would be a likely person to work up a trade for us. I have got South America going pretty well & the West Indies & am working on Mexico & ~~the~~ Central America.[4] Who is working the 'Manchester district' for the London people? Every thing has been dull here for the last few months owing to the Presidential Election but they are now picking up & the prospect looks brighter for the winter. I was very much obliged to you for report of Breckon[5] it gave us a hint although I believe he is good He has already paid a bill for us at Elliot Bros[6] of £90.17— & accepted our drafts which I put into Drexel Morgan & Cos[7] hands to collect.

All well at home. I have got some help to run this thing & am spending a great deal of my time on a new electric Sewing machine.[8] I expect to use up some 8 or 9 months on it but if it is a success we shall be all right. Yours as ever

Charley.

ALS (letterpress copy), NjWOE, Lbk. 3:19 (*TAEM* 28:507).

1. This letterbook contains Batchelor's correspondence as agent of Edison's Electric Pen and Duplicating Press Co., which business he conducted at the office in New York.

2. Not found.

3. The party interested in Belgium is unidentified. A copy in Edison's hand of a cable to Ireland, dated 27 October 1876, states "Sell Belgium." DF (*TAEM* 13:985).

4. This may refer to Vesey Butler, who was headquartered in Havana, Cuba. See Doc. 830.

5. Not found.

6. Electrical instrument makers in London. See *TAEB* 2:176, 188.

7. Drexel, Morgan & Co., formed in 1871, brought together the New York–based Morgan and Philadelphia-based Drexel banking houses. With Drexel & Co. in Philadelphia, J. S. Morgan & Co. in London, and Drexel, Harjes & Co. in Paris, it constituted one of the most powerful banking operations in the world (Carosso 1987, 136–74). See Frederic Ireland to TAE, 9 Oct. 1876 (DF [*TAEM* 13:978]), regarding these acceptances.

8. Either Ezra Gilliland or Robert Henry might have been assisting Batchelor. On the sewing machine see Doc. 802.

[Menlo Park,] November 1 1876[1]

*Notebook Entry: Miscellaneous*

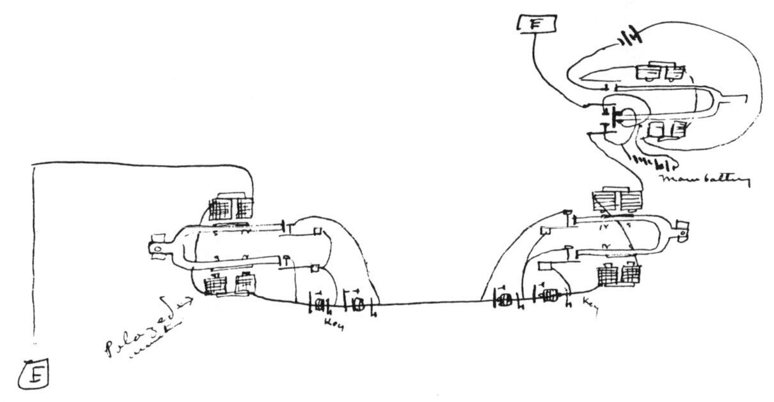


You Can Duplex This & make .6.[2]

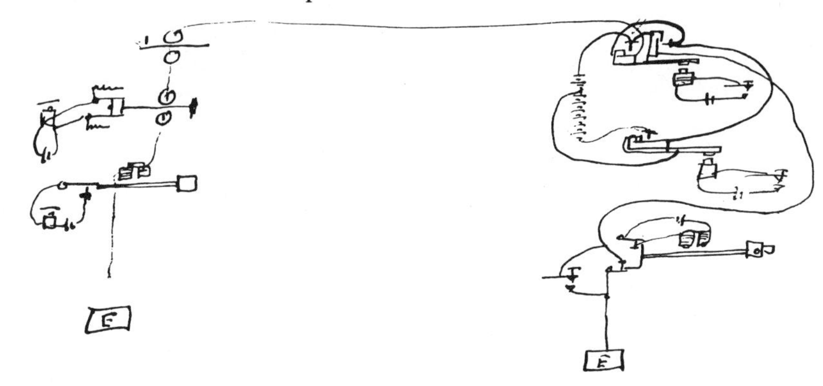


1 message by reversals 1 [message][a] by increase & decrease 1 [message][a] by acoustic Vibrations[3b]

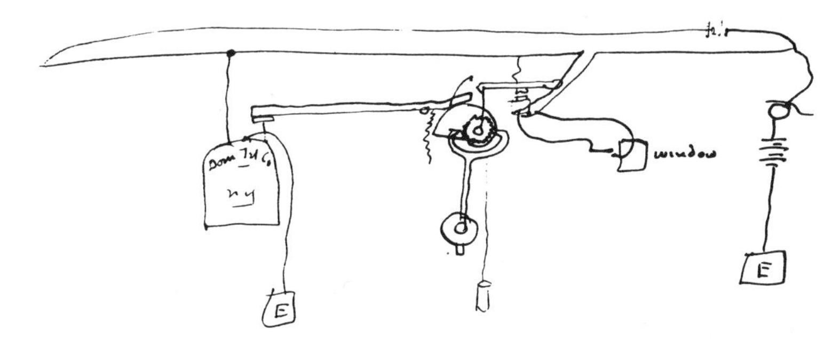


with[c] burglar alarm circuit closers on windows of a house contact short circuits magnet & that releases button pushing mechanism which is so arranged that the cord is just long nuf to depress button 4 times more or less=[4]

Tower bell striker for Fire alarm[d] Telegh.[5]

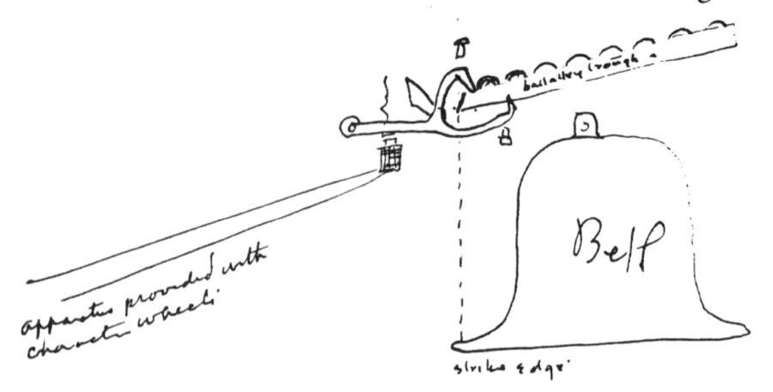


Cannon balls released at proper intervals by a magnets

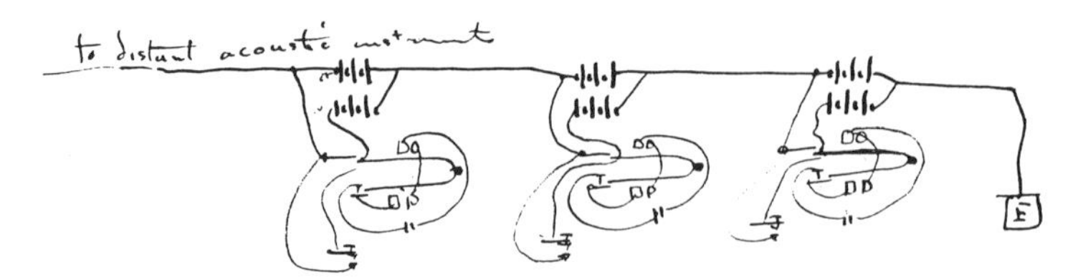


Acoustic[6] Transmitters each making different note

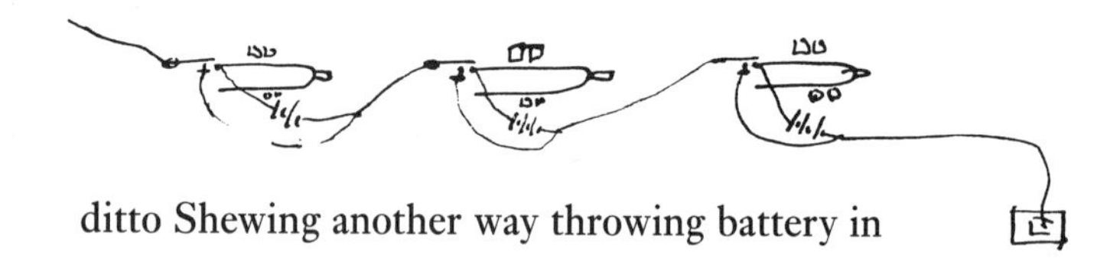

ditto Shewing another way throwing battery in

Throws in a neutralizing battery[7]  to right neutralizes  to left doubles strength= Acoustic System=

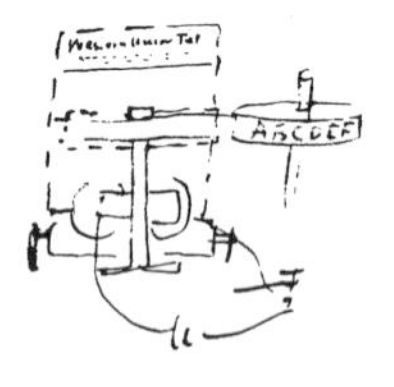

Combination Printer method[8] of pasting strips on WU Blanx  Every time you print letter it puts paste on or you can have a paste roller=

T. A. Edison

Chas Batchelor
James Adams

X, NjWOE, Lab., Cat. 997:35 (*TAEM* 3:369). Document multiply signed and dated. [a]Ditto marks in original. [b]"Haskins interference" added later in right margin. [c]Obscured overwritten letters. [d]Interlined above.

1. This document marks the end of Edison's attention to acoustic transfer telegraphy. Figure labels in the drawing below are "Polarzed armature," "key," "key," and "main battery." The armature for each of the four electromagnets driving the two lower tuning forks is a bar magnet (i.e., a polarized armature) attached to the tuning fork and labeled "n" and "s." The apparatus at top right would put the main battery rapidly into and out of the circuit without ever breaking continuity.

2. In this circuit Edison adds acoustic instruments to his basic diplex (and quadruplex) plan (Docs. 348, 472, 512, and 513) in an attempt to achieve a sextuplex circuit, much as he added his electromotograph the previous year (Docs. 583 and 611). While in theory this third type of signal should neither interfere with nor be disturbed by the other two types of signals, it was by no means obvious whether this could work in practice (see headnote, p. 280 and Doc. 877 n. 3). In the receivers, shown at left, a complex polarized relay (center) responds to increases and decreases in signal strength, replacing the neutral relay originally used. This had become typical in quadruplex practice by this time. Docs. 512 and 513 n. 2.; Preece and Sivewright 1891, 198.

3. Figure labels are "Dom Tel Co NY" and "window." Burglar alarms were becoming a common feature of district telegraph companies (Greer 1979, 25–27; Israel 1992, 111). Edison had drawn similar designs for a Domestic Telegraph Co. burglar alarm on 10 October (Cat. 997:26, Lab. [*TAEM* 3:365]).

4. In sophisticated district telegraph systems the signal transmitted indicated which of several services the subscriber wanted (messenger, medical help, police, firefighters, etc.).

5. Figure labels are "apparatus provided with character wheels," "ball alley trough," "Bell," and "strike edge." Edison probably intended this design for the Domestic Telegraph Co. fire alarm system. The intervals at which the cannon balls dropped were determined by the signal transmitted from the fire alarm box and would indicate the box's location. The signal also operated an apparatus (not shown; attached to the two wires at lower left) to provide a printed record of the signal.

6. Figure label is "to distant acoustic instruments." In this and the following two acoustic multiple telegraph sketches, as in the first design in this notebook entry, Edison devised signaling arrangements that maintain the line circuit's continuity. Here each transmitting key interrupts the regular pulses of its tuning fork by short-circuiting a battery; in the next design the key adds a battery to the line.

7. Figure labels are "same" and "ditto." What Edison refers to as right and left movements would be up and down in the drawing.

8. Figure labels are "Western Union Tel" and "ABCDEF." The Phelps combination printer typed its message on a paper tape. See *TAEB* 1:62 n. 1.

**–809–**

*Notebook Entry: Telegraph Recorder/Repeater*

[Menlo Park,] Nov 2 1876

I propose now to emboss the messages for transmission by having the blank stationary on a raidiating spiral block the blank being held down by another block with raidiating spiral groove cut clear through

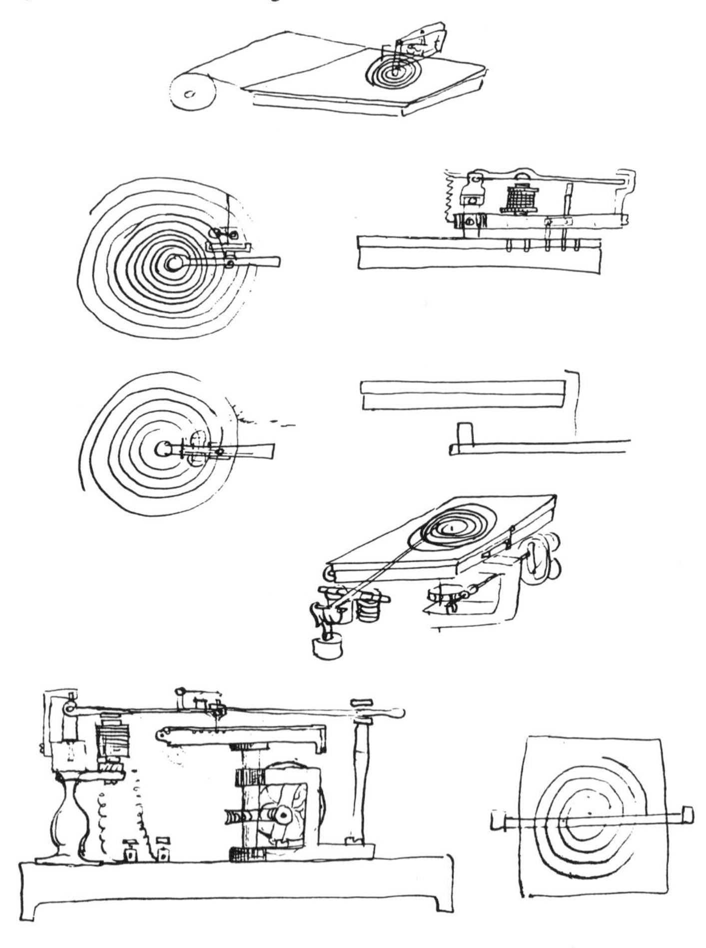

Perforating Embossed dots & dashes cut of[f] leaving a perforation doubled up

Perforating[1]

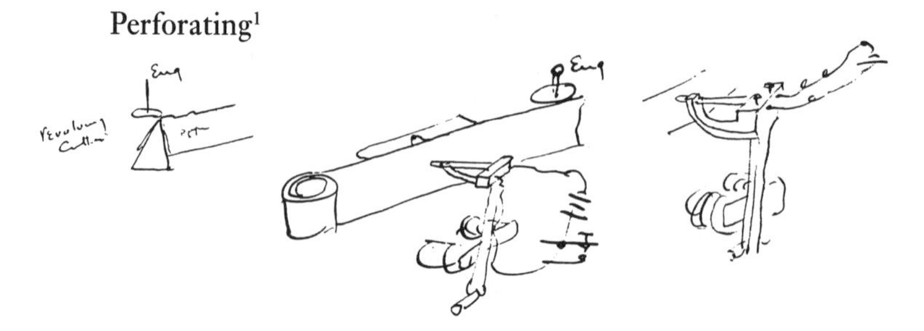

doubled up platina wire kept white hot by a battery and connected to a magnet operated by a relay or key or direct on a Telgh line, the closing of the magnet serving to bring white hot platina up to paper & burns thro it a hole the length of which is determined by the continuously moved paper, & length of closing of the magnet=

Wheatstone perforator[2a]

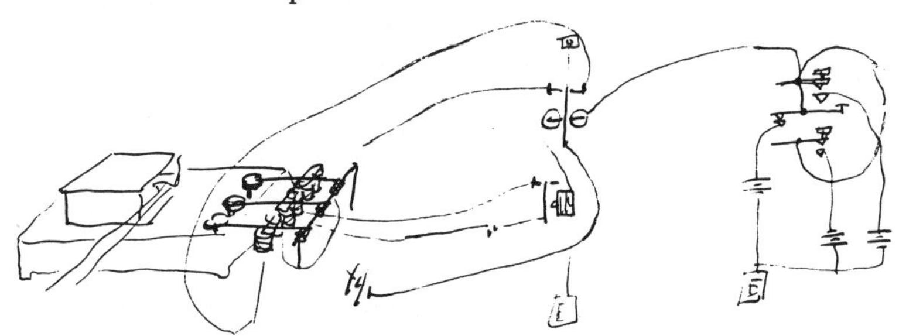

for working it over a Telgh line.

Embossing from a Continuous Roll—

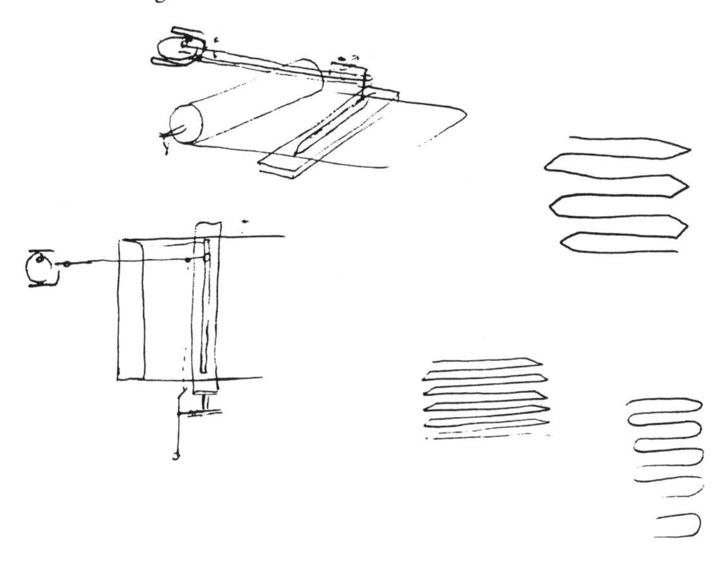

T A Edison James Adams

X, NjWOE, Lab., Cat. 997:37 (*TAEM* 3:370). Document multiply signed and dated. [a]"Nov 22 1876 T.A.E" in margin.

1. Figure labels are "revolving cutter," "Eng[ine]," and "paper"; and "Eng." Edison sketched similar ideas in Docs. 135C and 151A.

2. Edison had worked intermittently with the British automatic telegraph system of Charles Wheatstone (see, for example, Docs. 348 n. 2 and 457). Here he is using a polarized relay and a standard relay to send three signals that operate a Wheatstone perforator at a remote location.

[Menlo Park,] Nov 3rd 1876

*Notebook Entry: Chemistry*

Discovery[1a]

I have three bottles containing in a 100 cc HO, 1 5 & 20 gramme solution of Tannate of Iron I find after many weeks that a thick deposit has formed on the side of the bottles facing the light. I do not know ~~w~~how long it took to bring this about. on the 1 gramme solution it is thin & clean on the 5 gramme from some cause (pbly bottles been exposed all round) the deposit is all round but on the 20 grame it is sharp where exposed & very thick but not so clean as on the 1 grame solution= This fact can probably be used in photography and its allies, for printing etc= Note 10 gram sol Tannic a[cid] shews same but it dont adhere to g~~a~~lass; The 1 gram sol d[on']t sho=[b]

Phenomenon[a]— After many weeks there appears a crop of crystal both at the bottom of the bottle & f~~o~~loating on the solution. These crystals are of amber color and apparently square 1/32 sharp edged and are very transparent= I never saw such crystals from crystalizable chemicals= The solution is 1 gramme of Acetate Strychinine in 100 cc Rain water.[2c]

Phenomenon 14 gramme sol acetate urinum 100 cc HO[d] crystalizes only on side of bottle exposed to the sun—other side in difused daylight= These crystals creep up on side bottle & adhere 1/4 inch from surface of the Liquid= Same with 8 gramme 100 cc HO proto-acetate Copper—creeps up 1/2 inch deposits large crystals in quantity= Same with Cyanide Mercury 20 gramme sol= Same with acetate morphin 5 gramme solution. Cant say how long th~~is~~ese crystalizations[e] takes= pby 1 month at least= also formate copper in 4 10 & 20 gramme also oxalate ammonia in 20 gramme sol & many others= Sulphate anilin 10 gramm [---][f] shews greatest 1 [notg?][g] 5 same[h] crystalizes from inch of solution in bottom 4 inches above the surface of the Solution tho this may be due to pouring some sol out—but think not 1 & 5 gm of sul aniline turn Red but 20 gm doesnt=[i]

phenomenon= I notice in [Goltah?][g] ammonia 10 & 20 gramme solutions & in other sols which shew a fungus growth ie.[j] cotton like stuff that when it settles it settles at the bottom & back furtherst from the Sun—

Phenomenon= 5 gramm Gum Myrrh in 1 gramme Caustic Potash sol 100 cc a sponge like cloud rises up in the solution in that part of the Sol towards the sun= it doesnt adhere to sides bottle or apparently touch it—its clear on other side=

Phenomenon— Noticed that certain bottles & sols gave

streaks of colors out of 300 only 5 gave them distinct & these were colorless solutions. 1 gram sol Caustic Pot gave it good also oxalate Pot= changed bottles & this appeared to diminish it. very poor bottles may be something in it— try clean glass= (Oxalate Ammonia 5 grm (caustic Baryta 1 gramme (Stannate soda 5 gram) 500 milegms Caustic Strontia=

T. A. Edison

X, NjWOE, Lab., Cat. 996:67 (*TAEM* 3:312). Document multiply signed and dated. [a]Underlined twice. [b]"Note . . . sho=" crowded onto line; followed by centered horizontal line. [c]Followed by centered horizontal line. [d]"100 cc HO" written in margin. [e]Interlined above. [f]Canceled; "10 gramm [---]" interlined above. [g]Illegible. [h]"1 . . . same" interlined above. [i]"1 & . . . doesnt=" in margin. [j]Circled.

1. Edison stocked his laboratory shelves with hundreds of bottles of prepared chemical solutions. Sunlight and age had apparently ruined many of them. He continued his observations in Vol. 8:69 (Lab. [*TAEM* 3:501]) and published several of them (Docs. 813 and 845).

2. According to the next entry in this notebook, Edison used rainwater in many of his chemical solutions. Cat. 996:67, Lab. (*TAEM* 3:313).

## -811-

*From George Ward*

[New York,] Nov 4th [1876]

Dear Edison

I got your note all right.[1] Siemens[2] expects to return to NY on or about the 10th when I have no doubt he will avail himself of inspecting your many interesting objects.[3] I understand Preece and Fischer are not likely to come out before the spring.[4] I know the latter very well indeed and I shall not fail to let them know of your institution. I intended doing this if you had not mentioned it. I dare say I shall see something of them.

I shall try & come with Siemens.[5]

In haste. Yours, Truly,

G G Ward[6]

ALS, NjWOE, DF (*TAEM* 13:781).

1. Not found.

2. Charles William (Carl Wilhelm) Siemens, F.R.S., later knighted, was the fourth of eight adult brothers, five of whom became eminent figures in industrial, engineering, and scientific endeavors in the second half of the nineteenth century in Britain, Germany, and Russia. William made significant contributions in several fields of technology and science and headed a London firm that became a prominent producer and installer of submarine telegraph cables, among other electrical products. (This firm was one portion of the complex of family enterprises

led by the oldest brother, Ernst Werner von Siemens [see *TAEB* 1:82 n. 4].) William had been chosen as one of the judges for the Centennial Exhibition in Philadelphia and had come to the U.S. at the start of October. Pole 1888, esp. 9–15 and 280–81; Siemens 1968, 278–79.

In *TAEB* 1:82 n. 4, Johann Georg Siemens, a co-founder of the Berlin firm of Siemens and Halske, is incorrectly identified as one of Werner's brothers; he was a cousin. Siemens 1968, 52; Weiher and Goetzler 1977, 11, 25.

3. The Centennial Exhibition in Philadelphia closed on 10 November 1876.

4. See Doc. 803 n. 7.

5. No record of a visit by William Siemens to Menlo Park has been located in the U.S., U.K., or B.R.D. However, a member of the family and firm did visit the following summer (see Docs. 977 and 987). William spent 13–14 November visiting the Western Union and Atlantic and Pacific main offices in New York and there saw Edison's quadruplex and automatic telegraphs in regular operation and saw Gray's "electro-harmonic" (i.e., acoustic) telegraph apparatus as well. He left by ship on Wednesday, 15 November. *Telegr.* 12 (1876): 281. (This report, as was usual in the *Telegrapher* during these years, credited Gray with his devices but only mentioned Edison disparagingly.)

6. George Ward was American manager of the Direct United States Cable Co. (Reid 1886, 526). Siemens's firm had manufactured and laid that company's cable as its first major project in submarine telegraphy. Pole 1888, 206–10; Bright 1974, 128–29.

## -812-

*Notebook Entry: Battery and Copying Machine*

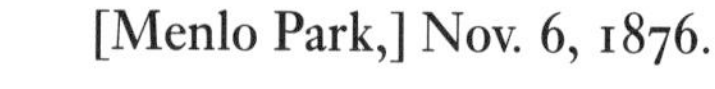
[Menlo Park,] Nov. 6, 1876.

Ga~~v~~lvanic battery[1]

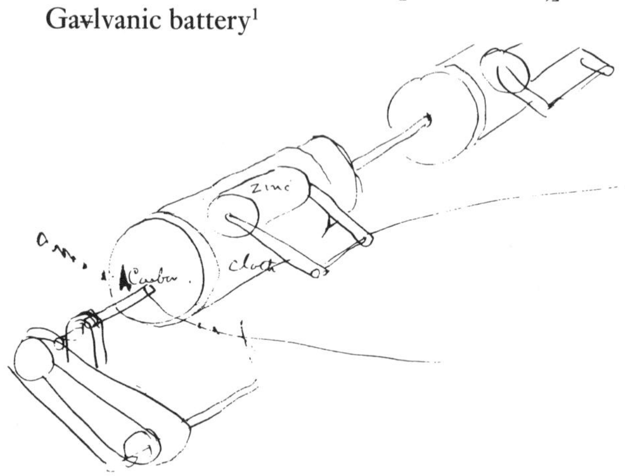


Cloth on Carbon Roller[a] saturated with exciting solution & a salt Hg. rotated. Zinc roller= or vice versa, Carbon roller & zinc with cloth on= object semi dry battery—no polarization

Copying machine.

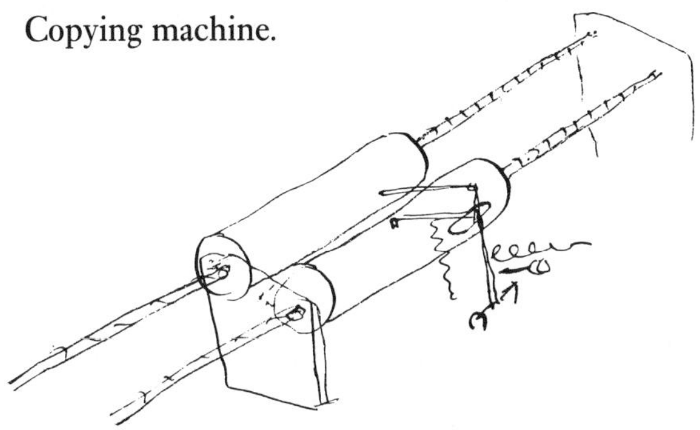

I propose to write the letter etc with an ink that will give great frictional surface; place it on a drum provided with an E.M.G lever on the extreme end of which is a fountain pen opposite another roller with the paper on for receiving the copy  the ink marks by their friction move the EMG lever cause it to bring the fountain pen against the other paper at every mark  the Cylinders revolve together[b] & are fed endwise by any suitable device such as a ratchet, worm-screw. the whole thing may be rotated by pen Electric Engine, clockwork handpower foot, or any source of power ~~ie~~[c] or tuning fork set in motion by magnet=[2] I can use a stylus & put carbonized paper[3] ~~on~~ over[d] second paper & roll, could use several EMG levers on one roll[e] & thus expediate copying—

T. A. Edison  Chas Batchelor
James Adams

X, NjWOE, Lab., Cat. 997:39 (*TAEM* 3:371). Document multiply signed and dated. [a]"on Carbon Roller" interlined above. [b]Added in margin. [c]Circled. [d]Interlined above. [e]"on one roll" interlined above.

1. Figure labels are "Carbon," "cloth," and "zinc."

2. On 10 November 1877 Edison sketched an alternative version of this electromotograph copying machine with electrical connections (NS-77-002, Lab. [*TAEM* 7:350]). An undated but more finished drawing of the 10 November device, without electrical connections, is in NS-77-002, Lab. (*TAEM* 7:445).

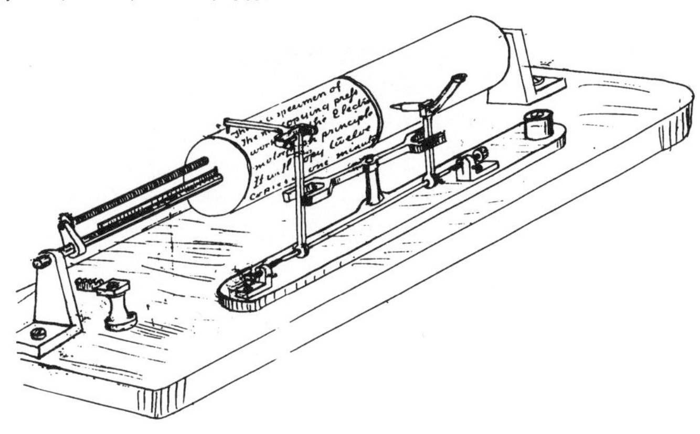

*Edison conceived this design for an electromotograph copying machine about 10 November 1876.*

3. Carbonized paper was a common material in the laboratory. See Doc. 829 n. 11.

-813-

*Article in the* American Chemist

MENLO PARK, N.J., Nov. 10, 1876.[a]

LABORATORY NOTES[1]

BY T. A. EDISON.

1. HARD RUBBER or vulcanite, placed for several weeks in nitrobenzol, becomes soft and pliable like leather, and easily broken.

2. The vapor of chloral hydrate is a solvent of cellulose. I have found the corks of bottles containing the crystals eaten away to the depth of a quarter of an inch, the cork being resolved into a black semi-liquid. Certain kinds of tissue paper are partially dissolved in time, if thrown in a bottle containing the crystals.[2]

3. A very difficult substance to dissolve is gum copal. I have found that aniline oil dissolves it with great facility.

4. Hyposulphite of soda is apparently soluble to a considerable extent in spirits of turpentine. Large crystals of "hypo" melt down to a liquid after several weeks, and if the bottle be shaken, partially disappear. The turpentine smell nearly disappears.

5. The vapors of iodine, in the course of several months, will penetrate deeply into lumps of beeswax.

6. If to a solution of bisulphide of carbon there be added twice its bulk of potassic hydrate in sticks, and the bottle be well sealed, the whole will, in two months, become an intense reddish, syrupy liquid, with scarcely any free bisulphide of carbon.

7. Some substances in solution form crystals or deposits on the sides of the bottles containing them generally above the water line. Among such solution in 100 c.c. of rain water may be mentioned a 14-gramme solution of acetate of uranium, 8-gramme d[itt]o. of protoacetate of copper, 5-gramme do. of acetate of morphine, 10-gramme do. of formate of copper, 20-gramme do. of tannate of iron. These deposits invariably take place on that part of the bottle most *exposed to light.* This phenomenon may be due to heat, but deposits or films occur in some solutions *within the liquid* as well as above it—especially noticeable with tannate of iron, the film of which adheres strongly to glass.

PD, *Am. Chem.* 7 (1876): 127. [a]Place and date not that of publication.

1. These notes were the product of various experiments undertaken by Edison and his staff or the results of observations made of chemical solutions prepared for experiments. See, for example, Docs. 791 and 810 and Cat. 996:41–47, Lab. (*TAEM* 3:299–302). These published notes are continued in Doc. 845.

2. Edison intermittently investigated the solubility, absorbency, and permeability of variously treated paper, and its constituent cellulose, both as related to automatic telegraphy and his electric pen copying system and as topics in their own right, over several years; see e.g. Docs. 579, 586, 655, and 915 n. 2.

## -814-

*From George Walter*

Phila Nov 13/76

Dear Sir

I have a prospect of selling yr Patents[1] for India, for Ital~~a~~y, for Spain, for Queensland, for Netherlands, for Austria, for Russia at least I have parties here at work on each of those countries also Norway & Sweden

I have strong parties considering my proposition for the purchase of the United States, But I guess the price will be too high for them[2] I have talked up very strong since I have been here with all these parties and I am satisfied that I shall make some sales for you before long.[3]

If any of these parties call upon you or Mr Batchellor say to them that I am authorised to sell the Patents for you and that all arrangements made through me will be fulfilled by you & oblige Truly Yrs

Geo B. Walter

⟨OK Edison⟩

ALS, NjWOE, DF (*TAEM* 13:1052).

1. That is, for the electric pen.

2. The next week (24 November) Walter wrote to Edison that the commission Edison had offered for the United States sale (five percent) was too low; Walters asked for ten percent. DF (*TAEM* 13:1053).

3. In January, Edison wrote to Walter about royalty terms for the sale of the copying system, and also discussed the possibility of beginning manufacture of the apparatus (excepting the pen itself) in Austria (4 Jan. 1877, DF [*TAEM* 28:532]). See also Doc. 892.

-815-

*From Frederic Ireland*

[London,] Novr 16 1876[a]

My dear Sir

Pneumatic Pen[b]

Our Patent Laws are so unsatisfactory that—rather than risk a law suit—we have consented to buy up the invention— It certainly works wonderfully well and with a little improvement, would do quite as well as the "Electric"— the cost of production was very Small and had we not come to terms the opposition would have been very Serious.

There is an effort being made to get the Pneumatic patented on the Continent— had you not better authorize me— or someone—on your behalf—to oppose this—or to come to terms with the inventor

Please send us an Embroidery Pen.[1]

—Press.—

I enclose you sketch of this—[2] Frame A has a groove in it— B has a projecting band of India Rubber— the two fasten together by two hooks and the paper is held firm at all points.— However thin the paper used may be it cannot "sag" and it cannot tear away as it is apt to do where it is pierced by the pins

Ink

Our new ink is working very well and—as a close assimilation to writing Ink is an immense improvement.

Batteries

These also are working well— the fact of their portability is alone—an immense—advantage—but independently of that they are lasting well— We are putting them to the severest tests and we quite believe[c] what is said of them—i.e. that they will keep in good working order—without renewal—for many months.[3]

I think we had better send you samples of the battery[4]— press & Ink.— Yours truly

Frederic Ireland

ALS, NjWOE, DF (*TAEM* 13:989). Letterhead of Electric Writing Co. [a]"187" preprinted. [b]Followed by centered horizontal line. [c]Obscured overwritten letters.

1. See Doc. 800.

2. Not found.

3. In a letter of 27 November, Ireland said that this battery—the Fuller—could still write after six weeks of use, unlike other batteries he had tried, which weakened after half an hour's writing (DF [*TAEM* 13:992]). See also Doc. 804 n. 6; and Niaudet 1884, 223.

4. On 1 December, Ireland wrote Edison that the batteries would be on a steamer to New York departing the next day. DF (*TAEM* 13:996).

**–816–**

*From Alfred Nelson*

New York, Nov 24 1876[a]

Dear Sir

The examination of witnesses is progressing in the Gamewell case vs Domestic Tel Co.[1] and Mr Pope[2] is being examined on behalf of the former Co= He is very voluminous and on a cross examination we shall need some one who can match him and by general consent you are looked on as the proper person— I have told Mr Wheeler[3] our counsel that I presumed you would appear if requested to do so, as I know how tenacious you were on the question of not having infringed on Gamewell's Patents in securing your own, as expressed to me, and the study necessary to make this certain would qualify you very fully to prove the fact of noninterference— Please say whether or not you will come tomorrow if possible[4] Respectfully

Alfred Nelson V.P.

ALS, NjWOE, DF (*TAEM* 13:1239). Letterhead of Domestic Telegraph Co. [a]"New York," and "187" preprinted.

1. Although infringement suits tried in New York City would have been heard by the Circuit Court for the Southern District of New York, this case could not be found listed in the court docket and no record appears to exist.

2. Franklin Pope, Edison's former business partner, was a prominent electrical engineer and inventor who was at this time in charge of Western Union's patent department. *TAEB* 1:115 n. 1, 226; *DAB*, s.v. "Pope, Franklin Leonard."

3. Everett Wheeler (1840–1925) was a prominent New York attorney who represented Atlantic and Pacific Telegraph Co. in the Quadruplex Cases. He later authored a reminiscence of his association with Edison. *DAB*, s.v. "Wheeler, Everett Pepperrel"; Wheeler 1927.

4. Edison's response is unknown.

**–817–**

*And Robert Gilliland Agreement with Western Electric Manufacturing Co.*

November 28, 1876[a]

This[b] agreement entered into this twenty-eighth day of November, eighteen hundred and seventy-six, by and between Thos. A Edison, of Menlo Park, County of Middlesex, and State of New Jersey, and Robert Gilliland,[1] of Hudson, Lenewee County, State of Michigan, parties of the first part, and the Western Electric Manufacturing Company,[2] of Chicago, Cook County, State of Illinois, party of the second part;

And WHEREAS the parties of the first part are the sole owners of all the right, title, and interest in a certain patent granted by the United States, August 8th, 1876, and numbered one hundred and eighty thousand eight hundred and fifty-seven, and also a certain patent granted by the Dominion of Canada, Sept. 6th, 1876, and numbered six thousand five hundred and eight, for an electrical pen and duplicating press for Autographic Printing;

And WHEREAS the said parties of the first part are in possession of a certain lot of special tools for the manufacture of the machinery under such letters patent, and exhibit of which is hereto annexed and marked Exhibit A,[3] and have established certain agencies for the sale of such manufactured articles in the United States and the Dominion of Canada;[4]

And WHEREAS the said party of the second part is desirous of acquiring the business of the said parties of the first part, and the sole right to manufacture and sell said Duplicating Apparatus under said patents within said countries for a period of three (3) years or more, and are also desirous of acquiring the said special tools to enable them to economically and expeditiously manufacture the said apparatus;

Now it is hereby agreed by and between the parties hereto as follows:

The said parties of the first part do hereby transfer to the party of the second part the business as above mentioned, and the sole right to manufacture and sell within the United States and the Dominion of Canada, and nowhere else, the articles covered by the above recited patents; and also transfer and sell the said special tools upon the following terms and conditions—namely:

That the said party of the second part shall pay to the parties of the first part five (5) dollars royalty on each and every Duplicating Apparatus as covered by said patents and which may be sold within said countries, and fifteen per cent on the price of all parts and supplies as fixed and shown in Exhibit B hereto annexed,[5] sold in such countries, which payment shall continue for a period of three (3) years from the date of this Agreement, and fifteen hundred dollars for the said special tools in three (3) equal annual payments of five hundred (500) dollars each, the first of such annual payments to be made on the delivery of said tools.[6]

The said party of the second part does hereby guarantee and bind itself that the royalty paid by it shall not be less than two hundred and fifty (250) dollars in each and every month during the said three (3) years on complete Apparatus.

It is further agreed that if six months before the expiration of this three (3) years, said party of the second part shall notify the said parties of the first part that they desire to continue this contract for a further period of three years or more, it shall be extended accordingly for that period.

It is further agreed that, in case the sales of said Duplicating Apparatus within the said countries (excepting those for exportation) shall reach fifteen hundred (1500) in any one year, then the said royalty of five (5) dollars shall be reduced to four (4) dollars in that year, and in every year in which that number is sold. If the sales shall reach twenty-five hundred (2500) in any one year (those for exportation excepted) then the royalty shall be reduced to three (3)[c] dollars.

It is further agreed that, upon the first day of January in each and every year, that the said party of the second part shall make full and true returns to the said parties of the first part, under oath, and that the said royalties shall be paid monthly to the said parties of the first part in the proportion of seven tenths 7/10 to said Edison,[7] and three tenths 3/10 to said Gilliland.

Should the party of the second part desire to discontinue the manufacture and sale of such patented articles after the expiration of the three (3) years they shall return said special tools in as good condition as when received, for which they are to be paid back the price—viz: fifteen hundred dollars.[8]

And WHEREAS the said party of the second part is desirous of manufacturing and selling to the said Edison complete machines under such patents for his foreign trade and to furnish the owners of the foreign patents with such machines and the parts thereof;[9]

Now it is hereby agreed that the said party of the second part shall deliver to the said Edison, in New York or Jersey City, all the machines that he may require to supply his foreign orders for the sum of twelve dollars and fifty cents, currency, for each and every duplicating apparatus (packed) and the parts thereof at prices shewn in Exhibit C hereto appended,[10] during the continuance of this contract;

And WHEREAS the said Edison has disposed of the patent for the said Duplicating Apparatus for the Kingdom of Great Britian to one John Robt. Breckon, of Sunderland, England, and his assigns, and has agreed to furnish him with as many duplicating apparatus up to thirty (30) per week for a period of four (4) years from June 29th, 1876, at a fixed price of two pounds nine shillings and six pence sterling for each and every duplicating apparatus, the same to be paid for in London by his acceptancies of ninety (90) days drafts, bearing date of delivery of such goods in New York or Jersey City, to any steamship company named by said Breckon.

Now, therefore, the said party of the second part agrees to furnish to the said John Robt. Breckon all the Duplicating Apparatus, or parts thereof, the prices of which are shewn in Exhibit D hereto appended,[11] necessary to carry out the said contract between the said Edison and said Breckon as long as said Breckon shall promptly pay such drafts. And the said party of the second part further agrees that the last mentioned machines shall be packed well and delivered F.O.B. in New York or Jersey City.

And WHEREAS the said Edison has patented said invention in foreign countries other than Great Britian and the Dominion of Canada, and has not yet disposed of his said patent rights in such foreign countries;

Now it is hereby agreed by the party of the second part that it will not sell any of the said patented Duplicating Apparatus, for exportation to foreign parts, other than the Dominion of Canada and Great Britian until the said Edison shall authorize it in writing signed by him so to do. The party of the second part shall have the benefit and advantage, if it shall elect to accept the same, of any further contract or contracts that may be hereafter made by the said Edison for the supply from the United States of the said patented Duplicating Apparatus. The license hereby granted is to extend to all reissues and renewals of the said patents, or either of them, and similar license or licenses shall be granted to the party hereto of the second part by the said T. A. Edison for the use of any invention or inventions of any addition to or improvement or improvements upon the said invention, such additional license or licenses to be granted upon fair and equitable terms. And the said parties of the first part further agree to prosecute all infringers upon the said patent and bear all the expenses of such prosecution, for the purpose of protecting said party of the second part in the full and exclusive enjoyment of said patent.[12]

And finally it is agreed that the provisions of this agreement shall go into effect on and after December fifteenth (15) 1876.

In witness whereof, the said Thos. A. Edison and Robert Gilliland have hereunto affixed their hands and seals; and the Western Electric Manufacturing Company has caused these presents to be signed by its President and its seal to be affixed and attested by its Secretary.

(Signed) Thos. A. EDISON. (Signed) ROBERT GILLILAND.

THE WESTERN ELECTRIC MANUFACTURING CO.,

By (Signed) ANSON STAGER, Pres't.[13]

Attest: (Signed) E. M. BARTON, Sec'y.[14]

TD (transcript), NjWOE, Miller (*TAEM* 28:1019). [a]Date taken from text, form altered. [b]"(COPY)" in top margin. [c]Interlined above.

1. Robert Gilliland, Ezra's father, owned a three-tenths interest in Edison's autographic printing (i.e., electric pen system) patents. See Doc. 724.

2. Western Electric was a major electrical manufacturer in which Western Union owned a one-third interest. See *TAEB* 1:402 n. 5.

3. Not found.

4. See, for example, *TAEB* 2:598 n. 11.

5. Not found.

6. The first $500 installment was the only money received until the end of February 1877, when the first royalty check arrived. Western Electric paid $3,779.56 in royalties in 1877 and just over $2,000 by September 1878, when Edison's account record ends. Batchelor to Gilliland, 31 Jan., 21 Feb. 1877, Lbk. 10:68, 81 (*TAEM* 93:77, 87); Cat. 1185:88, Accts. (*TAEM* 22:595).

7. Edison shared the profits from the electric pen system with Batchelor and Adams. See Doc. 637.

8. On 27 April 1877 Edison and Gilliland signed an agreement with George Bliss and Charles Holland granting Bliss and Holland manufacturing rights if Western Electric discontinued its manufacture of the pens. Miller (*TAEM* 28:1044).

9. See Doc. 892.

10. Not found; but see Doc. 892, Schedule B.

11. Not found.

12. The term infringement here referred to any legal challenge or illegal manufacturing of the electric pen system that threatened Western Electric's exclusive manufacturing rights. See, for example, Doc. 900 n. 8.

13. Anson Stager (1825–1885) entered telegraphy as an operator in 1846. He was general superintendent of Western Union Telegraph Co. in the 1850s and during the Civil War also served as head of the Union military telegraph, attaining the rank of brigadier-general. After the war he declined the general superintendency of Western Union, choosing instead to head the company's newly created Central Division. In 1878 Stager was elected a Western Union vice president. In 1872 he helped organize Western Electric Manufacturing Co. and served as its president until 1884. *DAB*, s.v. "Stager, Anson."

14. Enos Barton (1842–1916) became a telegraph operator in 1859. In 1869 he organized the telegraph manufacturing firm of Shawk & Barton, which soon became Gray & Barton when Elisha Gray bought out George Shawk's interest. Anson Stager also became a partner in the firm and the three men formed Western Electric in 1872, with Barton serving as secretary. Barton became company president in 1886. *NCAB* 30:383.

**-818-**

*To William Orton*

[Menlo Park,] ⟨Dec 4/76⟩[a]

Dear Sir

Before you was taken ill you agreed to allow me to draw on account of Acoustic for four weeks. I drew for three only. may

I have the balance. I would not ask it only I need it, to pay my "Intellectuals" Yours

T A Edison

ALS, NjWOE, DF (*TAEM* 13:1259). [a]Dated by Orton.

**-819-**

*To George Prescott* [1]

⟨Dec 4/76⟩[2a]

I have something that will allow one man to work one end of a Quadruplex:=[3] no change in present system= all morse: will you cooperate with me so that I can introduce it= It will save the Co a large sum by reducing the number of employees:[4]

AL, DSI-NMAH, WUTAE. [a]Dated (with initials) by William Orton.

1. George Prescott was chief electrician of the Western Union Telegraph Co. See *TAEB* 1:258 n. 4.

2. This note and Doc. 820 were written on the same sheet of paper.

3. Edison is referring to his embossing translator. Doc. 857; also see Docs. 789, 806, and 809.

4. On 28 November, Norman Miller had written Edison a letter saying that he had "had some talk with Mr. O[rton]. about your economic and he says 'that sounds like business, but don't print it in the newspapers.'" Miller offered to set up a meeting between Edison and Orton, but Orton was sick for much of the fall and there is no evidence that they had met by the time of this letter. DF (*TAEM* 13:1257).

**-820-**

*George Prescott to William Orton*

[New York,] ⟨Dec 4/76⟩[1a]

Hon. William Orton President

I have said to Mr. Edison, verbally, that I would assist him in every possible way to bring out and introduce upon our lines any improvements which he might devise. I told him I had never accepted, or agreed to accept, any pecuniary interest in any invention designed for the use of the Company, with the single exception of the Duplex & Quadruplex, & had not done so in these cases until he had experimented thirteen months upon our lines, and had had full use of our workshop & apparatus and had then informed me that he could accomplish nothing without my personal assistance, and personal interest in the invention. That I had not taken such part—accepted such an interest—until I had submitted the proposition to you and received your assent—and that now I expected, and had no doubt of receiving proper compensation for that interest, but that I should never accept another under any circumstances.[2]

I would, however, assist him in developing his invention for the Company to the full extent of my ability, and as fully as if I had a primary interest in its success. Mr. Edison seemed to feel satisfied and pleased with my assurance of help and said he would bring his apparatus here for development. I offered him a room for experimenting and all assistance necessary. He said he had just come from Gould & they wanted him to testify in their case.[3] I said I hoped he would do so—that the truth could do us no harm—and the more we had of it the better. Yours truly

Geo B. Prescott

ALS, DSI-NMAH, WUTAE. [a]Dated (with initials) by William Orton.

1. This letter and Doc. 819 were written on the same sheet of paper.

2. See Docs. 432, 445, 451, 466, 509, 519, 535, 603, 694, and 727.

3. *Atlantic & Pacific v. Prescott & others,* Quad. 70, 71, 73 (*TAEM* 9:288–10:797).

**-821-**

*From Pitt Edison*

Sarnia, Ont., Dec 4th 1876[a]

Dea Bro

Our Suits came off Last week I send you copy of each[1] the Wm Stewart & Co claim he levied on our lots which ~~we~~ will give us 15 months time at 7 per cent which which is less intrest than we could get the mony for so that disposes of No 1 the next claim is Guy Kimbals[2] I fixed with him this[b] way I paid him $50.00 down and am to pay $25.00 a week untill it is paid so that fixes No 2 for I can pay it out of the earnings of the road the F C Harrington[3] Judmnt is what we owed the Savings bank I see him Saturday also H M McMorran[4] the Vice Prest of the bank thay will give 6 months on thair amt so that will arrange No 3 we have been suid on 4 small accts last week[b] which I will put in a Stay for 5 months the notes at the first Nat bank Mr Barnum[5] told me to pay $10.00 or $15.00 each month on the principal and the intrest all would be OK so you see with this to take care off and to look to this & the Sarnia Road and to watch the opposition you may immagine I have something to do although thay write that I do not attend to buisness so you see something must be done within the next 3 or 4 months and in my opinion one of the following first to sell out if posable second to buy out if posable Third to lease the new road or to leas our road to them if we can agree fifth which I think would be the boss to rais Money if posable and buy up all the notes and claims of the old road and force the paymt as soon as posable bid it in[b] and build arround to Gratiot & go for them  it needs about 4000.00 to bring this gents to Tirm  if you could come out here with 4000 you would fix this thing up in some way in less than 48 hours[6]

AL, NjWOE, DF (*TAEM* 13:1079). Letterhead of Sarnia Street Railway Co. [a]"Sarnia, Ont.," and "187" preprinted. [b]Interlined above.

1. Pitt enclosed a newspaper clipping listing the proceedings of the November term of the Circuit Court. Included were Wm. Stewart & Co. vs. the Port Huron and Gratiot Street Railway Co., judgment for $1,148.71; Guy Kimball vs. the Port Huron and Gratiot Street Railway Co., judgment for $306.07; and Charles F. Harrington vs. the Port Huron and Gratiot Street Railway Co., Wm. Wastell, and Wm. P. Edison, judgment for $578.63. DF (*TAEM* 13:1080).

2. Guy Kimball was a Port Huron dealer in flour, feed, and seeds. *History of St. Clair County* 1883, 581.

3. C. F. Harrington was cashier and later vice president of the Port Huron Savings Bank. Jenks 1912, 429.

4. Henry McMorran, a wholesale and retail grocer and ship chandler, was vice president and later president of the Port Huron Savings Bank. Endlich 1981, 101; Jenks 1912, 429.

5. H. G. Barnum was cashier and later president of the First National Exchange Bank. Jenks 1912, 428.

6. On 12 December 1876 Pitt acknowledged the receipt of Edison's $600 to pay Harrington (see note 3) and also noted that the men connected with the City Railroad were about to make Edison an offer to settle the street railroad business. On 27 December Pitt wrote that they were ready to make a formal proposition. He indicated that the City Railroad people wanted to "call both roads even including real estate," but noted that the Gratiot Street Railroad had "4 cars and ten Horses" while City Railroad had "two devlish poor cars and three horses" and were hiring three horses. In this letter he again urged Edison to "come out here and you can make a trade in one day  it would be to your advantage to do so" as the City Railroad people were "red hot for to do something right away." DF (*TAEM* 13:1082–84).

**-822-**

*From Josiah Reiff*

New York, Decr 8th 1876[a]

Strictly Confidential

Dear Edison

I desire to see you as soon as possible & alone. I am seriously contemplating going into bankruptcy, to rid myself of the load that is crushing me & I want your advice in several matters.

I must avoid[b] anything that will injuriously affect our joint interest, (yours & mine)[c] but I must do something. O[1] seems to promise fairly with his lips,[b] but for some mysterious reason, either of over confidence or insincerity, he simply delays. If after a careful review of the situation, it is deemed betr, I shall go through the form of legal relief, but my moral obligation to pay every dollar I owe, still remains absolute, if I am ever able. When will you be over? Truly

J. C Reiff

You better destroy this after reading[d] I sent copy Bill[2] to G.B.P[3] today.

ALS, NjWOE, DF (*TAEM* 13:1175). Letterhead of J.C. Reiff. [a]"New York," and "187" preprinted. [b]Obscured overwritten letters. [c]"(yours & mine)" interlined above. [d]On two lines spanned by brace.

1. Perhaps William Orton. See Doc. 823.
2. Probably the bill of complaint in *Harrington v. A&P.*
3. George Prescott.

**-823-**

*From Robert Russell*

N.Y. 8th December 1876

Confidential

Dear Sir/

Having been informed that you have recently made several new and important inventions in Telegraphy[1] I beg leave to offer for your patient and careful consideration the following suggestions.

1. If you submit these new inventions to Mr Orton with the understanding that he may control them if he pleases disconnected from your prior inventions in Automatic and Quadruplex telegraphy you will seriously damage your own interests as well as those of your friends.

2. Mr Orton has recently felt well disposed to adopt our theory that the true way to impair the strength of the A.&P. and to prevent the further acquisition of prestige by them is to make an arrangement with you and your party whereby the Western Union may hold the title to your Automatic and quadruplex patents in connection with your more recent inventions.

3. It is a question of policy which they have to consider. An improvement in their own present mode of doing business is of course of some importance to them, but it is not so important to them as the weakening of their great enemy,[2] and especially the prevention of the acquisition by that enemy of the great strength they would obtain by coming to terms with Mr Reiff and your party[3] and thereby holding your automatic and quadruplex inventions as the basis of further improvements in Automatic and Quadruplex telegraphy.[4]

4. It is true you might show the Western Union how to improve the present system but if the A.&P. should then acquire your Automatic and Quadruplex inventions[5] they might have afterwards important improvements invented and thus be able to maintain a strong competition with the W.U. That is what they want to prevent.

5. Our true plan is to couple your former inventions with those now contemplated by you, not to dispose of them separately to the W.U. By coupling them together you will have in your own hands the control of the negotiation and we should so apportion the total price paid as to secure the larger part for you individually, for your newer inventions

6. If on the other hand you should endeavor to dispose of the new things separately to the W.U. you would have to wait a long while before you would receive any considerable payment on account of them— the payment would depend upon future experiments and trials.

And by separating the newer inventions from the former ones you would disparage the Automatic and Quadruplex and cool the ardor of the W.U. in the pursuit of the policy—which they have (or recently had) in view viz the deprival of their enemy of their prestige derived from the supposed control of your Automatic and Quadruplex systems.

It appears clear to my mind that you have everything to gain by holding the inventions to be disposed of as a unit and that great loss and damage would result to you if you were to adopt the opposite policy Yours truly

R W. Russell

P.S. We must take care to prevent as much as possible the disparagement of your prior inventions seeing that your reputation is at stake This letter has been written without conference with Mr Reiff

ALS, NjWOE, DF (*TAEM* 13:1176).

1. Edison had filed patent applications for his acoustic transfer telegraph system in August (U.S. Pats. 185,507, 200,993, and 235,142) and for the perforating telegraph recorder/repeater in November (U.S. Pat. 200,994).

2. Jay Gould's Atlantic and Pacific Telegraph Co. was Western Union's primary competitor.

3. That is, the American Automatic Telegraph Co., of which Russell was company trustee and in which he held 100 shares of stock (Doc. 676; Articles of Association and By-Laws, 31 Aug. 1875, DF [*TAEM* 13:489]). Russell had previously acted as Edison's confidential adviser and attorney (see Docs. 577 nn. 2–4 and 734 n. 6; and Russell's undated memorandum [probably written earlier in 1876], DF [*TAEM* 13:1194]).

4. This assumes the validity and legal vindication of the claims of Edison and George Harrington to patent rights for Edison's quadruplex and automatic telegraph technologies. Those claims, however, were contested in pending litigation, primarily *Atlantic & Pacific v. Prescott & others* and *Harrington v. A&P.* See *TAEB* 2, app. 3; and Docs. 740 n. 5 and 750 n. 5.

5. That is, if they won *Atlantic & Pacific v. Prescott & others.*

**-824-**

*Notebook Entry: Chemistry*

[Menlo Park,] Dec 10 1876.

5 below Zero—

No fire in Laboratory fearful cold great wind storm nearly blown Laboratory down had brace it with 20 poles (Telgh)[1] Record of phenomenon of freezing of different chemical solutions=

100 Milgrms Citrate Quinine, in 100 cc HO, Frozen solid, in 4 oz bottles transparent cracked beautifully in centre there is a egg shaped peice which slooks like Cotton. The surface towards Cork is very uneven sunk in place & raised in others ¼ inch

1 gram Camphoric Acid in 100 cc HO= raised in places on top ½ inch= greatly expanded bottle busted=

1 gramme— & 400 m.g. Cyanide Silver & 100 cc HO. shoots formed by gas= spread out from centre in curves ½ bending upwards & ½ downwards

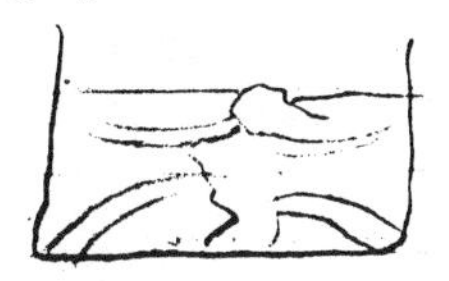

500 mg. Hydrate strontia same.

Chloride Zinc 1 gramme 100 cc— same on end of shoots have a globule=

1 gramm Sulphate Soda Expansion burst botttle.

20 grammes pyrogallic acid= Beautiful Brown scintillations of crystals on side of bottle

Frozen Ferridcyanide of potassium looks deep yellow=

1 gramme gum Guiacum[2] has fuzzy looking elongated egg in centre bottle with innumerable shoots running from it

20 grms Nitrate lead throws up a buttion ½ inch high in centre on surface of solution

Picrotoxin[a] 500 m.g.= white opaque column in centre & clear [~~and~~?][b] otherwise looks as if ice had thrown the picrotoxin with great force= this appear to ~~e~~illustrate[c] that phenomenon strongly=

Iodine 1 grm in HO 100 cc shows fuzzy Column in centre

*The Menlo Park laboratory, showing telegraph poles bracing the walls.*

~~w~~colored but the ice around it is clear[c] this is like picrotoxin

Ammonic Chloride crystals of Ice similra to the salt of the Am Chl cross-hatched

X, NjWOE, Lab., Cat. 996:71 (*TAEM* 3:314). [a]Multiply underlined. [b]Canceled. [c]Obscured overwritten letters.

1. A photograph dated 1878 shows such braces. Pretzer 1989, 19.
2. Guaiacum.

## –825–

*Notebook Entry: Duplicating Ink*

[Menlo Park,] Dec 13th 1876

Edisons Duplicating Ink

On the 1st day of December 1876 or thereabout Edison compounded the following ingredients to form a new copying ink:— Aniline Violet, Alcohol and Gum dextrine in the proportions as mentioned below. We find it superior to anything in the market it being able to do the following:—

A copy can be taken on letter press paper & from such copy numerous copies can be taken on letter or other paper. A copy can be taken from such copy at any future time—[1]

The following are the proportions to manufacture this ink:—

No 1 Solution (Stand.)[a]

2 galls & 1 pint of Alcohol
5 lb 5 oz of Aniline Violet
4 galls & ½ pint of Water

No 2 (Stand. Solution)[a]

20¼ galls Hot water
21½ lbs of Gum dextrine

1 Gall of No 1 to 4 galls of No 2 and stir well before bottling

| | |
|---|---|
| Thos. A. Edison | Chas Batchelor |
| E. H. Johnson | James Adams |

X, NjWOE, Lab., Cat. 997:44 (*TAEM* 3:374). [a]This heading appears at right of list, which is spanned by brace.

1. Batchelor's scrapbook contains a printed sheet of directions for using Edison's duplicating ink, which he hand-dated 13 December 1876. Cat. 1240, item 6, Batchelor (*TAEM* 94:7); see Doc. 831 n.4.

**-826-**

*From George Gouraud*

New York[1] 15 Dec 76

My dear Edison

Ref'g to our conversation of a few days since when I suggested to you the idea of your discovering some means of detecting sewer gases in the atmosphere of dwellings or other rooms and thus preventing an enormous percentage of illness derived from such an insidious enemy and which you said you felt no doubt of being able to accomplish— referring that interview I now wish to repeat what I then said that in my opinion you wd render a lasting and inestimable service to the whole sewered world comprising a pretty large proportion of ~~civili~~ existing civilization.

The subject is at this moment brought particularly "home to me by the fact that I have recently several times detected the presence of abominable ~~odorers~~ odors in my bed room & trace them unmistakably to the wash stand drain pipe & yet I can never get anybody to the spot in time to corroborate my statement; consequently my landlord believing his plumbing to be perfect does nothing in the premises, & I am obliged for the sake of safety keep water standing in the basin, and yet sometimes upon my letting it out I get a puff of the vilest air imaginable—a poison which must impregnate the room until the air is completely changed. And thus is the health of my family endangered.

Here is a great field for you, & honor enough to me, the credit of having put your great genius on the track.

Do it & be blessed of all men

Geo. E Gouraud[2]

Written with the English ink GEG

ALS, NjWOE, DF (*TAEM* 13:786).

1. Gouraud addressed this from the Albany Hotel.
2. George Gouraud had represented the Automatic Telegraph Co.

in London and subsequently became Edison's principal agent in Great Britain. See *TAEB* 1:280 n. 7.

**–827–**

*Uriah Painter to Josiah Reiff*

WASHINGTON, D.C. DEC 24 1876[a]

Dr Rff.

A short note from Edison dated 12-23, rec'd to-day declines to do as I suggested—[1] Says he is well pleased with his situation & pleasant relations to WU—& such is life! Truly

U H Painter

ALS, NjWOE, DF (*TAEM* 13:1180). [a]Place and date from *Philadelphia Inquirer* hand stamp.

1. On 20 December, Painter had written to Reiff that he had heard nothing from Edison in reply to an earlier letter, which probably contained the suggestions to which Painter refers. Reiff forwarded that note to Edison, and Edison apparently wrote to Painter. Reiff noted to Edison on the back of this letter, "Within is his interpretation of your letter." (DF [*TAEM* 13:1178–79, 1181]). Neither Painter's earlier letter nor Edison's answer has been found.

**–828–**

*Edison's Electric Pen and Duplicating Press Co. Account*

[New York?,][1] Dec 27th 1876

Up to this date we have[a] sold (205) two hundred & five presses complete and quite some extras amounting to altogether to $3,880.28 for which we paid $2717 98/100 leaving a balance of

| | | |
|---|---|---|
| ~~Net~~ profit of sales | | $1162.30 |
| Cash advanced etc | | 420.70 |
| | | 1583 00 to account for |
| Expenses | $216.77 | |
| Brecons[2] D for (Due) | 857.79 | |
| Cash in hand | 217.05 | |
| Paper on stock | 3.44 | |
| Fogg & Co[3] | 16.00 | |
| E T Gilliland | 48.40 | |
| Lefferts[b] | 3 | |
| T A Edison | 220.55 | |
| | $158~~0~~3.00[a] | $ 1583.00 |

I have taken $450– & divided it as follows[4]

| | | |
|---|---|---|
| Edison | $327.60 | |
| Batchelor | 110.2~~0~~5 | |
| Adams | 12.15[c] | $450.00[d] |

AD, NjWOE, DF (*TAEM* 13:943). Written by Batchelor on letterhead of Charles Batchelor. [a]Obscured overwritten character. [b]"Lefferts" and "3" interlined. [c]Underlined twice. [d]Preceded by brace spanning last three entries.

1. Batchelor crossed out the preprinted "NEWARK, N.J." He probably wrote this account in the New York office of the company.

2. John Breckon.

3. H. Fogg and Co., New York importers, were agents for Edison's electric pen and press in China and Japan. Wilson 1877, 447; see Doc. 892, Schedule C.

4. See Doc. 637 regarding the division of electric pen royalties.

**-829-**

*Charles Batchelor to Thomas Batchelor*

[Menlo Park,] Dec 28th 1876.

Dear Tom.

Yours of the 6th received[1] Glad to hear we are all in the same mind in regard to Father.[2] I think it would be a great deal better for him if he could be induced to give up and lead a quiet life at home.

I am not afraid that anyone will get ahead at present in regard to running sewing machines by electricity. It is a very difficult subject to handle. During the last five years I have made more than 50 different applications of electric engines (every one of different principle) to running small machinery and I also know from experience that all the patents taken out during that time for improvements in such engines are worth nothing, although in some cases there has been an immense amount of talk spent over them, in fact all the improvements made on "Electric Engines" during the last ten years you can value by calling it one, whereas to make it a successful motor you want one hundred. It is a well known fact that we do not get in practice more than 1/10 part of the power given out by a battery now if we can find some means whereby we can utilize the whole why then you can talk about Electric Motors, and until this is done or partly so, or some new principle found out Electric Motors will be a failure.[3] Of course this does not apply to motors of very small power such as the pen and those used in Telegraphic printing Insts. but for sewing machines, printing presses, etc. I have been working on a new principle for the last 3 months and I must confess I have not met with very much success although I have succeeded in making a Wilcox & Gibbs machine stitch 130 stitches per minute through 8 thicknesses of cloth right along through the day with 3 cells of ordinary Bunsen battery.[4] This of course is nothing great, but I do not know and cannot tell for certain for some short time whether I have the principle applied right. The power derived from an electromagnet is the most deceptive thing possible, and in this country has been the means through which some big frauds have been perpetrated.[5]

I am glad you are connected with us and hope you'll make a little out of it.[6] Is the Mr Chas Wilson[7] the gentlemen who used to live near St Luke's Church? Much obliged for the names of his friends. I just sent 12 to Dunedin New Zealand today.[8] Mr Ireland has sent us 2 batteries out by Steamer and they have arrived but I have not got them yet.[9] We have started a new enterprise here in the shape of "the American Novelty Co"[10] which we got up with the idea of working a great many of the "little things" that we should otherwise pass over in our experiments; many of them when worked are very valuable.[11] It is a regular organized company of 50,000 shares I think it will be very successful I have another very good thing which I have already secured, but have not had time to do anything more than the first experiments It is a new Holtz machine for the production of statical electricity[12] After it is out I believe I can sell one to every college & professor in the world. It is a new application. Yours

Charley.

ALS (letterpress copy), NjWOE, Batchelor, Cat. 1238:57 (*TAEM* 93:68).

1. Not found.

2. Batchelor's father had apparently suffered some sort of attack earlier in the fall and Charles had advised him to give up business and offered to help out financially if this proved necessary. He also offered to pay his way to Menlo Park for a period of recuperation. Batchelor to James Batchelor, 9 Nov. 1876, Cat. 1238:55, Batchelor (*TAEM* 93:67).

3. On electric motors see *TAEB* 1:293 n. 3.

4. See Doc. 802.

5. For example, the Paine motor. See *TAEB* 1:230 n. 2.

6. This probably refers to the electric pen. Batchelor had offered his brother the general agency for Great Britain, but had to withdraw the offer after Edison sold the rights to John Breckon. Charles Batchelor to Thomas Batchelor, 29 Mar. and 9 May 1876, Cat. 1238:46, 51, Batchelor (*TAEM* 93:59, 63).

7. Unidentified.

8. No other record exists of this order for electric pens.

9. See Doc. 815.

10. The American Novelty Co. was incorporated in New York on 28 November 1876, with Edison as president and Edward Johnson as secretary and general manager. American Novelty Co. stock certificate, 15 Dec. 1876, DF (*TAEM* 13:791); Cat. 1240, item 4, Batchelor (*TAEM* 94:7).

11. Edison later testified that the American Novelty Co. was formed to "work off some of the small inventions which I was making. The manufacture of a great many articles from carbon was one of these inventions. Among other things was an electrical sheepshearing machine, an electrical drill, an electrical engraving machine, ribbon mucilage and other inventions which I cannot now remember. They were small things. I think he [Johnson] had a list of about twenty." P. 3040, *Sawyer and Man v. Edison* (*TAEM* 48:27).

Edward Johnson testified that the object of the American Novelty Co. "was to acquire numerous inventions of Mr. Edison, Mr. Batchelor, Mr. Adams, my own, and others, and to put them upon the market. Several such were acquired, namely, Edison's duplicating ink, Edison's battery carbons, Edison's jeweller's engraving machines and others which I cannot now recall" (p. 105, *Sawyer and Man v. Edison* [*TAEM* 46:235]). At least one more product can be identified—Batchelor's "Office Door attachment," which indicated whether the occupant was in or out and, if out, the time of return (Cat. 1307:59; Cat. 1233:19; both Batchelor [*TAEM* 90:644, 61]).

Edison also described the experiments leading to the American Novelty Co.

> I carbonized paper in the summer of 1876. Such paper was to be used for battery carbons, for non-conductors of heat, and articles were to be made in different shapes and carbonized. A great many articles were so made and carbonized from paper. Sheets of carbonized paper were used for electrical resistance about that time. Strips of cardboard or Bristol board, about a quarter of an inch wide and five inches long were placed in gas tubes and carbonized by placing the same in a furnace and heating the tube to a white heat. The strips were packed in the tube one upon the other and the interstices were filled with charcoal powder. Sheets of tissue paper were laid in iron boxes, fifty to a hundred deep, on the top of which was laid a weight of metal so that the carbon would remain straight after being carbonized. Also sheets of thick Bristol board several inches square were carbonized under strain to keep them straight. Some experiments were also made to carbonize small crucibles made out of Bristol board.
>
> The experiments were quite extensive. My intention was to go into the business of making carbon wire for various purposes, electrical and chemical, for electric lighting and batteries. A company called the American Novelty Company was to handle the goods in connection with a gentleman named James, who is now dead. Mr. Charles Batchelor, and I believe Mr. E. H. Johnson, saw many of the experiments. Mr. Adams, one of my assistants, now dead, helped me in the experiments. I would mention that we also carbonized wood made up in various shapes, as well as paper. [Pp. 3006–7, *Sawyer and Man v. Edison* (*TAEM* 48:10)]

In his testimony Edison listed the following items to be manufactured of carbonized paper: "Battery carbons, strips of carbon for electrical resistances, dishes for use in making chemical reactions and tissue paper as a non-conducting packing." He said they also "made four or five crucibles, a great many flat sheets and a half a pound to a pound of carbonized tissue paper" over the course of two months during the summer. P. 3041, *Sawyer and Man v. Edison* (*TAEM* 48:27); see also Charles Batchelor's testimony, pp. 3164–69, ibid. (*TAEM* 48:92).

12. Wilhelm Holtz's machine, which produced static electricity by induction, employed two glass plates (Atkinson 1890, 726–29). Apparently an important feature of the new Holtz machine mentioned by Batchelor was that it did not use glass. In a 4 November 1876 letter to Robert Spice, professor of chemistry at Brooklyn High School, regarding Spice's joining him and a man named Benjamin in developing such a machine, Batchelor noted that if experiments proved satisfactory the machines could be "made a great deal cheaper than glass" (Cat. 1238:54, Batchelor [*TAEM* 93:66]). A clipping from the October 1876 issue of the *American Chemist*, kept in one of Batchelor's scrapbooks, noted a new form of Holtz machine developed in Prussia that employed ebonite (hard rubber) instead of glass plates (Cat. 1240, item 30, Batchelor [*TAEM* 94:15]).

**-830-**

*Charles Batchelor to Vesey Butler*[1]

[Menlo Park,] Dec 28th 1876

Dear Sir,

Your letter of Dec 2[2nd?][a] received and contents noted. The goods shall be shipped immediately though not with large presses. When we can get these large presses your troubles will be over for then all your orders can be filled promptly I have already got your orders down first in the list for large presses so your people cannot buy in New-York till after yours are shipped If we have any such applicants I shall tell them that I have shipped you a lot and it would be better to get from you.

In this case I also ship you two bottles of writing fluid,[2] (invented by Edison but the property of the American Novelty Co) which I want you to try, and see what you think of it; and if you like it to introduce in Cuba. from a single writing with an ordinary pen you can take copies on letter or other paper up to 20 good and then lots of letter press copies after. I think it is good where they only want a few say 4 to 10, and I sell a great deal of it I will send you further particulars in regard to it shortly Respectfully Yours

Chas Batchelor

ALS (letterpress copy), NjWOE, Lbk. 3:44 (*TAEM* 28:526). [a]Illegible.

1. Nothing is known of Vesey Butler beyond his association with Edison. The familiarity shown in correspondence indicates that he knew Edison rather well, perhaps as a telegraph operator. In 1876 he was selling Edison's electric pen copying system. He continued his association with Edison through 1880, promoting the inventor's electrical inventions in Cuba. See *TAEM-G1*, s.v. "Butler, Vesey F."

2. That is, Edison's duplicating ink.

-831-

*From Edward Johnson*

[New York,] 12/30/76.

My Dear Alva—

Saw Batch today—[1] thot. you would be in— told him about success at Direct Cable[2] how it was obtained &c—but he thot. you would like to have it direct from me—on paper—. So I set about it in the am & worked till 3 pm before we got it satisfactory—

Then did it by means of cloths obtained from the Erie RW[3] ofs—[--][a] the occupants of which I astonished by an Exhibition of the working of the Ink—[4] We took 4 sheets of cloth down to cable ofs— ran em thro. the wringer— Had men make very light copies

Put em thro the wringer  Dried em—and took em up stairs to Letter Press— There put em thro the following treatment with immense success—

Placed on Counter

1st an Oil Board
2d a cloth—
3rd a sheet of tissue
4th the Dry Copy—
5th an Oil Board
~~6th a sheet of tissue~~
6th a cloth
7th Tissue
8th Dry Copy
9th Oil Board

& so on till we had 4 complete sandwiches of this character= put em in Press give em one min. & took em out before an appreciative & applauding audience— Ward[5] being the Enthusiast— Did[b] the same thing over again 4 times with the same copies & getting the same results— Whoop-la—being the tone of the meeting[c]

Explained to Ward how by going down to Ann St. & buying a stout 2d hand Press long enough to take 2 sheets on a page—

getting 4 long stout pieces of Card Board—
50 Cloths same length
50 Oil Boards " "
a small wash tub—with Cheap wringer attached

He could make—25 Sandwiches (50 messages) Put em in press— then proceed to make another 25—by which time magnificent copies would be found in Press—& not a seconds time lost in waiting on em— Even Bucknor[6] tumbled to this— The whole party saw the point— Ward ordered the ½

*A label for Johnson's ribbon mucilage.*

Doz I gave him sent to Rye Beach by that night Express—& is waiting for me to Deliver him some in half pint Bottles—[7] give him the price for the same & pay me the money—

Trot out em ½ pints & your prices for same

Come over on Tuesday to attend a meeting—

Hurry up the mucilage[8] Sutterlin[9] wants put out Canvassers on em at once says more in it than in the Ink.

Happy New Year

E. H. Johnson

ALS, NjWOE, DF (*TAEM* 13:944). [a]Canceled. [b]Obscured overwritten letters. [c]Followed by centered, horizontal line.

1. Batchelor spent his days at the electric pen office at 41 Dey St., New York.

2. The Direct United States Cable Co.; see *TAEB* 2:566 n. 2.

3. The Erie Railroad Co.

4. A label for the duplicating ink gives the following directions for its use.

> Keep Bottle closed when not using. Don't use a pen dipped in Other Ink. For LETTER PAPER COPIES; Copy from original on tissue sheet; Sponge a Letter Sheet both sides, blot surplus water from it; Remove original and put Letter sheet Exactly in its place on Tissue Copy. When this tissue copy ceases to yield other copies, take a Second impression from Original.
>
> For COPYING FROM OLD COPY; Take a BLOTTER; put a tissue sheet upon it, thoroughly SPONGE put INKED side of copy upon tissue sheet; put under good pressure One minute. To COPY A NUMBER AT ONE TIME; put in wet muslin cloths in place of blotters and do not sponge. TO TAKE THESE COPIES ON LETTER PAPER; Sponge the Letter sheets in addition to use of cloths. [Cat. 1144, Scraps. (*TAEM* 27:338)]

5. George Ward.

6. Unidentified.

7. See Doc. 836.

8. Johnson had invented a ribbon gum mucilage, which was marketed by the American Novelty Co. Cat. 1240, items 8, 27, 35, Batchelor (*TAEM* 94:8, 14, 17)

9. American Novelty Co. circulars identify J. E. Suitterlin as manager of the company's stationery department. Cat. 1144, Scraps. (*TAEM* 27:338).

## –832–

*Draft to Joseph Hickson*

New York[1] Dec 1876[2]

Joseph Hickson[3] General Manager G.T.R

Parties inimical to the interest of the old street railway between Fort Gratiot and ~~p~~Port Huron have been scheming to have your company eject the road from the yard at that point.

From motives of spite against my brother and with hopes that they might succeed in ruining the old road through the G.T.R. by ejectment they have built a cheap and straggling road to run in competition with it. My brother Wm P Edison was the pioneer of the old road as well as the one running from Point Edwards to Sarnia and has worked in harmony with the G.T.R Cos representatives for many years and as yet without profit on either road. The rumors set afloat By the parties above mentioned of ejectment etc. are a course of constant fear and expense.

The road from Pt Edwards to Sarnia has a right of way (with (certain restrictions) through the Cos grounds for a period of 10 years, signed by yourself.[4]

If not detrimental to the interest of your Co would you sign a similar right of way through the yard on the other side of the river  Yours Respy

Thos. A. Edison

ADfS, NjWOE, DF (*TAEM* 13:1087).

1. Edison gave "41 Dey St" (the Electric Pen business office) as his return address.

2. Another, undated draft is in DF (*TAEM* 13:1085).

3. Joseph Hickson became general manager of the Grand Trunk Railroad in 1874. Currie 1957, 142.

4. The agreement, dated 9 December 1874, is in DF (*TAEM* 13:29).

## -833-

*Draft to William Orton*

[Menlo Park, December 1876?][1]

Dear Sir:

Some six months ago while experimenting with the system of Acoustic Telegraphy called for by my contract with the Co[2] &[a] described in the French work of J Baille,[3] on which system both Mr Gray of Chicago & Bell of Boston are also working, I was led into another principle of Multiplex transmission[4] which promised great results and although not called for by my contract, I thought should be worked up and secured to the Co. and I have been continuously engaged in perfecting the same; and have reached a point where I am able to send and receive eight different messages simultaneously over a single wire, with Morse apparatus, and the same has been covered by patents & caveats[5] setting forth every ~~imaginable~~ modification that I could conceive; the copies of such patents and caveats are hereto annexed= These ~~pats &~~[b] papers ~~c~~all of which have been assigned to the Company[6c] contain many devices which I have been unable to test & it will require some time to perfect the system= At the present time I can only work between New York and Philidilphia, on longer line the phenomenon of the Static Charge causes mutilation of the signals and I have been unable up to the present time to devise a compensation ~~p~~ sufficiently powerful to counteract this effect beyond a distance of 100 miles= This apparatus upon an artificial line, has been shewn to Mr Smith assist to Mr Jno Van Horn[7] ~~&~~ who[b] I believe has made a report to you~~r~~ ~~I~~ my assistants[d] having sent & Recd 8 messages simultaneously on the artificial wire in his presence;[8] These Instruments I have laid ~~to~~ aside, and am now engaged in experimenting ~~up~~ with a modification shewn in the patent papers with hopes of obtaining the compensation spoken of= It is only a question of time when the system will be perfected as the principle I am sure is correct, but you will readly allow that it is a very complicated subject and requires some elaborate contortions of the brain to manipulate thes signals ~~over th~~ and eradicate the defects ~~and~~ hence the experimental labor is exceedingly tedious= While I have been engaged on this side issue, I have ~~had~~ caused Mr Murray to make me a complete set of Acoustic apparatus upon the principle set forth in the work of J Baille & called for by my contract; I now propose to take hold of the subject again having gained much experience since I ~~dr~~ ceased experimenting with that principle. I annex hereto copies of patents & caveats which are in the patent office covering this principle= Two claims in these papers have been conceeded to Mr Gray & Bell ~~& he ha~~ they being prior inventors & Mr Gray has conceeded me one, that one being substantially ~~his~~ his Receiving[e] system as shewn in the Centennial=[9] It is not possible for either Mr Gray or Bell to obtain a base patent in this principle, and the only troublesome claim Mr Gray could obtain would be a method of manipulating the currents by transmittors ~~which~~ on which he has two plans; ~~but~~ one of which I had applied for & have now conceeded him but I have 4 other methods ~~entirely~~ of effecting the same object which are shewn in the annexed papers and Mr Gray has filed no papers shewing these.[10] I have no doubt of ultimate success with both systems, but the whole subject is of such a nature that quick results are impossible. I have drawn I think since last — thirty two hundred dollars[b] & have expended ~~considerably~~ half as much[f] more ~~than this~~ of my own funds; ~~the latter It is~~ which I shall not call on the Co to pay except in the event of success; ~~If you~~

~~think that the showing is good,~~ if you desire me to continue the experimenting It would be more satisfactory to reduce the $200. to $100. and arrange so[g] I can draw it regularly; because at the end of each week if the experimenting has not been satisfactory to myself, I find it impossible to screw up my courage to the point of asking for the $200. & consequently have to suffer— ~~Whereas if~~ & when[h] it has been satisfactory ~~you a~~ I do not always find you in & then I have to suffer= Your Respy

T. A. Edison

ADfS, NjWOE, DF (*TAEM* 13:1261). [a]"called . . . &" interlined above. [b]Interlined above. [c]"eall . . . Company" interlined above; "of which" interlined above that. [d]"my assistants" interlined above. [e]"his Receiving" interlined above. [f]"half as much" interlined above. [g]Obscured overwritten letter. [h]"& when" interlined above.

1. As the text makes clear, this document was written sometime following the resolution of interferences between two of Edison's acoustic telegraph applications and patents of Elisha Gray and Alexander Graham Bell, which occurred in mid-December 1876. Porter, Lowrey, Soren and Stone to TAE, 12 Dec. 1876, DF (*TAEM* 13:1055).

2. Doc. 695.

3. Jean Baptiste Alexandre Baille, *Wonders of Electricity,* tr. from the French with numerous additions by Dr. John W. Armstrong (New York: Scribner, Armstrong, and Co., 1872). See Doc. 671.

4. The acoustic transfer telegraph. See headnote, p. 27.

5. One caveat (Doc. 754) and three patent applications—Edison Cases 122, 124, and 125 (U.S. Pats. 185,507, 200,993, and 235,142)—covered the acoustic transfer system. Edison may also have considered a fourth application, Case 126 (U.S. Pat. 200,032), to have been related.

6. The caveat and Case 122 (U.S. Pat. 185,507) were assigned on 16 September 1876, and Cases 124–25 (U.S. Pats. 200,993 and 235,142) were assigned on 7 October 1876. "Letters Patent granted to Thomas A. Edison for, or relating to 'Telegraphy,'" WU Coll.

7. Smith is otherwise unidentified; see Docs. 786 and 788.

8. This test probably took place in September (see Doc. 788); no report has been found.

9. According to the judges, one type of receiving equipment exhibited by Gray was "founded on the remarkable property . . . discovered by Mr. Edison . . . in his electromotograph." U.S. Centennial Commission 1880, 454.

10. Four of Edison's pending patent applications involved claims at issue with those in patents and applications of Gray and Bell. Case 116, later issued as U.S. Patent 198,089, was placed in interference with Gray in July 1876. Edison resolved the interference by conceding the second claim to Gray and substituting devices not shown by Gray in the first claim. However, the wording of the claims continued to pose a problem and the patent did not issue until December 1877, after Edison erased all but the first claim and substituted two new claims. Case 118, later issued as U.S. Patent 198,087, was originally placed in interference with both Gray and Bell in June 1876, but this was amended to an inter-

ference with only Gray in July. In that case Edison dropped his first claim and Gray conceded Edison's seventh claim. Both of these patents issued in December 1877, as did Case 115 (U.S. Pat. 198,088). Edison originally intended Case 115 to go into interference with Bell, but he amended the contested claim in October 1877. One other application disputed with Gray, Case 117 (U.S. Patent 186,330), was amended in December 1876 to avoid interference with Gray and issued in January. Pat. Apps. 186,330 and 198,087–89.

# -4- January–March 1877

Edison began the new year with three extensive series of chemical experiments. The first grew out of new etheric force experiments involving an electrochemical cell. As Edison substituted various solutions in the cell he investigated in turn the effect on the etheric spark, the polarization of the electrodes, etheric spark again, and the possibility of using the sound of bursting bubbles evolved at the electrodes to replace a telegraph sounder. In the second series Edison attempted to measure, over a period of three weeks, the relative capillary action of numerous solutions on a paper strip. The third series of experiments, which lasted eight days, was related to the behavior of mercury submerged in electrolytic solutions and was prompted by Edison's observations of the Fuller battery he had recently received from England. In mid-February Edison briefly returned to experiments with the etheric force, studying its decomposing power and trying again to determine whether it exhibited polarity.

In February and March the staff constructed a number of different telephone devices. Most of the transmitters varied the resistance of the circuit. Some switched discrete resistances in and out and some varied the resistance continuously; most of the latter employed carbon, either as a coating or as a solid body. In one case Edison employed condensers to vary the electrical condition of the line. The receivers the staff used were generally simple electromagnets moving diaphragms, but in mid-March they devised their first electromotograph receiver. They carried out tests both in the laboratory and over a line to New York. At the end of March two men showed an interest in "taking hold of" the "singing telegraph," but declined after an unsuccessful demonstration at the Western Union office.[1]

During this period Edison developed other projects as well. In January he designed and built his embossing recorder/repeater, executing a patent early the next month. Western Union found the instrument sufficiently promising to authorize the construction of six machines. At the end of March he turned to the problem of sending three Morse messages simultaneously in both directions on one wire—his "sextuplex" system.

Edison's and Western Union's mutual concerns over their contractual relations led to a new agreement under which Edison received a guarantee of one hundred dollars per week toward his laboratory expenses and Western Union received "sole right within the United States to all his inventions and improvements capable to be used on land lines of telegraph or upon cables."[2] Edison entered this arrangement despite entreaties from Uriah Painter, Robert Russell, and Josiah Reiff to maintain his independence. Edison was willing to give Western Union these rights because he had finally abandoned hope for a satisfactory, timely settlement of the questions of patent rights and compensation from Jay Gould and the Atlantic and Pacific Telegraph Company. In February, Western Union filed a broad countersuit in the quadruplex litigation—*Western Union v. Harrington & others*—contesting various aspects of the ownership and validity of Edison's quadruplex telegraph patents.[3]

Urged by George Bliss—who assumed charge of the electric pen business in America—and the British promoters of the pen, Edison and Batchelor applied themselves to the problem of a faster press to use with electric pen stencils. In the second half of January they devised, and Batchelor built, a press with a rotating drum and automatic paper feed. Although they noted experimental speeds of 2,000 copies per hour and apparently built a patent model, they did not apply for a patent and dropped development by early March. Competing copying technologies also prompted laboratory work. A British pneumatic perforating pen, not yet patented in the United States, led Edison to file a patent application for one of his own design. In America the papyrograph was competing for the commercial market with the electric pen and Edison spent some time investigating similar copying methods.

The Western Electric Manufacturing Company began producing the electric pen under the contract signed in November.[4] Bliss was in New York from mid-January to the end of February, when he and Batchelor traveled to Chicago to settle business affairs there. Western Electric initially had problems filling orders and for several months Batchelor and a number of pen agents complained about the manufacturing quality. While Batchelor continued to run the overseas business, appointing an agent in New Zealand and advertising in several countries, he and Edison began discussions with Bliss and Anson Stager concerning a transfer of that part of the business as well.[5] On 13 March, Batchelor moved the New York electric pen office from 41 Dey St. to 20 New Church St.[6]

In January, Edison, Batchelor, and Edward Johnson assigned to the American Novelty Company the products they had developed for it—Edison's duplicating ink, Batchelor's door indicator, and Johnson's ribbon mucilage.[7] Although Johnson's only employee, J. E. Suitterlin, left in mid-February, Johnson appointed several canvassers and agents to sell the company's products. Edison considered other inventions for the company, such as small engines for a variety of uses, but never developed any of them. One minor project that Edison did briefly pursue was a method of waterproofing paper, undertaken for a manufacturer of paper barrels.

Personal account records reveal that the Edisons had domestic help by January. Alice Stilwell, Mary's sister, was also apparently living with them. Edison's childhood friend Tony Bronk left, and Edison hired someone in his place.[8] He invited his father in January to come to Menlo Park to help around the laboratory, and in late March Samuel Edison arrived for a visit.[9] Edison himself traveled to Port Huron in early March to effect the merger of the Port Huron and Gratiot Street Railway Co. with the opposition line. The machinations surrounding the street railway business so disgusted Edison that he referred to Port Huron as "that god damn hole of a [place] which contains most despicable remenants of the human race that can be found on the earth."[10]

1. Cat. 1233:80, 89, Batchelor (*TAEM* 90:93, 97).

2. Doc. 876.

3. Western Union filed this countersuit in New York state court on 21 February 1877, asserting its claims on Edison's quadruplex patent rights against Edison and George Harrington, the Atlantic and Pacific Telegraph Co., Samuel Mills, Josiah Reiff, William Seyfert, William Palmer, Henry Dallet, Jr., Augustus Ward, Robert Russell, the American Automatic Telegraph Co., and George Prescott. Western Union asked that the defendants acknowledge Western Union's ownership of

Edison's quadruplex inventions and show cause why they should not all be enjoined from any sale of the inventions and from pursuing any suit or patent office proceeding regarding them. Quad. TLC 1–2 (*TAEM* 10:800–908); *TAEB* 2:796 n. 8.

4. Doc. 817.

5. Batchelor recorded some of his electric pen business activities in his diary. Cat. 1233, Batchelor (*TAEM* 90:52).

6. Cat. 1233:72, Batchelor (*TAEM* 90:89).

7. Cat. 1233:19, 30, Batchelor (*TAEM* 90:62, 68); see Doc. 829 n. 11. Some accounts for this period, listed as the "Edison Manufacturing Co.," are in Cat. 1184:151–55, Accts. (*TAEM* 21:853–55); see also Charley Edison's account book, PN-77-12-18, Accts. (*TAEM* 20:65).

8. Cat. 1213:7–8, Accts. (*TAEM* 20:8–9).

9. Cat. 1233:80, Batchelor (*TAEM* 90:93).

10. Doc. 849.

## –834–

*From Edward Johnson*

N York Jan 3/76[1877]

My Dear Edison

You have evidently come to the conclusion that I am a burden to you and that something is necessary to be done to rid yourself of it—a la Gilliland—[1]

I understand you make quite frequent allusions to my improvidence[a] &c when mention is incidentally made of monies you have advanced me—[2]

As I do not propose to[a] occupy any such position in relation to you, or any one else—I ask you as a special favor to prepare for me from your Books & Memoranda a Memoranda[a] of all Monies loaned me or paid me in any way shape or form & to Include the following—[3]

| | |
|---|---|
| Loaned at the Date of Quad sale[4] | $200. |
| Bill paid clothier at A&P ofs | 60. |
| Money loaned some time subsequent to A&P—sale—[5] | 50. |
| 20 Shares Domestic stock[6] | 100. |
| Odd times loaned me $25.[7] of which I have no mem. Say 4 times[b] | 100. |
| Check Sent me July[a] 28th[8] | 100.[a] |
| 5 Months Board—say[9] | 125. |

Being a total of $735. to which add everything that may have escaped my memory and every thing[a] you have paid me or for me during my stay at Menlo Park—[10]

I want this for purpose of record & entry in my Private Book solely—in order that I may as soon as possible make proper returns for it— I am not intending[a] to dramatise for the moment at all— I simply express my obligations for what you have done for me— Put it on record—I promise you now to pay it all back with interest the moment I can do so—either from a realization upon Automatic or some interest that you may have given me—

And in re—to the future I hereby agree to look elsewhere for my living. In consideration for all of which I ask only that you take me out of the catagory of those who have fed upon your sustainence & hindered rather than promoted your Welfare, at least when you are discussing me with others— I cannot afford to have this sort of stigma precede me in my efforts to get along in the world Meantime Reiff is going to try & raise me enough money on the 56 shares of my Colorado stock[11] to pay my board for one month—during which time I shall make constant effort to get some one to make me an advance on my stock—or hapily to get this thing under way so as to bring me at least a partial living.[12]

You know by experience what the cost & inconvenience of keeping one branch of my family is— You can guess readily enough therefore how the row in Phila[13] originated. & you may only ask yourself how you would feel—to understand my feelings tonight—supplemented as these circumstances are—by the conviction that you as well as others have finally concluded that "no hope is left in me" & that while not exactly one[a] to be plainly told so—yet one whom it were safer to expect Evil things of & the part of wisdom to seperate from

Nothing however shall cause me to foolishly quarrel with you because you happen to hold an opinion of me different from my own— Yours Very Truly

E. H. Johnson

ALS, NjWOE, DF (*TAEM* 14:3). Letterhead of American Novelty Co. [a]Obscured overwritten letters. [b]"Say 4 times" interlined above.

1. This same day Ezra Gilliland returned to Edison the machinery, stock, and remaining promissory notes that Gilliland had obtained in July 1875 when he formed Gilliland & Co. *TAEB* 2:544 n. 1; list of notes and agreement attached to mortgage, 2 and 3 Jan. 1877, Miller (*TAEM* 28:984–85).

2. On 9 January, apparently in response to a letter from Edison, Johnson wrote:

> Everything I do or say is not an appeal for money, hence my letter will bear some other than the interpretation you evidently put upon it. When you say I am not just in expecting you to give me money when you had none—objection to your criticisms of me

and an appeal from them is not an appeal for money [DF (*TAEM* 14:13)]

3. For Edison's accounts with Johnson, see PN-75-01-05 and Cat. 1185:200, both Accts. (*TAEM* 20:29, 31–33, 36; 22:646).

4. On 7 January 1875, three days after the sale of quadruplex rights to Jay Gould, Edison had loaned Johnson $200. See Doc. 565 n. 4.

5. Gould purchased the Automatic Telegraph Co. for Atlantic and Pacific Telegraph Co. on 16 April 1875 (Doc. 561). An 1875 account book entry lists the following amounts paid to Johnson after that date: $10 on 5 and 12 August, $5 on 19 August and 6 September, $15 on 8 September, and $10 on 2 November (Cat. 1185:200, Accts. [*TAEM* 22:646]).

6. Johnson's 20 shares of stock sold for $100, $50 in cash and $50 in notes. Undated statement, DF (*TAEM* 13:532); Doc. 739 n. 1.

7. On 21 December, Johnson used 50 of his shares in the American Novelty Co. as collateral for a $25 loan from Edison. DF (*TAEM* 13:790); see also note 5.

8. This is shown in Edison's account book under the date of 26 July. Cat. 1185:200, Accts. (*TAEM* 22:646).

9. With money earned from the American Novelty Co., Johnson was able to pay $20 of this amount in February. Johnson to TAE, 8 Feb. 1877, DF (*TAEM* 14:23).

10. Notebook entries show that Johnson was in Menlo Park between 3 August and 14 September 1876. Vol. 8:334, Lab. (*TAEM* 3:764); Doc. 792.

11. Probably stock he had in connection with William Palmer's Denver & Rio Grande Railroad. See *TAEB* 1:279 n. 1, 505 n. 13.

12. This probably refers to the American Novelty Co., of which Johnson was secretary and general manager. For the year 1877 Johnson was to receive a salary of $1,000 and 10% of all company profits above that amount; if the company's profits were less than $1,000 his salary would be whatever profit accrued. Charles Batchelor to Johnson, 23 Jan. 1877, Cat. 1238:65, Batchelor (*TAEM* 93:75).

13. Johnson was in Philadelphia during the Centennial, but the subject of this reference is unknown.

**-835-**

*From Pitt Edison*

Sarnia, Ont., Jan 4th 1877[a]

Dear Bro

You will find Reports for the Sarnia road also the amt it owes Itimised[1b] I see J P Sanborns letter sent you was dated Dec 22th and the (Drugist WW) writes you on the 24th[2] now if you watch this RR matter you will see that Wastell is to sicken you while they make you proppositions Sanborns tells you we have been ordered out of the GT Yard[3] I have not seen or heard of any such order neither can I find out if such an order has been given Wastell would hold[b] up both hands for such to be for him and W Wiley of GTR his bosum freind has bin working for it for Two years Wastell is the man that wants it done for in my opinion he ownes the most of the new road[4] and would be willing to bust the old road for the benefit of the new

he tells you we wore short paying the men off some $12.00 but he dont tell you that week we pd Guy Kimball on Judgmnt $25.00 also straw & hay that will last a month he speaks of the Omnibus Earnings to be sure but it was in 1865 in the Great Oil Excitemt here when we could not carry the Pasingers at 50 cts each and it cost $10.00 for a horse & cutter to go to Lake Port he tells you we have a note to renew for 1100.00 which is only $1050.00 it was redused $10.00 on the principbale and the intrest pd for 30 days $11.00 making $21.00 and it was paid out of the St Car mony he says he knowes you ha~~s~~ve ben keep in the dark and you know nothing abot the facts and that you have had false representations made to you and lies ben manufacturd which he knows is a damd lie and he is the only man that that doing that buisness than in the next line he tells you (but sir I tell you the damd road is[b] eleven years old half worn out Horses worn out cars half used up &c and he hears that the Sarnia & Gratiot Roads are to vacate the GTR grounds whats he to say about the Sarnia road he only tells you this to sicken you you want to know of his letter which is lies ~~w~~and which isent the shortest way that I can answer it that his whole letter is a damd lie the whole of it ~~if~~ and should you come out here I will prove it to you to his teeth it may be that he may succeed in getting the GTR to order us out of the yard but if such shoud be the case it is through him that it is done for ~~I~~ his tool Wiley is and has ben trying but I think the GTR has got all they want on thair hands to take care of the Engine drivers and see that thay have no more strikes dam these fellows (Al) if you could come out here with $4000.00 dollars or make them believe you had it it[d] would make a differance of 4000.00 with you in the trade all they want is to let them know that you mean buisness and they will come down to your tearms you will see ~~W~~by Wastells letter that he makes no propposition that if ~~you~~we had to go out of the GTR yard to build around or to protect ourselves but his whole hobby is to sell out or give our road to the Opposition I look apon him as the only enemey that the Gratiot road[d] has got today and you tell him so with about $4000.00 in you pocket and he is gone for whin he makes up his mind that nothing can be done with you and that you dont scare he will try and have all the claims pressed and try and have the road sold out—and thay will bid it in and thare is whare it wants ~~the~~ some Greenbacks and if thay find you have got[b] the money to pay the depts and build arroud if nessesary or make them beleieve so which is just as good thay will gracefuly come[b] and see you and than you can dictater termes to them not them to you   we have got the best of this thing if we can stay with Wastell long enouf for I think he is getting near the bottom of his pocket   you will find all that J P Sanborn is doing he is doing for Wastell   I dont think that Sanborn cares a cent about either roads once[d] sicken Wastell and the bottom falls out of the new road in one day and you see by his corespondence that he dont draw ~~s~~over one inch[e] of water to the foot   he is a bull headed Englishman that hates to give up   Should you write him as you and him had the largest intrest in the old road you thought it best to take care of the Gratiot road and build around and let the new road run thair road and we would run ours and you get his reply he would advise you before doing that to give our road to them (write such a letter and try him

The Item of advertisng on the Sarnia Road is for putting[b] sighns on the cars for one year   I also let the Gratiot road for the same price which you will see in January 1877 Report   This is something we have never had before

WW tells you we have no legal rights in the Park   why should he say so he knows better for we have a lease from the Sec of war authorised by act of Congress[5] which he knows has never been Repealed but I will in my Offical Capasity take care of Park buisness as fore as St RR is conserned   if you make any arrangments whareby you let the controll go out of your hands I would let them state how much Gratiot Stock thay would take for thair road clear of Dept and than let the old road pay her depts   it is certain the old road with her surplus of Stock & Real Estate is worth her dept more than the new road   I will send the notes as soon as we can have a meeting   also the old road dept

WPE

(How is Chas)

Dont forget the Car in Sarnia and the little peice of Land

ALS, NjWOE, DF (*TAEM* 14:520). Letterhead of Sarnia Street Railway Co. [a]"Sarnia, Ont.," and "187" preprinted. [b]Obscured overwritten letters. [c]Another "&" appears above. [d]Interlined above. [e]"ch" interlined above.

1. Not found.
2. "WW" is William Wastell. Edison apparently sent both letters to Pitt; neither has been found.
3. See Doc. 832.
4. According to Jenks (n.d., 8), when the City Railroad Co. was organized in 1873 F. H. Vanderburgh owned 55 shares, while Wastell and John Cole each owned 50.
5. Pitt is referring to the right-of-way through the Fort Gratiot military reservation. Statement of Secretary of War, 14 Mar. 1874, DF (*TAEM* 13:1113).

-836-

*From Edward Johnson*

N York 1/5/77

My Dear Edison

We must have some mucilage— The Demand is for that while[a] the work & Expenses are on the Ink— Give us some mucilage to pay for the Ink We are getting out circulars—[1] Ad. in Herald[2] brought nothing but customers—no agents who could take hold of Ink—but one who has 50 men—who will take hold of mucilage—

Come to my ofs tomorrow[3]—& write out a form[b] contract for your things—[4] Russell will read it over—make it OK— &c will put ~~int~~ in final shape for you to sign.

We copied for Ward[5] today 2 Days business & have to copy another tomorrow— Meantime they are using a fresh bottle of Ink— Concluded that they must have got some foreign stuff in the Ink they were using— Yours

EH.J.

ALS, NjWOE, DF (*TAEM* 14:9). [a]"while" overwrites a dash. [b]Interlined above.

1. Copies of circulars for Edison's duplicating ink and Johnson's ribbon mucilage are in Cat. 1204, items 24, 33, 35, 49, Batchelor (*TAEM* 94:13, 16–17, 20).
2. Not found.
3. The American Novelty Co. was located at 52 Broadway, New York, the same address as Josiah Reiff's office. See Doc. 952 n. 2.
4. Not found.
5. See Doc. 831. On 9 January, Ward ordered an additional 5½ pints for Rye Beach, and a month later he ordered half a gross of Johnson's ribbon mucilage (Johnson to TAE, 9 Jan. and 8 Feb. 1877, DF [*TAEM* 14:13, 23]).

*Advertising circulars for Edison's duplicating ink and Johnson's ribbon mucilage.*

24 THE

# AMERICAN NOVELTY COMPANY

CIRCULAR.

*52 Broadway, New York, January 10, 1877.*

NEW INVENTIONS.

## EDISON'S DUPLICATING INK.

This Duplicating or Transfer Ink, designed for the multiplication of copies by means of the common Letter Press, has properties entirely distinct from any other in the market. It will do the following:

First. One original writing will yield in the ordinary Letter Press, without other appliances than those used in ordinary copying, 20 to 30 copies upon Tissue Paper.

Second. It will yield from 10 to 20 copies upon LETTER PAPER—an entirely new feature; these copies being re-transferred from a Tissue Copy, are right side up and readable from the face of the paper and not *through* it.

Third. It will also yield copies either from the original or from the copy at any time subsequent to the original writing, thus making it practicable for Lawyers, Merchants, Brokers, Reporters, Insurance and Real Estate Agents, Clergymen, and the Professional Business Community generally, at any time to duplicate any paper which they may have on record. Also enabling the recipient of a letter written with this ink, to obtain from such letter a number of copies (fac-similes) without the labor of re-writing; also giving the writer of a letter a copy, in addition to his letter-book record for filing in his pigeon hole with matter of the same subject, thus keeping a complete record of the whole matter at hand for convenience of reference; this saves the annoyance of hunting through a promiscuous letter-book for correspondence upon any given matter. The property of transferring upon hard paper (letter paper,) being an entirely new one, and possible with no other Ink, creates for this a new field, and one, which, in view of the simplicity of the means by which these novel results are obtained, must immediately be occupied by it.

PRICE.

| | | |
|---|---|---|
| No. 1 Bottles, | | .50 |
| No. 2 " | (Half-pints,) | $1.50 |

FOR SALE BY

---

THE AMERICAN NOVELTY COMPANY,

**52 Broadway, New York,**

49 *Proprietors*

# Johnson's Ribbon Mucilage.

THIS IS A NEAT, HANDY, AND VERY EFFICIENT SUBSTITUTE FOR LIQUID MUCILAGE.

A long narrow tape of paper is gummed on *both sides* with the same material as used on postage stamps, put up in a neat box having in its edge a slot from which projects the end of the ribbon, or tape. When a piece is required, the requisite length is drawn out and broken off. The gumming of both sides, of course, makes it practically a ribbon of gum—or mucilage—and to use it for any purpose whatever as a substitute for mucilage it is only necessary to moisten it on the tongue and insert between the papers which it is desired to fasten together.

It is superior to liquid mucilage in the following particulars:

1st.—It is always neat and handy.

2d.—It contains only the moisture gathered by contact with the tongue, and will not therefore cause the paper to crumple as does the liquid mucilage, which is necessarily largely composed of water.

3d.—It will adhere more quickly and firmly, there being less water to evaporate than in the case of liquid mucilage.

4th.—It is always ready for use, and is not subject to different degrees of consistency and an almost universal unreadiness for use.

5th.—It acts so quickly, and is so neat, that it fills some wants which mucilage cannot; as, for instance, tacking together a number of sheets of a written document, for which brass tacks, or pins, are now used.

6th.—For numberless trifling purposes, requiring haste, of not sufficient importance to warrant the slow application of a mucilage brush and pot, this ever ready and handy material will make its value felt.

7th.—No matter why! But where and when did anyone ever see an orderly, well behaved mucilage pot and brush in a private family? And who has not had painful experience of such absenteeism? Now the remedy for this distressing state of things is clearly in something which, while cheap and always ready, can be thrown into the sewing machine drawer, "my lady's" toilet case, the housekeeper's pocket, or the chambermaid's mysterious receptacle; one box in each for convenience of access.

A moment's thought and a personal examination and trial of the ribbon mucilage will take the glamour of jest off the 7th recommendation and clothe it in the habiliments of sober truth, when it will be readily seen an immense unoccupied field is open for this valuable invention to enter.

**GIVE IT A TRIAL.**

Price per Dozen, $1.50. 40 per cent. Discount to Trade.

-837-

*Notebook Entry: Miscellaneous*

[Menlo Park,] Jany 8 1876~~6~~7

Proposed application of small electric engines, some of which I have already experimented with, and am now engaged in experimenting with.[1]

Flying Bird, composed of hollow cores pen magnet, paper wings and self vibrating break, with a homeopathic vial attached containing zinc highly amalgamated & carbon—using strong bichromate of potash. The idea being that it will fly a 1000 or more feet high & a considerable distance, according to the strength of the battery & its lasting qualitiles. Object. a pleasing Scientific Toy & perhaps it might be used for carrying communications short distance with proper wind~~er~~=

Electric Shear for cutting cloth, paper, tin foil[a] etc. consists of a rapidly revolving[a] sharp edged disk against & a little below a straight edged presser foot=

An electric engine, poratable like Electric Pen[b] with vertial ~~&~~revolving shaft holding a diamond pointed tool— it may vibrate[a] up & down like ~~on~~the pen—& the Diamond act like The Sand Blast= for engraving on glass[2]

An Electric pen on a pantograph— also The Engraving Electric Engine.=[3]

A portable electric engine, with double needle one shoveing thread through cloth & the other passing through & hooking it pulling up a loop for embroidery

An electric engine with worm & worm wheel for revolving the Limes of a Coleman light[4] Also for Revolving show goods in windows Also for working Morse Registers

Ideas

Dental plugger—[5]

Make a ribbon of glycerine & aniline for clephane[6] for type writter for duplicating= Try Chl Zinc Dextrin & Aniline—which is pbly best=

Get up a bath for Carbon paper & sell this to users of paper let them make their own

Electric Vibrater or Engine to give a down forward & upward motion to a graver for Engraving

Pneumatic Pen

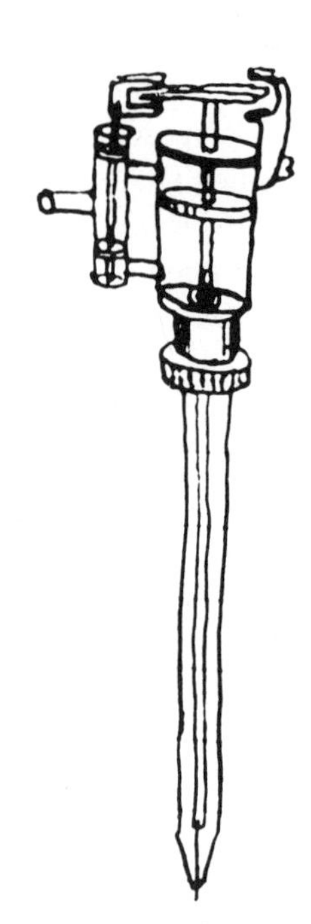

Batchelor Adams

X, NjWOE, Lab., Cat. 997:45 (*TAEM* 3:374). [a]Obscured overwritten letters. [b]"poratable . . . Pen" interlined above.

1. The following day Edison described several other ideas (Cat. 997:47, Lab. [*TAEM* 3:38]), including a variation on the copying machine design shown in Doc. 812, portable tool attachments, and a horse clipper (see also Doc. 994). He also drew a new stand for the electric pen. These inventions were probably intended for the American Novelty Co. (see Doc. 829 n. 11). He continued this line of thought on 10 January, suggesting among other things that electric engines could run ventilators, a battery depolarizer, a music box, a clock, and a telegraph ink recorder, and noting that John Kruesi was making an electric engine (NS-77-001, Lab. [*TAEM* 7:417]).

2. For Edison's subsequent work on this idea see Doc. 940.

3. For Edison's earlier design of this idea see Doc. 784.

4. On 15 November 1876 Edison described electric engine designs to replace the clockwork fan blowers used in chimneyless kerosene lamps (Cat. 997:40–41, Lab. [*TAEM* 3:372]). See also Doc. 579 for Edison's earlier proposal regarding a chimneyless lamp.

5. Edison may have become interested in dental pluggers as a result of a patent search conducted on this technology in relation to his electric pen. Lemuel Serrell to TAE, 9 Aug. 1876, DF (*TAEM* 13:1043); patent search notes, NS-Undated-001, Lab. (*TAEM* 8:13–19).

6. Probably James Clephane, electric pen agent for the District of Columbia. See *TAEB* 2:598 n. 11.

-838-

*From Edward Johnson*

N York Jan 10/77.

My Dear Sir:

I am in receipt of a communication from U.H.P. at Washington intimating his ability to negotiate a quick cash sale of the new embosser[1] to the A.&P. Co. & asking me if I can secure from you an option on it for 30 Days for that purpose Says if it will do what I claim—(and I was very moderate) it can be done— He mentions $100,000 as the price probably obtainable—What do you think—would it not be a good thing to give them a chance at it?— I doubt not in view of the present status of instrumentation they are solicitious of securing something outright.[2]

I call your attention to this because I have before thought it was the part of wisdom to get a proposition from both parties— Of course any from A&P should be on a basis of cash (C.O.D.) while from W.U.—your Royalty plan[3] is probably the best. Truly Yours

E. H. Johnson

ALS, NjWOE, DF (*TAEM* 14:16). Letterhead of American Novelty Co.

1. Doc. 857.
2. See Doc. 760 n. 2 regarding the apparatus on Atlantic and Pacific Telegraph Co. lines.
3. See Doc. 695.

-839-

*From Uriah Painter*

WASHINGTON, D.C. JAN 10 1877[a]

My Dr E

I enclose copy of the act you desire—[1]

There are two sides, if not more to history of A&P & auto[2]—without going into past you must admit

First That A&P have put about $200,000 into auto—

Second—The present outlay has not been warranted by its utility

Third—If they are not shown how to make it save them money more than it does now, they will not pay another cent out on it or grieve at its total loss—

Fourth—That if you or Reiff or both can get a reasonable sum of money out of it by you expending a little time & labor on it with A&P then it would be better to do it, than to abandon it to be devoured by lawyers & thrown into disuse by newer things—

Fifth.—That with two telegph companies to bid on your future inventions they will bring you more money than with but one Shylock in the field[3] who has publicly & privately for years heralded you as "a thief & pretender"

Now put a price on your new perforator & repeater EHJ writes me off[4] & let me make a raise for all hands on it— Dont make a d——d fool of yourself by sacrificing it Y[ours]

U H Painter

ALS, NjWOE, DF (*TAEM* 14:636). [a]Date and place from *Philadelphia Inquirer* hand stamp.

1. Probably the congressional act of 8 February 1859 granting right-of-way through the military reservation at Ft. Gratiot for a railroad to Port Huron. 76-013, DF (*TAEM* 13:1107).

2. Legal action about this lasted from 1876 to 1913. See *TAEB* 2:469 n. 2; *Harrington v. A&P.*

3. That is, Western Union.

4. Should be "of."

## -840-

*Notebook Entry: Miscellaneous*

[Menlo Park,] Jan 11[–12] 1877

Etheric[1]

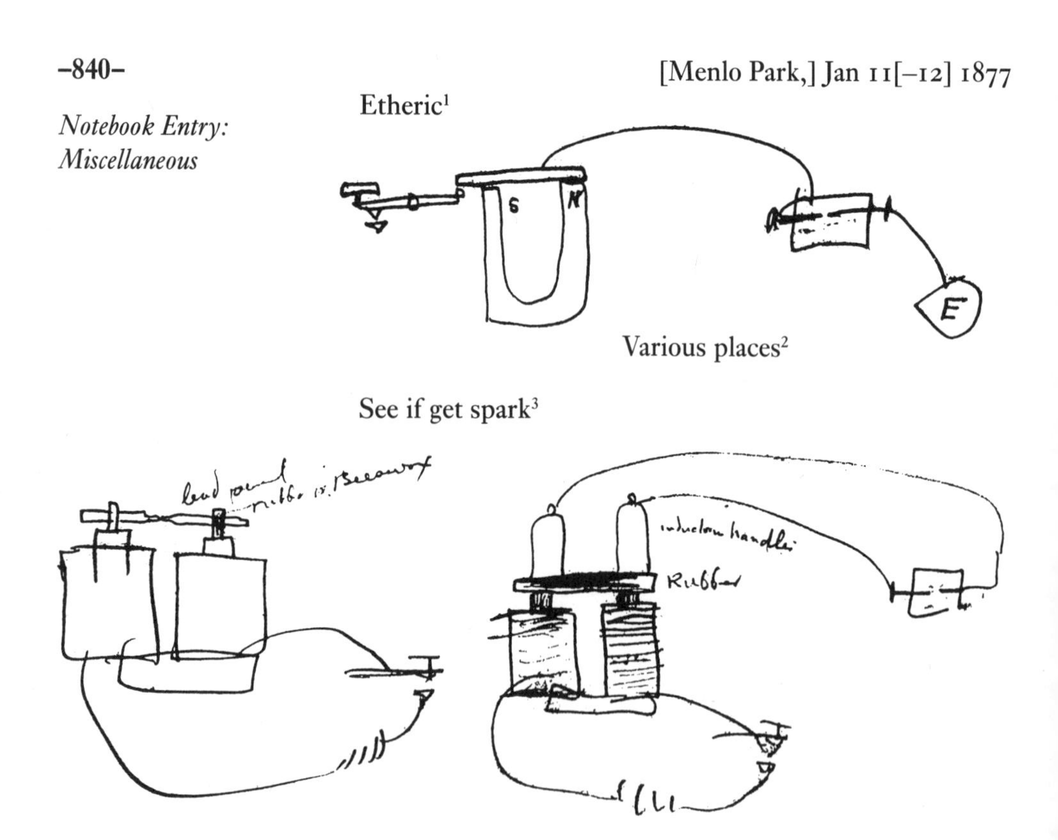


Various places[2]

See if get spark[3]

Try different Solutions[4]

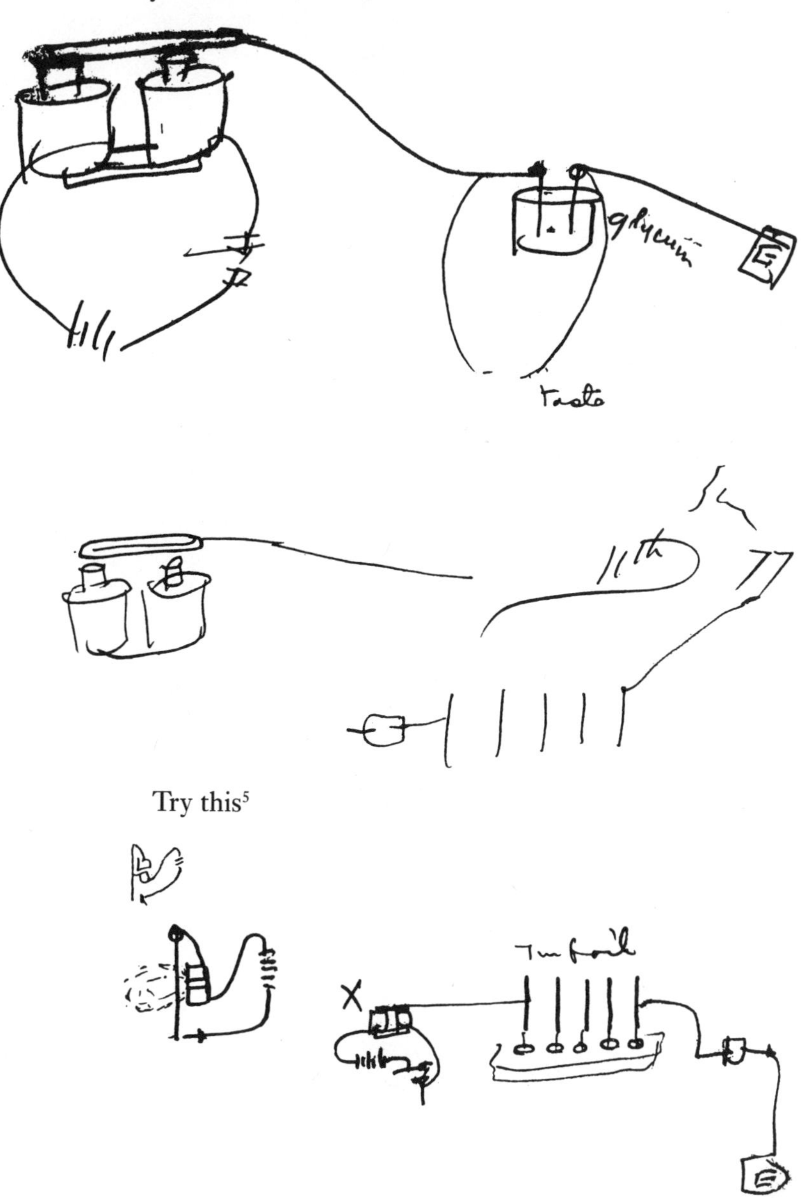


A .X. try magnets of very high & low resistance with high & low R batteries to see if tension increase or dec ascertained by moving tinfoil plates[6]

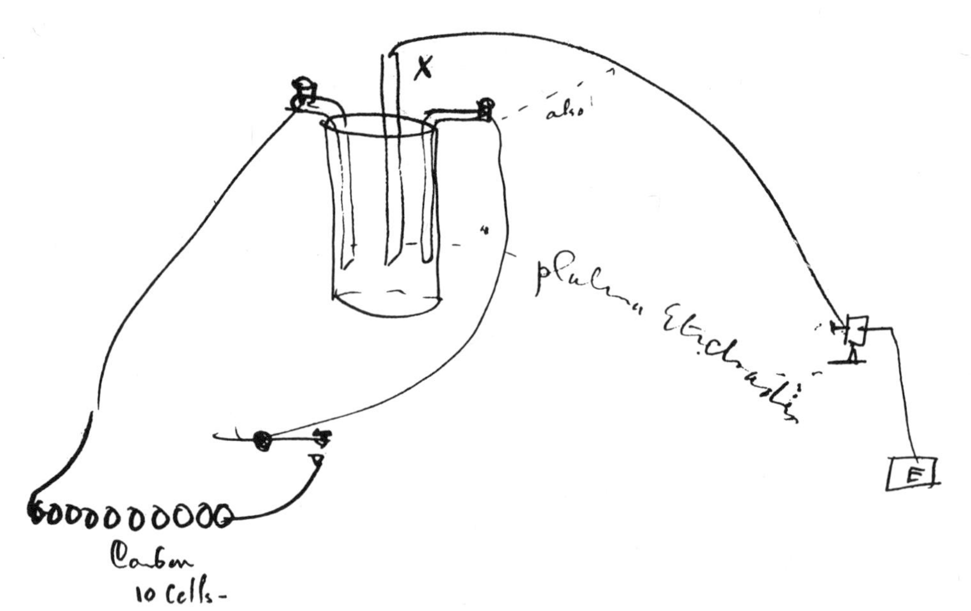


Place[7] in the cell different Solutions[a] moving X to one side or the other or taking it out altogether & connecting Etheriscope direct ~~on~~to one of the Electrodes =[8]

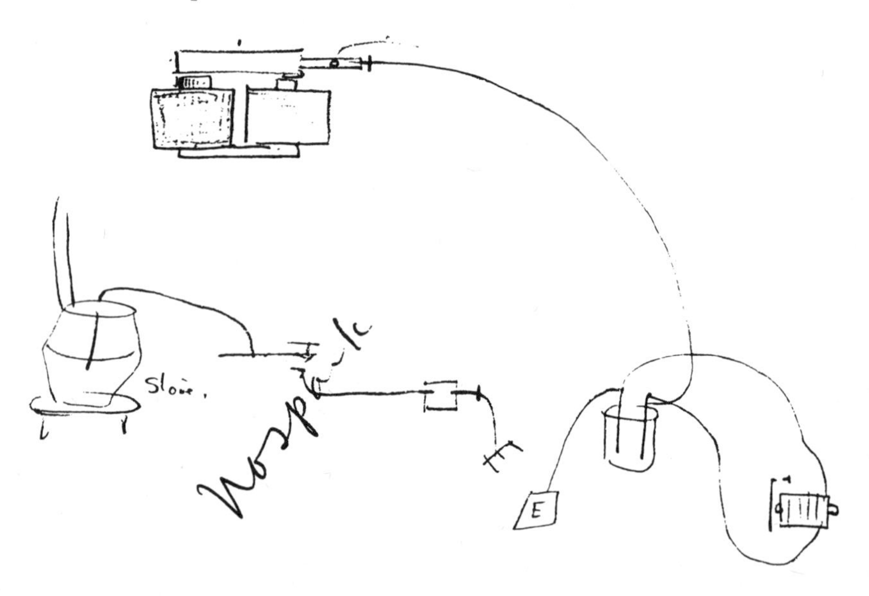


The Etheric spark may be a molecular vibration set up in wire by Electric Spark.

Carbon points Electric Light  Key brings wire into Light also[9]

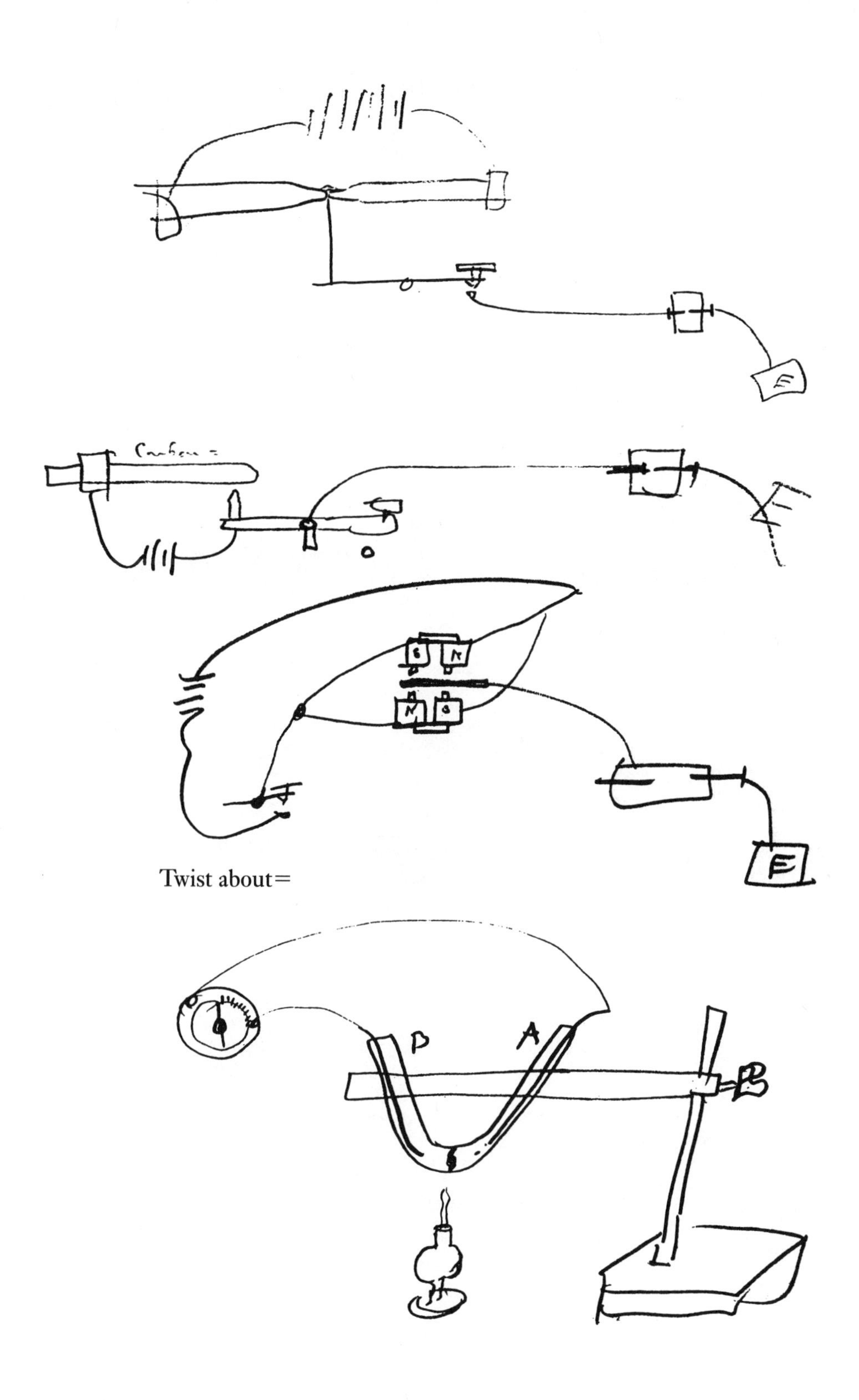


Twist about=

Bent tube, Platina Electrodes  1 Sol in one prong another kind in other poured same time  Chalk partition or spongne on End wire which can be pulled out every experiment & squeezed—

This page is a failure[10]

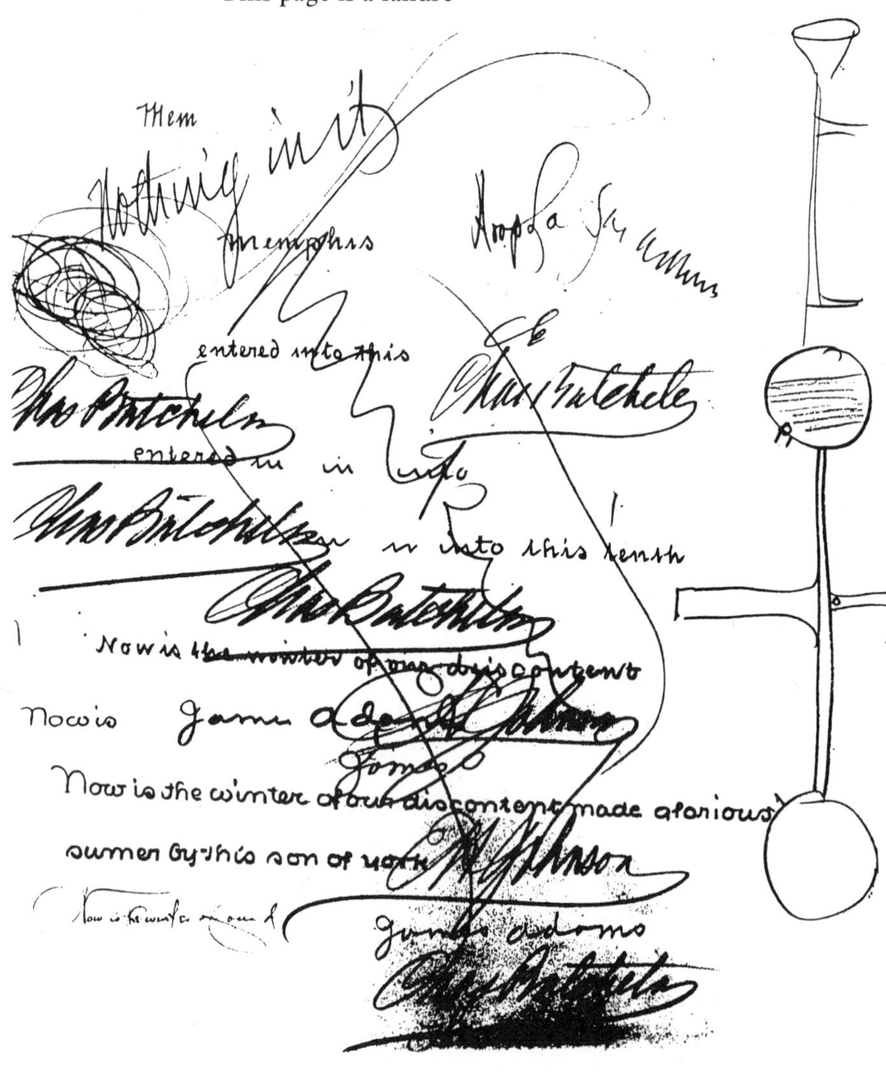


Shunt etheriscope with 200 ohms mag dont effect brilliancy of spark[11]

Get spark with it shunted thus when I put my hand at .X.[b]

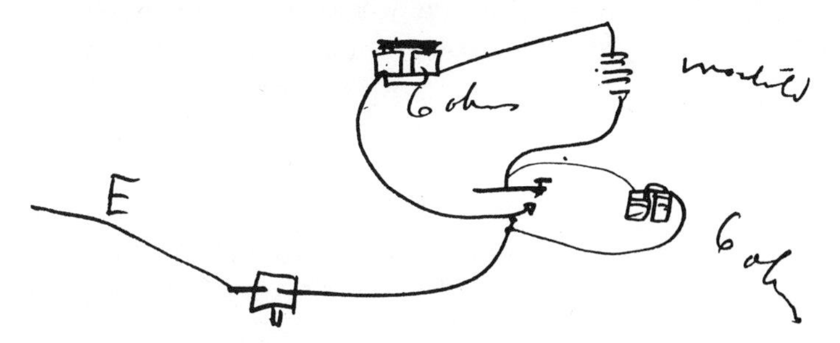


get nothing[12b]

I have ~~nothi~~ noticed with the carbon points that no matter how powerful the spark is there must be a chain of carbon particles touching each other= If I seperate the points a little distance and blow very hard in them thereby cleaning them of particles I cannot obtain[c] the spark until I have brought points in contact and withdraw them, the spark is weak though & I do not get it with full brilliancy until I have brought the points together & worked them so as to grind some of the carbon particles off= when considerable have been ground the points may be seperated 1/32 of an inch the spark[d] getting weaker & weaker but if a gentle blow is given the spark disappears & can ~~b~~not be made to appear until the points are again brought together

Also while sparks are passing between the two points, another set of points will give sparks also of equall brilliancy without diminishing those in the 1st points= Its probable that ½ a doz or doz points might be made to give sparks.[13b]

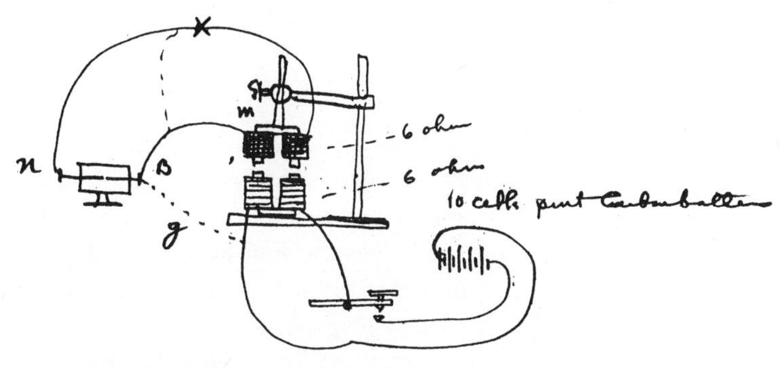


In contact quite brilliant spark observed on points of the etheroscope. spark~~s~~ gradually dies out until when inch away

get it exceedingly[e] feint  yet the shock is violent on the tongue when inserted in circuit at X. if now Magnet M be detached from the dark box & the dotted line wire g placed in connection the spark is at once brilliant ~~almost~~ equal~~l~~ to[c] that from m when it is laying on the cores of the energized magnet only seperated by a piece of paper yet no shock can be perceived at g= The ground ~~mag or~~ was connected & disconnected from n with no great material diff  get just as much spark whe~~n~~ther X wire is disconnected or connected when M is inch away

When M[c] is in actual contact with cores of other magnet, & its two wires taken from dark box & put together and a single wire run to .B. thus

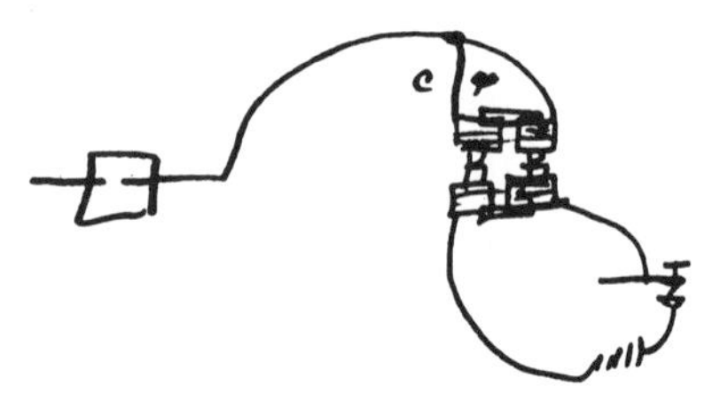


get just as much spark  by disconnecting c get same  this is when this connected etheric from old magnet=

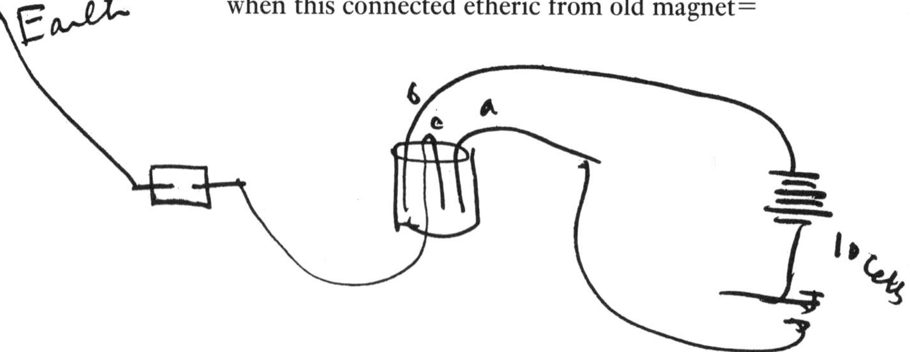


Platina Electrodes.[14] c electrode being connected to the Etheroscope= by very fine manipulation of points got several sparks but that was during ½ hour manipulation=

Bichromate Potash in cell[15] 10 grammes in 100 cc HO,= moved c towards a. & .b, couldnt say ~~whether~~[c] where c was when got sparks=

8 gramme Sol  No 2. Borate of Soda in cell. Couldnt get a spark  10 minutes manipulation[f]

40 gramme Sol  Sulphate Magnesia  Couldnt get a spark 10 minutes manipulation[f]

20. Grammes in 100 cc HO. Pyrogallic Acid= Nothing after 10 minutes manipulation Resistance very high Could charge magnet with pyro in ckt strong enough to get etheric in usual manner=[b]

20 gramme Bicarb Potash No spark=

I have noticed that the gas given off on the electrode connected to the zinc pole of the battery ~~ie~~[g] in the BiCarb Pot which is pbly Hydrogen is composed of very fine bubbles & of great number wherase on the other pole the bubbles are larger & much fewer of them—but they shoot up to the top of the ~~water~~ solution[e] with greater velocity= Is it not possible to deduce from this the size of the H & Oxygen atoms, ie[g] relative size=[16] is it that the specific gravity of the infinitely fine Oxygen bubbles being ~~sine~~ so much greater than the H bubbles retain their hold on the electrode ~~being~~& thus allowing a number to coalece & run into each other this would account for the difference in size.

=I notice today that aluminum with carbon in sol of sal ammoniac there was formed an immense number of gas bubbles one series apparently superimposed on the other & appeared 1/16 inch thick they appear to cling to that metal with remarkable tenacity they may be slid up & down on the aluminum with a point under the solution without detaching them in fact it is very hard to detach them no amount of shaking will clear the plate of bubbles— this probably accounts for the tremendous polarizing properties of the metal as shewn in EMG experiments.[b]

It was Carbonic Acid gas given off on the Electrode connected to Coke= it put a light out=[b]

5 gramme Licorice— Nothing= forms gas on electrode connected to Carbon which collects in a mass in the thick solution & comes to the top in lumps like the curds of a precipitate=[b]

8 gramme BiCarb Soda= nothing= Bubbles on both appear to be larger than with BiCarb Potash, but that was a 20 gramme sol, that would seem to prove that the weaker a solution or the smaller the amount of decomposition the larger & slower the Bubbles were formed & detached, hence this might explain battery polarization to a certain extent

Nitrate Potash 20 grammes No spark=

Notice that electrode connected[c] to carbon gives small[c] bubbles which run into exceedingly large ones they only coming to top & slowly they h~~an~~ug Electrode persistently after circuit open on Electrode Connected to Zinc, Bubbles smaller ~~to~~dont run into each other in the least ~~ie~~[g] apparently have no affinity=[h] on opening the circuit the[c] bubbles gradually rise & it takes sometime before the plate is clear,= after ckt open 5 minutes bubbles come out of the platina quite copously  if it is cleaned of bubbles by shaking they immediately commence to pour out of it= This is not the case with the Electrode connected to the carbon which remains full of large bubbles  I now shake it, it becomes clean  in 2 minutes it collects some large bubbles but they do not rise  ~~th~~ the other Electrode still continues to pour out ~~co~~ bubbles ~~f~~But not so copiously=[b]

20 grammes gum arabic  nothing=

The ~~e~~bubbles formed in this solution leave the electrode with Extreme slowness hence it wouldnt be a good depolarizer for a battery if any one had the insanity to use it for that purpose—

Discovery[a]

In decomposing with the electrodes (platina) & the 10 cells carbon battery a 20 gramme sol of Bromide of Potassium in 100 cc HO ~~We heard a Sound every time the key was closed and could read morse~~[e] ~~by it~~ it occurred to me seeing the great volume of gas arising to form "Votex rings" like[c] the smoke rings to carry out this idea I bent ~~the~~ a[e] No 20 copper wire in form of a loop thus  on inserting it passable rings[c] were made, when the loop was immersed a proper distance in the solution= too deep breaking them up before they reached the top— while watching these rings heard a sound[h] every time closed key= After attention was called to it found it was very loud and could read morse by it 10 & 15 feet away= found that by putting ear near cell that could read morse through resistance of 1000 ohms[h]= found that it made a difference in size of ring  straight wire dont give sound  a wire bent at right angles gives it nearly as loud as ring=

found that placing ring in particular part of solution got it louder= [--][i] Keeping the key continuously closed the sound was a vibrating one, like a tuning fork making about 40 per second— by pushing loop down lower in Liquid & in fact up & down could get different tone= tried peice iron wire on end of copper wire in a loop= it was fine wire= gave it very good= straight platina wire nothing  ditto 1/8 inch wide sheet aluminum  The iron wire after being in use a minute commenced[c] to give off bubbles after circuit ~~closed~~ opened &

continued to do so for 5 or 6 minutes until took it out  probably the absorbed Hydrogen—

Bromide Cadmium gives off no gass= Cadmium deposition on Hydrogen electrode & Bromine set free on Oxygen= perhaps Cadmium & Zinc might be used for a battery, Brom Cadmium used instead of Sulphate of Cu in Callaud battery=[b]

Phosphate ammonia 5 grammes  Curious= with bent loop Copper wire= it works quite plain ~~but~~ giving a vibrating sound like Brom Pot= but on opening key there is a short after sound[h] like the frying in lard in a frying pan  I reversed the battery so that Ox would come on the copper loop  heard nothing of the regular signals but the lard frying pan noise was there & heard only after opening=[b]

Chlorate of Potash 8 grammes= Very lardy not sharp only hear it when loop is near top of water— We notice that H leave loop much sooner than O leaves platina & get full quicker  this explains why H is best for producing a mark in chem prepared paper for Auto Telegraphs ~~in~~=[j]

1 gramme Formate of copper= No gas—deposits Cu on Loop[b]

1 Gramme Citrate of Potash  Best yet= Very strong can hear it myself 2 feet away= the lower the loop is immersed the greater the sounder  if brought near (1 inch) top of liquid hear nothing= its like lard frying but clear & sharp gives beautiful morse= but dont know if its as delicate as Bromide Kali= ~~if~~ought to have a 10 gramme or more solution=

There is something curious ~~appa~~ about the looped wire= if you use a Rubber Coated wire just leaving slight bit copper out to secure copper fine wire loop to you can hardly feel it= if ~~us~~you use a No 20 bare wire ~~you~~ with loop its not good, but if you have the No 2 wire covered loosely with cotton & bare ~~nar~~ near[e] tip & turned to a loop its loud again= fine iron wire works well= the loop is not necessary as it straightened it out & it worked well though the sound appeared to be different=[k]

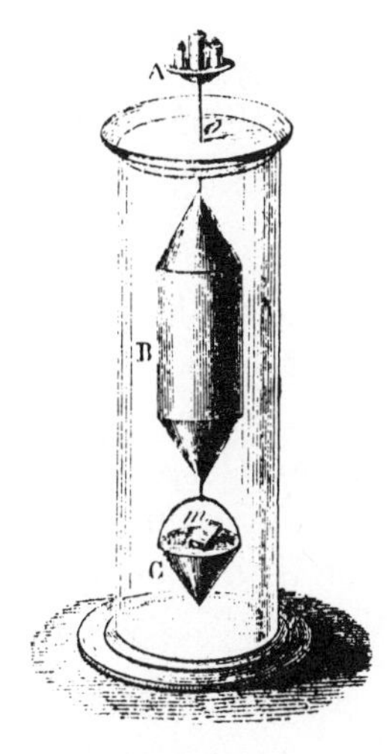

*A nineteenth-century hydrometer.*

2 grammes Tartarate Soda  Equal if not the best[b]

Tried a deep hydrometer[17] glass full of salt water— ~~I~~immersed electrodes deep  Jim could hear it faintly but I couldnt hence no good not because I couldt hear it but because if I couldnt it wasnt worth a damn[b]

Tannic Acid 10 gramms  Hear it myself with ear to cell but not good= think reason is due to high Resistance of Tannic acid

Sulphate Zinc 20 grammes gives a higher note & very nice & sharp  its loud too but alas the iron & Car soon deposits Zinc & then the sweet music stops=[b]

Chromic Acid—1[c] gram  Can just hear it  nix good[b]

It is probable that if an electrode be made throughly insulated except at the end & that be provided with a copper ~~tup~~ tube ~~with~~ closed at the end & the outside ~~i~~entirely insulated & a very small hole drilled in tube so that gas is formed inside and must escape by this orifice that it will produce a greater noise and give any note required

Thus

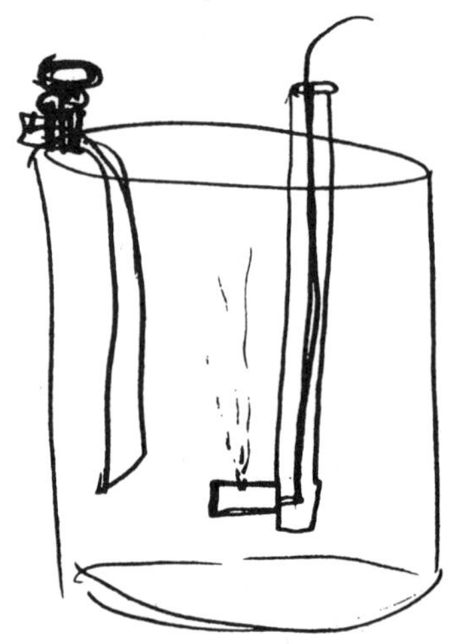

Sulphuric Acid & HO  Very strong pbly strongest  dont hear it on 1000 ohm though= perhaps electrode bad=

Hydrochloric Acid & HO, First Class=

New Discovery

If a peice of carbon points for electric light be held in the hand and inserted in the liquid one electrode being platina & connected to the carbon of the battery & the other being a peice of No 2 lose covered copper with an iron ~~loop~~ loop and the carbon allowed to touch the platina it will when entirely disconnected from it & as[e] far away as can be give a spark by touching the iron wire[h] of the loop under water[l]

Quit work 6 am=[18]

X, NjWOE, Lab., Vol. 8:1 (*TAEM* 3:432). Document multiply dated, once as 1876. [a]"different Solutions" underlined twice. [b]Followed by centered horizontal line. [c]Obscured overwritten letters. [d]"the spark" interlined above. [e]Interlined above. [f]Followed by horizontal line across page. [g]Circled. [h]Multiply underlined. [i]Canceled. [j]Followed by "get over" to indicate page turn. [k]Followed by horizontal line across page and "flap over" to indicate page turn. [l]Multiply underlined; followed by "over" to indicate page turn.

1. Since Edison's etheric force experiments in the summer of 1876 (Docs. 764 and 766), the British electrical scientist Silvanus Thompson (with Frederick Guthrie) had published a major investigation in September (Thompson 1876) and Edison also had performed further experiments (Cat. 1172:70, Lab. [*TAEM* 3:314]; see Doc. 764 n. 10). A clipping about Thompson's experiments is in Cat. 1031:13, Scraps. (*TAEM* 27:739).

2. That is, connect the wire leading to the etheroscope to different points on the bar.

3. Figure labels are "lead pencil" and "rubber or Beeswax"; and "induction handles" and "Rubber."

4. Figure labels are "glycerin" and "Taste."

5. Figure label is "tin foil."

6. Cf. Doc. 701.

7. Figure labels are "also," "platina electrodes," and "Carbon 10 cells—."

8. Figure labels are "stove" and "no spark."

9. Label on lower figure is "Carbon=."

10. Edison and his co-workers did a lot of this sort of doodling in their notebooks.

11. Figure labels are "6 ohm" and "10 cells."

12. Figure labels are "6 ohms," "insulated," and "6 ohms."

13. Figure labels are (clockwise) "x," "m," "6 ohm," "6 ohm," "10 cells pint carbon batteries," "g," "B," and "n."

14. Figure labels are "Earth," "b," "c," "a," and "10 cells."

15. Edison had recently received samples of a powerful bichromate of potash battery—the Fuller batteries Frederic Ireland had sent him at the start of December 1876. See Docs. 804 n. 5 and 815.

16. Determining the sizes of atoms or molecules from various phenomena exhibited by gases, liquids, and surfaces was a recurring topic in physical science in the second half of the nineteenth century (Brush 1976, 75–78, 201; Maxwell 1873b; Thomson 1869–70). Edison was aware of at least Maxwell's article on this subject but also had treated it as a jocular matter (see Docs. 370 n. 3 and 418).

17. A hydrometer measures the specific gravity of a liquid, often using a weighted tube in a larger glass. Atkinson 1890, 105–13.

18. When Edison returned to these experiments at 2 P.M. he wrote

> The idea has occurred to me seeing the great volume of gas given off by electrolysis in proper solutions and its suddeness that by the use a proper cylindr & valve that more power could be obtained by electrolysis than through the intervention of magnetism= [Vol. 8:29, Lab. (*TAEM* 3:460)]

These experiments continued for two more days. After forty pages of experiments on aqueous solutions, the staff began experiments on alcohol solutions. Edison soon noted "that it is useless to go further in the high resistance solutions  We are now going into experimenting with HCl & HO, with different shapes of electrodes jars etc etc" (Vol. 8:74, Lab. [*TAEM* 3:505]).However, no such series of experiments appears; the next entry in the notebook is Doc. 842.

**-841-**

*From John Sanborn*

Port Huron Jany 16th 1877

Dear Sir

Not having heard from you in reply to my last letter which I conclude you have not received I telegraphed you last night and am just in receipt of your reply.[1] As I had taken hold of the matter not at the request of any one here, but to see ~~that~~ if I could not do something to bring about a settlement between those who ought to be neighbors in a business like way, I thought I would telegraph you as I did—and I did it without consultation with any one. What I meant to convey is this. 1st To consolidate the two roads, at an equal value for each as proposed in the proposition sent you. 2d. The reorganized Company to assume the entire debt of the old Company, and an amount of the new (or City RR) Company twenty five hundred dollars less than the amount of the old road indebtedness In other words if the old road should owe $5,000– dollars then the reorganized Company to assume that amount and $2,500– dollars of the new RR the balance of the new RRs indebtedness to be paid by their old stockholders the real difference between the papers I sent you and your proposition is as I understand it say $5,000 Now this would be dividing it, and in case you say so, I will try and see what I can do— Is[n']t it better to do this if it can be done and get this thing which I know must be an annoyance to you as it certainly is to me put[a] on a fair basis—where there will be some prospect of its being paying property. I wish you could spare the time to come out here if you can I will take hold with you and do all I can. please let me hear from you by letter or telegram.[2]

Yours truly

Jno P Sanborn

ALS, NjWOE, DF (*TAEM* 14:526). [a]Obscured overwritten letters.

1. Sanborn may be referring to the letter Pitt Edison discusses in Doc. 835. Sanborn's telegram to Edison has not been found.

2. On 27 January Sanborn acknowledged Edison's answer and indicated that he would "take hold and see what I can do" to "get it fixed up" (DF [*TAEM* 14:527]). Two days later the stockholders of the Port Huron and Gratiot Street Railway and those of the City Railroad agreed to consolidate their interests into a new company known as the Port Huron Railway Co. Each stockholder received shares in proportion to those held in the prior companies; Edison received 158 shares. The new corporation assumed $4,000 of indebtedness of each prior road with the stockholders of the old roads responsible for any remaining debt. What appears to be a draft copy of this agreement is in DF (*TAEM* 14:530) on Josiah Reiff's letterhead. That copy includes the names of all stockholders and their shares except for Edison. Also see draft agreement, March 1877, DF (*TAEM* 14:538); and Jenks n.d., 14.

**–842–**

*Notebook Entry: Introduction to Mercury Experiments*

[Menlo Park, January 16, 1877?][1]

Experiments with globule of mercury placed in flat dish covered with different solutions=[2] 2 cells carbon battery platina electrode thick[3]

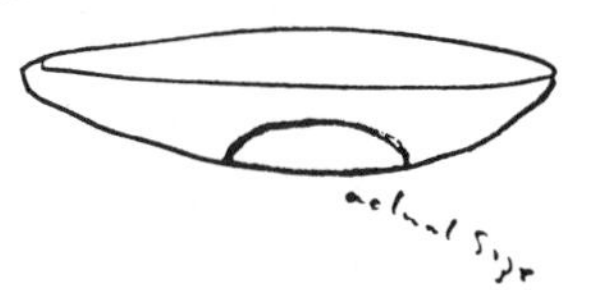

mercury not very pure=[a]

I believe that friction of surface keeps the mercury globular & that if globule forms ~~one p~~ the Zinc pole[4] hydrogen is set f[re]e[b] and this acts to lubricate like the E.M.G the surface hence it spreads out= I notice [that?][b] if it f[or]m[b] the O pole[5] it becomes oxidized ~~& friction is~~ & ceases to be mercury so to speak but becom[es][b] an alloy of an oxide & mercury hence lessening of the globule & flattening out= after its been flattened it may be brought to globule state by causing it to be the Zinc pole= the H reducing the oxide= I notice that when it is the Zinc pole the globule moves freely on bottom of dish but if the O pole it sticks somewhat doesnt have free movement:[6]

X, NjWOE, Lab., Vol. 8:76–77 (*TAEM* 3:506–7). [a]Followed by centered horizontal line. [b]Document damaged.

1. The twenty-page set of mercury experiments that follows this introduction ends with the note "Wednesday 16 1877= 2 am." However, 16 January 1877 was a Tuesday. It seems more likely that at 2 o'clock in the morning Edison got the day right and mistook the date, in which case this introduction was written on Tuesday the 16th. Moreover, Edison resumed these experiments on 17 January at 8 P.M. Vol. 8:97, Lab. (*TAEM* 3:528).

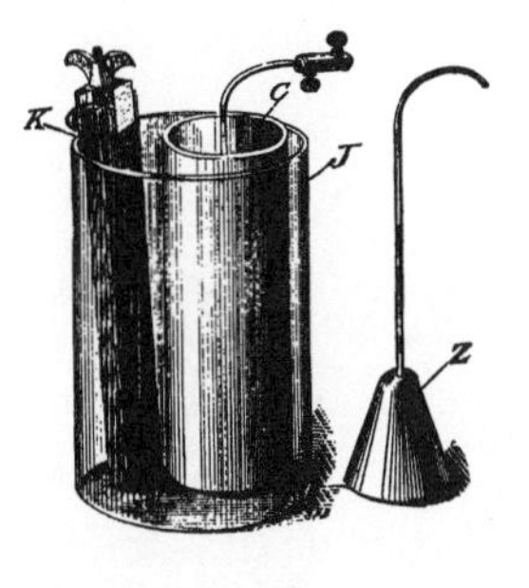

*In the Fuller battery, zinc electrode* **Z** *touched a small pool of mercury inside porous cup* **C.**

2. Fuller batteries, samples of which Edison had recently received from Frederic Ireland and which he had apparently begun to use in his experiments, contained mercury under dilute sulphuric acid, one of two electrolytic fluids employed (Docs. 804 n. 5 and 840 n. 15; Niaudet 1884, 218–20; Maver 1892, 17–18). English, French, and German investigators during the preceding several years had begun to subject the special physical and chemical characteristics of mercury drops in an electric cell to scientific scrutiny. Researchers included people with whose work Edison was familiar, such as Cromwell Varley, and Hermann Helmholtz. Sir Charles Wheatstone had recently "produced motion in globule of mercury enclosed in a glass tube by . . . oxidation at one end of its surface and deoxidation at the opposite end," and "devised a telegraph on this principle" ("Society of Telegraph Engineers," *Sci. Am. Suppl.* 57 [1877]: 908). No practical telegraph resulted, but such work eventually opened a significant field of study within electrochemistry (Stock and Orna 1989, 339–401).

3. Figure label is "actual size."

4. That is, the cathode.

5. That is, the anode.

6. This entry begins more than eighty pages of notes on several series of experiments carried out during the following week on the behavior of mercury drops submerged in various solutions and subjected to an electric current. That material ends with Doc. 846. See Vol. 8:97, 126, 136, 183, Lab. (*TAEM* 3:528, 558, 568, 616).

**-843-**

*Notebook Entry: Autographic Printing*[1]

[Menlo Park,] Jan 17th 1877

Edisons Autographic power press.[2]

Fig 31

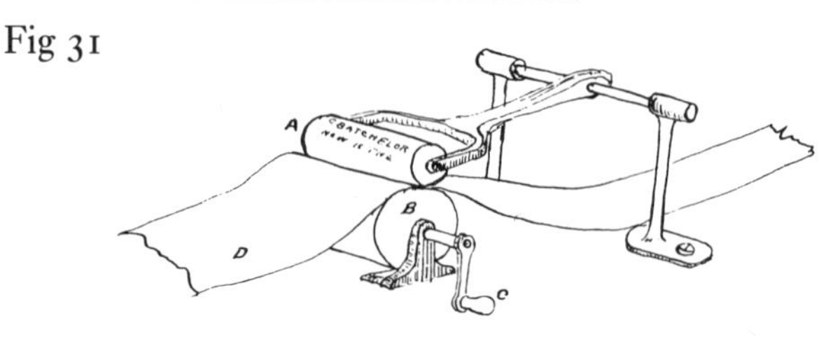


I made a machine like sketch to prove the practicability of working on a rotatory press.[3] The stencil is fastened on the felt roller A by lapping it round and is held tight by bands on the end. The roller B is turned by handle C and at each revolution of A a copy is left on the endless band D. This worked well but wanted great pressure and we now think that the stencil ought to be placed on roller B and to have three or four rollers placed like[a] Fig 32. with considerable pressure on them this being the same as rolling over the stencil four times.

Fig 32.

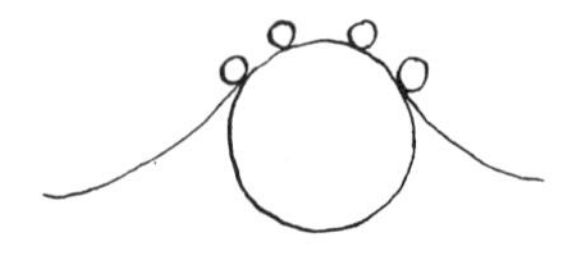

Chas Batchelor

X, NjWOE, Batchelor, Cat. 1317:31 (*TAEM* 90:672). Written by Batchelor. [a]Obscured overwritten letters.

1. This entry is continued in Doc. 852.

2. Edison had apparently proposed a power press for use with electric pen stencils sometime earlier (Frederic Ireland to TAE, 27 Nov. 1876, DF [*TAEM* 13:992]). Drawings dated 6 November 1876 show press designs that would print two copies at once with a very different printing mechanism (NS-76-001, Lab., Supp. III [*TAEM* 162:529]).

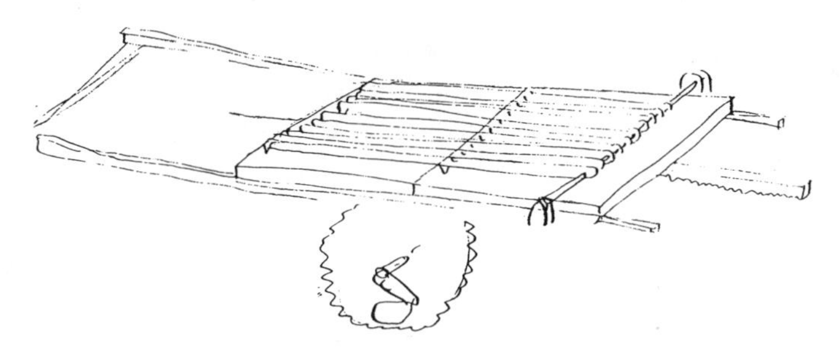

*An autographic press design that would make copies from two stencils at once.*

3. This entry reflects Batchelor's work on the press (Cat. 1233:17, Batchelor [*TAEM* 90:60]). At midnight on the morning of 18 January, Edison stopped his mercury globule experiments (see Doc. 842) to work on the press, beginning with the design shown here in figure 31 (Vol. 8:128, Lab. [*TAEM* 3:560]). Several pages of notebook drawings follow, ending with a sketch similar to figure 33 in Doc. 852 (Vol. 8:130–35, Lab. [*TAEM* 3:562–67]).

This marked the beginning of nearly two months' concerted effort to develop a power press. A time record for Batchelor's and John Kruesi's work on a power press is in a notebook whose cover Batchelor labeled "Edison's Duplicating Press Rotary Jan. 19th 1877" (Rotary Press Notebook 1, Lab., Supp. III [*TAEM* 162:265]; Batchelor noted his work in Cat. 1233:17, 22–25, 32–33, 35–36, 38–39, 49–50, Batchelor [*TAEM* 90:61, 64–65, 69–72, 77–78]). This and another notebook contain rough drawings of various press designs from January and February, which are summarized by Batchelor in the notebook containing this document and Docs. 852 and 854. A sample identified as "First specimen of rotary press Jan 23" was probably made on the press described in this entry (Cat. 1240, item 36, Batchelor [*TAEM* 94:18]).

## –844–

*Technical Note: Telephony*

[Menlo Park,] Jany 20 18~~6~~7

Speaking telegraph

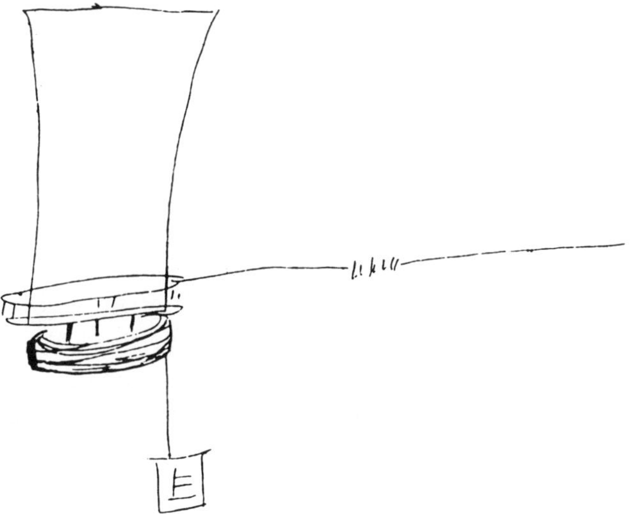

three Platina points dont seem to work any better than one point[1] get it very good on a western union relay through my teeth.[2] but I think that I could get it better if I had an adjustment to it for Charley cant Hold it steady enough[3]

X (photographic transcript), NjWOE, TI 2, Edison's Exhibit 5-11 (*TAEM* 11:224). Written by James Adams.

1. Edison identified this as the first of his experiments with a telephone transmitter that varied the resistance of the circuit by changing the pressure on carbon rather than changing the amount of carbon included in the circuit. Edison recounted that something early in 1877 led him to recall his 1873 observation that the resistance of a mass of carbon changed in response to pressure (see Doc. 351). He had already used plumbago in his telephonic experiments, and he had just used carbon as well in experiments on etheric force and electric arcs, in which passage of a current was facilitated by pushing and grinding carbon contacts together rather than simply having them touch (Docs. 799 and 840).

As was usual in his telephone designs, Edison avoided the approaches of Reis and Bell, rejecting breaks in circuit continuity as well as reliance upon voice power to create the necessary signal current. The short vertical marks in the middle of the line extending to the right in the drawing represent the battery; the platina point or points in the experiment were attached to the diaphragm on the base of the tube and extended down to a dish containing "loose carbon." Edison said that this initial arrangement produced adequate volume but only poor intelligibility. Edison's testimony, TI 1:46–47 (*TAEM* 11:44).

2. It was possible to roughly test a transmitter by sending its signal through a telegraph relay, some part of which an experimenter held against or between his teeth. This technique was convenient for Edison because of his partial hearing loss and sometimes was used by others as well. Edison's testimony, TI 1:46 (*TAEM* 11:44).

3. In the initial trials Charles Edison apparently held the combination of tube, diaphragm, and points above the dish of carbon. "An adjustment" would have involved a fixed stand or frame with an adjusting screw to gradually change and firmly maintain the distance of the tube and diaphragm above the dish and thus control the initial pressure of the point or points on the carbon.

## -845-

*Article in the* American Chemist

MENLO PARK, N.J., Jan. 21, 1877.[a]

LABORATORY NOTES.[1]

BY THOS. A. EDISON.

*No.* 8.—Certain kinds of tissue-paper, if cut in strips and laid upon the palm of the hand, will curl and uncurl at regular intervals for hours. On the hands of some persons it cannot be made to move.

*No.* 9.—Many bottles containing substances in solution show upon the side placed nearest to the wall, and the least exposed to the light, a bow formed of drops, which extends from the surface of the liquid on one side nearly to the cork, and thence to the surface of the liquid on the other side. The smaller the amount of liquid the higher the bow, which is generally sharp and well defined.

*No.* 10.—Crystals of protosulphate of iron will turn white only in that part of the bottle exposed to light.

*No.* 11.—The smoke from burning camphor, when inhaled, has a most powerful effect upon the lungs. It causes a violent cough and spitting of blood almost immediately, and the effects are greater than with sulphurous acid, chloride of bromine, etc.

PD, *Am. Chem.* 7 (1877): 356. [a]Place and date not those of publication.

1. Notes 1–7 are in Doc. 813. The original observations of notes 8–10 from this series are in Cat. 996:69, 73, Lab. (*TAEM* 3:313, 315). Drafts for further notes (through 22) are in Menlo Park Period, Undated Notes and Drawings; NS-Undated-002; and NS-Undated-003; all Lab. (*TAEM* 45:128–31, 137–39; 8:169–70, 178–79).

**-846-**

*Notebook Entry: Mercury Experiments and Miscellaneous*

[Menlo Park, January 23, 1877?][1]

Desultory Experiment on Mercury Globules=
Carbon plate with hollow cut in it

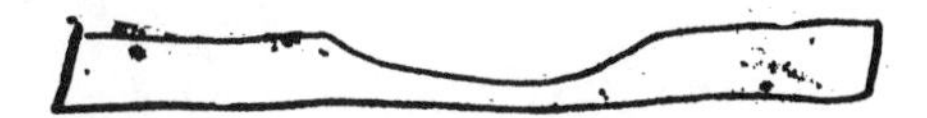

The Hg in 20 grm sol Caustic Soda without Electrodes keeps up a surface current little whirlpools they have continued for 10 minutes & will pbly continue long time nothing very strange ocurs with Hydrochloric acid nothing strange=[a]

I now have a bent U tube.[2]

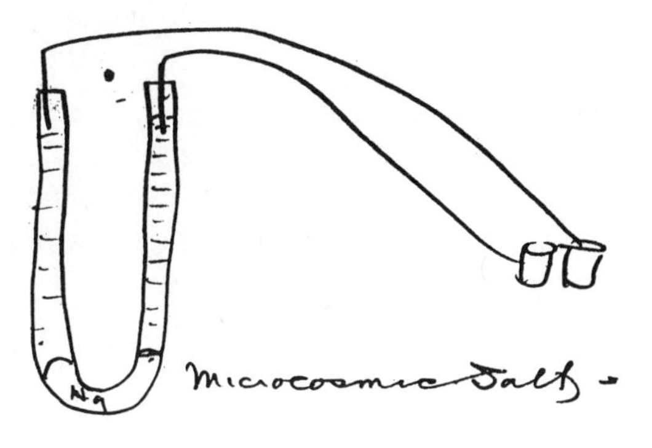

Hg moves towards Oxygen Electrode about 1/32 of an inch on closing key=[3] It vibrates too easy to work morse well= diameter bore tube about 3/16=

I notice that the water passes freely from one tube to the other between the glass & the Hg[a]

I have now made another capilliary tube

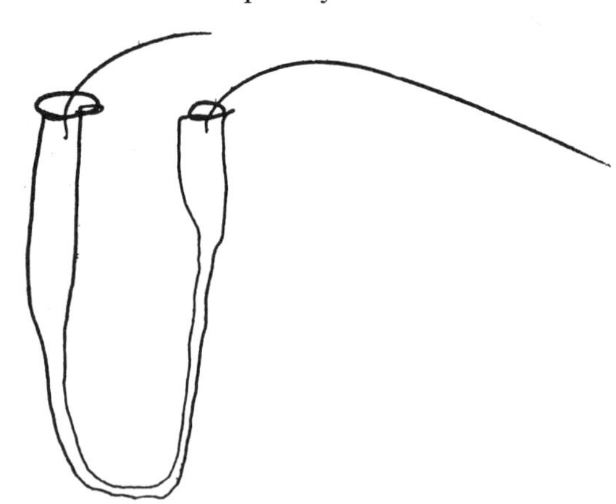

with considerable Hg & Copper Electrodes it scarcely moves  with platina moves much better but not so well as in flat dish  I got less Hg in now[a]

The tube I drew didnt answer well= gone back to U tube 3/16 bore small Hg in bottom & $SO_3$ & HO, dilute.=

Thus[4]

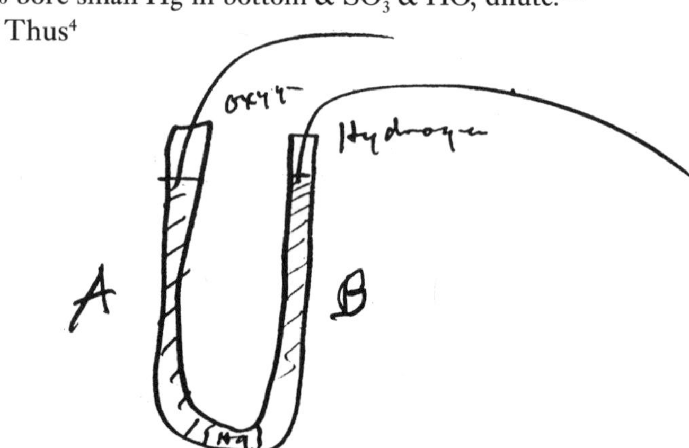


Platina Electrodes  2 cells Carbon battery through 10 000 ohms  water raises in A 1/200 inch & lowers same rate in B. it responds good to Morse=[a]

With this U tube & with Hg thus[5]

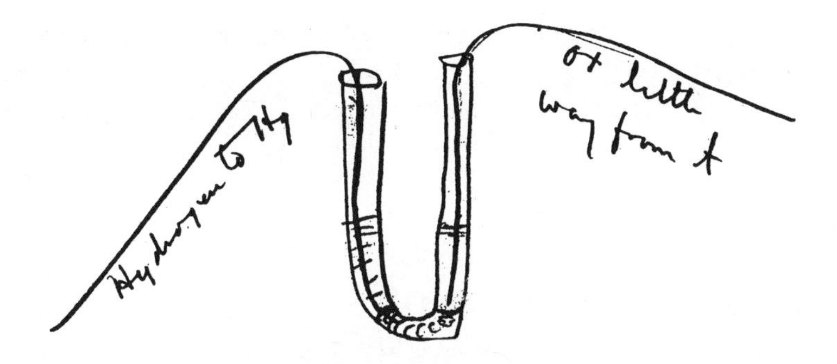


water moves up in ~~H~~O & down in H. ~~1/200~~ 1/364 to 1/32 inch quite sharp through 24 000[b] ohms with 1 cell Bunsen  it dont work at all with reverse currents  the reverse current or O to Hg sticks it & it dont recover for several minutes ~~mistake~~[a]

~~By keeping key closed o~~[c] ~~10 see~~ If it had a capilliary bore it would move a long distance=$SO_3$ & water dont act anywhere near so well besides this aint pure Hg there's lots Zinc in it[a]

I find that with this tube 3/16 bore that with 2 cells carbon it moves 1/64 through 36 000 ohms'= it moves sharply & plainly through 70 000 ohms'=1/100. of inch=with magnifying glass magnifying double it moves plainly about 1/300 inch through 70 000 ohms with one[b] cell carbon battery= this above with 5 gram solution phosphate manganese

Been trying $SO_3$= & 40 grams Sulphate Magnesia= but theres no doubt but 5 gram phos manganese is best=. I find that when the Hydrogen wire is made in a loop thus & dipped in Hg in ben~~t~~d of tube and drawn up as far as Hg will adhere without breaking that the water rises & falls a greater distance hence mkg it more sensitive. This puts Hg on a tension=

Senstiveness, strange to say appears to increase as the .O. wire is ~~put~~ just put in liquid=[a]

Note= Acetate Aniline, & solution protochloride palladium in test ~~tube if heated~~ tube. two solutions dont mix— if heated, a large bubble 1/4 diam comes up from bottom solution occasionally with a thump, but if peice ~~pa~~Biblus paper be stuck in enormous amount gas[b] comes up and liquid boils= This shows that two liquids one above other or liquids that boil by jerks can be prevented by a biblus paper, or other substance connecting bottom with top.[a]

Mercury Continued= Experiment

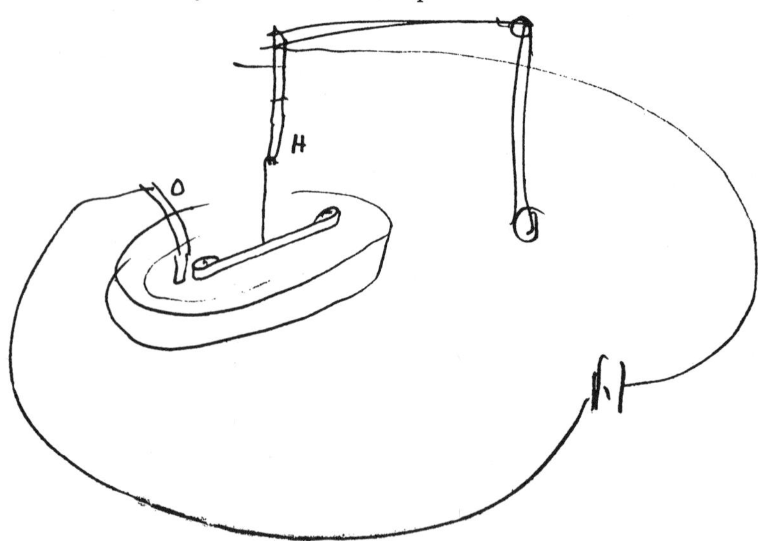


Hg in loops ~~w~~attracted slightly with Sulphate Magnesia— not attracted with solution 20 grms Caustic Soda= with 5 grms Phosphate Manganese not apparently attracted.=

bent thus[6]

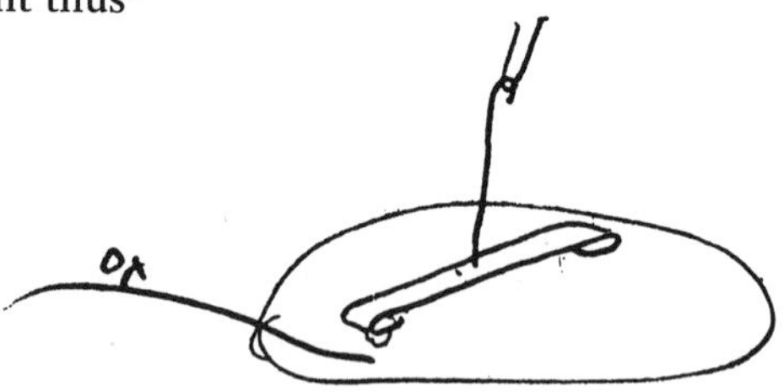


it teters, ie[d] when ckt closed the Hg in loop is squeezed so to speak out & reaches to bottom dish & on bkg ckt is drawn up by amalgmation capilliary=

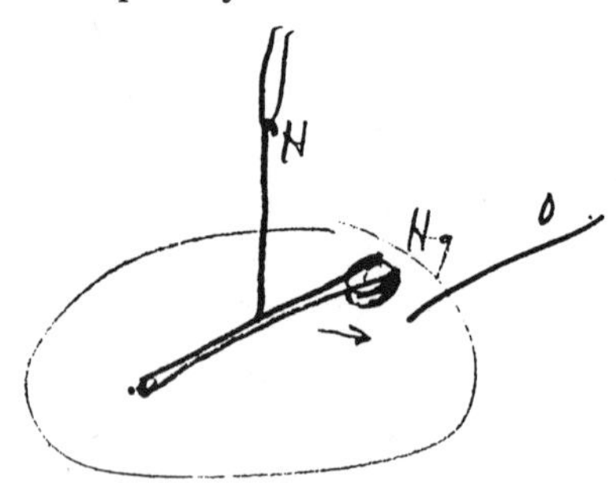

5 grms phos manganese on 1800 ohms 2 cells its attracted by .O. 1/16 inch= the globule laying on bottom dish but [~~16~~?][e] 3000 ohms' moves 1/64 6000 ohms 1/200= 2 cells Carbon battery[7]

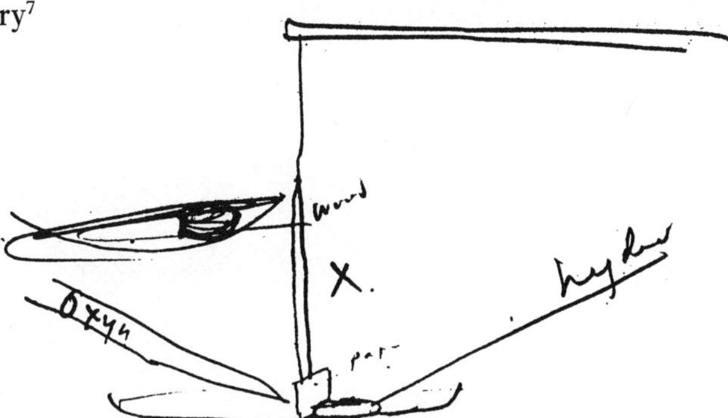


With 5 gramme solution phosphate manganese paper moves 1/64 inch through 12 000 ohms'=

Very Sensitive=[8]

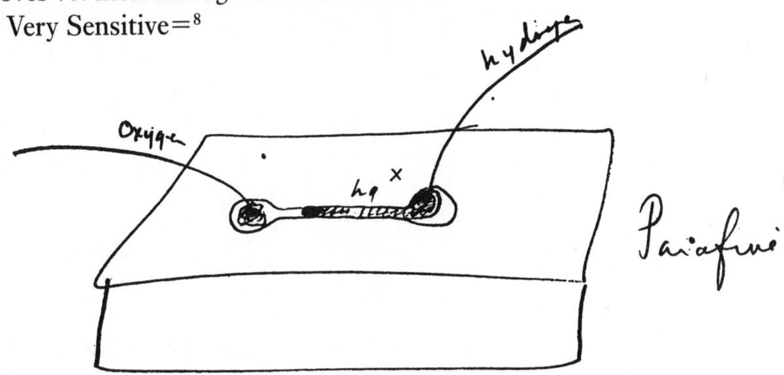


groove cut in parafin about 1/8 wide & about 1/8 deep conical this with rounded part at each end with 5 gram solution of Phos Manganese; end of hg .X. is attracted towards .O. with 2 cells Carbon 1/64 inch through ~~6~~70,000 ohms & 1/100 with 1 cell through same resistance= Theres scarcely any solution in bottom cant see any just moist pbly if too much solution not so sensitive. The long tail of the Hg must be manipulated for sometimes I can get no movement.

I propose to make a Mirror Electrometer=[9]

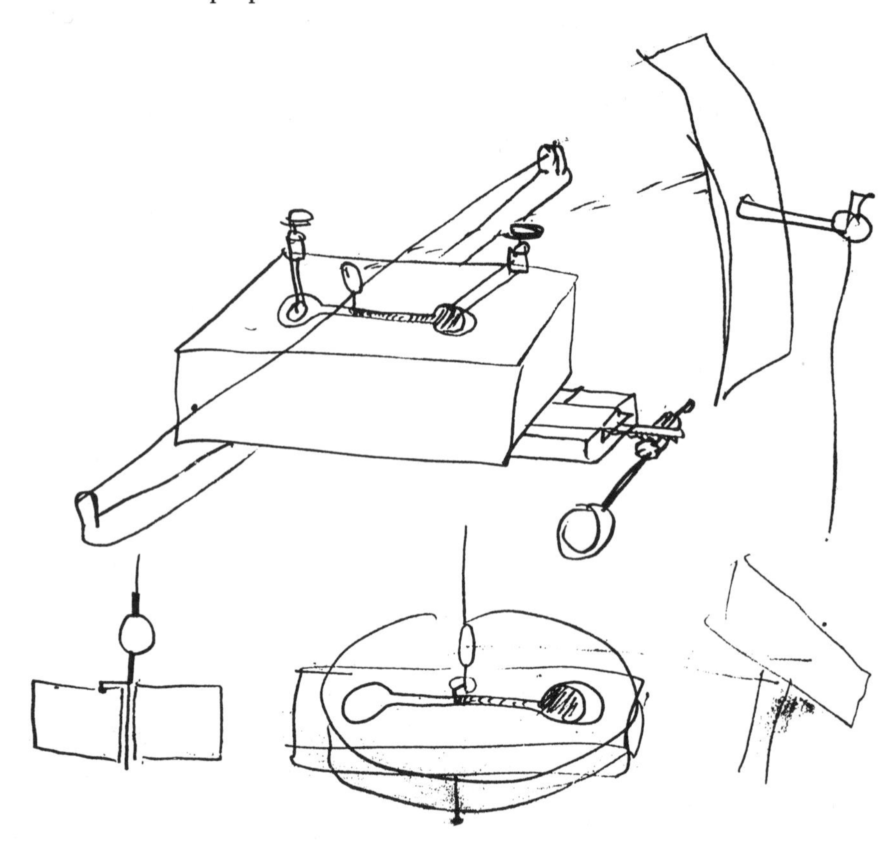

X, NjWOE, Lab., Vol. 8:198–208 (*TAEM* 3:632–42). [a]Followed by centered horizontal line. [b]Multiply underlined. [c]Interlined above. [d]Circled. [e]Canceled.

1. This entry begins on the sixteenth page of a tablet whose cover is dated 23 January and labeled "Electrolysis Mercury Globules continued" (Vol. 8, following p. 182, Lab. [*TAEM* 3:616]). The last several pages of the tablet, following this entry, relate to other subjects, and there is no evidence that Edison continued experimenting with mercury in electrolytic cells past this date.

2. Figure labels are "Hg" (in tube) and "Microcosmic Salt=," which is a phosphate of sodium and ammonia ($NaNH_3PO_4{\cdot}4H_2O$). The name is an allusion to the human as microcosm, the salt having been first extracted from human urine.

3. Not shown in diagram.

4. Figure labels are "oxygen," "hydrogen," "A," "B," and "Hg."

5. Figure labels are "Hydrogen to Hg" and "Ox little way from it."

6. Figure label is "ox."

7. Figure labels are "Oxyg," "wood," "X," "paper," and "hydro."

8. Figure labels are "oxygen," "hg," "x," "hydrogen," and "Parafine—."

9. An electrometer measures electrical charge. In Edison's first

sketch, movement of the mercury in the groove would move the lower end of the vertical stick attached to the small round mirror at center. The curved scale for reading the reflected spot of light is at right. In the last, large sketch the mirror turns from side to side.

**–847–**

*Notebook Entry: Automatic Telegraphy*

[Menlo Park,] Jany 25th 1877

I think the best solution found yet with Platina pen is standard[a] Aniline oil Acetic Acid. Small amount of Sulphuric Acid and about 10 times its bulk of water and add to this a little iodide Potass which makes the solution very delicate the ~~Aniline~~ Iodide[a] mark will fade away and there will remain a greenish yellow mark which is Permanent[1]

X, NjWOE, Lab., Vol. 8:172 (*TAEM* 3:605). Written by James Adams. [a]Interlined above.

1. This reflects Edison's continuing search for a sensitive, permanent recording solution for his automatic and domestic telegraph systems.

**–848–**

*From Jay Gould*

[New York,] Jany 26, 1877

Dr Sir

Your letter of Jany 20th received—[1]

I have always felt very kindly toward you & assisted you financially when but for it you would have been ruined at least so you told me— How have you requited me for that kindness? I leave your own conscience to answer

Genl Eckert is absent when he returns I[a] will lay your complaints before him— I am certain he would not intentally cheat you or do you a wrong certainly I will not be a party to any Your

Jay Gould

ALS, NjWOE, DF (*TAEM* 14:755). Monogram letterhead of Jay Gould. [a]Obscured overwritten letter.

1. Not found.

**–849–**

*To Samuel Edison*

New York, Jany 29, 187~~6~~7[a]

Dear Father,

You better come on. Tony[1] who came on here has left and I have no[--- ----- ----- -----][b] and if you will come on I will have everything fixed for you— and you will have a nice easy job= I havent a dead beat around me= No one at the house= ~~if you~~ Unlike Pitt my wife does not nor never can control me and you can have anything that I have= Im not poor by any means: I dont want you to stay in that god damn hole of a Port Huron which contains the most despicable remenants of the human race that can be found on the earth= I want you to come here & settle down. if you wish I will give you money enough to go to Florida, Tampa Bay etc. on your arrival, & your salary will be sufficient to keep you in fine style— Your son

T.A.E.

ALS, MiDbEI, EP&RI. Letterhead of Edison's Electrical Pen and Duplicating Press. [a]"New York," and "187" preprinted. [b]Document damaged.

1. The previous day Anthony (Tony) Bronk, whom Edison had known as a youth in Port Huron, left the laboratory. He had been working there since 22 December 1876 for one dollar a day without board. *TAEB* 1:29 n. 4; Cat. 1213:7–8, Accts. (*TAEM* 20:8–9).

**-850-**

*Draft to William Orton*

[Menlo Park, c. January 29, 1877][1]

~~I make the following proposition which I believe will be advantageous both to the Co and to myself.~~

I have ~~a~~ now thoroughly completed & stocked my[a] laboratory at Menlo Park N.J. 26 miles from New York on the Penn. R.R. The building is[b] 25 × 100 .&. 2 stories fitted with every kind of apparatus for scientific research= I have in the Laboratory a machine shop run by a 5 horse power engine. The machinery is of the finest description. I employ three workmen two of whom have been in my Employ for five years and have much experience.[2] I have also two assistants who have been with me 5 and 7 years, respectively ~~with~~ both of which are very expert.[3] At present the cost of running my machine shop including coal kerosene & labor ~~ais~~ ~~for~~ about, 15 per day or 100 per week; at present I have no source of income which will warrant continuing my machine shop[c] and I ~~am~~ shall be[d] compelled to close ~~my machine shop~~ it,[e] ~~without~~ unless[e] I am able to provide funds for continuing the same and keep my skilled workmen the loss of which would seriously cripple ~~me in~~ me[e] I therefore make the following[f] proposition to the Co to wit:—that if it will ~~enter into a contrac~~[g] pay the running expenses of my shop ie[h] $100 per week, I will enter into a contract for a period of time to be agreed upon giving them every invention that I can make during that time which is applicable to ~~th Land Telegraphy~~ commercial telegraphy in the US[i] ~~and adaptable~~ ~~applicable~~[e] ~~on~~ [---][j] ~~to~~ & the Cos ~~needs~~ system,[e] they ~~company~~ paying me a small royalty on ~~every~~ such inventions as they may[e] adopt and find of value to them This royalty will be the[e] ~~my~~ source of income, ~~and will pay me for my time~~ of myself and 2 assistants, for time and talents which ~~of course will be entirely dependant upon the value of the inventions to the Co. adopted & found useful to the Co.~~ will be large or small according to the value of the invention to the Co. ~~The CoValue I have invested in~~ my laboratory ~~h~~with machinery & apparatus have cost[k] about $40,000, ~~and I~~ and if the Co desire to enter into such a contract[l] I suggest that Mr. Prescott be sent to ~~look at it and learn its completeness~~ investigate ~~my unusual~~ the unusual facilities which I have[m] for perfecting any kind of Telegraphic[e] invention. Yours

T. A. Edison

ADfS, NjWOE, DF (*TAEM* 14:226). [a]"now . . . my" interlined above. [b]"The . . . is" interlined above. [c]"which . . . shop" interlined above. [d]"shall be" interlined above. [e]Interlined above. [f]"therefore . . . following" interlined above. [g]"~~enter~~ . . . ~~contrac~~" interlined above. [h]Circled. [i]"commercial . . . US" interlined above. [j]Canceled; "~~on~~ [---]" interlined above "~~applicable~~". [k]"~~h~~with . . . cost" interlined. [l]"and . . . contract" interlined above. [m]"investigate . . . have" interlined above.

1. This draft is the foundation for Edison's 22 March contract with Western Union (Doc. 876). On 29 January, Charles Batchelor wrote in his diary that "Edison saw Orton and broached the subject of getting $100— per week to pay Expenses at Laboratory" (Cat. 1233:29, Batchelor [*TAEM* 90:67]). Edison had apparently been considering the terms of his relationship with Western Union for some time, as indicated by letters from Robert Russell and Uriah Painter and his earlier letter to Orton (Docs. 823, 827, and 833).

2. Account records show entries for Charles Wurth and John Kruesi, both of whom had been with Edison for several years; the other workman may have been Tony Bronk. *TAEB* 2:519 n. 2; 633 n. 6; Cat. 1213:7–12, Accts. (*TAEM* 20:8–11).

3. James Adams and Charles Batchelor.

## -851-

*From Uriah Painter*

WASHINGTON, D.C. JAN 31 1877[a]

My Dr E

You remind me of "the boy stood on the burning deck when all but he had fled"!!![1] Much obliged for note,[2] I'll realize on it & use for "commonweal." You said you would see what

R[eiff] would take & let me know— I am satisfied that nothing can be got out of A&P. for Auto[b] except you go & set it up on its legs & make it work— Now the thing to do, is to take advantage of the situation & get an option on all the rest of interests[c] at its present value: (which is awfully low-d~~on~~ewn, about where Little sold—) then put in the past with your genius to bring it up, & ~~then~~ we can get something that will do somebody some good! Let Reiff settle with his creditors on the basis of what there is to-day, & what is made to-morrow beyond what you & I need we can settle ~~it~~ on the Phila girl:—[3] So he will have a home to go to, which he has not now—& is not likely to have if allowed to have his own way— The advice you gave him one night two years ago when we were in Newark was, to "go & bust," & "make fresh start":—was sound then, & is sound now—

You do not put a price on your new perforating repeater! Is it in condition where you can do so? Yours Truly

U H. Painter

Come over here & stay a day or two it will pay you! UH.P

ALS, NjWOE, DF (*TAEM* 14:641). [a]Place and date from *Philadelphia Inquirer* hand stamp. [b]"for Auto" interlined above. [c]"of interests" interlined above.

1. From Felicia Dorothea Hemans's poem "Casabianca."
2. Not found.
3. Unidentified.

**-852-**

*Notebook Entry: Autographic Printing*[1]

[Menlo Park,] Feb 1st, 1877

Edisons Autographic Power press.

Our next endeavour was to turn a cylinder and cut a piece from the surface so that the stencil could be fastened to it (as in Fig 33) and have a roller running on top which acts in place of a press bed and the ink roller placed inside which when not running on the stencil is distributing on the inside surface of the cylinder.

Fig 33

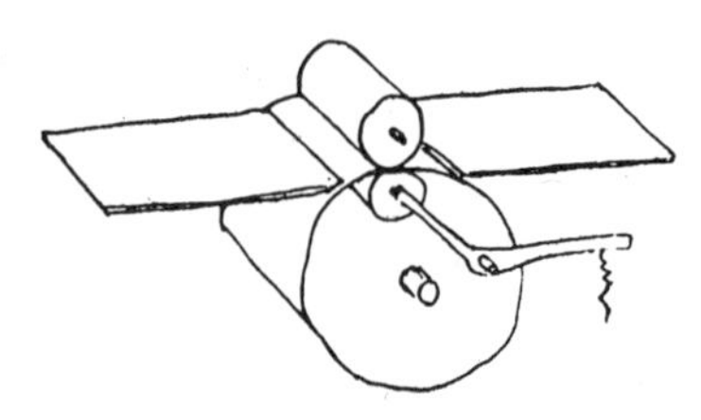

We got fair results from this device. The paper was fed by two springs fastened on the cylinder which were lifted at the right time by the cam on the feed plate and after raising camming out and falling back in position it cam down on the paper and held it tight on to the cylinder  this device shown better in Fig 34.[2]

Fig 34

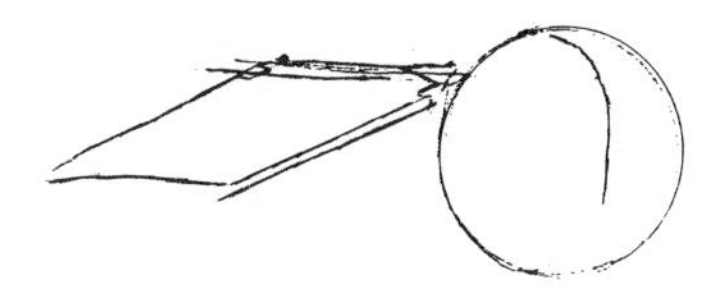

It was completed January  1877[3]

We next devised the means thus:—

Fig 35

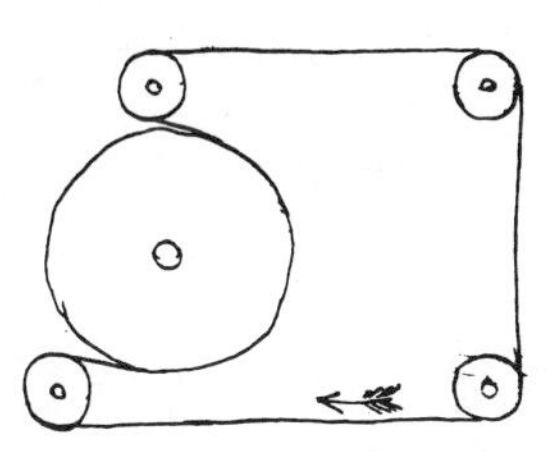

We used an endless band running as in Fig. 35 over the cylinder A and the four rollers B C D E. Ink rollers F G H run on the outside of the cylinder and are independent of each other. The stripper I turns the paper over the top band which it travels along depositing over the end of machine  This plan we have concluded is so far the best  we are using rubber cloth as a band, we have tried brass and copper sheet[4]

X, NjWOE, Batchelor, Cat. 1317:31 (*TAEM* 90:672). Written by Charles Batchelor.

1. This entry continues Doc. 843 and is continued in Doc. 854.
2. Neither Figure 34 nor 35 was completed.
3. Edison drew sketches of this design on 26 and 27 January (Rotary Press Notebooks 1 and 2, Lab., Supp. III [*TAEM* 162:268–80, 309–19]) and a refined sketch on 29 January (NS-77-002, Lab., Supp. III [*TAEM* 162:544]). The press was completed by 31 January, when it produced a number of samples (Rotary Press Notebook 1, Lab., Supp. III [*TAEM* 162:283]).
4. In Rotary Press Notebook 1, Lab., Supp. III [*TAEM* 162:285]) Batchelor wrote "How would a Copper foil band do  The Brass foil band appears to work well." He also noted that "It may be that in putting steel spring on the sides of Endless band that the cloth will stretch more than the steel and consequently the cloth will always be baggy. We must prevent the paper from blurring when going round a curve."

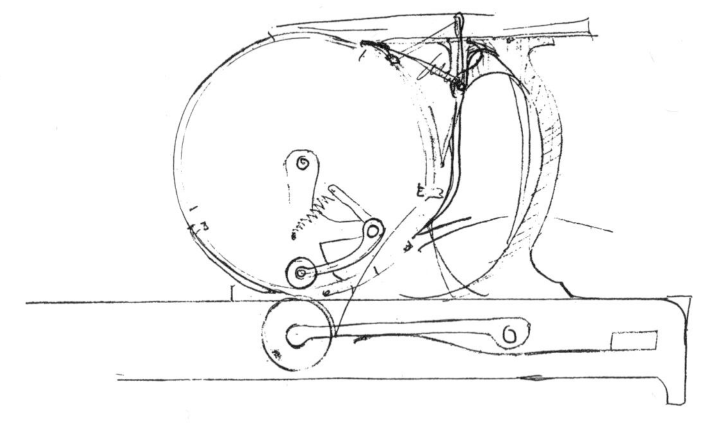

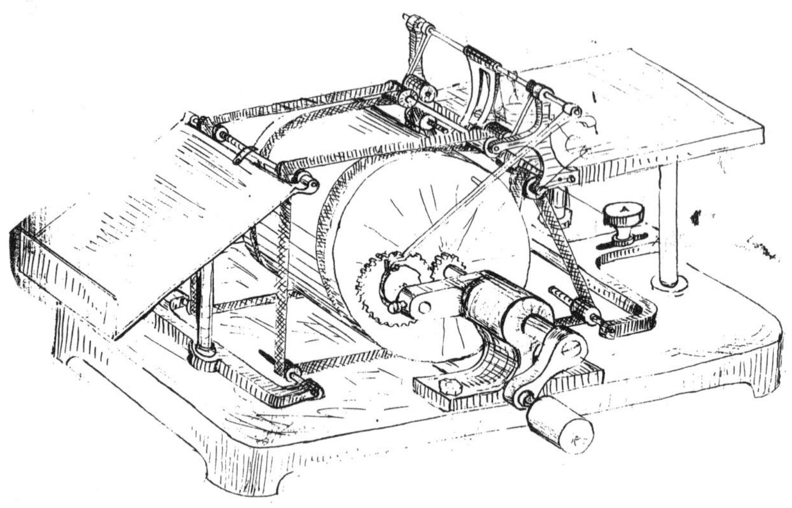

*The rotary press with automatic paper feed. The rough sketch shows the placement of the inking rollers at the bottom of the cylinder.*

-853-

*To Jay Gould*[1]

MENLO PARK, N.J., Feby. 2/77.

DEAR SIR:—

In reality it is not the desire of Mr. Eckert that I should be connected with the A.&P. no matter what he may say to you to the contrary.[2] It is not necessary that a fact should be forced down one's throat with a crow-bar, there are gentler methods. For instance, keep an impatient man like myself waiting 3 weeks' to decide about removing a partition, causing me to wait for weeks, daily for small sums due me, which the treasurer said was in bank for that purpose—doubting my honesty in a four dollar transaction, refusing to pay for improvements put on machinery in course of construction, urging the necessity of producing certain things which when finished was laid aside & pay refused. Refusing to grant money to conduct experiments & then report that I do nothing for the Co. & 1001 things which acting cumulatively was a long unbroken disappointment to me.

I have never complained but once or twice to you in these two years. You know that I was led to expect that I should be paid for the 5 years labor & money expended on the automatic, you told me so yourself on the day you bought the Quadruplex.[3] I supposed I should be furnished funds by the Co. to go on and perfect the system—introduce the Roman Letter apparatus and bring the instrumentation of the A.&P. to a high state of perfection. The result has just been the opposite. I have been made to suffer for the personal spite of Mr. Eckert against others and my presence at the A&P office not proving agreeable I had no other recourse than to take hold of some experimental work for the Western Union.[4] I must live Mr. Gould and if the agents between me and you are not gifted with perception enough to deal with an inventor and his peculiar business in a different way than with a business man, then you must blame your agents not me. I am the last person in the World to go back on a person who has befriended me. Although I have every reason to strike back at Eckert for his mean treatment of myself through the introduction of economizing apparatus in the Western Union. I have refrained from doing so, because it would or might cause loss to you, the work so far that I have produced is of a purely scientific character and of no immediate value to the W.U. but some day I must be compelled to sell something to keep the pot boiling. How can it be expected that I should go back on Mr. Reiff.

For years he has assisted me and as regards obligations is in the same position as yourself. I could perhaps have been paid for my interest in Automatic—I needed it—the temptation was great—and yet I did not take it, nor never would unless Mr. Reiff got something. Most of the stories floated by Mr. Eckert and told you about Reiff are false & he knows them to be false, and you having been alienated from Reiff have no means of proving their falsity. The whole trouble arose early in the negotiations, Eckert being jealous that Reiff would be connected with the Co. A Tallow candle if it were given its choice would not surround itself with gas jets. Yours truly,

⟨This is a true copy of Letter sent today—how does it sound?⟩

PL (transcript), NjBaFAR, *Harrington v. A&P*, Box 17B, Complainant's Exhibit 4, 2:185.

1. A draft of this letter appears in DF (*TAEM* 14:749). Edison sent this copy to Josiah Reiff, who introduced it as an exhibit in *Harrington v. A&P.*

2. On Edison's relationship with Atlantic and Pacific, see *TAEB* 2, chaps. 6–11 and app. 3.

3. That is, 4 January 1875; see Doc. 526.

4. That is, acoustic telegraphy; see *TAEB* 2:524.

**-854-**

*Notebook Entry: Autographic Printing*[1]

[Menlo Park,] Feb 2nd 1877.

Automatic Feed for Autographic Rotary Press.

We consider that the press would be of very little consequence without a self feed and therefore put on the one shewn in Fig 36

Fig. 36

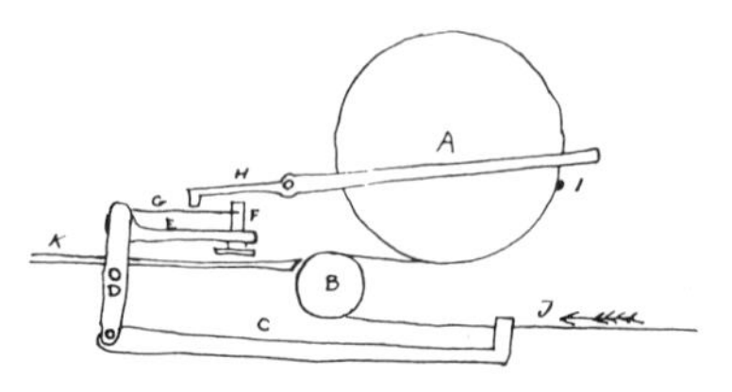


A is the cylinder B the feed roller C the feed lever. E picks up the paper H is a lever moved down by the pin I on the band. As the band J moves in the direction of the arrow projections on its sides move the lever C and another on the other side and gives the lever D one movement backward and forward for every revolution of the band. F is a pin held up by the spring G off the paper. This pin is furnished with a piece of rubber on its end which is heated occasionally to make it adhere to the papers. The pin I on the belt moves the lever H down on to the spring G and thereby pressing the rubber on to the paper it lifts one sheet up and waits till lever C carries it over to the drum A and takes it round. This works admirably.[2]

X, NjWOE, Batchelor, Cat. 1317:32 (*TAEM* 90:673). Written by Charles Batchelor.

1. This entry continues Doc. 852.

2. Batchelor noted in his diary on 7 February that he had "Finished Rotary press experiments having got a self feed and all necessaries"; the following day he reported "Worked at night on Rotary press got the experiments all tried and it now remains to design the instrument" (Cat. 1233:38–39, Batchelor [*TAEM* 90:72]). A sample dated 7 February in Batchelor's scrapbook indicates that the press "will duplicate letters at a rate of 2000 per hour" (Cat. 1240, item 48, Batchelor [*TAEM* 94:20]). Notes and drawings related to work on the rotary press through early March are in Rotary Press Notebooks 1 and 2, Lab., Supp. III

[*TAEM* 162:258–327]. An entry in notebook 2 notes that a wooden model, presumably for the Patent Office, was begun on 12 February but apparently no patent application was filed.

## –855–

*Notebook Entry: Electric Sparks*

[Menlo Park, February 2, 1877][1]

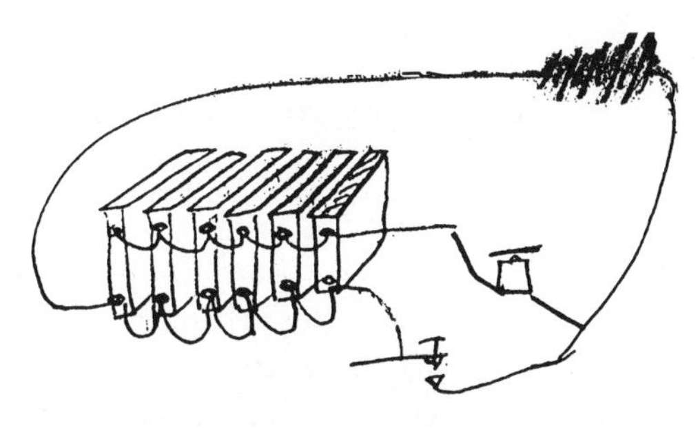

Callaud battery 70 cells not first class 5 Phelps Condensers M[icro]F[arads]. 1.9 1.6 2.5 1.6 1.6 Capacity & the red condenser= on closing ~~battery~~ Key a current (momentary) closes Sounder— (Bunnell)[2] on opening an opposite Current closes Sounder= this is correct as on closing ~~the battery is~~ a route is found for discharging the Condenser on opening both poles of the battery are free to charge the condenser=

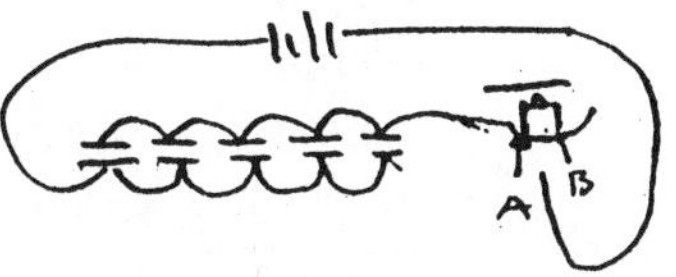


Phenomenon on closing on .A & not through the Sounder the snap[a] spark takes place with [s]cintillation[b] only on closing whereas on B it does not scintilate and you get the steady Electric Arc without scintillations only on opening—[c] ? why— perhaps self induction of the magnet=

~~We a~~[d] We are now adding Condensers and battery. got 110 cells= with hand motograph I notice that with Copper wire aluminium gives brighest sparks & that & Lead gives Loudest Cracks= With iron the spark shew Scintilations ~~farther apart than~~ closer together ~~thus~~ & more abundant & longer= appear thus

While Aluminium shows less abundant and the~~ir~~re are mixed sparks some appear like iron & others are thick & bright, while iron are thin & yellow= These bright Scintilations are thus the ends being turned down

The iron also have ie[e] some of the scintilations a forked end=

Nickel Zinc Copper Tin Platina give much less scintilations, Copper to Copper scarcely any.

Lead & Iron, give much finer than iron to iron= and the forked ends have more prongs= ~~it~~ In some cases the scintilations are not seen until an inch or so away there is apparently a break in em=

Phenomenon[c] if a peice of metallic silicum[3] be ~~b~~placed between two metallic or carbon poles & manipulated so as to get it red hot, A continuous[f] Electric light[c] with snaps is obtained. ~~it is~~ I ~~E~~ have the condensers as connected in first part of book[4] I have no doubte but a peice of silcum would serve to keep the Carbon points of a Common Electric Light going constantly= Lithium does same thing but only for a time until its volatilized which is not the case with silicum Boron gets very hot & keeps hot if small grain be used but does not act so well as silicum.

Of all metals Magnesium Ribbon gives best scintilations but they dont shoot out so far as iron[5]

Putting Parafine on metal & using copper wire to stick through it great snap & flash of ~~w~~fire, ¼ dramatic looks wicked & like an explosion of parafine in near vacinity of parafine=

Chloride Lime Salts Dry=[g] nothing Extra
Camphor acts like parafine probably explosion only effect of confinement of spark—
Starch Sul Copper n.g
Nitrate Potash n.g
Oil Sweet almond ng
Brom Pot ng
Salt pretty good.
Nitrate Ammonia Salt Electrolysied to brown powder

Oxalic A Ng
Coca butter acts similar to parafine

Putting in Rheostats with 200 or 400 ohms in wires leading to EMG kills spark=

X, NjWOE, Lab., Vol. 8:233–38 (*TAEM* 3:658–63). [a]Interlined above. [b]Document damaged. [c]Underlined twice. [d]"~~We a~~" interlined above. [e]Circled. [f]"Laboratory note" in top margin. [g]"Salts Dry=" interlined above.

1. This is the first entry in a tablet whose cover is dated 2–14 February 1877.

2. Jesse Bunnell invented and manufactured telegraph instruments. See *TAEB* 2:104 n. 8.

3. That is, silicium, the name given to metallic silicon by Humphry Davy. *OED*, s.v. "Silicium."

4. That is, the first drawing in this entry. This text is on the fourth page of the tablet.

5. Edison prepared a version of the results about the electric light and scintillations, together with a comment on disintegration of paper in a sodium manganate solution, as a draft for another series of published laboratory notes (NS-Undated-003, Lab. [*TAEM* 8:178–79]). Cf. Docs. 813 and 845, and NS-Undated-002, Lab. (*TAEM* 8:167–70); see also Doc. 988.

## –856–

*From Josiah Reiff*

N.Y. Feby 3/77

My dear Edison

To say I am delighted with the letter to J.G.[1] a copy of which you send me, would but illy express my appreciation[2] I can only wish it had been[a] dated many months ago.

You have done yourself credit & me a great service.

It will in my judgment bring things to a focus—

You evidently recd another letter from J.G.[3] probably he wrote it after receiving mine.

Regarding the stock, I must see you first. Suffice it to say, ~~t~~when the matter reaches a point where it will do any good to decide, you & I will be found ready to act together

AL (fragment?), NjWOE, DF (*TAEM* 14:644). [a]Interlined above.

1. Doc. 853.

2. The same day, Reiff sent Edison a telegram: "Reads like a novel unanswerably true. must have good effect." DF (*TAEM* 14:643).

3. Doc. 848.

## TRANSLATING EMBOSSER  Doc. 857

In September, Edison had conceived a recorder that would automatically repeat Morse telegraph messages.[1] At the end of October he had begun to consider using a machine that embossed the signal spirally on a piece of paper, even as he executed a patent application for an instrument that recorded by perforating paper tape.[2] These recorders were designed to record an outgoing message as the operator sent it, enabling the automatic, rapid retransmission of the same message later on other lines. This was particularly desirable for press copy, which consisted of long messages requiring the most skilled operators to transmit and receive. The incoming, high-speed message recorded by the embosser at each receiving station would then be transcribed at a slower speed by an operator using a standard sounder.[3] Although Edison's initial design for the spiral embosser used only a single revolving plate, he may have begun to consider a two-plate machine by early December, when he wrote George Prescott about the invention.[4]

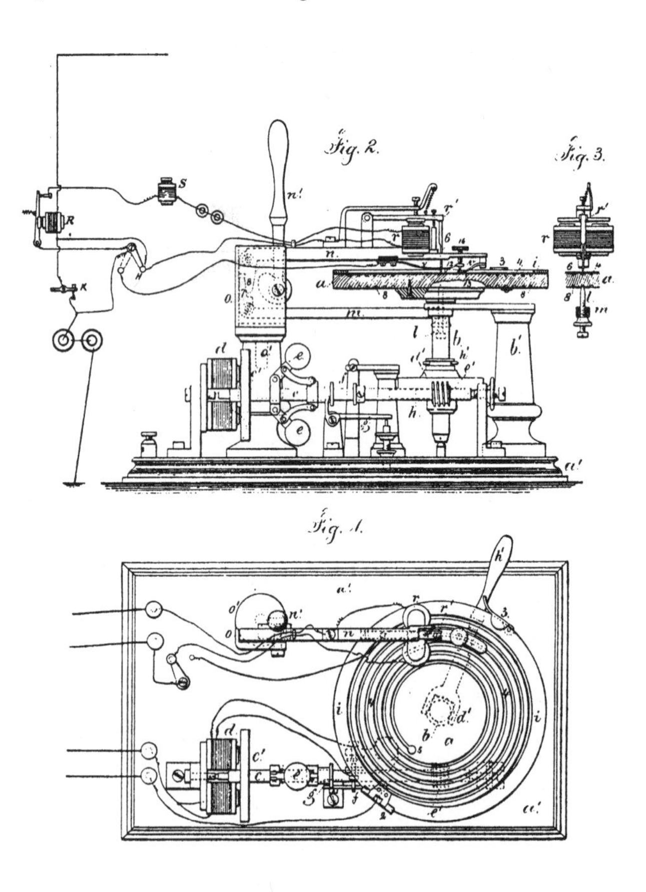


*Patent drawings for the translating embosser.*

Having two plates would allow an operator to switch the incoming message without interruption to the second piece of paper when the first was full. By January Edison was having such a "2 plate embosser" built, and by 8 February the laboratory staff made their first test of the new instrument.[5] The patent application Edison executed on 3 February included both single and double plate designs.[6]

Edison took the embosser to Western Union headquarters on 14 March, and on 30 March William Orton gave him permission to have Joseph Murray make some more.[7] On 2 April, Murray began to make six instruments, which he finished three months later.[8] During this time other refinements were made in the design.[9] On 2 July Edison and Charles Batchelor tested the instruments at Murray's shop in Newark; the following day they took them to the Western Union office where they were demonstrated for several Western Union officials, including Orton and Prescott. Also in attendance were British electricians William Preece and Henry Fischer. According to Batchelor this demonstration led them to find "considerable bugs amongst which the connections were wrong."[10] Although there is no record of further work on the translating embosser after 5 July, when Edison proposed some changes to the instrument and Batchelor spent time at the Western Union offices fixing the machines there,[11] the staff may well have worked on it periodically up to the end of August when Edison wrote William Orton, "which shall it be= more Telephone, or the Embosser!"[12]

1. Doc. 789.

2. Edison executed U.S. Patent 200,994 on 30 October 1876. See Docs. 806 and 809.

3. One function of these devices—recording a fast message for slower transcription—was the same as that of Edison's "first invention," his "practice instrument" (Doc. 10), invented in 1865 to help him increase his facility at receiving Morse transmissions. He had described that instrument in a 1 September 1874 *Operator* article, "A Novel Device" (p. 1).

4. Doc. 819.

5. Entries of 5, 6, 15, and 30 Jan. 1877, Cat. 1213:21–23, Accts. (*TAEM* 20:12–13); "Sample of Edisons new Embosser taken Feb. 8th 1877," Cat. 1240, item 70, Batchelor (*TAEM* 94:26). The embosser may be what "highly delighted" George Prescott when he was at the laboratory on 9 February (Cat. 1233:40, Batchelor [*TAEM* 90:73]).

6. U.S. Patent 213,554.

7. Cat. 1233:73, 89, Batchelor (*TAEM* 90:89, 97).

8. Cat. 1233:90, 92, Batchelor (*TAEM* 90:98, 99); Murray's bill, 2 May 1877, 77-017, DF (*TAEM* 14:665).

9. Cat. 1307:62–63, Batchelor (*TAEM* 90:647–48); Vol. 11:194, Lab. (*TAEM* 3:1004).

10. Cat. 1233:183–84, Batchelor (*TAEM* 90:144–45). On 3 April Batchelor had drawn the electrical connections for the six embossers being made at Murray's and, probably as a result of this demonstration, later indicated that these were "Wrong" (Cat. 1307:60, Batchelor [*TAEM* 90:645]).

*Charles Batchelor's initial (incorrect) wiring diagram for the six embossing recorder/repeaters made by Joseph Murray.*

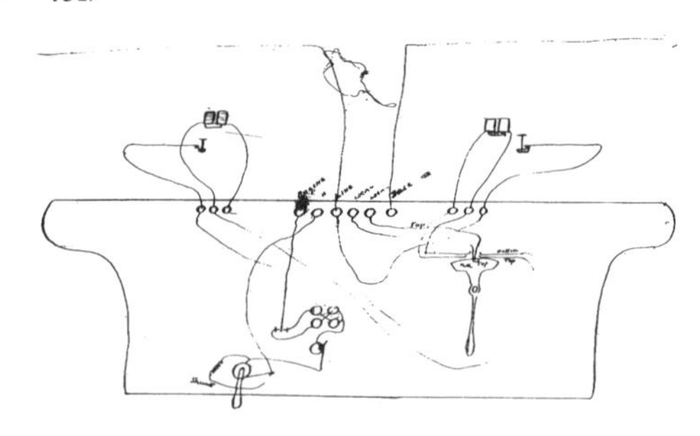

11. Doc. 955; Cat. 1233:186, Batchelor (*TAEM* 90:146).

12. Doc. 1030.

## –857–

[Menlo Park, February 3, 1877?][2]

*Production Model: Telegraph Recorder/ Repeater*[1]

M (60 cm × 22 cm × 20 cm), NjWOE, Cat. 100.

1. See headnote above.
2. Date of patent execution.

**-858-** [Menlo Park,] Feby 5, 1877

*Notebook Entry: Chemistry*

Capilliary Experiment with Alcohol & water in different proportions, small bottle & auto paper scale[1]

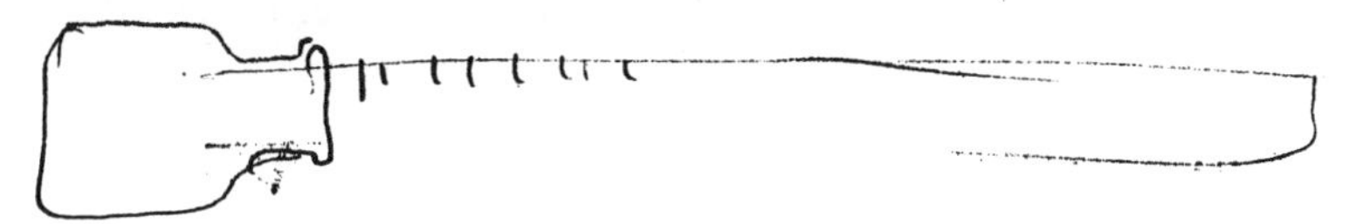

All placed at Zero thus allowing 3/8 inch in Solution= after 2 hours[a]

| | | |
|---|---|---|
| No 1 | 8½ | most alcohol |
| No 2 | 10½ | less " |
| No 3 | 11 | less |
| No 4 | 9 | less |
| No 5 | 9½ | less |

Unreliable[a]

With Capilliary tube Composed of one prong of Geo Little glass Capilliary U Tube=[2] Alcohol runs up a certain distance while HO a much greater distance & the distance is in proportion in which Al & HO are mixed I should think this would be a good thing for measuring percentages of Al & other liquids in HO

I find that with sesquichl Fe Sul Zinc—~~per~~[b] mangnate K, in Capilliary tube the Column is not ~~e~~affected by magnetism at all, possibly some liquids might but doubtful=[b]

I have an idea[c] for Crystalization that might be of some value for some purpose thus

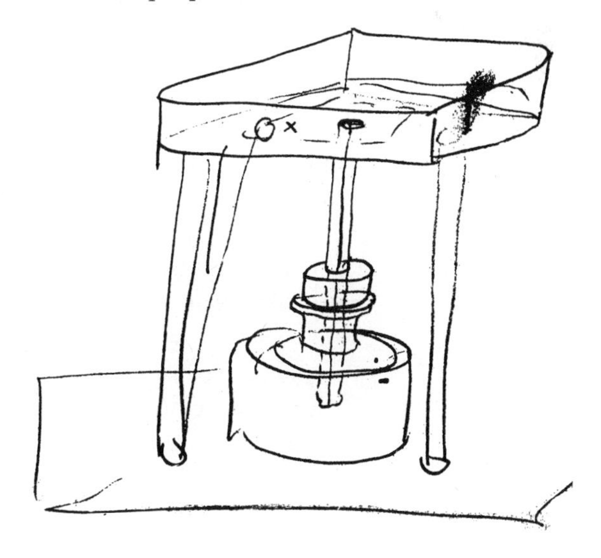

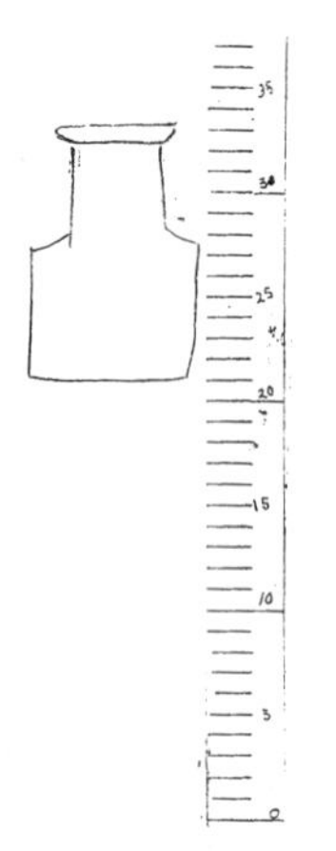
A ruled paper strip and a scale drawing of the ink bottle used in the capillary experiments.

Capilliary bore of a tube runs liquid from bottle up into flat dish onto a peice ~~p~~Biblus paper & there being rapid evaporation ~~er~~efficient crystalation takes place until dish is full  I see this takes place in several solutions with paper strips.

X, NjWOE, Lab., Vol. 8:114–15 (*TAEM* 3:547–48). [a]Written to right of list. [b]Followed by centered horizontal line. [c]Obscured overwritten letters.

1. On 14 January the staff had begun a series of experiments on "the relative capillary force" of a number of oils. They filled small ink bottles with the individual oils and placed a paper scale upright in each bottle, recording the rise of oil in the paper over the next two days and then again on 3 February. Edison observed that several oils dissolved paper and noted that the capillary action of acid in paper might be used in the manufacture of nitrocellulose. Following the experiments in this entry, Edison and Adams tried to find a substance that would increase the cellulose-dissolving property of oil of thyme. Vol. 8:108–13, 116–19, Lab. (*TAEM* 3:540–46, 549–52).

2. What constitutes a "Geo[rge] Little" capillary tube is unknown.

## –859–

*From Amasa Mason*

New York, Feby 8th 1877

Dear Sir

I have Sent you the copy book & ink as suggested which please use in your own way—[1]

I have written Mr Thomson who is now making the paper Barrels, asking to have some packages sent you and if in any way a compound can be made of small cost—by the use of which the packages will be made liquid proof it will be quite valuable—[2]

In the matter of the "Electric pantagraph"[3] a sample of work from which I showed you— You might quietly look in at 1½ Park place—where the Ingersoll Drill[4] office is—and see it at work. I am quite pleased with it—and somewhat disposed to buy it—but I should highly estimate your judgment in the matter— Should you go in, it would perhaps be as well not to use my name— One point I particularly want to know—And that is—if any liquid can be made, which being used as an ink—or to produce drawings and forms upon paper—would cause the magnet to act as the pantagraph pointer as the pointer passed over the tracing— I mean some material which could be made in a liquid form and which when put upon papers would break the circuit— Now coppertypes are used—or any metal which will act—but this involves the necessity of actually producing the form, from which the transfer engraving is made

I fear I am a little muddy about this—but your quick intelligence will doubtless comprehend it— I shall be out of Town until next week Monday— After that when you are in the city—I shall be glad to know it, and see you. Yours truly

Amasa Mason[5]

ALS, NjWOE, DF (*TAEM* 14:45). Letterhead of Seamless and Dove-Tailed Paper Barrels.

1. Mason and James James had been at the laboratory on 1 February. Cat. 1233:33, Batchelor (*TAEM* 90:69).

2. See Doc. 915.

3. A pantograph is an instrument for duplicating handwriting or drawings. No information could be found concerning this electric pantograph. Edison himself had considered attaching his electric pen to a pantograph (Doc. 837).

4. The Ingersoll mining rock drill had been recently introduced. For a description and drawings of the drill see André 1877, 33, plate XXX.

5. See Doc. 741 n. 2.

**-860-**

*Technical Note: Telephony*

[Menlo Park,] Feby 9 1877

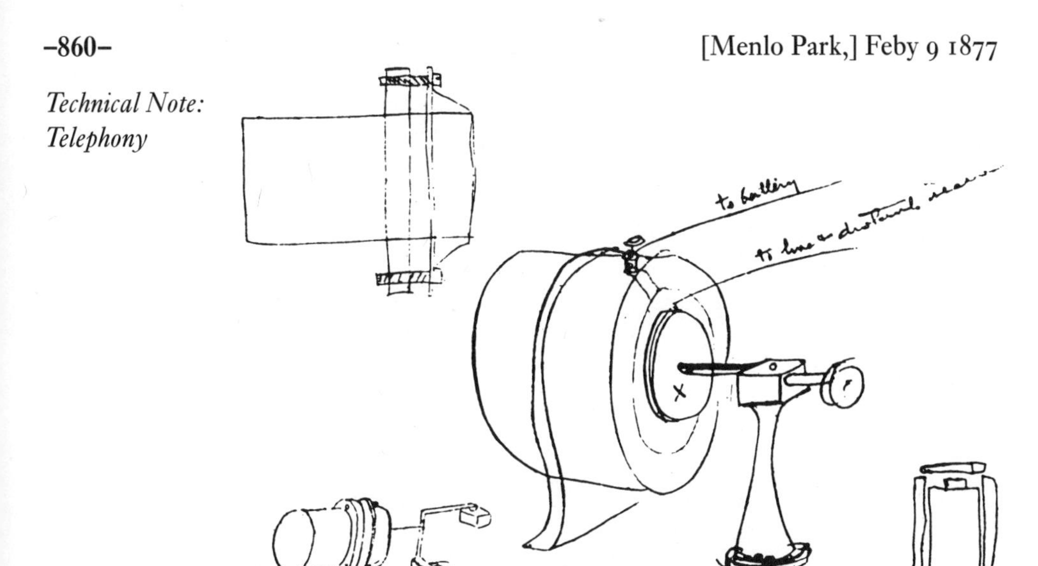


X[1] is a ~~p~~ Disk of hard rubber coated over with plumbago=[2] on the parchment[3] is a peice of tin foil facing the disk  on speaking this foil connects with the plumbago & closes ckt  if low notes a greater surface connects & [----][a] resistance is thereby reduced  if high notes but little touches[b] & current weak

The edg[e of][c] the disk is blackleaded & wire wound around it

T A Edison | Chas Batchelor[4]
Adams

X (photographic transcript), NjWOE, TI 2, Edison's Exhibit 8-11 (*TAEM* 11:226). [a]Canceled. [b]Illegible. [c]Incomplete transcript.

1. Figure labels are "X," "to battery," and "to line & distant receiver."

2. Edison had used rubber coated with carbon in his transmitter four months earlier (see Doc. 799). He identified this drawing in testimony as "the original sketch from which the instrument shown in my application 130 was made" (TI 1:48 [*TAEM* 11:45]). Case 130 eventually became U.S. Patent 474,230. Both Edison and Batchelor testified that the instrument worked; Edison remembered its transmitting "articulate speech with such a degree of success that it was capable for business purposes" (TI 1:130, 231 [*TAEM* 11:86, 92]). At the top of the page is a small table listing the hours spent (presumably) on this instrument by Kruesi and Batchelor on 11 and 12 February. Exhibit 28-11 (TI 2 [*TAEM* 11:242]) shows a similar instrument, finished on 21 March 1877, in which the current flows from one side of the carbon-coated disk to the other.

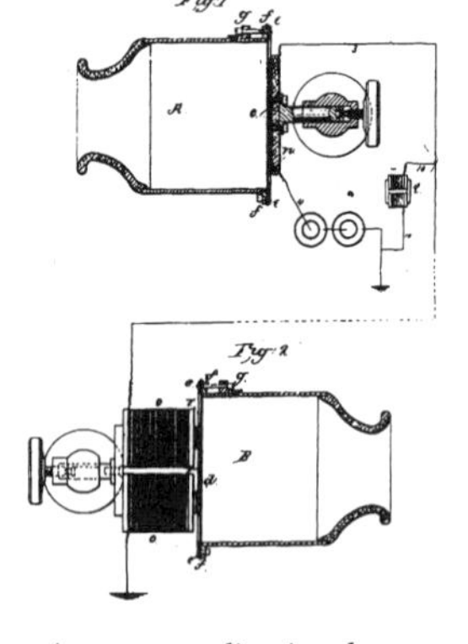

*A patent application drawing of a telephone transmitter (top) in which the diaphragm pressed against a hard rubber disk coated with plumbago.*

The sketch at upper left illustrates an arrangement Edison used to adjust the tension of the parchment diaphragm, which is also discussed in Case 130 (TI 1:130; 2:3–4 [*TAEM* 11:86, 185–86]). The sketch at lower left appears to show a sliding contact attached directly to the diaphragm. The meaning of the sketch at right is unclear.

3. Batchelor had bought parchment and rubber in New York for the telephone experiments two days earlier. Cat. 1233:38, Batchelor (*TAEM* 90:72).

4. This marks the beginning of Batchelor's work on the telephone. Cat. 1233:40, Batchelor (*TAEM* 90:73).

**-861-**

*From George Bliss*

New York.[1] February 10-th, 1877.

Dear Sir:—

We send you a check to-day for the goods turned over in New York.[2]

Ed Gilliland[3] writes me, urging the payment of all of the indebtedness to you.

He also states that a letter has been written to his father,[4] requesting that such an order be sent us.

Thus far the old gentleman has written us nothing, except that we are authorized to make one more payment. He talks about going to New York with me on my next trip. I expect to start about the middle of next week.

We shall take the responsibility of paying you soon the $650 bill, as that will still leave enough due to protect Gilliland.

I have, however, written him, urging that some final decision be made without delay.

The papyrograph[5] people are giving us a lively fight all over the country, and I have been compelled to take the pens turned out thus far to fill orders, as a great deal depends on our filling orders promptly.

The Western Electric are rapidly increasing their capacity, which I hope for months to come, beginning with March 1-st, will be at least two hundred pens per month.

This ought to enable us to fill all our orders and get a sufficient stock on hand to do a fair trade.

If you will only give me a press suited for rapid work, and for which I can charge a good round price, it will be a good many dollars in your pocket and help business immensely. Please don't let up on it until you arrive at perfection.

The general[6] is going to New York next week also; so we shall have a good chance at him there, and I hope the foreign pen business will be put on a basis satisfactory to you.[7] Respectfully,

Geo. H. Bliss Gen. Agent.[8]

TLS, NjWOE, DF (*TAEM* 14:369). Letterhead of Western Electric Manufacturing Co. Typed in upper case.

1. Bliss had been in New York in mid-January and stayed through February, when he and Batchelor left for Chicago with Edison, who was going to Port Huron. Cat. 1233:15, 17, 53, 57, 59, Batchelor (*TAEM* 90:60, 61, 79, 81, 82).

2. This refers to the transfer of machinery in accordance with terms of the contract with Western Electric (Doc. 817).

3. Ezra Gilliland was known as Ed.

4. Robert Gilliland.

5. On the papyrograph see *TAEB* 2:483 n. 1.

6. Anson Stager.

7. On 17 January, Edison and Batchelor had discussed with Bliss their "plan of forming a company of the Foreign Interest of Electric pen, . . . and he thought very favorably of it." On 26 February, Bliss and Stager came to Menlo Park "where Edison shewed [Stager] the ropes" and they discussed "the royalty on the Foreign Electric pen [and] concluded that three dollars was enough and about right." Cat. 1233:17, 57, Batchelor (*TAEM* 90:61, 81).

8. George H. Bliss (1840–1900) entered telegraphy in his teens as an operator with the Illinois and Mississippi Telegraph Co. He was in charge of the company's Chicago office in 1860 when he left to become an operator with the Chicago and Northwestern Railroad, which appointed him superintendent of its telegraph lines in 1865. In 1868 he formed the telegraph manufacturing firm Bliss, Tillotson & Co. in Chicago, which in 1873 became George H. Bliss & Co. That year Bliss also resigned his position as superintendent of telegraph lines for the Chi-

cago and Northwestern Railroad. In 1875 Bliss sold out to Western Electric Manufacturing Co. and became the company's general agent. Reid 1879, 234; "Bliss, Tillotson & Co.," *J. Teleg.* 1 (1868): 4; "Telegraph Manufacturers," *Telegr. Supplement,* 21 Aug. 1869, 1; *J. Teleg.* 6 (1873): 137; advertisement, *Telegr.* 11 (1875): 126.

**–862–**

*Charles Batchelor to Robert Gilliland*

Menlo Park, N.J. Feb 11th 1877

Dear Sir,

Yours of the 8th received.[1] If you could possibly let me have two or three hundred on that note it would oblige me and help me out during the time you mention. It is now two months since we turned over the goods to the Western Electric Co and Mr Bliss assured Edison & myself that the whole of the money would be paid before the end of January, on this I of course made promises to our creditors, some of whom now talk pretty rough. Later we received a letter from Mr Bliss to the effect that the money would have been paid but they had not got your signature to a certain paper which they sent to Edison to sign and then to you regarding the payment of the money to me. Edison now has a letter from Bliss dated Feb 6th[2] in which he says you are disposed to have another small payment made and that you want the balance to rest a few days. Why cannot this money be paid at once so that I can get rid of this disagreeable task of putting people off that come after their money? Mr Bliss then said that they should remit for the stock turned over in New York which is a very small item but as yet we have seen nothing of it.

In regard to the $2165^{74}/_{100}$ Cr to Gilliland & Co[3] How could I put it in any other way? I had overpaid them that amount and I never had anything turned over to the company to compensate for it. Yours respectfully

Chas Batchelor

ALS (letterpress copy), NjWOE, Batchelor, Cat. 1238:77 (*TAEM* 93:85).

1. Batchelor noted in his diary that he received a "Letter from Robert Gilliland saying he could not pay note for 60 or 90 days." The note was for 12 months. On 16 February, Batchelor noted that he had "Received note of Robert Gilliland back unpaid," and five days later that he had "Recd check for $300— from R Gilliland to be credited on note of $1200—." Cat. 1233:31, 41, 47, 52, Batchelor (*TAEM* 90:68, 73, 76, 79).

2. Not found.

3. There are no accounting records regarding this overpayment.

**–863–**

*Notebook Entry: Telephony*

[Menlo Park,] Feb 12th 1877.

Speaking Telegraph.

Edison thought that the speaking telegraph of Bell was very imperfect, seeing that it could only be used on very short lines, and he maintained that if we could by any means get the resistance of the circuit increased and decreased by the raising or lowering of your voice, it could be used on long lines.

I therefore made two instruments; a transmitter and receiver. The principle of the transmitter was to make the vibration of a membrane work a roller along a lead pencil mark or other high resistance, altering the resistance of the circuit every time the membrane is excited. This was done in the following manner:—as shewn in Fig 37[1]

Fig 37

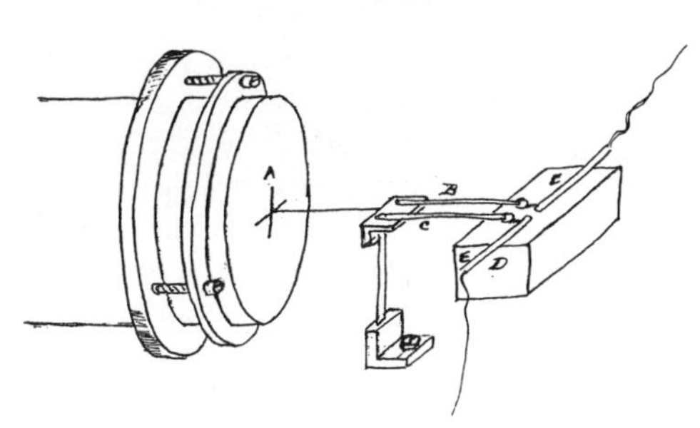


A is the membrane, B & C are fine springs with Platinum springs or rollers on their end  E & F are two pieces of metal used as connections  The lead pencil was rubbed on stone D between the rollers and the connecting pieces E & F  This however did not give us the desired result as all we could get was a mumbling sound.

Another plan we tried was a band across the diaphragm with projecting pins which operated on the springs B Fig. 38.

38

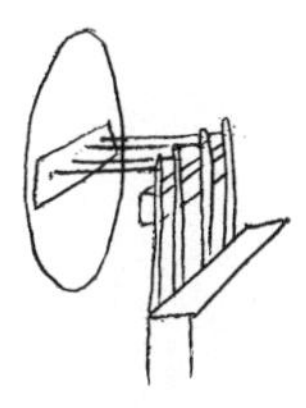

The idea being to get the articulation by cutting out or putting in resistance.[2] Our receiver for these instruments was merely a stretched diaphragm with an armature on the centre and a magnet adjustable to it as in fig 39.

39

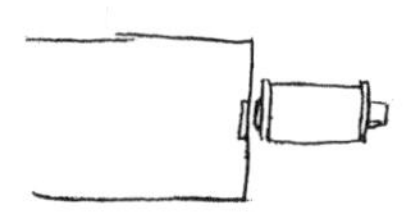

Our next transmitter was a disc of black lead in front (and adjustable) of the diaphragm.[3] This has been so far good. We have used a hard rubber diaphragm covered with blacklead. As the diaphragm vibrates it touches in more places and reduces the resistance. With this apparatus we have already been able to distinguish clearly (known) sentences well between New York and Menlo Park.[4]

X, NjWOE, Batchelor, Cat. 1317:33 (*TAEM* 90:673). Written by Charles Batchelor.

1. Charles Batchelor later testified that this design was a "method of increasing and decreasing the resistance of a current in a closed circuit by the movement of a diaghragm moving two springs, with rollers on their ends, on a film of plumbago or other conducting material, which film was part of the circuit, the movement of the diaphragm increasing the length of the film in circuit." In testing the transmitter the staff used a receiver consisting of "an electro magnet in front of an iron diaphragm placed in front of a resonant chamber, with an iron diaphragm on its end." Batchelor also commented that he had made the instrument himself and that it transmitted the human voice, but that he thought work on this design had begun previous to 6 February. A drawing of that date shows such a transmitter in a circuit. However, in a diary entry of 12 February, Batchelor drew a picture of this transmitter (and the receiver in figure 39) and said that he "Stayed [in Menlo Park] and worked all day on a new talking telegraph." The following day, he wrote "Worked all day on Talking Telegraph . . . not much good as yet very little encouragement from instruments." Batchelor's testimony, TI

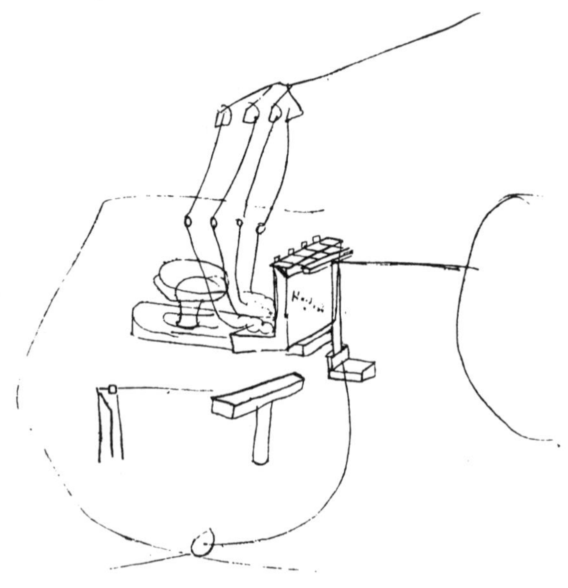

*A telephone transmitter design that uses a four-pronged contact instead of four individual pins to cut resistances in and out of the circuit.*

2:230–31; Edison's Exhibit 6-11, TI 2 (*TAEM* 11:92, 225); Cat. 1233:43–44, Batchelor (*TAEM* 90:74–75).

2. Two pages of drawings of 11 February (Exhibits 9-11 and 10-11), as well as an instrument introduced into the Telephone Interferences, are related to the work described here. They are similar to Batchelor's Figure 38 but have a four-pronged contact attached to the diaphragm rather than four individual pins. According to Batchelor's testimony, "They illustrate the principle of cutting in and out resistance or batteries by means of a diaphragm or springs, in order to accomplish the rise and fall of tension of current in a circuit which always remains closed." Each spring is attached to a different resistance; for example, the ones in Exhibit 10-11 were labeled "1500," "1000," "800," and "500." TI 1:231 (*TAEM* 11:92); Edison's Exhibits 9-11–10-11, Exhibit Instrument 10-11, all TI 2 (*TAEM* 11:227–28, 647).

3. This is the first time that either Edison or one of his associates mentioned using a disk of solid carbon against the diaphragm in order to vary the resistance. In previous experiments, the carbon was layered on a surface such as a rubber disk or cylinder (see Docs. 799 and 860) or in an unpacked, unshaped form in a container (see Doc. 844 and Edison's testimony, TI 1:46 [*TAEM* 11:44]).

4. Notes from about this time made during telephone experiments (probably between Menlo Park and New York) are in Vol. 8:230–32; also see Vol. 8:123–24; both Lab. (*TAEM* 3:652–56, 555–56).

**-864-**

*Notebook Entry: Introduction to Etheric Force Experiments*

[Menlo Park, February 14, 1877?][1]

Etheric Force[2]

With 6 sheets of suspend foil on glass rods each foil ~~having~~ being 12 × 8 and the total distan 4 feet get spark quite often in Etheriscope with 10 ohm spool connected in place of ground wire by grounding A makes no difference in getting spark at Etheriscope if connected thus.[3]

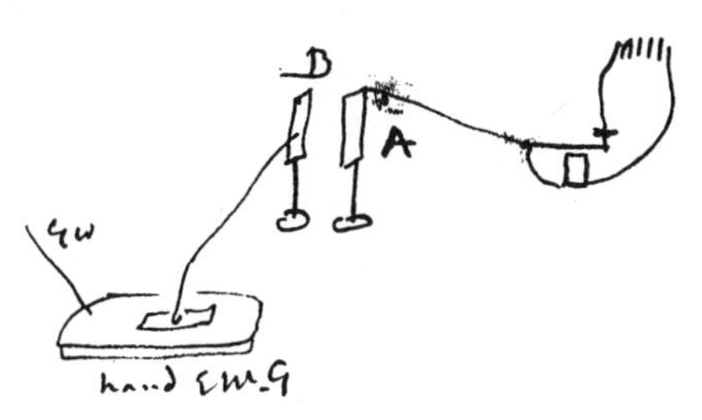

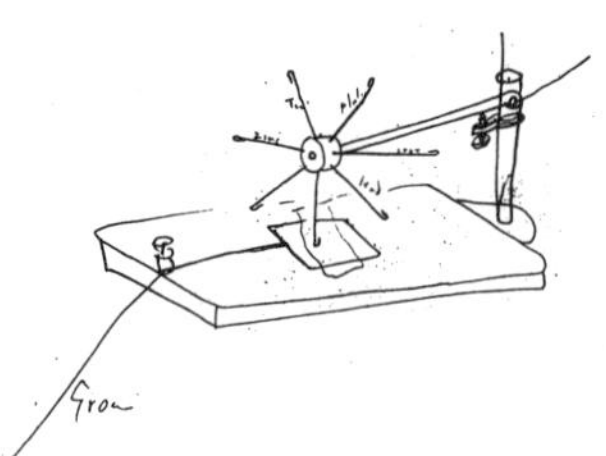

*The "Machine for Etherolysis" Edison and Adams used in etheric force experiments.*

With Iodide Pot 20 grm to 100 cc get no[a] spots shewing decomposition ~~probably from Escape of Electricity~~ but appears to be feint tinge though this may be due to pressure of platina Loop—but when platina connected direct to .A. get good spots in 2 minutes. the spark on iron with platina shews scintillations ¼ to ½ inch long=. When platina to .B sparks is bright with no scintillations   I believe connecting to B filters the E from the true Etheric

I now propose to submit all the HO solutions to the decomposing [power] of the Etheric by allowing platina point connected both to A & then to .B to rest on strip of paper for several minutes.[4]

X, NjWOE, Lab., Vol. 8:239 (*TAEM* 3:664). [a]Interlined above.

1. This notebook entry follows Doc. 855 in a notebook whose cover is dated 2–14 February 1877. It appears to be related to notes on etheric force experiments made by James Adams on 14 and 15 February 1877 (Vol. 8: 174–82, Lab. [*TAEM* 3:607–15]).

2. This setup is similar to one Edison used in a series of etheric force experiments begun on 11 January 1877 (Doc. 840). Figure labels are "magnet" (at far lower left), "G[round] W[ire]," "GW," "A," and "10   1½ carbon."

3. Figure labels are "GW," "hand EMG," "B," and "A."

4. In these experiments Edison was attempting to test the decomposing power of the etheric spark and used a microscope in order to see faint discolorations (Vol. 8:240–55, Lab. [*TAEM* 3:665–80]). After conducting some experiments on ferrocyanide of potassium Edison noted that "Jim [Adams] tries the passsage of current from 2 cells but gets nothing on one side & a blue coloration on other= evidently the Etheric decompostion is not the same as the Electrical from 2 cells at least" (Vol. 8:240, Lab. [*TAEM* 3:665]). The series of tests in which Adams tried different "kinds of electricity," and which he recorded in another notebook, were apparently intended to test this hypothesis (Vol. 8:174–76, Lab. [*TAEM* 3:607–9]). Probably on 15 February, Edison and Adams continued their etheric force experiments by using a "Machine for Etherolysis" to conduct a series of experiments which Adams labeled "trying to decompose different chemicals with different metals by Etheric force" in his notebook entry of that date (Vol. 8:246, 176–82, Lab. [*TAEM* 3:671, 609–15]; see also Vol. 12:43, Lab. [*TAEM* 4:24]).

Among Edison's other etheric experiments was one in which he used a quadruplex reversing sounder to reverse the etheric currents. After making this experiment he noted that it was a "Very important experiment proving that Etheric Currents are not reversed." Vol. 8:244, Lab. (*TAEM* 3:669).

**–865–**

*From Edward Johnson*

NYork 2/15/77.

My Dear Edison,

I think Suitterlin[1] is about to withdraw. He has not been at the ofs for any practical purpose for a week He has said nothing to me nor I to him but the above is my conviction—& I am glad of it. As with only the 2 articles[2] we have on hand & the strong probability that we shall have no more for some time, the thing wont overwork me—& I'll save his com.

This is the result I suppose of my having no money I presume when he found such was the case he concluded it would not be safe for him to invest much & hence his withdrawal—However I'm glad of it for more reasons than one, & hasten to inform you. Pending the receipt of some funds in the treasury I am compelled to simply make use of such channels as come to hand I can't go out for them. That is I cant ad. for Agents Canvassers &c The 2 or 3 that I have[3] out give indications of doing well & I should like to put an ad in the Herald for a week for Travelling Agents to sell on Com. & for City Canvassers. I find these 2 classes are the only ones who accomplish anything—

Another thing which operates bad for me is this— All good sold to the trade are on 30 Days so orders for the time being simply nil me of my stock & bring me no returns.

Nothing from Detroit yet. I have withdrawn the ad. from the stationer & from Harpers—& shall put no more in until we change our name which I strongly recommend doing at once— I have proposed what I think a good one & Reiff says "tip top"— Viz.

The
Electro Chemical and Mechanical Novelty Company[4]

Thomas A. Edison — E. H. Johnson
Prest. — Secy & Treas & Gen

52 Bdwy NYork /77[a]

How does it look?

Theres no question but such a name will give us prestige while our present one only Excites suspicion.

Come in Yours

E H Johnson

ALS, NjWOE, DF (*TAEM* 14:37). Letterhead of American Novelty Co. [a]Followed by flourished horizontal line.

1. According to the letterhead on this letter, J. E. Suitterlin was the manager of the American Novelty Co.'s stationery department. Bat-

chelor had noted the previous day, "Suit[t]erlin about to give up the Stat. Dept of Amn Nov Co." Cat. 1233:45, Batchelor (*TAEM* 90:75).

2. That is, Edison's duplicating ink and Johnson's ribbon mucilage.

3. See Johnson to T. A. Bailey, 27 Jan. 1877, and Johnson to TAE, 31 Jan. 1877, DF (*TAEM* 14:18, 21).

4. The Electro Chemical Manufacturing Co. was listed in the next edition of the New York City directory with Edison as president, Johnson as secretary, and a capital stock of $50,000. Wilson 1877 (City Register), 31.

**–866–** [Menlo Park,] Feby 1~~9~~7 1877

*Technical Note: Telephony*

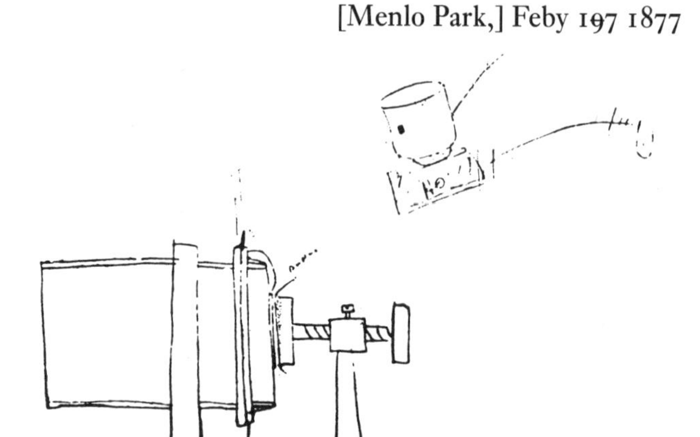

in this case[1] the disk is of metal & on it is a piece of moistened paper moistened with a chemical solution suitable for rapid polarization & deliquescent the tin foil, (or the diaphram may be of brass aluminum etc) faces the paper which is connected to the line on contact ckt is closed.[2]

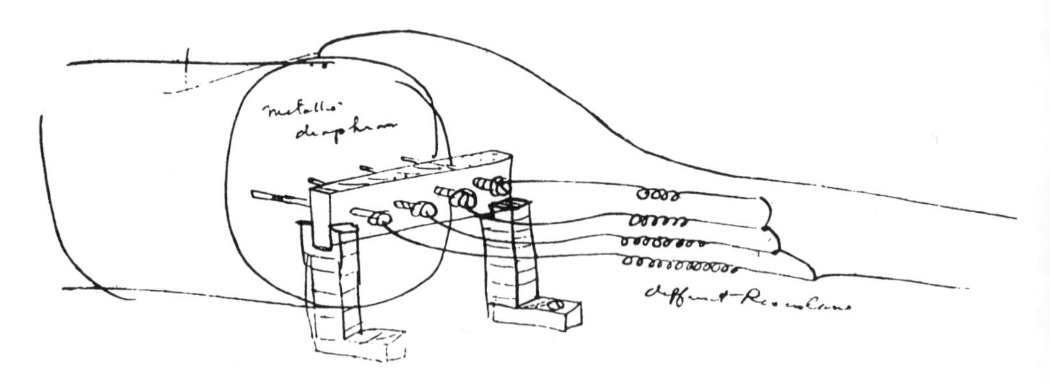

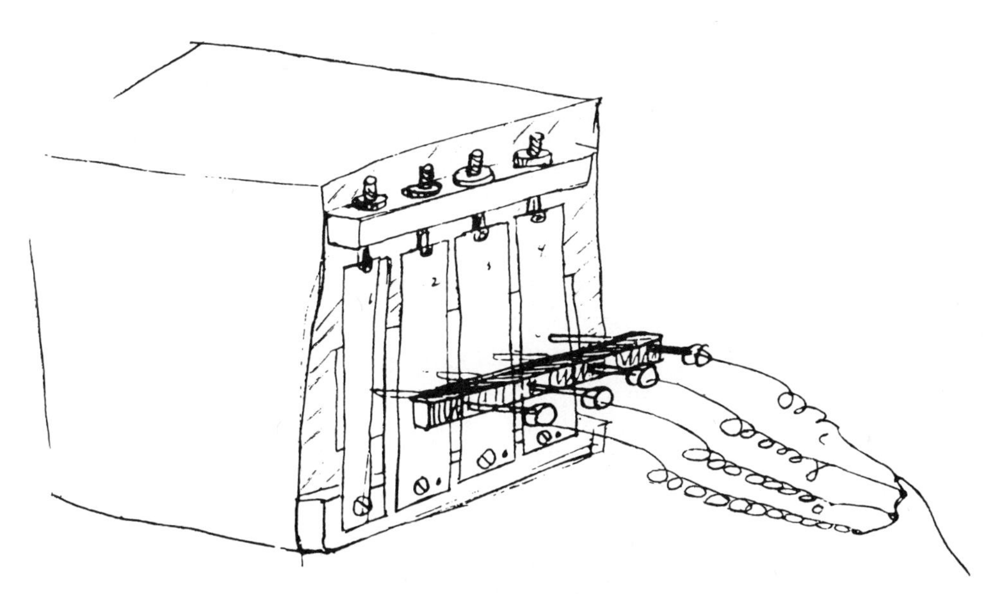

1 2 3 & 4 are strips of metal of membranes tuned by the screws to different rates of vibration[3] the centres of these springs are in front of a hole in the box on speaking different reeds act on the contact points & give [---]f[a] vibrations & strengths of cur[ren]t[a] by and of the resistances C, C[4]

TAE Chas Batchelor
Adams

X (photographic transcript), NjWOE, TI 2, Edison's Exhibits 15-11, 14-11 (*TAEM* 11:230, 229). Document multiply signed and dated. [a]Illegible.

1. Figure label in the large drawing is "paper." Edison tesified that this design used wet paper in place of the plumbago used in Doc. 860 to vary the circuit resistance (TI 1:48 [*TAEM* 11:45]). He mentions this idea again in Doc. 973.

The label of the small drawing is "HO" (meaning "water"); this drawing appears to be a variant of a water telephone.

2. Figure labels for the following sketch are "metallic diaphram" and "different Resistances." As it vibrates, the metal diaphragm makes contact with one or several of the contact points positioned behind it. Each contact point has a different-valued, separate resistance connected to it. As a result vibrations of varying intensity or in varying areas of the diaphragm change the circuit resistance. Greater contact (corresponding to higher volume) lowers the line resistance, increasing the signal strength. For an earlier experiment in which resistances were switched in and out of the circuit see figure 38 in Doc. 863; for a later example see Edison's Exhibit 41-11 (TI 2 [*TAEM* 11:251]), which was a patent model sketch for an application executed on 16 July 1877 (U.S. Pat. 208,299). On 8 December 1877 Edison executed another patent application that covered the switching in and out of resistances to transmit speech (U.S. Pat. 203,013).

3. Edison had used various configurations of tuned bars, reeds, and resonators in his acoustic telegraph research for some time (see, for example, Doc. 708). In this case he tuned the reeds with tension.

4. Edison introduced this sketch during his testimony in order to illustrate his efforts to vary resistance of a telephone circuit by having multiple resistors and contacts (there was no testimony regarding the other, similar sketch in this document). Edison also introduced the instrument based on this sketch, which had eight reeds instead of four (Edison's Exhibit Instrument 14-11, TI 2:521 [*TAEM* 11:649]), describing it this way:

> It consists of a number of thin strips of metal, their tension being adjustable, the whole forming a sort of a diaphragm placed in front of a box in such a manner that speaking into the box would set the strips in vibration.
>
> Platina points upon the end of contact screws formed one electrode, and the strips the other electrode.
>
> A contact screw was adjusted in front of the strips; each strip was provided with a contact screw. All of the screws were connected through resistance with the circuit. This was an attempt to vary the resistance of a circuit by a multiplicity of contacts. [TI 1:98 (*TAEM* 11:70)]

On 20 February, Batchelor wrote that he "Worked all night on Talking telegraph made the 8 reed instrument" (Cat. 1233:51, Batchelor [*TAEM* 90:78]).

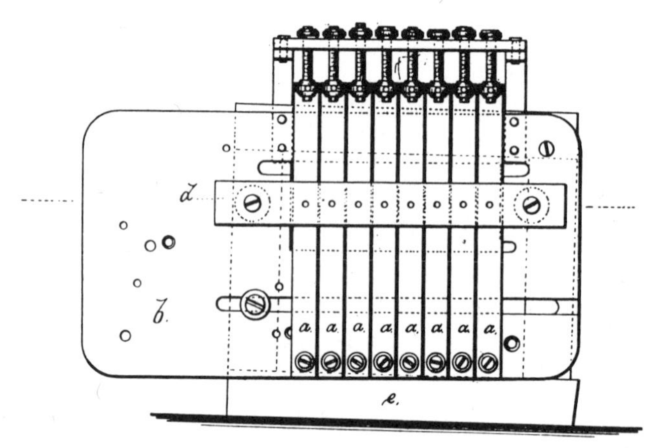


*Charles Batchelor made this eight-reed instrument, which used multiple contacts to vary the circuit resistance.*

## –867–

*To Jay Gould*[1]

Menlo Feby 19 1877

J. Gould Esq

it has occurred to me that it would be a good policy on the part of the A&P. to sell the Quad to the WU for $37 000. They would give it because the royalties amount to more than that sum & it has become of vital importance to them whereas it appears to only have a negative value to A&P can a trade be made Yours

TAE

ALS (copy?), NjWOE, DF (*TAEM* 14:755).

1. It is not known whether Edison sent this message to Gould.

## -868-

*Technical Note: Telephony*

[Menlo Park,] Feby 19 1877

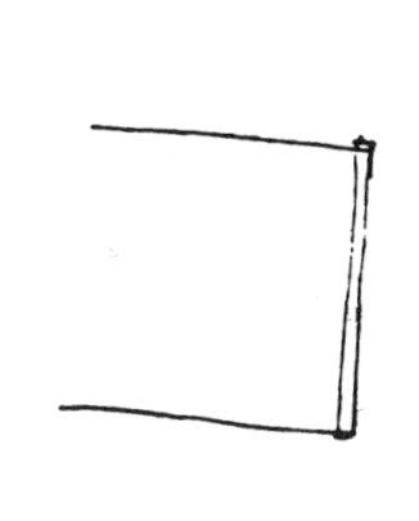

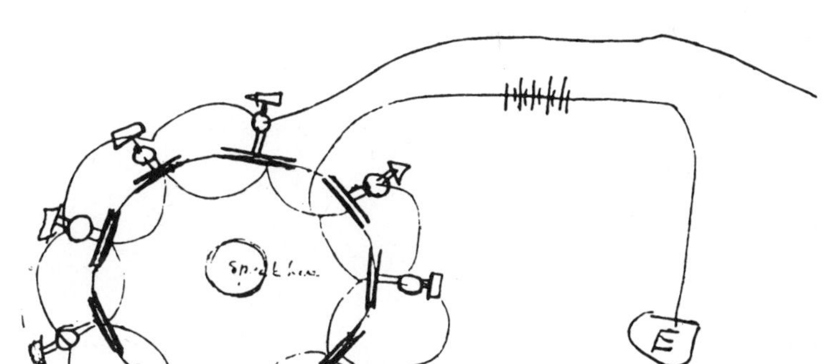

A condenser which charges & discharges by the voice ie[a] by bringing the metallized diaphragms together ie[a] close a battery charge rushes in & when the go[b] way it discharges these currents acting in electromagnet on the receiver set its diaphragm vibrating in unison[1]

T A Edison

Chas Batchelor
Adams

X (photographic transcript), NjWOE, TI 2, Edison's Exhibit 17-11 (*TAEM* 11:232). [a]Circled. [b]Obscured overwritten letter.

1. Each of the nine condensers in this instrument (viewed from the top) has one plate mounted as a diaphragm in the wall of a circular chamber; the other plate is mounted outside the chamber on a screw that adjusts the gap between the plates. The inner plates are all connected to one terminal of a grounded battery; the outer plates are connected to the line. Speaking into the chamber through the hole marked "Speak here" vibrates the inner condenser plates, varying the gap between them and the outer plates. This motion changes the condensers' static capacity and thus the electrical condition of the line. These changes in the line excite a telephone receiver at a distant station. This is, in fact, essentially the way a present-day condenser microphone works, although the equipment available to Edison was not sensitive enough for the purpose.

Edison or one of his lab associates sketched another, possibly earlier version of a condenser telephone. This undated sketch, submitted during the telephone interferences, may date from about 12 February, judging from other drawings on the page that are related to Doc. 863. The geometry and circuitry of this undated condenser telephone design are much simpler, but it would function similarly by varying the distance between the plates of a charged condenser to alter the line current to a receiver. Edison's Exhibit 80-15, TI 2 (*TAEM* 11:618).

Prescott 1879 (546–47) notes that Edison produced another condenser telephone on 10 December 1877. This version apparently used separated condenser plates stacked in a line directly behind the transmitter diaphragm. No documents related to this design have been found.

In 1875 Edison had filed two caveats in which he used a circular arrangement of condensers in series to store signals for the automatic and high speed telegraph systems. See Docs. 540 and 557.

**–869–**

*From Uriah Painter*

WASHINGTON, D.C. FEB 20 1877[a]

My Dr E.

I never said you never lost anything in Auto! Who says I did? Hurry up the letter from Port Huron so I can get the action of present Secy of War![1] I see A&P is putting on its War paint for WU![2] Sorry you are not at its head as you ought to be & would have been if JCR had taken my advice! Yours

U H Painter

ALS, NjWOE, DF (*TAEM* 14:49). [a]Place and date from *Philadelphia Inquirer* hand stamp.

1. This is in reference to the right-of-way through the Fort Gratiot military reservation. See Doc. 835.

2. Painter is probably referring to decisions by the Atlantic and Pacific Telegraph Co. to extend its lines and to further reduce its rates. "Important Action of the A. and P. Directors," *Operator,* 1 Mar. 1877, 6.

**–870–**

*Charles Batchelor to James Batchelor*

Menlo Park Feb 27 1877

Dear Jim

Yours enclosing cartes of the twins to hand. Please accept our thanks I enclose in this one of our baby. Emma has got quite a big girl over three years old now and this one is no chicken[1] they are both in very good health and very fat; we are thankful to say the Doctor makes very little out of us

The Pen business is assuming large proportions here far exceeding our expectations. We have given up manufacturing them ourselves, that and the selling in this country and Canada being done by the Western Electric Mfg Co of Chicago. Edison and I still hold control of all foreign countries and we are now making an arrangement with some prominent men here to take the foreign interest and work it up properly paying us a royalty on every press sold and a percentage on all the extra parts, we do this because the French and Austrian patent laws oblige us to manufacture in those countries inside of 12 months after date of patent issue. Their intention is to make their European headquarters at Paris with branch offices in all the other large cities The South American trade will be supplied from New York and the China Japan and Australian trade from San Francisco[2] I leave here for Chicago tonight on this business stopping at Buffalo & Detroit.

We have lots of new things coming out and if we can get this arranged all right I shall be free to go into them Yours

Charley

ALS (letterpress copy), NjWOE, Batchelor, Cat. 1238:87 (*TAEM* 92:93).

1. Batchelor and his wife, Rosanna, had two daughters, Emma (b. 7 Apr. 1874) and Rosa (b. 5 [or 3] Feb. 1876). In *TAEB* 2, relying on 1920 passports issued to the Batchelor children, the editors noted that both of the girls born in the 1870s died in infancy (pp. 557 n. 7, 760 n. 6). In fact the 1880 U.S. Census record sheets indicate that the daughters falsified their ages on the passports (claiming birth years of 1881 and 1883), perhaps in order to maintain their mother's dignity, as she claimed a birth year of 1856 (she was at least seven and perhaps twelve years older). Batchelor family genealogy, NjWOE; passports, Batchelor; U.S. Bureau of the Census 1952, roll T4.

2. On 3 January 1877 Batchelor had appointed Montagu Pym electric pen agent for New Zealand (Batchelor to Pym, 3 Jan. 1877; Power of Attorney, 3 Jan. 1877; both DF [*TAEM* 28:529, 535]). H. Fogg and Co. had been acting as electric pen agents in China and Japan since February 1876 (Doc. 892, Schedule C).

**-871-**

*Charles Batchelor Diary Entry*

[Port Huron, Mich.,] FRIDAY, MARCH 2, 1877.[1a]

Went over with Al & Pitt to Sarnia to make arrangements about the control of the Horse RR there. Edison said if he could get back the money that he put in there he would be satisfied. This proposal would have been taken immediately if Pitt had not suddenly remembered that he had put something into the road before Edison had taken the shares and he said he did not understand it that way he then proposed that the Symington bros should lend Edison $1500 for 2 years at 10 per cent and in return Edison to give them note secured by

Stock (90 shares) which they were to hold & vote but which Edison was to get the dividends on Edison then left $1000 with them to pay his share of an assessment they were to make to pay off all indebtedness & commence~~ment~~ to pay dividends immediately

Port Huron is a very pretty little place with about 10,000 inhabitants. A good theatre, a handsome large post office and Custom house covers a large territory.

AD, NjWOE, Batchelor, Cat. 1233:61 (*TAEM* 90:83). [a]"Friday, March 2, 1877." preprinted.

1. Edison and Batchelor had intended to leave for Port Huron on 27 February, but "Edison forgot his certificate of RR Stock & could not go." After dining with Josiah Reiff and George Bliss and attending *Round the Clock* at the Olympic Theatre with Reiff, Edison returned to Menlo Park and Batchelor spent the night at the Astor House. After breakfast with Reiff the next morning they took an 11 A.M. train. They stopped in Albany to view the unfinished State House and left for Buffalo at 8:15 P.M., arriving at 7:00 the next morning. From Buffalo they took the Grand Trunk Railroad across Ontario, arriving at Sarnia at 4:30 P.M., where the train was ferried across the St. Clair River to Port Huron. They spent that night at Pitt's house. Batchelor left the next evening to settle accounts with the Western Electric Manufacturing Co. in Chicago, returning home on 8 March. Cat. 1233:58–67, Batchelor (*TAEM* 90:82–86).

Edison remained in Port Huron, where he, William Wastell, and William Stewart—the stockholders of the Port Huron and Gratiot Street Railway—drafted an agreement to settle the indebtedness of the road as per the 29 January agreement of consolidation with the City Railroad (agreement of March 1877, DF [*TAEM* 14:539]). Edison returned to Menlo Park on 12 March (Cat. 1233:71, Batchelor [*TAEM* 90:88]).

**-872-**

*From J. L. Thomson*

New York March 10th 1877.

My Esteemed Sir

Your letter of Feby 26th came to the New York office[1] I was In Syracuse I shall remain here this summer[2] Your sample Waterproof Paper undoubtably would be an Excelent thing & I should be pleased to have Enough of it for a test can the paper or material you treat with be deoderised It must be Sweet & Pure free from oder &c.[3]

In my my mind what would be more valuble to the Success of the "Pkgs" for Licquids would be some durible Cement for Out Side & Inside a heavy coat Elastic If possible and very hard & durible something like Assfelt & Sand Something after the Combination of Pavements used in Paris only of

Light color for in Side & a color Similar to the out Side of Pkgs for have for the Out Side. Any such thing that would make the Bbls" Strong Stiff & Strong would be valuble Even for Out Side of Our presant Dry Bbls.

The Trouble now with Our full Bbls for flour Sugar & all heavy weights the Bbls is to light & not Strong Enough for rough handling; Any cheep Cement that can be put on to accomplish this result will be very valuble.

Any Samples of Packages or of the Board will be Sent you If you desire It. for Dry Packages Keep in view cheepness & only for Out Side finish—but for licquids an Expense of Even 50c to $1. for full Bbls Can stand it.

Please keep me posted and when you desire it I will visitt you.

I saw Mr. [Amasa] Mason on the 8th at his House in Williamsburg he was quite well we discussed your ability quite freely he has great confidence in your being able to acomplish what I want— Yours most Respectfully

J L Thomson

ALS, NjWOE, DF (*TAEM* 14:51).

1. This letter has not been found. Edison was probably replying to Thomson's letter of 15 February 1877 indicating that the company desired a treatment for the inside of its barrels (DF [*TAEM* 14:48]). Unfortunately, an earlier letter sent by Thomson that apparently provided a fuller explanation of the company's needs is also lost.

2. The factory of Thomson's New York Paper Barrel Co. was in Syracuse; the office was in New York. See letterhead, Thomson to TAE, c. Feb. 1877, DF (*TAEM* 14:50).

3. In a letter to Edison of 28 February (DF [*TAEM* 14:49]), the secretary of the New York Paper Barrel Co. described the company's requirements in this regard, noting that a "cheap paper for lining our packages, that will accomplish what you state, could probably be made to answer if it possesses no offensive odor, or will impart none. Used largely for prepared Cereals, refined Sugar, etc. the package must be entirely inodorous."

**-873-**

*Notebook Entry: Telephony*

[Menlo Park,] March 18th, 1877

Speaking Telegraph

We made a transmitter[1] in which the diaphragm struck against two 2[a] discs of Plumbago fastened to springs this seemed a little better

We also found that the words appeared plainer when received on a reed fastened at both ends, to which was attached a sounding tube[2]

39

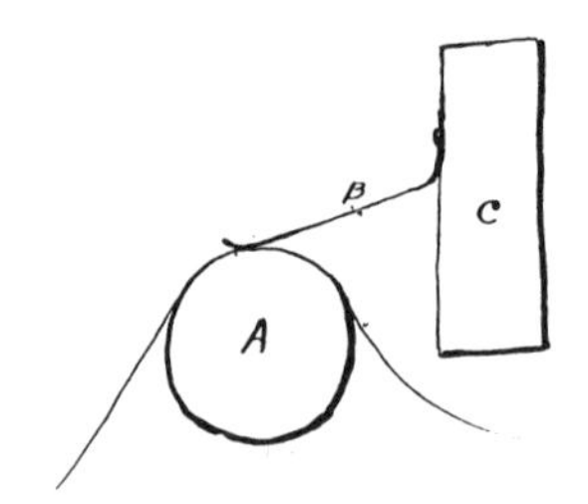


We also find that the best way to receive and hear the words is on the Electromotograph principle in which the current is passed through a spring B during the time it is pressing on a strip of chemically prepared paper carried along by the drum A C is a resonant box to increase the effect[3]

X, NjWOE, Batchelor, Cat. 1317:34 (*TAEM* 90:674). Written by Charles Batchelor. [a]Circled.

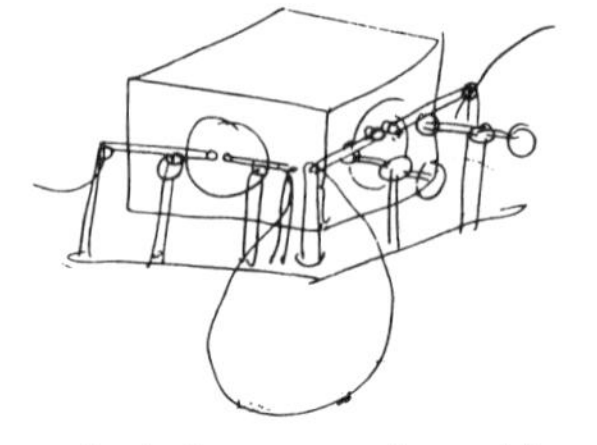

*A telephone transmitter with a pair of plumbago disks held by springs against each of two diaphragms.*

1. A sketch of 25 March 1877 (Edison's Exhibit 29-11, TI 2 [*TAEM* 11:243]) shows an instrument that appears to match this description, except that there are two diaphragms, each in contact with a pair of plumbago disks at the end of springs. The diaphragms' changing pressure on the carbon varied the resistance in the circuit, which was wired in series from a battery (not shown), through the springs, plumbago, and diaphragms, to the line. Edison indicated later that in such multiple-diaphragm transmitters "the tension [of the diaphragms was] various, so as to respond to different sound waves" (Doc. 973).

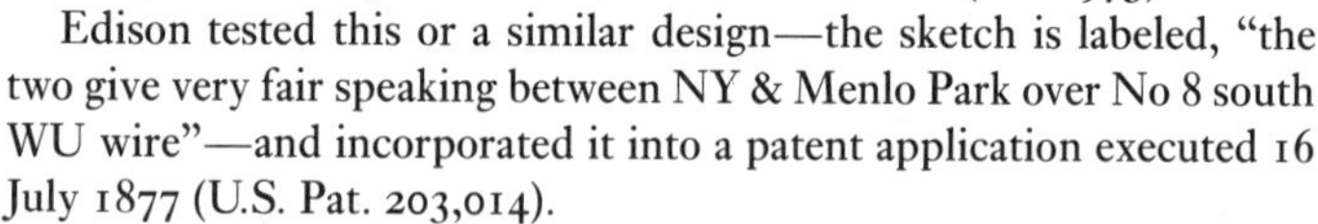

Edison tested this or a similar design—the sketch is labeled, "the two give very fair speaking between NY & Menlo Park over No 8 south WU wire"—and incorporated it into a patent application executed 16 July 1877 (U.S. Pat. 203,014).

The next day Batchelor noted that he "got good reading through the duplex plumbago points & a large tuning fork fastened at both ends" (Cat. 1233:85, Batchelor [*TAEM* 90:95]). Edison and his associates sketched other designs using two plumbago disks supported by springs in late March and early April 1877 (Edison's Exhibits 32-11, 49-11, 51¼-11, 51½-11, 56-11, 87-11 [*TAEM* 11:246, 257–60, 272]). During this same period they also used the two springs with platinum contacts in place of carbon disks to modify the current (see Doc. 866 n. 2).

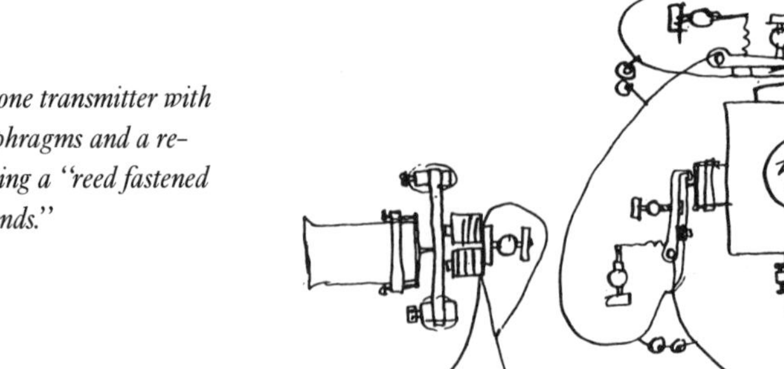


*A telephone transmitter with four diaphragms and a receiver using a "reed fastened at both ends."*

*A telephone receiver (right) using "a reed fastened at both ends to which was attached a sounding tube."*

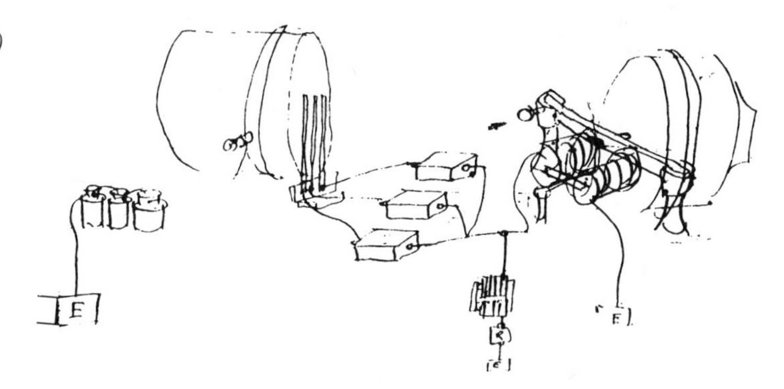

2. Such a receiver appeared in a 26 March sketch and a 29 March drawing for a model (apparently for an unfiled patent application). Edison's Exhibits 30-11, 39-11 (*TAEM* 11:244, 249).

3. This is the first evidence of Edison's use of the electromotograph as a telephone receiver.

**-874-**

*Technical Note: Telephony*

[Menlo Park,] Mch 20/77[1]

Speaking Telgh

[A][2]

[B]

[C][3]

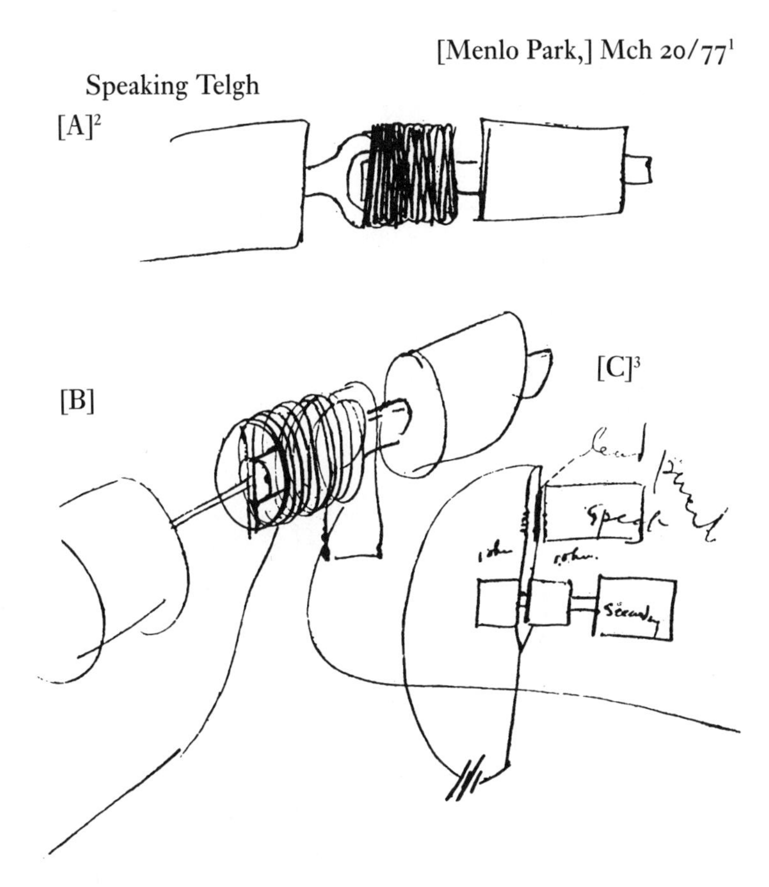

[D]

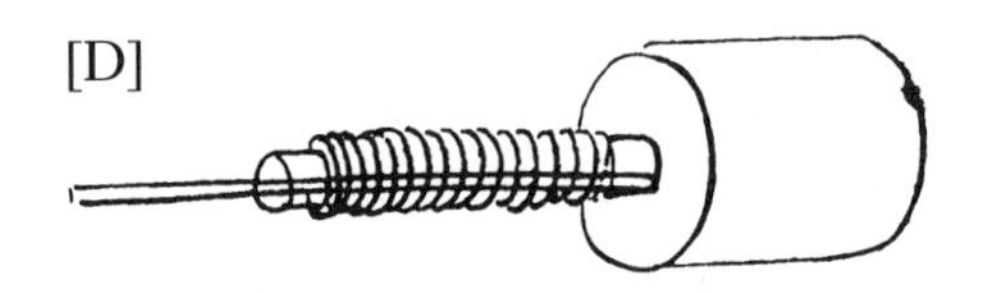

Short ckt primary wires ~~in~~ of indctn coil[4]

[E]

[F][5]

[G][6]

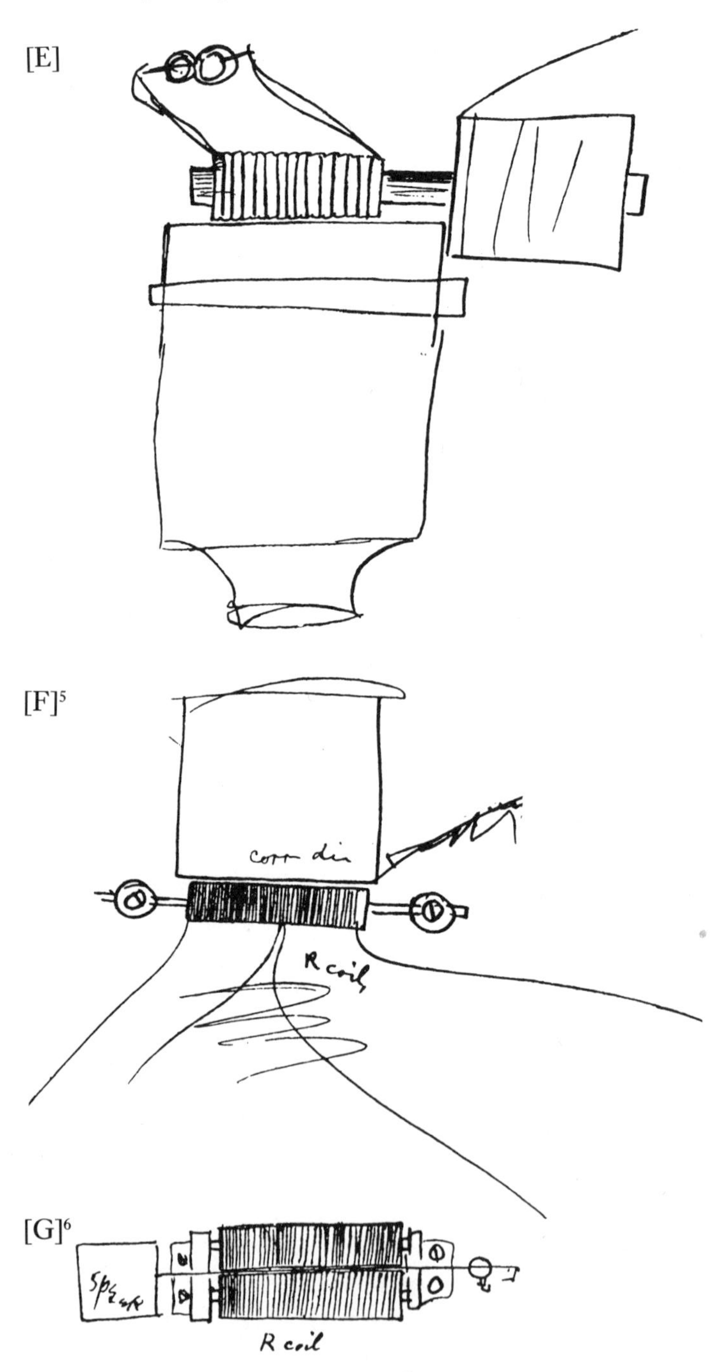

Rubber cylinders covered with bare G[erman] Silver wire & vibrating shaft

[H][7]

[I][8]

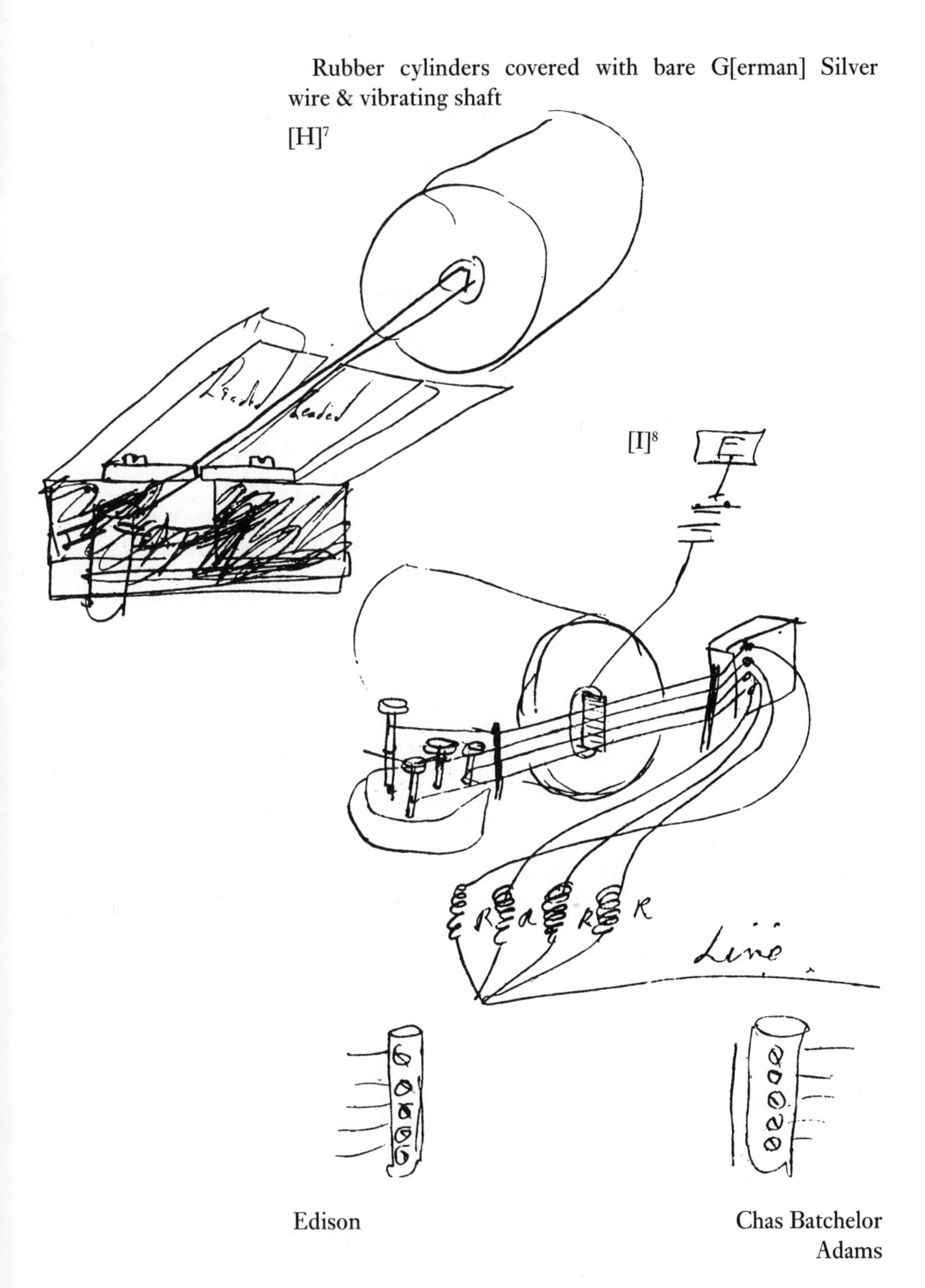


Edison

Chas Batchelor
Adams

X (photographic transcript), NjWOE, TI 2, Edison's Exhibits 20-11, 25-11, 21-11–23-11 (*TAEM* 11:235, 239, 236–38). Document multiply signed and dated.

1. The previous day, Batchelor recorded in his diary, "Worked all night on talking telegraph  Al & C[harley].P.E. in New York and Jim and I at laboratory  during the day they had a great many people to see it  Orton, Prescot, [Alfred] Brown, etc." His 20 March entry says, "Worked all night on speaker telegraph  Edison myself and Jim at laboratory and Charley at W.U. office. put on sheet iron diaphragm on receiver." Cat. 1233:78–79, Batchelor (*TAEM* 90:92).

2. In drawings A and B the primary coil of an induction coil is wound around a frame attached to the diaphragm of a speaking tube (left). The core of the secondary coil (right) extends into the open primary coil. A permanent current passing through the secondary to the receiver creates a magnetic field around the core; when the primary coil vibrates with the diaphragm, it induces a changing current in the secondary.

3. Figure labels are "lead pencil," "speak," "1 ohm," "1 ohm," and "secondary." In this sketch the changing pressure or degree of contact between the lead pencil and diaphragm changes the resistance in one branch of the parallel wires, which together constitute the primary of an induction coil. The idea of short-circuiting a high resistance to transmit human speech, which Edison also explored in drawings E–H, is the basis of Edison's U.S. Patent 203,013 (executed 8 December 1877).

4. In this design the vibrating diaphragm short circuits more or less of the bare primary coil, causing fluctuations in the permanent current in the secondary coil.

5. Figure labels are "copper dia[phragm]" and "R[esistance] coil." Here Edison has dispensed with the induction coil and passes the current directly through the bare wires of the resistance.

6. Figure labels are "Speak" and "R coil." In this design the diaphragm is attached to a long flexible conductor that passes between two high-resistance coils and is clamped at the other end. Vibrations of the diaphragm cause the conductor to touch the coils, short-circuiting them and changing the resistance of the circuit.

7. Figure labels are "Leaded" twice. In this drawing two blocks coated with plumbago have replaced the coils shown in drawing G.

8. In this design the vertical contact attached to the diaphragm touches the strings variously as the diaphragm vibrates, cutting in and out the four resistance coils (**RRRR**).

## –875–

*From Frank Whipple*

Port Huron, Mich., March 22d 1877[a]

Dear Sir

Your note left with B. H. Welton[1] was not handed me till Tuesday evening[2] and my time has been fully occupied from that time so that I have been unable to get this statement out sooner.[3]

Mr. Wastell has not handed me in an exact statement of his account with the Ry Co. but showed me the Books he has kept[b] which disclose a balance due the road of about $22.00

I enclose with this a statement showing all the indebtedness of the road that I can discover deducting the $500.00 paid W.P.

From this comes $3500 to be assumed by the consolidated company and the balance to be adjusted by you Wastell & Stewart

Mr. Stewart went as far West as Chicago & returned here being gone but four days on his trip[b] to California— The road is running averaging about $18.00 per day now & the snow is deeper now than ever seen before in this section— It is however melting rapidly to-day—

I send this today without waiting for Wastells detailed statement because that will delay another day & I presume you want to hear from me Truly Yours

Frank Whipple

ALS, NjWOE, DF (*TAEM* 14:535). Letterhead of Whipple & Potter, Attorneys at Law. [a]"Port Huron, Mich.," and "187" preprinted. [b]Obscured overwritten letters.

1. B. H. Welton was a Port Huron grocer who is identified in *History of St. Clair County* 1883 (602) as having at one point been engaged in the "railroad ticket business." A statement of indebtedness of the Port Huron and Gratiot Street Railway Co. dated 12 March 1877 shows $524.21 owed Welton. 77-016, DF (*TAEM* 14:537).

2. Whipple wrote this on a Thursday.

3. A statement in Whipple's hand listing the company's debts as of 12 March 1877—which may be either the one enclosed by Whipple with this letter or the schedule of indebtedness referred to in the draft agreement of March 1877—shows a total debt of $5,872.09, with Edison holding promissory notes worth $1,104.51 (DF [*TAEM* 14:537]).

**-876-**

*Agreement with Western Union Telegraph Co.*

[New York,] March 22, 1877[a]

This Memorandum of Agreement entered into this twenty second day of March 1877 by and between the Western Union Telegraph Company of New York and Thomas A. Edison of Menlo Park Middlesex County New Jersey.

Witnesseth

Whereas the said Edison being the owner of an experimental laboratory at Menlo Park New Jersey provided with all necessary facilities for perfecting new inventions applicable to the business of the said Telegraph Company and being himself an expert in devising and perfecting such new inventions is desirous of obtaining the aid and co-operation of the Telegraph Company as hereinafter provided.

And whereas The Telegraph Company is desirous of securing for its benefit the skill and experience of said Edison and the sole right within the United States to all his inventions and improvements capable to be used on land lines of telegraph or upon cables within the United States which may be made during the period covered by this agreement.

Therefore be it agreed:

First: This Agreement is to continue for five years from the date hereof unless sooner terminated as hereinafter provided The Company will assume a portion of the cost of experimenting with and producing such inventions and improvements and will pay to the said Edison in aid of such experimentation subject to the provisions hereinafter stated the sum of One hundred and fifty dollars per week during the period covered by this contract, the same to be used by the said Edison in payment of laboratory expenses incurred in perfecting inventions applicable to land lines of telegraph or cables within the United States.[1]

Second. Whenever any such invention or improvement shall be made by said Edison the same shall be deemed to have been made for and shall belong to the Telegraph Company and the said Edison through the Attorney hereinafter named shall promptly file in The Patent Office sufficient caveats to protect the same and shall thereafter duly and promptly file his application for Letters Patent thereon and all re-issues thereof with request and direction that the same be issued to said Company as sole assignee and shall at all times and from time to time execute such other applications assignment or assignments or other papers as may be requisite or desirable to vest the sole title to the monopoly of such invention or improvement and all letters patent therefor and all re-issues thereof exclusively in the Company.

Third. Immediately upon the complete issue of letters patent as above provided for any invention improvement or device and after three months satisfactory trial by the Company of the same it shall be the duty of the parties to agree upon a royalty to be paid to said Edison upon each and every article or machine covered by such Letters Patent and used by the Company; and in case they are not able to agree within three months from date of such Letters Patent and the completion of such satisfactory trial the same shall on the written demand of either party be submitted to arbitration by two persons one to be chosen by each party and they being unable to agree they shall select a third and the decision of such arbitration shall be final

All royalties shall be fixed at a rate per annum payable monthly and shall continue during the life of the patent subject to temporary cessation as the use of such articles or any of them may be suspended by the Company and subject to complete cessation as hereinafter provided.

Fourth: In case the Company after such satisfactory trial shall elect not to put in use any invention improvement or device covered by any certain Letters Patent it may retain the title to said respective Letters Patent paying the round sum of Two hundred and fifty dollars in full compensation for each said Letters Patent or may re-assign the same to said Edison in which latter case all charges and expenses incurred by the Company in respect to such letters patent or the fair proportion of general expense for solicitors and Patent office fees applicable thereto shall be re-imbursed to the Company by said Edison. If the Company shall elect to discontinue the use of any patented article or device which has been after such satisfactory trial adopted for use in business on its lines or in its offices it may do so retaining the title to the Letters Patent on paying said Edison so much as will with the royalties already paid thereunder amount to Two hundred and fifty dollars.

Provided However that this stipulation shall not apply to any case in which the Company shall acquire or attempt to exercise the right to accomplish the same work as is done by the article or device whose use it has abandoned or is about to abandon through the use of any device or invention of another person, but shall apply only in cases where the Company elects to abandon the use of such patented article or device invented by Edison without substituting and does not afterwards substitute in its place any other patented article or device of equivalent function.

The Company shall have the right at any time after payment by royalties thereon or otherwise the sum of Two hundred and fifty dollars to abandon suspend or limit the use of any article or device of Edison whenever it shall have acquired the right to use any equivalent article or device not an infringement of any of said Edisons inventions or Letters Patent thereon and which is of general superiority or which shall be preferable to such article or device of Edison in respect to economy or convenience in manufacture or use; and in case of any abandonment suspension or limitation of use under the conditions in the last sentence recited, it shall be the right of said Edison in the manner and under all the conditions herein provided as to other arbitrators to call for a similar arbitration to fix certain commutation for royalties for said invention not exceeding in any event the sum of Five thousand dollars and in determining such sum the arbitrators shall consider whether such invention of said Edison was in any way by suggestion or otherwise the cause of the invention which is being adopted or used in its place it being the intention of the parties to provide for the payment of a commutation sum as last above mentioned only in cases where the Edison invention being put out of use has clearly been the efficient parent or cause of or has suggested the making of the other and superior invention then being put in use in its place, and it being also their intention (in view of the labor and skill which is often required to perfect inventions which are afterwards easily improvable by much less labor and skill) that in case of a prior invention by Edison which has lead to such improvements by others that he shall be suitably compensated therefor even though subsequent improvements may render his inventions less valuable. If after satisfactory trial and being put in use for business such use of an article or device shall be once abandoned and afterward resumed and in like manner whenever thc Company shall put in use any article or device covered by any Letters Patent for which it has paid the lump sum of Two hundred and fifty dollars, as first hereinabove provided the payment of royalties shall then begin or be resumed as the case may be and there shall be credited on such royalties all sums already paid for commutation or otherwise on such invention. In like manner should the Company be dissatisfied with the royalties fixed by agreement or by the arbitration it may at any time after the payment of the sum of Two hundred and fifty dollars first above named in gross or in royalties elect to discontinue working under any of such patents and thereupon after one months notice in writing to said Edison all royalties thereunder shall cease subject however to all the provisions above made in cases where such discontinuance shall be followed by the substitution of other patented means to accomplish the same object In case at any time the use of any article or device shall by reason of the patenting of any invention of equivalent use by another person become of less value to the Telegraph Company it shall have the right to require the appointment of arbitrators as above provided to fix a reduced royalty to be paid to said Edison thereupon. In fixing royalty or commutation prices to be paid the arbitrators chosen shall take into consideration the economy in manufacturing and in the use of such invention; and the value of the monopoly thereof as well as the sums already paid to said Edison for weekly expenses or for other inventions of which the article or device then being under consideration is an improvement on or modification or which are a necessary or convenient part of any apparatus or method in connection with which the article or device under consideration is to be used.[2]

Fifth The Company may at any time after the expiration of one year reduce the weekly sum above provided for to One hundred dollars and may in like manner determine this agreement altogether by giving three months notice to that effect.

Sixth. All expenses for the taking out of patents and solicitors and counsel fees in connection therewith shall be borne by the Telegraph Company subject to reimbursement in certain cases as above provided. And in any case where after issue of Letters Patent thereon, the Company are put to legal expenses to defend the priority of invention or patentability of any such article or device the said Edison shall contribute to such expense by waiving one half of the royalties on the same until such reductions have amounted to one half of such expense.

Seventh: The said Edison hereby appoints Grosvenor P. Lowrey his attorney irrevocable with full power of substitution and revocation to represent him in all matters before the Patent Office or Department of the Interior or Courts of law in respect to any inventions or patents above referred to or any legal proceedings thereunder or in respect thereto hereby covenanting to execute and deliver from time to time all special authorities or letters of Attorney to said Lowrey for this purpose which may be required of him by the Company.

Eighth: It shall be the duty of the Company to keep strict accounts of all matters covered by this agreement and to permit said Edison at all proper times to inspect such accounts and otherwise to take account of the number of articles machines or devices being used by the Company under this agreement.

Ninth. This agreement shall not be construed to apply to any improvements upon the system of chemical telegraphy sometimes called Automatic or fast telegraphy and shall take effect from March 1st, 1877.

In Witness whereof the parties to these presents have hereunto set their hands and seals the day and year first above written.

The Western Union Telegraph Co. By Willm Orton
President

A. R. Brewer[3] Secy Thos. A. Edison

DS, NjWOE, Miller (*TAEM* 28:1029). Notarization omitted. [a]Date taken from text, form altered.

1. In Doc. 850 Edison had asked that the figure be set at $100 per week for his shop expenses and account records show that this is what he received. The extra fifty dollars was apparently intended for Josiah Reiff and was included at William Orton's suggestion as part of a proposed settlement of outstanding claims by the American Automatic Telegraph Co. on Edison's quadruplex patents. There was to have been an agreement between Western Union and American Automatic concerning this settlement, and Reiff urged Edison not to sign his contract with Western Union until this second agreement was finalized. No final settlement on the quadruplex was ever reached and Reiff never received his fifty dollars per week. Reiff to TAE, 13 Apr. 1880; TAE to Reiff, c. 13 Apr. 1880; Reiff to Norvin Green, 17 Apr. 1880; Reiff to TAE, 12 May 1880; Reiff to TAE, 22 Mar. 1882; all DF (*TAEM* 55:540, 544, 588; 63:678).

2. This paragraph is probably a response to Edison's experience with Gerritt Smith and the quadruplex. See *TAEB* 2:295 n. 4, 353 n. 3, 354 n. 5, 359 nn. 2 and 4.

3. Abijah Brewer was secretary of Western Union. Reid 1886, 690–91.

## SEXTUPLEX TELEGRAPHY Docs. 808, 877, 881, 884, 897, 898, 901, 903, 905, 910, 916, 919, 927, 935, 938, and 957

With Edison's designs of 1873 and 1874 quadruplex telegraphy had become a practical reality, and the possibility of conveying still more simultaneous messages on a single wire attracted increased attention. Edison was aware of the various approaches attempted by others—even before his successful quadruplex—and soon tried to surpass his previous accomplishment.[1] During 1875 and 1876 he worked with acoustic telegraphy and then with his acoustic transfer system. In the spring of 1877, having secured financing for his shop from Western Union, he undertook a renewed campaign to create a practical sextuplex telegraph system.[2]

Edison focused his attention on sending more than two messages in one direction, assuming that conventional duplexing techniques would allow any system to work in both directions at once.[3] Each receiving instrument had to respond accurately to only the signals made by its corresponding distant transmitter, remaining effectively undisturbed by the activity of all other instruments. Edison had succeeded in quadruplex telegraphy by varying the strength of one signal and the current polarity of the other.[4] Now, in 1877, he experimented with a large number of systems, most of which combined aspects of his quadruplex designs with one of three other approaches: acoustic instruments, acoustic transfer circuits, and signals of different strengths.[5]

Edison's first sextuplex systems used a normal quadruplex arrangement combined with devices for acoustic telegraphy. He tried to use messages of different current strengths in various ways, receiving them with combinations of specially designed relays;[6] he tried having one receiver respond to signal currents of intermediate strengths while another responded only to pulses of weaker or stronger current; and he used his acoustic transfer system to switch transmitters and receivers on and off the line many times each second. In searching for techniques that would prevent signal interference and allow duplexing, Edison drew upon many aspects of his multiple telegraphy drawings and caveats from the last third of 1874 and the start of 1875.[7]

Each of the general approaches could be used in theory to send any number of simultaneous signals over a single wire and could be duplexed for an equal number of messages in the opposite direction. Nevertheless, none had yet been found commercially practicable.[8] Despite some laboratory success, Edison failed to devise a satisfactory sextuplex system, and after three months of work and four patent applications he essentially abandoned the subject.[9]

1. Doc. 579.

2. Edison occasionally called it "sextruplex" or "sexduplex." Edison also considered multiple telegraph systems of higher capacity, envisioning as many as 24 messages at once on a single wire. NS-Undated-005, Lab. (*TAEM* 8:373); see also the many sketches in NS-77-004, Lab. (*TAEM* 7:457–713 passim).

3. These techniques—using differentially wound relay coils or bridge circuit arrangements, together with artificial lines and other devices—neutralized the effects of outgoing signals on local receiving apparatus while allowing incoming signals to act. However, in practice it was difficult to make such combinations work without additional circuit elements and very careful adjustment. See *TAEB* 1:31–32; and Docs. 50, 283, and 285.

4. All of Edison's commercial quadruplex circuits and most of his experimental designs were like this. For the uniqueness of his approach and illustrations of the difficulty of combining the two signal types see Doc. 348 (esp. nn. 3, 9, 16, 21) and Prescott 1877, 834.

5. Edison had flirted briefly with six-message systems in mid-1875 and again in the fall of 1876 (Docs. 583, 611, and 808). His new contract with Western Union (Doc. 876) may have affected the timing and topic of his work, but about this time Edison also conducted a literature review of earlier multiple-telegraph alternatives in preparation for his Quadruplex Case testimony (NS-Undated-005, Lab. [*TAEM* 8:432–36, 444]). The most common diplex plan during the previous two decades

had employed currents of very different strengths (Prescott 1877, 825–32, 835–38).

6. The moving multiple-contact levers on many of those relays were key elements in Cases 138, 139, and 140 (U.S. Pats. 452,913, 512,872, and 453,601).

7. See Docs. 485, 496, 498, 507, 512, and 531–34, and the materials referred to in their notes.

8. Despite the assurances of British electrician Oliver Heaviside, the differential-strength method did not work well even with two signals (Heaviside 1873; Prescott 1877, 832–34). One of Elisha Gray's harmonic systems did handle four messages each way in some tests, but it was never used commercially by Western Union (Gray 1977, 86; Prescott 1877, 872–80). A system based on time-sharing (like Edison's acoustic transfer), later widely used, was just being introduced in France during these years (Prescott 1877, 862–66; King 1862b, 308–15; Guillemin 1977, 642–56).

9. The applications were Cases 133, 138, 139, and 140. Edison recorded some preparatory revisions for the last three between 30 April and 26 May (see NS-77-004, Lab. [*TAEM* 7:660–69]). Although his primary goal was a commercially practical sextuplex system, the patents were designed broadly enough to foreclose competitors' chances of producing alternative systems. Cases 138–40 had tortuous histories, going through interferences and numerous objections and alterations before issuing in the 1890s. See *Edison v. Dickerson.*

## -877-

*Patent Model Specification: Multiple Telegraphy*[1]

[Menlo Park,] March 23 1877

⟨G. M. Phelps Esqr Supt Please make this Model. G. B. Prescott Electrician⟩[2a]

Model Sextruplex[3]

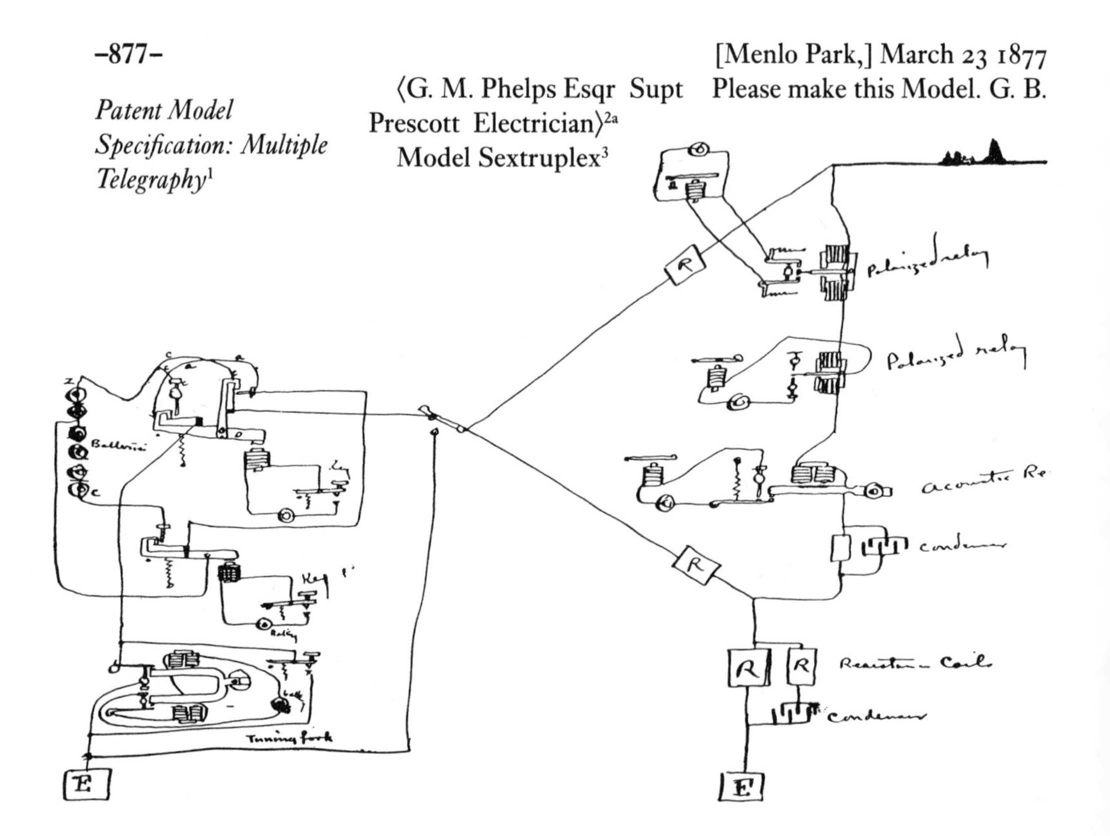

wooden batteries about this size[4]

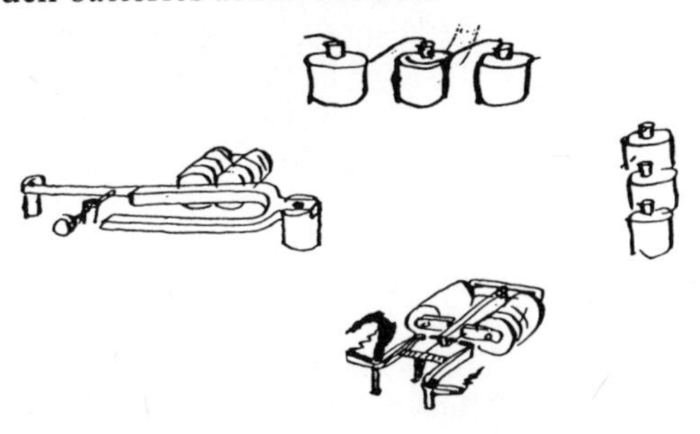

Model for Acoustic Quadruplex[5]

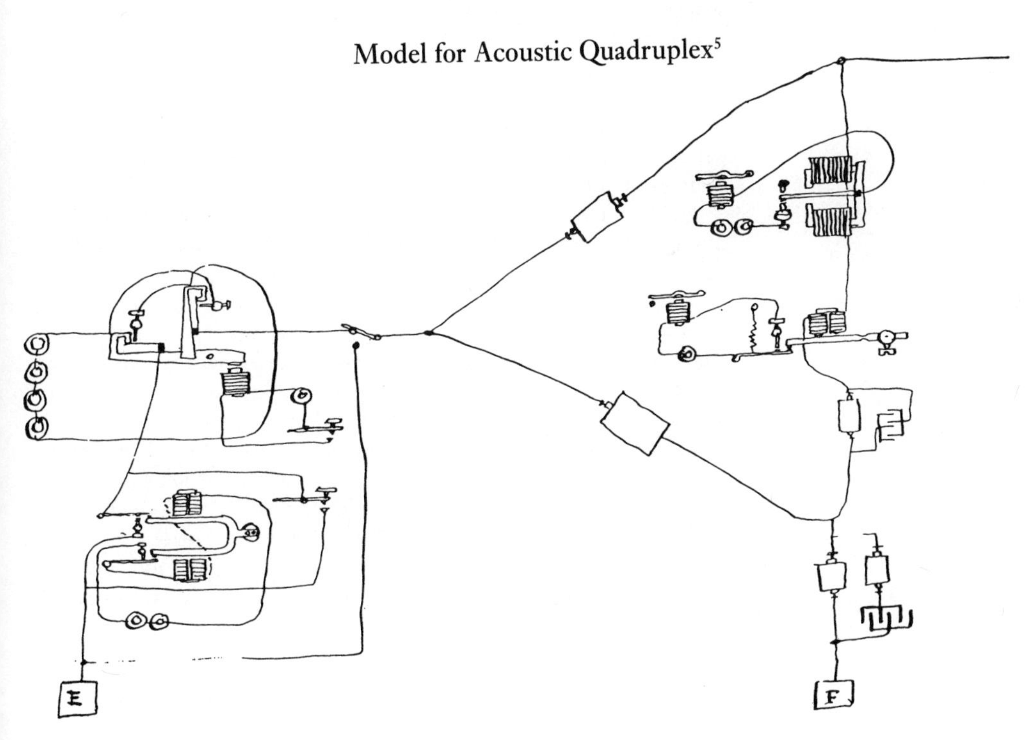


T A Edison

X, NjWAT, Box 92 02 163 01. Document multiply signed and dated. [a]Marginalia written by Prescott.

1. See headnote above.

2. The day after signing his contract with Western Union (Doc. 876), Edison took these patent model drawings to George Prescott at the company's New York office. The models were for patent applications covering multiple telegraph designs Edison had been considering for some time (see Doc. 808). Edison also gave Prescott a specification for acoustic reed instruments for tests (Box 92 02 163 01, NjWAT). Prescott forwarded the drawings to George Phelps, who ran Western Union's manufacturing shop in New York; Phelps later annotated the pages "Delivered & charged March 29."

The same day Edison made other drawings to help in the drafting of his first telephone patent application (Case 130). See Edison's Exhibit, TI 2:508 (*TAEM* 11:633).

3. Figure labels on the left are: "Batteries," "Key," "Key," "Battery," "battery," "Tuning forks," and several individual letters for wires and batteries; on the right are: "Polarized relay," "Polarized relay," "Acoustic Reed," "Condenser," "Resistance Coils," and "Condenser." The drawings for these two models, with another circuit drawn the same day (NS-77-004, Lab. [*TAEM* 7:658]), are the first dated evidence of Edison's work on sextuplex telegraphy since November (Doc. 808). The first arrangement is very similar to the November sketch, adding an acoustic telegraph transmitter ("tuning fork" at left) and receiver ("Acoustic Reed" at right) to a bridge version of Edison's quadruplex system, although the earlier design called for a reed transmitter rather than a tuning fork. Edison did not execute the patent application that included this model until May 8 (Case 133; U.S. Patent 217,781). The patent model is at the Edison Institute, Dearborn, Mich. (Acc. 29.1980.1375).

4. The originals are about 1 cm by 1.5 cm. The other items are the tuning fork transmitter (at lower left in the larger drawing) and the polarized relay (at upper right).

5. This design is the basis for Case 132, which eventuated in U.S. Patent 377,374. Although it shows acoustic telegraph instruments coupled with apparatus that work by reversing the polarity of a circuit, the application also covered the substitution of instruments that used increase/decrease of current strength in place of the reversal devices.

Edison executed and filed Cases 132 and 133 together. The patent examiner initially rejected the applications, one objection being that there was at most only one patentable design involved. In a 10 October 1877 letter, Serrell indicated that working such combinations practically was not straightforward and that Edison had spent a good deal of effort on these designs before making the applications. He tried to make the significance of each clear.

> Edison no doubt was the first to discover that the rise and fall of tension or reversing instruments could be used in a harmonic telegraph without either interfering with the other and to arrange his instruments for this object: He had considerable difficulty before this was accomplished: In case 132, this feature is brought out, and the broader ground claimed: Having gone so far he next sought to bring together all three systems, rise and fall of tension, reversing polarity, and harmonic vibrations, this is the subject of the present application No 133, and the claim in the present case is limited to this point.
>
> It appeared best to introduce these in separate applications making the present tributary to 132, instead of crowding too much into that case. [Pat. App. 217,781]

See also Edison's undated memorandum about his early attempts to work acoustic and Morse instruments together (77-015, DF [*TAEM* 14:518]).

**–878–**

*Charles Batchelor Diary Entry*

[Menlo Park,] SUNDAY, MARCH 25, 1877.[a]

Worked in Laboratory and finished my speaking telegraph receiver made an iron disc also a thin rubber one.

Edison got a new theory in regard to magnets  he thinks that if a single magnet will give magnetism at the end of one foot then put in more magnets & project so much farther on this principle

I shall try it; he thinks also that as the magnetic field works like this magnet shows

and as Sprague[1] says a long one magnetizes like this

showing it travels further then he thinks that a number of magnets placed like the first figure will give intensity. He also thinks that it follows the same law as the bridge for Elec.[b]

AD, NjWOE, Batchelor, Cat. 1233:84 (*TAEM* 90:95). [a]"SUNDAY, MARCH 25, 1877." preprinted. [b]Obscured overwritten letters.

1. See Sprague 1875, 325–27.

**–879–**

*From John Breckon*

Sunderland. March 29 1877[a]

Dear Sir,

I learn, through Mr Ireland, that you have transferred the manufacture of the Electric Pens to the Western Electric Manufactg Co at Chicago.

I cannot agree to the transfer of the Contract between you & me—dated June 29th 1876—without further information.[1] I am quite satisfied with you and the Contract as it stands—but until I can be satisfied that the Company mentioned on the other side will perform that Contract to my entire satisfaction I must hold you responsible.

In this you will see nothing but what is prudent and reasonable— I shall be glad to meet your wishes in every thing that is fair & businesslike.

The last order for 100 pens has come forward very slowly—can you not cause them to be supplied more promptly in future. Mr. Ireland tells me that Mr Batchelor has called attention to a draft for £52.10.0—it has never reached me—or I should at once have accepted & sent it forward to you.

If you will draw upon me, according to contract, and add interest at say 5 pr cent pr annum I will accept & return it to you by first mail   Yours truly

J. R Breckon

ALS, NjWOE, DF (*TAEM* 14:374). [a]"Sunderland." and "18" preprinted.

1. See Doc. 779 n. 2.

# –5– April–June 1877

Edison and the laboratory staff—minus machinist Charles Wurth, who left in April[1]—worked intensively on two projects under his new contract with Western Union—telephone technology and his sextuplex system of multiple telegraphy. By the end of June, Edison had a musical telephone instrument that had been publicly exhibited, he had developed a new research program for a telephone transmitter, and he was coming to the conclusion that his sextuplex telegraph was a failure. He and his staff also briefly investigated a few other avenues of research and, in late June, considered some variant designs for his embossing recorder, six of which Joseph Murray was making.

The telephone research involved improvements to both transmitters and receivers. Earlier work with carbon as a medium for varying resistance led to the development in April of a "pressure relay," an electromagnet that varied the current in a local circuit by squeezing a block of plumbago between its core and armature. In late May, having settled on plumbago as the best form of carbon to use in the telephone transmitter, the staff started extensive tests to determine the best material for binding it into a block. At the same time, they were testing various diaphragm materials and found that mica in the place of a metallic diaphragm reduced unwanted vocal harmonics. They also tried to enhance the transmitter's ability to convey sibilant sounds, testing several different mouthpiece configurations and circuit designs.

Laboratory work on the receiver produced refinements of the electromotograph instrument first devised in March. At the end of April, Edison demonstrated this "musical telephone" at the Newark Opera House for a week, transmitting several pieces of music to an electromotograph receiver.[2] In May, Charles Edison and electric pen agent George Caldwell attempted to exhibit the musical telephone commercially but within a short time gave up the enterprise. News of the instrument prompted an inquiry from Boston telegraph entrepreneur Peter Dowd, who would later become associated with Edison's telephones through the Gold and Stock Telegraph Company. The next month Edward Johnson arranged for a demonstration to be held in Philadelphia on the Fourth of July. The staff continued their development work and by mid-June had designed a combined transmitter/receiver that used the electromotograph.

At the end of March, Edison had started development of a sextuplex telegraph system, and the staff spent much time on the project during the next three months. Although this work initially involved searching for a way to use acoustic instruments as a third "side" for the quadruplex, Edison soon was exploring six-way transmission on principles entirely different from either his standard four-message or acoustic telegraph systems. He filed four sextuplex patent applications in May and early June. He also considered some of his new ideas as bases for alternative quadruplex methods, filing one such patent application in May and another in August. Western Union gave Edison the use of a New York–Boston line to try the sextuplex in early June, but the tests seem to have proved unsatisfactory, and a month later he dropped this line of research.

The manufacture and sale of the electric pen moved entirely from Edison's and Batchelor's hands. The Western Electric Manufacturing Company was already producing and marketing the pen domestically. George Bliss, who managed the North American market, and his associate Charles Holland assumed responsibility for the foreign pen business in late April, leaving Batchelor with "nothing whatever to do for it except receive [his] share of Royalty."[3] Ezra Gilliland, who had been the general agent for the pen in the eastern states, left the company in June. In Britain, "the . . . public—with its usual stupidity—did not take kindly to the apparatus as it was," according to Frederic Ireland, who outfitted the pen with a new battery and press. Edison executed two alternative copying pen patent applications in April, and he and Batchelor briefly investigated other copying technologies in June and July, but that marked the end of their interest in the subject.

The American Novelty Company, which had earlier seemed promising as an outlet for Edison's minor inventions, failed. The company changed its name to the Electro-Chemical Manufacturing Company, hoping to generate more interest in its products, but George Bliss and Charles Holland took over the duplicating ink business in early June, leaving the company with little but its new name.[4] Edison briefly continued his research into a waterproofing compound for paper containers and, at the request of a glass ornamenting concern, worked on modifying his electric pen to etch metal foil laid on glass. He also had Batchelor and Kruesi work up an address label machine at the end of April.[5]

Edison spent more than a week in New York City at the end of April testifying about his quadruplex inventions and patent rights in *Atlantic & Pacific v. Prescott & others.* The British telegraph engineer William Preece, who witnessed some of Edison's testimony concerning George Prescott, declared that he "would not for £50,000 have my name bespattered as Prescott's was."[6] (He also quoted William Orton as having once told him that Edison had "a vacuum where his conscience ought to be.")[7] Preece visited Menlo Park on 18 May and spent the day "experimenting and examining apparatus,"[8] enjoying himself enough to return on three subsequent occasions before leaving for England on 4 July.[9]

Samuel Edison, who had come to Menlo Park in late March, left on 2 July.[10] The staff's pet bear "got loose at Laboratory" on 6 April and, Charles Batchelor noted, "we caught him and afterwards killed him."[11] Finally, two small personal details of Edison's life appeared in print. A letter in a local newspaper described Edison's compassion for a seriously ill tramp:[12]

> Mr. Edison coming to the station to take a train for New-York, learning of the circumstances, gave Mr. Stryker a dollar to pay for a telegram to the Overseer of the Poor, whose duty it is to attend to such matters, and to use the remainder of the funds for the benefit of the tramp.

And an item in the *Operator* noted that "T. A. Edison is gray as a badger, and rapidly growing old."[13]

1. "Charles Nicholaus Wurth," Pioneers Bio.
2. See Doc. 889 n. 1.
3. Doc. 922.
4. On 14 April, Batchelor noted "a conversation with Johnson, Edison & [James] James about selling James some stock in the E.C. Mfg

Co we agreed to let him have (4500) forty five hundred shares for (1500) fifteen hundred dollars he will think about it." On 21 May, Batchelor "Saw James who wanted to advance money to carry on the carbon experiments but I told him we could not take it in that way, if he put his money into the company and the company advanced it it would be all right." Cat. 1233:104, 141, Batchelor (*TAEM* 90:105, 123).

5. Cat. 1233:120, 122, Batchelor (*TAEM* 90:113, 114); Cat. 1171:74, Lab. (*TAEM* 6:294); U.S. Pat. 217,781.

6. Quoted in Baker 1976, 157.

7. Quoted in ibid.

8. Ibid., 162.

9. See Doc. 976.

10. Cat. 1233:183, Batchelor (*TAEM* 90:144).

11. Cat. 1233:96, Batchelor (*TAEM* 90:101).

12. New Brunswick *Weekly Fredonian*, 7 June 1877, Cat. 1240, item 183, Batchelor (*TAEM* 94:56).

13. "Echoes from 197," 15 June 1877, p. 9; cf. Docs. 77 and 89.

**–880–**

*Technical Note: Telephony*

[Menlo Park,] April 1 1877[1]

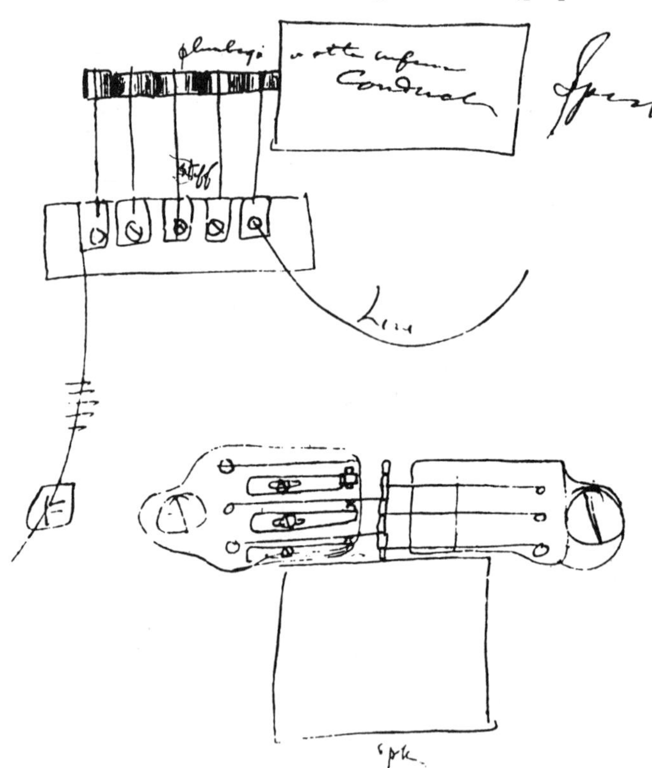

T A Edison

Chas Batchelor
Adams

X (photographic transcript), NjWOE, TI 2, Edison's Exhibit 42-11 (*TAEM* 11:252).

1. Figure labels are "Plumbago or other inferior conductor," "stiff," "Line," "Speak"; and "spk" on lower drawing. Edison later testified that he made a telephone incorporating the design shown in the first sketch. He described it as

> a telephone having a diaphragm; in front of this diaphragm were several springs similar to those shown in patent No. 203,014. On the extreme end of these springs were brass cups or shells; in these shells were cylinders of plumbago; the ends of the cylinders were all adjusted so as to be in contact with each other, and with the diaphragm of the speaking instrument, the circuit passing up one spring to the extremity holding the plumbago passed through it to the next plumbago, and so on through the rest of the cylinders to the spring furthest from the diaphragm, thence to the other pole of the circuit; the lower end of these springs were secured to blocks of brass, provided with slots, and these were all secured to a block of insulating material by screws, which passed through the slots, which admitted of adjustment of the springs up or down and to or from the diaphragm or the adjacent springs. I have a drawing of this instrument; it was made and used, and worked well. This was done within one or two days, I think, of April 1, 1877. [TI 1:50 (*TAEM* 11:46)]

Batchelor recalled that he and James Adams made and experimented with the device as a transmitter in a telephone circuit within the next two or three days and that it "transmitted ordinary speech perfectly." He also described its construction.

> The springs were made of steel, about 2 inches long and 1-50 of an inch thick, and about 3-8 of an inch long; the bottom end of these springs was fastened to brass about 5-16 of an inch square and 3-4 of an inch long, with an elongated hole in them, so that the spring could be screwed to an insulating bar and adjusted to position. The other end of the spring carried a brass sleeve projecting about 1-8 of an inch from each side, in which was pressed tightly a hard pressed button of plumbago. This was afterwards filed down on each side so as to project about 1-16 from each end of the sleeve. There were five of these springs in the instrument. The diaphragm of the instrument was 3 and 1-2 inches and the speaking case was turned from wood. The diaphragm being held to the case by a brass ring. This whole apparatus was placed on a base with the springs upright. The electric circuit passed in at the bottom of the spring nearest the diaphragm, thence through all the plumbagos and passed out at the bottom of the furtherest spring from the diaphragm. The springs themselves were made by Adams. [TI 1:248–49 (*TAEM* 11:101)]

The springs were introduced as evidence in the telephone interferences (TI 2:519 [*TAEM* 11:647]), but Batchelor noted that "most of the parts of that instrument have been used in other experiments, only a few of them being able to be found at present." He went on to explain that, "After an experiment is tried in the laboratory it is usual with us to use as much of the instrument as can be used in making the next experiment" (TI 1:249 [*TAEM* 11:101]).

A close variation of this device is shown in Edison's Exhibit 99-11, TI 2 (*TAEM* 11:274) and also in Prescott 1879 (534).

The next year Edison used this device as evidence in a priority dispute with inventor David Hughes over the invention of the microphone.

The lower sketch is similar to a drawing of the following day (Edison's Exhibit 43-11, TI 2 [*TAEM* 11:253]). That drawing shows a series of contacts switching resistances in and out of the main circuit. At the end of May, Edison made an instrument that used batteries in place of the resistances (Edison's Exhibit 106-11, TI 2 [*TAEM* 11:278]).

**-881-**

[Menlo Park,] April 4 1877

*Technical Note: Multiple Telegraphy*[1]

Quadruplex
Increase & decrease with acoustic

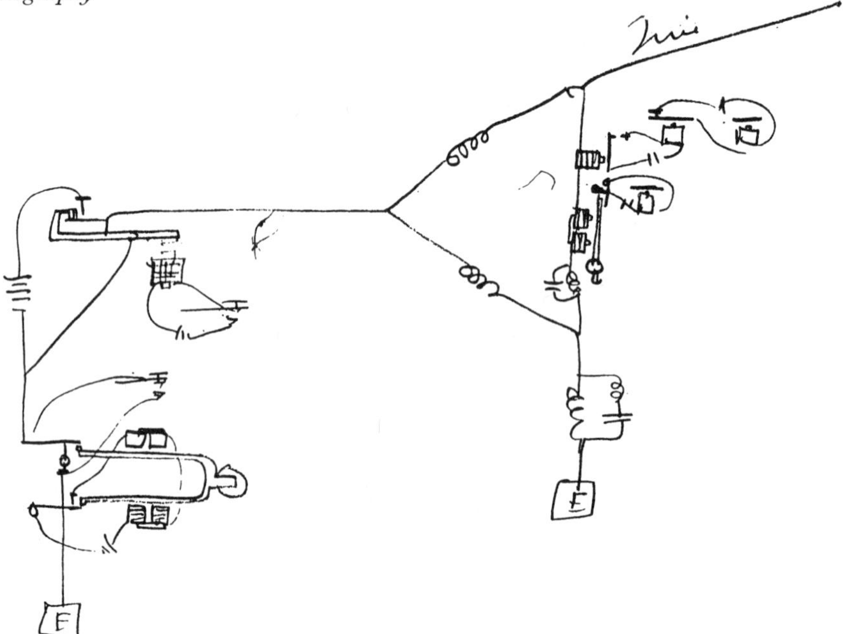

Reversals with acoustic[2]

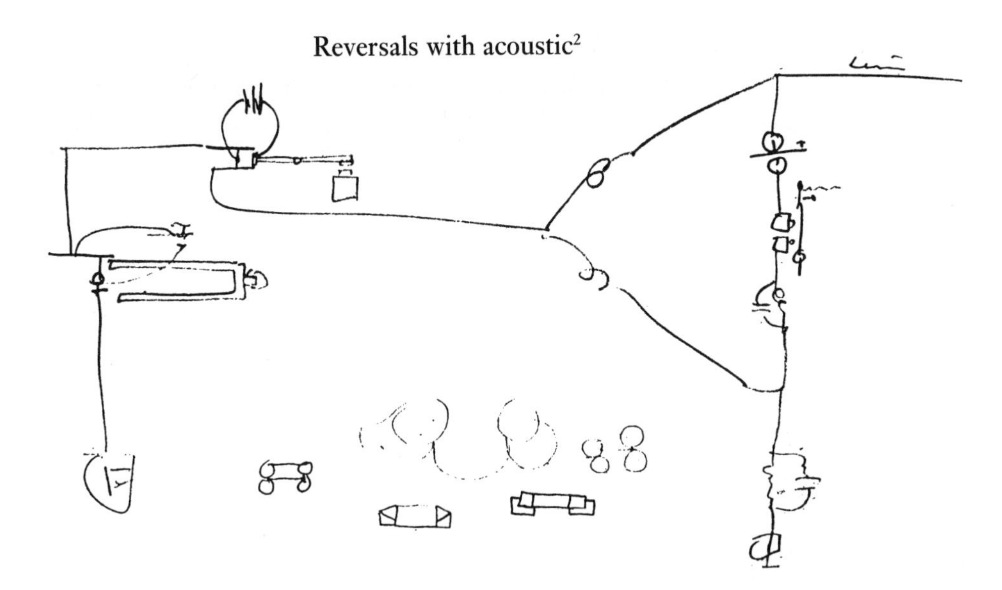

Sextuplex

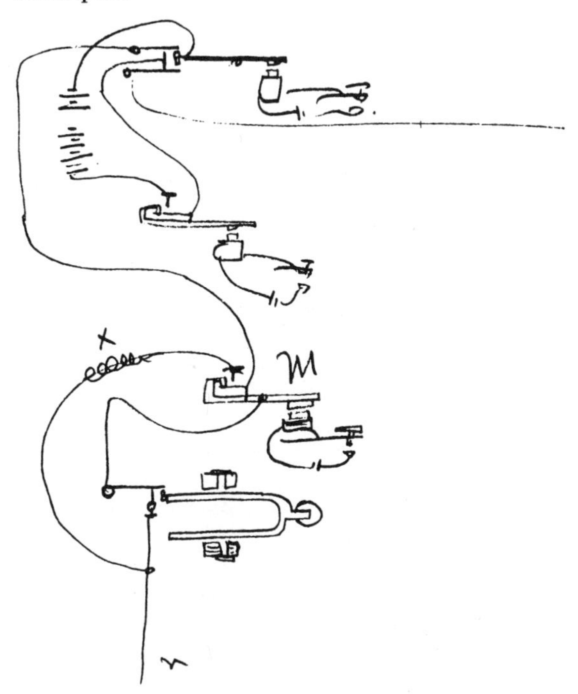


Object of x is to cause even signalling because the vibrations weaken the current and when they cease it strengthens it hence when I stop I throw in a resistance=.

Octuplex[3] is made by inserting another tuning fork at n before it is connected to the earth & this is connected exactly like the one shewn but has a different vibrating rate and other forks may be inserted as long as they can be arranged so as not to interfere with the regular polarized & increased current Morse instruments  If interrupting the continuity is bad & it is not found requisite to use m it may be arranged thus[4]

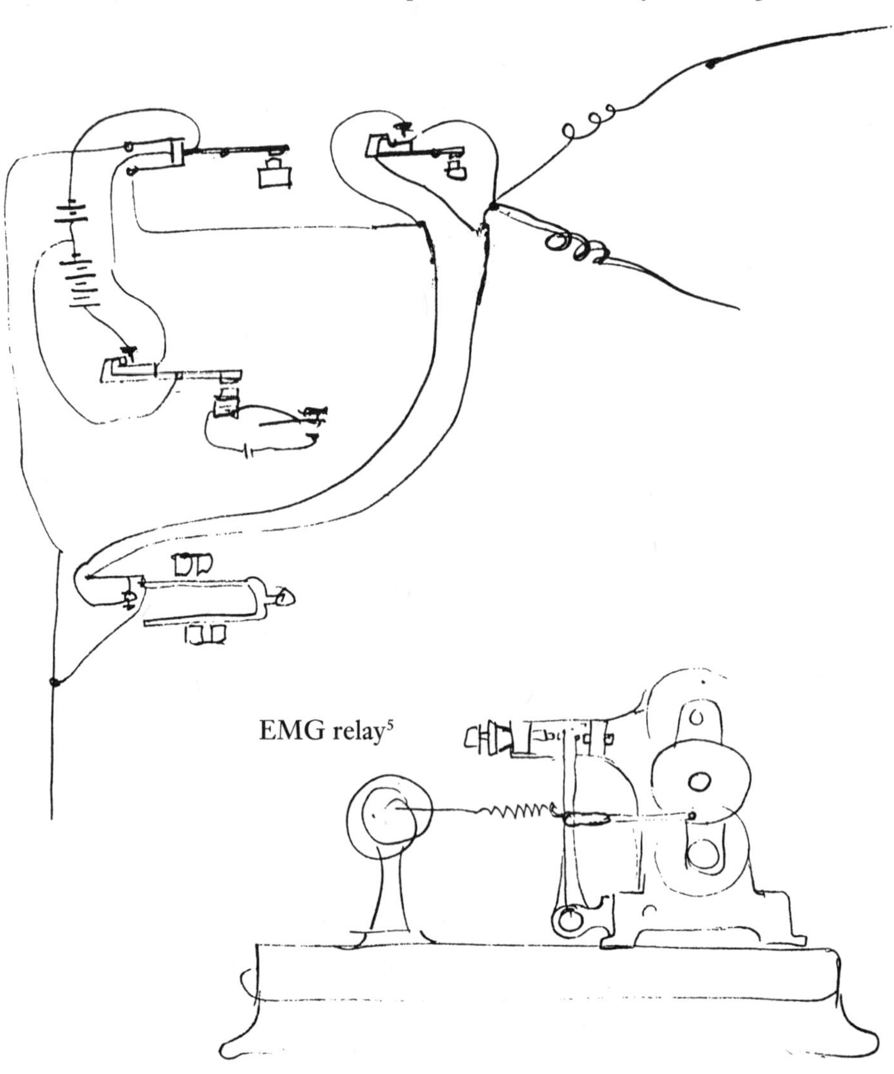

EMG relay[5]

Differential Sextuplex

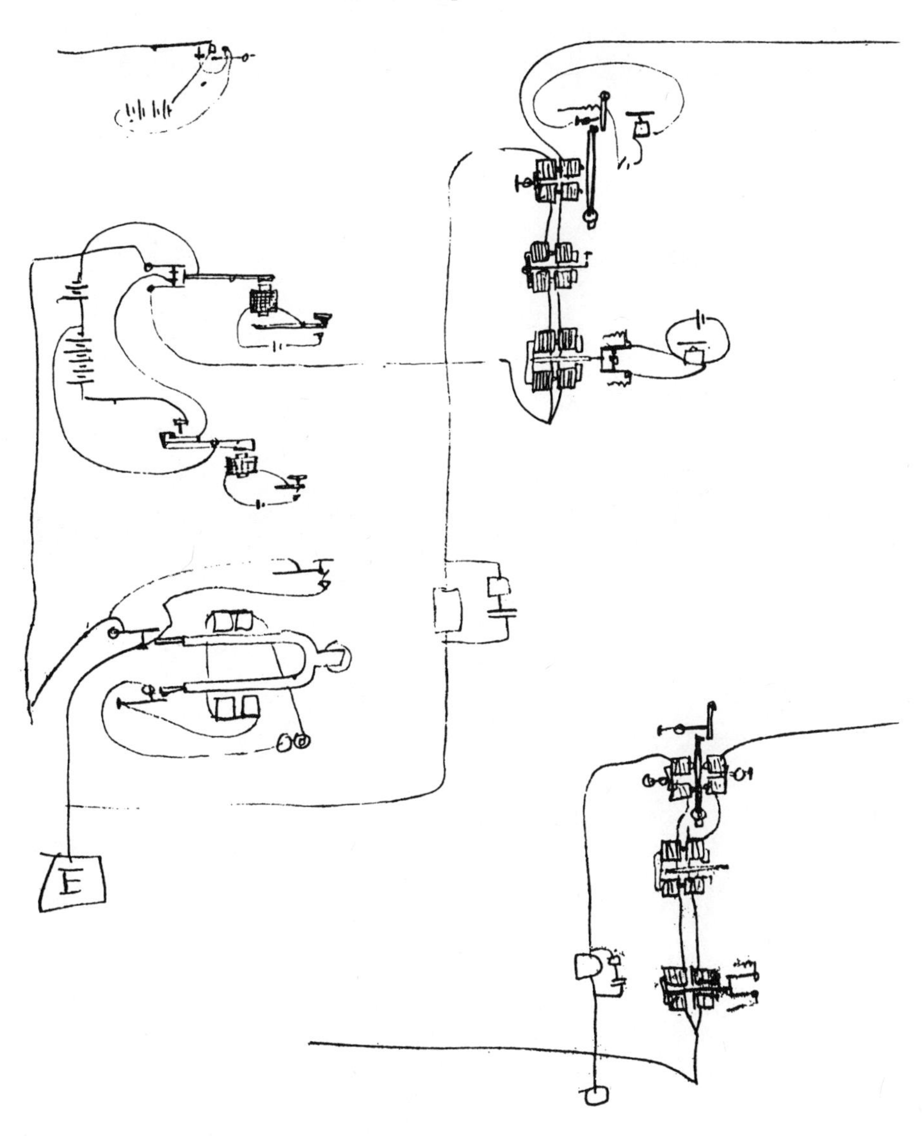

Reed worked by induction[6]

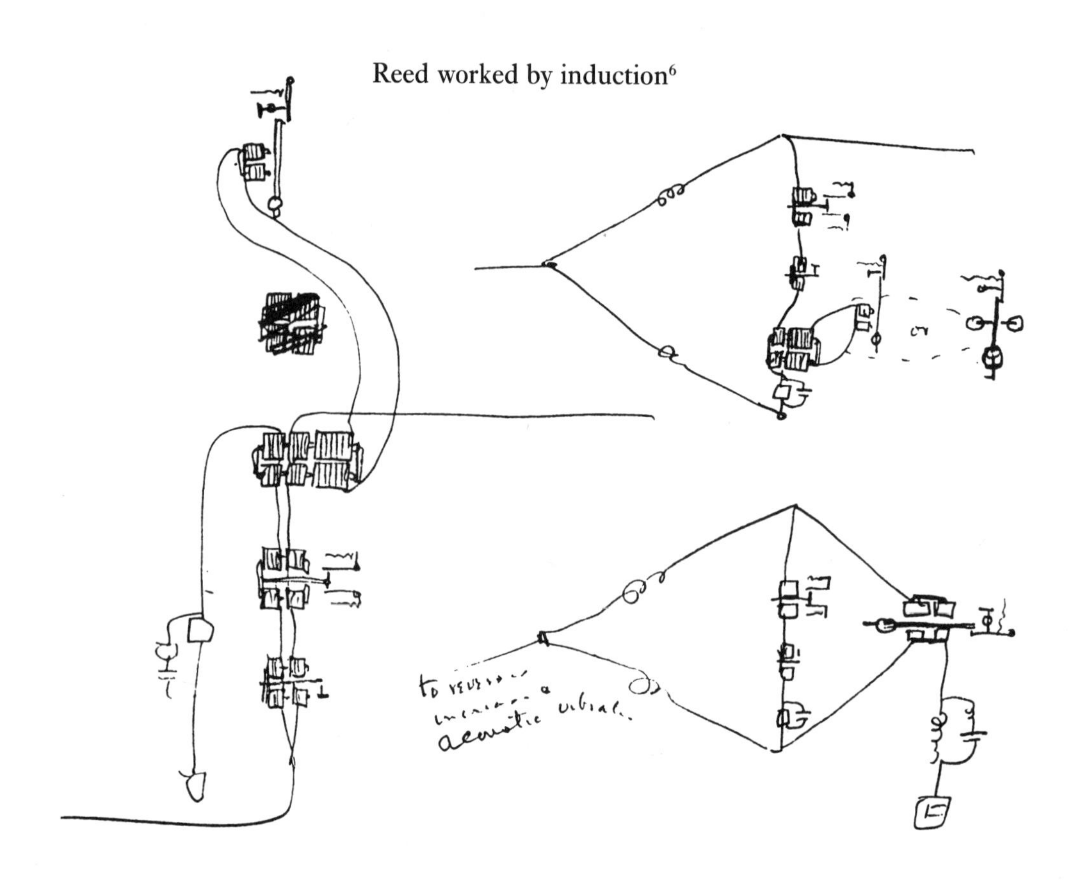

Quadruplex & Sextuplex devices[7]

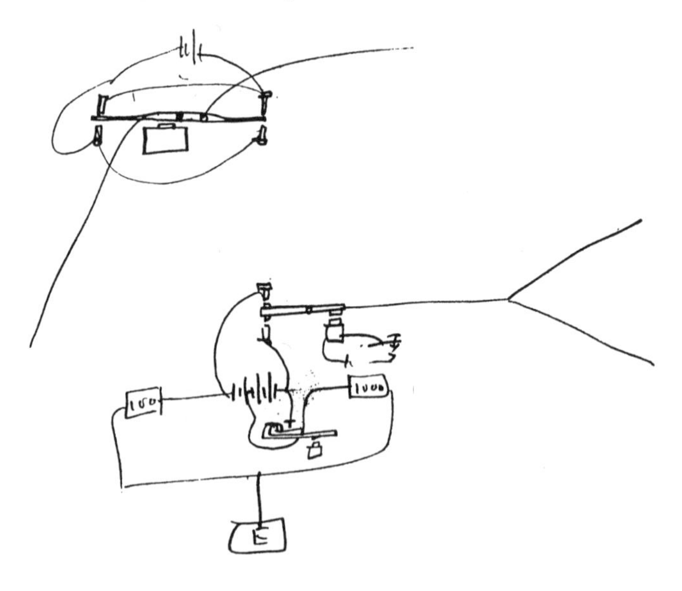

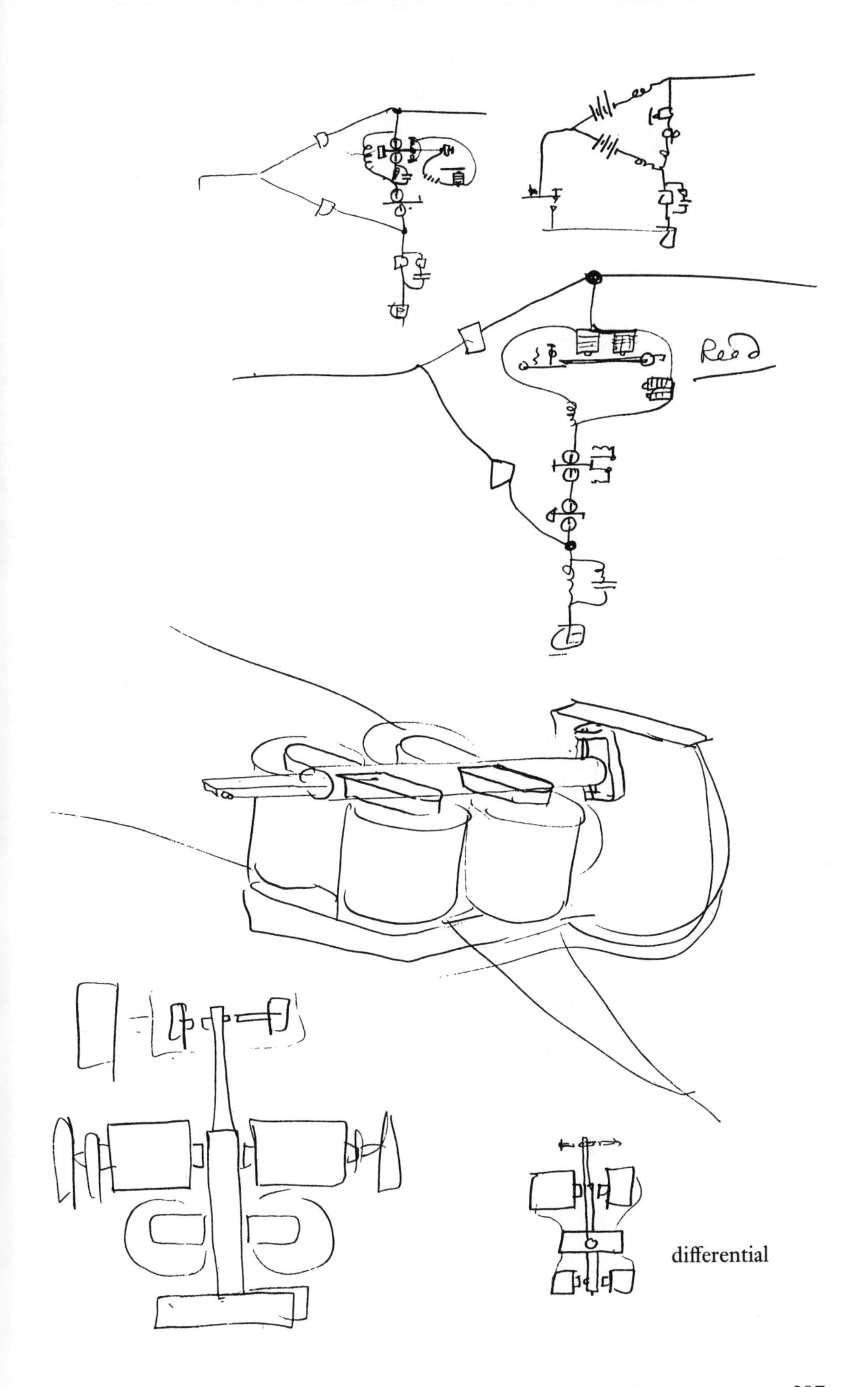
Reed
differential

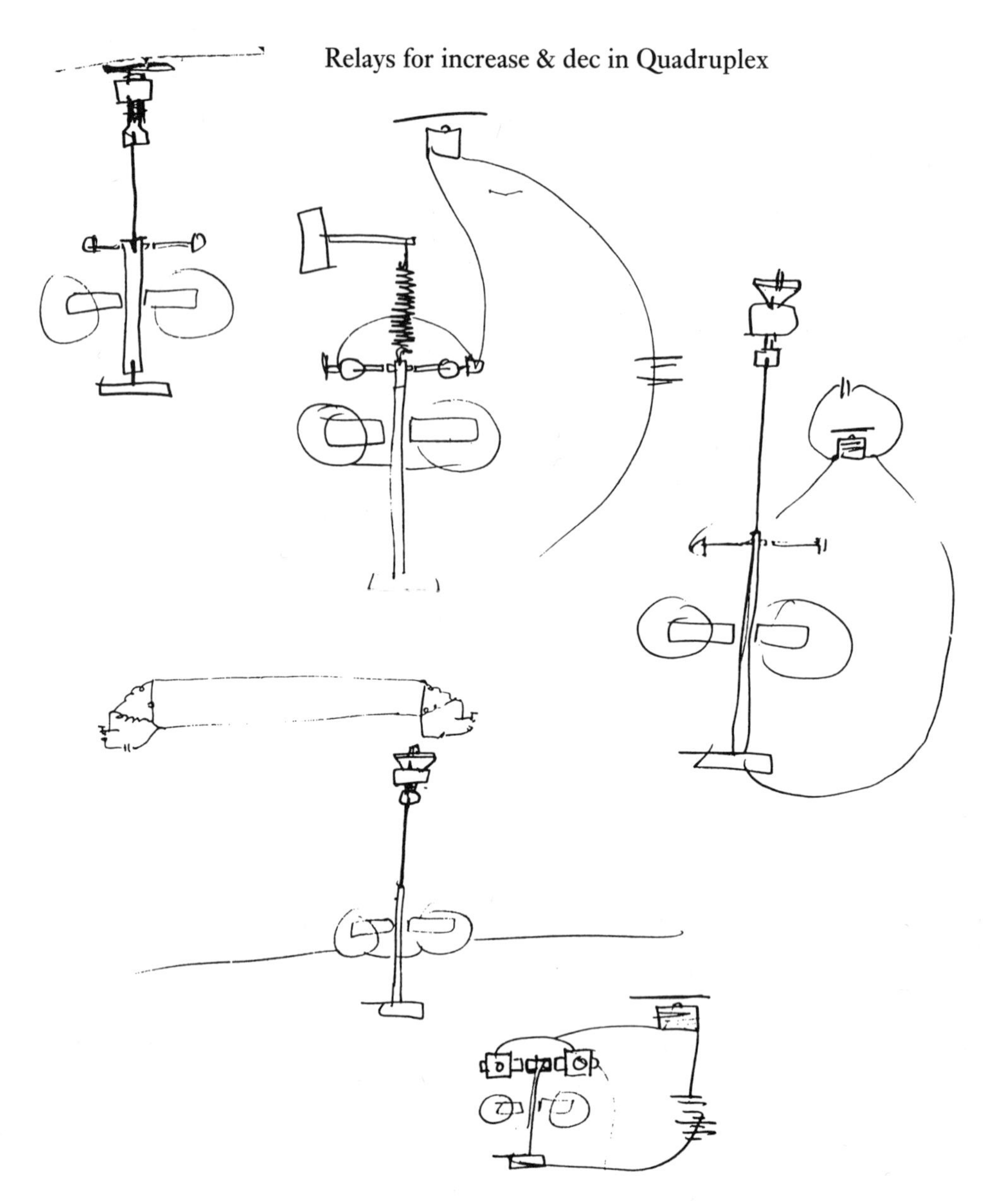


Plumbago coal or manganese blk ox point[8]

T A Edison

X, NjWOE, Lab., NS-77-004 (*TAEM* 7:464, 460–61, 459, 458, 463, 462, 465). Document multiply signed and dated.

1. See headnote, p. 280.
2. Figure label in preceding and following drawings is "Line."
3. See Docs. 773 and 795. In the fall of 1876 Elisha Gray had exhibited his octruplex system between New York and Philadelphia. Gray 1977, 86.
4. The drawing that follows this text is on a separate page but ap-

pears to be the modification described in the text because it preserves the continuity of the incoming signal.

5. One week later Edison drew a motor-driven electromotograph coupled to a reed and another "wkd by induction." NS-77-004, Lab. (*TAEM* 7:475).

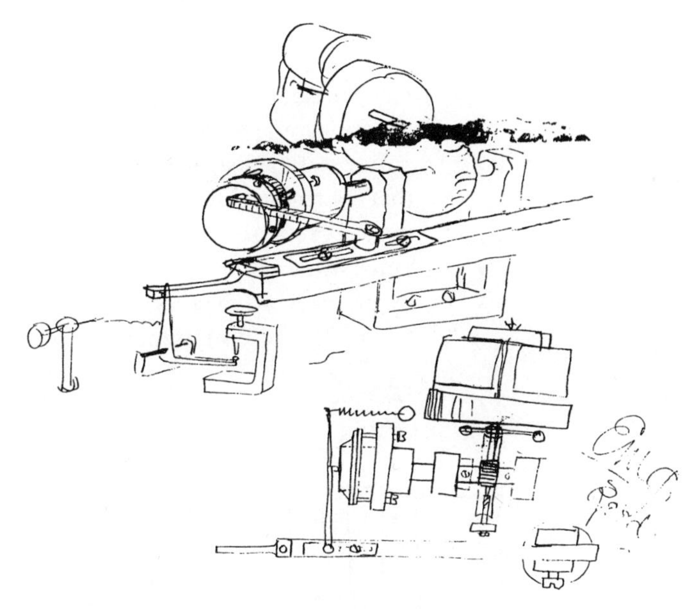

*Edison's idea for sextuplex relay incorporating a motor-driven electromotograph and an acoustic reed.*

6. Figure label is "to reverser increase & acoustic vibrator."

7. The resistances (rectangles) in the second sketch are each labeled "1000."

8. Figure label is "small cell intensity battery." See Doc. 882.

**-882-**

*Technical Note: Telephony*

[Menlo Park,] April 4 1877[1]

Speaking Telegh=

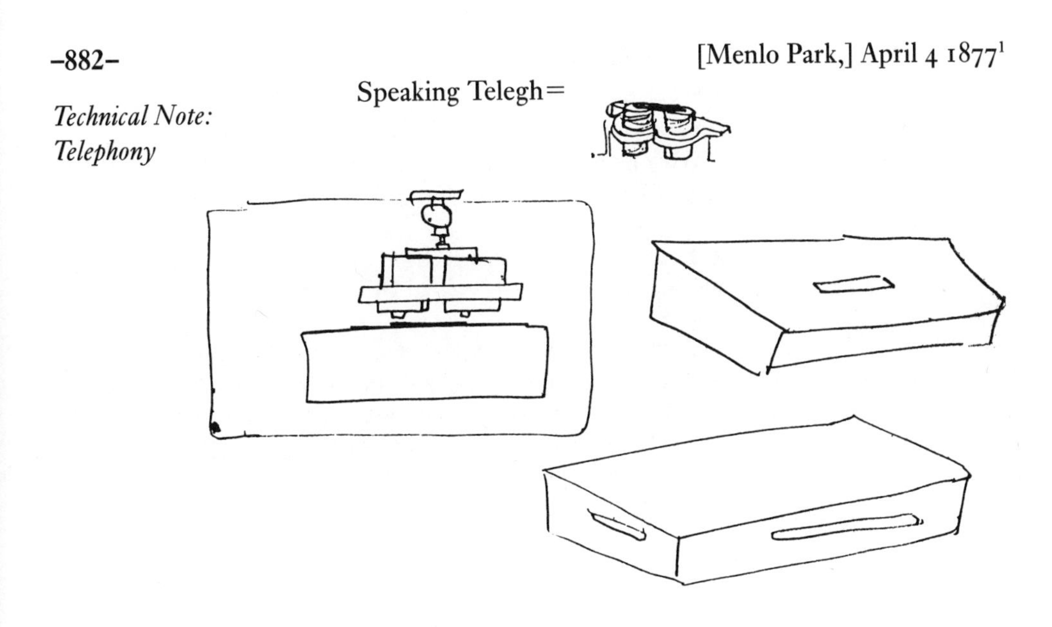

We can get everything perfect except the lisps & hissing parts of speach such as 'Sh' in shall= get only .o. in coach[2] I propose in addition to the regular diaphram to have an additional diaphram[a] adjusted very delicately to [resp]ond[b] to these lisping or hissing sounds= I propose that it either cut in and out resistance or act in the same manner as the regular diaphram—[3] I have just been listing Substances[a] for imperfect contact points.[4] I found that good ~~B~~Lumps of black oxide of Manganese ~~have~~ Anthracite & Bituminous Coal only give a moderate resistance somewhat higher than plumbago and that enormous difference of resistance is obtained ~~on~~ by Varying the pressure for instance Anthracite Coal having a resistance with light pressure of 1700 ohms is reduced to 300 ohms by pressure, Manganese Oxide, from 1500 to 600. I propose to employ these substances as Contact points=

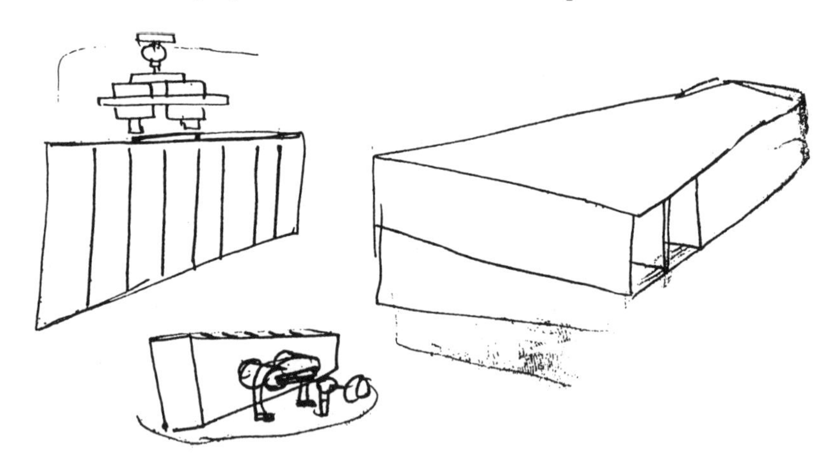

~~B~~Combination of resonant boxes in one: ie[c] 1 large box & several partitions to reinforce the sound communicated to back of larger box

T A Edison

James Adams
Chas Batchelor

X, NjWOE, Lab., Vol. 11:50 (*TAEM* 3:933). Document multiply signed and dated. [a]Obscured overwritten letters. [b]Document damaged. [c]Circled.

1. The previous night Marcus Hussey had brought his cornet to the laboratory and played it over the wires to New York. Batchelor noted that "we got it very perfect." Cat. 1233:93, Batchelor (*TAEM* 90:99).

2. This is the first mention of problems with the transmission and reproduction of specific vocal sounds.

3. In U.S. Patent 203,014, executed 16 July 1877, Edison claimed the

use of a separate diaphragm designed to respond to hissing speech sounds by switching a resistor in and out of the circuit.

4. The list of substances mentioned here has not been located. See Doc. 887 n. 2.

**-883-**

*From George Bliss*

Chicago. Ill.[a] Apl 9. 1877

Dear Sir:

I expected to have started for the east the first of this week but unexpected matters have detained me here and it will be some days before I can leave.

About the Foreign Pen business I have been unable to force matters to a conclusion.

Gen'l Stager is adverse to putting in any money unless some one who has money to invest also can be found who is willing to invest & go to Europe to represent the interest there.

The man must also be a first class business man.

The Gen'l then does not care to put in a large sum of money but he will give his influence & the backing of the W.E.M. Co. to such a move. This of course is a good deal but as matters stand I do not feel like going outside of the Gen'l & the W.E. people for the means to carry out this project.

Several of my friends have expressed a willingness to go in but before anything can be closed I must have a written proposition open for 30 days. This cannot be arranged till we meet as I have some new ideas on the subject.

Now if you have a good opportunity to arrange with other parties for Europe my advice is to complete the same.

I have been extremely anxious to control this Pen business but the obstacles which have come up unforeseen make me unwilling to stand in the way of a friends success.

You will see me the last of this or the first of next week I hope. Respectfully

Geo H. Bliss Gen'l Man.

ALS, NjWOE, DF (*TAEM* 14:376). Letterhead of Electric Pen. [a]"Chicago. Ill." preprinted.

[Menlo Park,] April 10, 1877—[2]

*Technical Note: Multiple Telegraphy*[1]

Sextruplex

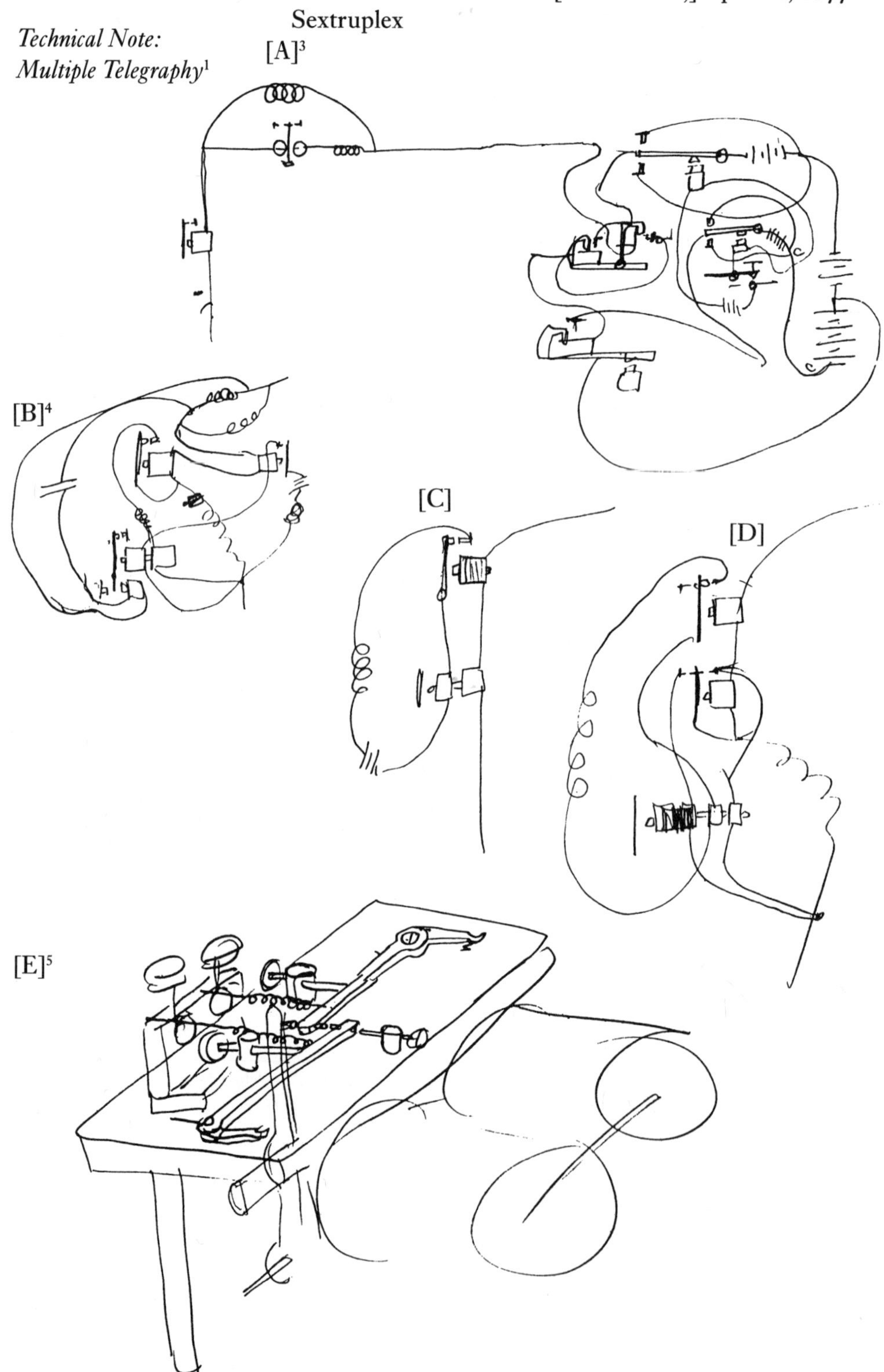

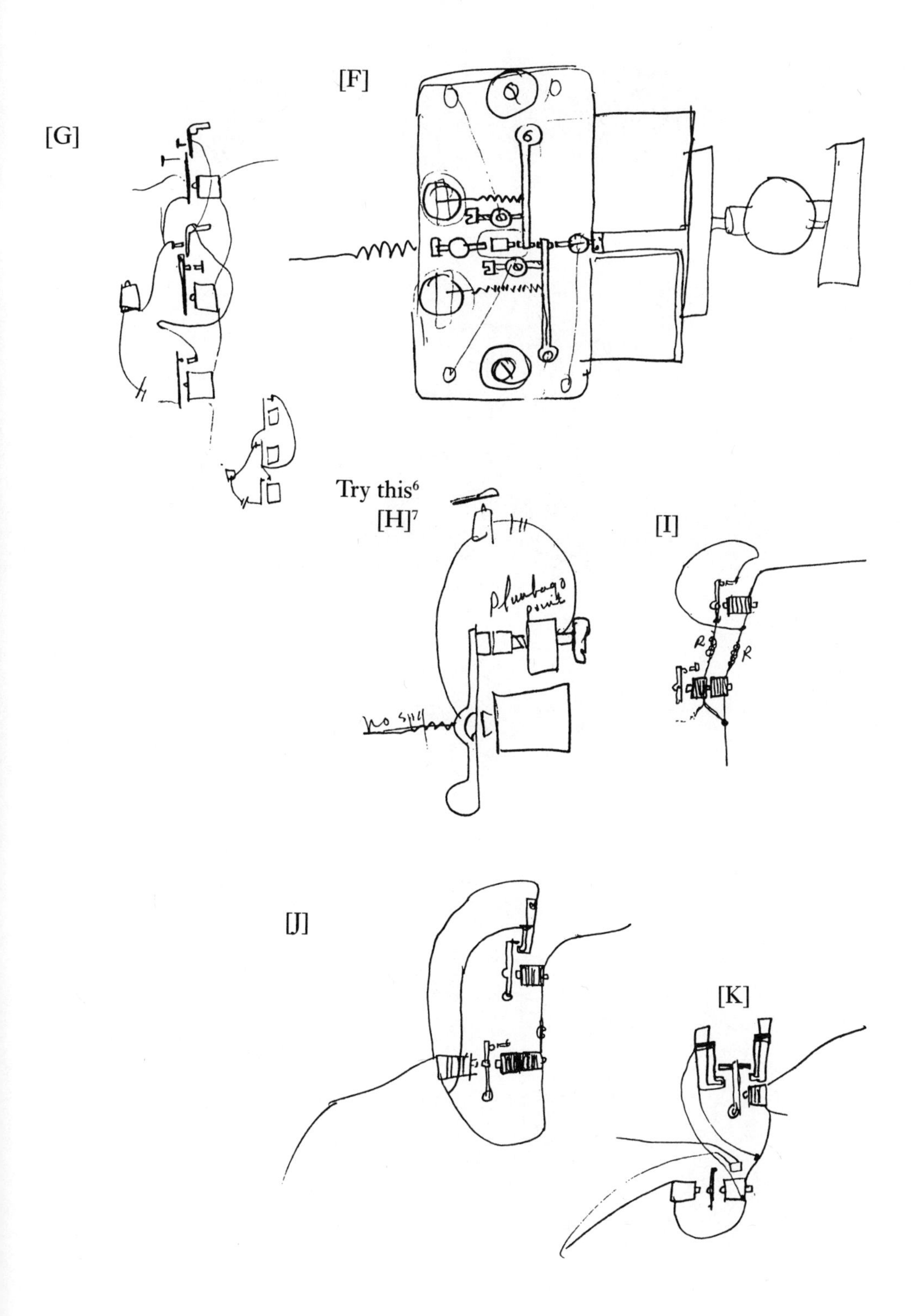
[F]
[G]
Try this[6]
[H][7]
plumbago point
no spg
[I]
R
R
[J]
[K]

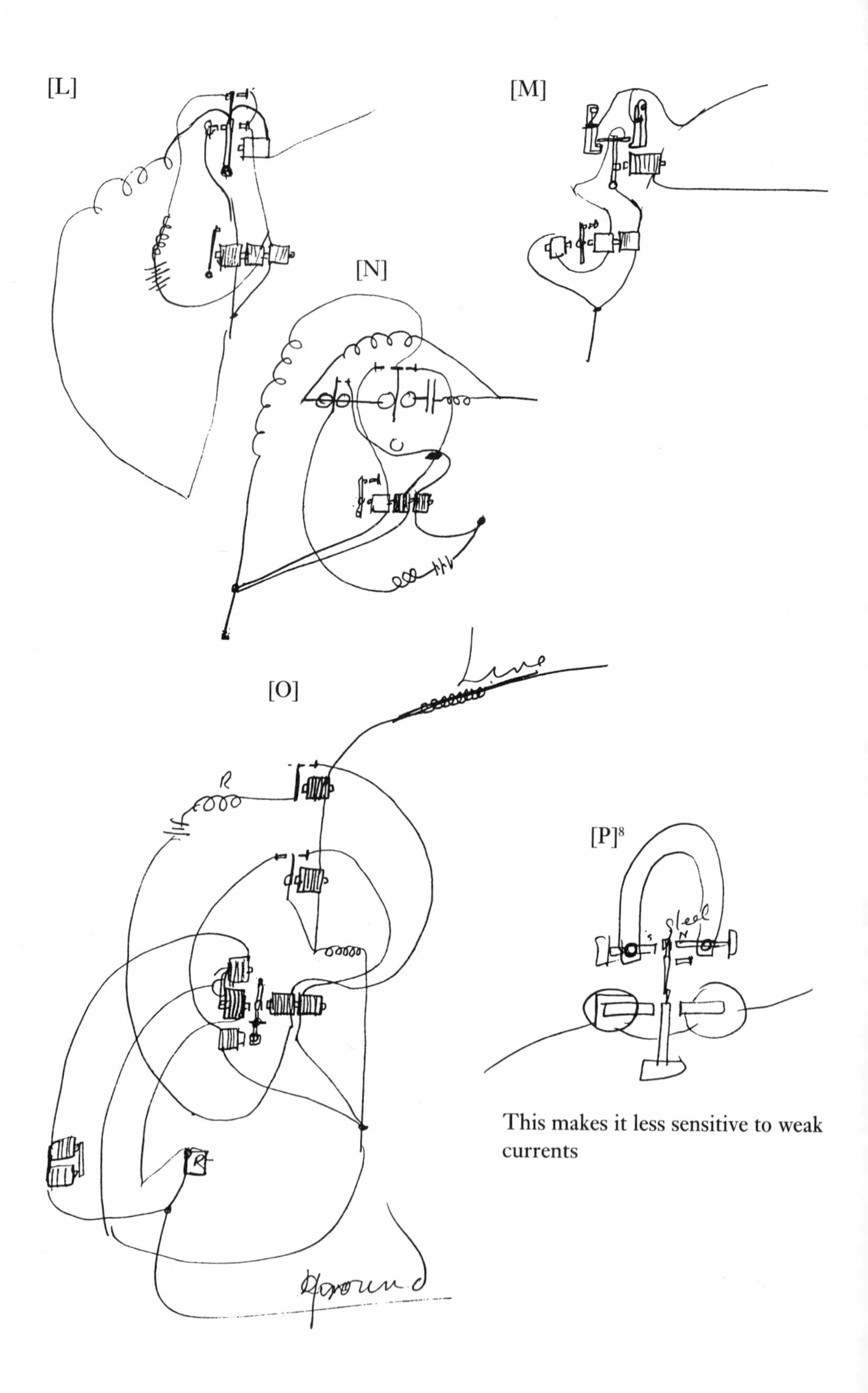


This makes it less sensitive to weak currents

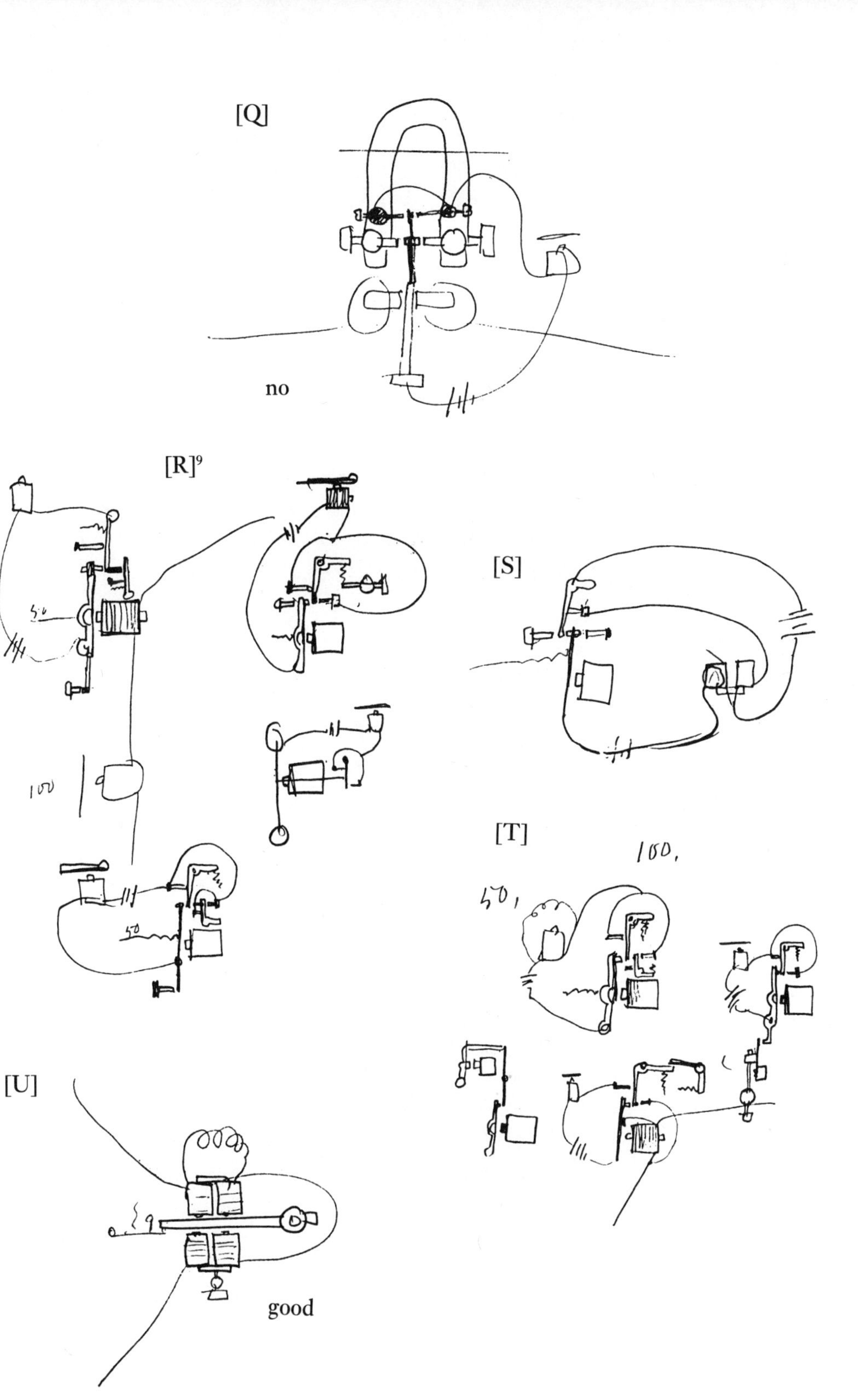
[Q]
no
[R][9]
50
100
50
[S]
[T]
100,
50,
[U]
good

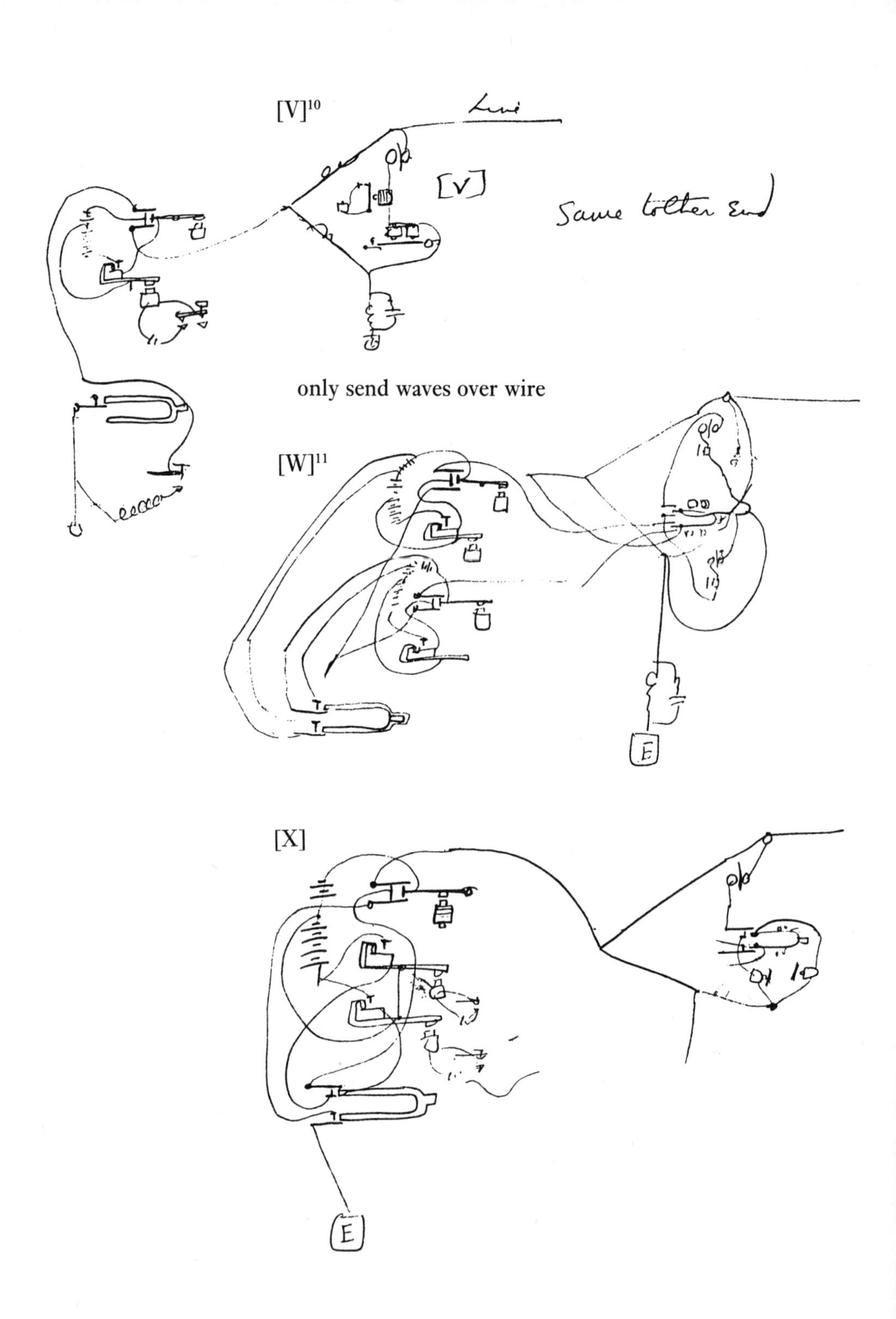
[V][10]
Line
[V]
Same to other end
only send waves over wire
[W][11]
E
[X]
E

[Y]

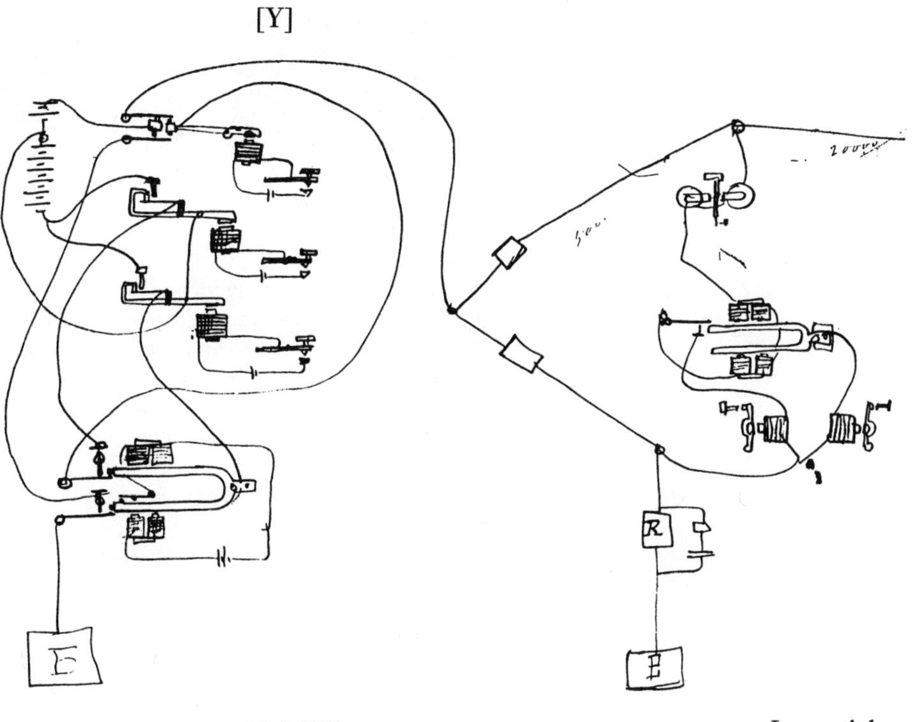

T A Edison

James Adams[a]
Chas Batchelor

X, NjWOE, Lab., NS-77-004 (*TAEM* 7:467, 468, 466, 470, 473, 471, 474, 472). Document multiply signed and dated. [a]"the shit" appears on one page under Adams's signature; "the [gut?]" appears under his signature on another.

1. See headnote, p. 280.

2. Edison continued his experiments and tests with the sextuplex during the following weeks. On 22 April, Batchelor noted that "we made sextuplex work very successfully." Cat. 1233:112, Batchelor (*TAEM* 90:109).

3. The right side of this sketch is a transmitting circuit, with only one key shown (at center).

4. Sketches B, C, and D are ideas for receiving circuits. These drawings, like most others in this document, focus on particular parts or aspects of sextuplex design, and their exact operation is uncertain without such information as current intensity, spring strengths, and circuit arrangements.

5. The instrument design shown in drawings E and F—a relay that could activate or cut out different receiving circuits depending on the strength of the incoming current—was central to much of Edison's sextuplex work (F is a top view of the relay). In the diagram (p. 308), **A** is the end of the armature lever (seen more fully in drawing E) for magnet **MM**; the magnet is in the main line circuit. A local circuit (not shown) is connected to **A,** which is held back against contact point **B** by spring **S″** when no current runs through **MM.** A weak current passing over

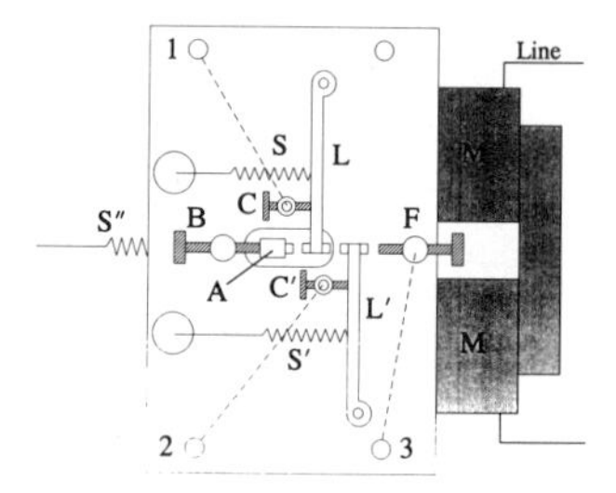


the main line and through **MM** will pull **A** against lever **L**, which is held back against **C** by spring **S**. Current can then pass from **A** through **L** and **C** to binding post (terminal) **1**. A stronger current through **MM** will pull **A** hard enough to overcome **S**, and move **L** to touch **L′**. Then current will pass through **A**, the points at the end of **L**, **L′**, and **C′** to binding post **2**. Finally, a still stronger single current (or a combination of the two weaker currents) will pull **A**, **L**, and **L′** against **F**, and the current will flow to binding post **3**. Edison tried many variations on this design over the next three months (for example, drawings R, S, and T in this document), with different mechanical and electrical arrangements depending on how he set up the local and line circuits.

6. This text appears in the midst of sketches E, F, and H; it is unclear to which it refers.

7. Figure labels are "plumbago point" and "no sp[rin]g." See Doc. 885.

8. This horseshoe magnet is an alternative to the double-magnet devices in Doc. 881 (immediately above the text "differential"). Figure labels are "S," "steel," and "N."

9. The numeric labels ("50," "100") on these sketches indicate the relative strengths of the currents to which the relays respond.

10. Figure labels are "line" and "same tother end." The tuning fork at lower left is an acoustic transmitter. The rapid pulses it sends are interrupted by the key just below it and are received by the (horizontal) reed at center right.

11. In sketches V, W, and X, Edison uses his acoustic transfer technology. In each case the transmitters are on the left (with a tuning fork at the bottom) and the receiving circuit is on the right. Sketch V has two polarity-reversing transmitters and two that increase and decrease the current, and actually appears to be an octruplex design. Sketches W and X have only one pole-changing transmitter each (at top), and the tuning fork switches the other two transmitters alternately on and off the line. See Doc. 754, fig. 8.

**–885–** [Menlo Park,] April 10 1877

*Technical Note: Relay and Telephony*

Pressure Relay[1]

manganese bl[ac]k ox[ide] or other inferior conducting material may be used[2]

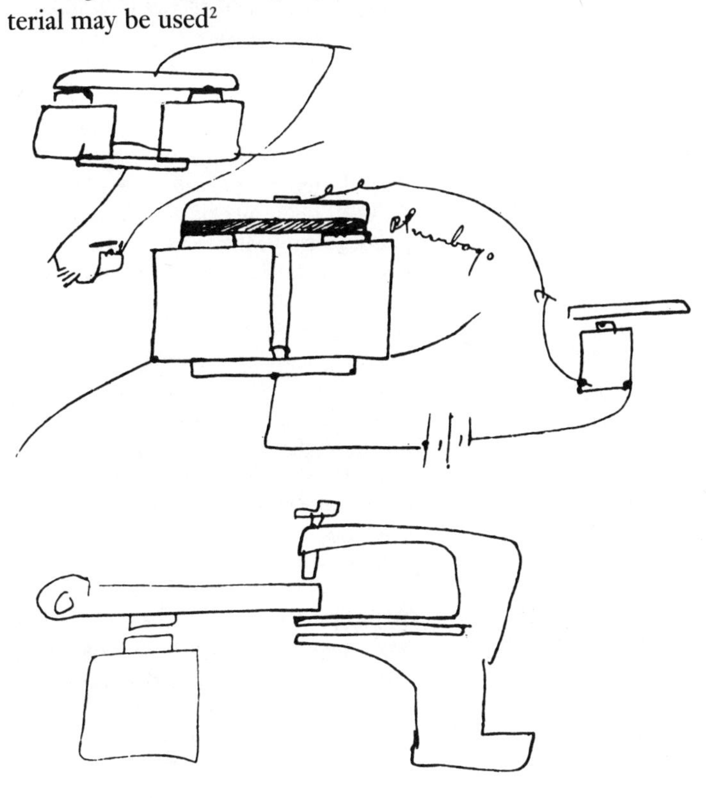


Spkg Telegh wkg by pressure Relay Principle
Batch & Jim Making it[3]

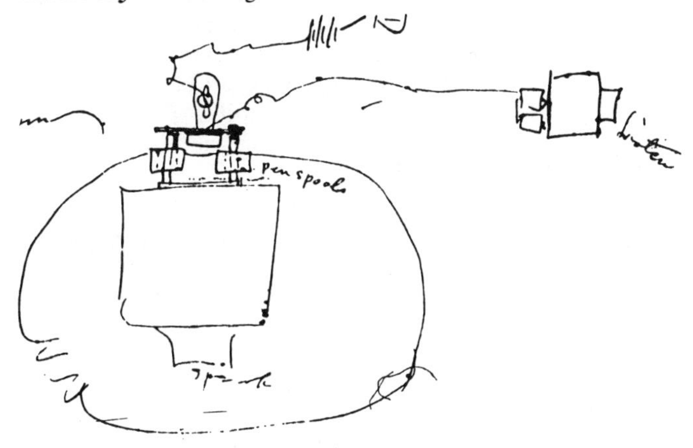


T A Edison

Chas Batchelor
James Adams

X (photographic transcript), NjWOE, TI 2, Edison's Exhibits 58-11, 57-11 (*TAEM* 11:262, 261). Document multiply signed and dated.

1. These sketches are the first that show the pressure relay. Six days earlier Edison had made a list of substances whose resistance varied under pressure (see Doc. 882), and on 16 April the shop delivered a finished instrument (Edison's Exhibit 60-11, TI 2 [*TAEM* 11:263]; see also Doc. 887). Although Batchelor stated in testimony that he thought they "had it working much longer than two months when it was illustrated in the *Journal of the Telegraph,* June 1, 1877" (TI 1:261 [*TAEM* 11:107]), these documents indicate otherwise.

The significance of the pressure relay lay in its potential as a telephone relay: it could transfer into a second circuit the fluctuating, relative strength of an incoming signal (although even as a telegraph relay this device had the advantage of needing no adjustment for varying line conditions). As the main current passed through the electromagnet coils, the changing attractive force on the armature variably compressed the plumbago. This altered the resistance in the local circuit, which passed through the electromagnet cores, plumbago, and armature. The pressure relay was used "on a regular telegraph line to Washington" from the laboratory, but whether with telephone or telegraph instruments is not clear (Batchelor's testimony, TI 1:232 [*TAEM* 11:93]). Several drawings of the relay, dated 10 May 1877, constitute Edison's Exhibit 86-11 (TI 2 [*TAEM* 11:271]).

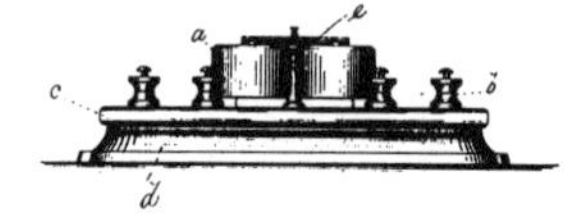

*A drawing for the Telephone Interferences of Edison's pressure relay.*

Apparently six of these instruments were made (Edison's testimony, TI 1:98 [*TAEM* 11:70]). None survives, but one with two armatures like those pictured here (the first with a solid bar of plumbago, the other with plumbago only over the electromagnet cores) was entered as an exhibit in the Telephone Interferences (Edison's Exhibit 58-11, TI 2 [*TAEM* 11:651]). The 1 June 1877 *Journal of the Telegraph* article ("Edison's Pressure Relay") referred to by Batchelor was entered in evidence as well (TI 2:629 [*TAEM* 11:731]).

Edison executed a patent application for the pressure relay on 29 July 1880 (U.S. Pat. 434,585); it issued 19 August 1890 after substantial alteration of the claims. Although Edison had earlier patented a device that passed a current through two needles that approached and receded from each other in a liquid (U.S. Pat. 141,777), it functioned solely as a switch, not a variable resistance.

2. Figure label is "plumbago." The meaning of the bottom sketch is not clear; Batchelor could not identify it in testimony. TI 1:261 (*TAEM* 11:107).

3. Figure labels are (clockwise from bottom) "speak," "pen spools," and "Listen." This design is illustrated and discussed in Prescott 1879 (530). For a slightly different version of this configuration, see Doc. 889. A copy of this drawing is in Vol. 11:57, Lab. (*TAEM* 3:937).

**-886-**

*From Josiah Reiff*

New York, Apl 132th 1877[a]

Confidential[b]

My dear T. A.

I have offered everything possible to meet the views of L[owrey]. who of influence O[rton]. but he (L) has insisted only within past few days, that we shall absolutely transfer as prerequisite, "Case H," leaving the contest still to go on as to whether Case H & 99 are the same etc—& whether we have the title etc as against Prescott.[1]

I of course proposed to agree to deliver, but as it was necessary to get the signatures of all our people for an actual[c] transfer it was impracticable unless I could shew them they were to get ~~it~~ something sure.[2] But they insist on having it now say we have nothing else they care for & that there is no way we can assist them— They also admit that case H involves the whole case & hence their anxiety to get the actual title but without real consideration— The fact is they gained almost all they wanted when they induced you to sign your contract for the future, without insisting as a consideration that all our matters certainly with Quad should be fixed—

I feared it & told you so. I see no other course than to shew our case in court in the A&P suit against WU—[3] They will then see that L is not altogether right & they will want to make the arrangement with us. I have sincerely tried to come together but they simply spurn us for the past, being in possession of Quad & also having your cooperation which they value very highly. I will not give up yet but you will see my difficulty— Yrs

JCR

They will respect R.W.R[ussell]. before many days.

ALS, NjWOE, DF (*TAEM* 14:650). Letterhead of J. C. Reiff. [a]"New York," "187" preprinted. [b]In top left margin. [c]Obscured overwritten letters.

1. See *TAEB* 1:556–57, 2:254–55.

2. This refers to the continuing American Automatic Telegraph Co. interest in the quadruplex. See Doc. 876 n. 1.

3. *Atlantic & Pacific v. Prescott & others,* filed a year previously in a New York state court. Quad. 70, 71, 73 (*TAEM* 9:288–10:797).

## -887-

*Notebook Entry: Relay*

[Menlo Park,] April 16th 1877

Chemical Repeating Magnet

Edison made a relay[1] on the principle of the difference of resistance on Plumbago[2] when under pressure of different degrees.

Fig. 40

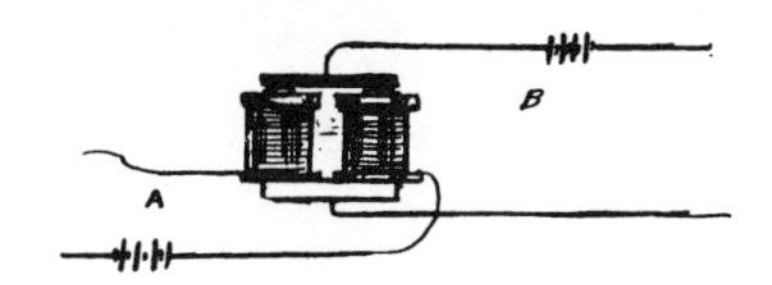

By taking a magnet (Electro) as in Fig 40 and placing the small cakes of Plumbago on the cores and letting the armature rest on these. When circuit A is open the armature does not press down and consequently circuit B is such a resistance that it is practically open, but when A is closed the magnet pulls the armature down with force, and the circuit B is worked with a force that is at all times regulated by the circuit A.[3]

X, NjWOE, Batchelor, Cat. 1317:34 (*TAEM* 90:674). Written by Charles Batchelor.

1. The shop delivered a pressure relay on this date. Edison's Exhibit 60-11, TI 2 (*TAEM* 11:263).

2. On this same date Edison's Exhibit 61-11 (TI 2 [*TAEM* 11:264]) shows a list of substances besides plumbago that Adams tested with the pressure relay in a telephone circuit.

3. A laboratory sketch indicates that Edison contemplated or actually experimented with the pressure relay on this day. NS-77-004, Lab. (*TAEM* 7:484).

## -888-

*Technical Note: Telephony*

[Menlo Park,] April 18 1877

Emg Spkg Tel Receiver Emg[1]

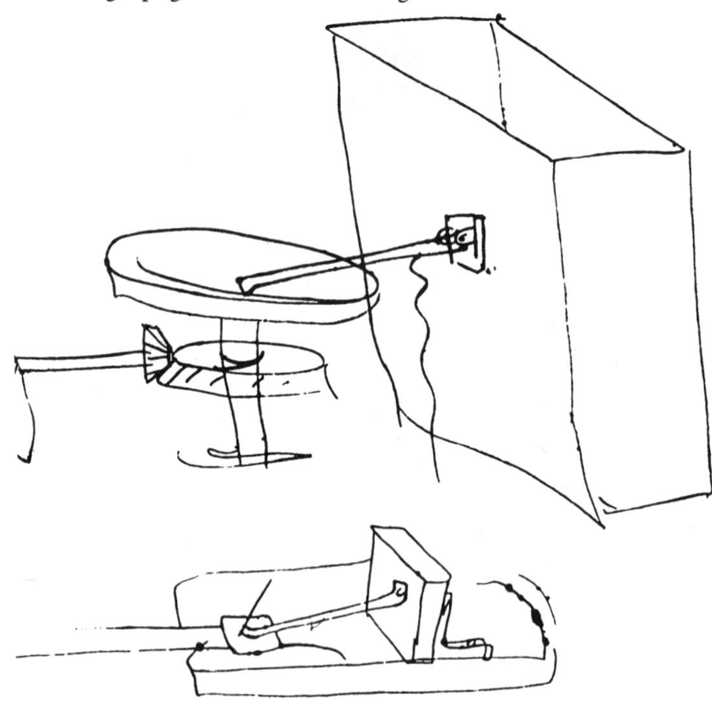

T A Edison

James Adams
Chas Batchelor

X (photographic transcript), NjWOE, TI 2, Edison's Exhibit 62-11 (*TAEM* 11:265).

1. In this electromotograph telephone receiver a lever rests near the edge of a rotating horizontal disk that is coated with a chemical.

In the second sketch, the rotating disk arrangement is replaced by Edison's standard chemically treated paper tape.

Batchelor's diary entry for this day noted "that the Electromotograph principle applied to singing telegraph makes it much better." Cat. 1233:108, Batchelor (*TAEM* 90:107).

**–889–** [Menlo Park,] April 19 1877

*Technical Note: Telephony*

Spkg Telgh[1]

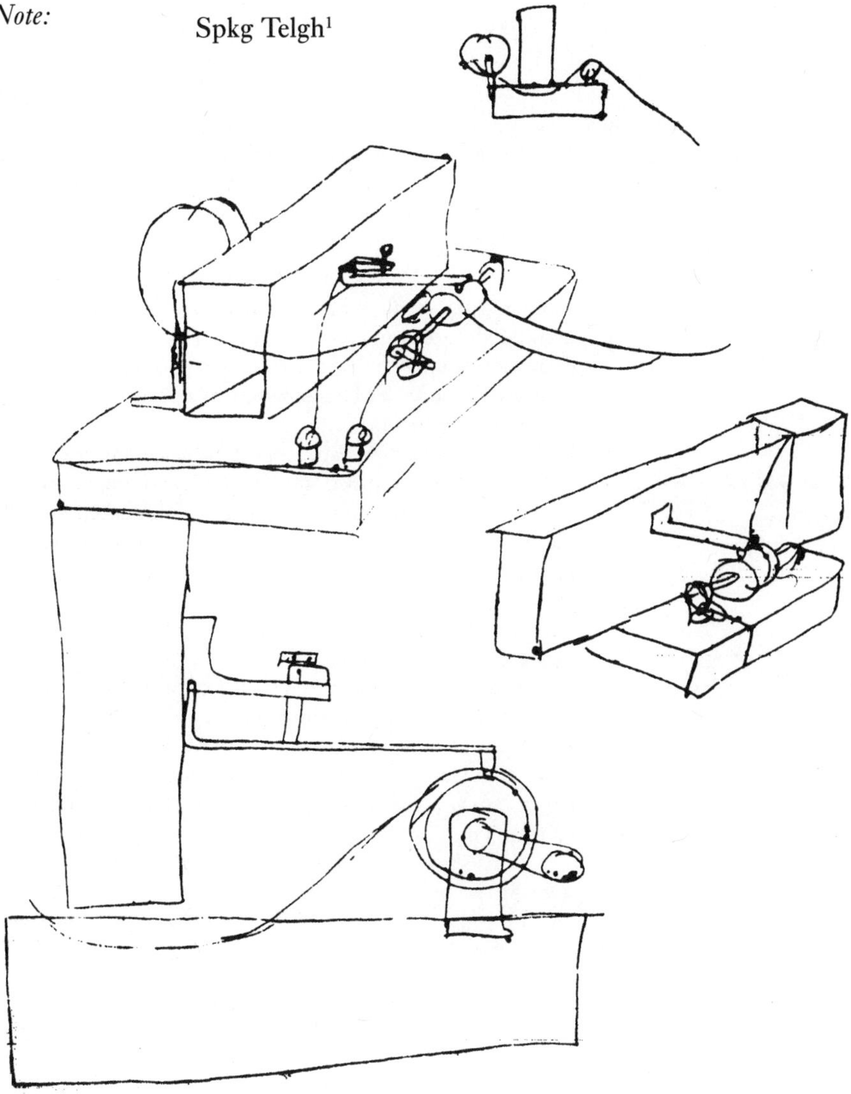

Speaker
Try this[2]

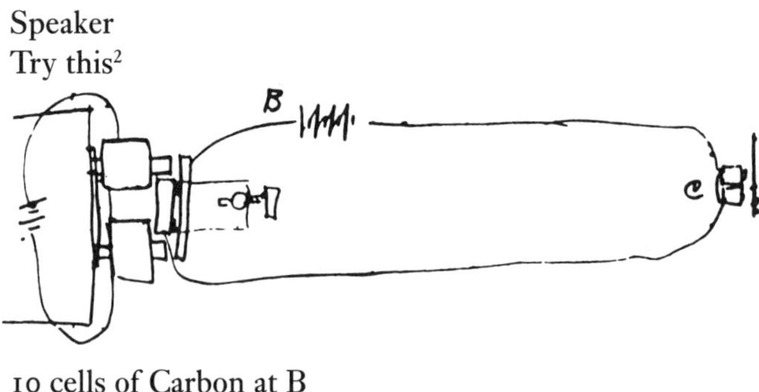


10 cells of Carbon at B
20 " " " " "
30 " " " " "

Make Motograph [re]ceiver[a] & try at C
T A Edison

Chas Batchelor
James Adams

X (photographic transcript), NjWOE, TI 2, Edison's Exhibits 65-11, 71-11 (*TAEM* 11:266–67). Document multiply signed and dated. "Spkg Telgh" and following drawings in Edison's hand; "Speaker . . . C" in Batchelor's hand. [a]Obscured by ink blot.

1. Although Edison labeled almost all of his telephone drawings "speaking telegraph," the electromotograph receiver shown here was used publicly only to receive music. On 28 April, Edison began a week of musical telephone demonstrations at the Newark Opera House that included the successful transmission of a duet and several compositions for wind instruments.

According to a newspaper report of the demonstration, "The transmitting apparatus consists simply of a long tube, having one end covered with a thin sheet-brass diaphragm, which is kept tight by a stretching ring. In the center of the brass diaphragm is soldered a thin disk of platina, and immediately in front of this disk is an adjustable platina-pointed screw secured to a rigid pillar." *Newark Daily Advertiser,* 2 and 4 May 1877; *New York Daily Tribune,* 6 May 1877; all Cat. 1240, items 147–48, 150, Batchelor (*TAEM* 94:46–47).

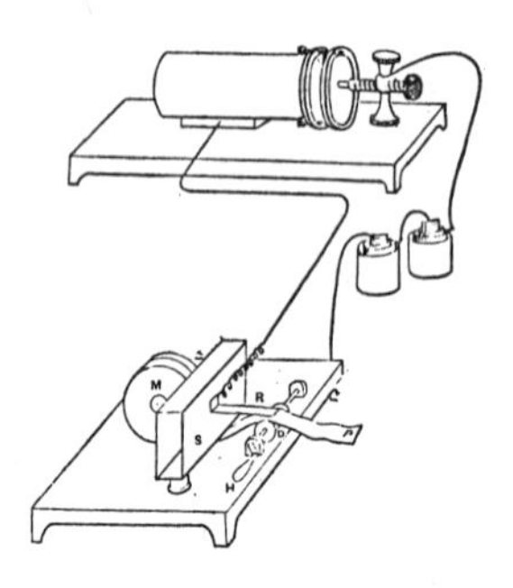

*A newspaper illustration of the Edison musical telephone on display at the Newark Opera House.*

On 19 April, Edison drew a sketch of the electromotograph receiver that appears to be the instrument used in the demonstration and wrote John Kruesi's name next to it (Vol. 11:63, Lab. [*TAEM* 3:939]). A shop order of the same day shows an electromotograph receiver with a much larger resonant box; it was delivered on 28 April (Vol. 11:70, 74, Lab. [*TAEM* 3:947, 952]).

2. This shows a variation on the pressure relay used in a telephone transmitter.

**–890–**

*From Albert Chandler*

[New York,] Monday Apr. 23. [1877][1]

As you address me Rev.[2] I address you, Prof.—or, to be more affectionate,

Dear Professor:—

I had an opportunity to-day for the first time, to present your suggestion for payment on a/c of salary, to Mr. G[ould].—[3]

He asked me to say to you that he will be very busy for two or three days, now,—but that, afterwards he would like to see you personally and have a good long talk.—[4] I think you would find such an interview useful.— In a hurry.—

Chandler

ALS, NjWOE, DF (*TAEM* 13:1203).

1. The reference to Edison's salary indicates that this letter could only have been sent in 1876 or 1877; 23 April fell on a Monday in 1877. Edison was officially appointed electrician of Atlantic and Pacific Telegraph Co. in June 1875 (see Doc. 585) and Western Union gained control of Atlantic and Pacific in August 1877.

2. Reference unknown.

3. Nothing is known of Edison's "suggestion."

4. It is unknown if this meeting occurred.

**–891–**

*Technical Note: Telephony*

[Menlo Park,] April 23— 1877

Spkg Telegraph

We get the best speaking from shield brass diaphram about 2 inches diam with ¾ inch surface round point of pressed black lead fastened to a 3/64 thick steel spring with screw with 1½ inches of centre of plumbago[1] & Receive on Emg ie[a] Resonant box with Sp[rin]g in centre Leading to & resting upon Chem paper moved by drum & belt & wheel   The singing with rigid platina point instead of plumbago is simply perfect= It is quite probably that for speaking Telegh something can be found better than plumbago[2]

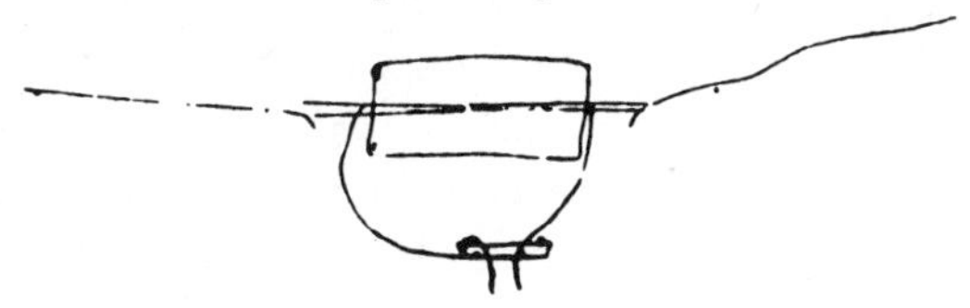

T A Edison

Chas Batchelor
James Adams

X (photographic transcript), NjWOE, TI 2, Edison's Exhibit 78-11 (*TAEM* 11:269). [a]Circled.

1. This transmitter was perhaps similar to one Edison sketched the same day (Vol. 11: 81, Lab. [*TAEM* 3:957]).

By "1½ inches of center of plumbago" Edison may have meant that he coated the center of the diaphragm.

2. The following, apparently unrelated sketch appears at the bottom of the page.

## -892-

*And Charles Batchelor and James Adams Agreement with Charles Holland and George Bliss*

April 24, 1877[a]

This Agreement[1] entered into this 24th day of April, 1877, by and between Thomas A. Edison, Charles Batchelor, and James Adams all of Menlo Park, Middlesex County, State of New Jersey, parties of the first part; and Chas E Holland[2] of Hancock, Houghton Co, State of Michigan, and George H. Bliss of Chicago, Cook County, State of Illinois parties of the Second part witnesseth that:—

Whereas, the said parties of the first part are the sole owners of all the right title and interest of and in certain patents granted in the Kingdom of Austria October the sixth 1876 and numbered (26935), in the Kingdom of Italy September 15th 1876 and numbered (8733), in the Kingdom of Belgium May 31st 1876 and numbered 39502 and in the Republic of France July 10 1876 and numbered (112,719) for an Electric Pen and duplicating press, with accessories and

Whereas the said parties of the first part have established a trade in such articles through certain agencies in various countries of the world and in some countries where patents cannot be obtained and,

Whereas the said parties of the second part are desirous of obtaining the business, goodwill, and rights already established by the said parties of the first part and the sole right to manufacture and sell, under the patents already granted in the countries aforesaid and under the patents to be obtained in other countries, for their sole use and benefit for a period of five 5[b] years from date hereof

Therefore be it Agreed that for and in consideration of the sum of one 1[b] dollar paid by the parties of the second part to each of the said parties of the first part, the receipt of which is hereby acknowledged, and for other valuable considerations hereinafter mentioned, the said parties of the first part do hereby transfer to the parties of the second part the business goodwill and rights as now established, and the sole right to manufacture and sell within all countries except the United States of America, Canada, and the Kingdom of Gt Britain and Ireland; the articles covered by the above recited patents

It is also agreed that the said parties of the second part shall pay to the parties of the first part 3[b] three dollars royalty on each and every complete machine sold within, or ~~sent~~ to, any country (whether a patent exists in such country or not) except the aforesaid United States, Canada and Great Britain and Ireland, and proportionately on the parts thereof, and also 15[b] fifteen per cent of the selling price of all supplies connected therewith, which payment shall continue for a period of 5[b] five years from the date of this agreement

It is also agreed that the said part~~y~~ies of the second part shall guarantee and bind themselves and do hereby guarantee and bind themselves to use diligence and energy in introducing the said apparatus in the various parts of the world

It is further agreed that if six months first previous to the expiration of the five years, the said party of the second part shall notify the said parties of the first part that they desire to continue this contract they may do so upon the same terms, providing they give such notice in each and every year they desire its continuance

It is also agreed that should the parties of the second part desire at any time to discontinue this contract they may do so by giving to the parties of the first part six months previous notice and retransferring the business, good will, and rights, herein specified.

It is also agreed that upon the first day of January in each and every year the said parties of the second part shall make full and true returns to the parties of the first part under oath, in the usual manner, and that the said royalty shall after the expiration of four months after date hereof be paid monthly to the said parties of the first part in the proportion of 67½[b] sixty seven and one half per cent to Edison, 22½[b] twenty two and one half per cent to Batchelor and 10[b] ten per cent to Adams

It is also agreed that the said parties of the first part take out patents in such other countries as is necessary for protecting the trade of the party of the second part

It is also agreed that the said parties of the second part will cause to be manufactured in the Republic of France and Kingdom of Austria before October the first 1877 the said apparatus as called for by the law under which the patents in such countries were granted, and will place on sale in the Kingdom of Belgium the said apparatus before the first day of June 1877 so as to comply with the law regarding the working of inventions under the Belgian patent law

It is also agreed that should the said parties of the second part desire to purchase outright the patent for the said Electric pen and duplicating press in any foreign country they may do so at any time during the continuance of this contract for the sums named in Schedule A annexed, which sums must be paid in gold and upon payment of which all royalty on apparatus or supplies sold in such countries shall cease

It is further agreed that all improvements upon the apparatus covered by the above recited patents shall be included in this contract, and any Duplicating Apparatus which the said parties of the first part may invent and which is capable of performing the same work as the Electric pen and which would be a competetor if introduced shall also be included in this contract subject however to an equitable arrangement of Royalty should the new system or any part of it be adopted in place of the Electric pen and Duplicating press or any part thereof.

It is also agreed that this contract is not transferable on the said parties of the second part without the written consent of the said Edison.

It is further agreed that the said Edison shall transfer and he does hereby transfer during the continuance of this contract the priviledges granted to him by the Western Electric Manufacturing company under a contract dated 1876[3] of purchasing each complete duplicating apparatus at $12^{50}/100^{b}$ twelve dollars and fifty cents each and the parts of such apparatus at the price named in a schedule to such contract marked C a copy of which is hereto annexed and marked Schedule B

It is further agreed that the said parties of the first part will accept $1^{75}/100^{b}$ one dollar and seventy five cents as full royalty on each and every complete apparatus sold to Messrs Fogg and Co for shipment to China and Japan during two $2^{b}$ years from the date hereof, providing that the said parties of the second part are compelled to carry out an agreement entered into by & between the Edison Electrical pen and Duplicating press Company and H. Fogg and Co dated Feb 1st 1876 a copy of which is hereto annexed and marked schedule C.[4]

It is also agreed that in case a new arrangement with the Messrs Fogg and Co. is entered into whereby they release their rights under said contract then the said parties of the first part shall receive their full royalty of $3^{b}$ three dollars as in all other sales.

In witness whereof all the said parties of the first and second part have hereunto set their hands and seals

Thos. A. Edison[c] Chas Batchelor[c]
James Adams[c]
Witnessed by John Kruesi Chas. P. Edison[d]
Chas E. Holland[c] Geo. H. Bliss[c]
Witnesses to the signatures[e] E. T. Gilliland G. A Mason[5]

Schedule A:—

For the French Patent (7000) Seven thousand dollars
For the Austrian Patent (5000) Five thousand dollars
For the Prussian Patent (4000) Four thousand dollars
For the Russian Patent (4000) Four thousand dollars
For the Italian Patent (3000) Three thousand dollars
For the Australian Patent (5000) Five thousand dollars
For the Spanish Patent (3000) Three thousand dollars
For the India Patent (5000) Five thousand dollars
For the Belgian Patent (3000) Three thousand dollars.[f]

Schedule B.

Prices of parts and supplies

| | |
|---|---|
| 7 x 11 press | $3.50 |
| Roller | .80 |
| Pen | 4.00 |
| Battery | 3.84 |
| Pen stand | .25 |
| Pen cord | .14 |
| Watch oil | .04 |
| Screwdriver | .04 |
| File | .10 |
| Pair Zincs | .16 |
| Bottle of Ink | .26[f] |

Schedule C:—[6]

Feb 1st 1876.

Messrs H. Fogg and Co
Gentlemen,

We hereby appoint you our agents for the sale of Edisons Electrical Pen and Duplicating press in China and Japan for a period of 3[b] three years, after the first shipment has arrived at its destination.

We shall protect you from intrusion on your territory as far as lays in our power so to do. We agree on the first order to deliver them in any part of New York for the sum of 16[b] sixteen dollars each and that in the event of the machines not being sold in China or Japan or elsewhere we will take them back at the same price, and pay you the freight and charges, provided they be returned in good order and that we shall have received the money within 30[b] thirty days after delivery in New York City

If these terms are satisfactory to you please send order and where to deliver  Respectfully yours

Edison Electrical Pen and Duplicating Press Co
(Sd) Chas Batchelor  Genl Manager.

DS (letterpress copy), NjWOE, Miller (*TAEM* 28:1038). Written by Batchelor; signatures are not letterpress copy. [a]Date taken from text, form altered. [b]Circled. [c]Followed by representation of a seal. [d]Brace spans signatures of Edison, Batchelor, and Adams, indicating they were witnessed by Kruesi and Charles Edison. [e]Brace spans signatures of Holland and Bliss, indicating they were witnessed by Gilliland and Mason. [f]Followed by centered, horizontal line.

1. Edison used a printed copy of the Articles of Association and By-Laws of the American Automatic Telegraph Co. to make what appears to be an earlier draft of an agreement regarding his foreign electric pen patents. 75-012, DF (*TAEM* 13:489).

2. Nothing is known of Charles Holland apart from his association with George Bliss in promoting Edison's electric pen.

3. Doc. 817, dated 28 November 1876.

4. See also *TAEM-G1*, s.vv. "Fogg, (W. H.) & Co."

5. Nothing is known of G. A. Mason apart from his role as general manager of the American Telephone Co., Ltd., of London in the mid-1880s. *TAEM-G2*, s.v. "Mason, G. A."

6. A somewhat more detailed letter of 28 January 1876, appointing Fogg and Co. agents for China and Japan, is in Lbk. 2:22 (*TAEM* 28:359).

**-893-**

*To John Breckon*

[New York,] April 25 [187]7

Dear Sir,

Your last note in regard to our contract received.[1] I shall hold myself responsible and shall see that our contract is carried out to the letter.

I have placed the manufacture of the 'Pen' into the Western Electric Co hands simply because they have so much better facilities for turning out large quantities.

The delay experienced in shipping the last lot of pens was unavoidable and cannot possibly occur again.[2]

I have made some improvements lately and have a new pen operates[a] simpler and not likely to get out of order  I am having a few made and shall send you one[3]

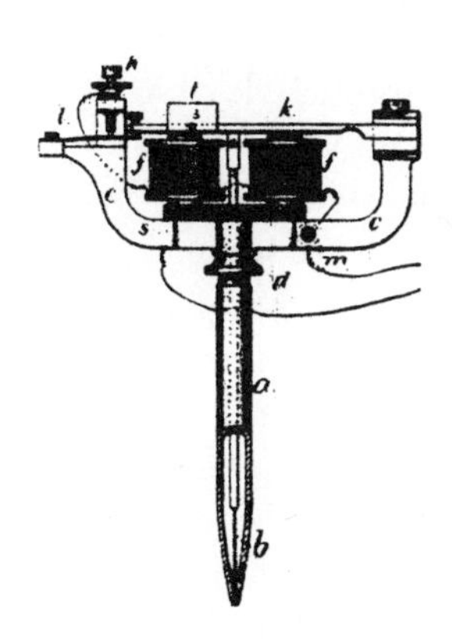

*A patent drawing of an electric pen driven by a vibrating reed.*

I have not received your telegram in regard to shipment. We have notice of 50 pens 50 stands & 50 Rollers on the way here for London and the others to follow before the month is out

Hoping you will not be inconvenienced by this arrangment[4] I am Very respectfully Yours

Thos A Edison per Batchelor

L (letterpress copy), NjWOE, Lbk. 3:178 (*TAEM* 28:644). [a]Obscured overwritten letters.

1. Doc. 879.

2. The delay was probably caused by production problems resulting from the change in manufacturers. On 12 March, after discovering "that there had been no platina on any of the batteries sent out by W E Mfg Co," Charles Batchelor had written Western Electric to complain about the poor quality of their output, which necessitated repairs at the Menlo Park laboratory. At the beginning of April he wrote that although the quality had improved there were still problems, and he had spent two nights fixing pens for one of Breckon's orders. Batchelor therefore suggested a number of alterations intended to improve manufacturing. Although still finding problems at the beginning of May, Batchelor indicated that the "work generally is greatly improving." Cat. 1233:71, 92–93, Batchelor (*TAEM* 90:88, 99); Batchelor to George Bliss and Western Electric, 12 Mar. and 4 Apr. 1877, Lbk. 3:97, 136 (*TAEM* 28:572, 609); Batchelor to Bliss, 6 Apr. 1877, and Batchelor to Western Electric, 2 May 1877, Cat. 1238:93, 108, Batchelor (*TAEM* 93:96, 107).

3. Besides working to develop a rotary power press (see Docs. 843, 852, and 854), Edison also applied for patents on several improvements, although there is no evidence that these were ever introduced commercially. The patents included a pen worked by a universal joint attached through a shaft to a foot-pedal or other motive force (U.S. Pat. 203,329) and a pen driven by air or water (U.S. Pat. 205,370). The simple pen design mentioned here is probably a pen driven by a vibrating reed (U.S. Pat. 196,747, executed on 18 April), which Batchelor drew in his diary for 7 April, when he noted, "Kruesi started on Reed pen He was to make 2 one for Bliss and the other for Patent office" (Cat. 1233:97, Batchelor [*TAEM* 90:101]). In the reed pen Edison moved the weight of the electromagnets from the side of the pen to the top. The competing, pneumatic pen was of this design and Breckon's assistant Frederic Ireland thought it an important improvement. Ireland to TAE, 11 May 1877, DF (*TAEM* 14:390).

4. On 10 May, Breckon responded that if Edison would see that Western Electric provided pens "with as good a finish upon them as those we received from you a few months ago that we shall be perfectly satisfied." DF (*TAEM* 14:388).

–894–

*From Pitt Edison*

[Port Huron,] May 1st 1877

Dear Bro

I have not received any letter from you, since you left here[1] did you get a letter from me the RR chaps are having a little trouble among themselves. Just at present thay have not Elected a Supt yet Cole Wastell & Beard are all after it and it is hard for anyone of them to get a majority of the Board of Directors but thay will have to do something soom for the road is getting out of repair and some one will have to be[a] putting it in order soon thay have not made the[b] connection of the two roads yet[2] thay are running the new road down to the Huron House[3] and the old road from the huron House to C[hicago]&L[ake] H[uron] Depot and change passengers at the office.

have thay sent you any statements of the earnings & disbursements let me know what thay are if thay have[4] I am Keeping a record of the changes in any property thay dispose of Wastell made me a propposition to get him elected Supt and than for me to do the work and draw the pay but I hardley think that me[c] him could get along smoothly after so much trouble and I told him so but he has no confidence in his ability to handle the road[5] if I could look over thair books I could keep you posted and had any stock I could attend thair Directors meetings and know all that was agoing on Stewart will not attend any meetings so thay have it all thair own way the Sarnia road is picking up I will send you a statement up to first of May[6]

Now Al I owe J D Carleton about thirteen hundred dollars and he has got 2000. in stock and he is a crowding me on it I went to Symington and asked him if he would lend me the amt on the same trm that he~~d~~ did the $1500. that was two years and he said if I would give him 3000 in stock he would now what I want is for you to send me 1000 stock and I will let them have[a] the 3000 take a rec[eip]t for it in your name and send it to you than you will hold thair rectns for $10,000.00 in stock the stock with the 1000 you wi~~l~~ll hold will more than pay the amt before it is due and that will make me feel easy for I would rather owe Symington $3000.00 than to owe J D Carleton $3.00 I would here less about it now Al dont fail to do this for I want to get out of his clutches now and forever I have got rid of Stewart and now I want to get rid of Carleton and than I think I can sleep easy so dont fail to send stock and I will send you their rect for the 3000 so no more at present how is Charley and dad and all the rest of the folks we are all well Y—

WPE

ALS, NjWOE, DF (*TAEM* 14:548). [a]Interlined above. [b]Obscured overwritten letters. [c]Followed by "over" to indicate page turn.

1. See Doc. 871.

2. That is, the lines of the old Port Huron and Gratiot Street Railway and the City Railroad.

3. A large hotel in Port Huron. *History of St. Clair County* 1883, 492.

4. Edison began to receive regular statements of earnings and disbursement in May 1877. DF 77-016, passim (*TAEM* 14:519–634).

5. According to Jenks (n.d., 14), John Sanborn was elected president and Pitt Edison was elected superintendent, filling that role until he retired in 1883, at which time the Edison interest in the road was sold.

6. Not found.

**-895-**

*From Josiah Reiff*

[New York,] 9:20 am May 3/77

Dear Tom—

I have a certified copy of the Welch paper,[1] which I want to show you, at once. It contains acknowledgements & promises which you have probably forgotten— Still I think it is absolutely worthless— yet WU have doubtless considered it & probably own it—

The evident intent now is to either to win or break the Patent[2] by proving (or trying to do so) that is nothing new in anything you have done—

This would not only be safe & profitable for WU, but nuts to[3] Ashley, Pope & all the others ~~y~~who have envied your ability & failed in their own efforts to discover or invent

This morning I feel as much interest in preserving your reputation as an Electrician as in getting money thru the suit—

Friend Dickinson[4] saw he had touched on tender ground & hence told you yesterday he did not want to reflect on your ability, or belittle the Patents etc—[5]

It is too late— ~~t~~He has shewn his hand— B.F.B[utler].[6] is not yet through with him

Will be at my office until ¼ of 11— Trly

J.C.R

ALS, NjWOE, DF (*TAEM* 14:667).

1. Doc. 61.

2. This refers to Case 99 (Doc. 472), which was the main quadruplex

patent application and the subject of Edward Dickerson's cross-examination of Edison on 2 May. Quad. 70.7, pp. 360–82 (*TAEM* 9:547–58).

3. That is, a source of pleasure to. *OED*, s.v. "Nut."

4. Edward Dickerson, one of the attorneys for Western Union in *Atlantic & Pacific v. Prescott & others.* Quad. 70, 71, 73 (*TAEM* 9:288–10:797).

5. Reiff is probably referring to Dickerson's statement to Edison, "I am not asking you about the merits of your invention; I have no doubt but that it is a new and valuable one." Quad. 70.7, p. 361 (*TAEM* 9:547).

6. The controversial congressman, lawyer, and former general Benjamin Butler was one of the attorneys for Jay Gould's Atlantic and Pacific Telegraph Co. in its suit against Western Union over control of patent rights for Edison's quadruplex telegraph designs (see note 4). In previous years he had appeared on Edison's behalf in related actions. See *TAEB* 2:374, 491, 806.

## -896-

*From William Orton*

New York May 3d 1877[a]

I want to see you tomorrow (Friday) ten o'clock without fail. I sent for you only a minute after you had left.[1]

Wm Orton

L (telegram), NjWOE, DF (*TAEM* 14:54). Message form of Western Union Telegraph Co. [a]"187" preprinted.

1. Edison had just completed his primary testimony (begun on 26 April) in *Atlantic & Pacific v. Prescott & others* at the New York Superior Court (Quad. 70.7, pp. 221–430 [*TAEM* 9:477–583]). It is not known whether Edison saw Orton the next day.

## -897-

*Technical Note: Multiple Telegraphy*[1]

[Menlo Park,] May 3rd 1877[2]

Sextuplex

Plans of working 3 messages in the same direction for ~~Qu~~Sextuplex working=[3]

| | |
|---|---|
| 1 Reversals[a] | Polarized relay 50 perm |
| 2 Increase & d[ecrease] | Common " 75 inc |
| 3 Increase & d[b] | Common " 150 " |
| Reversals[a] | Polarized relay ~~50~~75 perm |
| Increase & d | Com " 100 inc |
| Acoustic—[b] | Acoustic " breaks total |
| Reversals[a] | Polarized relay 50 perm |
| Very Small reversals of the 1st long reversals | " or acoustic same |
| Increase & d[b] | Com Relay[c] |

| | | |
|---|---|---|
| 1 Increase & d | Com relay | 75 |
| 2 Increase & d & Ex[4] relay[d] | Com relay | 150 |
| 3 Acoustic[b] | Acoustic | 50 perm |

3 increases 50 — 1500 — 200 —

T A Edison

Chas Batchelor
James Adams

X, NjWOE, Lab., NS-77-004 (*TAEM* 7:521). [a]Followed by horizontal line leading to "Polarized". [b]This and preceding two lines spanned by brace in left margin. [c]Followed by centered horizontal line. [d]Followed by brace spanning "Com relay 75" and "Com relay 150".

1. See headnote, p. 280.
2. On this day he also sketched multiple-contact polarized relays for his sextuplex system, and he drew various circuit arrangements employing multiple-contact levers. NS-77-004, Lab. (*TAEM* 7:522–26, 663).

*The work-order drawing for a multiple-contact polarized sextuplex relay made in the laboratory machine shop.*

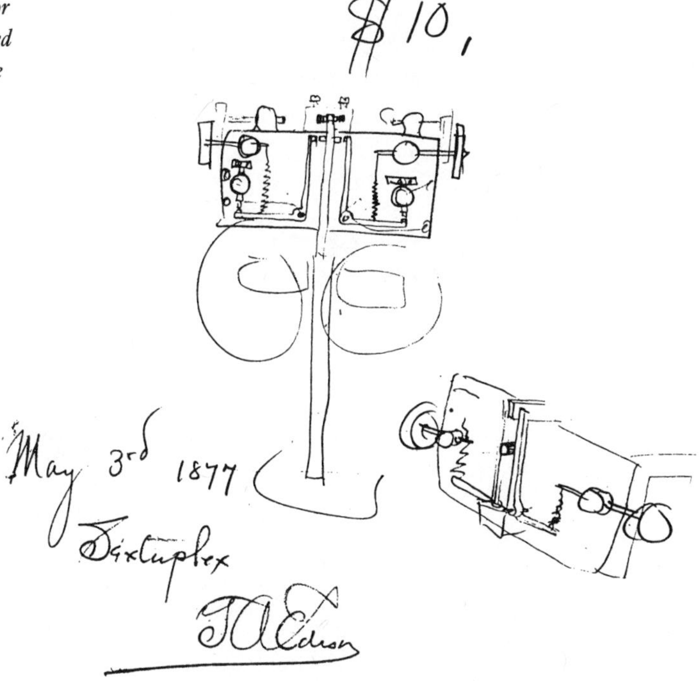


3. In the rightmost column the numbers "50," "75," "100," and "150" refer to relative current strength; "perm" indicates a current that is always on the line except for instantaneous interruptions when its polarity changes; "inc" means increase above the permanent current; and "breaks total" indicates that the acoustic transmitter is interrupting rather than modulating a current.
4. Perhaps "Extra," as in a bug trap.

**-898-** [Menlo Park,] May 6 1877

*Technical Note: Multiple Telegraphy*[1]

Improvement on the Quad[2]

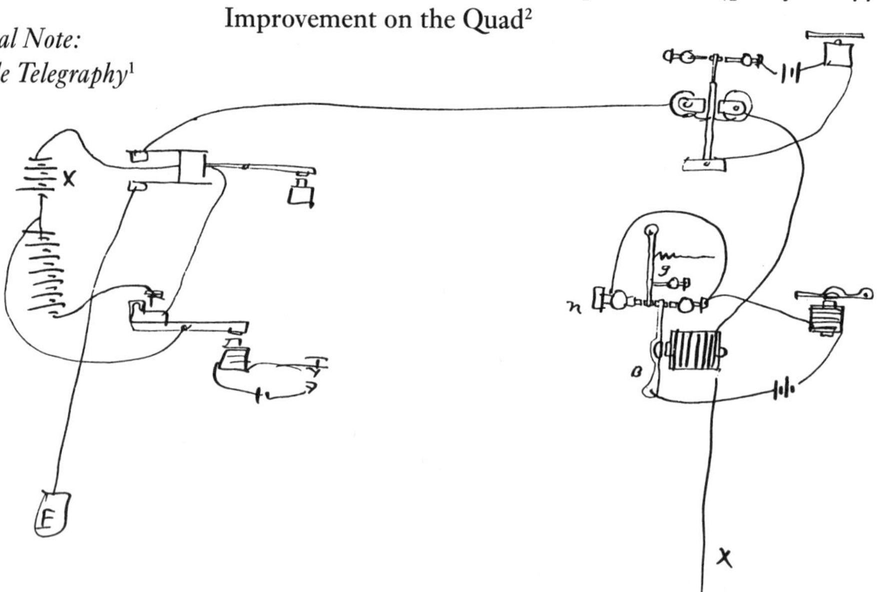


The permanent Current from X ~~is kept~~ keeps B from contact with .N. by reason of a constant magnetism but at the moment of reversal all magnetism disappears and B touches n through g. thus going from point to point keeps sounder closed=

Sextuplex[3]

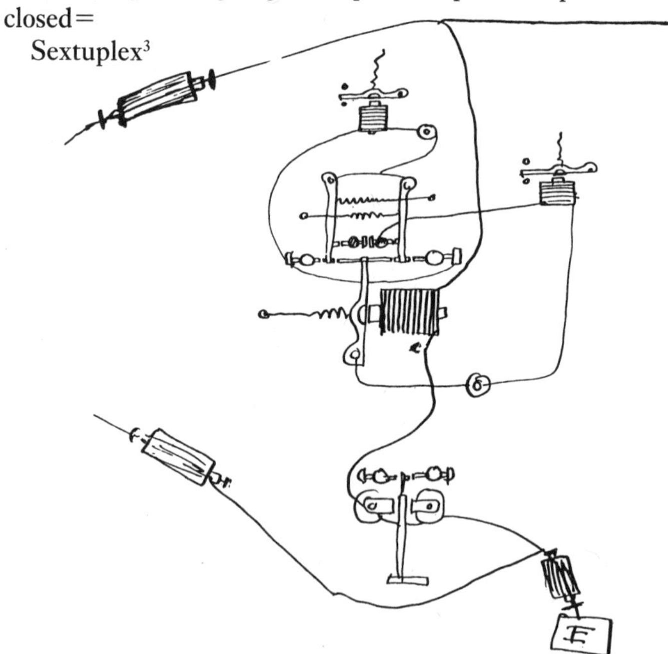


T A Edison James Adams

X, NjWOE, Supp. II, 1877 (*TAEM* 97:581, 580). Document multiply signed and dated.

1. See headnote, p. 280.

2. Here Edison appears to be adapting his sextuplex multiple-contact lever as a quadruplex bug trap, although the advantage of this arrangement is not clear. The same day he drew other versions of multiple-lever contacts. NS-74-004, Lab. (*TAEM* 7:527–31).

3. This drawing shows receiving instruments arranged in a bridge circuit. Different current strengths (and springs) would pull or release the armature lever of the central electromagnet to make or break various connections with the two levers above it. The design here is incomplete but was probably intended to operate in a manner similar to the sextuplex circuit in his U.S. Patent 452,913 (executed 31 May), in which one of the signals was made by either of two levels of current while another signal was made by either a much weaker or the strongest level of current. However, the relay in the center instead could be the basis for yet another alternative bug trap for the quadruplex, if the two sounders shown worked jointly to control another instrument.

**-899-**

*From Frederic Ireland*

[London,][a] May 7. 1877

My dear Sir

I have had very little time or opportunity to write you lately for I have been so much from home— have been so busy in forming the E W. Co into to Electric Writing Co Limited[1] and have had so much worry with respect to the change of apparatus that I have hardly known which way to turn—

The fact is the British public—with its usual stupidity—did not take kindly to the apparatus as it was and it has had to be practically revolutionized— I sent you two of the Fuller batteries—[2] How do you like them?— with us they answer spendidly lasting two or three months and then only requiring that the inner jar should be emptied and refilled with water and a little acid:—

I cannot tell how you travellers can possibly get on with the open batteries— We now cover our jars with an india rubber cap so that they are perfectly portable without any chance of spilling

—Our present press—after giving us a lot of trouble is, I think a great success—costs us 17/–[3] would you like to have one?—

I see that some new press has been invented in America—should like to have particulars

French Patent

I have worked hard at this but the unsettled state of Continental politics—which as you see has ended in War[4]—has clogged my operations and prevented my friends from coming to a decision—

I feel that I have now got England satisfactorily organized and if you are inclined to let me work France on a Royalty for two or three years on condition that I complete purchase at the end of the time I will take up the organization there— My experience here will be of immense service and I shall have every inducement to complete as otherwise all my labor would simply pass to you— Let me hear from you on this early.

American Novelties Co

I have hardly had time even to think of this— The idea seems good— do you care to have anything from this side— I have two patents that would pay—

Waiting for your reply I am Yours faithfully

Fredc Ireland

I dont think the new pens quite as good as yours.

ALS, NjWOE, DF (*TAEM* 14:386). Letterhead of Electric Writing Co. [a]"9 New Broad Street, E.C." printed above.

1. That is, the company was incorporated.
2. See Doc. 815 n. 4.
3. Shillings; about $4.25 at an assumed exchange rate of $5.00/£1.
4. The Russo-Turkish war of 1877–78.

## –900–

*Charles Batchelor to Robert Gilliland*

[Menlo Park,] May 8th [187]7

Dear Sir

Yours of the 3rd to hand.[1] Enclosed find statement[2] of your account with the Edison Electrical Pen and Duplicating press Co.. Dec 30th, as you see, it was $\$203^{44}/_{100}$, but Ed[3] has had on your account $\$27^{79}/_{100}$ and he has also accepted for you Barnard's[4] bill of $\$48^{45}/_{100}$ which of course leaves a balance of $\$127^{20}/_{100}$. The company owes me about $\$90^{00}/_{100}$ owing to Edison giving me instructions to pay one of Ed's bills for castings for $193— he also told me to pay Ed's bills whenever I could, to save him trouble. I told Ed that your account was to be paid next, and as you see $\$76^{24}/_{100}$ has been paid him and if it had not have been for this note[5] of his the rest would have been paid. The money for the old bills comes in slow but I suppose it is pretty sure.

In regard to the royalty from the W.E. Mfg. Co. it ought to have been paid promptly. When Bliss was here last, Edison got out of patience & told him about it, & he had them send him $200—then, and he promised the rest immediately he got home.[6] If you speak about it I think they would settle immediately.

Edison is continually getting up new things connected with the apparatus. We have just brought ~~up~~ out[a] a new pen, which will cost only about 1/3 of the cost of the present one, and numerous other improvements, which are being patented to secure the 'pen complete.[7]

The infringements of Stuart and Hix are in a fair way of being speedily knocked on the head.[8] My time is spent freely on all this sort of work, which as you know accrues for your benefit instead of mine, and therefore I think you ought to try and fix up my note[9] at as early a date as possible for I can assure you I need the money badly.

Edison and I went up to the Papyrograph[10] office today, as so much had been said about it, & we investigated it thoroughly. Edison wrote a sample as a test (of course they did not know who we were) and he wrote one line very heavy and the next line very small and light and the consequence was that the man tried to get a copy off it, but could not get a good one although it took him 29 minutes to manipulate it before it was ready.

We have an ink now that you write on ordinary paper with, and after writing you can use it in the press as a stencil, and roll the roller over[b] it and take as many copies as you please.[11] In talking over this sometimes Edison and myself have almost come to the conclusion to give away a bottle of this ink with every 'pen.' By that means you see when a man buys an 'Electric pen' he would also get a better process besides, than the Papyrograph What do you think of that idea? Yours Respy

Chas Batchelor.

P.S. Try and do something for me. C.B

ALS (letterpress copy), NjWOE, Batchelor, Cat. 1238:113 (*TAEM* 93:111). [a]Interlined above. [b]Obscured overwritten letters.

1. Not found.

2. Not found.

3. Ezra Gilliland.

4. William Barnard was the electric pen agent for Delaware, Maryland, and Virginia. *TAEB* 2:598 n. 11.

5. This may refer to one of the promissory notes held by Edison against Ezra Gilliland. See Doc. 834 n. 1.

6. The $200 check was credited on 6 May and another $250 was credited on 21 May. The total royalty due on 30 April for both March and April was $630.75. Cat. 1185:88, Accts. (*TAEM* 22:595).

7. See Doc. 893 n. 3.

8. On 12 May the Patent Office held that claim three of Edward Stewart's U.S. Patent 183,720 (issued in 1876) was in interference with one of Edison's applications. A ruling in Edison's favor on 12 December 1877 led to the issuance of U.S. Patent 203,329 (Pat. App. 203,329). Writing to Bliss on 26 February 1877 Batchelor noted that A. E. Hix's invention was an infringement of Edison's U.S. Patent 180,857, notwithstanding a letter of Hix's to the contrary (Cat. 1238:85, Batchelor [*TAEM* 93:89]). An undated note in Edison's hand indicates that Hix's invention was for a "cloth hand stamp" (Vol. 15:53, Lab. [*TAEM* 4:375]). On 10 September, Edison and Batchelor went into New York to give evidence in the Hix interference on "Hand stamp (Ink soaked pad on stencil sheet)" (Cat. 1233:253, Batchelor [*TAEM* 90:174]).

9. Nothing is known of this note.

10. The New York papyrograph office was W. F. Adams & Co. at 59 Murray St. See circular in Cat. 593 and letterhead in Cat. 1031:12, both Scraps. (*TAEM* 27:648, 738).

11. Batchelor spent 30 April and 2 May on copying experiments, and mentioned a "new composition" in diary entries of 4 and 7 May. There are also 30 April notes by Batchelor and Adams about a substitute for Edison's duplicating ink. Cat. 1233:120, 121, 124, 127, Batchelor (*TAEM* 90:113, 115, 116); Cat. 1317:37, Batchelor (*TAEM* 90:675); NS-77-002, Lab. (*TAEM* 7:431).

**–901–** [Menlo Park,] ~~S~~May 9 1877

*Technical Note: Multiple Telegraphy*[1]

Sextuplex[2]

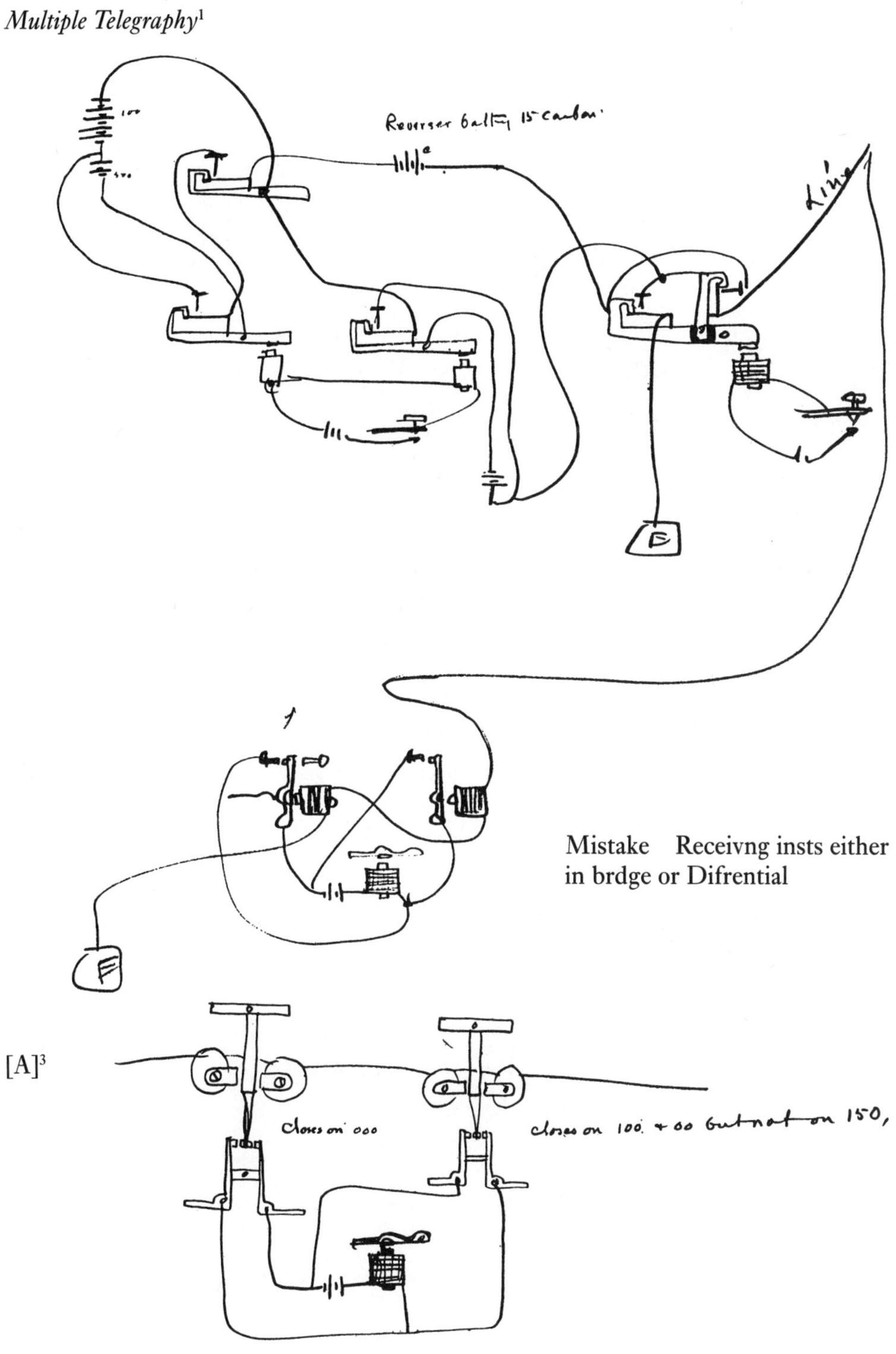


Mistake  Receivng insts either in brdge or Difrential

[A][3]

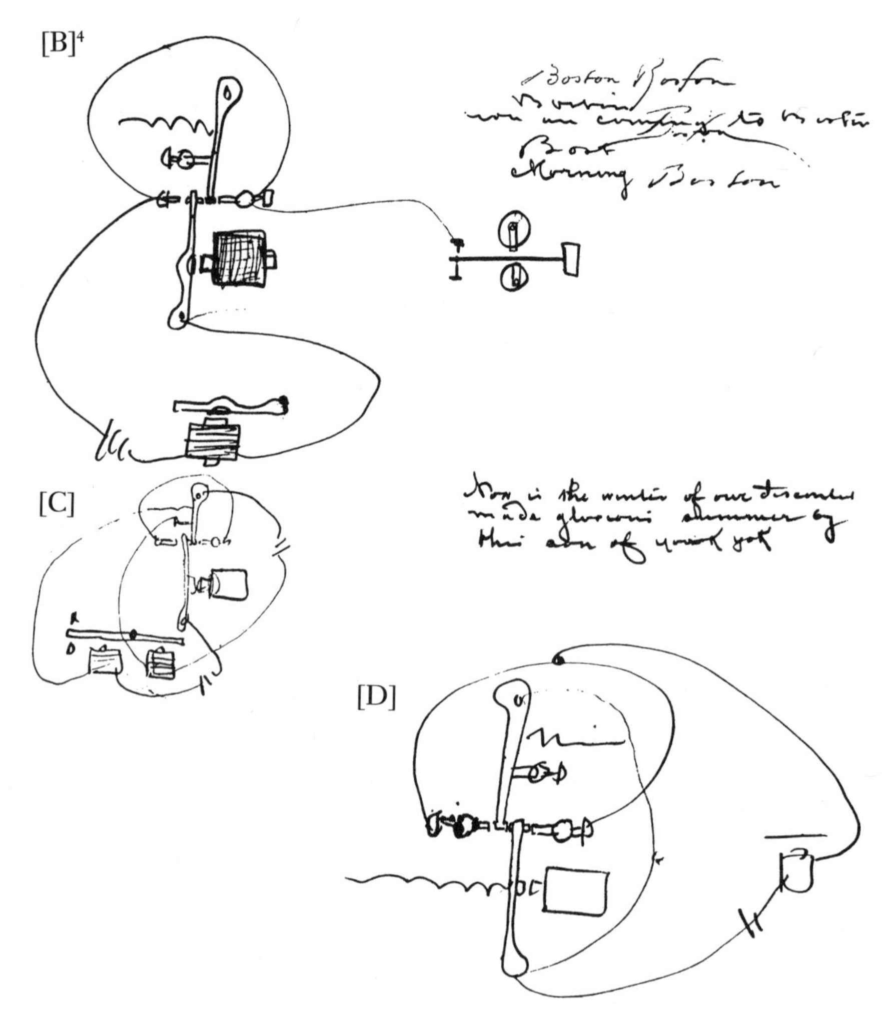


This[5] works about same as the single[a] lever in front[6] perhaps not quite so well.[7]

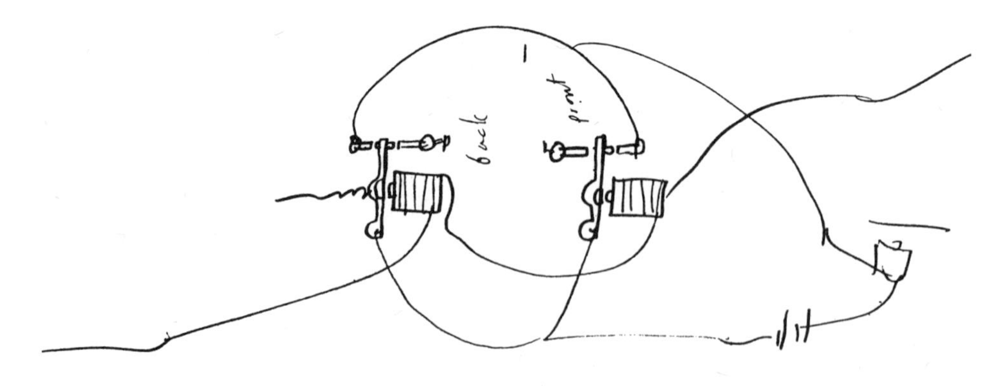


Perhaps this will replace= Works well. perhaps better than lever relays[8]

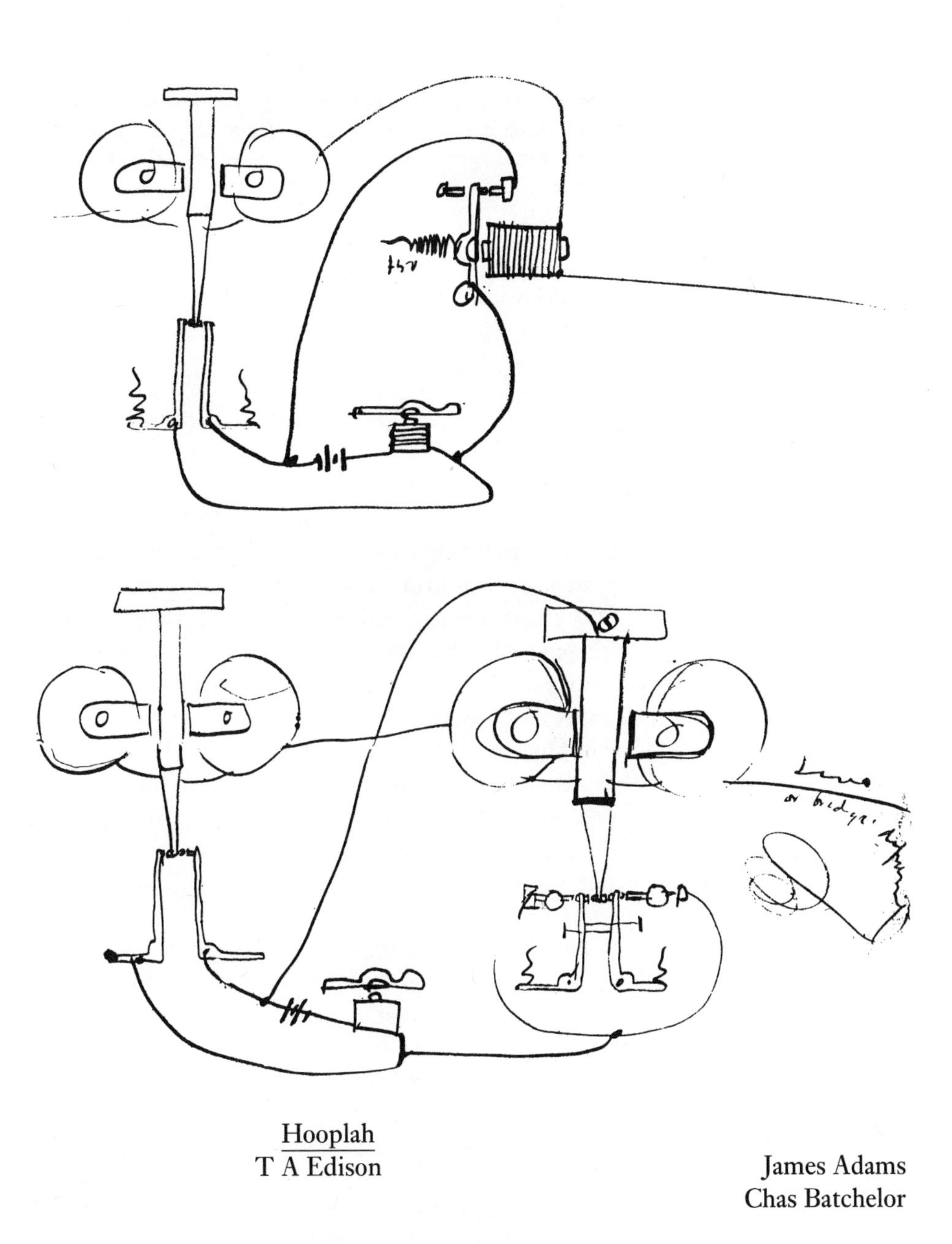

Hooplah
T A Edison

James Adams
Chas Batchelor

X, NjWOE, Lab., NS-77-004 (*TAEM* 7:538–41). Document multiply signed and dated. [a]Interlined above.

1. See headnote, p. 280.

2. Figure labels are "100" and "500" (on battery at upper left), "Reverser battery 15 carbon," and "Line." These drawings are representative of two days' work on multiple-contact relays (including working sketches for building one) and other aspects of sextuplex circuits. NS-77-004, Lab. (*TAEM* 7:532–37, 542–46).

3. Figure labels are "closes on 000" and "closes on 100 & 00 but not on 150."

4. The scribbling says, "Boston Boston Boston [----] are coming to Boston Boston Boston Morning Boston" and "Now is the winter of our discontent made glorious summer by this son of york yok."

5. Sketch D.

6. Sketch B.

7. Figure labels are "back [point]" and "front [point]."

8. Figure labels for the following drawings are "150" and "line or bridge differential."

**–902–**

*From Edward Johnson*

New York, May 10th 1877.[a]

My Dear Edison

The W.U. have attacked Reiff in a low mean cowardly manner—after Orton & Lowrey almost professing to Love him—They are producing all sorts of scraps & letters to destroy his creditability by putting upon them false constructions[1]

They have already made it appear that you & he have contradicted each other about some payments— Reiff was taken very sick in Phil= Sunday & only got back to N.Y. ~~last~~ yesterday= & went on the stand today in a miserable condition—to be subjected to this sort of thing—[2]

It becomes now your imperative duty to come over & see him & not longer absent yourself to the great advantage of the W.U. & consequent disadvantage of the man who has been a better friend to you than the W.U. Ever has or ever will be. ~~to y~~[3]

Your desire for neutrality is commendable within certain limits—beyond those limits you will be compelled to take sides whether you will it or not— Your absenting yourself at all times except when called for by some W.U. party has already placed you in their Ranks  If Reiff suffers at your hands you will never earn enough from the W.U. to compensate you for it— If it was simply a question of his losing or winning this suit[b]—that would be one thing—but it has got beyond that  Yours Very Truly

E. H. Johnson

ALS, NjWOE, DF (*TAEM* 14:669). Letterhead of Electro Chemical Manufacturing Co. [a]"New York," and "187" preprinted. [b]Obscured overwritten letters.

1. See Josiah Reiff's testimony, Quad. 70.7, pp. 439–87, 682–736 (*TAEM* 9:587–613, 711–38).

2. Reiff's testimony began on Friday, 4 May, but was suspended until 10 May because of his illness.

3. On either 4 or 11 May (the note is dated only "Friday") Norman Miller wrote Edison concerning the cross-examination of Reiff, noting

that “the papers introduced are such that you . . . would not like to be called to the stand to either identify dispute or explain— Keep away from court and every one—except [Western Union treasurer] Mr Rochester—if you come to City—” (DF [*TAEM* 14:740]). Nevertheless, on 12 May, Edison, Batchelor, and Johnson went to see Reiff (Cat. 1233:132, Batchelor [*TAEM* 90:119]).

**-903-**

*Technical Note: Multiple Telegraphy*[1]

[Menlo Park,] May 10 1877[2]

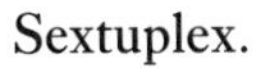

Sextuplex.

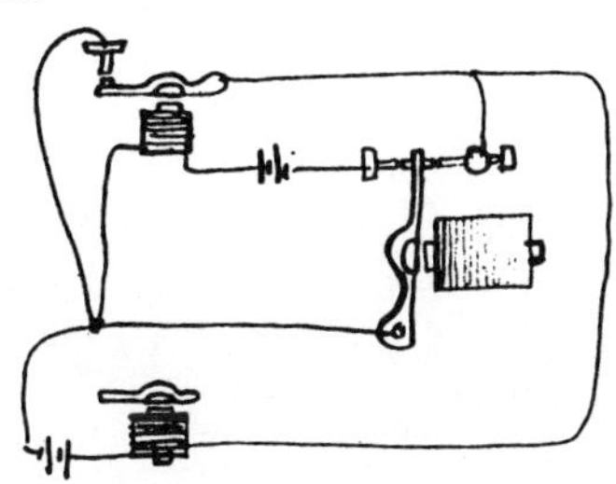

good bug Trap for reversals for Quad

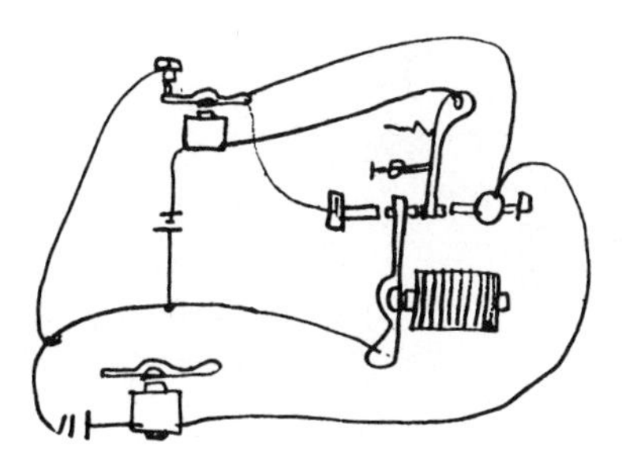

Bug trap for “Sexty”[3]

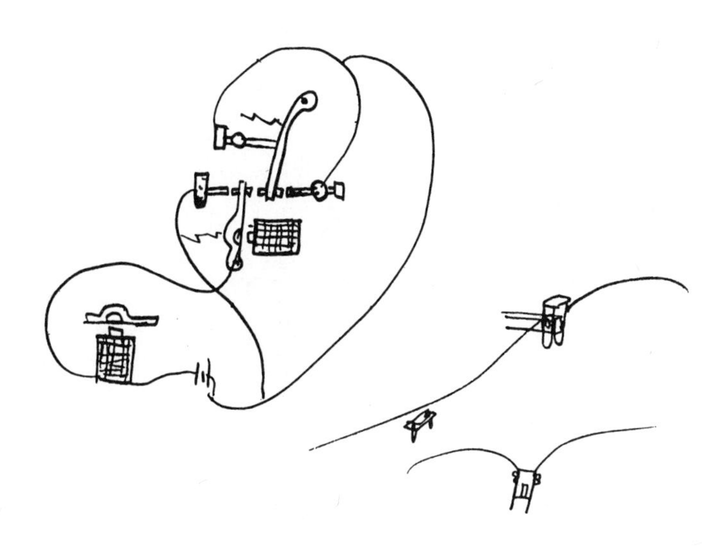

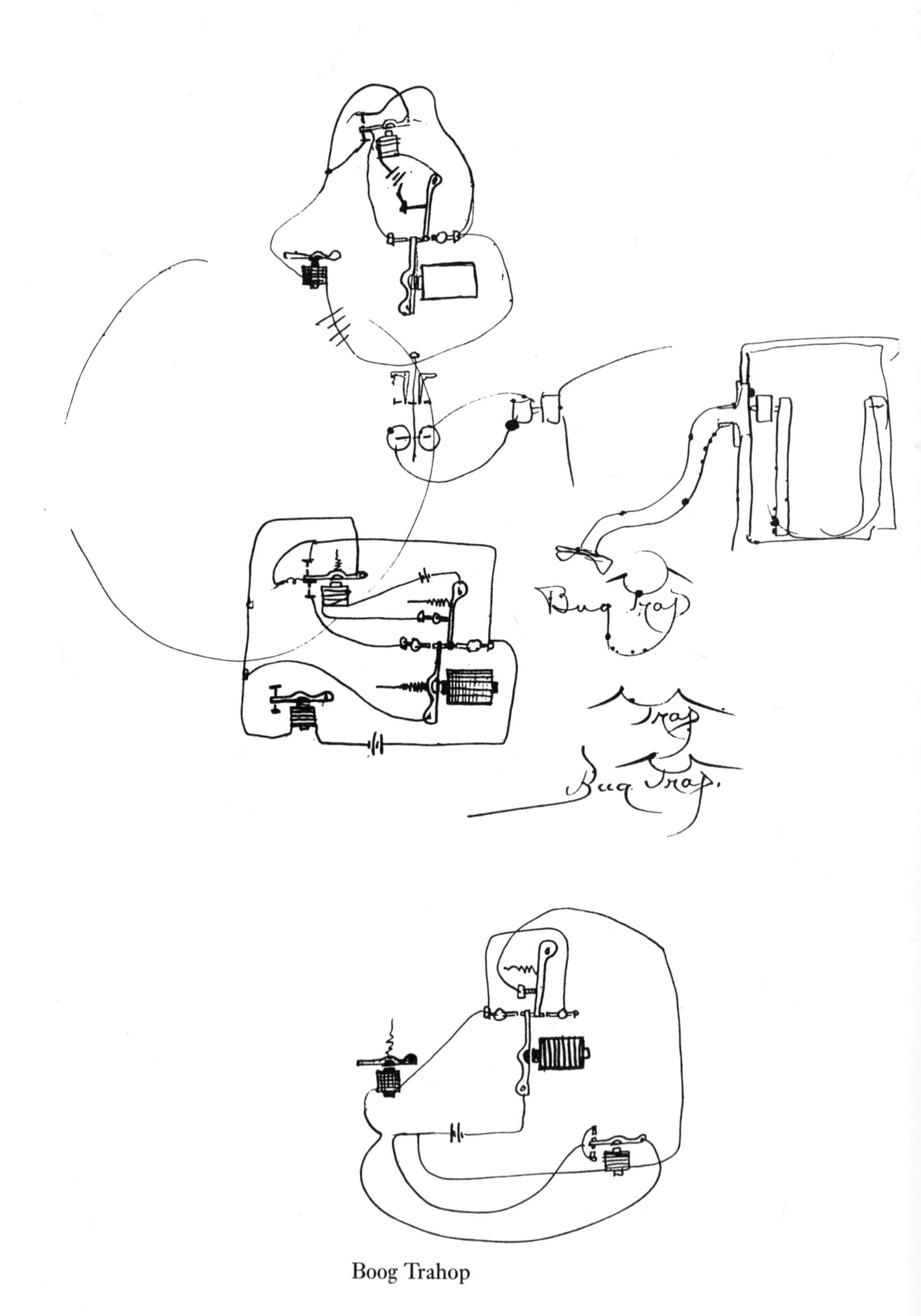


Boog Trahop

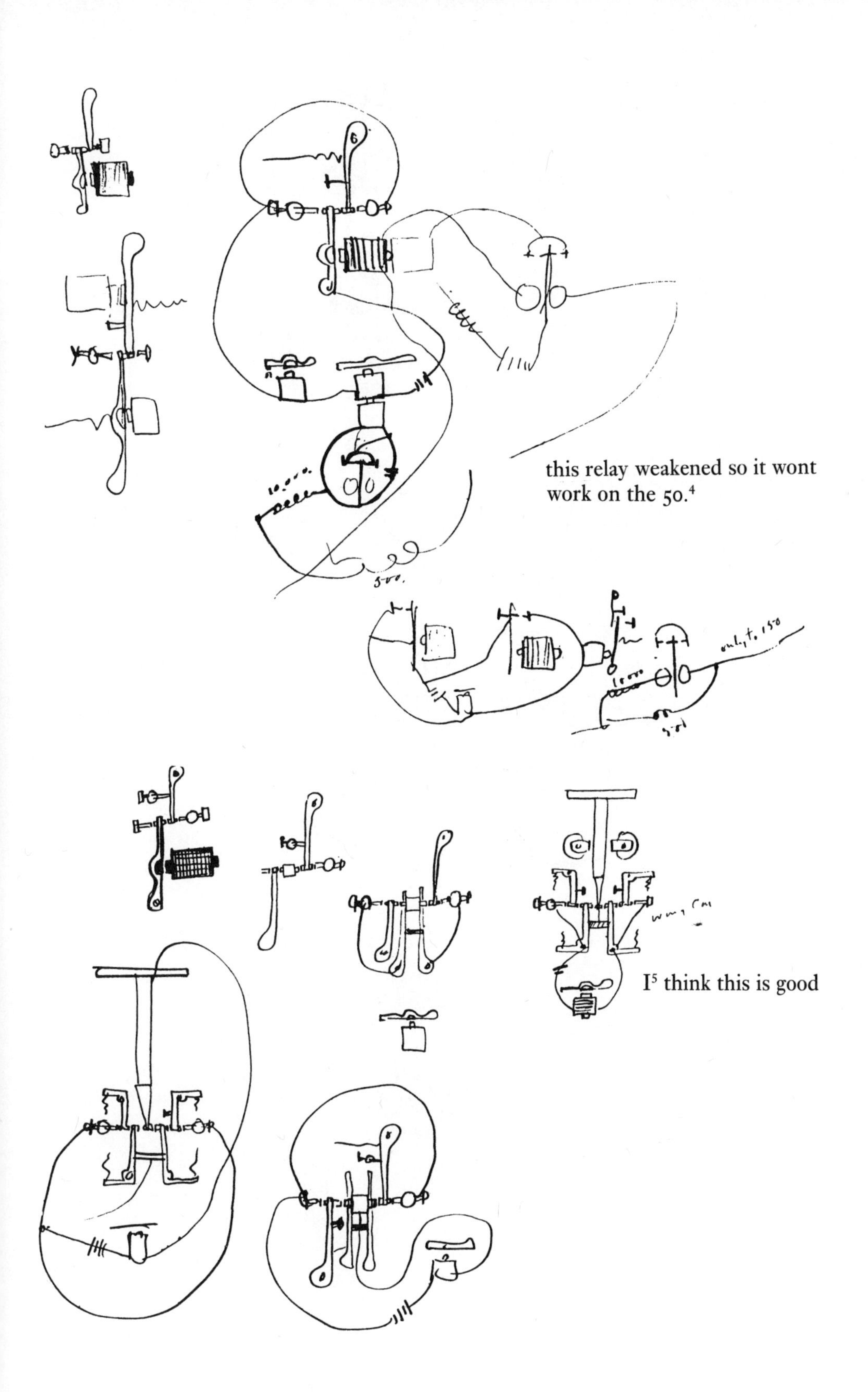
this relay weakened so it wont work on the 50.[4]
I[5] think this is good

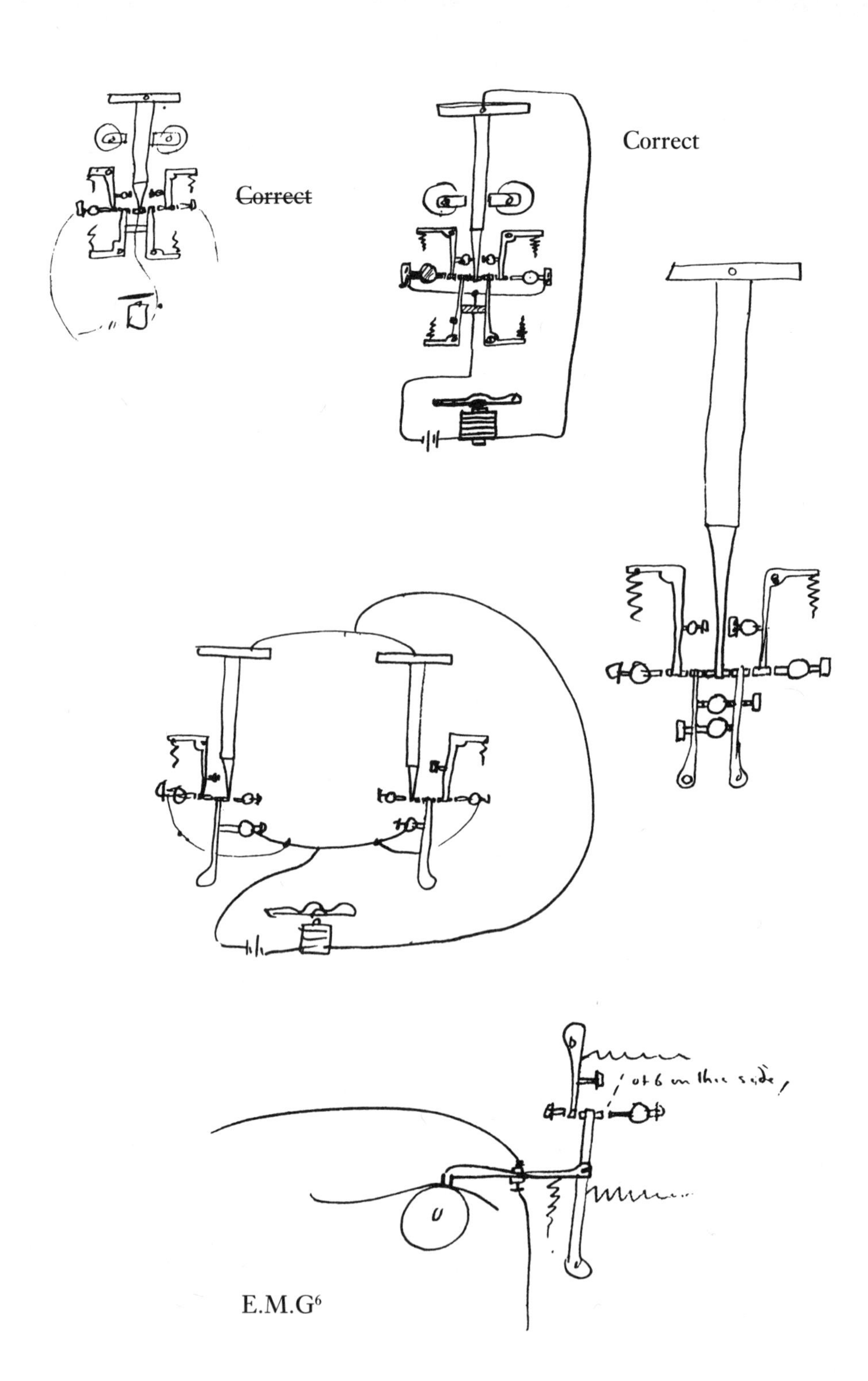
~~Correct~~
Correct
; or 6 on this side;
E.M.G[6]

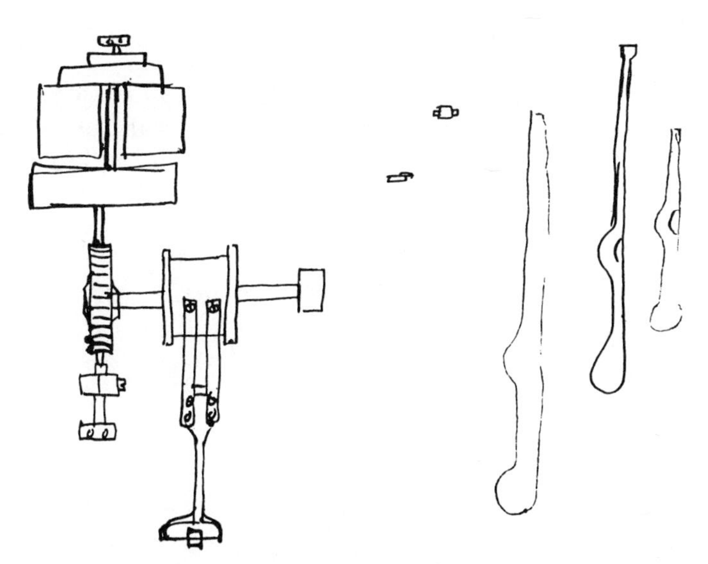

irregular dont work well=[7]

T A Edison

Chas Batchelor
James Adams

X, NjWOE, Lab., NS-77-004 (*TAEM* 7:549, 548, 558, 557, 559, 555). Document multiply signed and dated.

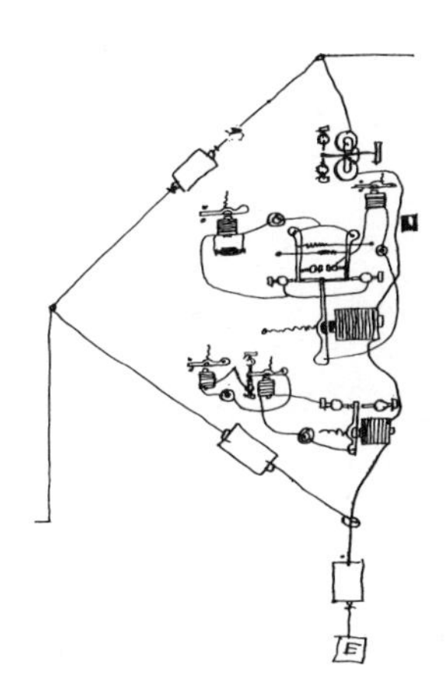

*A drawing of the receiving circuit for a sextuplex patent model.*

1. See headnote, p. 280.

2. The same day Edison drew other arrangements of continuity-preserving and reversing levers for sextuplex transmitters. NS-77-004, Lab. (*TAEM* 7:547, 551–52, 554, 556, 560).

3. The next day Edison drew bug trap variations that incorporated the principle of his pressure relay. NS-77-004, Lab. (*TAEM* 7:563).

4. Figure labels of drawing above are "500" and "10,000." Figure labels of drawing below are "10 000," "500," and "only to 150."

5. Figure label above is "wing con[nection]." Edison labeled a 30 April drawing of a sextuplex receiving circuit "wing for model." NS-77-004, Lab. (*TAEM* 7:662).

6. Figure label is "or b on this side." This refers to the upper vertical lever **b,** labeled at top. Drawings of 14 May show the electromotograph in a receiver circuit. NS-77-004, Lab. (*TAEM* 7:585).

7. This electromotograph design includes Edison's standard small motor (see Doc. 908 n. 5). Another drawing from 14 May shows a similar electromotograph design (NS-77-004, Lab. [*TAEM* 7:584]).

**–904–**

*From George Ward*

[New York,][1] 14 May 77.

Friend Edison

Preece was here this morning  he has fixed Friday next he wants a good day with you. What trains had we better start & return by. Tell me the times of all the trains. In haste. Yrs truly

G G Ward

ALS, NjWOE, DF (*TAEM* 14:55).

1. Ward wrote from 16 Broad St., the Direct U.S. Cable Co. office in New York. Wilson 1877, 340.

**–905–**

*Patent Model Specification: Multiple Telegraphy*[1]

[Menlo Park,] May 14 1877

Sextuplex[2]
Model for Phelps[3]

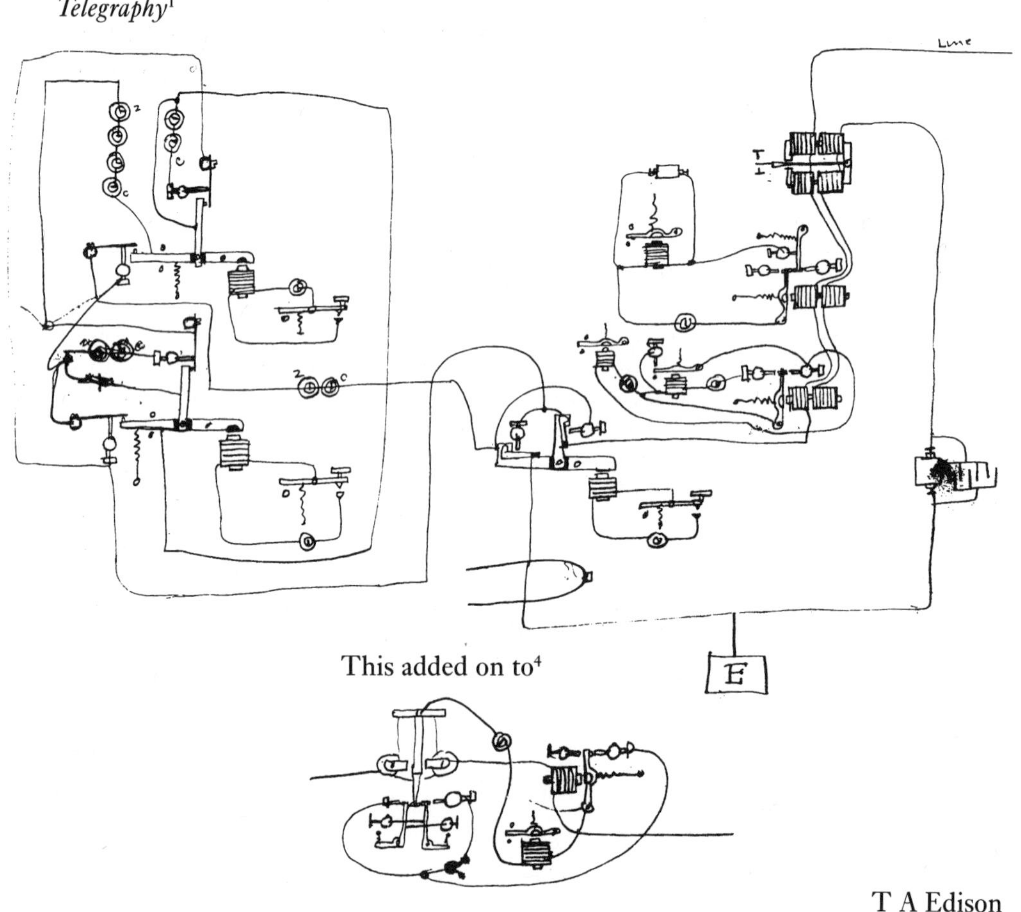


This added on to[4]

T A Edison

X, NjWOE, Lab., NS-77-004 (*TAEM* 7:583).

1. See headnote, p. 280.

2. This patent model drawing for Case 139, which shows one end of a sextuplex circuit with differentially wound relay coils, reflects much of Edison's work for several weeks around this date (see his almost identical 30 April sketch in NS-77-004, Lab. [*TAEM* 7:660]). A precise, complete explanation of the circuit is given in the patent (U.S. Pat. 512,872), but a general description follows.

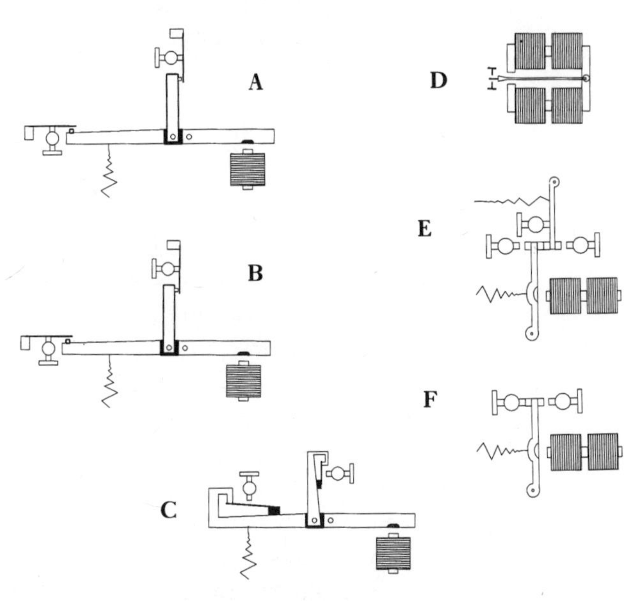


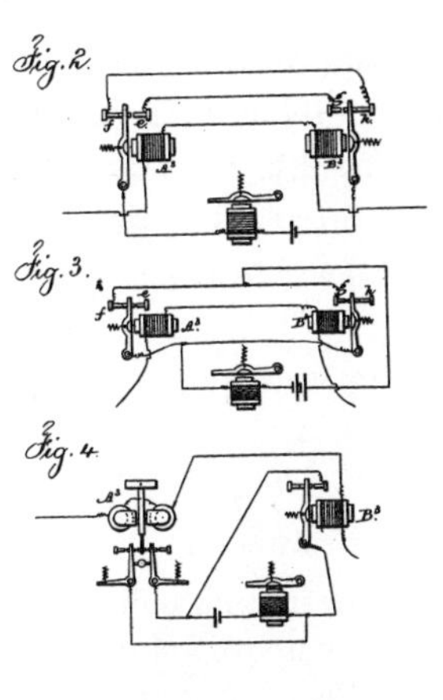

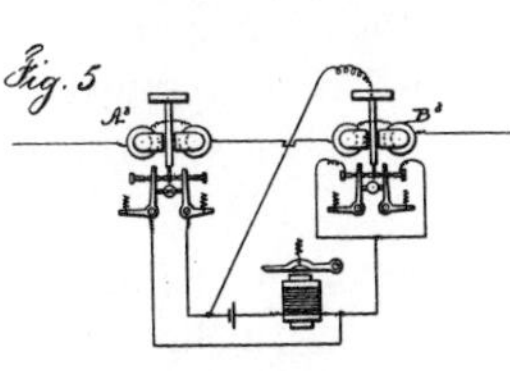

*Alternative designs for sextuplex receivers in U.S. Patent 452,913.*

Transmitter **C** (in the diagram) reverses the polarity of the entire system's base current, which is always on line. This affects the distant station's polarized relay corresponding to **D** (which lacks a sounder in the drawing). Transmitter **A** puts a powerful battery into the circuit, affecting the distant receiving relay corresponding to **F**, which will not respond to weaker signals. Transmitter **B** adds less power than **A** and affects the remote counterpart to **E**, which is sensitive enough to respond. The current reversals must not disrupt the other signals, the sensitive relay (**E**) must not respond to the strong signals, and the constant dynamic changes must not mutilate the signals on any of the relays. Relays such as **E**, with one or more extra contact levers, were important in many of Edison's sextuplex designs (see Doc. 884 n. 5). He had used them in quadruplex circuits as early as 1874 (Docs. 512 and 717).

3. The Western Union shop, run by George Phelps, was making the models for Edison's telegraph patent applications. See Doc. 877 n. 2; and *Edison v. Dickerson.*

4. According to the patent, the circuit shown in the main drawing was suitable for shorter lines but not practical for longer lines, where the static discharge effects were greater. Edison designed this arrangement, which became figure 2 in the patent, to replace receiver **E** (see note 2). It is one of a host of alternatives Edison considered during and after the preparation of the patents, examples of which can be found in technical notes ranging from 10 April through 7 July 1877 (NS-77-004, Lab.

[*TAEM* 7:466–691 passim]). U.S. Patent 452,913, also executed on 31 May, included several other alternatives to this design.

**–906–**

*To William Orton*

Menlo Park N.J. May 15 1877

Dear Sir

I replied that I had not perfected it, when perfected, it would be owned by the Gold and Stock Tel Co[1] Yours

Thos. A. Edison

ENCLOSURE[a]

Boston May 12th 1877

Dear Sir:

Within a few days I saw a notice in one of our newspapers that you had a new system of telephone called the Motograph. Now if you have perfected this apparatus to be as reliable as Bells telephone, I can sell like hot cakes, for the people of this city are ready to talk telephone with anybody. They are full of it and, although I know that Bells system is far from being perfect and is not calculated in its present form for practical use, yet he is talking freely with parties for putting it in and has already built several lines for the purpose. in fact he proposes to take possession of all the private telegraphing of this city and run a monopoly.

I have some sixty prvate lines of telegraph in this city and of course all my customers are interested in telephone. now I would like to hear from you as soon as possible and have no doubt I could sell a large number in Boston and vicinity I would like to put one in my office and show it to the public. let me hear from you. Respectfully yours

P. A. Dowd[2]

ALS, DSI-NMAH, WUTAE. [a]Enclosure is an ALS.

1. Edison's 14 May answer to Dowd (EP&RI) reads in full:

> I am working on the talking telephone but as yet it is not sufficiently perfect for introduction it is however more perfect than Bells. You need have no alarm about Bells monopoly as there are several things that he must discover before it will be at all practicable for every day [use?] when my apparatus is perfect you will be informed

2. Peter Dowd was a telegraph constructor in Boston who became agent for the Gold and Stock Telegraph Co.'s telephones, including Edison's. This led to his becoming the defendant in a major patent infringement case brought by the Bell Telephone Co. in 1878. *Boston Directory* 1877, 285; Bruce 1973, 263.

**-907-**

*From George Caldwell and Charles Edison*

New York May 15/77[1]

Sir

It is our intention to give the 1st Exhibition of the Edison Telephone at Newark on the 24th inst[2] and as soon thereaffter as possibble to go on the road and exhibit this telephone in every town where there is a possibilty of us receiving a pecuniary benefit for our venture

We intend to start within a week through Penna. on our way west. It is our intention to give a first class entertainment—that will both amuse and instruct the public. We are amply provided with Capital to properly carry on this enterprise independent of expectant sums we may realize from receipts from exhibitions. You are to find all the necessary instruments and supplies for a successful start hereafter we to furnish our own. we agree to pay you a royalty of twenty five dollars for each evening we give a exhibition to be paid daily or weekly through such parties or means as you may desire. We also agree to show six nights in every week if possible causes beyond our control may sometimes prevent us from showing every night.

We are to have the exclusive use of the telephone for Exhibition purposes for one year from date of performance we agreeing to give at least one performance per week[3]

Geo. W. Caldwell[4] Chas. P Edison

LS, NjWOE, DF (*TAEM* 14:772). Written by Caldwell.

1. Caldwell and Charles Batchelor attended a lecture by Alexander Graham Bell in New York on 17 May, and Batchelor attended another two days later. Batchelor said of the first one, "I had an opportunity to speak over the wire to New Brunswick but could not get it well I think it is no better than our own Cyrus W. Field apparently got it very well but I could not the singing was not anything so good as ours." Of the second he said "it was very poor indeed." Cat. 1233:137, 139, Batchelor (*TAEM* 90:121, 122); for newspaper accounts of the lectures, see Cat. 1240, items 162, 169–70, Batchelor (*TAEM* 94:151, 153–54).

2. Edison had exhibited his electromotograph telephone at the Newark Opera House at the end of April (see Doc. 889). On 9 May several gentlemen from Newark visited the laboratory and arranged to have Charley Edison demonstrate it before a scientific society in Newark. Charley held a rehearsal on 23 May "which was about to be a fizzle but [Batchelor] and Jim fixed it up." Batchelor spent the days after the Newark demonstration preparing instruments for Caldwell and Charley, who started for Reading, Pa., on 30 May (Cat. 1233:129, 143, 145–50, Batchelor [*TAEM* 90:117, 124, 125–28]).

3. Edison drafted and signed an agreement with Caldwell and Charley on 16 May 1877. DF (*TAEM* 14:773); Miller (*TAEM* 28:1048).

4. George Caldwell had assisted with the showing of the electric pen

at the Centennial Exhibition. Caldwell to French Commission to the Centennial Exhibition, 1876; U.S. Post Office to Caldwell, 12 Aug. 1876; both DF (*TAEM* 27:628, 641); electric pen pamphlet of George Caldwell, Agent, Centennial Exhibition (76-007, DF, Supp. III [*TAEM* 162:990]).

**-908-**

*Equipment Specification: Telephony*

[Menlo Park,] May 16 1877

Speaking Telegraph[1]
For G M Phelps to construct it.[2]

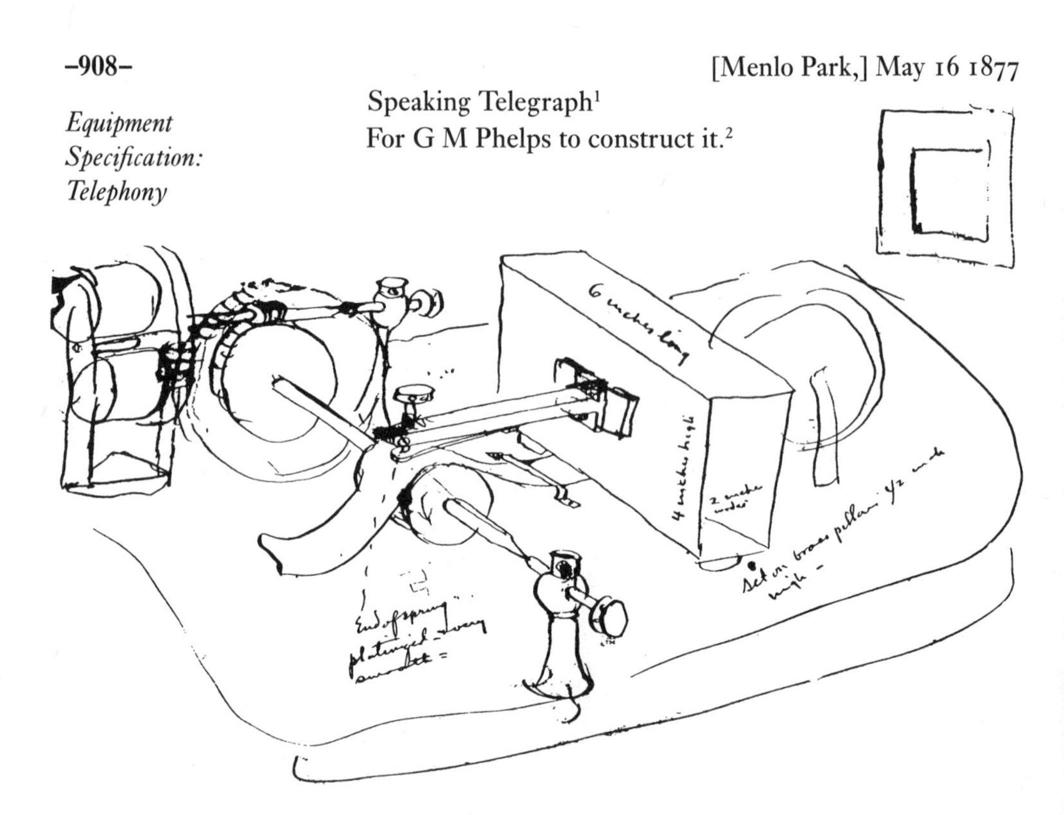


~~m~~box made of white spruce[3]—and glued together inside measurements[4] Wooden base with iron frame on bottom like relays= Surface of the drum roughened worm so cut as to cause wet paper drum to revolve either from 7 to 1 or 15 to 1. ie[a] the engine[5] revolving 7 or 15 times faster than the drum=[6]

Binding posts[7]

T A Edison

X, NjWAT, Box 92 02 163 01. [a]Circled.

1. This is a receiver. The file holding these drawings contains two other rough sketches of this instrument. On 24 May, Edison drew up specifications for a larger electromotograph receiver with a hand-crank

in place of the engine, and for a transmitter to be used with it (Doc. 917).

2. Figure labels are "End of spring platinized, & very smooth=" "6 inches long," "4 inches high," "2 inches wide," and "Set on brass pillars ½ inch high—." The documents were stamped at Phelps's factory on 17 May; a note on the principal drawing says, "Delivered at Mr. Prescotts office & charged June 11th."

3. White spruce is used for its resonant qualities in the tops of stringed instruments such as guitars.

4. Edison is referring to the box measurements on the drawing (this text is next to the box at upper right).

5. The engine, at far left, is the small electric motor design Edison had begun using in his 1871 printing telegraphs (*TAEB* 1:293 n. 3) and had used recently in his translating embosser (Doc. 857).

6. Edison drew the instrument with two gear wheels on the shaft (center right). It is unclear whether he meant Phelps to choose one or to design the instrument so that it could shift speeds.

7. Figure label is "to engine."

**–909–**

*Charles Batchelor Diary Entry*

[Menlo Park,] FRIDAY, MAY 18, 1877.[a]

Mr Preece Director of Telegraphs in England & Mr Ward Supt of Direct Cable here today had a splendid time they were highly pleased and are coming again.[1]

Preece very much interested in 'pressure relay' also in large tuning forks for splitting lines.[2] Mr Ward brought us our books of the Society of Telegraph Engineers

Worked on speaker at night

AD, NjWOE, Batchelor, Cat. 1233:138 (*TAEM* 90:122). [a]"FRIDAY, MAY 18, 1877." preprinted.

1. In a letter to his family, Preece described the trip to Menlo Park.

> Another blazing day which I spent at a place called Menlo Park with Edison—an ingenious electrician—experimenting and examining apparatus. He gave me for dinner *raw ham*! tea and iced water!! It is nearly 30 miles off. The railways here have no fences and they go bang through the streets of the towns. The whistles have the most horrid howls—more like an elephant's trumpet than anything else. The stations have no names and there are no porters about. Everyone has to look out for oneself. We nearly missed our station and as it was had to jump out while the train was moving. At level crossings the only notice put up is—'Look out for locomotive'—the *up* and *down* lines are reversed as compared with ours. They also drive on the opposite side to what we do. [Quoted in Baker 1976, 162]

Preece returned to the laboratory on three other occasions before embarking for England on 4 July (see Doc. 976; and Cat. 1233:176, Batchelor [*TAEM* 90:141]).

2. That is, acoustic transfer telegraphy.

*Technical Note: Multiple Telegraphy*[1]

Sextuplex

g[ues]s Il go back to the old auto shunt for clearing Quad Reversals[2]

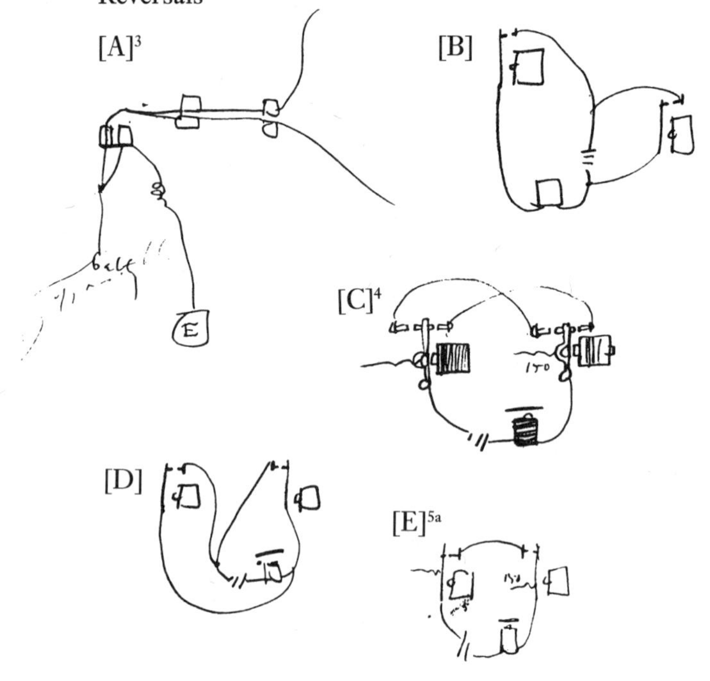


Compensation

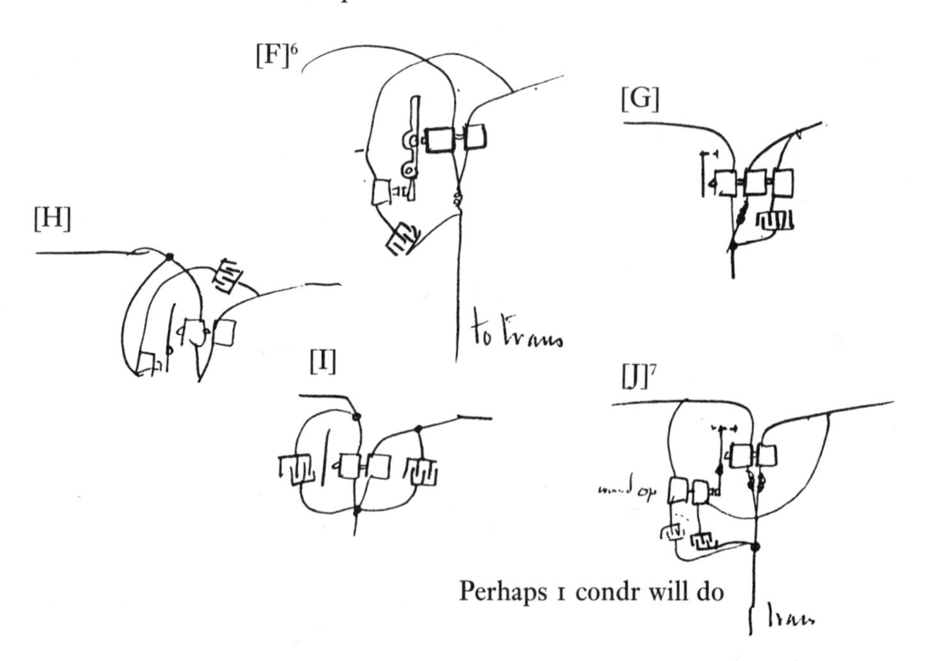


Perhaps 1 condr will do

[K]

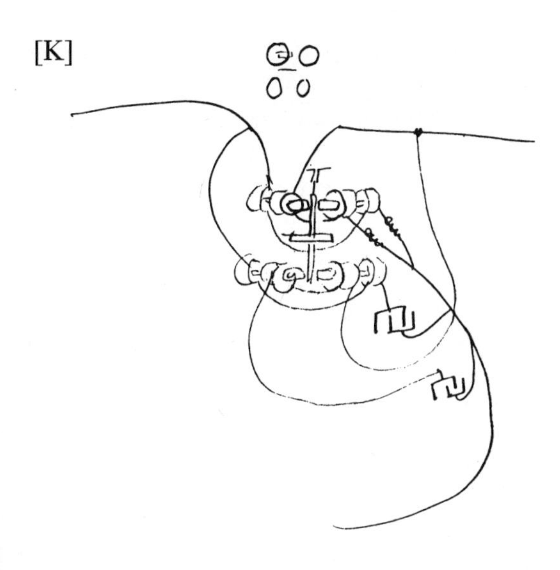

[L]

[M][8]

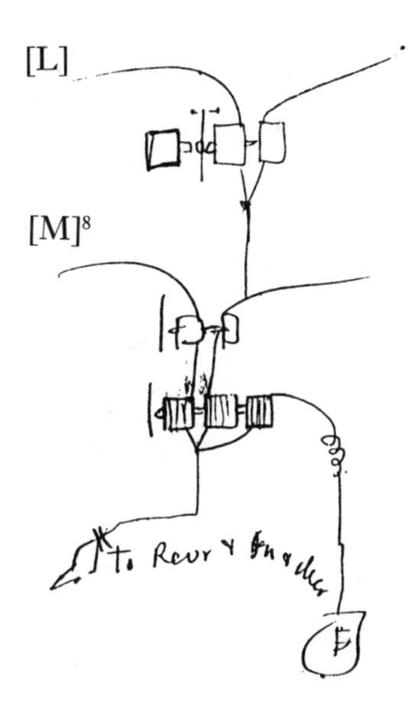

T A Edison

Charles Batchelor
James Adams

X, NjWOE, Lab., NS-77-004 (*TAEM* 7:590, 592). Document multiply signed and dated. [a]Doodled names, apparently by Edison, follow this drawing at bottom of page.

1. See headnote, p. 280.

2. These shunts appear in drawings F–K. Edison had discovered early in his automatic telegraph tests that a shunt around the receiving instrument greatly aided clear reception. See Docs. 317–19.

3. Figure labels (at lower left) are "Batty" and "Transmitter."

4. Figure label is "150." Edison had been working on this component of his sextuplex for several days (NS-77-004, Lab. [*TAEM* 7:570–89 passim]). It functioned as a replacement for the multicontact relay (see, e.g., Doc. 884). For a thorough description of its working, see U.S. Patent 452,913.

5. Figure label is "150."

6. Figure label is "to trans."

7. Figure labels are "wound op[posite]" and "trans."

8. Figure label is "to Rev[erse]r & ~~t~~In[crease] & decr[ease]"; that is, the transmitters.

**-911-**

*From Josiah Reiff*

New York, May 20th 1877[a]

Dear Edison.

You are needed as witness on Tuesday mng next @ 11 am & Genl Butler desires that you will come over early enough so as to meet him @ 5th Ave Hotel say about 9:30 to have a chat & ride down with him—[1]

Please let me know on Monday (if not in town today) that you receive this & will attend to it O.K.

I understand W.U. propose to put Farmer on stand Tuesday—[2]

Gerritt Smith[3] says 99[4] is not a working quadruplex with the differential or Bridge system of Stearns & the condenser as used by Stearns—[5] I guess the Edison Quad aint of much a/c[6] Eh? Yrs

J.C.R.

ALS, NjWOE, DF (*TAEM* 14:672). Letterhead of J. C. Reiff. [a]"New York," and "187" preprinted.

1. Edison was not recalled until Tuesday, 29 May. He had seen Butler the night of 16 May and did again before the 26th. Edison's testimony, Quad. 70.9, pp. 90–99 (*TAEM* 9:809–13); Cat. 1233:136, Batchelor (*TAEM* 90:121); Docs. 914 and 921.

2. Electrical inventor Moses Farmer (see *TAEB* 1:67) testified on Monday, 28 May. Quad. 71.1, pp. 356–80 (*TAEM* 10:185–97).

3. Gerritt Smith was the Western Union Telegraph Co. assistant electrician who helped Edison and George Prescott with quadruplex experiments and also received several patents on duplex and quadruplex telegraphy. *TAEB* 2:295 n. 4.

4. That is, Case 99 (Doc. 472).

5. See Smith's testimony, Quad. 71.1, pp. 298–305 (*TAEM* 10:156–59). Joseph B. Stearns (1831–1895) invented the first practical duplex telegraph (see Doc. 50 n. 3).

6. Account.

**–912–**

*From George Bliss*

Chicago May 23 1877.[a]

Dear Sir:

I am satisfied in my own mind that you are making a great mistake in the price at which the ink is being sold in this country.[1]

I have in my employ a first class man who was in the ink business three years.

He tells me there is a wide difference in the prices charged by Ink manufacturers.

The men who make the most money are those who put a first class price on their goods & stick to it.

The Public once becoming familiar with the goods wont be induced to buy a cheaper article though equally good.

There is an enormous profit to the retailer on ink.

For[~~ten~~?][b] instance an ink which costs by the gross 2.00 retails at 14.40 & other goods in nearly that proportion You will see from this that there is not enough inducement on your schedule to make dealers prefer your goods.

Again if sold largely by canvassers there is not margin enough to pay good men for taking hold of it.

The gent referred to says he has known ink makers to keep chemists at work long time trying to discover the ingredients & methods of Arnoldi[2] & other inks unsuccessfully

If your method of manufacture is kept well in hand he does not think there is much risk of successful imitation and at any rate your safety is in putting a good figure on the ink and paying no attention to cheap competitors.

We believe large amounts of your ink can be sold at first class prices.

My own idea would be to get up an outfit with the necessary ink, blotter, pens &c in good box & get 5.00 or 10.00 for it.[3]

I write this so you can give it consideration & we can talk it over when I come down.[4] Respectfully

Geo. H. Bliss Genl. Man.

ALS, NjWOE, DF (*TAEM* 14:396). Letterhead of Electric Pen & Duplicating Press. [a]"Chicago" and "1877." preprinted. [b]Canceled.

1. Edison was selling the ink in bottles of two sizes for $.50 and $1.00. After Bliss became general manager of Edison's duplicating ink business in the United States he raised the prices to $1.00 and $3.00. Edison's Duplicating Ink circulars: Cat. 1240, item 24, Batchelor (*TAEM* 94:13); and 77-011, DF, Supp. III (*TAEM* 162:1028).

2. Unidentified.

3. As general manager Bliss also sold the ink in three different outfits, each consisting of "a box, tray, plates, blotters, transfer paper, ink, sponge, and pen holder and pens." One outfit was for note paper, another for note and letter paper, and a third for note, letter, and legal paper. Edison's Duplicating Ink circular. 77-011, DF, Supp. III (*TAEM* 162:1030).

4. At the time Bliss wrote this he and Charles Holland had acquired foreign rights to the duplicating ink from Edison's Electro Chemical Manufacturing Co. (agreement of 5 May 1877, Miller [*TAEM* 28:1045]). On 4 June 1877 the Electro Chemical Manufacturing Co. transferred the entire duplicating ink business to Bliss and Holland, with Bliss becoming general manager of the business (Miller [*TAEM* 28:1051]; Edison's Duplicating Ink circular, 77-011, DF, Supp. III [*TAEM* 162:1021]).

## -913-

*From Norman Miller*

NEW YORK May 23 1877. morning[a]

Edison

I have been present at all of Mr. O's examinations. He has probably another full day before him—[1] I can see that Butler is not getting out what he hoped to do. So far you have been well protected, and unless <u>they</u> call you again to contradict Mr. O. you are in harmony on all important points.[2]

If you should go on the stand again I fear that some old and ugly affidavits of yours will come out that will do you discredit with the court.[3]

Reiff keeps his boy on the look out for you all the time.[4] Keep away from him. I will look for you at recess at Prescotts Office— Please be there if you come to the City.

Miller

ALS, NjWOE, DF (*TAEM* 14:680). Letterhead of Electric Pen and Duplicating Press. [a]"New York" and "187" preprinted.

1. Orton testified between 17 and 24 May. Quad. 71.1, pp. 107–294 (*TAEM* 10:58–154).

2. In a letter to Edison of 22 May, Reiff specifically requested that Edison come in to look at Orton's testimony. DF (*TAEM* 14:677).

3. E.g., "Edison's Affidavit," *TAEB* 2:806–15.

4. See Docs. 911 and 914.

**–914–**

*From Josiah Reiff*

May 24/77

My dear Edison

I learn you were in town yesterday & did not [go?][a] to see the Genl B. notwithstanding my Telegram to you.[1] If[b] you are in town today, please do not fail to see the Genl this evening— or tomorrow—any hour.[2] I understand WU intend to call you & the Genl wishes particularly to see you about some of your own testimony already given Yrs

J.C.R

ALS, NjWOE, DF (*TAEM* 14:682). [a]Illegible. [b]Obscured overwritten letters.

1. On 22 May, Reiff telegraphed Edison that he "must meet the party at Fifth Ave Hotel tonight sure—." In response James Adams wired back that Edison was sick. Reiff also wrote Edison on both 22 and 23 May requesting that he come in to meet with Butler. DF (*TAEM* 14:675–79).

2. Edison had seen Butler by 26 May. See Doc. 921.

**–915–**

*From J. L. Thomson*

New York, May 24th 1877[a]

My Dear Sir

I have this day sent you with this letter some Peices of Heavy board for Experiment  I was very sorry I could not of Seen you when I was out with Mr. James.[1]

If you can practically make glue Impervious by some chemical change I know of nothing so well addapted for treating cheeply this fiber. Still you experiment with what you are a mind to.[2] What I want is a perfectly Impervious material cheep & durible that will be practical to use for all styles of my Packages for [Liquids?].[b] I will try & see you when you have the Experiments where you would like my advise about them[3] Yours Respectfully

J L Thomson

ALS, NjWOE, DF (*TAEM* 14:62). Letterhead of New York Paper Barrel Co. [a]"New York," and "187" preprinted. [b]Illegible.

1. James James, who was associated with Amasa Mason; Mason also introduced Thomson to Edison. James and Thomson had visited Menlo Park on 17 May and spent the day with Batchelor discussing the waterproofing project. In late September Edison installed his telephones on a line to James's office at 72 Broadway. James later became interested in Edison's phonograph inventions. See Doc. 859 n. 1; Mason to Marshall Lefferts, 13 Apr. 1876, ML; Cat. 1233:137, 272, Batchelor (*TAEM* 90:121, 189); draft agreements, 1878, Miller (*TAEM* 28:1072, 1078).

2. Edison was apparently experimenting with the resin copal, which was used for varnishes and lacquers. Experimental notes in an unknown hand show attempts to make it soluble in solutions of various oils as well as solvents such as turpentine, kerosene, ether, and alcohol. One entry notes that "Caoutchouc [rubber] mixes with Soluble copal when heated to a very high heat it becomes vulcanized but when heated more it gets to a hard substance and I think would be very good for our purpose it would have to be pressed hot" (NS-Undated-002, Lab. [*TAEM* 8:145–50]). These notes may predate Edison's observation that aniline oil acted as a good solvent for gum copal (Doc. 813). A notebook entry of 11 May showing a thermostat employing gum copal probably postdates Edison's copal experiments (NS-77-004, Lab. [*TAEM* 7:566]).

3. There is no evidence of further work by Edison on this problem, nor are there further communications from Thomson.

**–916–**

*Technical Note: Multiple Telegraphy and Telephony*[1]

[Menlo Park,] May 24 1877

A principle of balancing in Du Quad & Sextuplex transmission without bridge or differential=[2]

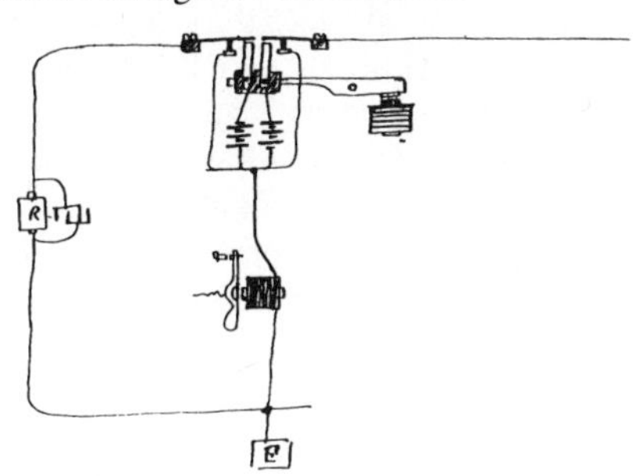

Iodide of Hg I find may replace plumbago in Talking Telgh— I notice decomp accompanied with slight[a] explosion on connecting 40 cells Callaud to hard pressed peice=

Sextuplex—[3]

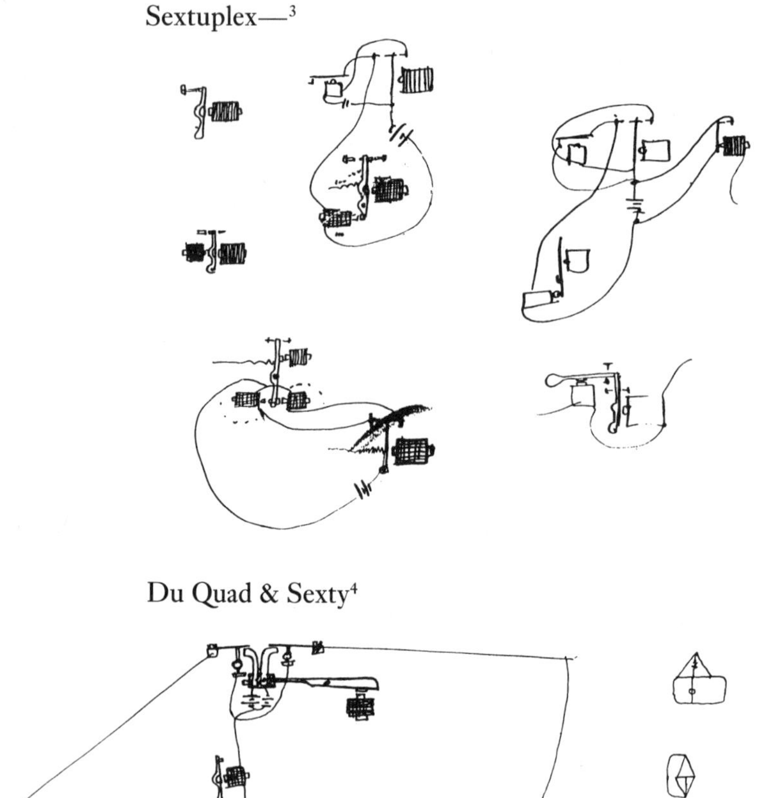

Du Quad & Sexty[4]

T A Edison

Chas Batchelor
James Adams

X, NjWOE, NS-77-004 (*TAEM* 7:601, 600). Document multiply signed and dated. [a]Interlined above.

1. See headnote, p. 280.

2. Standard methods of sending two messages in opposite directions involved either the bridge or differential methods (see *TAEB* 1:32, 530 n. 17). In this case Edison proposes isolating the receiving instrument from outgoing signals through an arrangement of batteries.

3. In these designs Edison is concerned with circuit designs that allow two messages to be sent in the same direction.

4. The following drawings show a bench test design and (perhaps) a schematic circuit analysis.

**–917–**

[Menlo Park,] May 24 1877

*Equipment Specification: Telephony*

Telephone[1]

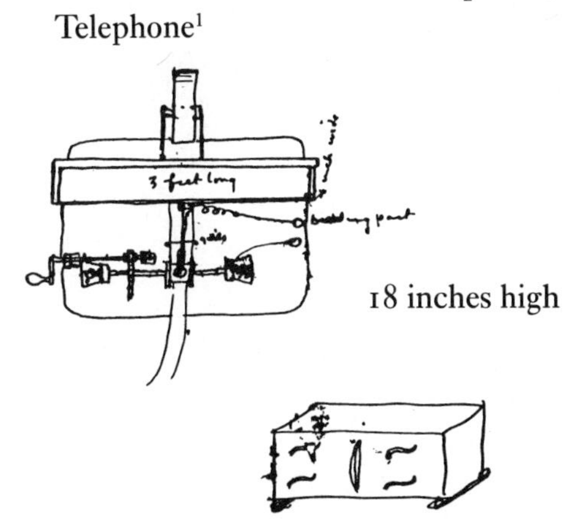

18 inches high

¾ stock spruce except front that ¼ inch with bridge in centre where EMG lever is fastened=[2]

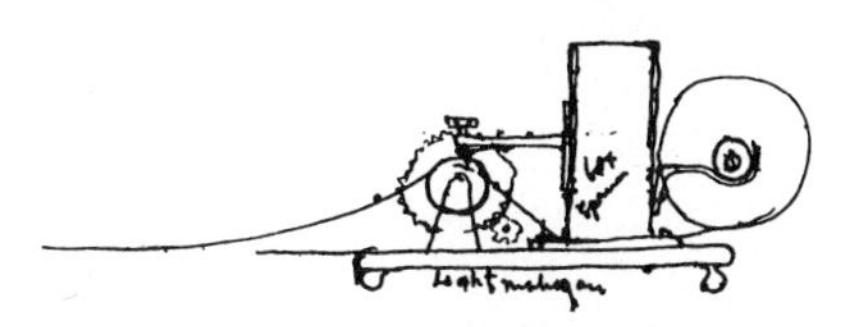

gear 5 to 1 or thereabouts even 10 to 1 or 3 to 1 would answer

Base Light Mahogany inch stock rounded edges also[3] = round legs ½ [--][a] inch high thus

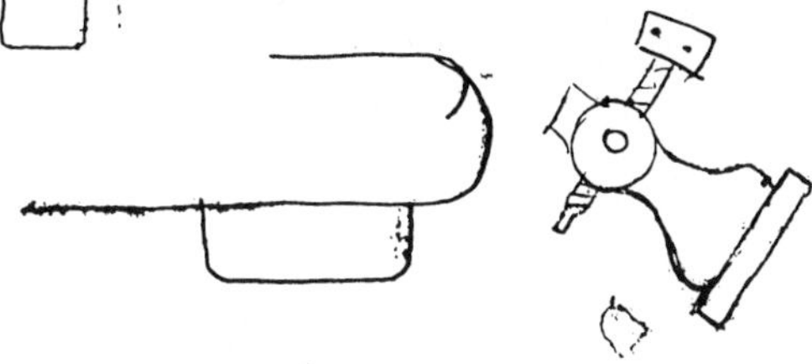

Transmitter for Telephone[4]

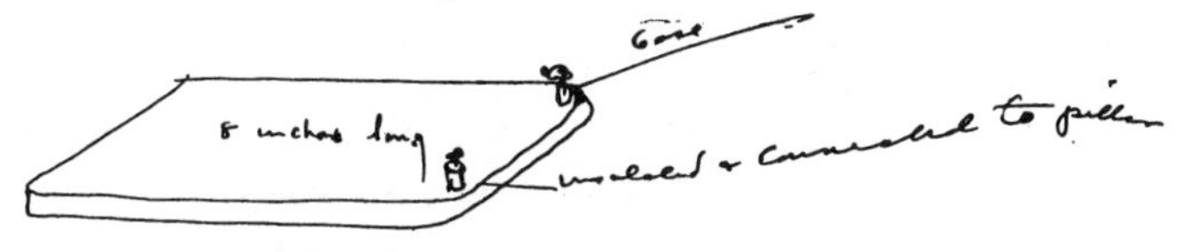

½ inch cast iron plan~~n~~ed on top[5]

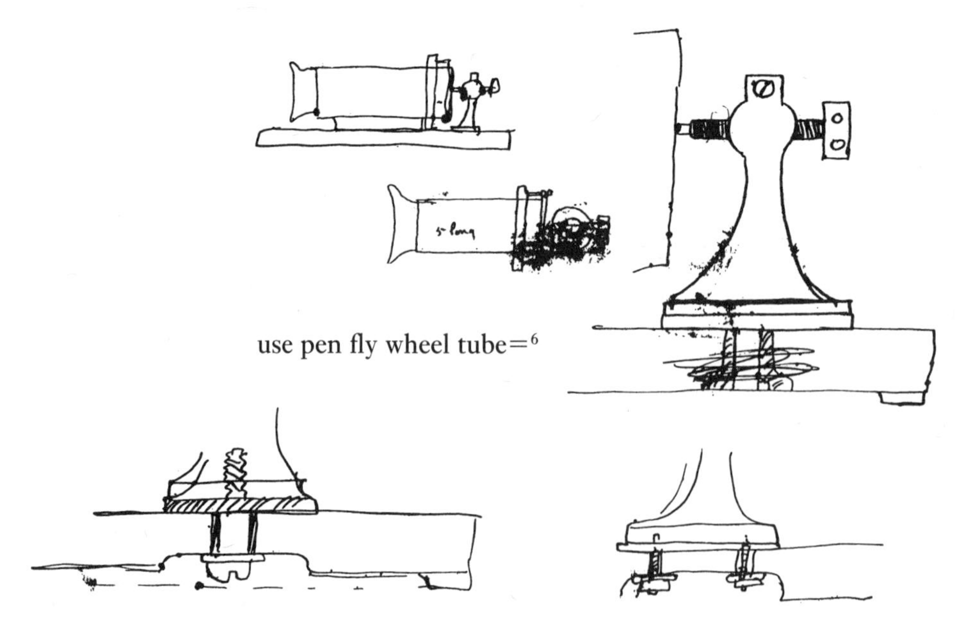

use pen fly wheel tube=[6]

T A Edison

Adams
Charles Batchelor

X, NjWOE, Lab., Vol. 11:97 (*TAEM* 3:971). Document multiply signed and dated. [a]Canceled.

1. Figure labels are "3 feet long," "guide," "4 inch wide," and "binding post."

2. Figure labels are "light mahogan" and "box spruce."

3. Figure label is "rounded."

4. Figure labels are "8 inches long," "base," and "insulated & connected to pillar."

5. Figure label is "5 long."

6. That is, make the tube from the stock used for the fly wheels on electric pen motors.

**–918–**

*From Frank Whipple*

Port Huron, Mich., May 25th 1877[a]

Dear Sir

Your note of the 21st rec'd[1] In order to make that balance sheet spoken of in it—it is absolutely necessary to know the amount of the note held by you against the company. I cannot make one as I suggested in my last letter to you[2] without that note or at least without ~~n~~knowing its amount date and rate per cent of interest.

The method I proposed was to save time but if the other suits you best send me the above data & I will prepare it[3]
Truly Yours

Frank Whipple

ALS, NjWOE, DF (*TAEM* 14:555). Letterhead of Whipple & Potter, Attorneys at Law. [a]"Port Huron, Mich.," and "187" preprinted.

1. Not found.

2. 17 May 1877, DF (*TAEM* 14:554).

3. Whipple subsequently sent Edison a statement (docketed by Edison as 15 June). It showed that Edison's share (160 of 302 shares) of the $2,439.42 debt assumed by the old Port Huron and Gratiot Street Railway amounted to $1,292.41. Edison held notes worth (with interest) $1,154.73, leaving due $137.68. DF (*TAEM* 14:556).

**-919-**

*Technical Note: Multiple Telegraphy*[1]

[Menlo Park,] May 25 1877

Sextuplex without reversals 3 increase & decreases[2]

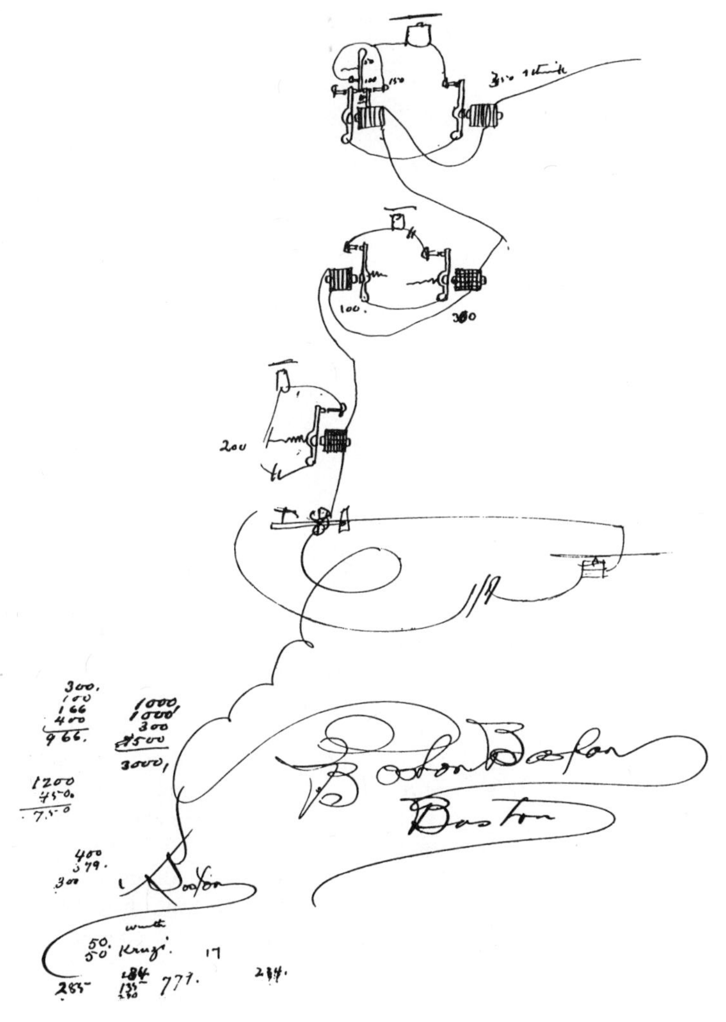


if this wud w[or]k cld send Reversals too & thus make it Octoplex[3] Hoop Lah How the art advances

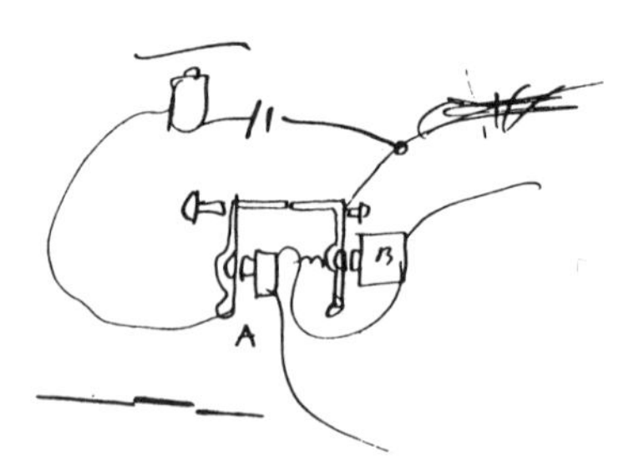


device for reversing thro A & B  B is adjusted to stay closed on the permanent current. A to closed on the increased it closes local through lever of B which at the moment of reversal flies back with the lever of A and thus prevents the Local circuit from being broken—[4]

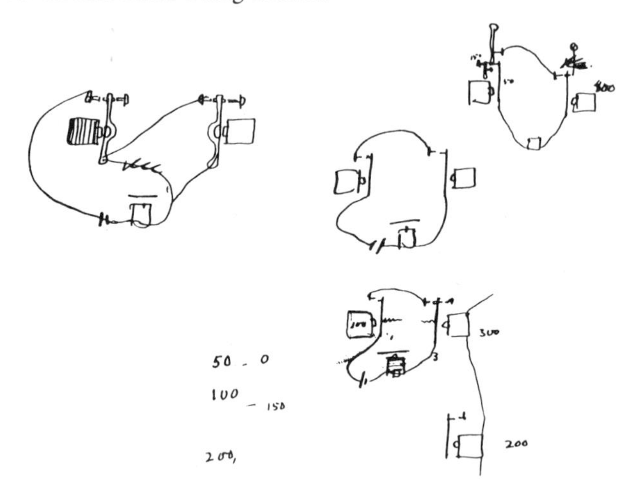


T A Edison

James Adams
Chas Batchelor

X, NjWOE, Lab., NS-77-004 (*TAEM* 7:602). Document multiply signed and dated.

1. See headnote, p. 280.

2. Figure labels are (top) "50," "100," "150," and "+350 I think"; (middle) "100" and "3500"; (lower) "200"; the names "Wurth" and "Kruzi" appear in the numbers and doodles at bottom. In this unfinished sketch of receiving arrangements for an octruplex telegraph, Edison tried to transmit three independent signals in the same direction using variations in current strength, with polarity reversals carrying the fourth signal (see note 3). All these receivers are independent, which means the transmitters would have been intricately interconnected. Earlier attempts in this vein had generally involved interaction among receivers instead. Prescott 1877, 827–32, 835–38; idem 1879, 364–70.

The receiver labeled "200" works with currents above a certain level and the corresponding transmitter would signal on its own using very strong current. Each of the other receivers, which respond to weaker signals, incorporates a second relay that stops its sounder from responding to stronger currents (the sounders are the blank rectangles, as distinct from the relay magnets' patterned coils). It is not clear that this particular arrangement would work. (The point of the calculations and other numbers at the bottom of the sketch is not known.)

Edison included a two-signal version of this plan in a sextuplex application executed six days later (Case 138; U.S. Pat. 452,913).

3. A polarized relay to receive reversal signals is shown in the circuit just below the relay labeled "200."

4. Figure labels in the following receiver sketches are (top right) "150," "50," and "[-]00"; and (lower right) "100," "1," "3," "300," and "200." The numbers at lower left ("50–0," "100–150," and "200.") apparently indicate ranges of signal strengths to which receiving apparatus would respond.

**–920–**

*Technical Note: Telephony*

[Menlo Park,] May 25 1877

Speaking Telegraph

A plumbago secured to platina Cup and faced with a platina Cap for the purpose of preserving it from abrasion & get the full effect of pressure[1]

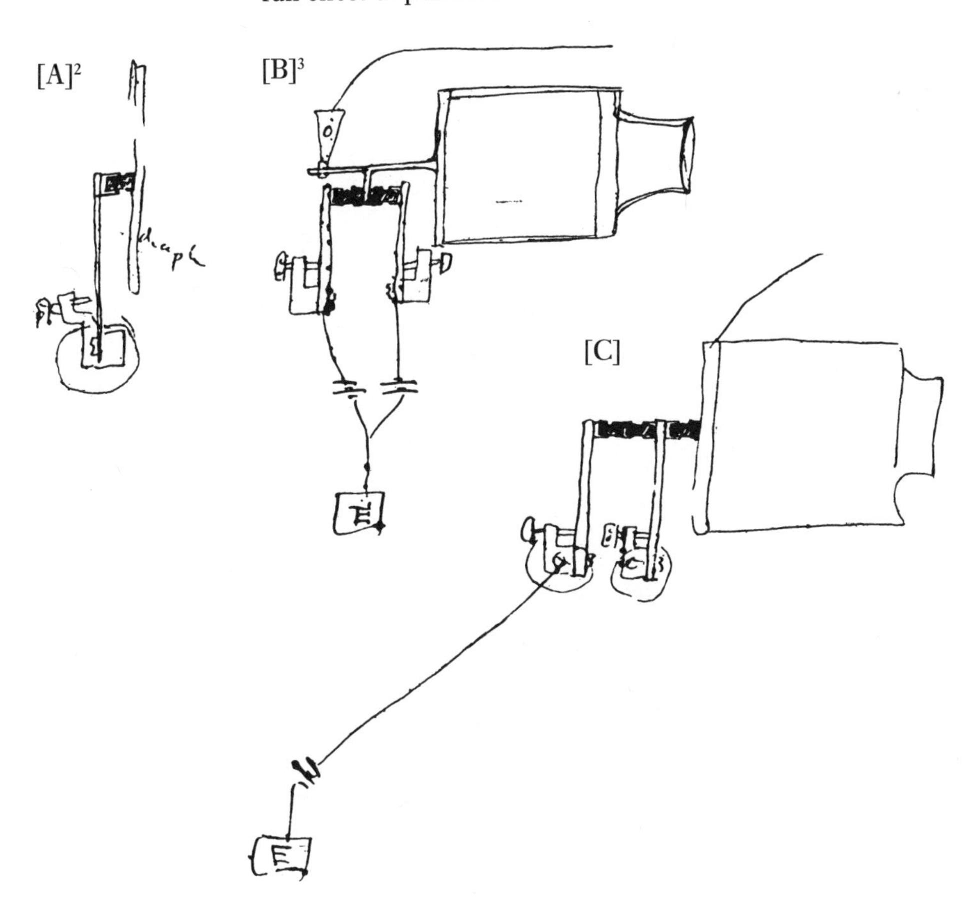

[D][4]

[E][5]

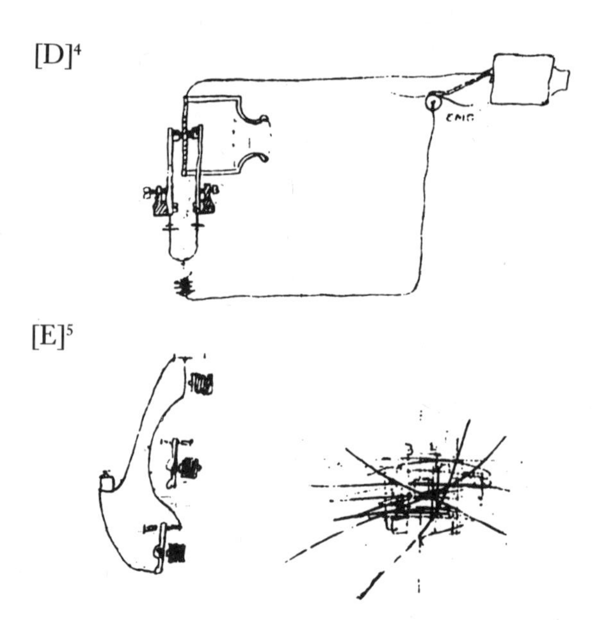

T. A. Edison

James Adams
Chas Batchelor

X (photographic transcript), NjWOE, TI 2, Edison's Exhibits 99-11–100-11 (*TAEM* 11:274–75). Document multiply signed and dated.

1. Edison introduced one of the platinum-cupped springs into evidence during the telephone interferences (Edison's Exhibit Instrument 100-11, TI 2 [*TAEM* 11:647]). In describing it, he stated, "I think the spring was made previous to May 25, 1877. It carried a carbon cylinder resting against a diaphragm. As a transmitter, the instrument of which Exhibit 100-11 is a part operated well" (TI 1:99 [*TAEM* 11:70]). Edison's Exhibit 100-11, to which the instrument corresponded, comprised drawings D and E in this document; D is the basis for Edison's patent application Case 141, executed on 9 July (U.S. Pat. 474,231).

The use of springs with plumbago contacts was a principal part of Edison's Case 141 and a major point later at issue in related intereference cases. On 29 May, Edison drew a full-size sketch of another platina-faced plumbago contact on a spring (Edison's Exhibit 111-11, TI 2 [*TAEM* 11:279]).

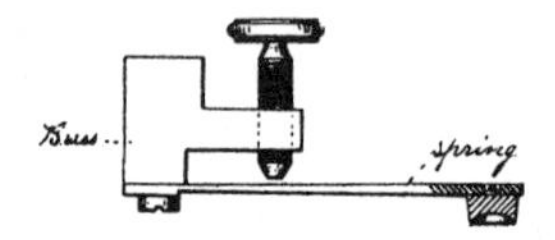

*Edison's Exhibit Instrument 100-11.*

2. Figure label is "diaph."

3. This drawing (like D below) closely resembles the circuit used in Case 141 (U.S. Pat. 474,231, figure 1). In these designs forward and backward movements of the diaphragm press separate carbon contacts, reducing resistance and thus increasing current from separate batteries connected with opposite poles to the earth so the signal current will alternate. When the diaphragm is at rest the batteries exactly cancel each other and no current flows.

4. Figure label is "EMG." Case 141 shows such a transmitter, with one of two diaphragm contacts on the inside of the speaking tube. The application also included Edison's electromotograph telephone receiver, as shown on the right of this sketch, with variations. Edison's Exhibits 101-11 and 102-11 (TI 2 [*TAEM* 11:276–77]) are measured

drawings also prepared on this day for an electromotograph receiver.

5. This appears to be an unfinished sextuplex receiving circuit.

**-921-**

*Technical Note: Telephony*

[Menlo Park,] May 26 1877.

Talking Telegraph

I find that it is absolutely essential to perfect the speaking Telegraph that the th sh ch s and other hissing sounds should be sent over the wire. now the diaphrams we have made do not respond to an appreciable extent to this hissing sounds=[1] now I think these hissing sounds are composed of vibrations and are notes having an exceedingly low rate of vibration pbly 10 per second & weak at[a] that. Now I propose to construct a tube & diaphram which will respond to them and this tube I propose to have made like a telescope so I can increase or decrease its length[2] so as to get a length of tube which will be in tune with these hisses & reinforce them and I propose to have this tube adjusted only to reinforce the hissing there being no trouble in getting the higher rates of vibration without the assistance of the Column of air= if necessary the higher notes can be made on another tube & diaphram The Talker talks into both tubes at once & both serve to close the battery=

free reed to respond to hissing consonants[3]

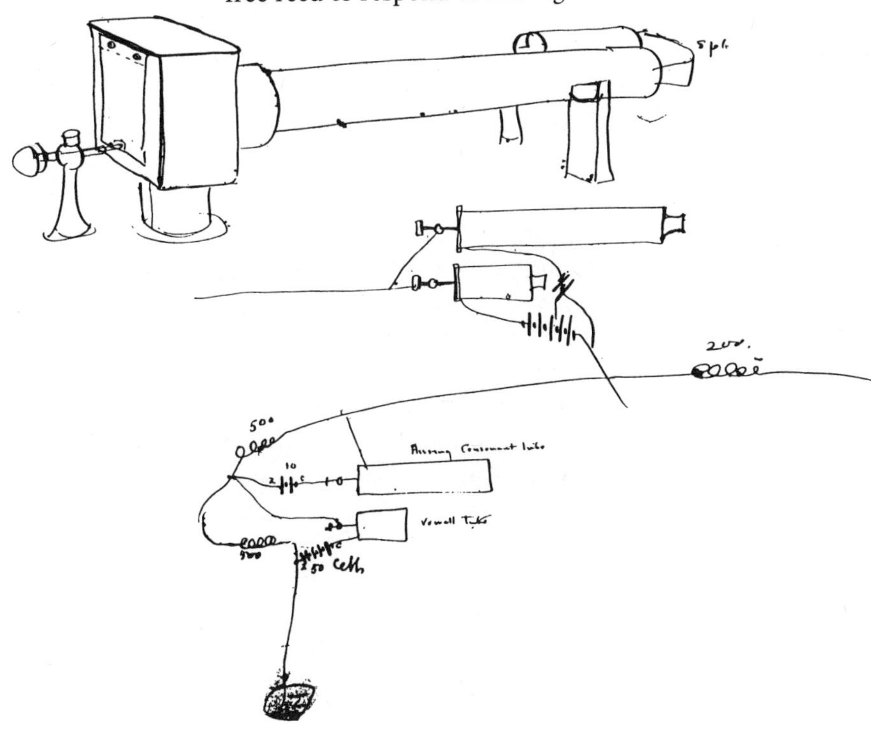

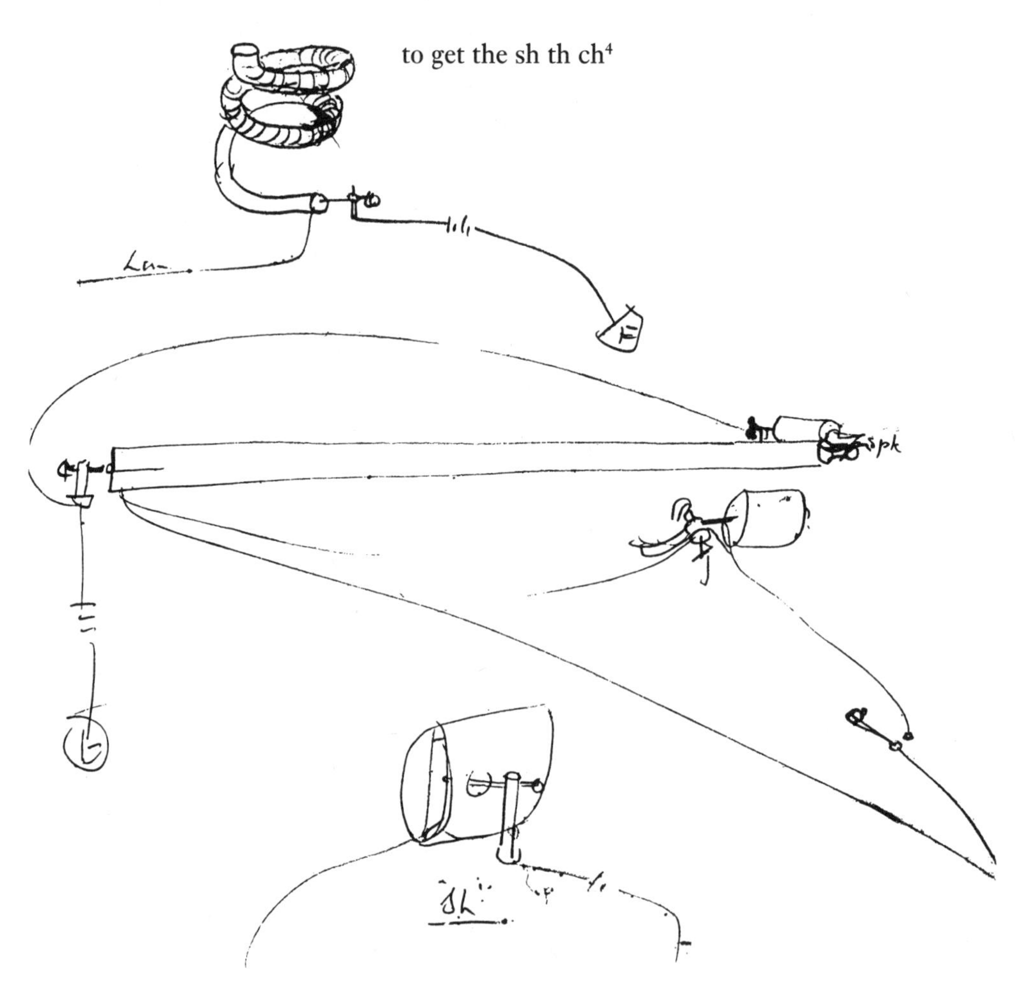


Keyboard Talking Telgh,

I propose to have a long shaft with wheels on having breaks (ie electrical) so arranged ~~th~~with a Key board that ~~is~~by depressing say the letters T H I S simultaneously that contact springs will one after another send the proper vibrations over the wire to cause the Emg & diaphram to speak plainly the word this=[5] by this means no difficulty will be had in obtaining the hissing consonants and as the breaks wh~~ie~~eels & contact springs may be arranged in any form and as many as required used the overtones harmonics of the parts of speech can easily be obtained turn this over in your mind Mr E & hoop it up

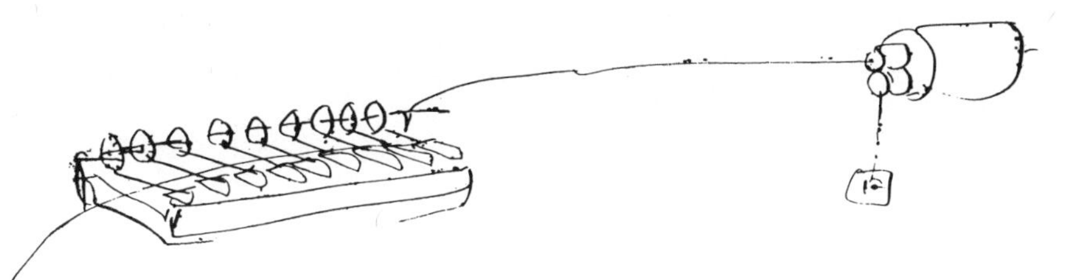

Speaking Telegh which prints automattically the speech[6]

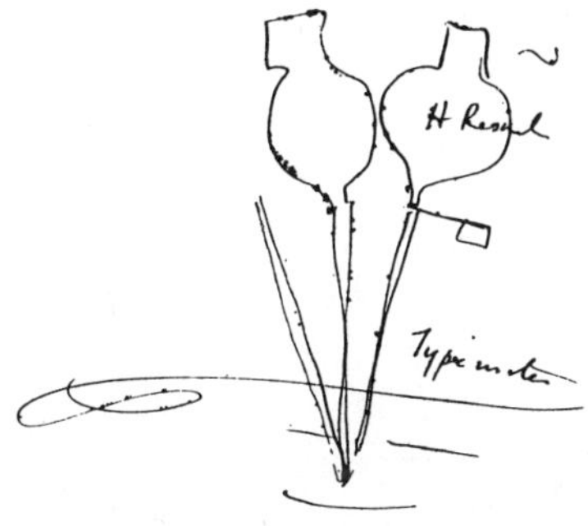

Each resonater is tuned to respond to any particular  it may be adjusted to open & close an Electric circuit & thus control the type writer or there may be bunched together 25 chemical recording points like my perforator and these p~~oi~~ressing on paper  the resonators will control which shall print the letters=[7] Ben Butler suggested strings at 5 ave Hotel in presence of Leonard[a] Myers & his stenographer= n[o] g[ood]=

T A Edison  Chas Batchelor
James Adams

X, NjWOE, Lab., Vol. 11:110, 105, 107, 109, 108 (*TAEM* 3:982, 977, 979, 981, 980). Document multiply signed and dated. [a]Obscured overwritten letters.

1. Accounts of Bell's early telephone demonstrations indicate that his telephones had difficulty in reproducing these speech sounds. Bell 1876, 8; Bruce 1973, 189; *Scientific American Supplement,* 10 Feb. 1877, Cat. 1240, item 59, Batchelor (*TAEM* 94:24).

2. See Doc. 708 (fig. 14).

3. Figure labels are "spk"; and (clockwise from upper right) "200," "Hissing Consonant tube," "vowell tube," "50 cells," "500," "10," and "500." Both batteries are labeled with their carbon poles toward the diaphragms and zinc poles toward the line.

4. Figure labels are "Line," "spk," and '"Sh."'

5. During his harmonic telegraph experiments conducted in 1875 and 1876 Edison developed "acoustic engines" that used break wheels to transmit high-frequency signals (Doc. 708 n. 13). Edison received U.S. Patent 198,087 (executed 9 May 1876) for one of these devices. In

1875 Elisha Gray had produced vowel sounds and groaning noises using break wheel cams on a shaft driven by a belt. Gray 1977, 52–53; see also Doc. 964.

6. Figure labels are "H[elmholtz] Resonator" and "Typewriter." Cf. claim 3 in Doc. 708.

7. Cf. Doc. 709 (fig. 9).

**-922-**

*Charles Batchelor to James Batchelor*

[Menlo Park,] May 28th [187]7

Dear Father.

Just a line to tell you we are all well. We are right in the middle of spring and having nice weather. The country looks beautiful. This is what they call Locust Year here as the 17 year locusts appear this year, and already there seems to be any quantity of them around. We find lots of snakes round here, we have killed no less than 10 Black snakes about 6 feet long already this spring. There is also good shooting in the woods here.[1] Your last letter was marked Newark; Menlo Park is 16 miles from Newark. We have a nice lot of vegetables planted this year. I hope you have got to like your change by this time & that its effect on your health is beneficial I know it will suit mother and the girls. We have now got the 'Electric Pen' fairly out on Royalty and in a very short time I shall have nothing whatever to do for it except receive my share of Royalty. We are now putting on the wires a machine which sends 6[a] six messages over the wire at the same time we call it the Sextuplex. We shall put it on royalty. Rosa unites with me in love to mother and girls & believe me Your affectionate Son

Chas Batchelor

ALS (letterpress copy), NjWOE, Batchelor, Cat. 1238:125 (*TAEM* 93:122). [a]Circled.

1. Batchelor's diary records occasional walks as well as other hunting expeditions in the woods. E.g., Cat. 1233:125, Batchelor (*TAEM* 94:115).

**-923-**

*From Thomas David*

Pittsburgh, May 29th 1877.[a]

Friend Edison—

I have been making some investigations in the varnish business, and am frank to say that I believe you can do better with your discovery by disposing of it to some of the larger concerns in the business East—[1] From what I have learned, and it is reliable information, one[b] wants to be near the seaboard to successfully compete with those already in the business—

I will tell you what I should do if it were mine; I should endeavor to make contracts with the larger manufacturers to furnish them with varnish at a certain price, leaving them to invest the capital necessary[c] to carry the stock the length of time varnish requires to be carried—

The party with whom I have been in[d] ~~in~~communication[b] ~~with~~ was in the business for a great many years— He says, however, that the process of manufacture was not hard, nor lengthy— however, if by your process it is done without fire, and is rapid there are decided advantages— The black color given the varnish results from burning, or it may come from the adulterations now largely in use—

How does all this compare with the information you have been able to get? Yours truly

T. B. A. David[2]

ALS, NjWOE, DF (*TAEM* 14:65). Letterhead of the Central District and Printing Telegraph Co. [a]"Pittsburgh," and "187" preprinted. [b]Obscured overwritten letters. [c]Interlined above. [d]Inserted.

1. For what is known of Edison's "discovery" see Docs. 872 and 915. David had visited Menlo Park on 10 May and discussed "the Damar waterproofing and varnishing & he agreed to go in & give half profit Electro Chemical Mfg Co & keep ½ profit to himself" (Cat. 1233:130, Batchelor [*TAEM* 90:118]). On 17 May, David had written Edison, saying, "It occurs to me to say, that if you have not taken out patent for that Copal discovery, you would be wise not to— It should be far easier to keep the thing secret, than to find parties who might infringe" (DF [*TAEM* 14:59]).

2. Thomas B. A. David (b. 1836) entered telegraphy as a messenger at the age of thirteen, becoming an operator three years later. He subsequently managed a telegraph office in Wheeling, Va. (later West Virginia), and served as a captain of the United States Military Telegraph there during the Civil War. After the war David returned to his home in Pittsburgh, Pa., where he served as superintendent of the Fourth District of the Western Union Telegraph Co. Central Division. Around 1869 he became associated with the Central District and Printing Telegraph Co. in Pittsburgh. Plum 1882, 2:145–46; Reid 1879, 563.

**–924–**

*Technical Note: Telephony*

[Menlo Park,] May 31st 1877

Speaking Telegraph    50 Cells Callaud Battery

Experiments to determine the availability of Plumbago mixed with different substances for a disc which by variable pressure shall give us variable resistance.[1] 5 gramme weight has piece of platina on it 9/16 diameter. The blocks are all 1 in thick

| | | 5 gramme weight | 1 lb & 5 gram | thickness |
|---|---|---|---|---|
| 1 | Pure Plumbago | 75 ohms | 75 ohm | 5/8 in |
| 2[a] | Plumbago 66:6. Gelatin33. | over 10,000 ohms | 200. | 1/2 |
| 3 | Plumbago 66. Gum Damar[2] & Linseed Oil 33[b] | 200 | 75 | 5/8 |
| On a second test of 2 | Plumbago 66 Gelatin | 2,000 | — | 1/2 |
| On a third test of 2 | Plumbago 66 & Gelatin 33 | 1400 | 400 | 1/2 |
| 4 | Plumbago 66 Plast. Paris 33 | over 10,000 | 100 | very sensitive!!![c] |
| Ditto Second test | | over 10,000 | 100 | 1/2 |
| | This has more than 10,000 probably 20,000 ohms | | | |
| 5 | Plumbago 2/3 Collodion 1/3 | 100 | 100 | 1/2 |
| 6— | Plumbago 1/2 Sulphur 1/2 | over[b] 10,000 | 2100 | 7/8 |
| 7— | Plumbago 2/3 Isinglass[3] Amn 1/3 Dissolved in Acetic Acid[d] | 400 | 100 | 5/8 |
| 8 | Plumbago 1/2 Sugar 1/2 | over[b] 10,000 | 100 | 1 in |
| 9 | Plumbago 2/3 Tragacanth 1/3 | 200 | 30 | 1/4 |
| 10 | Plumbago 2/3 Glue 1/3 | over[b] 10,000 | 400 | |

This is exceedingly variable sometimes ~~a very~~ 10 000 then goes to 1000

11 Plumbago 2/3 Sulphur 1/3

This has probably 100,000 ohms with 5 gm and about 10 000 ohms with 1 lb 5 gramme sometimes by a little pressure it can be brought down to 2 or 300 ohms but it is not reliable

12

T. A Edison[e]

Chas Batchelor
Jas Adams[e]

X (photographic transcript), NjWOE, TI 2, Edison's Exhibits 113-11–114-11 (*TAEM* 11:281). Written by Batchelor. Several numerical entries preceded and/or followed by dashes of varying length. [a]From here to "third test of 2" entry spanned by brace in left margin. [b]Interlined above. [c]Multiply underlined. [d]"Dissolved in Acetic Acid" interlined below. [e]Name written by Batchelor on first page; signed on second page.

1. This document marks the beginnings of a systematic search to find the best binder and form for plumbago cakes to use in the telephone transmitter. The extensive tests, almost all recorded by Charles Batchelor, continued into the summer. Although the surviving record is fragmentary, it does indicate the scope of the work. On 20 June, Edison and Batchelor listed 151 substances the staff had tested (the compounds in this document are not on that list). On 24 June the staff listed 7 more

(Edison's Exhibits 164-11–168-11, 172-11, TI 2 [*TAEM* 11:319–23, 325]). See Docs. 928, 937, 941, and 945.

2. Usually spelled "dammar," this is a tropical resin similar to copal. In April 1877 the staff had pressed this and several other substances into cakes in a one-inch-diameter die and had used it in their attempts to find a waterproof varnish. Cat. 1317:35; Cat. 1233:130; both Batchelor (*TAEM* 90:674, 113).

3. Isinglass is a fine gelatin usually made from the air bladders of fish (particularly sturgeon).

**-925-**

*From Thomas Clare*

Birmingham May 1877

Dear Sir

As you know I contracted with Mr. J. F. Gloyn to purchase your patent for the Electric Pen for Gr Britain but I never had the pleasure of having any correspondence with you or matters would have perhaps gone more pleasantly

I never knew the exact merits of the transactions when they originated. Mr. Gloyn I had known a considerable time & after he went to America he sent me one of your pens & said he had an offer of the patent from you & he wished me to join him in acquiring the English patent.

after many troubles in printing (as then I only had the first purple analine ink) I found I could print well with a different Ink & Roller & I cabled Mr. Gloyn to close an arrangement— which replied he had done— We passed many letters on the mode of dealing but I could not agree to his proposition but finally after agreeing with my friend Mr. J. R Breckon to join me in the purchase we sent over Mr. Ireland to settle with you which he reported he had done after considerable difficulty but as I could not obtain any satisfactory information from either side & the assignment was made according to arrangement I did not trouble about what had passed, but lately it occurred to me that representations had been made which might give you wrong impressions regarding me and my position as regards your patent here & I thought I would write you a few lines of explanation—

Before he went out (Mr. Ireland) we had a pow[er of?][a] attorney prepared, signed by myself & Mr Breckon authorizing him to act for us both— My [power?][a] was accidentally left behind but I had written fully to Mr. Gloyn & authorized him to permit the name of Mr Breckon alone to appear in the deed In that Mr. Ireland had sufficient authority & I had advised Mr Gloyn he would come out & would act for me & Mr Breckon— I heard subsequently he said he had nothing to do with me but had brought money to take up the patent & was acting for Mr. Breckon alone

To shew the truth of this Mr Ireland met me with Mr Breckon specially at Birm'm for final arrangements & we went to my Solicitors Mess Ryland, Martineau, & Carslake to have the regular papers drawn up It was Saturday & business greatly close at 2 pm but they staid till 4 oclock to complete this business & the document was sent to the American Consul for signature but he was absent & it had to be left till Monday & then followed a blunder so that Mr. Ireland did not get the power of Atty & went without it but it was really the only legal power by which he could act but Mr Gloyn being prepared dispensed with it & all you had to do was to complete your contract with Mr. Gloyn & receive your money—[1]

My connection with this business was very simple I recd copy of your contract to sell to Gloyn & his contract to sell to me— I agreed to purchase & then saw friends to join me but in the end agreed with Mr Breckon & I hold a moiety of the patent subject to Mr Breckons [moi---s?][b] claims—

There are so many false things said that I wished to shew you at some time how matters had really stood— Accidentally some time since Mr Ireland said you had offered him your foreign patents; I at once wrote Mr Breckon & said Mr Ireland going out to represent us as agent ought not to deal with matters connected for his own benefit & I should expect to participate in anything that took place but I have not heard if anything further has been done but I should be glad to know how this stands if you have offered the patents to above and what has been done

I may say matters have not been carried on here to my satisfaction Mr Ireland has been manager but I should think the business has been about a tenth of what an American man of business would have done here & as I understand has been done in your country—[2] It has been a disappointment to me as I expected good results for the pen is a wonderfully perfect instrument though the printing is faulty & slow but with energy & push I consider 10 times as much could have done if money money & energy had been used—

I was pleased to read of your improvement on the Telephone— The progress of science is marvellous in all directions

Your Country means to turn the tables on us & instead of we supplying you we have your various manufacturers coming in here of all kinds & they seem to be able to compete successfully— The production of Bessemer Steel has been marvellous the last year especially. I hold three fourths in Mushet's patent[3] which practically controlled the process but we did not play our cards well & only get a percentage [of?][a] what we ought to have got. Still we received Royal[ties?][a] last year nearly £5000 but at our first scale it would be ha[ve?][a] realized £250,000 Sterling for the one year— Bessemer [---][a] £2. put in & made an enormous fortune in ten Years, in whi[ch?][a] we ought to have participated but for our accidental loss [of?][a] the patent— I shall be glad at your convenie[nce?][a] to have a few lines— I am Dear Sir Yours truly

Thos. D. Clare

I exhibited "The Pen" at a Conversazione of the Royal Society,[4] where it was received with distinguished success. I also exhibited it at the Society of Telegraph Engineers two consecutive meetings,[5] & at the Society of Arts, the Royal Institution, Brm'm & Midland Institute 4 evenings, Iron & Steel Institute Leeds—

The Secretary to the Treasury wrote me for specimens of the work done by the pen & particulars as they are investigating the best mode of multiplying copies of writings &c—

It ought to have a large sale & will eventually but it requires very energetic management with no lack of funds & some one acquainted with & fitted for business— There is a press newly brought out which Mr Breckon says will print 1000 per hour good copies—

I know very little of it at present as I am rather at impass with him about certain arrangements—

ALS, NjWOE, DF (*TAEM* 14:397). [a]Document damaged. [b]Interlined above; illegible.

1. On 23 June 1876 Lemuel Serrell had advised Edison:

> The British Ass[ignmen]t is all right except that you should not guarantee to defend the patent. . . . You should have from Mr Gloyn a paper showing that you are relieved from any question on Clares rights, that he Gloyn alone has to carry out the arrangement with Clare so that Clare cannot come on you for not recognizing a right you know he is entitled to. [DF (*TAEM* 13:1032)]

The assignment has not been found, but Edison discussed it in a 29 June 1876 letter to Breckon. DF (*TAEM* 13:1035).

2. There were at least some sales of the pen. In mid-June, Oxford mathematics don Charles Dodgson—better known as Lewis Carroll—bought one through a stationer. He used it for drawings, multiple correspondence, and examination papers, and was quite taken with it. Cohen 1976.

3. Clare provided the financial backing for Robert Mushet, who devised the manganese process that made Bessemer's steel-making system practical. McHugh 1980, 148, 162, 183.

4. See Doc. 748. The Royal Society of London, the leading general scientific society in Great Britain, would occasionally host an informal social evening for members where some scientific or technical items of note might be displayed or demonstrated. These receptions, like similar ones held by other technical and scientific societies, were regularly reported in the scientific and technical press of the world.

5. A report of a conversazione of the Society of Telegraph Engineers held in January 1877 noted that Edison's electric pen, exhibited by the Electric Writing Co., was "one of the most popular objects in the whole exhibition." *Sci. Am. Supplement* 3 (1877): 909; see also Doc. 748 n. 7.

**–926–**

*Article in the* Journal of the Telegraph

New York, June 1, 1877

EDISON'S PRESSURE RELAY.

MR. EDISON has recently invented an ingenious and novel relay instrument, based upon an entirely new principle. He takes advantage of the property which plumbago possess of decreasing its resistance enormously under slight pressure. Thin discs of that material are placed upon the cupped poles of an electro-magnet, (see diagram), the coils of which have several hundred ohms resistance.

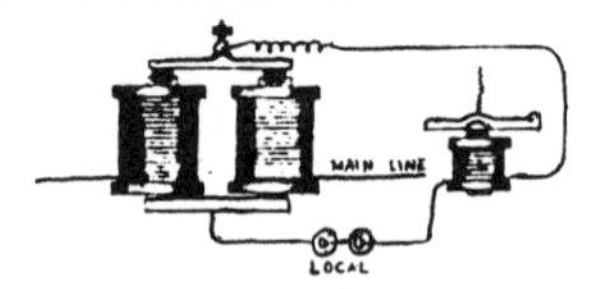


Upon the discs of plumbago is laid the armature which is provided with a binding post for clamping the local battery wire.

The cores of the magnet, the plumbago discs, and the armature are included in a local circuit, which also contains an ordinary sounder and several cells of bichromate battery. The relay magnet is inserted in the main line in the usual manner. The operation is as follows: When the main circuit is opened the attraction for the armature ceases, and the only pressure upon the plumbago discs is due to the weight of the armature itself. With this pressure only the resistance of the plumbago to the passage of the local current amounts to several hundred ohms; with this resistance in the local circuit the sounder remains open. If now the main circuit be closed, a powerful attraction is set up between the poles of the relay magnet and its armature, causing a great increase in the pressure upon the plumbago discs, and reducing its resistance from several hundred to several ohms, consequently the sounder closes. So far the result differs but little from the ordinary relay and sounder. But the great difference between this relay and those in common use, and its value, rests upon the fact that it repeats or translates from one circuit into another, the relative strengths on the first circuit. For instance, if a weak current circulates upon the line in which the relay magnet is inserted, the attraction for its armature will be small, the pressure upon the plumbago discs will be light, consequently a weak current will circulate within the second circuit; and on the contrary, if the current in the first circuit be strong, the pressure upon the plumbago discs will be increased, and in proportion will the current in the second circuit be increased. No adjustment is ever required. It is probably the only device yet invented which will allow of the translation of signals of *variable strengths,* from one circuit into another, by the use of batteries in the ordinary manner. This apparatus was designed by Mr. Edison for repeating acoustical vibrations of variable strengths in his speaking telegraph, a description of which we shall shortly publish.

PD, *J. Teleg.* 10 (1877): 163.

**-927-**

*Technical Note: Multiple Telegraphy*[1]

[Menlo Park,] June 1 1877.

Sextuplex[2]

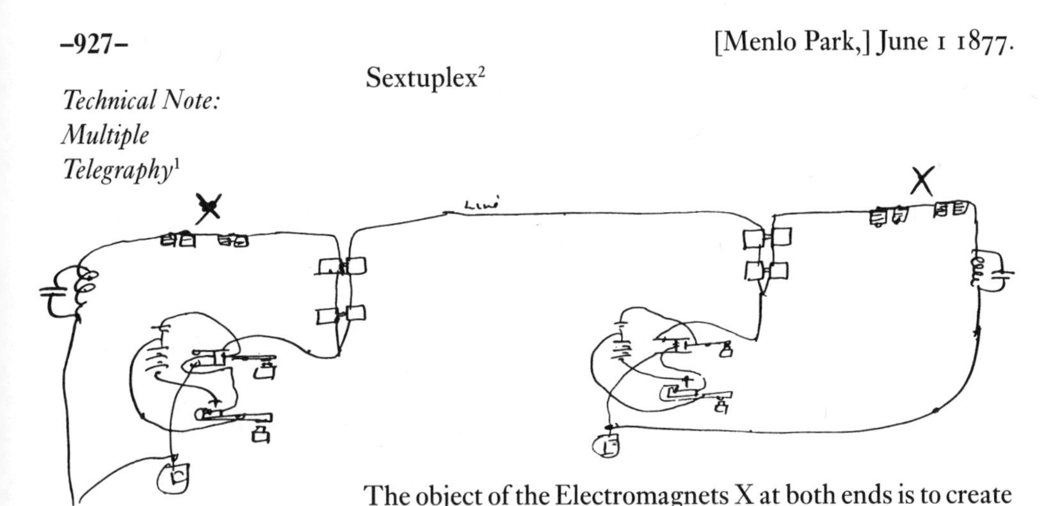


The object of the Electromagnets X at both ends is to create the same conditions upon the artificial as there is on the line, and it is impossible to obtain an accurate balance unless the 2 relays of X are inserted to counteract the effect of the 2 Receiving relays at the other[a] station & vice versa=

Pretty fair but reversals touch X

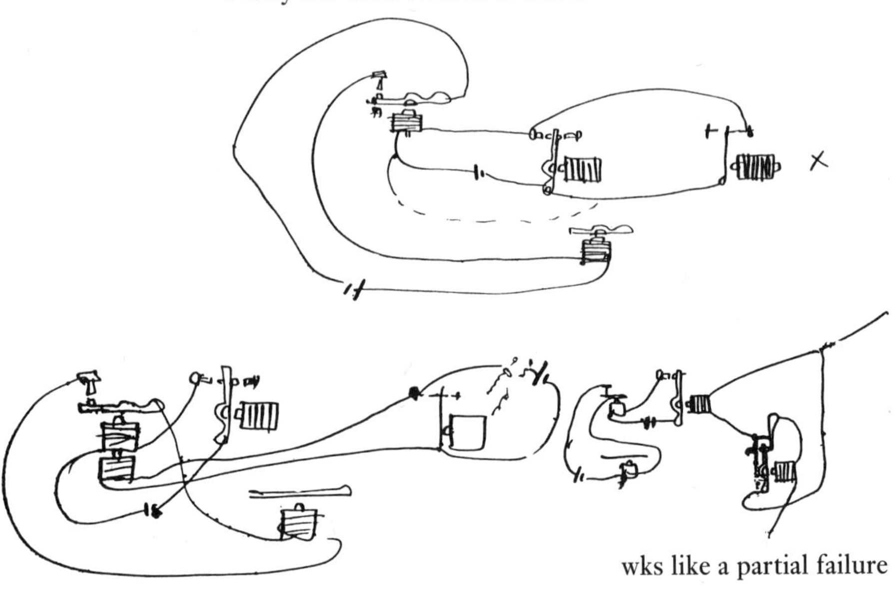


wks like a partial failure

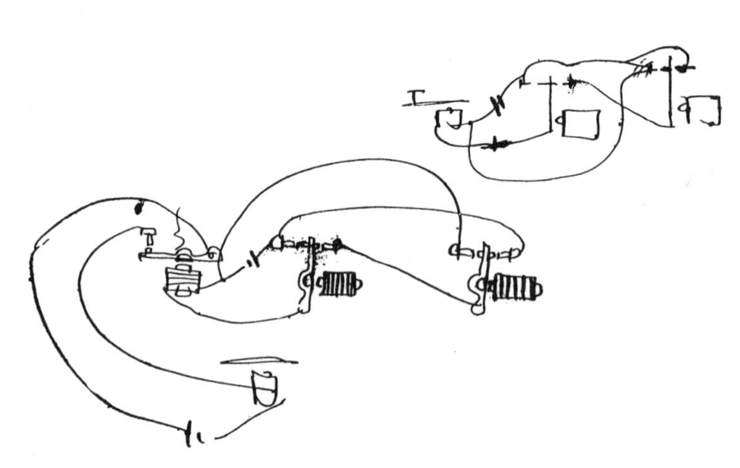

Dont better it much=

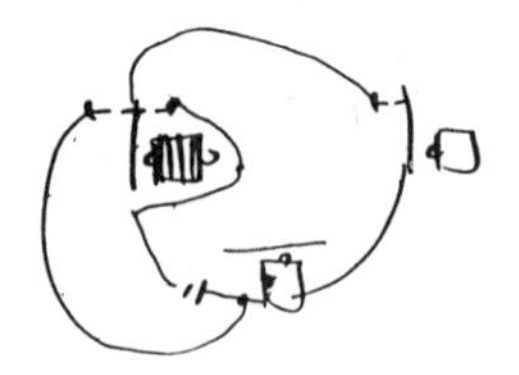

This works splendid  Scarcely a bug in it  Besides its very simple.[3]

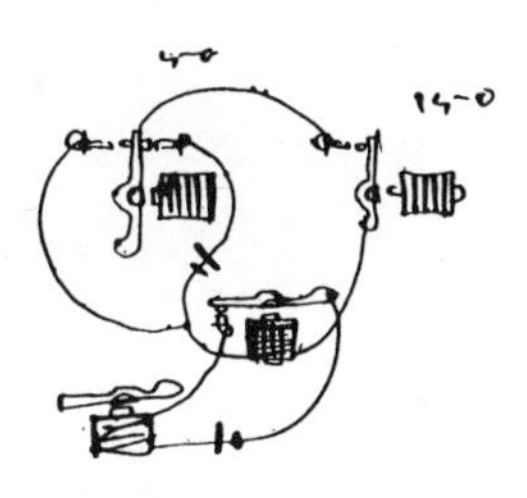

I think a repeating sounder better be put in as owing to 150 relay working sounder not shunted gives light writing whereas 50 shunting sounder his writing comes heavy  hence although no bugs the writing would be uneven by reason of different strengths producing the sound  the Repeating Sounder would obviate this  I put~~t~~ repeating sounder in but It gave bks  I then put in Duplex spg sounder & that worked better scarcely bug but its no great if any impvt on straight sounder with static off line 4500 ohms full condr  It works quite even & prompt with scarcely a bug=[4]

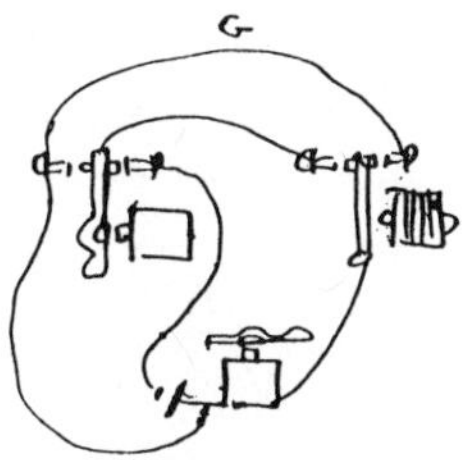


Putting .G. in dont appear to make it any better.=

Try this

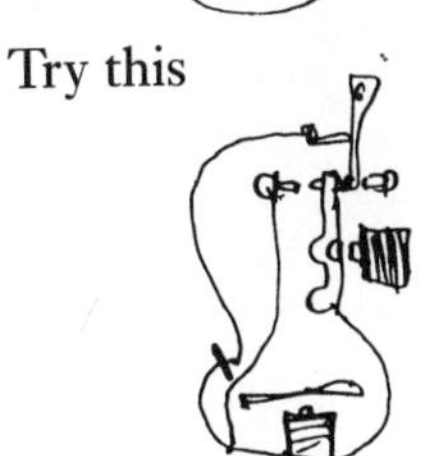

Not very good cant wk close enough adjstments without points sticking

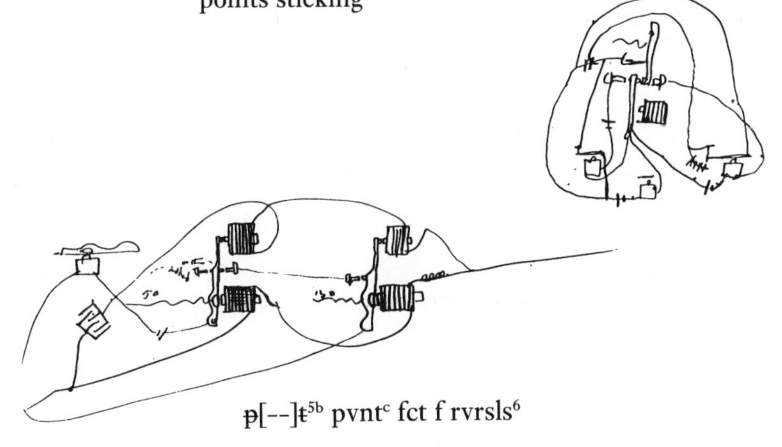

~~p~~[--]~~t~~[5b] pvnt[c] fct f rvrsls[6]

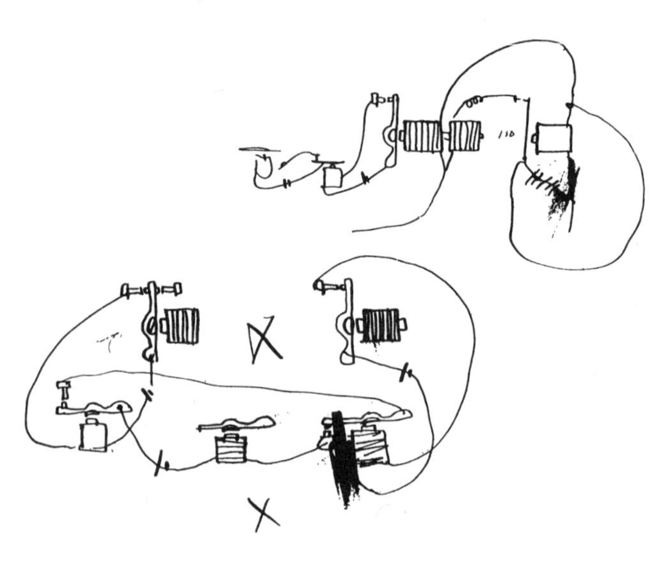

This I think works the best yet  the only bug noticed is when 50 man[7] is open  every few seconds the sounder gives a closure for an instant. There is some defect pbly in the manipulation of the current. Best bug trap yet for Sexty=[8]

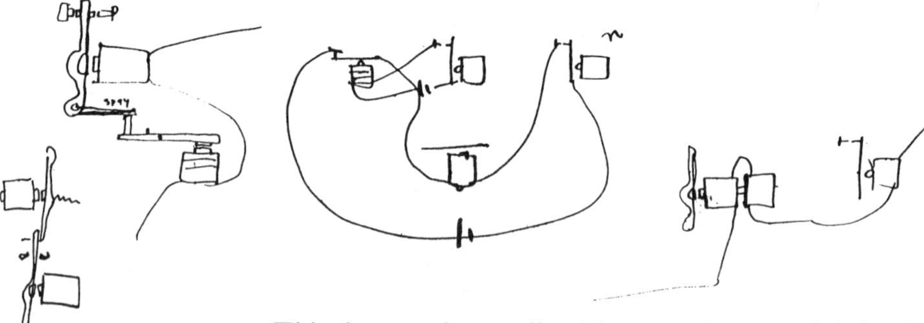

This dont work as well as X cos n wks too quick for the repeating sounder
good[9]

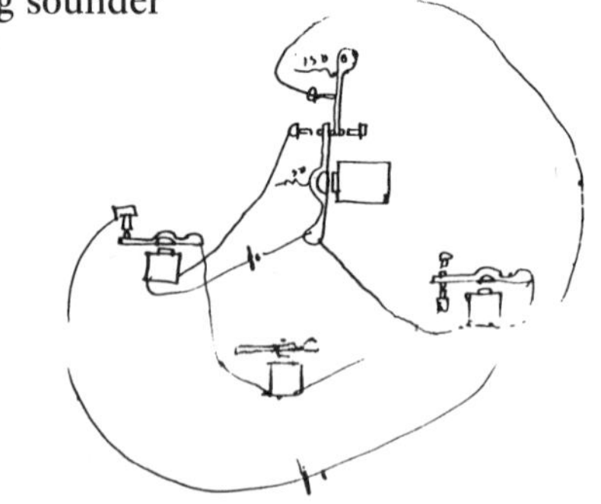

This dont appear to work very well  although the reversals dont effect No 2[10] 100 man's does

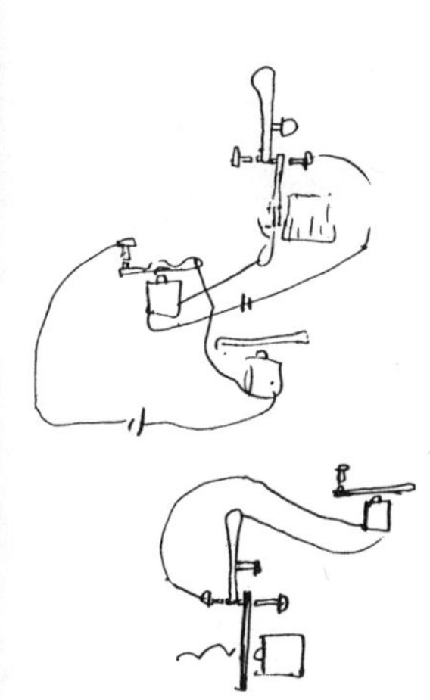

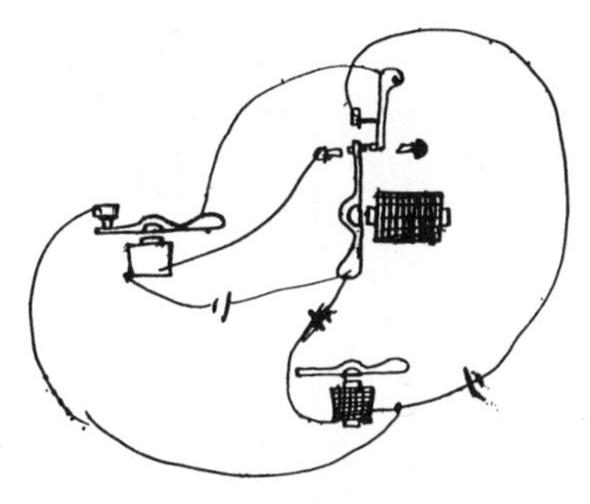

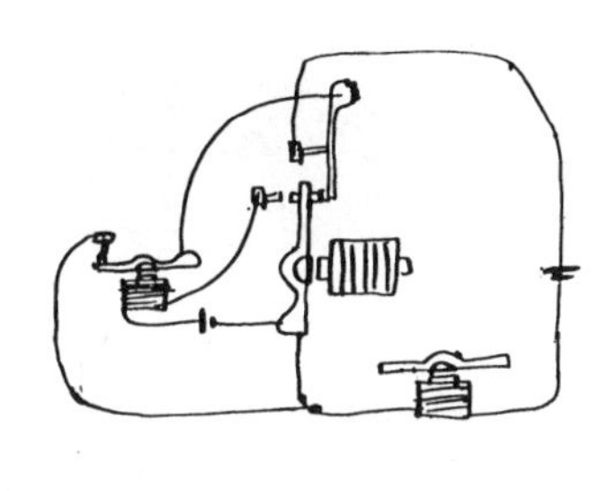

This appears to give a short wave to Sounder when 1 is open & No 2 100= after closing is opened= Reversals effect it also alone when 2 is closed=

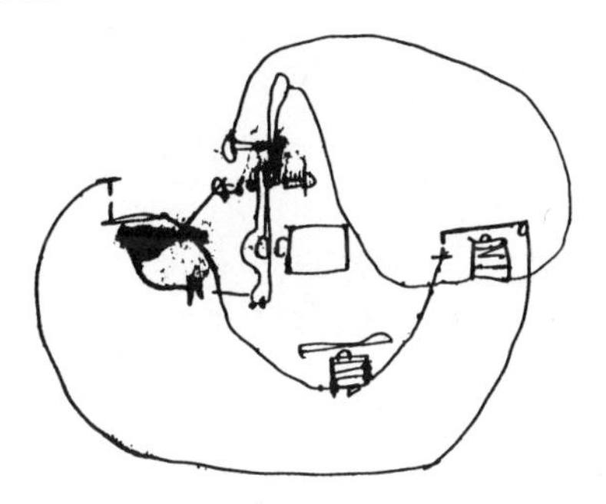

This is about same thing as X on preceding page[11] except only 1 Relay used[12]

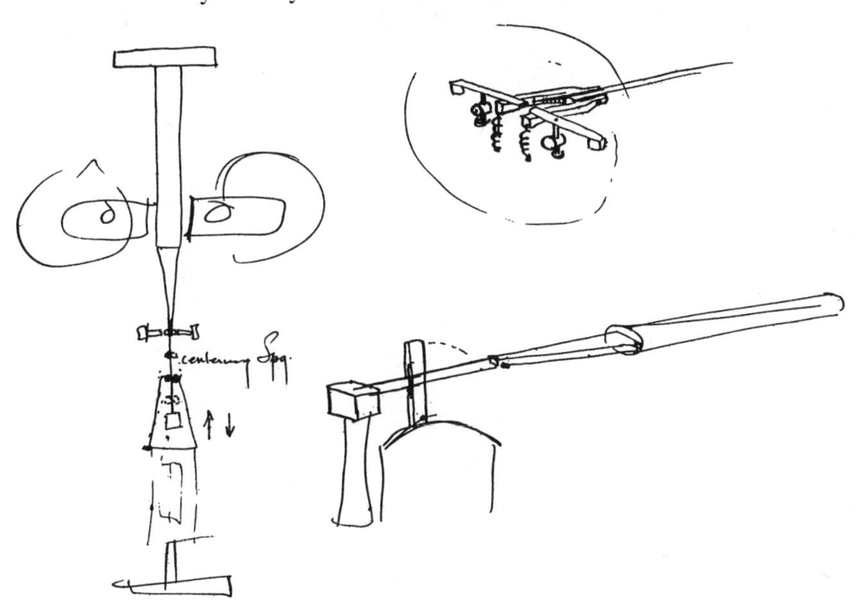

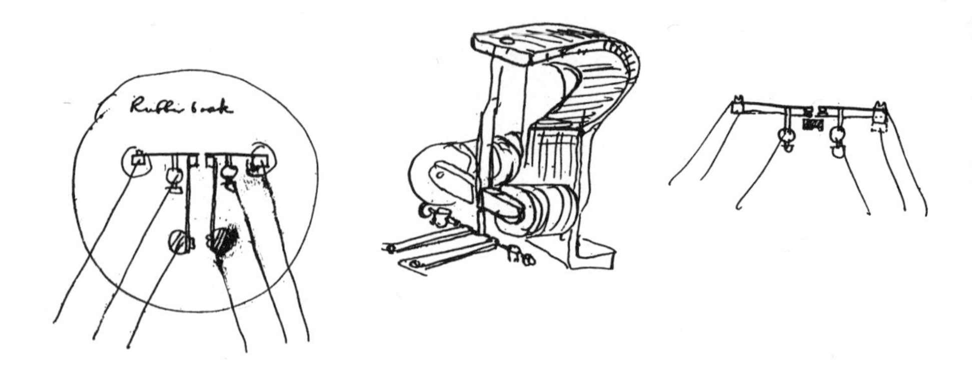

Reversing transmitters[13]

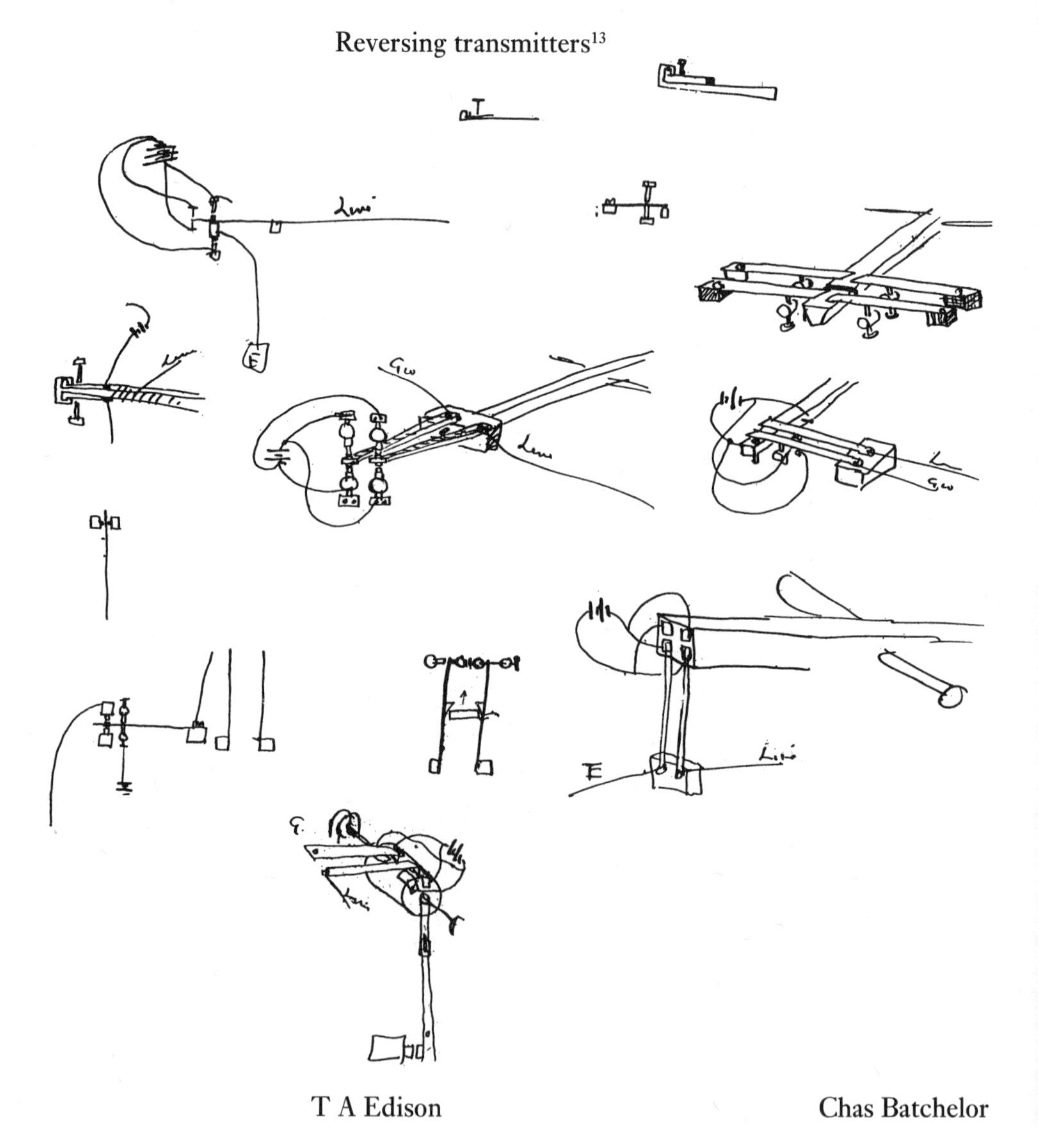

T A Edison Chas Batchelor

X, NjWOE, Lab., NS-77-004 (*TAEM* 7:614, 612, 613, 616, 615, 618, 617). Document multiply signed and dated. [a]Obscured overwritten letters. [b]Canceled. [c]Interlined above.

1. See headnote, p. 280
2. This drawing is labeled sextuplex but only a quadruplex circuit is shown.
3. Figure labels are "50" and "150."
4. After "straight sounder" the physical arrangement of the text does not clearly indicate either the referent or breaks in thought.
5. Figure labels above are "50," "or this added," and "150."
6. That is, "prevent effect of reversals." Figure label on following drawing is "150."
7. The "50 man" is the operator on the key that uses a signal strength of 50.
8. Figure labels are "spgy" and "n."
9. Figure labels are "150" and "50."
10. This and other similar designations are missing from the drawings.
11. That is, the drawing that precedes "This I think works the best yet."
12. Figure labels are "centering Spg." and "Rubber back."
13. Several of the following figures have the labels "G[round] w[ire]," "G," "E," and "Line." Edison drew configurations for other reversing transmitters around this time. See drawings of 18 and 19 May and 5 June 1877, NS-77-004, Lab. (*TAEM* 7:591, 595, 599).

**-928-**

*Technical Note: Telephony*

[Menlo Park,] June 2nd 1877

Speaking telegraph

Tried our new speaker made of cast iron cylinder with diaphragm 6 in on solid iron base and plumbago disc 1 in diameter but it evidently is not so good as former experiments with less diaphragms so

We put in ~~an~~ instrument with 1⅜ diaphragm and got better results

This with pure plumbago[1] was good[2]

" " Plumbago & sugar no better

Plumbago & Sulphur <u>no good at all</u>

Plumbago & Isinglass[3] best yet

We now took away cylinder from receiver & left the diaphragm standing & found we received better[4]

We now put back Plumbago & Isinglass in large diaphragm sender but no good at all—[5]

T. A. Edison[a] Chas Batchelor
James Adams

X (photographic transcript), NjWOE, TI 2, Edison's Exhibits 116-11, 115-11 (*TAEM* 11:284, 283). Written by Batchelor. [a]Written by Batchelor; re-signed "Edison" by Edison.

1. At the end of the month Batchelor noted that the laboratory now had "Pure Plumbago $1 per lb. previous plumbago used by us only 6¢ per lb greatly adulterated." Edison's Exhibits 172-11, 120-11, TI 2 (*TAEM* 11:325, 285) (Edison's Exhibit 120-11, of 26 June, is misdated 6 June).

2. Unlike the earlier resistance tests of plumbago mixtures (Doc. 924), here the staff evaluated the mixtures in an actual transmitter. In these experiments, which continued through 26 June (Edison's Exhibits 120-11–121-11, TI 2 [*TAEM* 11:285]; Edison's Exhibit 120-11 is misdated), the plumbago mixture was pressed into the shape of a cylinder with a small bump or "tit" on one end (as shown in Doc. 937). Edison drew plumbago in this shape on 7 June and noted additionally that "with the plumbago fine points the diaphragm ought to be faced with platina" (Vol. 11:127, Lab. [*TAEM* 11:992]). The cylinder was apparently held against the diaphragm by an adjustable screw; later the staff added a spring (Edison's Exhibits 181-11, 184-11, 187-11, TI 2 [*TAEM* 11:334, 337, 340]).

3. On 7 June, Edison again recorded the superiority of the plumbago-isinglass mixture in a lab notebook. NS-77-004, Lab. (*TAEM* 7:628).

4. The tests apparently used a Bell-type magneto receiver until 25 June, when Batchelor noted that "starting these experiments today we replaced the receiver with the wooden sounding box one [i.e., the electromotograph]." Edison's Exhibit 187-11, TI 2 (*TAEM* 11:340).

5. The next surviving record of these tests is Doc. 937.

## –929–

*From George Caldwell*

Phila June 3rd [1877]

T A Edison

As you are aware we have met with a failure financially[1] the business was outragous and up to the present time I am out nearly $300. I thought it best to come while I had money enough to pay the fares home. I have however not givin up the telephone, but will work it different I will now try to make arrangements for Exhibitions for churches for a certainty and in the fall take the road again Business was $26 in Reading expenses $76. but when we got to Lancaster 85 cents was the sum total in the house at 8 P.M. we didnt show there but got out of town as soon as possible. Charlie will tell you all the particulars I shall remain in Phila a few days and try and arrange for a chance to show the telephone.[2]

Caldwell

The telephone worked very successfully and gave great satisfaction we started out too late the weather is too hot for amusements—

ALS, NjWOE, DF (*TAEM* 14:778).

1. A flyer, prepared by electric pen, advertising one of their telephone concert programs is in Cat. 1240:175, Batchelor (*TAEM* 94:54).
2. Caldwell did not exhibit the telephone again. In July, when Edward Johnson began a series of exhibitions (see Doc. 967), Caldwell wrote to Edison asking why he had not been included and seeking instruments to exhibit overseas. Caldwell to TAE, 12 July 1877, DF (*TAEM* 14:816).

**-930-**

*From Josiah Reiff*

New York, June 4th 1877[a]

My dear Edison—

I know I am "a squezed orange," "an old worn out coat" or an equally valuable relic, still whenever you have an absolutely idle moment you might call down & see if I still live—

Notwithstanding Lowrey said[b] I intended to Blackmail Orton, You can be sure I wont attempt to blackmail you, so come along. Occasionally I have something to tell you—

I have been quite unwell & lame, but am better now & hope to spend an evening with you soon. Yours

JCR

ALS, NjWOE, DF (*TAEM* 14:689). Letterhead of J. C. Reiff. [a]"New York," and "187" preprinted. [b]Obscured overwritten letters.

**-931-**

*Notebook Entry: Electromagnetism*

[Menlo Park,] June 6th 1877

Induction of Magnets[1]

41

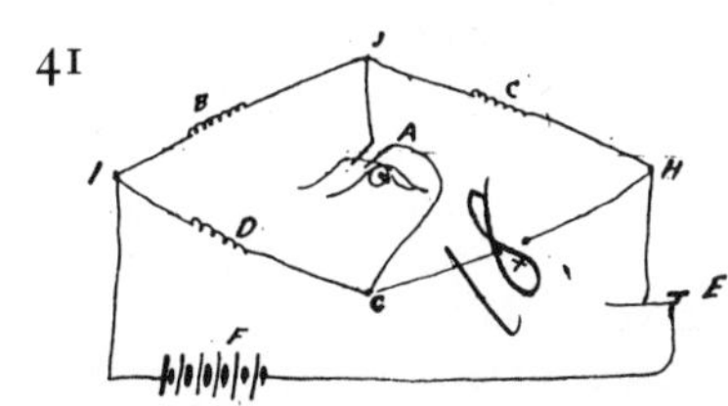


In the diagram 41 A is a drum carrying chemical paper (preferably Edison's solution of Ferridcyanide of Potassium and salt as it is exceedingly sensitive and permanent) B. C. D are resistances in three sides of a Wheatstone bridge B and D being equal and of high resistance and C being equal to the magnet to be tested which is placed at X E is a Morse Key and F a battery.

If X[a] were plain resistance there would be no mark on either pen on opening or closing E, the bridge being balanced, but being a magnet it sets up a current itself which circulates in circuit H, J, G, overbalancing the main current and obliging a part of it equal in volume to itself to pass through the bridge wire & in doing so it leaves its mark (either positive or negative) on one or other pen. On opening the key the magnet discharges in the same circuit HJG but in the opposite direction and leaves its direct mark on the other pen.

Thus with this device we have on closing, a mark produced by the exact equivalent of the current set up by the magnet and on opening we have the same magnet's direct mark for its own discharge.

These marks have pecularities which determine for us decisively the action of magnets during their charge and discharge  Thus when an ordinary 126 ohm relay is put in at X and the cores adjusted to touch the armature the closing of key or charging of magnet will give a mark like A and on opening one like B

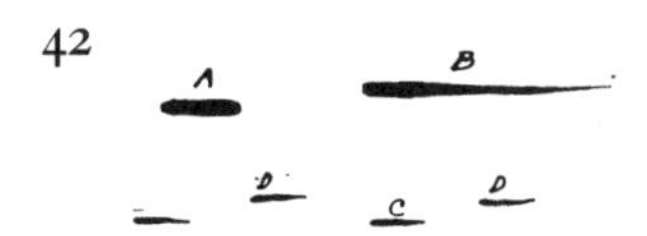


Now when the armature is adjusted ~~so~~a little away from the cores there is not so much difference in the marks and when the armature is entirely away the marks come equal for both opening and closing as at C, D, fig 42.[2]

X, NjWOE, Batchelor, Cat. 1317:38 (*TAEM* 90:676). Written by Charles Batchelor. [a]Obscured overwritten letter.

1. These experiments were apparently done in connection with sextuplex experiments. On the same day Edison proposed to use an induction magnet as a bug trap for the sextuplex by setting up an inductive current equal to that of the magnet in the main line as a means of bridging over the interval of no magnetism at the moment of reversal. NS-77-004, Lab. (*TAEM* 7:626).

2. A set of notes from 25 July 1877 describes further experiments with magnetic induction using this setup (NS-77-004, NS-Undated-005, both Lab. [*TAEM* 7:693; 8:229–31]). A Batchelor scrapbook contains two strips of chemical paper dated June 1877 that show the "difference of inductive effect on opening and closing a circuit when the armature is close up and far away" (Cat. 1240, item 194, Batchelor [*TAEM* 94:60]).

[Menlo Park,] Wednes. June 6th 1877.

*Notebook Entry: Telephony*

Speaking Telegraph

Our speaking telegraph as now improved is far plainer and better than Bell's. The apparatus at present consists of a speaker a receiver and a morse key and sounder[1]

The speaker is shown in Fig 42[2] and consists of a bent tube having a mouthpiece at one end and an adjustable diaphragm I[a] at the other. This diaphragm I when set in motion by a person speaking in the end of tube strikes agains[t][b] the point of adjusting post H this point is made of compressed plumbago and isinglass which has the property of altering the resistance of circuit when pressed with variable strength, as is done by the diaphragm when vibrated by talking[3] This vari~~ous~~able resistance in our circuit gives us the articulation of the wordsounds.

Our receiver is a special form of Edison's Electromotograph, the stylus being fixed to a resonant box. The arrangement of our circuit is thus:—

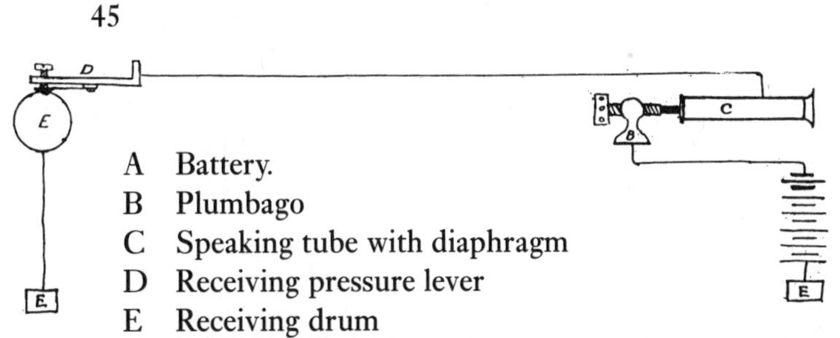


A Battery.
B Plumbago
C Speaking tube with diaphragm
D Receiving pressure lever
E Receiving drum

In the receiving end we find yet that Sulphate of Soda is the best solution to work with. We find that this tube does not give us such sounds as are accompanied by stress of air such as P, B, Sh, Th, etc and for this we have devised a new attachment as we believe that the stress of air carries the diaphragm forward and holds it from vibrating freely.[4]

46.

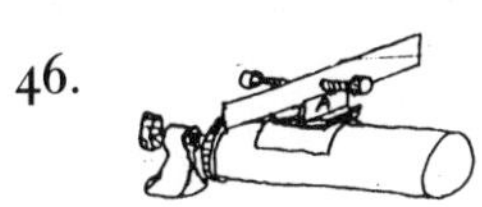

Our means for getting these Sh & Th vibrations is shown in 46 we cut a hole in the top of tube and put a pair of lips over it as at A above this and immediately between we stretch a piece of rubber or parchment or copper foil; when talking in the tube all notes that have a stress of air with them come up through the lips and vibrate the blade of copper like blowing on the edge of a blade of grass and the other sounds will not affect it by this means we got good results.

Chas Batchelor

X, NjWOE, Batchelor, Cat. 1317:39 (*TAEM* 90:676). Written by Batchelor. [a]Interlined above. [b]Page trimmed.

1. The transmitter and receiver are one unit; the Morse key and sounder can be seen in drawings from the following day (Edison's Exhibit 130-11, TI 2 [*TAEM* 11:288]). The several 7 June drawings of this instrument include alternative designs and electrical connections. There is also a Batchelor drawing of the pattern for the transmitter stand, which he may have given Joseph Murray to use in making the four transmitter-receivers Edison ordered from Murray on 11 June (Vol. 11:122–29, Lab. [*TAEM* 3:987–94]; Murray's Testimony, TI 1:362–63 [*TAEM* 11:158]). In two diary entries probably discussing this instrument, Batchelor wrote (on 6 June), "Worked on Speaker today. we now got it so that we consider it fit to be put on a line and shall now proceed to make 2 instruments alike to try it on a circuit"; and, two weeks later, "Took the Speaker Instrument for private lines to Murrays to make four" (Cat. 1233:157, 171, Batchelor [*TAEM* 90:131, 138]). The final design of this instrument (Doc. 962) was not completed until 10 July as Batchelor continued to work at the laboratory and at Joseph Murray's shop, designing alternatives for such components as the drive mechanism for the electromotograph paper feed (Vol. 11:138, Lab. [*TAEM* 3:997]; Cat. 1233:172–91, Batchelor [*TAEM* 90:139–48]; Edison's Exhibit 146-11, TI 2, and Murray's Exhibit Drawing, TI 2:608 [*TAEM* 11:301, 728]; see also headnote, p. 427).

*A double, combined transmitter–electromotograph receiver with a Morse key and sounder.*

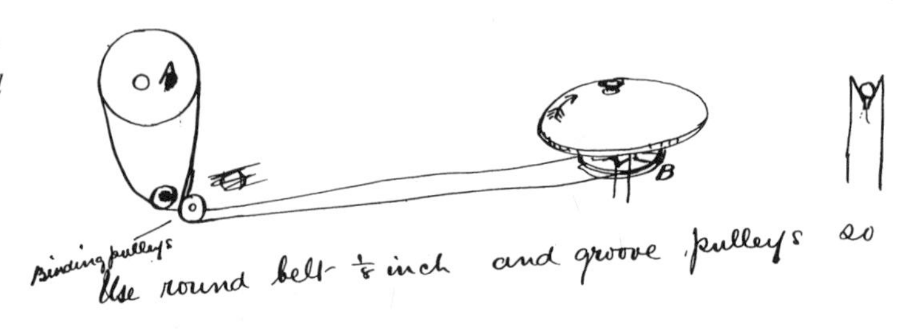


*Charles Batchelor's drawing of a design for the paper-feed mechanism of the combined transmitter–electromotograph receiver.*

2. That is, Fig. 44.
3. See Doc. 928.
4. See Doc. 921.

## –933–

*From Robert Morris*

[Boston,] June 8 1877[1a]

Edison

Heard you very distinctly with your pole changer at rest but cant anything on 3d side during your reversals[2]

Morris[3]

L (telegram), NjWOE, DF (*TAEM* 14:690). Message form of Western Union Telegraph Co. [a]"1877" preprinted.

1. This is one of two telegrams from Robert Morris on this date; he sent another the following day. Morris was in Boston to assist Edison in tests of his sextuplex system on Western Union wires. C. W. Henderson, chief operator in Boston, also appears to have been involved in the tests. Although instruments were left in both New York and Boston after 9 June, the tests did not resume and Edison abandoned work on the system in early July. Morris to TAE, 8 and 9 June 1877 and n.d.; C. W. H[enderson?] to TAE, 9 June 1877; C. W. H[enderson?] to Alfred Downer, 12 July 1877; Alfred Downer to TAE, n.d.; David Downer to Alfred Downer, n.d.; all DF (*TAEM* 14:690–91, 700, 745, 1024–25); "Echoes from 197," *Operator,* 15 June 1877, 9.

2. Undated drawings on the backs of Western Union forms show different arrangements of transmitting instruments, probably for these tests. Other drawings, from 26 May, are probably related to these tests, as are three pages of drawings made at the Western Union office on 6

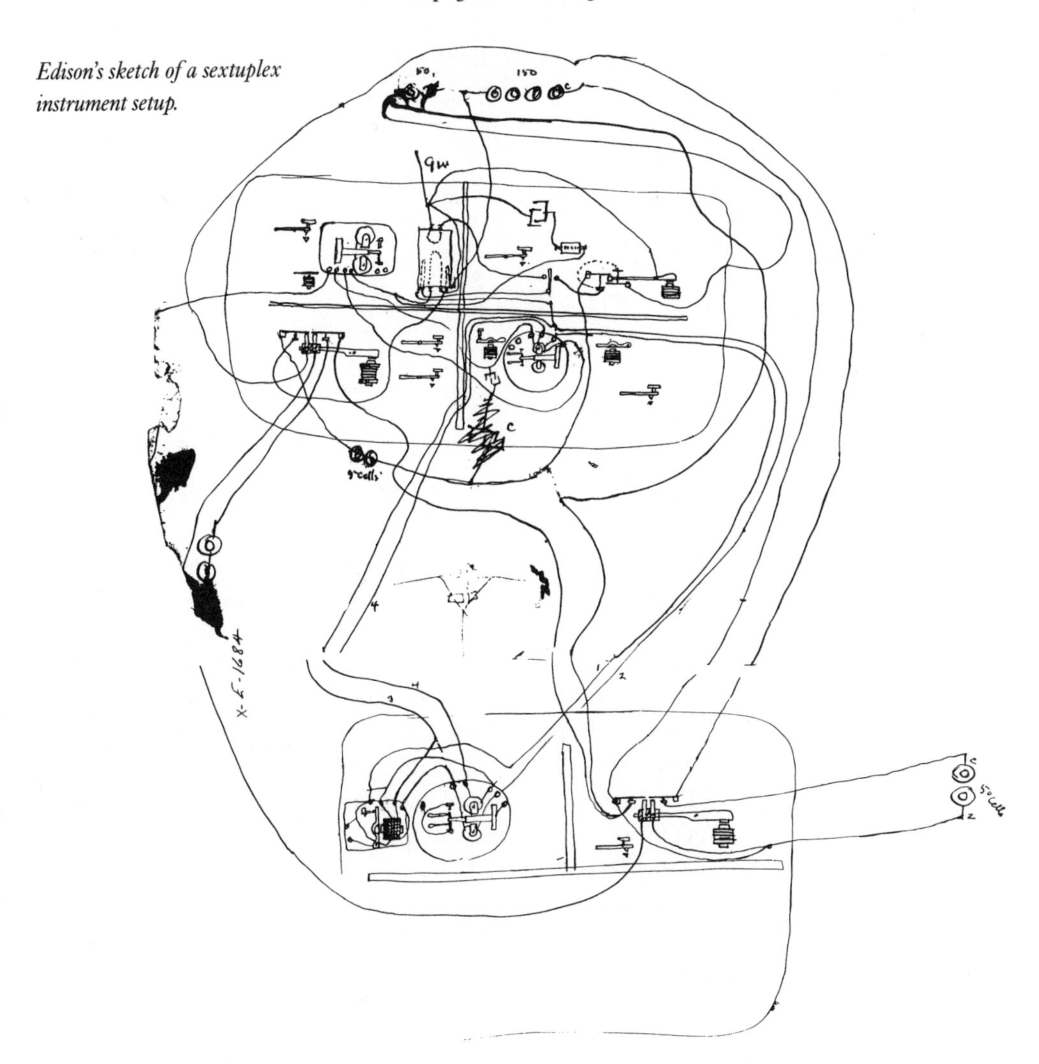


*Edison's sketch of a sextuplex instrument setup.*

and 8 June. There are also three pages of Edison's notes from the tests, written on the back of Western Union forms. NS-Undated-005, NS-77-004, both Lab. (*TAEM* 8:333, 335, 374; 7:667–68, 679–82]); Supp. II, 1877 (*TAEM* 97:583).

3. Robert Morris, senior chief and mechanical engineer of the operating room at Western Union headquarters in New York, had been sent to Boston on 4 June to assist in setting up and conducting the sextuplex tests. Reid 1886, 732; Stockton Griffin to TAE, 4 June 1877, DF (*TAEM* 14:689); "Echoes from 197," *Operator,* 15 June 1877, 9.

**–934–**

*Technical Note: Autographic Printing*

[Menlo Park,] June 10 1877

Autographic Printing
Autographic Stencil press—Continuous Roll printing

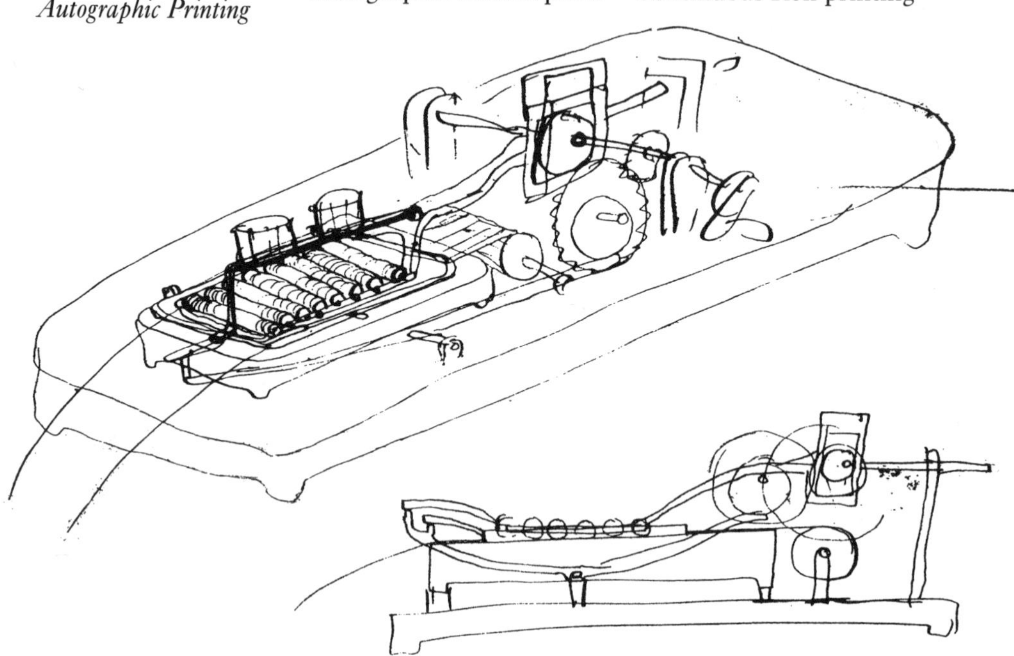

I have tried the stencil on a Gordon press[1] with a cloth but this dont give good printing= I propose to put a train of rollers which are to take the place of the ~~pr~~Regular roller on the Gordon press & roll instead of pressing=

I write on a high hard sized paper with arsenic [acid][a] or Acetate of Manganese. this di[ ][a] sizing & allows ink to pass through giv[ing?][a] many impressions=

I claim writing on proper paper with arsen[ic][a] acid or acetate of manganese & filing the swell~~i~~ed up writing with gum for the use of the blind=

By this arrangement you move the paper instead of the pen[2]

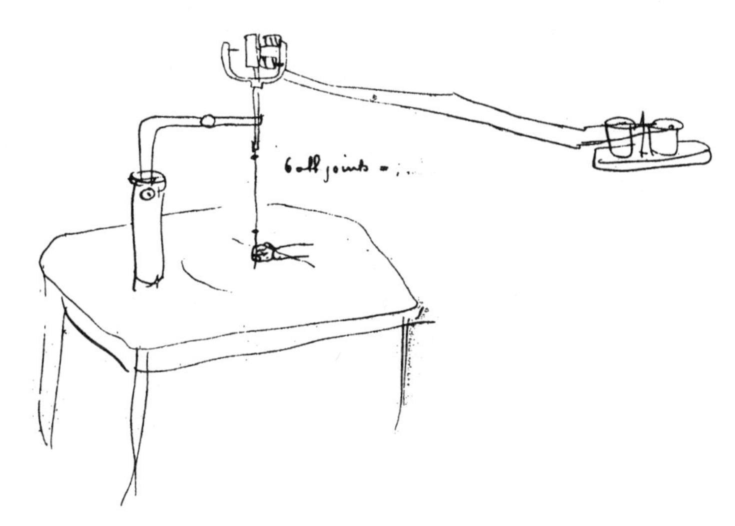

The pen may be arranged to rotate the shaft & on the tube at the top is the cam which serves to reciprocate the needle=

I propose to coat hard rubber with a conducting substance for static electricity & write on it with a point thus bearing the hard rubber I then propose to electrify it and hold it over powdered dry aniline or Lamp black & take impressions from it, or coat rubber with a conducting substance write on a peice of paper with a substance which when pressed on the rubber will take off the conducting substance & leave rubber bear ready for electrifying[3]

I just tried regular thick printers ink with a stencil I find that with rolling its too sticky but will work by pressure=

Make a vibrating pen vibrated with a magnet to write this way

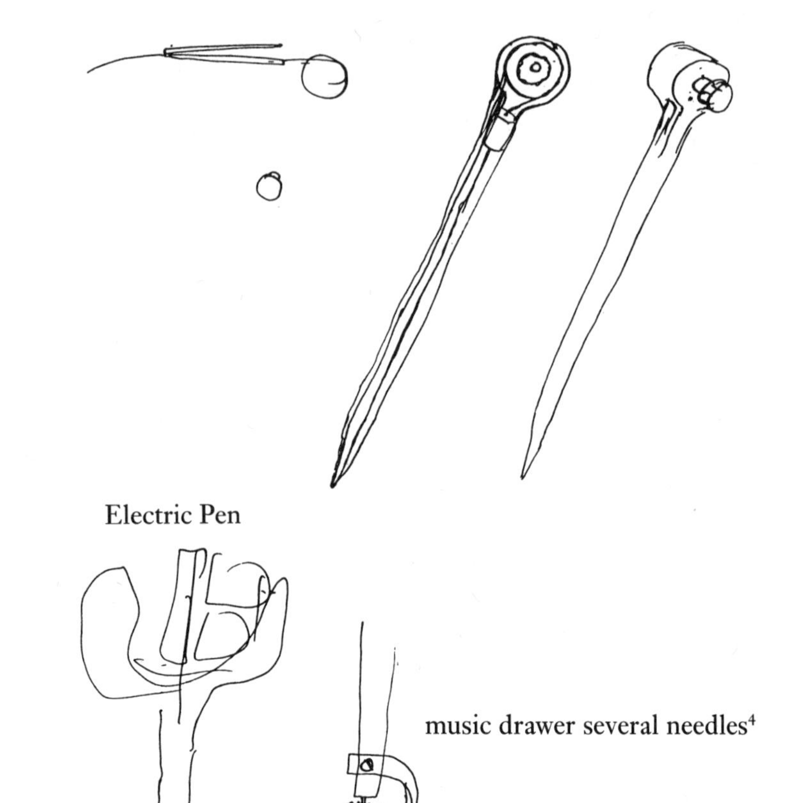

Electric Pen

music drawer several needles[4]

Pantograph with Electric pen

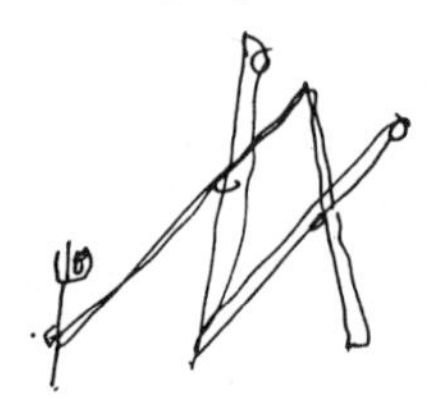

Works well.[5]

I have a new process for Duplicating and it works well  I take hard sized paper coat it with wax on one side, cold then write with a point. This scratches the wax away where you write. I then wet a pad with aniline ink & lay it on the wax side by wetting[b] the sheet  thus waxed on the other side, the ink comes through where the wax is gone & any no of dups can be taken

I propose to write with an amalgam of Mercury & sodium  lay this on a smooth iron plate, and it will amalgamate the iron where written.[c] I then[b] connect iron to zinc of a battery & moisten my paper in Ferridcyanide potassium salt Lay it on iron & over this another metal plate[b] connected to the carbon pole of a battery. the ~~am~~Mercury will prevent the iron reaction & I get white letters on a blue background

T A Edison James Adams

X, NjWOE, Lab., NS-77-002 (*TAEM* 7:432). Document multiply signed and dated. [a]Document damaged. [b]Obscured overwritten letters. [c]"where written" interlined above.

1. The Gordon press was one of a number of high-speed platen presses described and illustrated in Knight 1877 (2:1799–1800).

2. Figure label is "ball joints=."

3. Figure label is "hand stamp."

4. Figure label is "tube."

5. This comment may refer to the drawings or to the text that follows.

**–935–**

*Technical Note: Multiple Telegraphy*[1]

[Menlo Park,] June 11, 1877

Sextuplex=[2a]
Experiment No 1

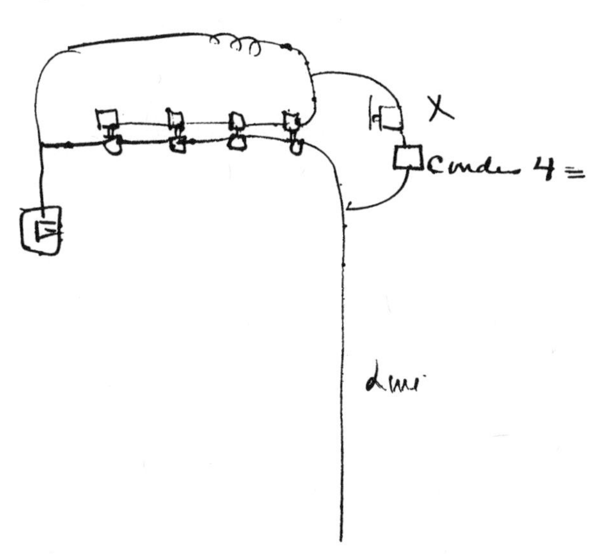


With 7 m[icro]f[arads]. I do not get any better kick on X than with 1¾ mf on 72 sheets[3] connected thus

No 2

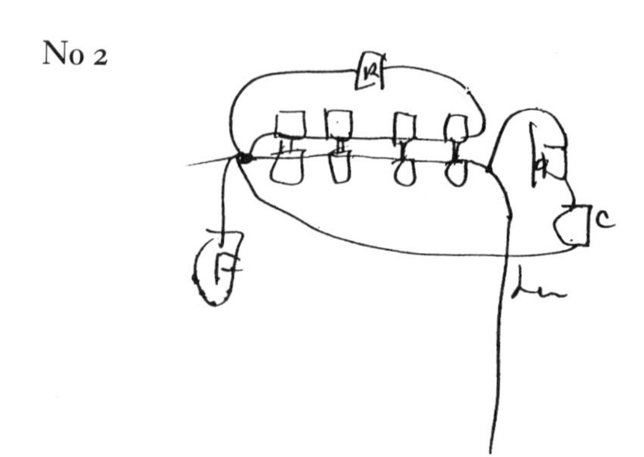


besides putting the Condenser as in No 1 it makes a bigger no magnetism time than in No. 2. therefore the Quad on Bosn is not good=

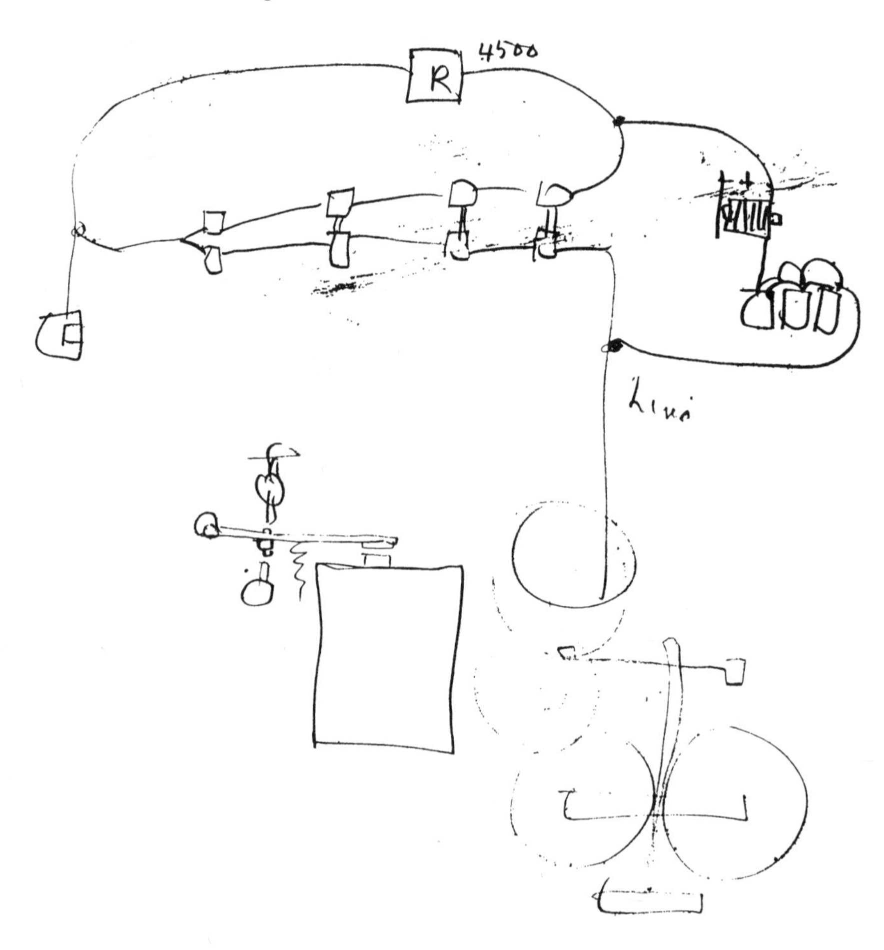

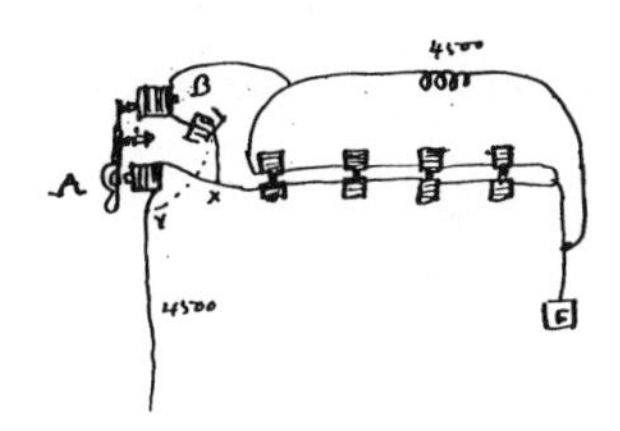


With the 7 mf condensers the effect on A is to cause writing to come heavier by reason of the slight magnetism thrown in B by the Condenser current on opening & quickening it on closing  I find that with end of condenser at .X. instead of X the writing is lighter. Therefore the heaviness cannot be attributed solely to the whifs of magnetism in .B. probably the discharge or effect of the Condensers on the magnet would be to make tailing and this would explain it together with the whifs of magnetism. the amount of Condenser charge or discharge ~~with~~ when operating by rise & fall of tension in exceedingly weak in comparison with that due to reversals perhaps $\frac{1}{40}$ to $\frac{1}{75}$th part as strong, even with No 2 of 150 cells No 1 the current barely moves the armature. I find that with the 50 cell No 1 the effect of permanent magnetism in the relays has very bad effect and with the top magnet arrangement I have the writing is rendered lighter when reversals are sent but you can get writing good at 40 or 50, in the front point even when rapidly reversed. Tried this the other night & it appeared to work when I had Condenser getting its current by a shunt around a 400 ohm short core relay but with the above arranged the relay .C being in the line & $D^4$ in the Condenser ckt the reversal shewed

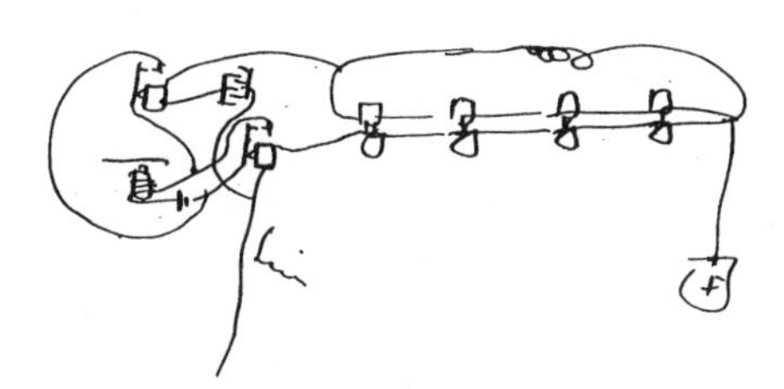

Reversing through a relay without opening[5]

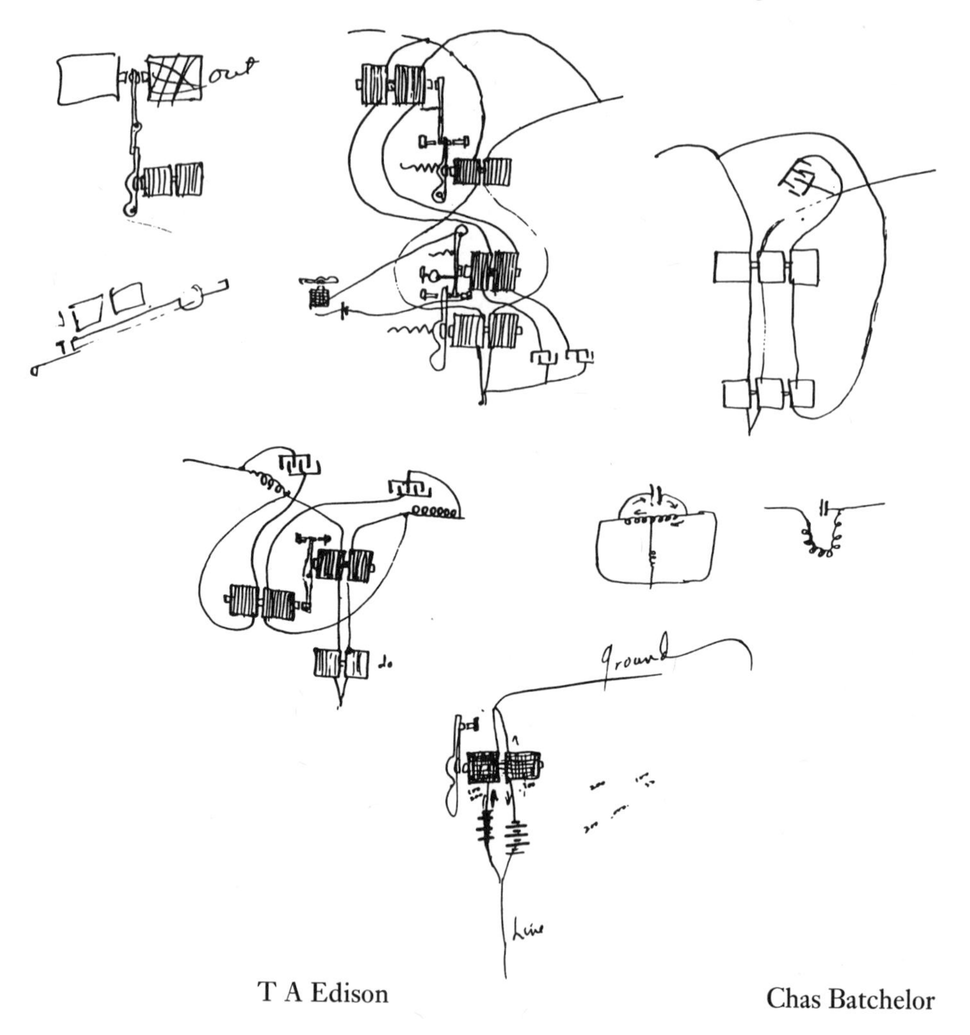


T A Edison Chas Batchelor

X, NjWOE, Lab., NS-77-004 (*TAEM* 7:639, 637, 636, 638). Document multiply signed and dated. [a]Separated from following text by horizontal line.

1. See headnote, p. 280.

2. Edison had found his quadruplex required condensers in service on regular telegraph lines. On 10 June, following the tests of 8 and 9 June, he first drew sextuplex circuits employing condensers. He experimented with such circuits over the next several days. NS-77-004, Lab. (*TAEM* 7:634–45).

3. See Doc. 768 n. 7.

4. No figure labels "C" or "D" appear in the drawing.

5. Figure labels are "out"; "do"; and, in the drawing with labels of "Ground" and "Line," "100" and "200" are below the left magnet coil, "300" below the right coil. Edison wrote the numbers "200," "100," "50," "200," and "000" to the right.

**–936–**

*From Edward Johnson*

New York, June 12th 1877.[a]

My Dear Edison

Phillips[1] Director of Telegraphs at the Centennial and now at the Permanent Exhibition in Philada came on to see you here— failing I took him in charge & after full talk came to this agreement— That if you will furnish me with a complete set of Telephone apparatus I will take it to Phila & give a series of concerts[2] under the ~~Centen P~~Auspices of the Exhibition folks. Phillips will furnish all the wire & Batterys needed—

Now the question is what will it cost to do so. This I could not tell him—but told him that I was just preparing to go to Phila for a few weeks & would be glad enough to engineer this thing for him at a moderate rate of compensation—

What will you charge me for the use of the Instruments for this purpose—that is to say what proportion of what I can get out of them would you want for your share— The auspices are good—as the Exhibition folks[b] will themselves make the arrangement & Prof Barker[3] will "do" the scientific—

If this should be a success one of the Interested parties will[b] want me to go elsewhere (say to the watering places)[4] with it under his patronage I think this is a first rate ~~way~~ opportunity[c] to bring this thing before the Public in the right manner I expect to see Prof Barker at the Stevens Institute tomorrow eve. & If you will advise me meantime that it will be agreeable I will ask him out to Menlo to see you (& it)

I want also to see the "Speaker"[5] as I am quite sure a good market can be had for it for Private Lines at once

Will you be in tomorrow

E H Johnson

ALS, NjWOE, DF (*TAEM* 14:781). Letterhead of J. C. Reiff. [a]"New York," and "187" preprinted. [b]Obscured overwritten letters. [c]Interlined above.

1. William Phillips displayed a printing telegraph (patented jointly with W. P. Phelps). A long-time telegrapher, Phillips was superintendent of the Philadelphia Fire Alarm Telegraph and general manager of the American District Telegraph Co. of Philadelphia. U.S. Centennial Commission 1880, 452; Reid 1879, 460; Obituary, *Elec. W.* 19 (1892): 391; American District Telegraph Co. of Philadelphia Minutebook, DSI-NMAH.

2. See Doc. 929 n. 2.

3. George Barker was professor of physics at the University of Pennsylvania (*TAEB* 2:328 n. 8). He and Johnson visited Menlo Park on 16 June (Cat. 1233:167, Batchelor [*TAEM* 90:136]; see also Edison's Exhibit 147-11, TI 2 [*TAEM* 11:302]).

4. That is, spas and health resorts.

5. Johnson is referring to the speaking (as opposed to the musical) telephone.

**-937-**

*Technical Note: Telephony*

[Menlo Park,] June 13 1877

Speaking Telegraph

Tests of pressed plumbago and other substances to see which is best for speaking.

The sticks pressed are this size[1]

Pure Plumbago— some buckling accompanies it when adjusted at any distance from diaphragm— call it 100—[2a] very difficult to understand an unknown sentence although it sounds well when you know the sentence before hand

~~15 grammes Plumbago & 2 grammes isinglass good about 100 110~~

⟨No 93⟩[3b] 15 Gr Plumbago & 6 Gramme Dextrine[4] about 110— Articulation pretty good

15 Gr Plumbago & 2 gram Coach Body Varnish articulation fair but inconstant about 108 ⟨No 90⟩[c]

⟨No 87⟩ 15 Gram of Plumbago and 2 gram Resin dissolved in alcohol good work about 115[d]

⟨No 84⟩[e] 15 Grm Plumbago & 2 grm Isinglass dissolved in Acetic Acid pretty good about 115 Still it seems to be inconstant[d]

We now put in another diaphragm made of shield brass with a small platina point in middle[5]

Unknown Plumbago mixture simply elegant[e] at least 125— It now becomes necessary to find out what this combination is We believe it is Plumbago with a large proportion of Isinglass or Gelatin

⟨No 94⟩ [7?][f] ~~Gram~~ 15— Plumbago & [-][f] 2 g. ~~R~~Soft Rubber dissolved in Bisulphide of Carbon very good at least 125 On this we got whistling and whispering perfectly

There is one difficulty yet and that is that the platina point on the diaphragm is too small, smaller than the tit on end of Plumbago This must be remedied[g]

⟨No 85⟩[e] 15 Plumbago & 2 Kaolin fair about 118 & seems to be a little variable[d]

⟨88⟩[e] 15 Plumbago & 2 Gelatin fair but only 115. It is not very loud[h]

⟨No 89⟩[e] 15 g. Plumbago & 2 g Tannin wheasy & rattling about 112[d]

⟨No 86.⟩ 15. G. Plumbago & 4 salt— —95— very rattly[d]

⟨No 91⟩[e] 15. G Plumbago & 2 Balsam Fir came rattly & then weakened right down— 104[d]

T A Edison Chas Batchelor
James Adams

X (photographic transcript), NjWOE, TI 2, Edison's Exhibits 134-11, 136-11, 135-11 (*TAEM* 11:290, 292, 291). Written by Batchelor; document multiply signed and dated. [a]"some . . . 100—" interlined above. [b]All marginalia written by Batchelor. [c]Followed by "over" to indicate page turn. [d]Followed by centered horizontal line. [e]Underlined twice. [f]Canceled. [g]Followed by broken double horizontal line. [h]Followed by horizontal line.

1. On the manuscript this drawing is 17 mm × 6 mm.

2. In the tests of plumbago mixtures, which continued through the month, the cylinders were evaluated for clarity and consistency of articulation and for the durability of the "tit" on the end of the cylinder. Edison's Exhibits 139-11–187-11 passim, TI 2 [*TAEM* 11:294–340]).

On 25 June the staff decided, "Hereafter we are going to test all for articulation by reading an article from Daily paper and if after the article is finished we get the sense of it we shall know how good our articulation is." Edison's Exhibit 186-11, TI 2 (*TAEM* 11:339).

3. The staff apparently assigned these numbers to the compounds in the next few days (see Doc. 941).

4. A gummy adhesive, often used in paper, textiles, syrup, and beer.

5. See Doc. 928 n. 2.

**-938-**

*Technical Note: Multiple Telegraphy*[1]

[Menlo Park,] June 14 1877

Sextuplex[2]

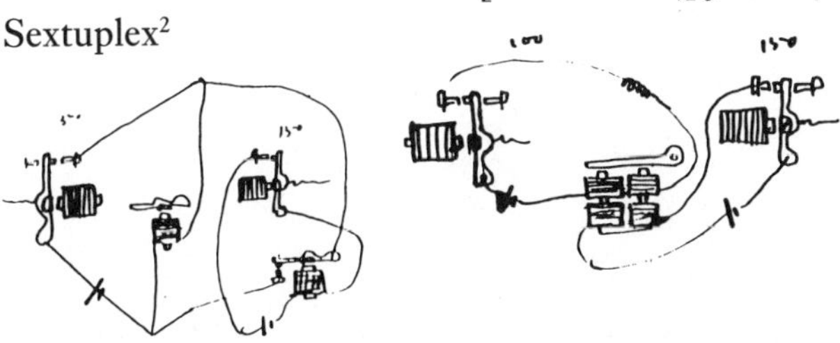

give 150 man[3] the advantage in strength in ~~m~~Sounder magnet so a little excess of his current will neutralize self magnetism of S[ounder]. although this may not be essential.[4]

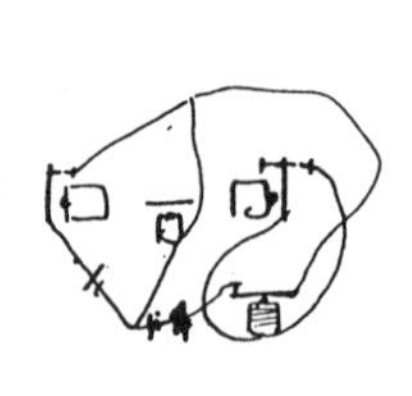

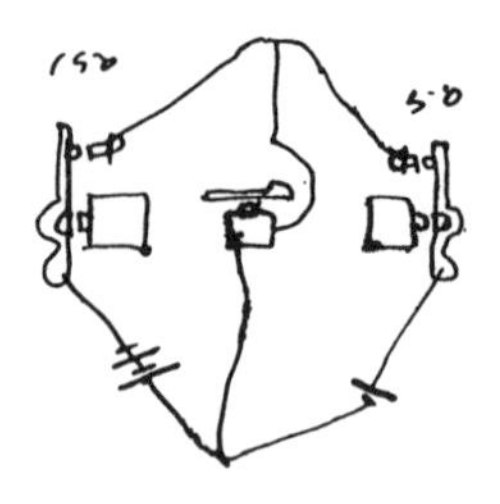

This works well still the ——— bug is in there but this time its on opening before twas on closing The trouble is caused by the 50 relay having a light adjustment the addition of the 150 causes a permanent set on the Cores so that when all current is taken off ~~that~~ the lever of 50 dont fly away as soon as that of the 150. if 150 mag is adjusted so its retractile spg is of the same tension as that of 50 then its cores are further away & act quick= I see this effect by the local spark on 50 points Theoretically both levers ought to leave at once but ~~the~~ & no spark appear but as a spark does appear & [stick?][a] on 50 point I infer that this lever is sluggish it should[5] be theoretically

I am testing for bugs with chemical recording instrument.[6]

~~responds to 100 or 150 but not to 50~~[7]

responds to 100 or 150[8]

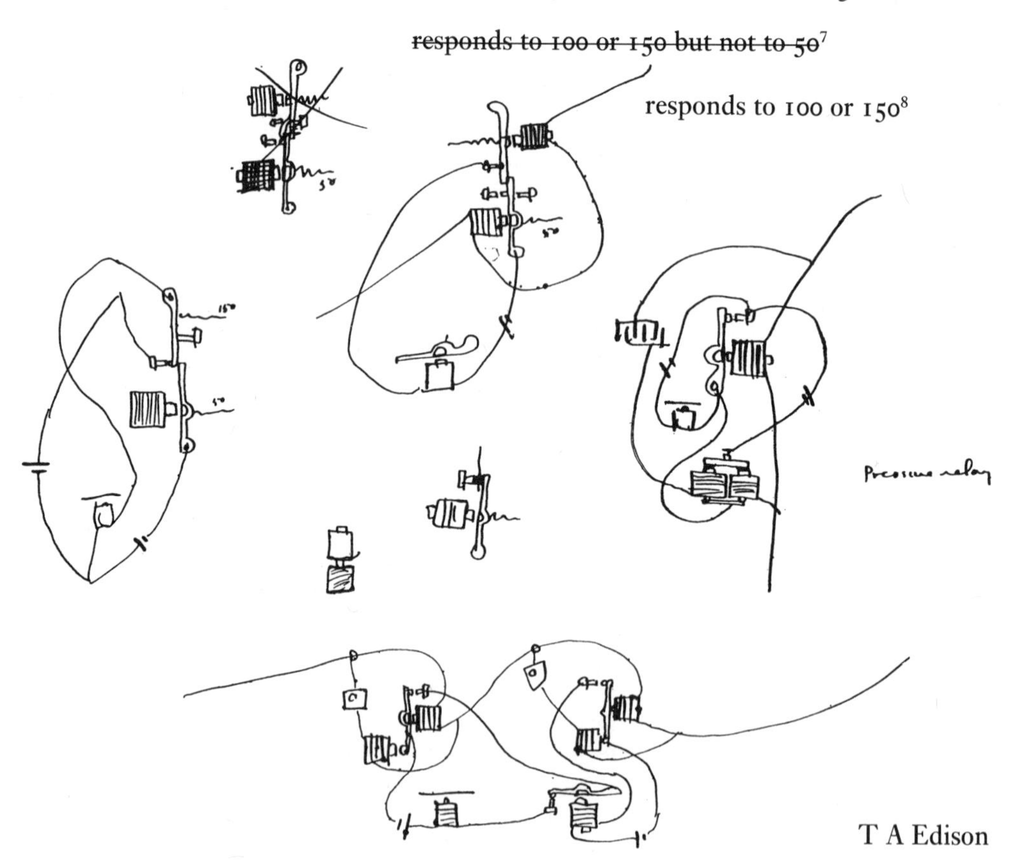

T A Edison

X, NjWOE, Lab., NS-77-004 (*TAEM* 7:646). Document multiply dated. [a]Illegible.

1. See headnote, p. 280.
2. Figure labels are "50," "150," "100," and "150."
3. That is, the operator on the relay marked "150."
4. Figure labels are "150" and "50."
5. Edison probably meant "shouldn't."
6. Edison often compared the strength and duration of electrical impulses by the marks they would make if passed through the chemical recorder of his automatic telegraph system (e.g., Doc. 931).
7. Figure labels above are "~~150~~" and "50."
8. Figure label of the drawing above is "50." Figure labels below are "150" and "50," and "Pressure relay."

## -939-

*Technical Note: Telephony*

[Menlo Park,] June 17, 1877

Speaking Telegraph

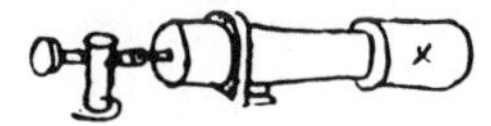

With the second tube & diaphram the talking comes in a higher key apparently an <u>Octave</u> higher & Adams says the sound given out when the diaphram strikes the plumbago is higher with speaking in the 2nd tube than when direct and the talking is not so good= by lengthening out the regular tube it appeared to come better but after careful tests the articulation was found to be inferior to the regular 8 inch tube=[1a]

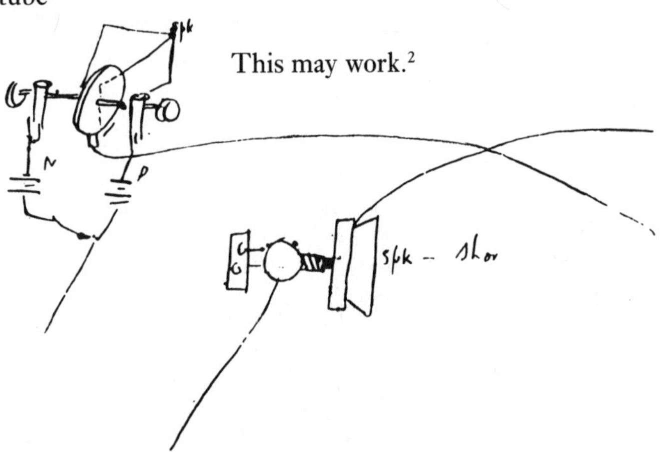

This may work.[2]

T A Edison James Adams

X (photographic transcript), NjWOE, TI 2, Edison's Exhibit 155-11 (*TAEM* 11:310). [a]Followed by centered horizontal line.

1. Batchelor recorded these experiments as well (Edison's Exhibit 147-11, TI 2 [*TAEM* 11:302]). He was also busy with the continuing tests of plumbago mixtures (Edison's Exhibits 147-11–154-11, TI 2 [*TAEM* 11:302–9]).

2. Figure labels are (top left) "spk" and (right) "spk—[shor?]." In the drawing at left the two sides of the transmitting diaphragm are insulated from each other. The diaphragm is placed at the end of a tapered resonant cavity and sends currents of opposite polarity (from the batteries labeled "N" and "P") as it vibrates between two contact points. Cf. Doc. 920 n. 3.

**-940-**

*From Henry Law*

New York, June 18th 1877.[a]

Dr Sir

I have thought yr Electric Needle might be turned to account in cutting of foil on Glass by using instead of a needle point one a little on the chisel order. it would need to cut a smooth cut, clear thro' the foil. I went to yr place in Church St\&\the Gent[1] who s~~a~~eemed to be in charge said he would give it a trial but he never did & I understand has now gone to Chicago. If you care to think of this & will make an appointment with me to meet you at yr place in this city at any day ~~or~~& hour most convenient to you I will call on you & shew you what we want[2] Yours truly

Henry A. Law.[3]

ALS, NjWOE, DF (*TAEM* 14:75). Letterhead of Glass Ornamenting Co. [a]"New York," and "1877." preprinted.

1. The Electric Pen Co. office was at 20 New Church St. Law may have been referring to Ezra Gilliland, who worked out of the office as general Eastern agent until leaving in mid-June. Electric pen circular, Cat. 593, Scraps. (*TAEM* 27:643); Charles Batchelor to Robert Gilliland, 21 Feb. 1877, Cat. 1238:81, Batchelor (*TAEM* 93:87); Doc. 952.

2. It is not known if they met, but Law did send Edison a piece of foiled glass (see Doc. 1130) and on 20 June Edison ordered an electric pen fixed with a "chisel for Tin foil glass" that was made the following day (NS-77-002, Lab. [*TAEM* 7:438]; Edison's Exhibit 169-11, TI 2 [*TAEM* 11:324]).

3. Little is known of Henry Law. He was probably associated with the New York Sand Blast Works, whose letterhead he used in his subsequent correspondence with Edison, and which made ornamental glass globes, signs, and other glass products. He may also have been associated with the Glass Ornamenting Co., whose letterhead he used for this letter.

*Technical Note: Telephony*

Speaking Telegraph

35[1] 4 Plumbago 500/MG Cyanide Copper. Strong but articulation about same as standard Plumbago 100[a] Tit clean Got Whistling & Whispering but no articulation[b]

27 4 Plumbago 1 500/ Cyanide Copper Articulation and strength good 129[a] get any thing on that Tit clean[b]

124 15 Plumbag 6 Flour. Very loud at first but after a little while working with it it went right down about 125[a] Tit clean—[b]

56 15 Plumbago 4 Carb Magnesia everything good articulation 129[c] Tit clean Whistling & Whispering good—[b]

41 4 Plumbago ~~2~~150/100 Iodine Articulation good 128[a] Tit clean Got Whispering & whistling ~~clear~~ well[b]

126 4 Plumbago 2 Camphor— Camphor in Alcohol[b] This started off quick but it dwindled away Articulation when it came was 125[a] Tit clean[b]

No 44 4 Plumbago 1 Iodine Articulation good about 128[a] Tit clean Whispering not very good It is a little inconstant[b]

No 71 4 Plumbago 2 Carbonate Magnesia Very good articulation 129—[a] Tit clean— This was very loud so much so that I had to have Jim speak low & then the articulation was beautiful I think perfectly constant The best tonight so far—[c]

No ~~89~~ 68 Composition ~~15~~ 4[d] Plumbago and ~~2 Tannin~~ 250 MG Bisulpuret Tin— Very good articulation about 128[a] Whispering good & whistling also Tit clean. I noticed that Tannin did not work well when we had it in before—[b]

No 125 Comp. 15 Plumbago 6 Dry Salt Very good articulation 128[a] Whistling & Whispering good Tit perfectly clean[b]

No. 89 Comp 15 Plumbago 2 Tannin— The articulation is good about 128[a] got several sentences conse~~q~~cutively unknown— Tit not altered; best whispering we have had a little rattly—[b]

149 Comp 15 Plumbago 10 Rubber in BiSulp. Current exceedingly weak due no doubt to the great resistance from such an amount of Rubber Articulation very good but so low it could be heard with difficulty[b]

141 Comp 15 Plumbago 4 Rubber in Sulphide Carb. Tit clean Good articulation about 126[a] Whispering distinct[b]

143 Comp 15 Plumbago 3 Rubber in Bisulph Very good I think best yet Articulation 129[a] best on ordinary talking very good on Whispering & whistling Tit clean[b]

144 Comp 15 Plumbago 6 Rubber Very good 128[a] Tit clean[b]

146 Comp 15 Plumbago 1 Rubber Very good 129[a] Articulation— comes best when talking low but can be made to come good in any talking tit burned a little. [b]

145 Comp 15 Plumbago— 8 Rubber tit a little burned Articulation very good equal to 128[f]

151 15 Plumbago 9 Rubber[g] Very constant articulation 129[a] Everything good Tit a little roughened This is the one Edison likes!!![h]

No 148 15 Plumbago 7 Rubber Constant articulation 129[c] Bully good[2] Tit very little altered—[b]

No. 61 ~~154~~ Plumbago 1 Caustic Magnesia Bully. Articulation good constant strong Tit not altered 128[f]

No 82. 4 Plumbago 1 $^{500}$/MG Caustic Magnesia Good articulation 129[c] good strong Whistling Whispering very good Tit not[i] altered[b]

80 4 Plumbago $^{250}$/MG Caustic Magnesia Good articulation 129[a] every thing good Whistling & Whispering good excellent Tit perfectly clean[b]

No. 94 15 Plumbago 2 Caoutchouc in BiSulph Tit a little roughened Very good about 128—[g]

No 129 4 Plumbago 1 $^{500}$/ Gelatin in $H_2O$ Not so good as best but 125[a] & generally good[b]

Edison Chas Batchelor
James Adams

X (photographic transcript), NjWOE, TI 2, Edison's Exhibits 156-11–161-11 (*TAEM* 11:311). Written by Batchelor; document multiply signed and dated. [a]Underlined twice. [b]Followed by centered horizontal line. [c]Multiply underlined. [d]Interlined above. [e]"250 . . . Tin" interlined above. [f]Underlined twice; followed by centered horizontal line. [g]"15 . . . Rubber" interlined above. [h]Multiply underlined; followed by centered horizontal line. [i]Obscured overwritten letters.

1. These numbers appear in a list of 20 June. Edison's Exhibits 164-11–168-11, TI 2 (*TAEM* 11:319).

2. According to Batchelor's 17 June diary entry, the staff "Worked on different plumbagos for speaking telegraph all night Found that the combination of Plumbago and Rubber also Plumbago and Caustic Magnesia are all excellent as also many others." Cat. 1233:168, Batchelor (*TAEM* 90:137).

**–942–**

*From H. L. R. Vandyck*

Jersey City, N.J., June 20 1877[a]

Dear Sir—

As I have been unable on account of business engagements and absence to call and see you—I send statement of our account—and check for balance due you— The matter against H C Fernando Rohe[1] is still pending will get it to trial as soon as possible.

The enclosed claim of D E Drake[2] was given to me by Elvin W Crane of 800 Broad St Newark Mr Drakes lawyer—as an offset to your claim. will not proceed further against Drake till you instruct me so to do.

Why do you not call at my office sometime when you are in the city Yours &c—

H L. R. Vandyck[3]

ENCLOSURES

[Jersey City,] June 20th 1877[b]

Statement.

| | | |
|---|---|---|
| Amount collected from W. H. Crossman & Son—[4] | | 159.50 |
| Amount collected from The Recording Steam Gauge Company—[5] | | 47.64 |
| Total Collection | | $207.14 |
| Collection fee on the above, 10%— | 20.71 | |
| Counsel fee &c. and services in the matter of the foreclosure of chattel mortgage given by E. T. Gilliland to Thomas A. Edison, to secure $3700.00, or thereabouts, and in the matter of the claim of Edison's E.P.&D.P. Co. vs. E. T. Gilliland[6] | 50.00 | |
| Case of Edison E.P.&D.P. Co. vs. D. E. Drake—Fee for services | 10.00 | |
| Amount retained as retaining fee to cover costs in case of Edison & Murray vs H. C. Fernando Rohe— | 25.00[c] | 105.71[c] |
| | | $101.43 |

Check, June 20th 1877, One hundred and one dollars and forty-three cents ($101.43)[d]

Newark N.J. Apl 12 1877[e]

Edisons Electrical Pen and Duplicating Press Co To D. E. Drake D[ebto]r[7]

| | |
|---|---|
| To loss to me for not furnishing Pens and Presses as wanted to fill my orders | $300.00 |
| To loss—difference on price of Pens and Presses. which company agreed to sell to me at $20.00 each and for[f] which they charged me $25. each. | 100.00 |
| To Commission due me for Pens and presses sold in my territory by the company befor my contracts with them expired | 150.00 |
| To wages for two weeks work at Union Square N.Y. at $25—per week commencing on Monday April 6, 1876 | 50.00 |
| To one Pen & Press left at shop in Ward St Newark N.J. that was moved to Menlo Park for which I paid Chas Batchelor | $20.00 |
| | $620.00 |

ALS, NjWOE, DF (*TAEM* 14:76). Letterhead of H. L. R. Vandyck. [a]"Jersey City, N.J.," and "187" preprinted. [b]Enclosure is a D, in another hand; date taken from text. [c]Underlined twice. [d]Obscured overwritten characters. [e]Enclosure is a D, in a third hand. [f]Interlined above.

1. Rohe had been involved in an attempt to set up a stock-quotation enterprise in Brazil using Edison's printers. *TAEB* 2:313.

2. D. E. Drake had been electric pen agent for the states of Connecticut, Maine, New Hampshire, Rhode Island, Vermont, and New York (except New York City and Brooklyn). On 22 January 1876 Charles Batchelor had notified Drake that his contracts as agent would be terminated in thirty days due to his failure to fulfill the average weekly sales for each of these states. *TAEB* 2:598 n. 11; agreements of 14 Sept. 1875, Miller (*TAEM* 28:351); Batchelor to Drake, Lbk. 2:13 (*TAEM* 28:987).

3. Vandyck's letterhead identifies him as a counselor at law. His business card is in Cat. 1031:13, Scraps. (*TAEM* 27:739).

4. W. H. Crossman & Bro. was a hardware dealer in New York. Wilson 1877, 289.

5. Edison and Murray manufactured items for the Recording Steam Gauge Co. in the fall of 1874. *TAEB* 2:228, 313.

6. See *TAEB* 2:544 n. 1.

7. That is, Drake claimed that the company owed him this money.

**-943-**

*From Josiah Reiff*

New York, June 26th 1877[a]

My dear Edison.

Your note duly recd.[1] I wrote accordingly E.H.J says the telephone did not meet him at New Brunswick, that he has telegraphed you & recd no reply—that Phillips has all the arrangements made for a first exhibition on July 4th & EH says he needs[b] to get everything in shape.

Will you ship it to him at once & write him to 20th & Poplar Phila? Butler begins tomorrow afternoon.[2] I believe it will result so that you & I will be benefited—if it does not God help me! What have I for all my money, years of labor & anxiety. It will simply Automatic which might have saved us, ruined by Quadruplex. Truly

J. C. Reiff

P.S. Can you not turn your attention a little to Cables, perhaps with your present knowledge a little effort might [-----][c] something better than Duplex.

ALS, NjWOE, DF (*TAEM* 14:695). Letterhead of J. C. Reiff. [a]"New York," and "187" preprinted. [b]Obscured overwritten letters. [c]Illegible.

1. Not found.

2. That is, Benjamin Butler's closing argument in *Atlantic & Pacific v. Prescott & others;* see Doc. 949 n. 1.

## -944-

*Technical Note: Telephony*

[Menlo Park,] June 26. 1877

Speaking Telegraph

Collodion film for a diaphram= Mica for ditto[1]

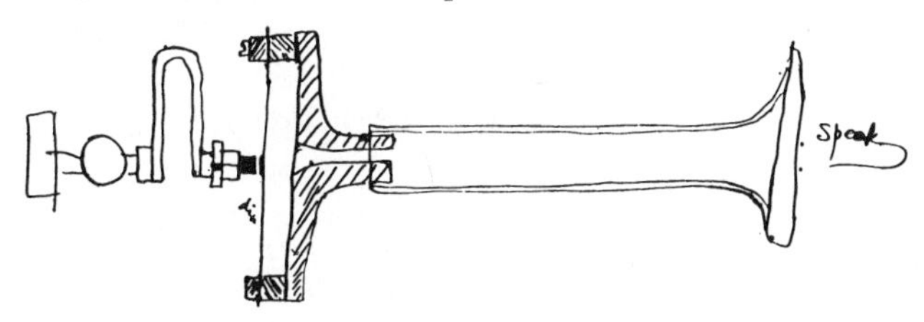


Having a small hole the amplitude of the vibration of the hissing Consonants are increased whereas the vowel etc Sounds are decreased which improves the articulation—

Just finished a Bell Magneto Telephone=

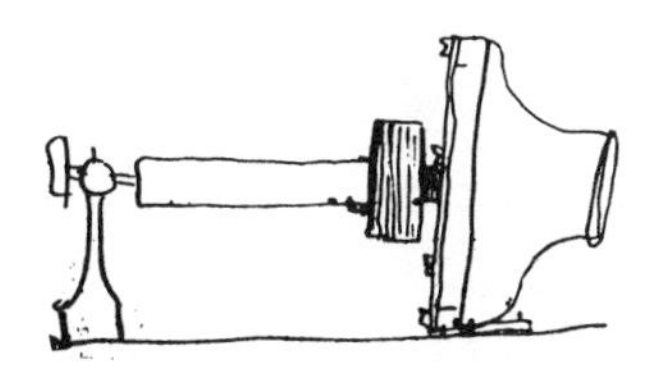

1/16 iron diaphram resistance magnet 250 ohms 1 inch long fork permanent magnet about 7 inches length lifts about 2 lbs iron I find that diaphram dont seem to vibrate but the case diaphram & all does & it is this that does the work & I believe this is the way Bell is fooled by not[a] securing his spkg devices rigidly=

We carried on Conversation with the pair it was so low I

couldnt hear but Batch & Jim by paying good could get it  ordinary conversation fair but not so good as ours. Reading out of a newspaper something strange & unknown couldnt catch a word in half column whereas with our you could get the general tenor of every thing that is read=

I propose to improve Bells as well as my own here goes![2]

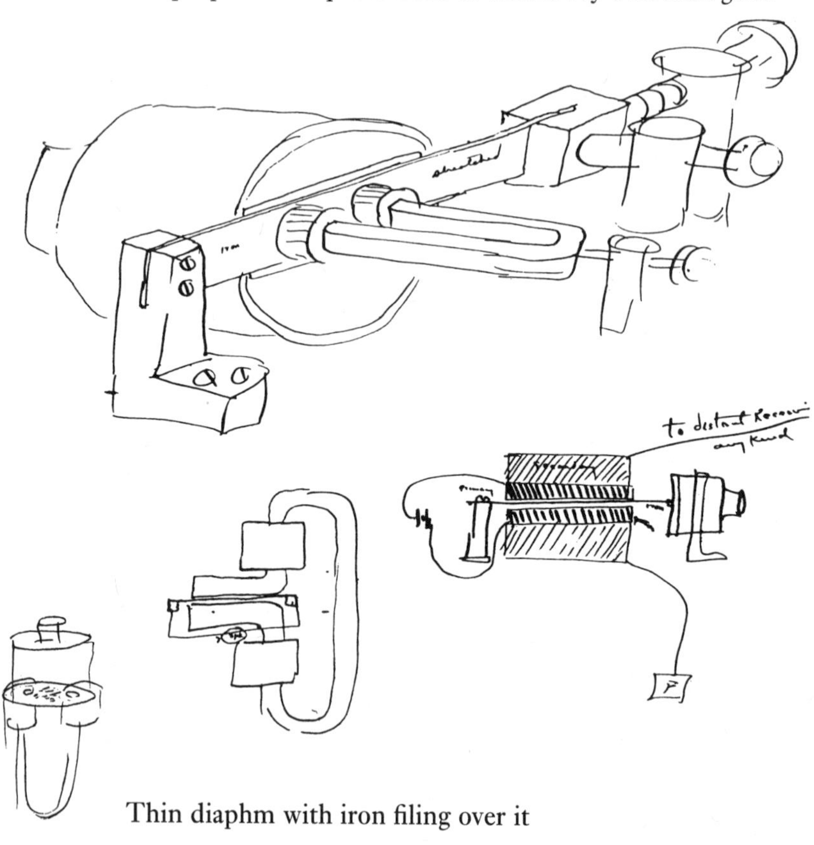


Thin diaphm with iron filing over it

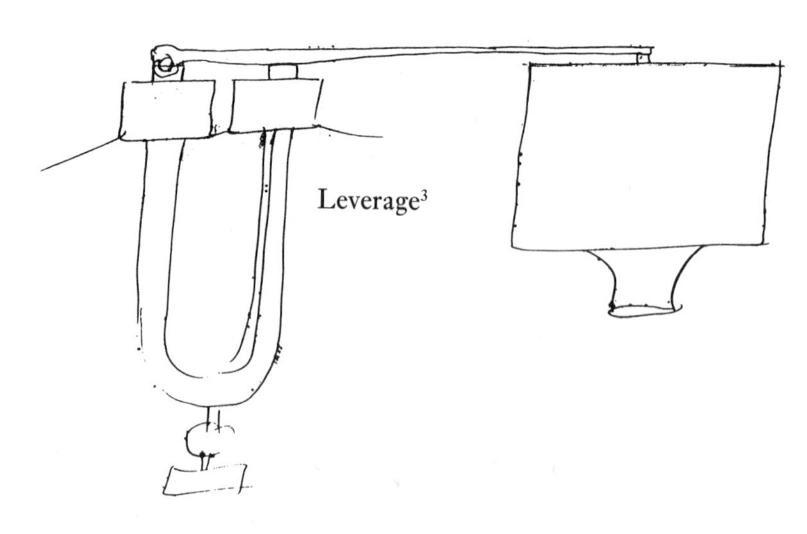

Leverage[3]

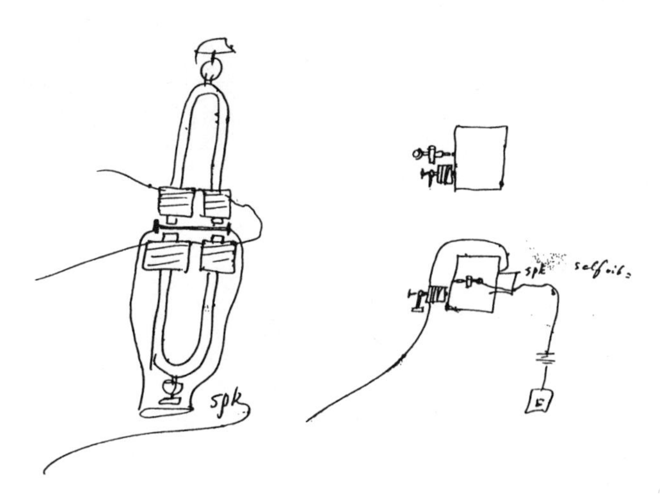


X, NjWOE, Lab., Vol. 11:190, 188, 193 (*TAEM* 3:1002, 1001, 1003).
[a]Interlined above.

1. Figure labels are "dia" and "speak." The previous day Batchelor and Edison had each sketched a transmitter configuration using a spring like the one shown here (at right). Edison's Exhibits 184-11, 187-11, TI 2 (*TAEM* 11:337, 340).

2. Figure labels are, top: "iron stretched"; right: "primary," "secondary," "~~prim~~," "iron," "to distant Receiver any kind"; and, bottom: "spk."

3. The following is apparently an amplifying receiver. The figure labels on the remaining sketches are "spk," "spk," and "self vib=."

**-945-**

*Technical Note: Telephony*

[Menlo Park,] June 26 1877

Speaking Telegraph

We are going to Mould Peroxide of Lead also Peroxide of Lead on the end of a plumbago stick=[1] a mixture of a lead & zinc amalg~~m~~am with plumbago=[2]

Things to be tried=

Platinum black=
Lead point=
Metallic points protoxidized=
Black finely divided mercury
Fused ~~S~~Conducting Salts such fused Chloride of Lead
Guttapercha mixed with various Conducting Substances=

we propose to make moulds to mould the plumbago in the following shapes for speaking[3]

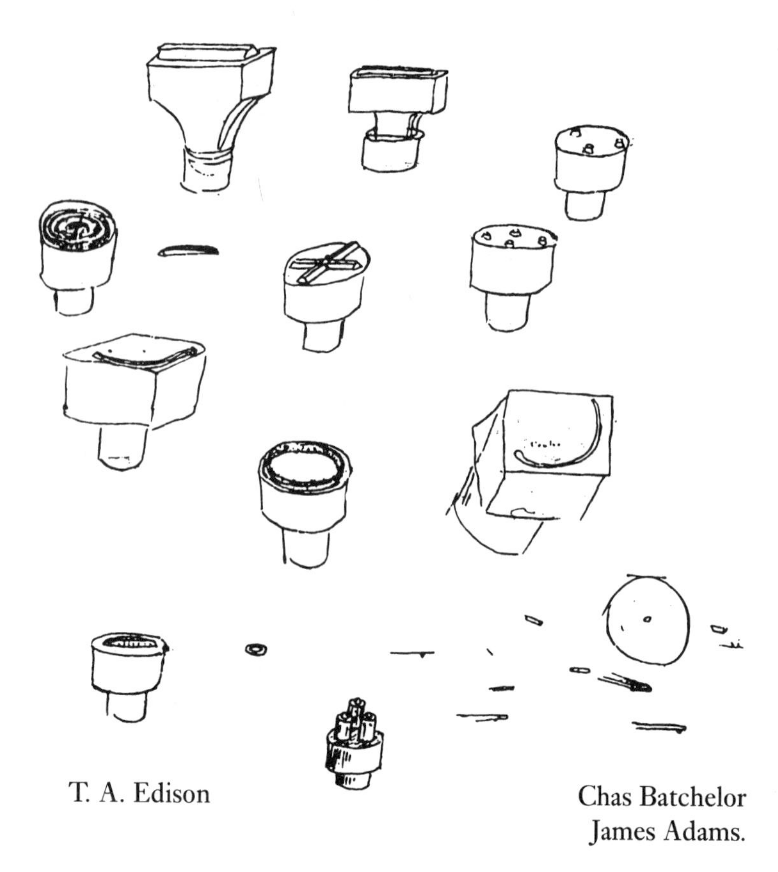

T. A. Edison

Chas Batchelor
James Adams.

X (photographic transcript), NjWOE, TI 2, Edison's Exhibits 192-11, 189-11 (*TAEM* 11:343, 341). Document multiply signed and dated.

1. A list of compounds from two days earlier includes "Plumbago 4 Gramm tit made of Peroxide Lead Body Plum." On this same day Batchelor noted, "Peroxide lead tit and plumbago shank No good." Edison's Exhibits 172-11, 121-11 (misdated), TI 2 (*TAEM* 11:325, 286).

2. On 25 June, Edison had written, "I find that although plumbago & the Conducting Oxides & other Compounds which conduct are good Lead & other metals & metaloids whose surface make bad Contact will act to give vibrations of variable strength= also the solid amalgmam." Edison's Exhibit 182-11, TI 2 (*TAEM* 11:335).

3. This word may be "sparking." The previous day Edison had proposed putting "several flexible contact points on a diaphragm" since "diaphragms probably vibrate in parts." Ibid.

**-946-**

*Technical Note: Autographic Printing*

[Menlo Park,] June 27th 1877

Autographic Printing[1]

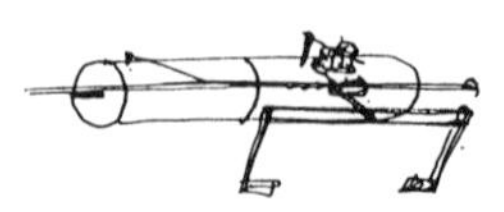

Write on paper with Collodon[2] Electrify the Mss by rubbing with dry substance= dust aniline dust over a flat surface with a sieve= put Collodion ~~sheet~~ Mss in frame & lower it w~~i~~hile electrified with 1/32 inch of the aniline it will attract the particles & these ~~m~~adhering to the Collodion Mss. ~~wi~~can be transferred to a wet sheet. The Collodion Mss is drying while you print from the transfer sheet & when dry re-electrify & go over same process—[3]

James Adams

X, NjWOE, Lab., NS-77-002 (*TAEM* 7:443).

1. This drawing appears to be related to the device shown in Doc. 812 n. 2.

2. Collodion was commonly used as a carrier for photosensitive materials in photography. Friedel 1983, 5, 91.

3. Edison described another duplication process two days later. NS-77-002, Lab. (*TAEM* 7:444).

## –947–

*Technical Note: Telegraph Recorder/Repeater*

[Menlo Park,] June 27. 1877

Translating Embosser[1]

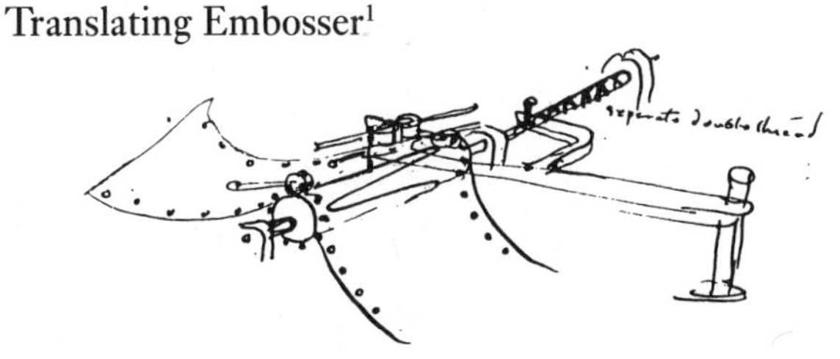

on the continuous roll embosser I propose for obtaining acurate registration to previously perforate holes either on both edges or on both edges & the centre or on the centre only and provide the rotating cylinder with pins to pass through the perforations & ensure the feed & registration

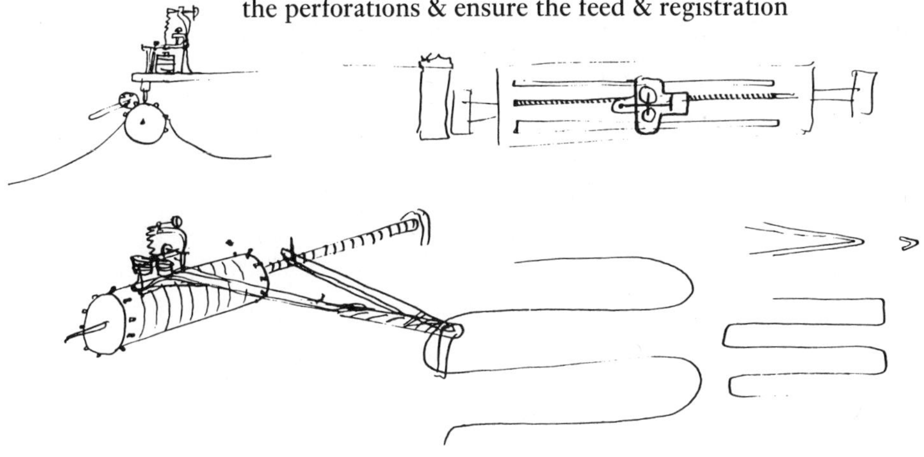

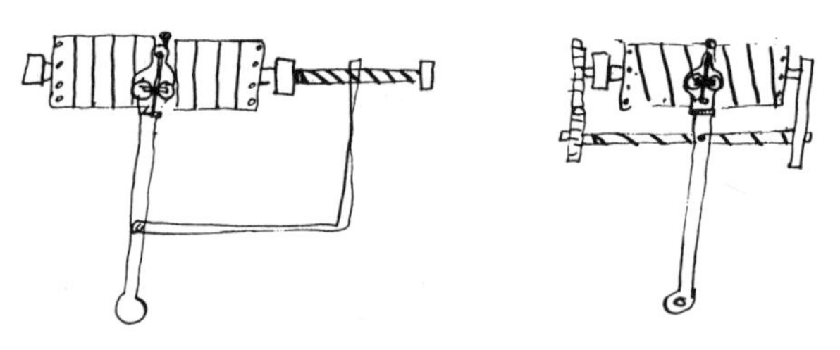

It may be possible that oiled indenting paper is preferable or that the paper should be parafined—shellacked or dipped in plaster paris water=[2]

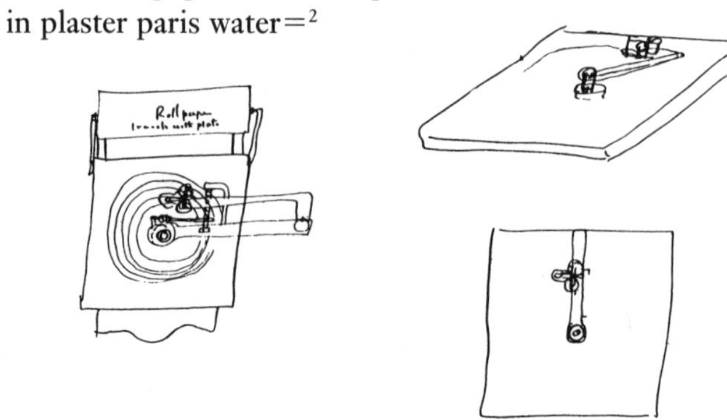


I propose in next translator to have the repeating points one line inside the embossing points, although it could be one line outside[3]

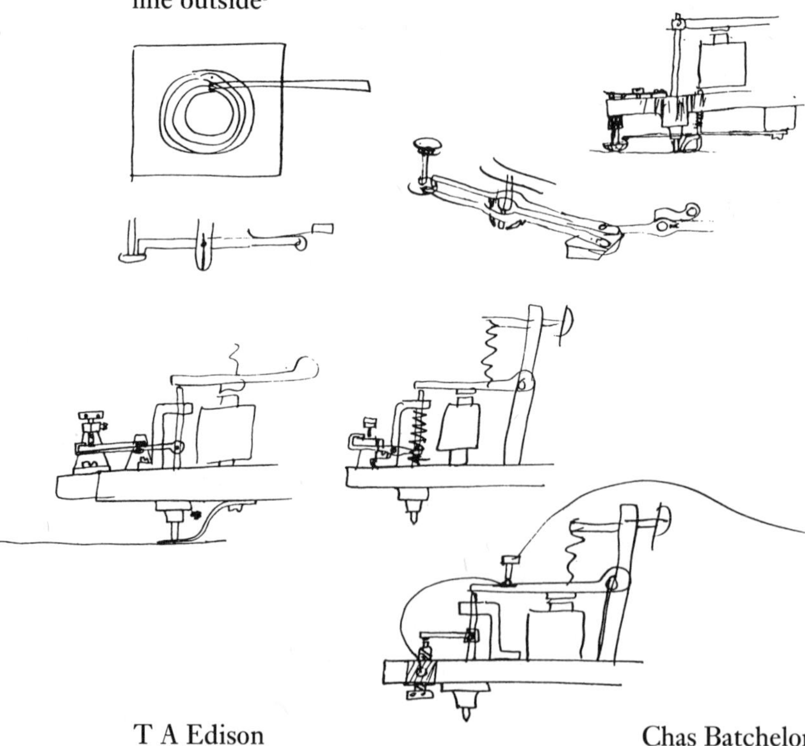

T A Edison

Chas Batchelor
James Adams

X, NjWOE, Lab., NS-77-004 (*TAEM* 7:696, 698–99). Continuity of subject matter links these separate scrapbook items. Document multiply signed and dated.

1. Figure label is "seperate double thread." Edison filed a provisional specification in the British Patent Office on 31 July 1877 for translating embossers recording either in "a volute line upon a correspondingly grooved disc" or "upon a sheet of paper in a zig-zag line, over a cylinder that is correspondingly grooved" and that included "register marks" to hold the paper or disc in place. Provisional Specification 2,927 (1877), Cat. 1321, Batchelor (*TAEM* 92:76).

2. Figure label is "Roll paper travels with plate."

3. On 26 June 1877 Batchelor recorded that

> Edison noticed a serious error in Embosser at Murrays today & that is that the repeating point following close behind the embossing pont and being in the same straight line from centre of swing lever will not follow ~~in the~~ exactly the marks made by embossing point. This we remedied by making the two points only 1/16 of an inch apart thus decreasing the error so as to make it hardly perceptible. [Vol. 11:194, Lab. (*TAEM* 3:1004)]

**–948–**

*Technical Note: Telephony*

[Menlo Park,] June 27/1877

Speaking Telegraph[1]

I find that weighting the centre of the diaphram gives better articulation therefore Batch is making 4 diaphrams of the same brass[2] & weighting the centre by soldering different thicknesses of brass[3]

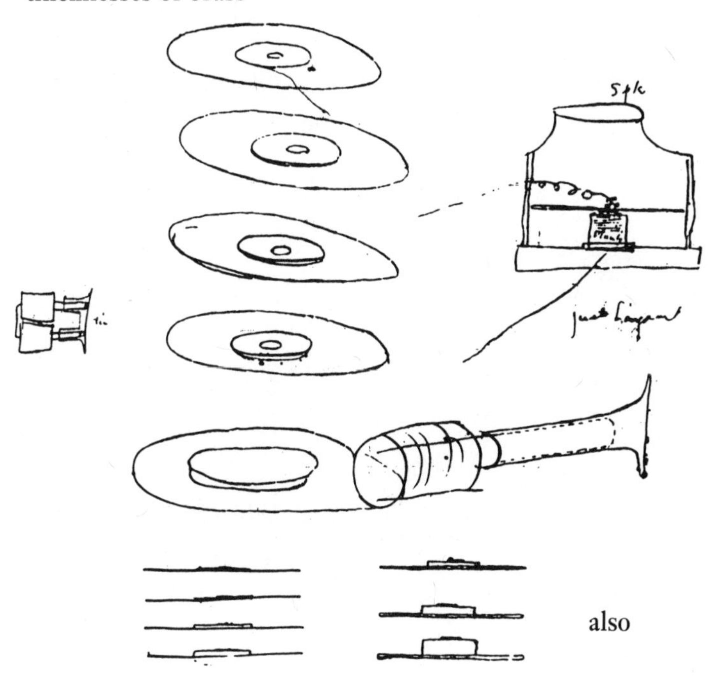

turn one thus also thus

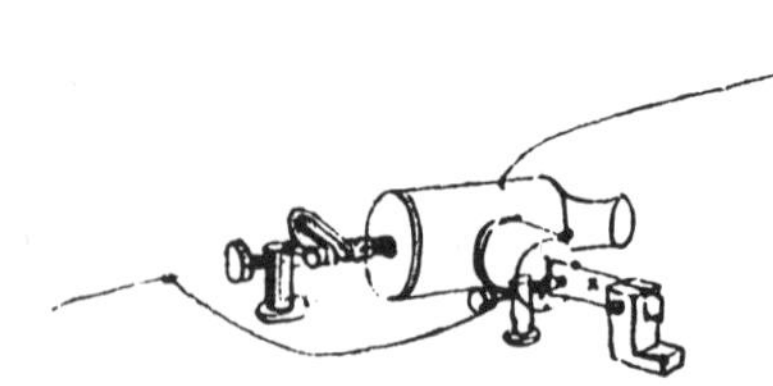

Screw in centre to put washer weights on to weight heavy diaphram

This leaf X is adjusted to respond to the hissing Consonants & hits on plumbago Connections may be made in various ways[4]

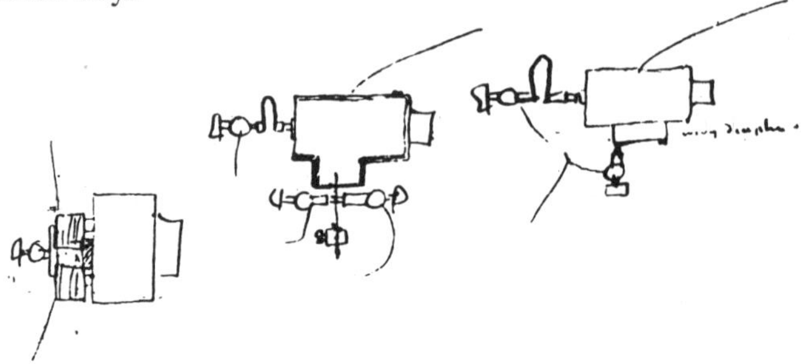

X is peice rubber which serves to dampen the diaphram receiving thus making articulation better by Lessening harmonics.

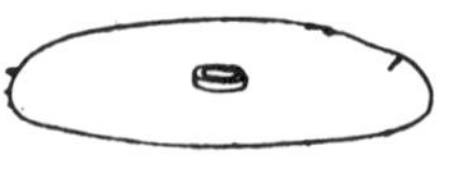

thin sardine tin[---][a] copper foil weighted & tuned to the hissing Consonants[5]

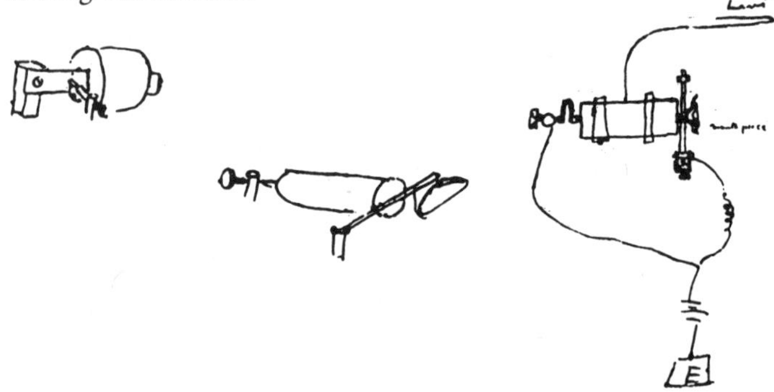

hissing Consonant vibrator in front of regular vibrator talker

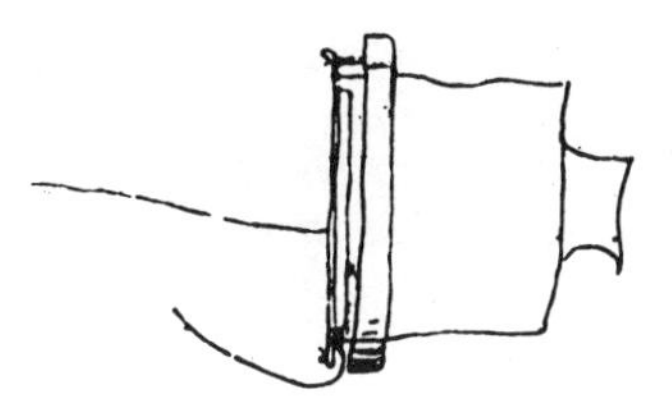

X plumbago  thick diaphram Loose hitting plumbago

Manometric speaker[6]

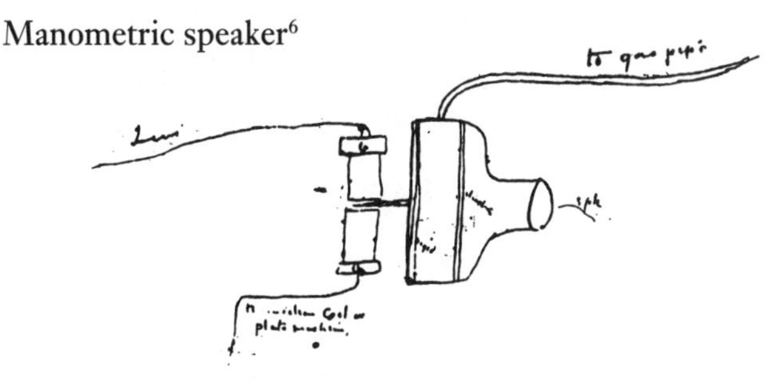

discharge through gas jet

Electromagnet with an iron disk laid on cores; this works beautifully & loud  good articulation & you press disk which is held slightly against cores by spring to your ear thus shutting out other noises;[7]

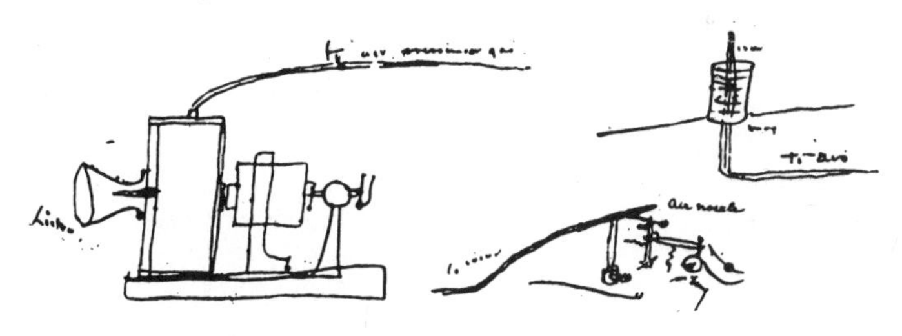

rubber metal or other diaphm[8]

Wind [Reservoirs?][a]  The E.M.Graph may replace the magnet

T A Edison

Charles Batchelor
James Adams

X (photographic transcript), NjWOE, TI 2, Edison's Exhibits 195-11–196-11, 198-11–199-11 (*TAEM* 11:344–45, 348–49). Document multiply signed and dated. [a]Illegible.

1. Edison had sketched another collection of ideas on 24 June. Edison's Exhibits 175-11, 177-11–179-11, TI 2 (*TAEM* 11:327, 330–32).

2. Batchelor's description of these diaphragms, also dated 27 June, is Edison Exhibit 197-11, TI 2 (*TAEM* 11:346).

3. Figure label of small drawing at left is "tin"; drawing at right is labeled "spk," "plumbg," and "just L[ays on?]."

4. Figure label is "wing diaphm."

5. Figure labels are "Line" and "mouth peice."

6. Figure labels are (clockwise from upper right) "to gas pipe," "spk," "diaphrm," "[rigid?]," "to induction Coil or plate machine," and "Line." The manometric flame used a burning gas jet influenced by sound hitting a diaphragm to illustrate the vibrations caused by different sounds. Atkinson 1890, 265–67; see also Doc. 964.

7. Figure labels are "Listen" and "to air reservoir or gas"; "to [sound?]," "air nozzle," and "Emg"; and "iron," "mag," and "to air."

8. This refers to the central diaphragm in the drawing at left.

**–949–**

*From Josiah Reiff*

New York, June 28th 1877[a]

Dear Edison.

Butler among other things said—

Edison has pushed himself forward into the foremost ranks of Electricians, if he himself is not the foremost of his time, & his name will go down with Henry in connection with Telgraphy.[1] Who can tell what good[b] God has in store for the world through the discoveries in Electrical Science of this man Edison if not held in Slavery by the WU. Yours

J C Reiff

Argument is closed— I am now ready for something else.

ALS, NjWOE, DF (*TAEM* 14:698). Letterhead of J. C. Reiff. [a]"New York," and "187" preprinted. [b]Interlined above.

1. Reiff is probably paraphrasing Benjamin Butler's closing argument (delivered this day) in *Atlantic & Pacific v. Prescott & others* (Quad. 70, 71, 73 [*TAEM* 9:288–10:797]). A *New York Tribune* article of 29 June describing Butler's argument (which is not in the printed court record) is in Cat. 1031:5, Scraps. (*TAEM* 27:735).

**–950–**

*From Edward Johnson*

Pha June 29/77.

My Dr Edison

Made a trial of Telephone last Eve. bet. 5th & Chest[nut].[1] & Centennial grounds 5 miles worked very loud & strong, but couldn't get coarse bugs out of it. Thinking they were caused by some defect in diaphram I came down & examined it & found Plat. point in ~~front~~ Diaphram loose—

Have had it riveted again—very little so as not to spoil vibratory power of brass,—& propose to try it again this P.M. I find it very handy to have a Rheostat at Rec. end to regulate the volume & purity of tone— I use gravity batty 75 cells works red hot—

I am going to cut down the block on which the wheel rests so as to put the brass arm lower down— It seems to me to want some sort of deadening its too sensitive—

They are perfectly satisfied with the quantity but want the quality of tone improved—that's all I got to do to make it a success in which case it will be extensively advertizsed & kept on the boards for some time. it is proposed to get a wire from A.&P & bring music consecutively from Balto. Washn & NYork— of course 1st trying it to see if it will carry—

Phillips was very much disappointed that I did not bring 3 or 4 Speakers— I did expect you would send me 2 & I write this to know if you wont send me another one at once— It is very risky business depending on one & it so delicate—besides we want a duet—at least—

If you will send this by 1st express I think I will be able to make this effort a financial success.

Write me a few lines making any suggestions that you think I haven't observed—

Send Everything in Care W J Phillips Supt Phila. Police & Fire Telegraphs Phila Yours

Johnson

ALS, NjWOE, DF (*TAEM* 14:785). Second leaf is letterhead of Police and Fire Alarm Telegraph, Office of the Superintendent.

1. The letterhead shows the office of the superintendent of the Police and Fire Alarm Telegraph to have been at the corner of 5th and Chestnut streets.

**-951-**

*Charles Batchelor to Robert Gilliland*

[Menlo Park,] June 30th [187]7

Dear Sir,

Yours enclosing $500.00 on account of note received and duly endorsed on back of same. It came just in time as we were entirely '<u>busted</u>' for 4th July. In regard to the royalty on Electric pen Edison has received his full amount up to 1st May somewhere about $1200— and they have just sent him a statement for the month of May in which his royalty is about $760— But I presume you have received one also.

Did Ed report on the condition of your shop?[1] I am communicating with Bliss about Daytons and Wheeler's account with the old pen Co and after I hear decidedly I will make a statement[2] at present it is in the same condition as on the last statement. I intended to go over the accounts with Ed before he left as he said he wanted to know something about them but he had to leave before I could get down there.

Edison thinks the prices are all right as long as the royalty keeps increasing.

Shall be glad to see you in the fall Respectfully yours
Chas Batchelor

P.S. Remember me to Ed.

ALS (letterpress copy), NjWOE, Batchelor, Cat. 1238:144 (*TAEM* 93:134).

1. On 12 June, Batchelor had written Robert Gilliland to say that he and Ezra Gilliland had inspected the old pen factory at Menlo Park and found that the reports of damage had been greatly exaggerated. He reported that while the windows were broken, the engine and boiler were in good shape except for broken pipes and the pump chamber, which he was repairing. The factory remained unused and windows were broken again in August and November, and tramps apparently used the building as well. Batchelor to Robert Gilliland, 12 June, 31 Aug., and 19 Nov. 1877, Cat. 1238:138, 202, 231, Batchelor (*TAEM* 93:130, 160, 168).

During the summer Edison arranged with the George Place Machine Agency for the sale of the factory equipment. Some pieces were purchased by sewing machine manufacturer Henry Stewart. *TAEM-G1*, s.vv. "Place (George) Machine Agency," "Stewart, Henry"; Cat. 1233:219, 220, 223, Batchelor (*TAEM* 90:162–64).

2. Dayton and Co. was electric pen agent in St. Louis and William Wheeler was electric pen agent in Chicago. On 18 August 1877 Batchelor wrote Robert Gilliland that both these accounts appeared to be "bad debts" (DF [*TAEM* 93:152]). The previous December, Bliss had told Edison that if their bills went unpaid he would not employ them as agents. However, Bliss subsequently made a new contract with Dayton and Co. and his lawyer advised not canceling Wheeler's contract. On Dayton and Co. see Lbk. 3:174 (*TAEM* 28:640) and stencil for circular, Cat. 593, Scraps. (*TAEM* 27:621). On Wheeler see Batchelor to Bliss, 6 Aug. 1877, Cat. 1238:186 (*TAEM* 93:149); correspondence from Batchelor (*TAEM-G1*, s.v. "Wheeler, W. F."); and his circular in Cat. 593, Scraps. (*TAEM* 27:667–69).

# –6– July–September 1877

The largest part of Edison's summer was devoted to the telephone. With Bell's telephone showing increasing commercial promise in Boston, Western Union became more concerned about establishing a foothold in the market. Although the company was also supporting Edison's work on his embossing recorder/repeater, when Edison asked company president William Orton, "Which shall it be= more Telephone, or Embosser," the answer was clearly "Telephone." In fact, Edison was kept so busy on telephone research that he had little time to pursue his newest and most astounding invention, the telephone recorder/repeater, or "phonograph." As was their custom, Edison and the staff continued to work late into the night. In mid-August night work became more comfortable with the installation of a gas-lighting system.

Edison encountered several formidable problems in developing a commercial telephone. Having settled on plumbago as the optimal type of carbon to use in his transmitter, he struggled throughout the summer to find the best way to use it in the instrument. Edison began the summer using the small cakes of plumbago mixtures he had been testing in June. Although Edison declared his telephone "perfected" (for the first time) on the morning of 17 July, he soon began to experiment with alternative arrangements of plumbago. He pressed plumbago-coated surfaces against the diaphragm as he had earlier and, at the end of July, developed a new approach—coating a fibrous substance with plumbago and packing it into a mass he called "fluff." Various forms of fluff and holders for it occupied the staff for the rest of the summer.

The telephone transmitter had other shortcomings as well.

Edison continued to substitute different materials as diaphragms to find one that was unaffected by moisture or heat from the speaker's breath. In late September he finally settled on a loose piece of mica over stretched rubber. The rubber, besides making the diaphragm responsive, also helped solve the problems of harmonics and sibilance, which had continued to plague the instrument. Before using rubber Edison had experimented with the use of extra speaking tubes and reeds that reinforced the sibilant vibrations.

Edison conducted line tests of his instruments throughout the summer. Thomas David, testing both Edison's and Bell's telephones in Pittsburgh, kept him apprised of the results. By September the articulation and loudness of Edison's transmitter had improved sufficiently to make it a practical instrument. (His earlier transmitter–electromotograph receiver design proved serviceable but too expensive.) In early September Edison arranged for the commercial introduction of his telephone system in Montreal, and in the middle of the month William Orton ordered 150 transmitters and receivers for the Gold and Stock Telegraph Company, a Western Union subsidiary, to use on private lines. Edison and others also considered the telephone's potential integration into district telegraph systems. The enterprising Edward Johnson was already successfully using Edison's musical telephone for well-publicized "concerts" at resorts such as Cape May, New Jersey, and Saratoga, New York. Edison also had Charles Batchelor build an electromotograph telephone that had a dulcimer as a sounding box for British electrician William Preece to demonstrate before scientific and technical societies.

As perceived by the telegraph industry, the telephone had a serious shortcoming in that it left only an oral message. On 17 July Edison designed a message recorder that would do for the telephone what his embossing repeater did for Morse telegraphy, only to realize hours later that he had conceived a means for permanently recording sound—the phonograph. Although he saw the potential of his new invention, the telephone research left him little time to experiment with it.

Intrigued by the development of the Jablochkoff arc lamp, two men involved in the electrical industry—George Field and Enos Barton—urged Edison to explore this exciting new electrical technology. Because arc lights were fit only for use outdoors or in very large interior spaces investigators considered ways to create smaller and less bright lamps—to "subdivide" the light. Edison briefly turned to this problem, first trying to dim the arc light and then using pieces of carbonized paper as incandescent "burners."

The laboratory in Menlo Park continued to attract important scientific visitors. In July, George Barker, Henry Draper, and William Wallace stopped in to see Edison on their way north from Philadelphia. Among the subjects they discussed was Draper's "discovery" of oxygen in the sun, a topic that led to correspondence between Edison and Draper about spectroscopy and Edison's "scintillations." Alexander Siemens, nephew of British electrical manufacturer and inventor William Siemens, also visited at the end of July. In September electrician Stephen Field and Dr. Cornelius Herz, on their way from San Francisco to Europe to acquire rights for the Jablochkoff arc light, witnessed some of Edison's own electric light experiments at the laboratory. George Bliss, Edward Johnson, Josiah Reiff, other business associates, and Western Union employees also made occasional visits to the laboratory.[1]

While in Menlo Park, Field and Herz acquired Edison's and George Prescott's rights to the quadruplex telegraph in Austria, Spain, France, and Belgium. Edison considered selling Field and Herz rights to his Continental telephone patents but did not do so.[2] At the same time Western Union electricians Gerritt Smith and George Hamilton were in England to introduce the quadruplex system on the Post Office lines, and George Bliss was working through his agent George Beetle in Paris to establish the electric pen business on the Continent.[3]

Several people tempted Edison with other ventures. Josiah Reiff (with Leonard Myers, one of the attorneys in the Quadruplex Case) encouraged Edison to develop apparatus to detect underground silver ore, something Edison had thought about in 1875.[4] James Hammond asked Edison for help in perfecting a typewriter. And G. A. Mason wanted Edison to develop his earlier idea for an electric horse clipper into a practical instrument.

Although the Port Huron street railway companies had merged in March, disputes over management plagued the new company and were a source of annoyance to Edison who received frequent letters from his brother Pitt and others connected to the road. Pitt was also involved in a dispute with other parties concerning the management of the Sarnia Street Railway Company, which led to negotiations to buy out Edison's share in that road. In early August, Edison sent Pitt's son Charley, who had been in Menlo Park, back to Port Huron to provide a more objective analysis of the railway situation there and to represent Edison's interest in the Sarnia Street Railway. Charley concluded these tasks by the end of the month but illness prevented his return to Menlo Park until the fall.

The summer was grueling for Edison, Batchelor, Adams, and the others at the laboratory as they strove to bring Edison's telephone to market. They often worked through the night in Menlo Park after testing apparatus in New York during the day, only to head back into the city with little or no sleep. In mid-September, Edison, Batchelor, Adams, and George Carman chartered a schooner for a three-day fishing expedition out in Raritan Bay.[5]

1. Cat. 1233:182–273, passim (*TAEM* 90:144–89).

2. George Bliss would also have been involved in this arrangement.

3. Bliss reported in mid-September that "they had sold 150 machines during the first four months on Foreign Pen." Cat. 1233:258, Batchelor (*TAEM* 90:182).

4. See Doc. 650.

5. The hectic pace of the summer can be followed in Batchelor's diary. The fishing trip is detailed on pages 254–56. Cat. 1233, Batchelor (*TAEM* 90:180–81).

**–952–**

*From Leonard Myers*

Phila. July 3, 1877.

Friend Edison

Mr. Larned[1] of 52 Broadway[2] Silver Island Mining Co. of Lake Superior[3] desires to see you at his office & I promised I would send you to him— It is said you have got a magician's wand and better than the witch hazel it will lean down & Kiss the earth when silver lies beneath— "4plex—why not?"[4]

I left at Reiff's office for you a copy of my Argument in the Quadruplex.[5] They all said you were a wonderful fellow[6]— neck and neck with Morse or still better, with Professor Henry. I did not quite say that, but I gave you full praise—

Now in return I want you to order from Washington & send to me one of your photographs such as I saw at Reiff's or even a smaller one.

Johnson I suppose is here fixing up your Telephone at the Exhibition grounds. Now "don't let's" read it Edison & Myers have it Prescott & Edison!

When you come to Philada. call on me at 125 South 7th st. Yours Very truly

Leonard Myers.[7]

ALS, NjWOE, DF (*TAEM* 14:84).

1. Unidentified. Josiah Reiff calls him "Learned" in Doc. 954.

2. Josiah Reiff's office was also at 52 Broadway, according to his letterhead. E.g., Reiff to Albert Chandler, 23 July 1877, DF (*TAEM* 14:701).

3. No information could be found regarding this company.

4. Myers is referring to Edison's notation in Doc. 285 (no. 14).

5. Quad. 73.9 (*TAEM* 10:508).

6. See the closing arguments in the Quadruplex case. Quad. 73 (*TAEM* 10:365–772).

7. Myers was a lawyer for the plaintiff in the Quadruplex Case. See *TAEB* 2:491 n. 3.

## –953–

*From Edward Johnson*

Pha July 5th 1877[a]

My Dr Edison.

All right send on the 2d speaker as soon as you can.

I have worked it up until it is in tip top condition made the change of position of vibrating arm on sounding board—Squared Everything up so as to make it reliable and will today get proper wires & substitute for the present ones—& have them properly tuned as you suggest

The speaker I have had a good deal of trouble with the plat[ina] point came out of Diaphram I had it reriveted but it came out again I then had a new one made somewhat larger—putting the plat on a piece of heavy copper wire turning it down so as to form a shoulder & then riveting it heavily on inside—resulting in a thoroughly good job & leaving Diaphram in better vibratory condition than before— The 3 screws which hold the Diaphram head in position worked so loose in the brass shoulder that they rattled loose & changed adjustment of Diaphram tension—so I set the threads up a trifle & fixed that— Finding that every voice requires a different adj. & that the fractional part of a hair adjustment made a difference I had 2 holes bored & tapped in end of base & a sheet of metal 8 inches high screwed on to stand upright in front of diaphram & marked off in a scale—a long wire or pointer set in the head of the adjusting screw thus—

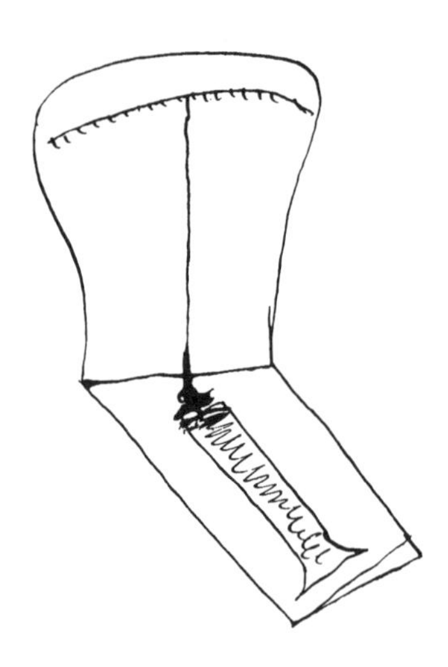

thats not very handsome but I guess youl divine what I mean

By this contrivance I am enabled to find the exact point on the scale a certain voice requires—& upon throwing it out & putting it back to that point I find the adj. right everytime— This saves annoying adjusting before an audience and enables me to move the adj. screw finer

I gave a preliminary rehearsal the night of the 3rd—having the Inst. in the immense auditorium— Everybody said it would simply be impossible to hear it owing to the huge dimensions of the Building You should have seen the surprize & heard the applause when the first tune came out It was clear & nice & ~~v~~extra ordinary loud— Several of the Exhibition folks went to the far side of the Building & heard it distinctly I kept it up 2½ hours with unwavering success—& at the close the managers pronounced themselves not only thoroughly satisfied but very agreeably disappointed and at once decided not to bring it out merely as an incidental thing but to make it a special feature of the Exhibition This necessitates the postponement of the Exhibition for 2 weeks—[1] The time meanwhile to be occupied in exciting a live interest among the proper classes This is to be done by inserting in the 15 Daily Papers in which the Ex. Co. advertizes—every day some new scientific note or squib about you & your career—[2] These 15 Daily papers are to talk of nothing but Edison & his inventions & discoveries during that time—all of course made to contribute to the Kindling of an interest in the "Electro Motograph Telephone" I am to furnish all the squibs—or rather the information upon which they are to be based so that it shall be reliable & have a real interest for Scientific people Hence I want you to give me every day or so often as you can some little Scientific note or observation that you have made that is new—like for instance the rapid reduction of Res. of Plumbago under pressure—, the effect of the passage of high notes through an electromagnet—, the effect through the motograph—, the peculiar property of the motograph in creating a dif of friction—, the method of transmitting graduated Battery currents—, &c &c— All of which will be duly credited to you for the immediate purpose of directing attention to your telephonic labors—

All this of course being done in a proper manner cannot fail to be advantageous to you All they ask that you may have objections to is this—the Exhibition is Established as an Industrial school— and it is the aim of the managers to get hold of every good thing they can and bring it before the public as emanating from the Industrial Palace of America of course your name will be constantly linked with every item—& in fact every effort made to establish a reputation for you individually—but [-----][b] rather as it were making the Industrial school your Labratory instead of Menlo Park— This[c] will call more attention to your works than anything else—& cannot but be beneficial to you so I have told them that I did not think you would object to their thus adverti~~z~~sing their Institution to the world if they took proper care to credit you individually with the works Inventions & discoveries you have wrought— This they say it is their interest to do as it will be their pleasure I think it the very finest kind of an opportunity for you to inform the world what you have been doing— These little squibs will be copied by papers the Country over as well as by papers abroad Hence I want nothing put in but what is sound & will reflect credit upon you The entire morale of the Permanent Exhibition—which by the way is much better than you think) is to be loaned to create a stir in scientific circles on this subject Barker is to do the talking when the time comes and it is confidently believed that the result will be a prolonged engagement of the telephone—

But the main idea is to advertize the Exhibition to the world—as the Grand Panoramic sheet upon which the Scientific Genius of America may show the world what they are doing I have sought to make this an opportunity to put you in your proper place in relation to Bell & Gray in the Telephonic Art.[3] I have Succeeded beyond my Expectations inasmuch as you will be given an opportunity to assume your proper place in relation to telegraphic inventions—etheric &

everything else— I am not limited to subjects, only to such as are yours—

You must heartily second me in this—& it is to get you to do so that I have so fully explained things generally Yours

Johnson

ALS, NjWOE, DF (*TAEM* 14:789). Letterhead of E. H. Johnson. [a]"187" preprinted. [b]Canceled. [c]"(now take back of Pages)" written at bottom of page.

1. The opening ceremonies of the Permanent Exhibition were held on 10 May 1877. During the four years after the close of the Centennial, the Permanent Exhibition Co. made many efforts to maintain a collection but the enterprise languished. Philadelphia merchants and manufacturers were generally indifferent to the enterprise, and the exhibition drew large crowds only at conventions, parades, receptions, and Fourth of July celebrations. In 1881 all further efforts to revive popular interest in it were abandoned, and in August 1881 the Main Building was sold. Scharf and Westcott 1884, 849.

2. See Doc. 974.

3. On Bell's and Gray's respective inventive work on the telephone, see Hounshell 1975.

**–954–**

*From Josiah Reiff*

[New York,] July 5/77

My Dear Edison—

I presume it is ordained I am not to see you anymore— After sending a good many notes without reply I finally was told to be here @ 4 PM—but no Edison.

I understand the translators[1] work well. J[ay]. G[ould]. will want to see you within a few days—

Can you not arrange to make an engagement for Learned[2] about the Silver Mine Experiment in Lake Superior— He is very anxious about it if you have any confidence—

Genl Butler & Leonard Myers both discussed the matter with Learned in connection with you—

Did you see Preece before he sailed? Did he say anything about Automatic or Quad for the other side?

Have you heard any more about [Pershire?][3a] & his Electric Pen? Have you seen it?

I leave for copies Russell' & Myers' briefs—[4]

If you wont come down to my office I shall have to come to Menlo for I dont propose to consider accounts closed with T. A Edison. Did you send telephone to EHJ? Y[ours]

JCR

ALS, NjWOE, DF (*TAEM* 14:86). [a]Obscured overwritten letters.

1. Embossing translators for automatic telegraphy.

2. Cf. Doc. 952.
3. Unidentified.
4. Robert Russell and Leonard Myers.

**–955–**

*Technical Note: Autographic Telegraphy and Telegraph Recorder/Repeater*

[Menlo Park,] July 5 1877.

Autographic Telegh Facsimile

For autographic Tel ~~em~~write your message by embossing on parafined paper & transmit by a spring Contact

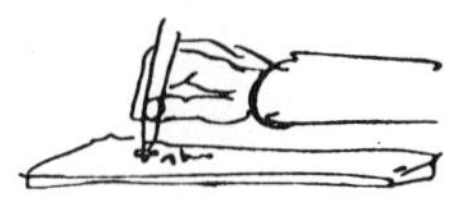

2 Write with an electric pen & then adapt the Wheatstone Auto plan of Jacquard[1] sending by vibrating the contact point ~~up und~~ against the paper by a tuning fork making several [hu]ndred[a] vibrations per second when no [sig]nals[a] occurs the needle cannot pass through the paper hence no contact & when it does occur it passes through & thus allows its spring to come in contact with its platina point closing the circuit=

3rd Write with a sticking ink broad nib pen dust over with plumbago & pass under a press or between rollers this will make it conducting writing use 2 pens to close ckt thus

in passing the mark the plumbago form a circuit between the pens

Embossing Translator

I propose to make[b] the grooves wider & to put repeating point one or 2 lines behind indenter & open Repeating points when necessary by extra lever or switch[2]

I oil my embossing paper

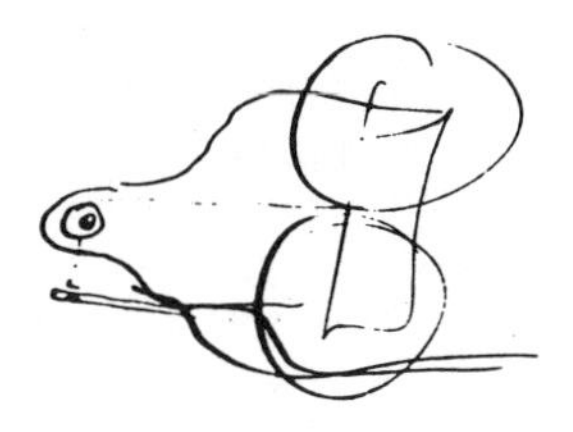

Machine oil I prefer= any oleagenous substance will answer even the gums will work by stiffening the paper & all semi hard materials which the[b] paper can be impregnated with will work= Instead of embossing in paper I can use thin Copper or other metallic foil, and I can arrange the circuit connections in such a manner that the foil will close circuit & the indentation will open by its point falling into it & ~~not~~ prevented from going to the Bottom by a limiting pin

The embossing point itself may be made to repeat the retactile spring being dispensed with the lever by its own weight causes the embossing point to lay on the paper where an indentation occurs it falls down & the lever striking a spring contact point closes the circuit[3]

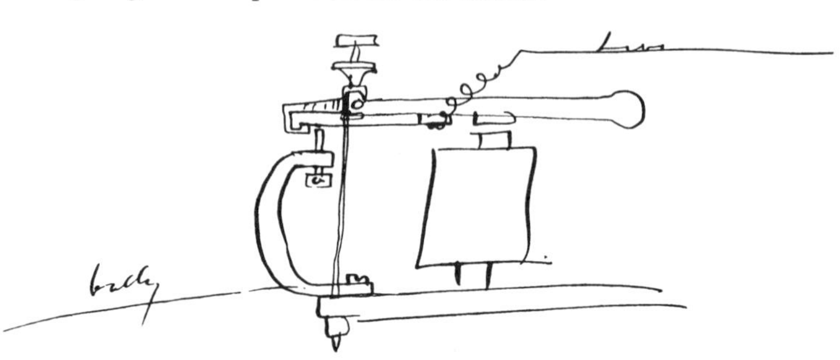

to obtain accurate repeating when it is undesirable to use the embossing point a hole may be drilled in its centre .& a needle passed down through it ths needle connected to the spring & act a circuit closer

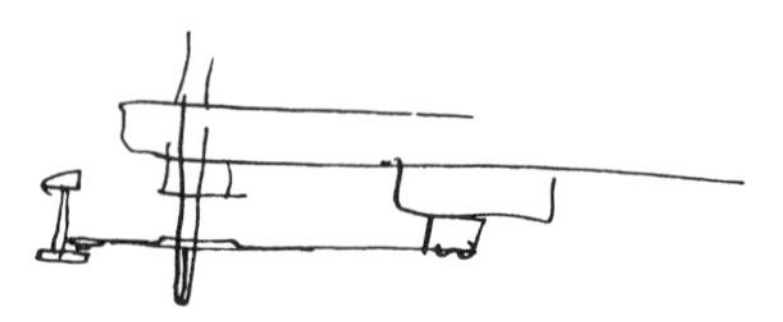

T A Edison James Adams

X, NjWOE, Lab., NS-77-004 (*TAEM* 7:701). Document multiply signed and dated. [a]Document damaged. [b]Obscured overwritten letters.

1. In the Jacquard loom (named for French inventor Joseph Marie Jacquard) the patterns of holes in a series of perforated cards dictated the design woven on the loom. In Charles Wheatstone's automatic telegraph, small metal rods pushed through the perforated message tape to make contact for transmission. Cardwell 1972, 119–21; Guillemin 1977, 627–32.
2. Figure labels are "now" and "[altered?]."
3. Figure labels are "batty" and "Line."

-956-

*To T. & J. S. Symington*[1]

Menlo Park 6 July 1877.[2]

Gents,

Your letter received.[3] As I am entirely unacquainted with the books of the road, and as no statement was enclosed in your letter, and no information good, bad, or indifferent has been received from my brother since I saw you it cannot be expected that I should grasp the full meaning of your letter. I have sent a copy of it to my brother for light on the subject. For your information I will state that I own and have paid for 180 shares at the rate of 67½ cents on the dollar gold [Originally?][a] I paid for 120 shares for which I hold receipts and return checks also I received 60 shares from M Fleming [held?][a] as collateral for the roads notes. I sent him $2300. Am currency through Vermylie & Co[4] New York for which I hold their receipt. There was returned to me by Mr Fleming the roads notes

Of the assessments on these 180 shares as you know 1000 was left in your hands which I believe was sufficient.[5] The only interest my brother has in this block of stock was the promise on my part that should it ever attain the dignity of par that he should have the difference between 67½ and that figure, and it was this that caused him to urge me to make a different bargain with you than I desired. Of the remainder of the Pt Huron stock I know nothing of who owns it, how much there is of it or if it has been paid for at the rate of 67½ or has paid the last assessment. Any other information on this subject I will be glad to give you. Yours

Thos. A. Edison

ALS (duplicating-ink copy), NjWOE, DF (*TAEM* 14:569). [a]Illegible.

1. T. & J. S. Symington ran a tailoring, dry goods, and carpet warehouse in Sarnia, Ontario. Letterhead, T. & J. S. Symington to TAE, 2 Mar. 1878, DF (*TAEM* 19:391).

2. An undated draft version precedes this document in DF (*TAEM* 14:566–68).

3. Probably the undated letter from the Symingtons to Edison in DF (*TAEM* 14:618–24).

4. Vermilye and Co. was a banking house at 18 Nassau St. in New York; William M. Vermilye is listed as a banker in the New York City directory for 1878. Wilson 1878, 1442.

5. See Doc. 871.

**–957–**

*Technical Note: Multiple Telegraphy*[1]

[Menlo Park,] July 7 1877

Quad & Sextuplex[2]

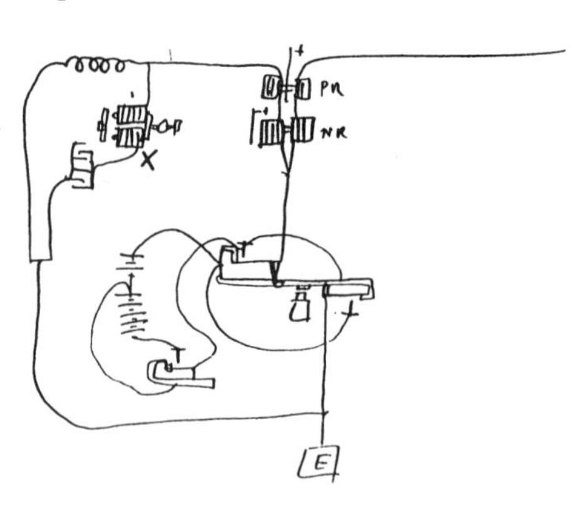


X[3] is an adjustable magnet to adjust the charge & discharge of the Condenser to meet that of the line which it does by its own self induction delaying both the charge & the discharge  The power of the self induction may be regulated by adjusting the cores to or from the set armature;

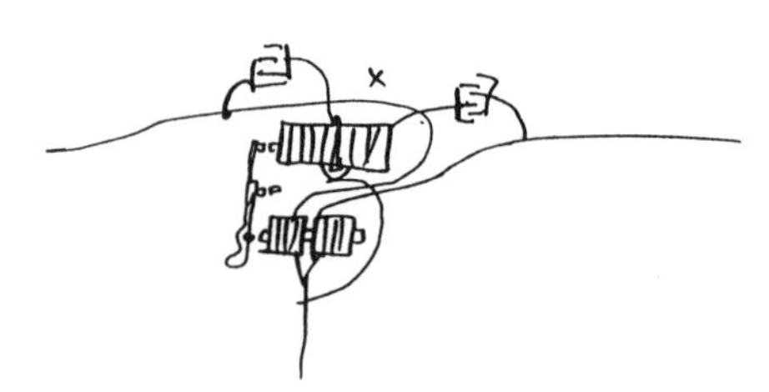


X long

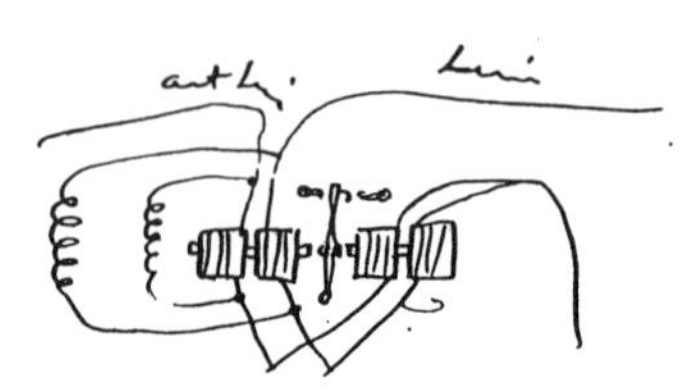

working by self induction[4]

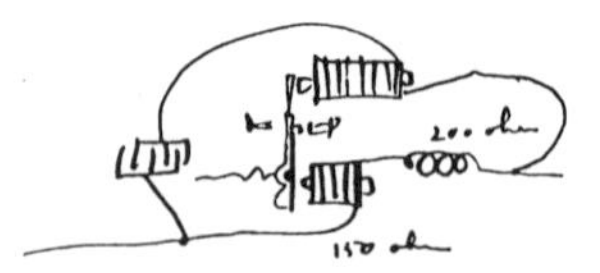

a relay you Can reverse through without opening=[5]

To get rid of the permanent magnetism or set in a magnet especially a polarzed relay used in Quad I propose to use Iron cores of layers of iron or wires=[6]

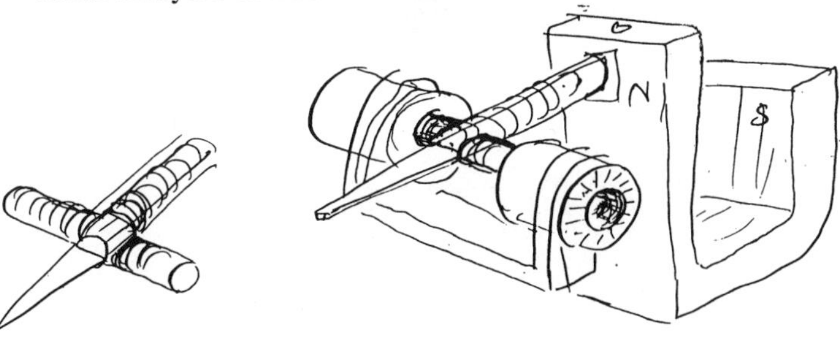


hollow spools to suck in light extensions of the polarzed tongue, passage of the current cannot leave any permanent magnetism in the wire helices as there is no iron to magnetise and if the tongues magnetism is increased or decreased it will make no difference as there is nothing to attract it or it to be attracted by[7]

T A Edison    Chas Batchelor

X, NjWOE, Lab., NS-77-004 (*TAEM* 7:656). Document multiply signed and dated.

1. See headnote, p. 280.

2. This document marks the end of Edison's principal work on sextuplex telegraphy, although there is a technical note dated 8 September 1877 that contains two drawings related to his earlier work (NS-77-004, Lab. [*TAEM* 7:705]). Edison apparently prepared notes for a sextuplex patent application during June (NS-77-004, Lab. [*TAEM* 7:635, 690]), but he never submitted it. In August, however, he used his sextuplex experience in drawing up a quadruplex patent application that described a circuit using two different current strengths (U.S. Pat. 420,594).

*A quadruplex circuit that works without current reversals.*

T. A. EDISON.
QUADRUPLEX TELEGRAPH.
No. 420,594.    Patented Feb. 4, 1890.
Fig. 1.
LINE

3. Figure labels are "X," "P[olarized] R[elay]," and "N[eutral] R[elay]."

4. Figure labels above are "art[ificial] line" and "line."

5. Figure labels above are "150 ohms" and "200 ohms."

6. Edison had drawn a number of designs using permanent magnets in a polarized relay on 5 July. NS-77-004, Lab. (*TAEM* 7:655).

7. On the normal operation of a polarized relay, see *TAEB* 1:38–39. At the end of this document is an unrelated, later set of notes on the resistance of plumbago.

**-958-**

*To Thomas David*

Menlo Park N.J. July 9 1877

Friend David,

Did I understand you to say when I saw you last that when the newspaper article was read to you that you had no idea of the nature of the article beforehand and did not follow the speaker with a duplicate newspaper=[1] why I ask this is that I remembered one of your remarks to the effect that it came so perfect that you even corrected an error and that made me think that you was following the speaker with a duplicate. I have mine to such a point that although it is difficult to understand a long sentence from a newspaper the subject being unknown to you that ordinary conversation is quite easy so much so that I do not have any doubt but that it will answer for private lines= it is very loud & no extraneous noises bother it between here & NYork on wire although sympathy[2] could be felt on the relay I have just finished a pair of Bells instruments little different from first ones[a] and I cannot do near as well with them[b] as with my own but perhaps I have made them wrong— Write me results you have attained so far— Yours

T A Edison

ALS, NjWOE, Supp. II, 1877 (*TAEM* 97:587). [a]"little . . . ones" interlined above. [b]"with them" interlined above.

1. See Doc. 963.

2. Inductive interference from nearby telegraph wires.

**-959-**

*To Leonard Myers*

Menlo Park July 9 1877

Friend Myers,

I received your argument, its very plain to me, and until I read the other side I should think the W.U. had'nt a ghost of a chance= A B Chandler spoke in very complimentary terms regarding it to me.= I went and saw the mine people. Their ore is peculiar, and it would require great labor in preliminary testing to establish an acurate method of testing lodes and deposits, but It can be done. Yours

T. A. Edison

ALS, MiDbEI, EP&RI.

**–960–**

*From Pitt Edison*

Port Huron July 9th 1877. [a]

Dear Bro

Pleas find the within letter which states what Mr Sanborn requires before making the loan to the company  I see nothing unfair in it, the road is being pushed for its debts and something must be don  I dont think that Mr Sanborn will interfere with the running of the road as long as it is run in the intrest of the stockholders and dos not depreciate his security and than in either case it would be to your interest to have him step in and stop it  every thing would of ben smooth were it not for Mr Cole and because thay would not consent to let him step in and plunder the road as he did the new road he comenced throughing mud at the present management which I think is running the road to the best interest of the Stockholders

WPE

This letter[1] was handed to me by Mr Voorheis  WPE

ENCLOSURE[b]

Port Huron Mich July 9th 1877

Dr Sir:

To get the loan which we had hoped to get before this time, it becomes necessary for a balance of the stock[c] to be transferred to ~~him~~ John P. Sanborn, who makes the loan, together with power of attorney, during lifetime of loan, as some of the stockholders refuse to guarantee payment of loan.[2] This seems necessary ~~because~~ to protect the company & Mr Sanborn, from outside influences, that may be made injurious to his loan or interests of the company. Wastell & Beard will transfer their stock and that will be sufficient, although others could be induced to do likewise.

Some of the stockholders would prefer to see the road sued, & sold, taking their chances on the sale to become purchasers.

Debtors are pressing the road, and the loan must be made forthwith.

Mr Sanborn will loan five thousand dollars, and that will take care of all pressing debts at present—and put the company in a shape where it can protect itself and the interests of all the stockholders.

This being seemingly the only alternative open to the[c] company, it seems necessary that the proposition named in the forepart of this letter should be made effective. As many as to whom this seems feasible, they[d] will take immediate steps to carry the same out, it is to be hoped.

The connection has been made, and the road is doing well, for the hard times. Yours Truly,

Geo. P. Voorheis[3]

ALS, NjWOE, DF (*TAEM* 14:570). Letterhead of Sarnia Street Railway Co. [a]"187" preprinted. [b]Enclosure is an ALS. [c]Obscured overwritten letters. [d]Interlined above.

1. The enclosed letter is addressed to "T. A. Edison Esq."

2. The loan was made after the stockholders signed an irrevocable five-year power of attorney giving James Beard and James Sanborn the management of the road. Power of attorney, 13 July 1877; Voorheis to TAE, 16 July 1877; both DF (*TAEM* 14:573, 576).

3. George Voorheis was an attorney with the Port Huron firm of Chadwick & Voorheis and secretary of the Port Huron Railway Co. Voorheis to TAE, 16 July and 10 Oct. 1877, DF (*TAEM* 14:577, 597).

**-961-**

*From Edward Johnson*

Pha July 10 1877[a]

My Dr E.

Since writing today I recd yours—[1] The record you send is very interesting—[2] I shall try setting the contact point to one side—leaving the present one in its position as a weight Its quite long & will probably be sufficient(?) I should much prefer however not making any change for fear of disabling the thing—& if you could only send me another I would not touch this one. Charlie is wrong— I think I've done more experimenting to get the Bugs out than they ever did[3]—& I find it makes a develish big difference what sort of voice you have on— I've got a scale for various voices— I use 60 cells of carbon on 5 miles of No 12 Iron wire— & I just tell you I make Rome howl You must remember I've got to ride over the Enormous acoustic difficulty of filling the Biggest Barn in the U.S.—the academy of Music would be a flea bite to it—& I've done it so as to be heard clear & sharp 500 feet away to do that & keep away the Bugs is no mean job—

Phillips is my right hand man & a Brick— He tells them every word I say about you is Gospel—that "you'r all my fancy painted you. You'r lovely & divine"—& they are willing to pay that the Public shall know it— You don't understand about the "Industrial Palace" Eh. Well it is simply this they want the world to think they deserve great credit for Introducing you to it—ie—putting you on exhibition so that the aforesaid World shall know what you are doing—that's all—simply want credit for preventing you from "hiding your light under a Bushel"— guess you can afford to take unlimited doses of that. But you can get a good ~~d~~Idea of all they want by reading a line in One of the Articles "under the auspices of the Permanent Exhibition" I'll send you papers daily—but they would be more interesting if you[b] would send me some Items— I see the N.Y. Sun[4] copied part of the Press article today I didn't revise that printed description you gave me—not thinking they meant to publish it— I'm sorry for this as its not very Clear—

Scrawl me some fodder Yours

Johnson

ALS, NjWOE, DF (*TAEM* 14:811). Letterhead of E. H. Johnson. [a]"187" preprinted. [b]Obscured overwritten letters.

1. The same day Johnson sent Edison clippings from that and the previous day's papers (Johnson to TAE, DF [*TAEM* 14:809]). Edison's letter has not been found.

2. Johnson appears to be discussing the embossing repeater at the beginning of this letter (until the word "Charlie"). A "record" would be the embossed paper disk.

3. See Doc. 929.

4. The *New York Sun* of 10 July 1877 (p. 4) published an article, "Mr. T. A. Edison's Telephone," credited to the *Philadelphia Press.* In his letter written earlier that day Johnson had included an article from the *Press* titled "Telephone" (DF [*TAEM* 14:809]).

## COMBINATION TELEPHONE TRANSMITTER–ELECTROMOTOGRAPH RECEIVER Doc. 962

Edison conceived a combination telephone transmitter–electromotograph receiver in early June 1877.[1] Although he gave Joseph Murray an order for four such instruments on 11 June, Charles Batchelor continued to work on the design until 10 July, when he finished the drawings for Murray, which Murray received on 13 July.[2] On 9 July, Edison also executed a patent application that included the combination instrument.[3] Murray finished the four instruments on 10 August and delivered them to Western Union headquarters at 197 Broadway, where he participated in an apparently successful test of the instruments. According to Murray "These machines were objected to on account of cost. There was no other objection made to them that I know of."[4]

One of these instruments (Doc. 962) was introduced into evidence during the Telephone Interferences.[5] At that time it no longer had a holder or case for the chemically treated paper.[6]

1. Doc. 932.

2. See Doc. 932 n. 1; Cat. 1233:191, 194, Batchelor (*TAEM* 90:148, 150). Batchelor's 7 July drawing of a wiring diagram for the instrument is in Vol. 12:4, Lab. (*TAEM* 4:2).

3. Edison's patent application (Case 141) was filed on 20 July 1877. It was later divided into two patents (U.S. Pats. 474,231 and 474,232), each of which described but did not claim the combination instrument. These patents did not issue until 1892. Pat. Apps. 474,231 and 474,232.

4. Murray recalled that the tests were made on a line to Philadelphia. Murray's testimony, TI 1:363 (*TAEM* 11:158).

5. This instrument may be Artifact CH-3439 in the AT&T Archives, Warren, N.J.

6. A Charles Batchelor drawing dated 17 June 1877 shows the paper holder and case. It also shows a redesigned mechanism for turning the electromotograph drum which is similar to the one actually used in the instrument at the AT&T Archives. Edison's Exhibit 146-11, TI 2 (*TAEM* 11:301).

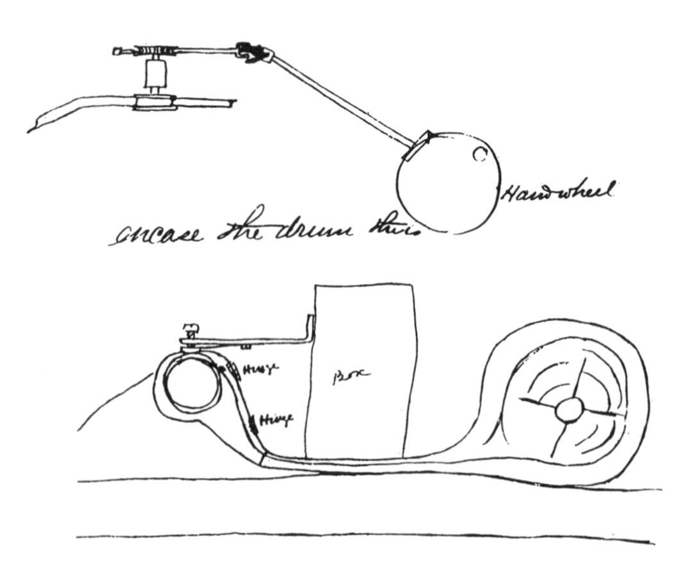


*Batchelor's sketches for the combination transmitter-receiver.*

–962–

[Menlo Park, July 10, 1877?]

*Experimental Model: Telephony*[1]

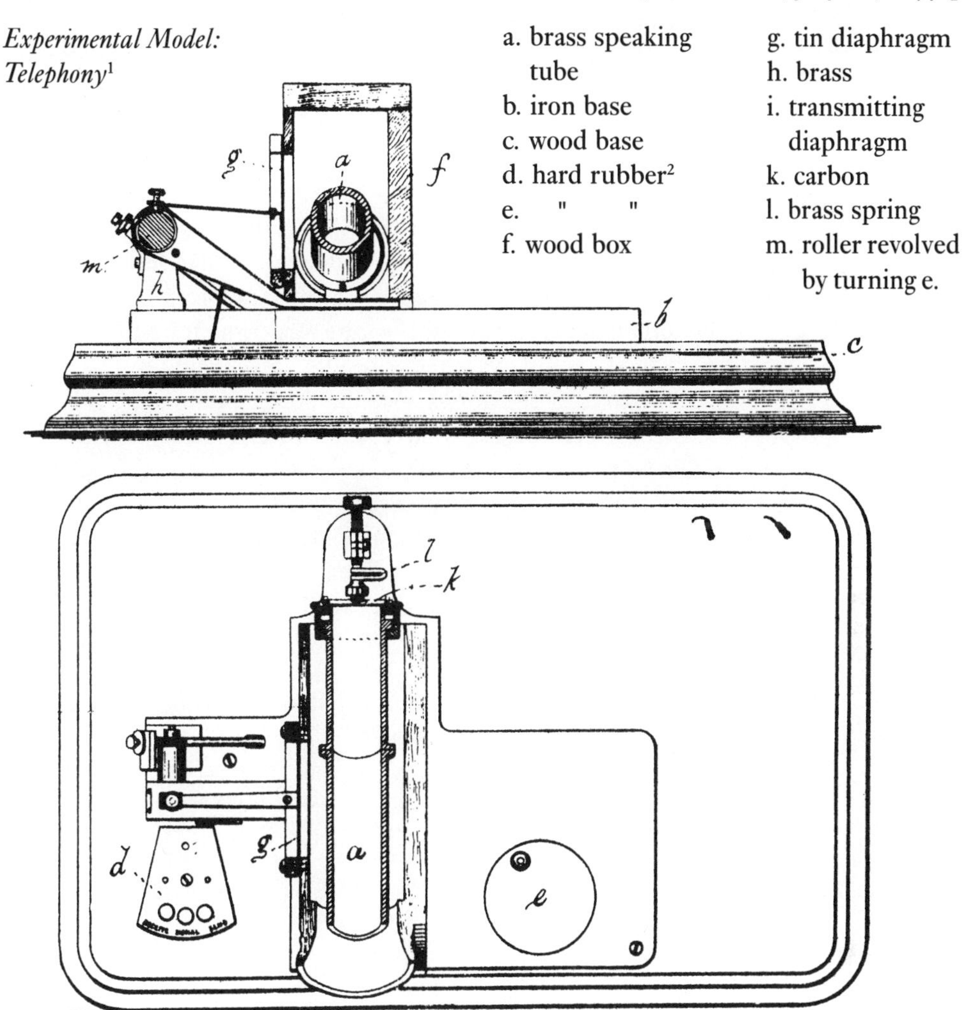


M (historic drawing) (56 cm × 36 cm × 23 cm), NjWOE, TI 2:525 (*TAEM* 11:653). These drawings were made from an instrument entered as an exhibit in the patent interference.

1. See headnote above.
2. The labels on **d** are "RECEIVE," "SIGNAL," and "SEND."

–963–

[Pittsburgh, July 11, 1877][1]

*From Thomas David*

Friend Edison—

I write on this side[2] in order to answer your questions <u>in order</u>— You misunderstood me; the party reading corrected himself— I spoke of it ~~of~~ as showing how readily all articulations were communicated— I was not following copy—

I am experimenting every day with Bell's. The greatest difficulty is the interference from other lines, and from earth currents—[3] If we attempt to use it on poles where there are a great many wires, we[a] have a[a] sound like frying fat which drowns the articulations— I yesterday tried the experiment on a single, independent wire, ten miles long— The articulations were perfect, but we had some trouble from the frying tho' mark you, the line was on seperate poles and away from other lines— we tried the same experiment on the same line the day before with better results, there[a] being less frying— I propose to try now a metalic circuit in amongst our other wires and see, to what extent we can overcome the ~~noise~~ frying.

Bell's will answer admirably out of cities, at least, on a line of a single wire— The articulations are so true that you readily make out who is talking to you— I want to go on and test yours as soon as you are ready. Yours truly

T. B. A. David

ALS, NjWOE, Supp. II, 1877 (*TAEM* 97:588). Written on back of Doc. 958. [a]Obscured overwritten letters.

1. In a letter of 10 June to Gardiner Hubbard, David described the telephone tests discussed in this letter as being done "yesterday" and "today." Box 1175 (Corp.), NjWAT.

2. David wrote this letter on the back of Doc. 958.

3. That is, currents caused by differences in the electrical potential of the earth along the length of the wire.

**–964–**

*Technical Note: Telephony*

[Menlo Park,] July 11 1877

Speaking Telegraph[1]

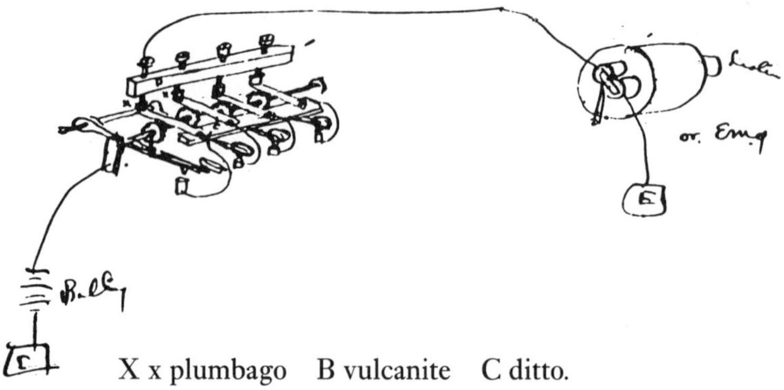

Line

or. Em'g

E

B. Ely

E

X x plumbago B vulcanite C ditto.

30 wheels each having sufficient no of teeth & of dif heighths to give proper vibration of spring & contact against plumbago & proper pressure by teeth being low & high to send letter over wire & have it reproduced= each key sends letter & words may be sent by proper depression of several letters[2]

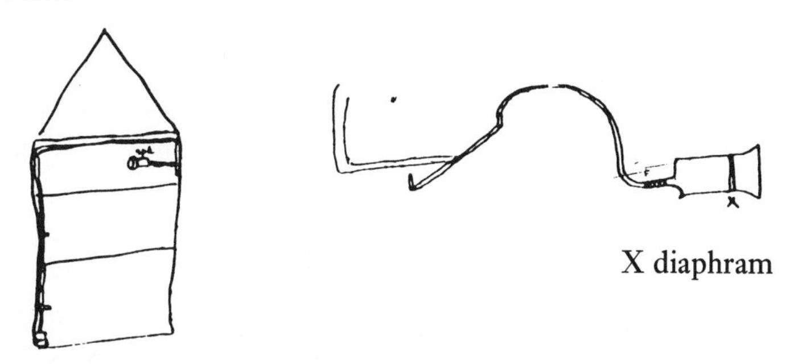

X diaphram

I propose to utilize the gas pipe system in factories houses etc to convey the human voice & allow of conversation being carried on ~~at~~ from any one point of the gas system to the other & even to great distances in different parts of a city especially applicable to fire alarm Telegh & I accomplish it by having an extra orifice fitted to the gas pipe similiar to that made by disconnecting a burner from the pipe— on this I either attach an extra tube ~~or f~~ of metal or a flexible tube the other end of which is attached to the metal tit of a tube the other end of which has a diaphram of such material as will prevent the gas escaping  thus speaking in one of these sets the whole of the gas in the tubes vibrating & if another person places one of these to his ear while another person is talking in another part of the building he will plainly hear what is said=[3]

Charley is now trying the experiment.[4]

T A Edison

Chas Batchelor
James Adams

X (photographic transcript) and X, NjWOE, TI 2, Edison's Exhibit 7-12; Lab., Vol. 12:8 (*TAEM* 11:356; 4:5). Document multiply signed and dated.

1. Figure labels are "Battery," "Listen," and "or Emg"; the two horizontal bars at left are labeled "B" and "C"; two points under **B** are each "x."

2. Figure labels are "spk" and "x."

3. In a 16 July note headed "Gas-Pipe Telephone" Edison described a telephone call bell that worked like this system (Edison's Exhibit 13-

12, TI 2 [*TAEM* 11:361]). There is also a description of this system in Prescott 1879 (548–49).

4. On 31 July, Charles Batchelor noted:

> We find that by putting a diaphragm on the gaspipe in a shop conversation can be carried on from one floor to another without in any way injuring or affecting the working of the gas. It will not go through the meter. [Cat. 1317:42, Batchelor (*TAEM* 90:678)]

**–965–**

*From John Breckon*

Sunderland 12th July 1877[a]

Dear Sir,

I have communicated with the Directors of the Electric Writing Co Ld & they will order one of the Electric pens for shading Scrolling & ruling music in their next requisition.[1]

I hope shortly to be able to say that the Directors of that Company will recognize the Western Electric Manufacturing Company in respect of the Contract entered into between you & me for the supply of Electric Pens.

I have been hoping to hear from you to say when the Press which is to print copies so rapidly may be expected in England.[2]

You also spoke about an improvement in the Electric Pen whereby liability to get out of order would be reduced— will you kindly say how soon the specimen pen will be delivered in London.

I am in correspondence with Mr. Ireland respecting your draft for £8.3.0 & hope in a day or two to have his reply to say that it is all in order—it shall then be accepted & forwarded to you without delay. Yours truly

J. R Breckon

ALS, NjWOE, DF (*TAEM* 14:410). [a]"Sunderland," and "18" preprinted.

1. See Doc. 934.

2. That is, the rotary press. The next week Breckon wrote again to ask for one, saying he believed "that sales of the Pen [were] being prevented in many quarters from want of a more rapid mode of printing." He was still waiting at the end of September. Breckon to TAE, 20 July and 27 Sept. 1877, DF (*TAEM* 14:411, 417).

**–966–**

*Technical Note: Telephony*

[Menlo Park,] July 16 1877

Speaking Telegraph[1]

Device for Transmitting the hissing consonants i.e. [a] the hissing parts of Speech[2]

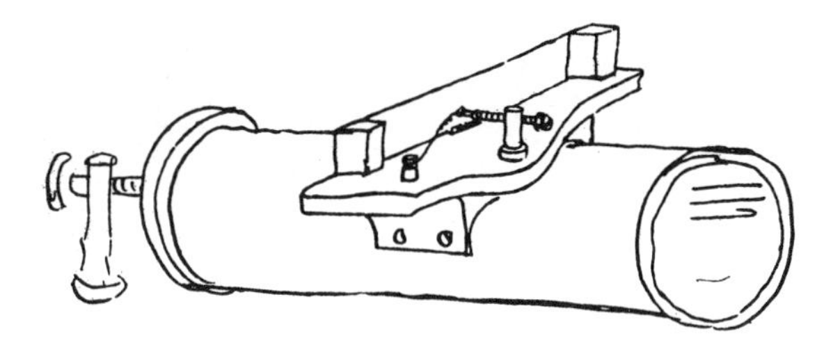

The leaf or reed fastened at both ends works when it is made little[b] stiffer & has one end[3] free over the slot=

It may be necessary to dispense with ~~a~~the rigid platina point & substitute a spring platina point so[b] as not to check vibrations of reed[4]

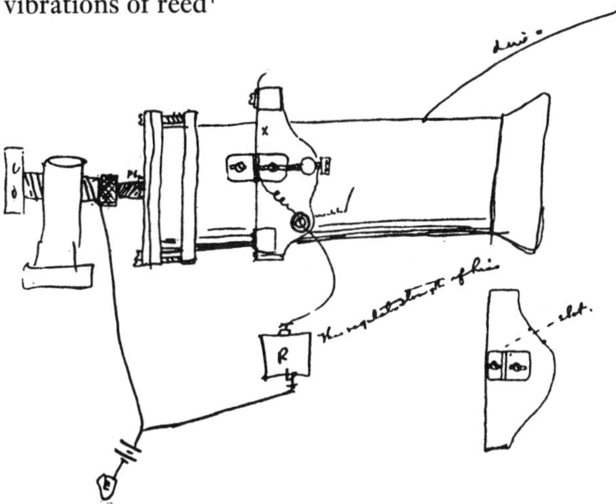

The front diaphram of mica gives all the musical sounds of talking & the leaf & slot gives the hissing Air Sounds. it works very fairly in first trial. Had good deal bother making it=

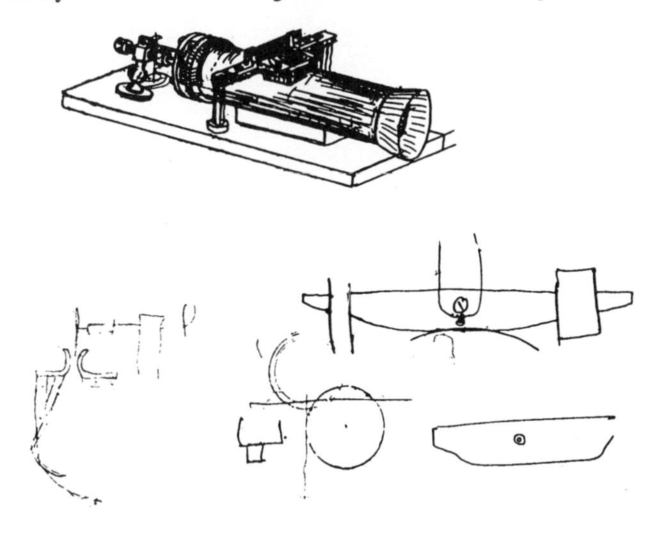

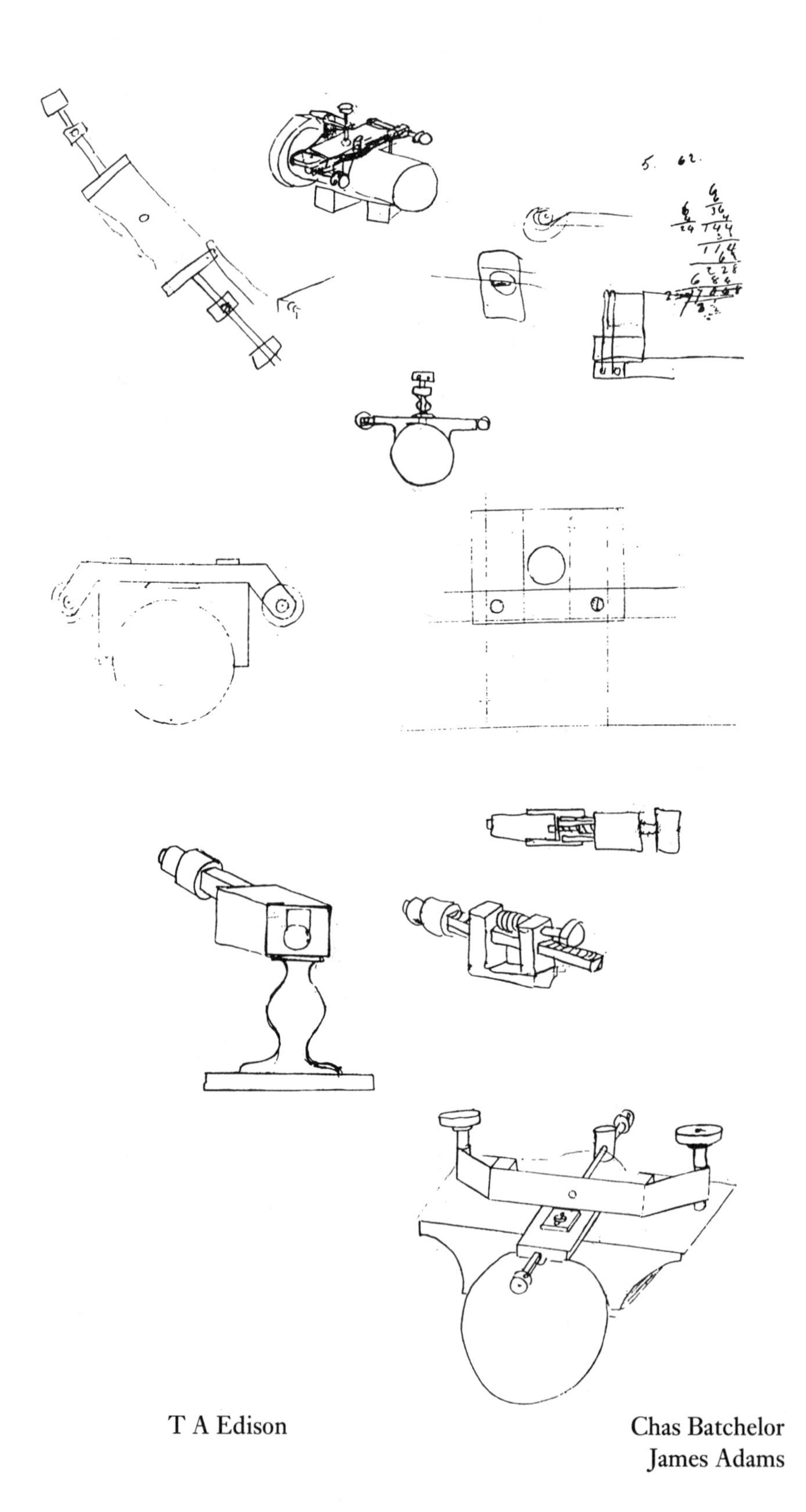

T A Edison

Chas Batchelor
James Adams

X (photographic transcript), NjWOE, TI 2, Edison's Exhibits 10-12-12-12, 14-12 (*TAEM* 11:358–60, 362). Document multiply signed and dated. [a]Circled. [b]Obscured overwritten letters.

1. On this day, Elisha Andrews wrote to Gardiner Hubbard:

> I went to see Mr Edison who give me a full exhibition of his speaking telephone. It has the following characteristics.
>
> 1st It gives forth any volume of sound, fills a Room.
>
> 2d The sounds are simply a combination of longer or shorter vibrations capable of any pitch from a whisper to shout, almost deafening.
>
> 3d Some sounds & words are not audible as sh.—ch.—ist th. Scythe theist scism & generally the hissing sounds, the word telegraph is plainly spoken. The words shoot out rapidly & with great force, but have a general lack of articulation. If Edison gets the articulation more perfect, which he is now working at, he can talk in thunder tones any distance.
>
> 4th He requires 6 to 10 cups of Battery & direct current. The mouth piece solid cast brass tube 1¾ to 2 in. Diam 10 to 12 inches long. Diaphragm is of platinum. The resistance 800 to 1000 Ohms or more, (sometimes) is obtained at the speaking end by a hard pressed Carbon (plumbago) terminal, solidly fixed behind the diaphragm, at the receiving end he now uses what he terms the Electro Motor, which is simply an arrangement to pass the current thus intensified through moistened paper kept slowly in motion. Sounds can be obtained with Diaphragm & Electro-Mag instead of the paper. If you would like any further information I can get it for you before I start for Buffalo— . . .
>
> The Western Union own all of Mr Edisons Telephone patents. He is virtually at work for them, & is now making 50 sets, for their private leased lines, in NY city & around. Altogether there are only about 650 private or leased lines, of the W.U controll, in the country & cities They will put their telephones on at once, in its present state. They work any distance, unaffected by other wires or cables & override any sounds around. They have 50 set about ready to put on. [Box 1067 (Corp.), NjWAT]

2. The first drawing of this device is from 14 July; three other pages of drawings dated 16 July show alternative designs for adjusting the reed. Edison's Exhibits 9-12, 11-12–12-12, 14-12, TI 2 (*TAEM* 11:357, 359–60, 362).

3. That is, the edge of the reed.

4. Figure labels are "Pbgo," "X," "insulated," "line=," "R," "This regulates strength of hiss," and "slot."

## –967–

*From Edward Johnson*

Phila. July 17th/77.

My Dr. Edison

I send you todays papers showing effect of rehearsal given last night.[1] it came very near being a failure owing primarily to my having but one transmitter but directly to the interference of a Musical Director who being at the Transmitting Station insisted upon trying a clarionette during the performance— Now as the air vibrations of that Inst. go in great part out of the vents on top—instead of out of the funnel the natural result was that we only received one half the notes—and further disaster was wrought by getting the Inst. all out of tune for the voices—it requiring some little time to get it back again— Nevertheless the day was saved by promptly remedying the trouble—& better times were promised at the 1st public concert which comes off tomorrow night—& for which I am praying you will furnish me with a 2nd transmitter I am risking everything on this one— I confidently expected you would send it on the 12th as you said in your Letter[2] & today could only send you a despatch to "Get it to me" somehow sure tomorrow am

Barker is quite enthuzed & talks right out—making however one of Bells absurd predictions for which I was sorry & would have[a] prevented If I could.—viz—about bringing European opera here by wire—[3] He will again lecture tonight before a large audience & I presume will be reported He will be accompanied by Draper[4]—& Wallace[5]—of Ansonia—The trio leave here Saturday for Menlo Park where they hope to see "much that will be of Interest to the Student of Science"—[6]

You will find the best a/c in the Ledger—Child's paper[7]—they sent for me to give them an accurate description, which they publish—making me say "solder" where I said "Rivet" but with that exception publishing me correctly—so you may criticize it

The Graphic[8] also sent for me ~~yesterday~~ today—& arranged with me to sketch the transmitting end this evening & the other end tomorrow evening when the audience is assembled—[9] This will be quite an Interesting Puff. Taylor[10] the art correspondent here says under such circumstances the sketch is a legitimate "news" production—but as it is also quite an[a] effective ad, it is customary for parties interested to take some papers I have agreed to take 100. The Exhib. folks I presume will take some—& Taylor wants you to take some, through me—as otherwise they dont go to his credit— terms $5. per 100. Will you take a few?—in order to make up the 300 that he says pays the cost of the Drawing—

The Drawing will be ready Thursday & I should know Thursday Early if you want any & how many as the No engaged I presume has to accompany the sketch to NYork to ensure its publication—

Taylor is very anxious to come to Menlo Park & sketch your

Laboratory as a business ad. Says owing to the agitation of your works & name at the moment the Co would make very liberal terms—& says that (in view of the fact that I wrote the "Press" article & the Ledger description) that the columns of the paper would be open for me to write anything about your establishment I chose —I told him I should set the matter before you & have you think on[a] & consider it—Please do [----][b]

Now as to the latest Idea of mechanicaly speaking the Letters of the alphabet.[11] Prof B. is delighted & says it looks as if you might reach by a short cut the end sought by scientists for ages, viz—the ascertainment 1st of what constitutes a vocal sound of a letter & 2—How to mechanically reproduce it— It was that Letter[12] which decided him to take his guests to Menlo Park—

He alluded in his speech to your Plumbago discovery & characterized it as the most Important & valuable discovery of your many extraordinary ones & otherwise eulogized it & you[13]— You will observe that he unequivocally pronounces your Telephone the best of all—as do the newspapers

The speakers must follow in the wake of all this Sure[c]

E H Johnson

ALS, NjWOE, DF (*TAEM* 14:819). [a]Obscured overwritten letters. [b]Canceled. [c]Multiply underlined.

1. The *Philadelphia Inquirer* for Tuesday, 17 July 1877, ran an article describing the rehearsal at the Permanent Exhibition. The transmitting apparatus was located at Fifth and Chestnut streets (see Doc. 950 n. 1).

2. Edison's letter in reply to Johnson's 10 July request for a second transmitter (DF [*TAEM* 14:809]) has not been found.

3. George Barker "explained the manner in which the telephone worked. He said that he considered the Edison telephone the most valuable of all the different inventions, and believed that the time would come when the Berlin opera would be listened to in Philadelphia." *Philadelphia Inquirer,* 17 July 1877, 3.

4. Henry Draper (1837–1882), professor of chemistry and natural science at the University of the City of New York (now New York University), was a pioneer in the application of photography to astronomy, especially for spectral analysis. *DSB,* s.v. "Draper, Henry."

5. William Wallace (1825–1904), British-born wire manufacturer and inventor, established Wallace & Sons at Ansonia, Conn., with his father in 1848. With electrical inventor Moses Farmer he constructed dynamos based on Farmer's patent design, which were used in combination with Wallace's own patent carbon-arc lamp. One of the Wallace-Farmer dynamos had been used to light the Centennial Exhibition. *DAB,* s.v. "Wallace, William."

6. That is, on 21 July, after Draper presented a paper on 20 July to the American Academy of Arts and Sciences in Philadelphia. Earlier in the month Barker had planned to meet Wallace and Draper in New

York and to stop at Menlo Park on the way to Philadelphia (Barker to Draper, 2 July 1877, HD). During their visit Edison showed them his speaking telegraph (which "worked beautifully") and conducted some "Experiments to obtain a recording solution for jump spark for use in determining rate of tuning forks etc" (Cat. 1233:202, Batchelor [*TAEM* 90:154]; Vol. 12:43, Lab. [*TAEM* 4:24]). Correspondence in the Barker and Draper collections indicates that Barker, Draper, Wallace, and Stevens Institute President Henry Morton were friends who had met over the years for scientific experiments and tests as well as social occasions (e.g., Barker to Draper, 10 Dec. 1875, 21 May 1876, 15 Aug. 1877; Henry Morton to Draper, 30 July 1877; all HD).

7. George Childs was publisher of the *Philadelphia Public Ledger. DAB,* s.v. "Childs, George W."

8. The *New York Daily Graphic.*

9. Neither this sketch nor any story about these exhibitions has been found in the *Graphic.*

10. Unidentified.

11. See Doc. 964.

12. Not found.

13. Edison believed that he was the first to discover the pressure dependency of carbon's resistance, which would be the basis for his later, controversial claim to have invented the microphone. However, various European engineers and scientists had noticed the relationship earlier (see, e.g., Du Moncel 1974, 144).

**–968–**

*Technical Note: Telephony and Ink*

[Menlo Park,] July 17, 1877

Speaking Telegraph[1]

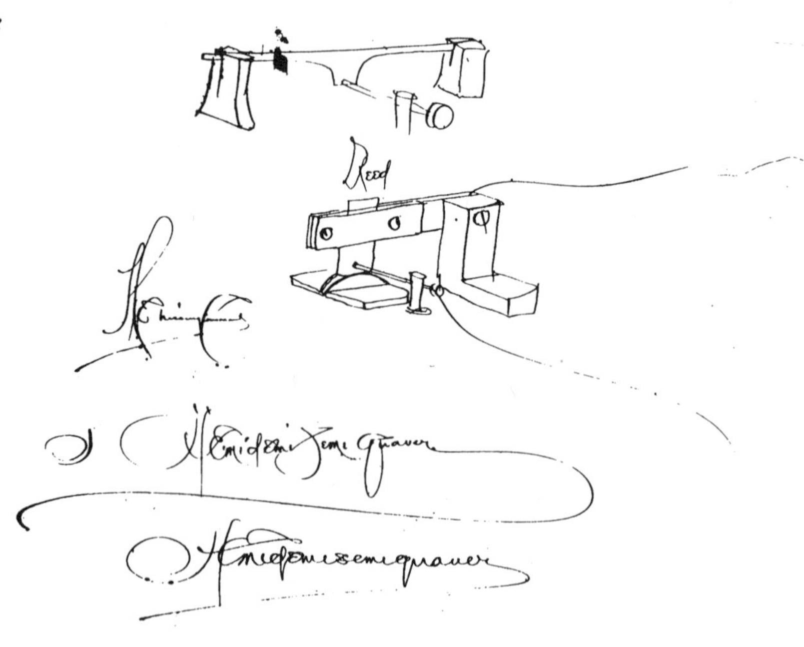

Glorious= Telephone perfected this morning 5 am= articulation perfect got ¼ column newspaper every word. had ricketty transmitter at that we are making it solid[2]

X Mica or Rubber Reed for the Hissing Consonants—[a]

Idea— Chalk Balls are soaked in a concentrated Alcoholic Solution of Aniline Violet or other A. color and dried & coated with Dextrine or other substance to keep them from staining hands these are enclosed in Pill boxes for transport by Mail for use in mkg ink place one of the Balls in ink[b] Bottle containing water shake & its ready for use= Perhaps Tragacanth or plaster paris would be better than chalk=[c]

Chas Batchelor James Adams

X and X (photographic transcript), NjWOE, Lab., Vol. 12:20–21; TI 2, Edison's Exhibit 22-12 (*TAEM* 4:11–12, 11:365). Document multiply signed and dated. [a]Followed by horizontal line across page. [b]Interlined above. [c]Paragraph canceled with large X.

1. Figure labels are "Reed," and "The hissing Consonants." The writing following the figure is "Hemidemisemiquaver" twice.

2. Figure labels below are "insulated" and "Hissing Consonants." The major part of the calligraphy below says, "The vibration of the oscillation," "Physcists and Sphynxes in majestical Mists The majestical myth which Physcists seek," "Protochloride," "The vibration," and "Protochloride of Lead."

## –969–

*Technical Note: Telephony*

[Menlo Park,] July 17 1877

Spkg Telephone

reproduced slow or fast by a copyist & written down[1] This can be applied telegraphically thus[2]

Telephone transmitter same as my Talker Transmitter

E

Sheet after received is sent to Copyist whole pass it in machine similar to that shewn on other page[3] & copied[a] at rate of 25 words per minute whereas it was sent at rate of 100 per minute thus Saving all skilled[a] oprs & 5 persons doing work of .8. Emg might be used instead of magnet to receive it might be done in other ways besides indenting—such as perforating with needle or by a friction ink=

Revolving plate two telephone tubes=
T A Edison                                   Chas Batchelor
James Adams

X (fragment?), NjWOE, Lab., Vol. 12:16 (*TAEM* 4:8). [a]Obscured overwritten letters.

1. Here Edison proposes a solution to the problem of recording telephone messages, which many in the telegraph industry thought essential for practical use of the telephone. Believing that the telephone would continue to be used by the public in much the same manner as the telegraph, with operators sending and recording messages that would then have to be delivered to the customer, they argued that

> inasmuch as the public must delegate to corporate bodies the carrying of their telegrams, and neither telegraph company or the public will receive one from the other *verbal* messages; they must be *written,* and this necessity neutralizes any advantages that might otherwise accrue to the telegraph company by reason of the increased capacity given a wire by the application of the telephone. [Johnson 1877, 10]

2. Figure label is "Telephone transmitter same as my Talker Transmitter."
3. This referent is unidentified.

**-970-**

*From George Field*

St Louis Mo July 18th 77

Dr Sir

Your favor of the 9th instant[1] has just reached me— by its date I observe that it was written the very day I left New York. I regret not having received it a few days earlier— I desired to see you for the purpose of calling your attention to the subject of Electric light—or the production of light by electricity  It appears that the invention of Paul Jablochkoff of Paris[2] demonstrates the possibility of utilizing this element for purposes of illumination—and I feel quite confident that if you will apply yourself to it that important results might follow. I shall be absent until about the 1st of Sept— On my return will call on you and talk this matter over— Yours Very Truly

Geo B Field[3]

ALS, NjWOE, DF (*TAEM* 14:95).

1. Not found.

2. Russian military engineer Paul Jablochkoff devised a carbon arc lighting system that had both a less intense light than other contemporary arc lamps and a regulating mechanism that allowed several to be placed in a single circuit. This opened the possibility of large-scale electric illumination for outdoors and large indoor spaces and provided a new impetus to the field of electric lighting (King 1962c, 393–96). Charles Batchelor's scrapbook contains a 30 June 1877 clipping from the British journal *Iron* describing and illustrating the Jablochkoff candle (Cat. 1240, item 207, Batchelor [*TAEM* 94:63]).

3. George Field was former president of the Gold and Stock Telegraph Co. See *TAEB* 1:153 n. 1.

## -971-

*Notebook Entry: Copying*

[Menlo Park,] July 18 1877.

Reduplication of copies

After vainly searching for a long time to find a process that shall be easier to work than Zuccato's papyrograph we have at last found a process.[1] Take an ordinary sheet of writing paper and cover it with collodion, now take a stylus and write your letter with it, this either breaks the surface of the collodion or rubs off a portion (we think the former) this sheet can now be used as a stencil and copies can be taken from it immediately.[2]

We find that Balsam Peru and Asphaltum Varnish answers better to prepare the paper with.

X, NjWOE, Batchelor, Cat. 1317:40 (*TAEM* 90:677). Written by Charles Batchelor.

1. What appear to be notes related to this search for an alternative to Zucatto's process are in Vol. 8:209–25, Lab. (*TAEM* 3:643–51).

2. On 29 June, Edison had described this process in another notebook. At that time he also noted that gums in alcohol or other substances that made the paper impervious to ink could be used in place of collodion. NS-77-002, Lab. (*TAEM* 7:444).

-972-

*Technical Note: Telephony and Phonograph*

[Menlo Park,] July 18 1877.

Speaking Telegraph

Improvement on Bells Magneto Telephone;[1] which improvement consists in reinforcing the hissing Consonants such[a] as S, T, V, P, C J etc. Thus

B enclosed at both ends.

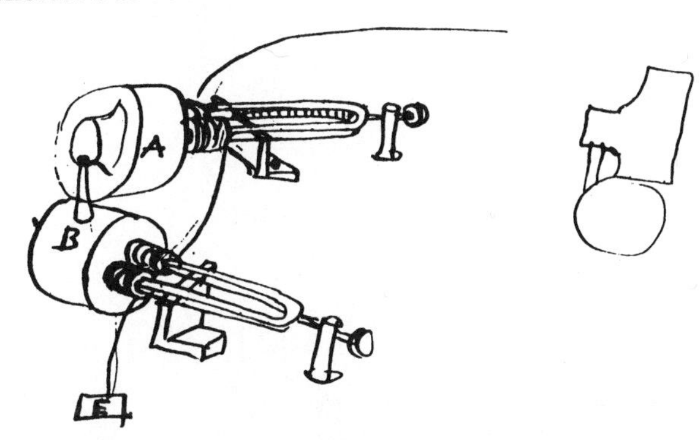


Conveying air to the second Telephone through tube ~~inn~~ in manner shewn sets diaphragm of B in powerful vibration at each wind sound or hiss  the vowel & musical sounds be obtained from .A.

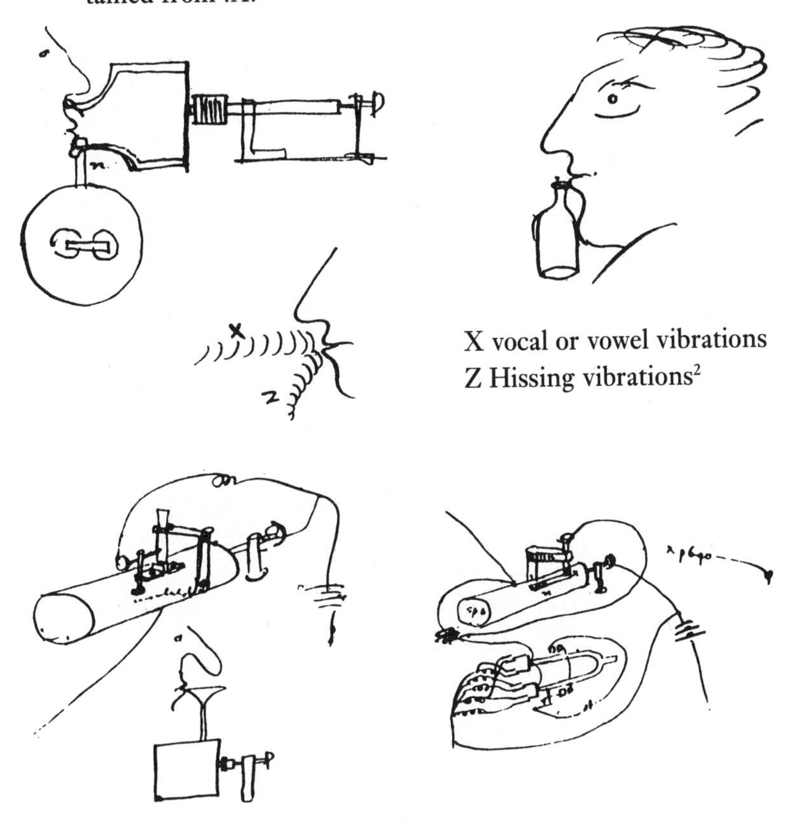

X vocal or vowel vibrations
Z Hissing vibrations[2]

A hiss produces a wind pressure in tube & this raises n with more or less pressure against the plumbago point X  this throws in a hiss sound produced by a tuning fork arranged to give a number of Contacts 1 after the other. [b]

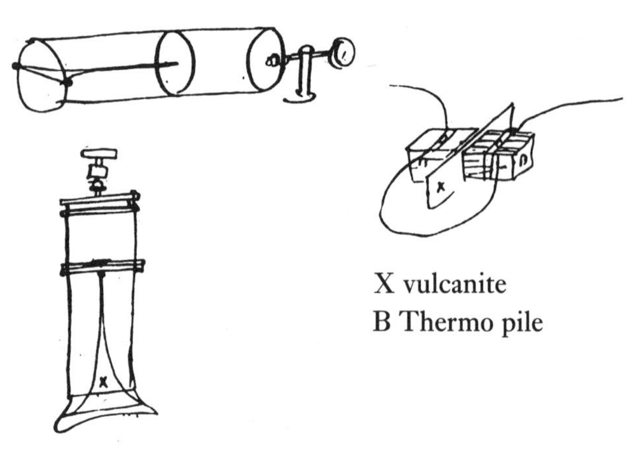

X vulcanite
B Thermo pile

X is a rubber membrane connected to the central diaphram and the edge being near or between the lips in the act of spkg it gets a vibration which is communicated to the central diapham & this in its turn sets the outer diaphragm vibrating hence the hissing consonants are reinforced & made to set the diaphram in motion  we just tried an experiment similar to this thus[3]

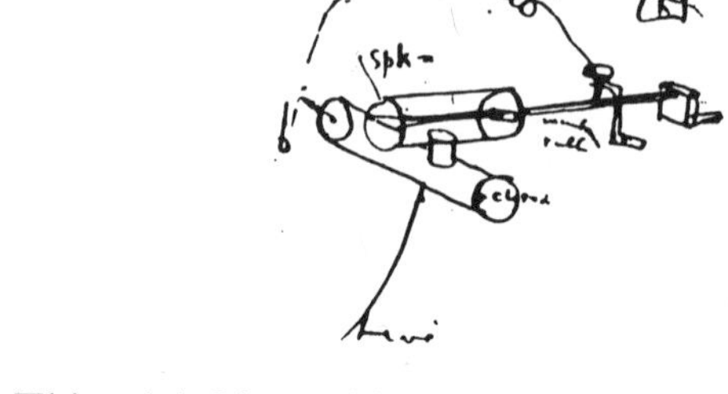


This wkd fair on hissing but it spoilt the regular vowell sounds on lower tube=

Just tried experiment with a diaphram having an embossing point & held against parafin paper moving rapidly  the ~~new~~ spkg vibrations are indented nicely & theres no doubt that I shall be able to store up & reproduce automatically at any future time the human voice perfectly[4]

T A Edison

Chas Batchelor
James Adams

X and X (photographic transcript), NjWOE, Lab., Vol. 12:26; TI 2, Edison's Exhibits 24-12–25-12 (*TAEM* 4:15; 11:366–67). Document multiply signed and dated. [a]Obscured overwritten letters. [b]Followed by row of closely spaced dots.

1. Bell's 1877 telephone transmitter did not use batteries to generate a line current. Instead, sound vibrations moved the metallic diaphragm in front of a permanent magnet, inducing a current in the line wire, which was wrapped around the magnet. Here Edison seems to be using long magnets (possibly permanently magnetized tuning forks) in an attempt to increase the strength of the signal, a serious problem with the Bell magneto telephone.

2. Figure labels are "insulated"; and "spk," "n," "X," and "X pbgo."

3. Figure labels are "spk=," "membrane rubber," "closed," and "line=." Two days later the staff further developed this idea of two chambers connected by a narrow tube to enhance sibilance. Edison's Exhibits 28-12, 32-12, TI 2 (*TAEM* 11:369, 371); Cat. 1233:201, Batchelor (*TAEM* 90:153).

4. Charles Batchelor later testified about these first sound-recording experiments (cf. App. 2).

> The first experiment, as I remember it, was made in this way: Mr. Edison had a telephone diaphragm mounted in a mouth-piece of rubber in his hand, and he was sounding notes in front of it and feeling the vibration of the center of the diaphragm with his finger. After amusing himself with this for some time, he turned round to me and he said: "Batch, if we had a point on this we could make a record on some material which we could afterwards pull under the point, and it would give us the speech back." I said, "Well, we can try it in a very few minutes," and I had a point put on the diaphragm in the center. This I had mounted on a grooved piece of wood that had been used for an old automatic telegraph. With this machine we got some of the old automatic telegraph paper, coated it over with wax, and I pulled it through the groove, while Mr. Edison talked to it. On pulling the paper through a second time, we both of us recognized that we had recorded the speech. We made quite a number of modifications of this the same night, and Mr. Edison immediately designed a machine which should be better adapted for giving us better talking.
>
> . . . The diaphragm holder was screwed down to the wood, but the screw at one side could be raised or lowered a little in order to bring the knife down for adjustment. . . .
>
> The shape of the channel at the bottom was perfectly flat. The adjustment, I think, was about a sixteenth of an inch, or thereabouts, and this was obviously necessary, as we could not tell exactly how thick our wax would be coated on the paper.
>
> The operation was: We put in the waxed paper, then adjusted the diaphragm so that the knife cut slightly into the wax, and then pulled it through, talking at the same time. Sometimes we adjusted the knife till it only just touched the wax. We adjusted this knife very many times in the first experiment; tried it in many different ways. I remember, when the wax was very thick, or when we had other devices, that I have lifted up the diaphragm as much as a thirty-second of an inch by putting washers under the screw-head, thus making an adjustment for a much thicker substance. [Pp.

586–87, *American Graphophone v. U.S. Phonograph* (*TAEM* 116:367)]

Edison's first public account of these experiments was in "A Marvelous Discovery," published in the *New York Sun* of 22 February 1878 (Cat. 1240, item 378, Batchelor [*TAEM* 94:115–16]). William Applebaugh, a telegraph and telephone company executive and friend of Edison's, related a story similar to Batchelor's in February 1878 ("Phonograph: A Machine that Talks and Sings," *Brooklyn Daily Eagle*, 26 Feb. 1878, Cat. 1240, item 385 [*TAEM* 94:117]). Other stories of the phonograph's genesis, more or less attributable to Edison, differ in many ways and often condense the time involved (e.g., App. 1.G24–26, Shaw 1878, and Lathrop 1890).

**–973–**

*British Provisional Patent Specification: Telephony and Phonograph*

Menlo Park, New Jersey, July [19,][1] 1877[a]

"IMPROVEMENT IN INSTRUMENTS FOR CONTROLING BY SOUND THE TRANSMISSION OF ELECTRIC CURRENTS, AND THE REPRODUCTION OF CORRESPONDING SOUNDS AT A DISTANCE."[2]

The vibrations of the atmosphere, which result from the human voice or from any musical instrument or otherwise, are made to act in increasing or lessening the electric force upon a line by opening or closing the circuit,[3] or increasing or lessening the intimacy of contact between conducting surfaces placed in the circuit at the receiving station; the electric action in one or more electro magnets causes a vibration in a tympan, or other instrument similar to a drum, and produces a sound, but this sound is greatly augmented by mechanical action. I have discovered that the friction of a point or surface that is in contact with a properly prepared and slowly moving surface, is very much increased or lessened by the strength of the electric wave passing at such point of contact, and from this variation in the friction a greater or less vibration is given to the mechanism or means that produce or develope the sound at the receiving station, thereby rendering clear and distinct the sound received that otherwise would not be audible.[4]

To carry out the peculiarities of my Invention under the varying conditions of use, I have devised several modifications of the transmitting, receiving, and intensifying devises employed in this sound telegraph; portions of the apparatus are interchangeably, available in transmitting or recording; others are adapted to local use; some are only available in transmitting, and others are only for receiving; and some portions of my improvement can be availed of to make a record of the atmospheric sound waves, or of the electric waves, or pulsations corresponding thereto or resulting therefrom.[5]

In one form of my apparatus the sound passes into a resonant box having one, two, or more tympans at its sides that are vibrated thereby, the tension of these being various, so as to respond to different sound waves, and the electric connections pass through all to one line or circuit, in which is a battery and the distant receiving instrument.[6] Circuit contact points are provided at one or both surfaces of the tympan or tympans;[7] the tympans are of parchment, foil, mica, sheet metal, or similar material.[8] I find platinum, foil, or mica to respond advantageously when the waves from the mouth are made to pass through a slot resembling the larynx placed within the resonant tube.[9] The contact points or surfaces are sometimes metallic, but plumbago, or similar semiconducting material, into which the tympan or diaphragm is brought more or less intimately into contact, or a point or pin thereon serves to lessen or increase the electricity passing at that point.[10]

In the receiving portion of the instrument the tympan is acted upon directly by an electro magnet, or through an armature,[11] or the tympan is provided with an arm extending out over a slowly moving surface or cylinder, and the electric current, passing at the point of contact, increases or lessens the friction, and produces the vibration of the arm and tympan in proportion to the difference of friction developed between the arm and moving surface by the passage and cessation of the current through the chemically prepared paper, preferably moistened with a salt of mercury and an alkali.

This feature is capable of very extended developement in telegraphy; the clearness and extent of sound produced by the receiving tympan exceeds anything heretofore attained in acoustic telegraphs.

The frictional surface may be any material which absorbs liquids, such as chalk or paper;[12] the transmitting points in all instances should be a poor conductor of electricity, such as cyanide of copper, peroxide of lead, or plumbago, which substance may be moulded with non-conductors.[13]

Where plumbago is used between the cores and armature of an electro magnet, and such plumbago is included in a local or relay circuit, the pulsations or rise and fall of electric tension will correspond to that in the coils of the electro magnet in consequence of the varying compressing action on the plumbago or similar material, which increases and decreases its resistance.[14]

A polarized arm, near a soft iron plate upon the tympan, will receive a motion from the vibration of the tympan, and make more or less intimate contact with the plumbago point without arresting the motion of such tympan.[15]

In the line the effect of the static charges and discharges are neutralized by condensers or by helices in shunts.[16] Where a rapidly revolving wheel is contiguous to the transmitting tympan, and there is a plate of platina on that tympan to which the circuit is connected, the rise and fall of electric tension will result accurately from the movement of the tympan, producing greater or less contact with the wheel; a spring may intervene between the wheel and tympan.[17]

A spring or reed, attached to a stud at one end, and to the tympan at the other, is vibrated or undulated by the sound waves acting on the tympan, and produces greater or less contact with pieces of tin foil surrounding it, which also are in the electric circuit or plumbago; contact points near the side of such spring effect the same object.[18] A column formed of disks of tin foil in a glass tube, pressed upon by the vibrations of the tympan, also are employed under some circumstances to regulate the electric tension in the circuit.[19]

The moving surface, whose friction is increased or decreased upon the passage and cessation of the current aforesaid, is employed in some instances to vibrate by the electric and mechanical action a frame of strings of different pitch to produce musical sounds, or to give motion to reeds or resonant boxes.[20]

The transmitting tympan is sometimes made of cloth, with plumbago pressed into its surfaces, and generally it is preferable to weight the centre of the tympan, and take the motion a little one side of the centre to prevent a flase[21] movement or rebound.[22] These sound transmitting and receiving instruments are applied in single or multipley telegraphs; the connections through the Wheatstone bridge,[23] or the induction coils, being used in the well known ways,[24] and the circuits being either single or relays.[25]

Where alluminium is introduced in the electric circuit in connection with a material containing an electrolite, the vibration of the tympan, and an extension from it to the material containing the electrolyte, causes a depolarization at the point of contact at every vibration, thus increasing and decreasing the strength of the current by depolarization.[26]

Where a slot is made in the diaphragm, and the airways impinge against the same, the vibration of the hissing sounds is augmented.[27]

The resonant chamber for convenience is made portable, so as to be presented to the mouth.[28]

In addition to the tympan for transmitting the vowel sounds, I employ another opening in the speaking tube or chamber, which opening may be increased or lessened; I stretch edgewise over this a thin reed of foil, which is set in vibration by all hissing consonants, and make contact with a plumbago point, or with a platina point, which is connected to the line through an adjustable resistance coil, or a self-acting reed giving a hiss may be thrown in and out of circuit by the movement of the foil reed.[29]

For transmitting and receiving letter by letter, I employ a shaft with 30 wheels, and contact springs resting upon them; the wheels are provided with teeth of such number and character that they will cause the springs to be vibrated against a plumbago point the necessary number of times, and with proper pressure to transmit the letter, which is rendered audible at the distant station by the magnet or frictional surface. Each wheel is controlled by a key of a key board.[30]

THOMAS ALVA EDISON

PD, NjWOE, Batchelor, Cat. 1321 (*TAEM* 92:56). [a]Place, month and year from printed docket.

1. This document was filed in London on 30 July. Unless it was cabled—an extremely expensive undertaking for a document of this length—it had to be mailed from New York at least ten days earlier (see Doc. 977 n. 4), and its drafting and copying had to take some time; its third paragraph could not have been completed until after Edison's realization on 18 July of the possibility of directly recording sound (Doc. 972) as well as telephone messages (Doc. 969). It was probably largely drafted earlier and then modified to include recent developments just before sending. Its content overlaps Edison's Cases 135–36 and 141, all of which were filed on 20 July, having been executed earlier in the month.

2. Edison filed a full specification on 30 January 1878 (signed on 24 December), making this British Patent 2,909 (1877). Edison executed and filed nine U.S. patent applications for telephones during 1877, five of which led to patents in the following year and four of which (along with one on acoustic telegraphy) were caught up in the Telephone Interferences. Much of the content of the earlier applications is briefly summarized in this document, and much of the later ones appears in the full specification in December. The only one of these that never issued as a patent was Case 145, for which see TI 2, following p. 24 (*TAEM* 11:198).

*Edison's Case 145, his only 1877 telephone patent application that failed completely.*

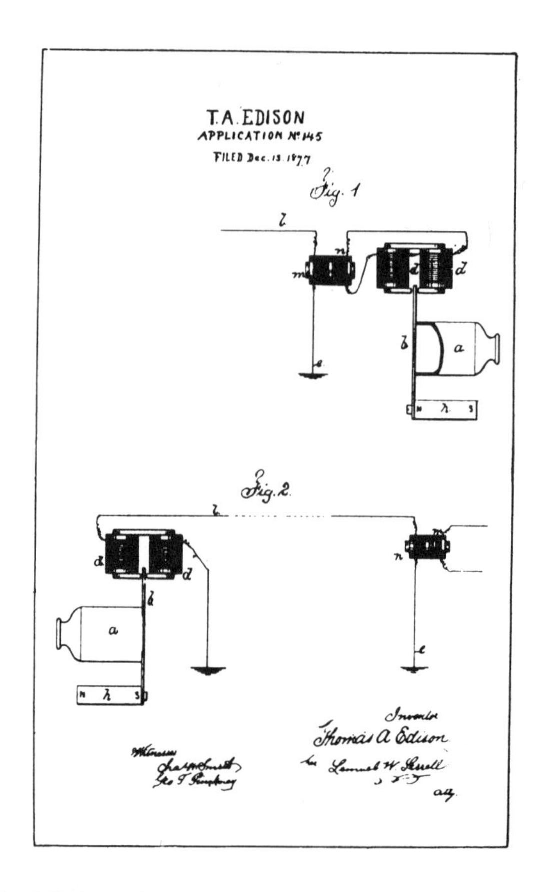


3. The full December specification insisted that such breaks in the circuit were only suitable for transmission of music, and that speech required a continuous circuit.

4. This refers to Edison's electromotograph telephone receiver. The U.S. patent application describing this was Case 141, executed on 9 July and later involved in the Telephone Interferences. It finally issued in 1892 as U.S. Patent 474,231. Edison included some drawings from that application in the full British specification in December. Edison's electromotograph effect had been exhibited in England but not patented there.

5. This is Edison's first public statement that he could record sound. The full specification in December included several detailed phonograph designs.

6. This refers to the designs described in Case 136, drafted in April or May and executed on 16 July, which became U.S. Patent 203,014. See Doc. 873 n. 2.

7. Contacts on both sides of the diaphragm also are described and shown in Case 141; see Doc. 920.

8. The staff started a series of diaphragm-material tests the next day (Doc. 975). See also Docs. 966 and 968.

9. See, for example, Edison's Exhibit 178-11, TI 2 (*TAEM* 11:331). Edison did not include this approach in the December specification.

10. Metallic contacts were successful in Edison's make-and-break musical telephone transmitter. Otherwise Edison's designs mostly relied upon carbon or other semi-conductors, using a principle of operation he described in Case 130. See Doc. 860 n. 2.

11. Edison described the operations of these arrangements in Cases 130 and 135. The latter, having been initially prepared in April or May and executed on 16 July, issued as U.S. Patent 208,299.

12. In extensive experiments the previous summer Edison had found chalk very good for electromotograph action. Docs. 780, 781, 783, and 787.

13. See Doc. 945.

14. This is Edison's pressure relay; see Doc. 885.

15. A discussion and illustration of this kind of telephone design using a pressure relay is in Prescott 1879 (530).

16. Edison describes such designs in Cases 135 and 136.

17. For this design, see Edison's Exhibits 179-11 and 1-12, TI 2 (*TAEM* 11:332, 352). Edison did not include it in the later specification or any U.S. application. Another inventor, Emile Berliner, applied for a patent on a related design on 4 June 1877 (U.S. Pat. 463,569).

18. See Doc. 874. Edison did not include these designs in any other patent application.

19. See Prescott 1879, 535–36. Edison did not include this in any other patent application.

20. See Doc. 987 n. 2. Edison did not include this in any other patent application.

21. Presumably "false" in original.

22. See Doc. 948.

23. No successful duplex telephone circuits were actually developed for many years. King 1962b, 332.

24. Edison included some arrangements for working telephones in local circuits with induction coils in Case 146 and in the December specification for this British patent.

25. No practical relay for telephone circuits existed.

26. See Doc. 866.

27. See, for example, Edison's Exhibit 177-11, TI 2 (*TAEM* 11:330); and Vol. 11:174, Lab. (*TAEM* 3:1000).

28. The principal drawings in the December specification showed hand-held transmitting and receiving apparatus. They are reproduced as the illustration accompanying Doc. 1030 n. 2.

29. See Doc. 966. Edison did not include this idea in any other patent application.

30. See Doc. 921. Edison did not introduce this concept in any other patent application.

## –974–

*Edward Johnson to Josiah Reiff*

Philad, July 20/77

My Dr. Rf.

Your postal recd.[1] It was brief but I suppose you thought it contained the Pith of what would interest me. Still I must insist that I would be somewhat gratified to have a more definite & specific reply to my interrogatory—put to both you & RWR[ussell]—as to the Status of the Telegraph Question— I presume something—however trifling—has been accomplished in the 4 weeks of my absence— Or is it transpiring that Edisons prediction is to be Verified & that we have only[a] been indulging in the construction of Castles in the air.

B.F.[2] was taken back July 4th—at $60. Per Month. (Blair[3] gets $80.)— I wrote DHB[4] remonstrating and shewing up some of the pettyness's of the Phila outfit[5] but it has borne no fruit unless some removals of incompetents within a few days are traceable to it— Of course I got no direct reply— They never do that sort of thing— I wonder upon what principles they conduct correspondence—

I am having a good time—professionally— my "Button holes" are ~~a~~fingered by gentry with Prefixes to their names—scientific of course— A genuine sensation has been created— I am told my hastily written article on Telephones ~~of~~ published in the "Press" couple weeks ago has been extensively copied by such papers as the Boston Advertizer The NYork Sun—the Cincin Comrcl &c The NYork Times Item is a palpable misrepresentation—[6]

Last nights Performance—the 1st Public Exhibition—is pronounced upon very favorably by todays papers[7] notably the Press—which alludes to it Editorially & styles Edison "One of the greatest inventors of the Age"— I send him copies of all most important papers.

The Exhibition last eve—was in Connection with a stereopticon display which combination I thought much more appropriate than to attempt as Gray & others did to, make it a musical performance—& thus have the product criticized as music—instead of a scientific display— The audience—consisted of a very nice class of our Philada people. considering the season. & was considerably over 5000 in numbers— They left no doubt either as to what they came to see—it was the Telephone— now as to hearing it—upon the announcement that the Telephone was about to speak, or sing there was a general move from the outer circles toward the[a] stage. This I caused to be interrupted by having them requested to await the 1st tune—before coming closer— the result was that of the 5000 people present not a score came a step nearer during the performance. In fact many who were near went back & many strolled off toward either nave until the audience was very loosely spread over almost the entire space between the

Roosevelt organ and the south centre-entrance door— Now with this vast concourse of People a moist atmosphere and almost an open air space—no walls to throw back the sound—Everyone was able to hear & enjoy the music— Can anyone do this with magnets for receiving instruments— If so his name isn't Gray—

As for Bell Barker told them last eve just what I did in the papers—that it was limited in capacity to produce sound— He went so far as to assert that it would never come into general use because it was impossible to hear it 6 feet away from it— His speech should have been reported— it was a description of Grays Bells & Edisons Telephones He dismissed Gray as a Toy & left him no where—& as for Bell—well he practically put him in the same boat— then he took up Edison & shewed what he had accomplished—& predicted Wonders of practical character as Edison was a man who if he wanted anything & it couldn't be found in the Laboratory of known things went to work & made Original discovery to supply it. (Hows that)— He then illustrated—by explaining the motograph & the Graphite discoveries—& their uses this to an audience including in its bosom many Profsers from all parts of the Country Prof. McGill[8]—I think from Boston— Draper NYork—Prof of Racine Wis. College[9]—& a host of others whose names I didn't catch— I was introduced to so many I lost all remembrance of them or their Post as soon as they got done talking— Chicago Tribune NYork Tribune, Louisville Courier & a host of smaller fry were on deck.

I got the 2d transmitter from Edison after performance began last night— It was fearfully disgusting He promised to send it on the 12th—& upon my telegraphing He sent it on 8 AM train the morning of the day of the concert— It failed to reach me—& though I turned Penna RR upside down I couldn't find it— they teleghd New Brunswick & Metuchen to know if there & if it had been were going to order an afternoon through train to stop & bring it on— It turned out Adams Express had it en route all day— It was a great disappointment though my good luck didn't otherwise desert me The old one worked with much better effect than on the rehearsal of monday last & the entertainment as such was an immense success, whe~~r~~n by the turn of a hair—it might have been the most ridiculous farce ever heard of. I now have 2 good Insts. he having sent me some Extra Diaphragms & I am preparing for another entertainment next Wednesday which will be the "Best I can do"—& at which we will all be disappointed if there are not 8 or 10,000 people— I got 3 cheers & a tiger Wednesday night—which was my first compensation—but I shall get 50$ Saturday—& each succeeding week as long as I can improve on it & keep the interest up. They will only give one Ex a week in order to prevent[a] its being made commonplace—& to give me time to work up something new. If I had Edisons fertility I suppose I might keep things 'Hot' all summer— nevertheless a 6000 audience, in mid summer the greater part of them going a distance of 4 miles & in a rain storm—isn't bad—.

The next epistle passing between No 243 North 12th & 52 Broadway I fancy will emanate from the latter address Truly Yours

Johnson

⟨Referred to TAE Splendid JCR⟩[b]

ALS, NjWOE, DF (*TAEM* 14:825). [a]Obscured overwritten letters. [b]Written by Reiff.

1. Not found.

2. Benjamin F. Johnson was the chief Atlantic and Pacific Telegraph Co. operator at the Centennial. *Operator,* 1 July 1876, 6–7; Gopsill 1875, 779.

3. Unidentified.

4. David Bates was general superintendent of the Atlantic Division of Atlantic and Pacific. Bates's testimony, 1:17, Box 17A, *Harrington v. A&P.*

5. That is, the Atlantic and Pacific office in Philadelphia.

6. Johnson is alluding to an article titled "Telephones and Their Different Uses," which appeared in the *Philadelphia Press,* 9 July 1877, p. 6.

7. Although this letter is dated 20 July, the newspaper reports appeared on 19 July, indicating that the concert was on 18 July. *Philadelphia Press,* 19 July 1877, p. 6; *Philadelphia Inquirer,* 19 July 1877, p. 8.

8. Unidentified.

9. Unidentified.

## –975–

*Technical Note: Telephony*

[Menlo Park,] July 20 1877

Speaking Telegh
Substances to be tried as diaphragms[1]

| | | |
|---|---|---|
| Boxwood— | Paper soaked various subs | Dammar[2] |
| Pine Wood | Chamois skin | Alblumized paper |
| Kruzi's Veneering | Kid | Wire gauze Coated or w Tissue ppr |
| Cedar Wood | Patent Leather | Parchment |
| Cork | | |

| | | |
|---|---|---|
| Celluloid | Bladder Skin Coated V. S. | Parchment paper |
| Collodion | Ivory | Bank note paper coated |
| Cardboard various sizes | Sheet Sulphur | Papyrograph paper |
| Glass | Vulcanized Fiber | Pressed Benzoic acid |
| Mica | Sheet Tin | Rawhide |
| Copper foil | Gelatin Coated | Balata[3] |
| Tin foil | Birch Bark | papyrus |
| Gutta Percha | Doe skin | Fish bladder |
| Rubber | Bamboo | Platina |
| Silk | Lemon skin dried | Phosphor bronze |
| Gutta Percha Covered with rubber | Tabacco Leaf | Aluminum bronze |
| Aluminum | Cabbage Leaf | Russia Leather[4] |
| Steel foil | Linen soaked in Collodion | White spruce shellac'd |
| Soft iron foil | Bone | oil cloth— |
| Brass Foil | Dried Linseed film | |
| German silver foil | Fish skin | |

T A Edison

Chas Batchelor
Adams

X (photographic transcript), NjWOE, TI 2, Edison's Exhibit 38-12 (*TAEM* 11:375).

1. Three other lists written on this date, two by Edison and one by Batchelor, contain items found in this list as well as additional items (Vol. 12:39, 42, Lab. [*TAEM* 4:21, 23]; Edison's Exhibit 14, *Edison Electric Light v. U.S. Electric Lighting,* Lit. [*TAEM* 46:271]; and Vol. 12:40, Lab. [*TAEM* 4:22]). On the same day Edison also noted unsuccessful tests with shield brass, pine wood, and bristol board, and limited success with cardboard (Edison's Exhibit 32-12, TI 2 [*TAEM* 11:371]).

2. A resin similar to copal. See Doc. 915 n. 2.

3. A gutta-percha-like substance derived from tropical trees of the sapodilla family.

4. A type of leather originally made in Russia but at this time largely made in Paris from goat and sheep skins. Also known as jucten, it is resistant to moisture. A description of the process for preparing Russia leather is in Knight 1876, s.v. "Russia Leather."

## -976-

*Charles Batchelor to Thomas Batchelor*

[Menlo Park,] July 21st [187]7

Dear Tom,

I have only today got the Directory[a] you sent me; so much for Red Tape in the Custom House. It is just the thing I wanted, exactly, in the pen business, but since I have gone out of that it will not be of so much use. I want you to let me know what you paid for it with expressage and I will send you the amount; as it was got for the Foreign Pen Co. and they must pay for it; you can make it in the form of a receipted bill & enclose in your next letter.

I have been exceedingly hard at work night and day on our speaking telegraph and at last we have got it in good shape and hope it will soon be in such a condition that it will bring in some money.

Two of the British Postal Telegraph Engineers have been in the States lat[ely?][b] Messrs Preece & Fischer, and Mr Preece spent four different days with us and we just showed him lots of things he did not know. He is a member of the Society of Telegraph Engineers of London of which both Edison and myself are members and he [purposes writing?][b] a paper for the Society [ ][c] [day?][b] [ ][c] I am making a [ ][c] [speaking telegraph machine?][b] which he is going to have exhibited before the British Association.

We have an infringement case in regard to the 'Pen' by a man named Trueman in Manchester, do you know anything about him?

There is a gentleman exhibiting our Singing Telephone at the Permanent Exhibition Philadelphia Pa I shall try and get some papers & send you

We have two or three splendid irons in the fire at present but they are all telegraphic The 'Pen business' is increasing very rapidly we have raised the price to $50 and $60 and they sell more now than they ever did— I send portrait of Emma, it is not very good it makes her look sickly. Rosa took her to Newark and had a large one taken of her & then she thought she might as well get a few to send away and I suppose she was tired when she stood for this.

Do the folks at Llandudno[1] get along nicely and is Father's health improving? Rosa desires to join me in sending kind love to Maria and the youngsters and believe me Your affectionate brother

Charley.

P.S. Dont fail to send bill because you know it does not come out of my pocket CB

ALS (letterpress copy), NjWOE, Batchelor, Cat. 1238:162 (*TAEM* 93:140). [a]Obscured overwritten letters. [b]Extremely faint. [c]Illegibly faint.

1. A town in northern Wales.

-977-

*From George Ward*

New York 23 July 1877[a]

Friend Edison

Thanks for your note.[1] I shall come with Siemens if possible & will let you Know a day beforehand.[2] I have just got the following from Preece.

"Tell Edison British Association August fifteenth dont fail with apparatus"—[3]

Send me any reply you want to give him. You ought to be shipping it soon to get it comfortably over there.[4] Yours very truly

Geo G Ward

ALS, NjWOE, DF (*TAEM* 14:100). Letterhead of the Direct United States Cable Co. [a]"New York" and "18" preprinted.

1. Edison had responded to a letter of 20 July from Ward proposing the visit discussed here and enclosing something from William or Alexander Siemens to the effect that Alexander would be able to make the visit that William had been unable to include in his U.S. visit the previous autumn. DF (*TAEM* 14:99); Doc. 811.

2. This was Alexander Siemens. He and Ward visited the laboratory on 27 July (Cat. 1233:208, Batchelor [*TAEM* 90:157]). Alexander, William Siemens's nephew and already his assistant for several years, became the managing director of the London firm some years after William's death and was president of the Institution of Civil Engineers, twice president of the Institution of Electrical Engineers, and secretary of the Royal Society (Appleyard 1939, 291; Pole 1888, 198; Siemens 1968, 246, 291, 298). According to a catalog card (item Z 307 Nr. 192) in the Siemens Museum in Munich, Alexander Siemens wrote a report later in 1877 (probably for William Siemens) about his visit, a German translation of which was sent to Werner Siemens in Berlin. However, neither the original nor the translation can be found.

3. On 26 July the staff "Designed Receiving instrument for Preece to show before the British Association." They built it over the next week and shipped it on 2 August. Cat. 1233:207–9, 214, Batchelor (*TAEM* 90:156–57, 160); see also Doc. 987 n. 2.

4. Although steamships had made record transatlantic crossings in less than eight days, they commonly took ten days or more. Tyler 1972, 370; "An American Telegrapher in England" (*Telegr.* 12 [1876]: 93), in which one J.C.B. noted that with fine weather the whole way a Cunard line steamer took eleven days to get from New York to Liverpool.

-978-

*From Josiah Reiff*

NY July 24/77

Dear T.A.

Have reported to Silver Mine people that you will advise them within a few days just what the Glass[1] will cost. Hurry Spice[2] up—

Let me know when you come in town next I wish you to see a man acquainted with the ores of Utah & Colorado & Explain to him your belief as to working low grade ores—we may interest him. He has been connected with the [Ei---d?][a] Mine.[3] In fact was W[ener?][a] Parks[4] Partner. Yrs

JCR

ALS, NjWOE, DF (*TAEM* 14:101). [a]Illegible.

1. Unidentified.
2. Robert Spice.
3. This mine has not been identified.
4. Unidentified.

## -979-

*From Edward Johnson*

Philada July 25/77

TA.E.

I sent you today a small Pkg of chloride of Potash I had contemplated sending it by mail Else I should have sent more of it— It will suffice however to Enable you to make a test—& (if you think it worth while—) to wet me a Roll or two & send on.[1] I have no means at all of wetting paper here or should not bother you. I want some more paper any way so may as well have a Roll or two of it— Please let it come forward Immediately so I will not get caught in a trap again.

Prof Barker suggested my making an attachment to the sounding Board of a Piano—& Immedy on this your letter came telling me to try a larger Dulcimer—[2] Not being able to find such a thing here—It struck me that the Piano was the same thing so I tried it— (I have since learned where I can find the Dulcimers & as soon as I get some money out of the concern I am going to see them & if the proper thing for you buy you the one promised— I wish you would meantime send me a description of the exact article you want—

The Piano test was partially successful—though by reason of my fearing to injure the Instrument I was compelled to make the attachment too close to the Edge of the ~~P~~sounding Board to get much volume of sound— It was however voted by everybody to be much sweeter— I have therefore made a proper fixture & attachment on the Board immediately in the Centre of the ~~P~~Instrument in front of the Key Board & just inside of the music rack so that it occupies about the same relative position it did on the old Box— I will test it tomorrow P.M. Could not do so today as I am moving my transmitting station out of Phillips office into a room he got for me across street— This strike[3] has made such a commotion around his ofs[a] that I have been unable to obtain a moments use of it for Experiment hence the change after tomorrow I will have full privacy & can work all the time—so send on your suggestions— I got but a minute or two to try Duet—but in that time I fancied one Inst cut the other out— are they not more liable to interfere placed in the manner you direct thus

than if ~~a~~in a straight ckt—thus

I will however be able to judge for myself tomorrow when I will try them fully both ways—

I was to have[a] had a Big time tonight but the mob has scared every body so that they are afraid to open the Building at night— beside no one would go out if they did— so it is postponed until next week[4] Meantime I want to make bid improvements so that the next show will be immense The Diaphragms you sent are an immense improvement on the Brass ones. Especially the one which was on the new transmitter— that was simply good for nix— The old Brass one however was pretty good—because thinner— I find the less resistance the metal offers of itself the better— The Brass foil and the mica are tip top—though in getting the mica on first I got a Bulge in it, which "set" it so as to make it difficult to vibrate I took that out & it is now immense I anticipate a tip top performance tomorrow PM on experimental Rehearsal— I found out as you thought I would—that such things as the Last Rose of Summer—came best— I have bought quite a lot of music which I have my wife & sister practicing on My wife plays them on the Piano thus giving my sister the air— who then sings them. Her voice has a peculiar vibration about it which adapts it especially to this thing so she has got the advantage of some better singers & gets furious encores Her Selection[a] for the next is

Last Rose of Summer
Won't you tell me why Robbin
Her Bright Smile haunts me still
5. oclock in the morning
Air from—Fra Diavolo[5]
& 2 or 3 others somewhat different in style.

Prof Barker has left us  I cannot yet say who will do the honors for us next time but no doubt we will find a proper party— I want it to be somebody whose name will have weight— I will take care that he is previously so fully enthuzed as to commit himself unhesitatingly when he gets warmed up—

I want to get hold of Draper—& think I will If I can keep this going till Barker gets back again— I see Draper has been immortalizing himself discovering Oxygen in the Sun. How did he proceed to find it out?—

Your last letter was simply astounding— I am already getting anxious to pay you a visit to see these things for myself. When you get me ready for the speaker I am going to make a new deal with these [Ex?][b] folks— Thus far I have had but little except advertizing. I mean to get not less than $250 per week for the speaker if perfect, & they will pay it—as they can put 10 000 nightly in that Building on it— then there will be something to divide—

The Graphic man was called off suddenly by this strike to go to Pittsburgh  It will probably so occupy them for some time to come as to exclude us, but that will all be over before we come to the speaker & then I mean to hook him, & of course will see that the Exhibit he makes is complete in all particulars even to the laboratory—

Prof Barker told me to tell you to use Iridium in place of Platinum as it fuses at 10 to 20 times greater heat— He uses it & is no longer troubled with oxydizing points[6]  I'd like to have it tried—as it is a great annoyance to us to keep points clean  they get dirty in one performance & spoil the last pieces— Give me a few minutes when at leisure at nights

Phillips is a Brick

EHJ

ALS, NjWOE, DF (*TAEM* 14:838). [a]Obscured overwritten letters. [b]Illegible.

1. Johnson is referring to chemically treated paper for the electromotograph telephone.

2. Edison's letter has not been found. For dulcimer telephone designs see Doc. 987 n. 2.

3. The railroad strike of 1877 began on 19 July in Pittsburgh and quickly spread to other cities including Philadelphia, Buffalo, and St. Louis. In Philadelphia rioters vandalized the Pennsylvania Railroad's telegraph lines, severing their communication with Pittsburgh, and Western Union had to supply wires. The "commotion" to which Johnson refers may have been the large crowds that gathered in West Phila-

delphia, the extra police precautions taken, and a fire deliberately set on an oil train. On 22 July, Mayor William Stokeley issued a proclamation appealing to the citizenry to "suppress outbreak and violence" in the city. *Philadelphia Inquirer,* 23 July 1877, p. 2; Bruce 1959, 194–96.

4. The telephone concert scheduled for 25 July had to be postponed until 2 August "on account of the public disquietude growing out of the riots." *Philadelphia Inquirer,* 25 July 1877, p. 2.

5. *Fra Diavolo* was a popular opera written by French composer Daniel François Esprit Auber in 1830.

6. That is, the contact points on the telephone transmitter.

**-980-**

*Charles Batchelor to Ezra Gilliland*

[Menlo Park,] July 26th. [18]77

Dear Ed,

Yours of 23rd[1] received, .O.K. no drawing enclosed but I understand everything from your description. I see you have tumbled to the 'flatdown box'[2] I am now making one which is going to be shown before the British Association in England on the very plan[3] there is no doubt it will work.[4] Transmitters 4″ long and size of 'pen wheel tubing' is all right. when you have finished your diaphragm rings take an impression on paper of one showing all the holes and send to me and I will make you a few diaphragms of different styles and send you.

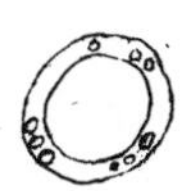

You will find them far superior to the ordinary thin copper or brass. We use ½ inch paper but the drum ought to be 1/16 wider, so as to have margin The diameter of drum does not matter but I make all one inch and I do not put the handle direct on wheel shaft so:

but I gear it up about 4 to ~~2~~1 so that you have to turn the handle 4 times to drum once so:—

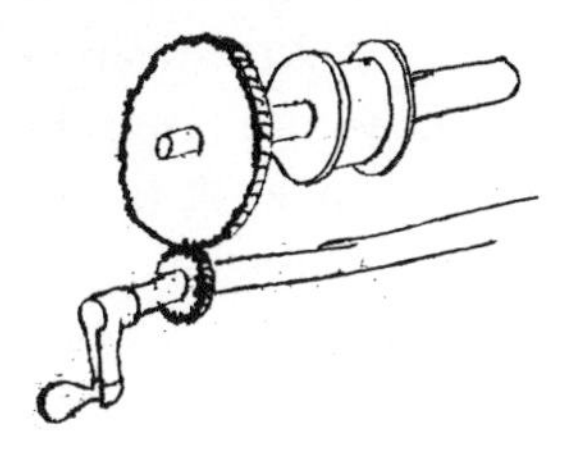

The drum will do nickelplated best & the surface of it is knurled straight across so as to make it rough ~~a~~ enough to carry the paper along by mere friction with the spring. the spring is tipped with platina and is this shape as at B and about ¼ inch wide

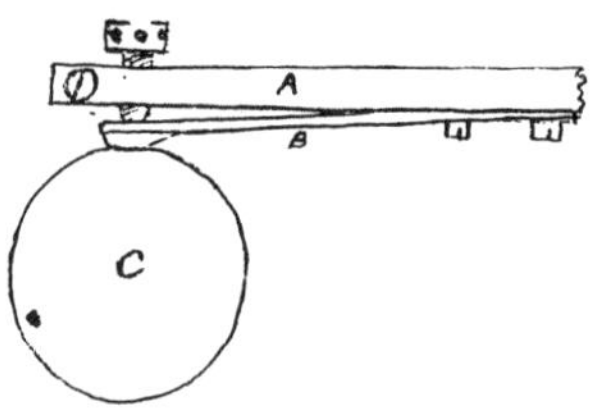


A is Brass lever fastened to sounding box and B is steel spring screwed on to the underneath side with a piece of brass faced with platina on the end working on drum C. The pressure of this spring on the knurled drum carries the paper along The solution we use is common soda water but I will wet up a box of paper and you can keep it in crocks. The paper reel[a] is the same as used by automatic for wet paper and is made this way

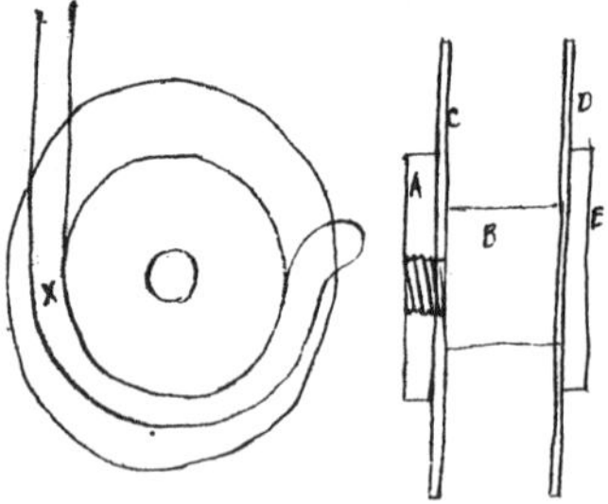


A is a nut same size as Head E C & D are rubber (hard) washers B is body where paper is wo[und][b] on A and C are same thickness as E & D. The reel revolves on the edges of nut and head as at X dont put it on centres as you want a little drag. I would not bother if I was you with the leather covered wheel as it is not reliable for showing. the paper is simple. In making the transmitter have it on a very solid base ~~as~~ I put them on iron bases but I suppose thick wood will do—

Send me print of rings and I will send you diaphragms with paper right away, as for price I guess we can afford that for 'Auld acquaintance sake' In the reports of its working I want you to give Edison due credit for its invention and send him copies of the papers, and send me an extra one for my own scrapbook.[5]

I see by your letter you have left the 'Pen'. what do you mean: left Bliss altogether or only got through at Cincinnati.[6]

We have had terrible hard work on the Speaking telegraph This last 5 or six weeks frequently working 2 nights together until we all had to knock off from want of sleep, but we have today got the best speaker in the country and we are making some for Gold & Stock Tel Co

Edison sends respects Yours

Chas Batchelor

ALS (letterpress copy), NjWOE, Batchelor, Cat. 1238:169 (*TAEM* 93:142). [a]Obscured overwritten letters. [b]Illegible.

1. Not found.

2. In a letter of 19 July, Batchelor had instructed Gilliland to make his own electromotograph receiver using an upright box. Cat. 1238:161, Batchelor (*TAEM* 93:139).

3. He shipped this instrument on 2 August. See Doc. 987 n. 2.

4. On 25 and 26 July, Batchelor made scale drawings of parts for the British Association machine. Cat 1240, items 213, 216, 218, Batchelor (*TAEM* 94:64–65).

5. Batchelor's scrapbook contains Cincinnati newspaper clippings from 15 and 19 August describing tests of Bell's telephone. Although Gilliland participated, Edison's telephone is not mentioned. Cat. 1240, items 226–27, Batchelor (*TAEM* 94:68).

6. According to the internal address on this letter (and Batchelor's other contemporary correspondence with Gilliland), Gilliland was living in Cincinnati.

**-981-**

*Technical Note: Telephony*

[Menlo Park,] July 26 1877[1]

Speaking Telephone

To obtain a very delicate Telephone I make the Diaphram of Mica place a very delicate Bow Spring upon it and coat a rubber strip with plumbago thus

Instead of Coating this strip with plumbago it may be coated with any inferior Conducting substance such a finely divided silver or other Metal sometimes used by Electrotypers

to Coat their nonconducting surfaces it is not necesssary that the Reed should be made as shewn I find that it may be free at one end that it may be of any flexible substance or that the plumbago may be put on any surface such as a surface of rubber hard or soft which is laid on felt[2]

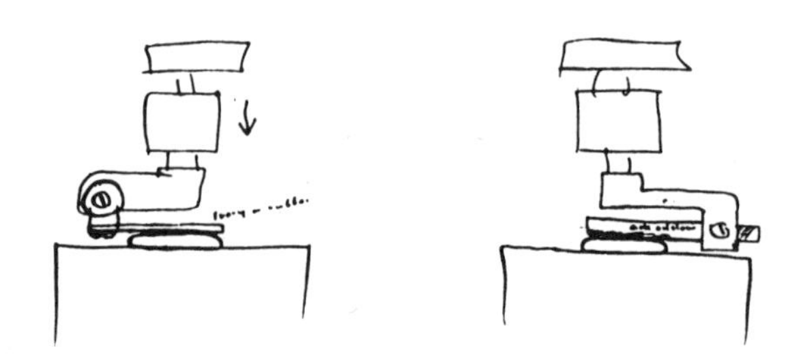

Arkansas oil stone or other hard Substance Coated with plumbago etc=[3]

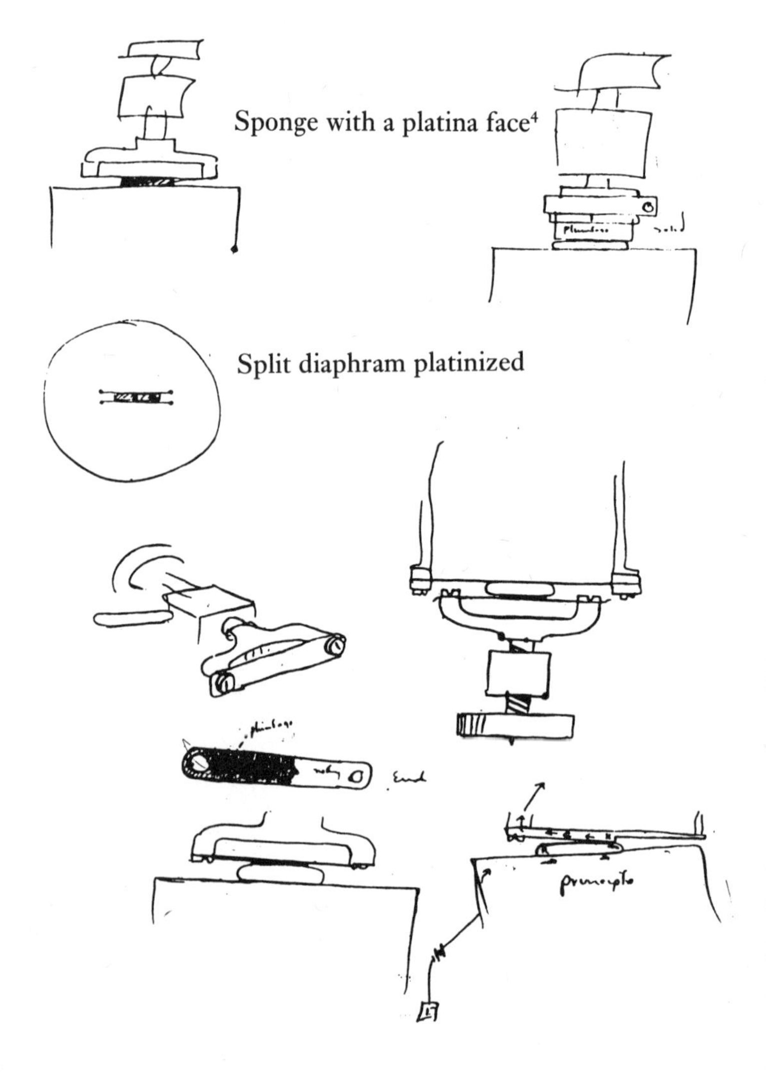

X[5] when in state rest is point of contact & current must pass through full resistance of plumbago Coating .C. when you talk the spring R makes more contact & the current does not pass through so much plumbago which being thin offers great resistance= The hisses come ok

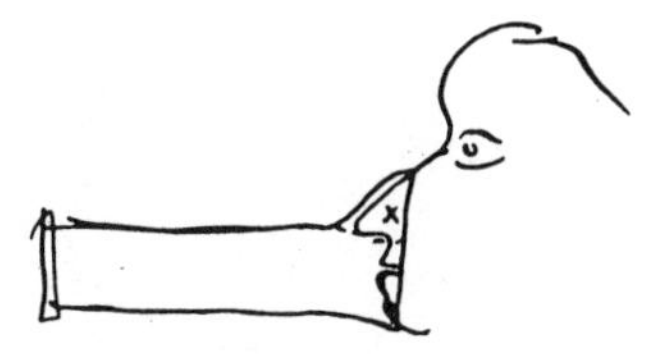


X is a nose piece serving to confine & render operative upon the diaphram the vibrations from the nose ie[a] the nasal sounds which is lost in Bell & ~~oth~~ Reuss Telephones & which are necessary for perfect articulation

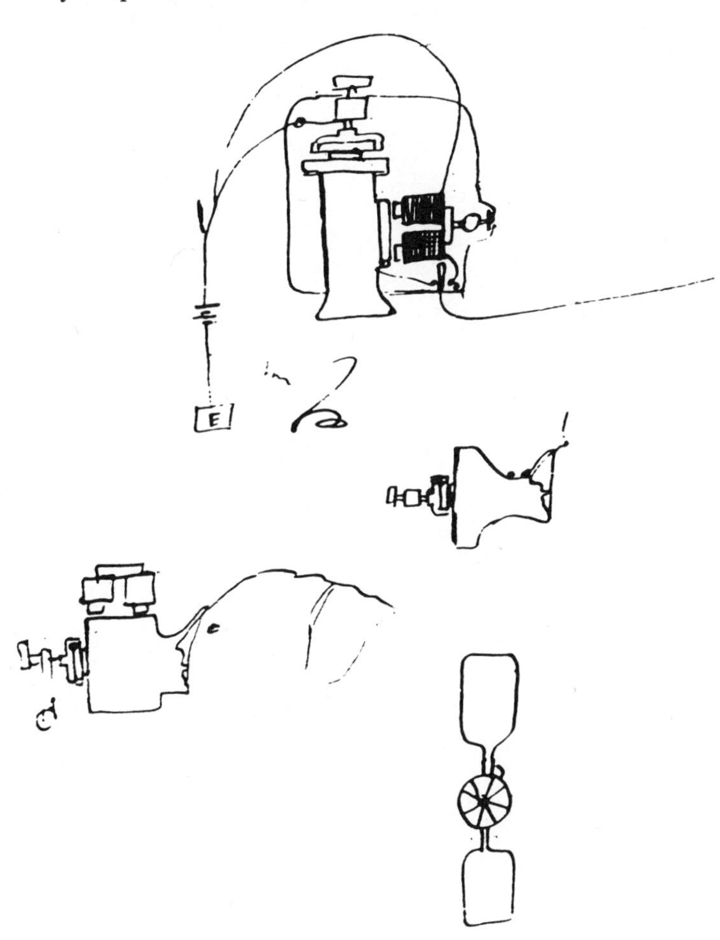

When several persons are to talk over one wire I provide each one with a Telephone although one or two can talk in the same tube

[A][6]

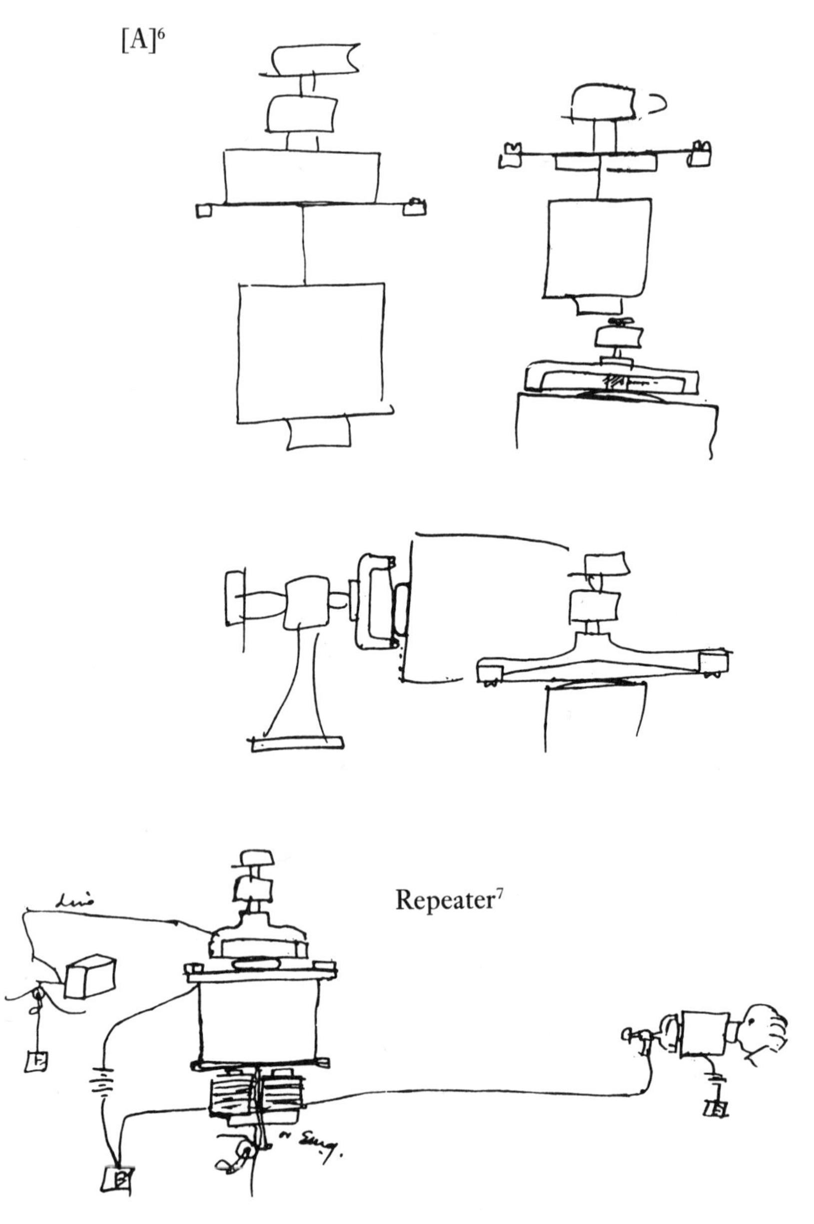

Repeater[7]

[B][8]

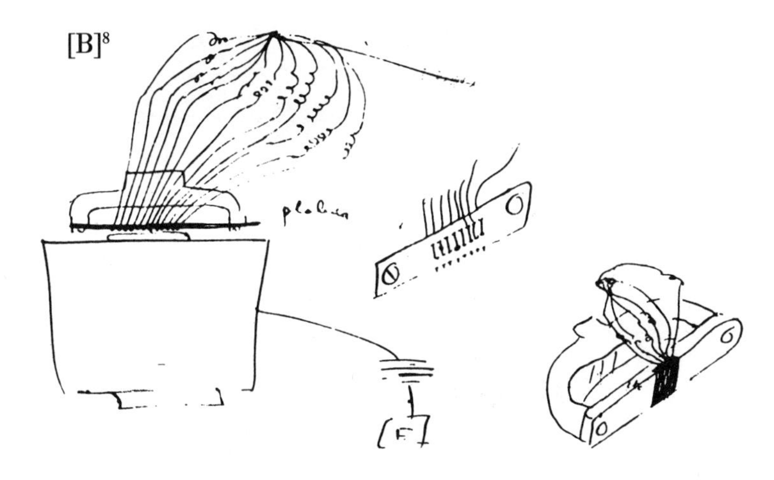

Platina sheet ends exposed put in slots in ivory or [such?][b] substances

[C][9]

[D][10]

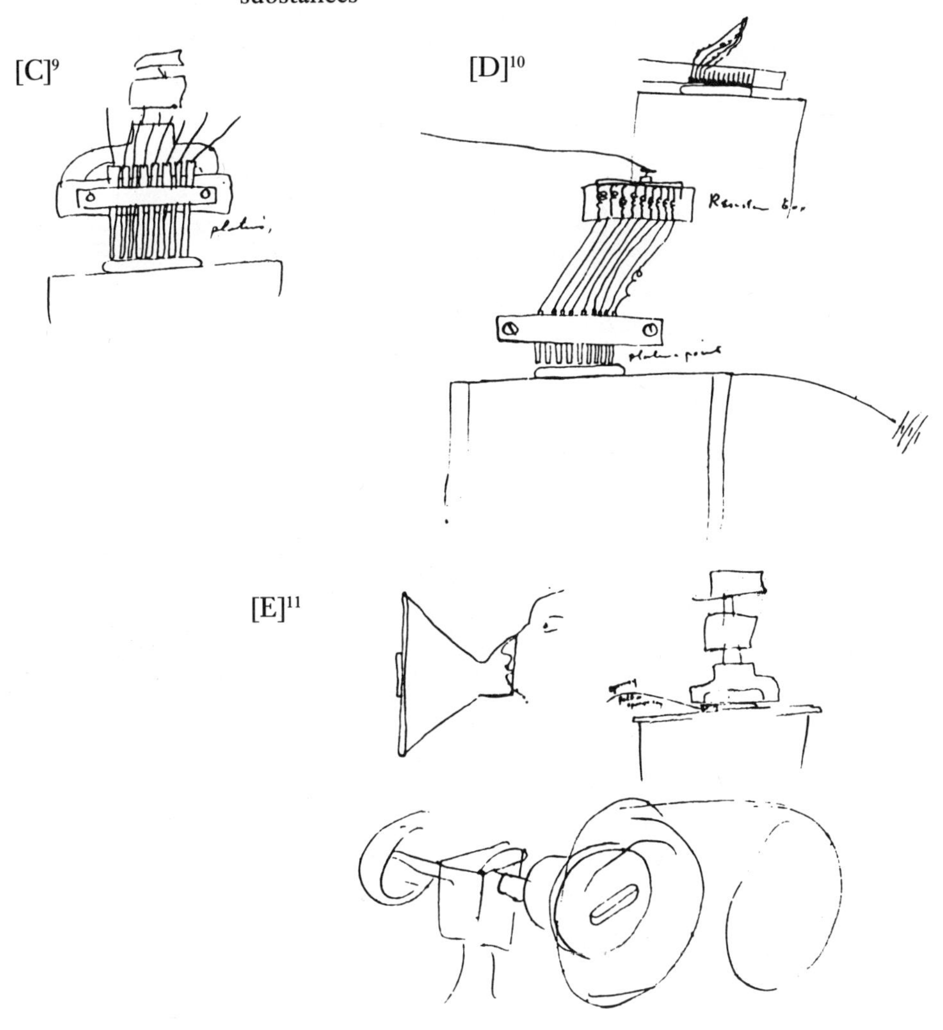

[E][11]

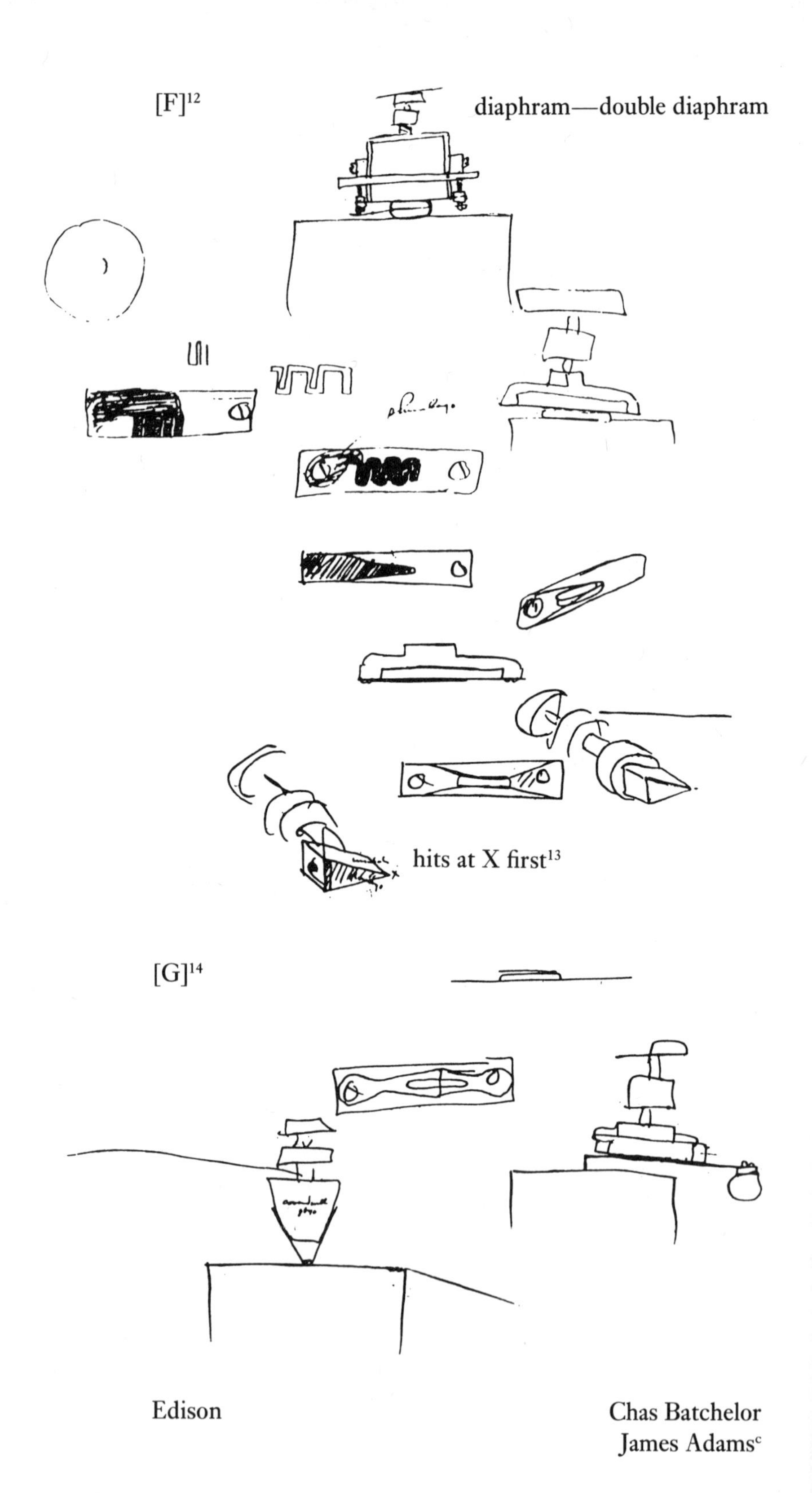
[F][12]
diaphram—double diaphram
hits at X first[13]
[G][14]
Edison
Chas Batchelor
James Adams[c]

X (photographic transcript), NjWOE, TI 2, Edison's Exhibits 58-12, 57-12, 59-12, 54-12, 61-12, 56-12, 55-12, 62-12 (*TAEM* 11:387, 386, 388, 383, 389, 385, 384, 390). Document multiply signed and dated. [a]Circled. [b]Illegible. [c]"S. Remark" in Edison's hand appears on one page.

1. The same day the laboratory conducted tests of the resistance of plumbago under the pressure of different weights. (Edison's Exhibits 51-12, 53-12, TI 2 [*TAEM* 11:381–82]). The Edison Institute in Dearborn, Mich., holds a device used in conducting such tests (Acc. 29.1980.42).

*A device for testing the resistance of plumbago mixtures under varying pressures.*

2. Figure label is "ivory or rubber."
3. Figure label is "ark oil stone."
4. Text connected by dotted line to dark rectangle in figure at left. Figure labels on second drawing are "plumbago" and "solid."
5. Figure labels on drawings above are (at left) "plumbago," "nothing," and "end"; and "R," "C," "X," and "Principle."
6. Figure label is "sponge."
7. Figure labels are "line" and "or EMG."
8. Figure label is "platina."
9. Figure label is "platina."
10. Figure labels are "Resistance box" and "platina point."
11. Figure label is "springy felt or sponge [---]."
12. Figure label is "plumbago."
13. Figure labels above are "insulated," "plumbago," and "X."
14. Figure label is "covered with pbgo."

**-982-**

*To William Orton*

Menlo Park N.J. July 27/77.

Wm Orton, Esq

Had some extra successful experiments with the Telephone last night= Read one column of a newspaper with the loss of only a few words.= also carried on a conversantion through 4000 ohms resistance without being within 10 feet of the speaking tube. The pair I am making is nearly finished, and I think Murrays[1] are well under way. Yours

T. A. Edison

ALS, DSI-NMAH, WUTAE.

1. See headnote, p. 427.

**–983–**

*Josiah Reiff to Albert Chandler*

New York, July 27th 1877[a]

Sir

Sincerly desiring accord & an adjustment in some form of our interests in Automatic Telegraphy & the Edison Patents, I am willing to labor regardless of any past misunderstanding or plans to secure that result.

I accept the situation & do not stop to discuss the causes that have produced the feeling that has existed on the failure to secure the results which I had anticipated=

I have unbounded faith in the Edison system & inventions & know that with Automatic Quadruplex & other improvements which ere now would have been in the control of A&.P but for the unfortunate lack of cooperation, great results would by this time have been obtained.[1]

Wiping out the past I am ready under any reasonable assurances of harmonious cooperation in Securing the best financial results from the joint facilities of Morse & Automatic to agree now to take for the Edison system & improvements whatever they shall prove themselves to be worth when a combined effort is made to test their capabilities.[2]

If the money, time & effort then prove to have been uselessly expended & the Patents worthless, that will be our misfortune & no one will be asked to pay an unjust value.

All I ask is hearty cooperation with it I am perfectly willing to share the fortunes of A&.P., taking the risk as to our own part & biding the demonstration.

The details should be easily agreed upon if this plan is heartily entered into.

Of more than $400,000 in greenbacks spent in experiment I have paid directly more than $150,000 much of it borrowed at severe sacrifice & which I can not now pay.

I have never personally received a single penny of benefit from my efforts, time or expenditure in connection with this business.

The money & Bonds loaned me by Mr Gould[3] were immediately exhausted in paying pressing Automatic debts for which I was personally responsible, or in returning loans long past due.

Some immediate help is needed, but if this general plan is assented to, I doubt not it will be redily granted.

Of course protection should be provided in case of possible arrangement, alliance or amalgamation with W.U. Tel. Co. Yours truly

Josiah C. Reiff

ALS, NjWOE, DF (*TAEM* 14:703). Letterhead of J. C. Reiff. [a]"New York," and "187" preprinted.

1. In testimony, Henry van Hoevenbergh and Edward Johnson referred to the use of the quadruplex and automatic systems together for higher speeds and greater efficiency. One of the improvements Johnson mentioned was the embossing transmitter. Quad. 70.6, pp. 524–26, 742–46 (*TAEM* 9:632–33, 742–44).

2. See Doc. 760 n. 2.

3. According to the Bill of Complaint in *Harrington v. A&P,* Gould, representing Atlantic and Pacific, loaned $35,000 to Automatic Telegraph to pay for rent, wages, other debts, and for the purchase from National Telegraph Co. (see Doc. 141) of that company's line between New York and Washington. Atlantic and Pacific responded that the money was part of the purchase price of the assets of Automatic Telegraph, including Edison's patents, but that there was no agreement to buy the line, even though Automatic Telegraph had given Atlantic and Pacific possession of its line and office. 1:40, 89, Box 17B, *Harrington v. A&P.*

**-984-**

*Charles Batchelor Diary Entry*

[Menlo Park,] MONDAY, JULY 30, 1877. [a]

Worked all day and night on Speaking telegraph. Got the articulation all perfect again by covering silk fibre with Plumbago and rolling into a lump and placing between a diaphragm with spring on and platina disc on screw:

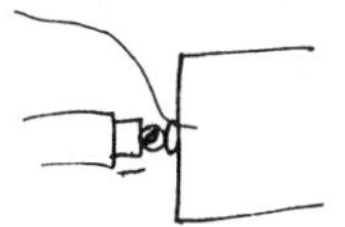

This we found so delicate that when you put this in circuit with a galvanometer so: — and put pressure on spring A so that it squeezes B the plumbago you can lessen the resistance so as to run the needle of galvanometer gently up and down the scale

AD, NjWOE, Batchelor, Cat. 1233:211 (*TAEM* 90:168). [a]"MONDAY, JULY 30, 1877." preprinted.

**-985-**

*From Josiah Reiff*

[New York,] July 31/77.

My dear TAE

It is stated positively today that some negotiation is progressing by which WU will compromise with G[ould] & A&P. Again it is put in shape of G going to control WU at this October election which is a reasonable expectation—[1]

Evidently something is doing outside or inside— Can you learn anything at WU? I will be absent tomorrow but will be here on Thursday, if you can tell me anything

AL, NjWOE, DF (*TAEM* 14:709).

1. The Western Union and Atlantic and Pacific telegraph companies would soon begin negotiating a consolidation. On 21 August the companies announced an agreement to pool their receipts, with Western Union to purchase a controlling interest in Atlantic and Pacific. Although Atlantic and Pacific remained nominally independent, after the Western Union purchase (on 11 September) it became effectively a subsidiary company. Jay Gould participated in the negotiations but had nothing to do with either company after the consolidation. However, he later formed the American Union Telegraph Co., which he used in a successful attempt to gain control of Western Union in 1881. Cat. 1240, items 231–34, Batchelor (*TAEM* 94:69–70); A&P Executive (1873–78): 201–19; A&P Stockholders (1875–88): 101–3; WU Executive (1875–78): 342–47, 359, 365, 385–95, 404–7; Reid 1886, 578–98.

**–986–**

*Notebook Entry: Telephony*

[Menlo Park,] July 31st 1877.

Speaking Telegraph

We find that our apparatus for getting the Sh, Th, and S, is not perfect and in experimenting we find that these sounds will vibrate a diaphragm when spoken across a tube same as speaking across the mouth of a bottle, so we constructed a speaker on this principle as shewn in Fig 47

47

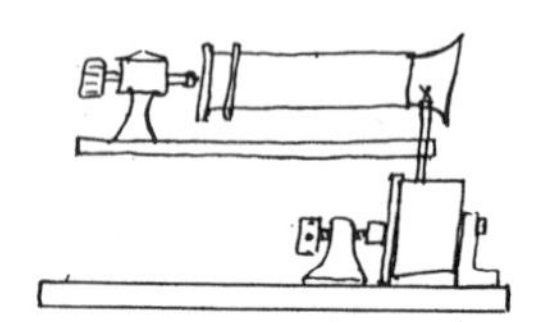

When you speak in the top tube you speak across the tube X and all 'ishing' sounds vibrate the lower diaphragm and make the talking more perfect.

We now began to experiment on different diaphragms and we have made them of, iron, steel, german silver, brass, mica, soft rubber hard rubber, paper, card, collodionised paper and many other materials. We have had different kinds of springs on them some of which are shown below the best of which we find is shewn at Fig 50

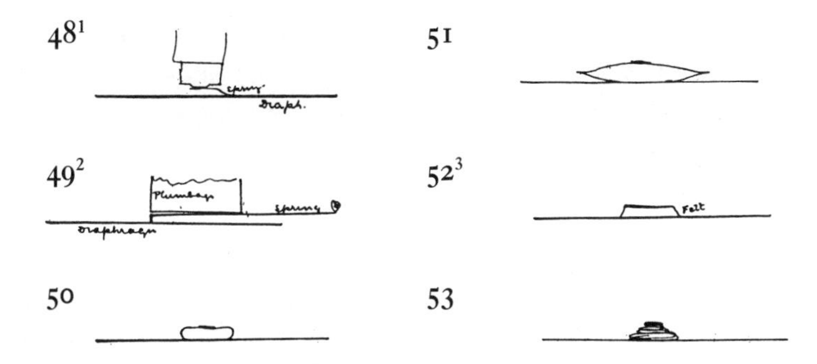


This must be made very delicate so that the diaphragm vibrates independently we have done some excellent work with this. We have now done away with the 'ish' tube and speak only in one tube the diaphragm being made to respond to the 'ish' sounds by cutting a slot in the mouth piece so that these sounds strike on the edge the stress of air passing down and out but causing a vibration which makes the diaphragm to respond.

Up to the present date we have used solid plumbago compositions but we find by testing on a galvanometer we find that we dont get the amount of variation in the resistance we require by slight pressure. Having found that we must not break the circuit but work entirely on differences of resistances as this stops all outside noises, bucklings etc. We take fine wool and rub it in plumbago and press into a small cake which of course is springy when this is put in between the spring on diaphragm and platina face of adjusting screw the talking is absolutely perfect and with this device I can and have taken a message of about 75 words through 1000 ohms perfectly without breaking of matter taken at random out of the newspaper. The plumbago is a little liable to shake out and consequently to deteriorate but we could make a band of it and continually expose a new place.

Chas Batchelor

X, NjWOE, Batchelor, Cat. 1317:41 (*TAEM* 90:677).

1. Figure labels are "Spring" and "Diaph."
2. Figure labels are "Diaphragm," "Plumbago," and "spring."
3. Figure label is "Felt."

## –987–

*To William Preece*

Menlo Park N.J. Aug 2nd 77

Friend Preece:

The musical Telephone goes on the White Star Steamer Germania Saturday—[1] There is a transmitter & Motograph

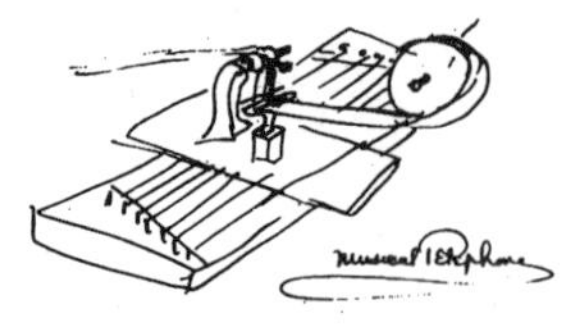

*An electromotograph telephone receiver built into a dulcimer.*

receiver. The dulcimer attachment is guess work,[2] and I tuned it myself and I dont know one note from another= You may be compelled to retune it=

If you think of working a long distance, ~~you~~ say 250 miles use large size wire, 100 cells Bunsen Bichromate with a shunt to earth at transmitting station wire a 1,000 ohm magnet having its keeper to poles this will Compensate for the static charge which greatly interferes with high notes owing to their rapidity & small time between each vibration to allow wire to discharge= Sing all tunes in high key= the low notes sound badly owing to rebound of diaphragm & loss of harmonics=. I think if you exhibit it before the B.A. 10 or 15 miles would be the most satisfactory= Have had two successful exhibitions ~~before~~ in[a] the Permanent Exhibition building Phila. 1st Exhibition there were 5000 people 2nd exhibition about 4000= I send you a slip from the paper=[3] The sounds were loud enough to be heard in every part of the main Centennial building= ~~the t~~ I am pleased to inform you that my speaking Telegraph is now absolutely perfect.[4] It talks as plain English as you or .I. and quite loud. The bark of a dog—jingling of keys clapping of hands—Breathing, yawning—Chirp of a Cricket goes through perfectly= 4 persons talked through it last night at once You would be and I am myself completely astonished the way it works= I can adjust it so sensitive and I have done so that a conversation carried on within 10 feet from the Telephone tube was heard at the distant [-------][b] end. [a] as a test one column of the N Y tribune was spoken through Telephone & copied with only the loss of a few proper names;[5] I have sent over my complete specifications to England and if you can arrange with some responsible house for its introduction on a royalty I should ~~b~~feel greatly obliged=[6] I have found a lot of new things since you left of which I will write soon[7] Yours

T A Edison

Ward was down the other day with young Siemens, but that was before I made the great improvement in articulation.[8] Will have him Come again & he can give you his impression—[9] Just as soon as the WU Co will allow me I will send a pair of speakers—[10]

ALS, UkLIEE, WHP. [a]Interlined above. [b]Canceled.

1. That is, 4 August; thus the instrument probably reached England around 15 August.

2. On the same day Batchelor wrote to Ezra Gilliland, "I finished

and shipped to England my telephone with box laid down today. I had 32 strings in it and it stood on legs 3 inches high it was polished mahogany a perfect beauty" (Cat. 1238:181, Batchelor [*TAEM* 93:148]; see also Doc. 979 and Cat. 1240, items 213, 216, 218, 219, Batchelor [*TAEM* 94:64]). There are several drawings of musical telephones dated 26 July but the design dates back at least to 18 May (Vol. 11:91, 12:63, Lab. [*TAEM* 3:966, 4:29]; Edison's Exhibit 176-11, TI 2 [*TAEM* 11:328]).

3. The *Philadelphia Press* for 19 July 1877 (p. 6) included a report on the first telephone concert at the Philadelphia Permanent Exhibition ("Exhibition Attractions: The Edison Telephone Concert").

4. This probably refers to the results recorded in Doc. 986.

5. See Doc. 982.

6. Edison's provisional British patent specification (Doc. 973) was filed in England on 30 July. If he refers here to manufacturing specifications, they have not been found. Commercial use of Edison's telephones did not begin in Britain until 1878.

7. Preece and Fischer had arrived in England on 14 July 1877 (*Operator,* 15 August 1877, 6).

8. On 31 July, Batchelor noted that although prior telephone devices had problems the newest one gave "absolutely perfect" speech (Doc. 986), as does Edison here, which may date their "great improvement."

9. It is not known whether Siemens made a second visit.

10. Edison's telephones, apart from research apparatus, were made for the Gold and Stock Telegraph Co., which was controlled by Western Union. See Docs. 980 and 982.

## –988–

*Technical Note: Electric Sparks*

[Menlo Park, c. August 2, 1877][1]

Watch spring on common semi[a] rusty iron

No. 1.[2] No 2[3]

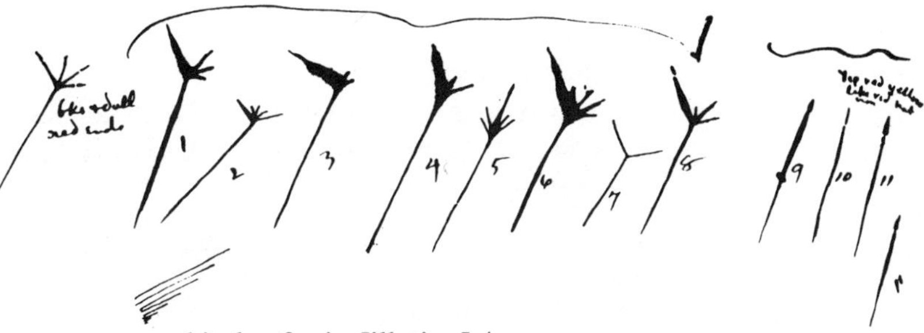

2 kinds of scint[illation]s[4]

Watch[b] spring steel, on aluminum[5]

Watch[b] spg & nickel

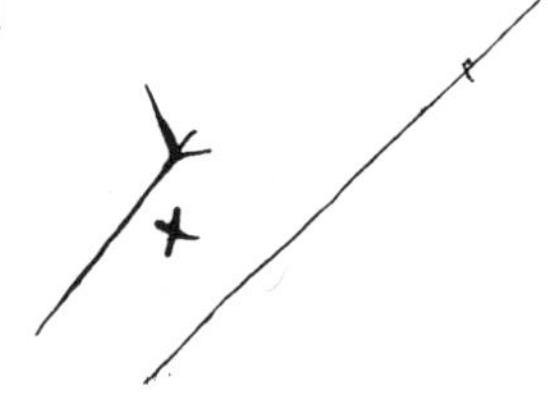

Straight even scint very pale scarcely see it shoots out far its like red hot iron just cooling

These fine lines shoot out horozontally dont see any above ang[l]e of 15.

X lines rise mostly above 45 and are apparently thicker than on iron. about 5 times as many scints on iron as on nickel.

No. 8 scint occurs also 11 but they are very weak.

watch spg on Zinc

I dont see any Zinc scint I notice that the forks of iron[c] are smaller although they shoot out apparently as far apparently as may scints as on iron=

Characteristic iron scints are mostly below 45 white Zinc scints are mostly above 45

watch[b] spg & Tin—

iron scints thin 8 & 11 also notice [an?][d] indistinct fine long nickel scint= but not so strong & long as nickel only got little tin on hand EMG have to try this expernt again

~~Iron~~[b] Watch spg & Platina.

Char[acter]istic

iron scints there bet[ween] dotted lines dull red x point white & bright

Charistics ~~iron forked scints have dull red points every scint fork~~[e] ~~i.e.~~[f] ~~if 3 forks every fork has a dull red end & a break slight this~~

See this on iron green didnt notice be4

d[itt]o[b] on Silver

Notice that scint shoots out & has no stalk you mosly see the[g] prongs appear ~~in~~an inch or so from the contact, noticed those that suddenly appear apparently have a doz prongs

This dull red

d[itt]o[b] Copper  dont notice anything peculiar theres same explosions or appearances as in Argent above

Lead & watch spg—

Seems to be more scints than on iron. the prongs are all short & do not spread out like on iron. the prongs seem spread most on Copper  Characteristic. nothing & then the appearance of a scint.

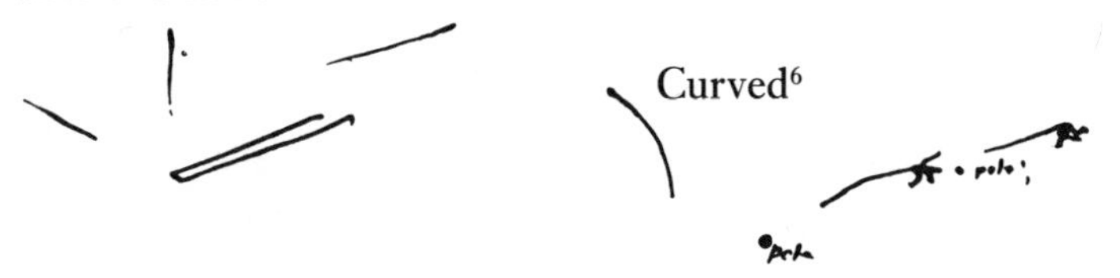


Copper[b] on iron. no forked scints yet if I lay a watch spring on iron & touch copper to it I get the forked Scints.= peculiarity scints of Cu on plate iron shoot out 9/10 of them below angle 3 deg[7]

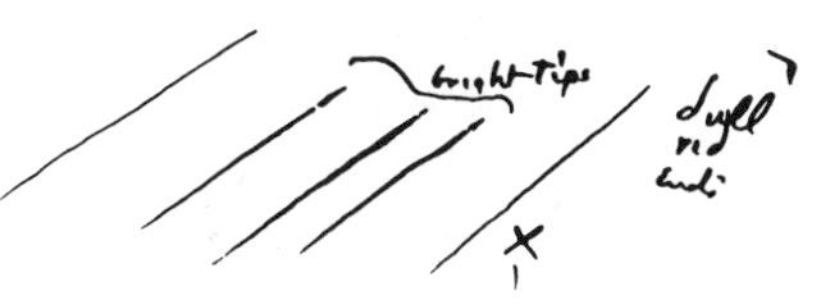

Copper[b] & Aluminum
aluminum scint & X[8] dull—
Cop[b] & Nickel—very dull scints but few

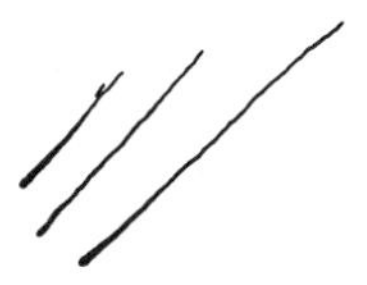

some of them at ½ nearest pole very dull last half brighter there is other peculiarities about these scints havent time to investigate

X, NjWOE, Lab., NS-Undated-001 (*TAEM* 8:8). [a]Interlined above. [b]Preceded by horizontal line across page. [c]"of iron" interlined above. [d]Illegible. [e]Underlined twice. [f]Circled. [g]Obscured overwritten letters.

1. Doc. 989 indicates that the experiments in this note only shortly preceded 3 August 1877. The beginning of this document was mistakenly included as an illustration for Doc. 690.

2. Figure label at left is "b[rea]ks & dull red ends."

3. Figure label is "tip red yellow like red hot iron."

4. These were curious, branching, rapidly varying sparks Edison observed in his experiments with electric spark discharges, arcs, and etheric force. He had been examining such scintillations since December 1875. See Docs. 690, 840, 855, and 864.

5. Figure label is "ought be even this is large & small."

6. Figure labels are "pole" and "pole."

7. Figure labels are "bright tips" and "dull red ends."

8. A line connects this "X" to X in the figure above.

## –989–

*To Henry Draper*

Menlo Park N.J. Aug 3 1877

Prof Draper.

When you were here[1] I did not understand about your discovery of oxygen in the sun.[2] I have just received my Sillimans Journal and I now see its great importance, its both a discovery of oxygen in the sun and in spectrum analysis.[3] I notice that the iron lines in the air spectrum[4] are slanting while the O lines[5] are perpindclr also that in my photo the two spectrums have not been matched the lower or air spectrum being a shade too far to the right. I also notice two dark lines in the large N. line next to Al. There appears to be a white line between the two Al lines which is composed of dots. I suppose these are due to the imperfections of the photo. Since you were here I have perfected the Speaking Telegraph so that it is now absolutely perfect and Talks as plain as you and I do.[6] The bark of a dog, jingling of keys, clapping hands, etc comes perfect. I have found one more phenomenon with the scintiltns[7] if two similar metals are used the scints go in all directions by for instance platina and Al. are used the Al scints shoot out horozontal and ~~p~~below angle of 45° while the platina scints shoot upwards between 45 and 90.[8] I have already 11 ~~k~~characteristic Scints=[9] when you are down this way again call and see me. Yours

Thos. A. Edison

ALS, NN, HD.

1. On 21 July; see Doc. 967 n. 6.

2. In a paper presented to the American Philosophical Society on 20 July, Draper claimed he had photographs of spectra that for the first time demonstrated the existence of oxygen in the sun. He also claimed that the photographs required changes in theories of how spectral lines are generated and in how spectrographic evidence should be interpreted (Draper 1877). His claims and theories were much controverted and eventually rejected (Plotkin 1972, chaps. 3–4; *DSB*, s.v. "Draper, Henry"). Draper had gone to some lengths to secure publication of his

work in several journals, and friends had worked on his behalf to ensure that the August issue of the *American Journal of Science and Arts* (popularly known as *Silliman's Journal*) appeared quickly (George Barker to Draper, 2 July 1877; P. Chase to Draper, 26 July 1877; Charles Young to Draper, 27 July 1877; all HD).

3. The light produced by incandescent gases can be dispersed with a prism or diffraction grating and the resulting spectra studied. The field of spectrum analysis primarily developed from the 1859–60 work of the German scientists Robert Bunsen and Gustav Kirchhoff (McGucken 1969, 1–34). Edison had been interested in spectroscopy for some time, acquiring a spectroscope in January 1875 and another in the spring of 1877 (*TAEB* 2:427 n. 2; Cat. 1233:148, Batchelor [*TAEM* 90:127]; bill of Asahel Eaton, 7 June 1877, 77-004, DF [*TAEM* 14:282]).

4. When in an incandescent gaseous state, every substance radiates light at a unique, characteristic set of wavelengths, which a spectroscope displays as separated bright bands or lines. Conversely, when light from a hotter source passes through a gas, the substances in the gas absorb radiation at their characteristic wavelengths and scatter the rest, resulting in dark lines across a continuous background spectrum. Draper's article was accompanied by a photograph showing a solar spectrum, with its many dark absorption lines, lined up above a bright-line spectrum of air heated by rapidly repeated electric sparks. The sparks passed between iron and aluminum electrodes, so the small amounts of those metals vaporized by the electricity added spectral lines identified with those elements to the air spectrum.

5. Spectral lines due to oxygen; the "N" and "Al" lines mentioned below refer to nitrogen and aluminum.

6. On the "perfection" of Edison's speaking telephone see Docs. 986 and 987 n. 8.

7. Edison's "scintillations" were significant to Draper because spectroscopic studies required generating electric sparks and arcs and carefully identifying all the light they produced.

8. See Doc. 990.

9. See Doc. 988.

**-990-**

*To Henry Draper*

Menlo Park N.J. Aug. 4. [1877]

Prof Draper.

In looking at your photo[1] again I notice that Aluminium lines in the air spectrum slant as well as the iron lines, but not nearly so much as the later. There may be some relation between this an the scintiltns thus[2]

I find that iron gives the greatest number of scints and Sodium the least this appears to correspond to the lines in the spectrum iron having the greatest number and sodium the least,[3] of the metals I have tried Do you think this is worth investigating. Yours

Thos. A. Edison

ALS, NN, HD.

1. That is, the photograph of spectra in Draper's article (see Doc. 989 n. 4).

2. Figure labels are "platina" and "aluminium."

3. This refers to the number of known lines in the spectra of various metallic elements. See, e.g., Roscoe 1870, 134ff, Plate I; and Schellen 1872, 102.

## -991-

*From Edward Johnson*

Philada Aug 4/77

T.A.E.

The second concert was, as you will see by paper sent you today,[1] more successful in the character of the music produced than was the first—but in point of numbers in attendance it was not so good. There were only about 2000 present— This was no doubt owing to the strike and to the fact that it rained steadily throughout the day until near the time for the concert— but it was too late then for many to come out.

I had the piano attachment made in the centre of the Inst. just inside the frame in front of Key Board— & it worked tip top— giving especially good volume & tone when the Pedal was pressed down thus leaving the strings free to vibrate— Nevertheless notwithstanding &c &c I am inclined to believe that the marked improvement was due to the better performance of that last Diaphragm you sent me. It beats all the others hollow—By the way— I could do nothing with the Mica. Its too stiff— The copper are the best though requiring very frequent readjustment—I had a happy thought yesterday morning— of course I was stupid not to have had it earlier— it would have saved me a heap of labor & have improved my performances wonderfully. It is this. Have a 2nd receiving Inst. made— Set it up in short circuit at Transmitting End & Let the operator adjust for each voice & each tune—then immediately switch into main ckt—say 77 & grind it out—thus obviating all necessity for adjusting from the Public end— & enabling me to have an adjustment exactly suited to each tune & voice— Having 2 Instruments you see one can be adjusted while the Other is singing—I am having ~~one m~~ a second crank & bar made, which I propose to attach to the Instrument which I will today buy for you, & use that outfit instead of the Piano— & use the old one for adjusting by—I am also having the plain copper Diaphragm made like the last one you sent me though I must make some improvement in the felt holding arrangement. It nearly played the D——l with me the Eve. of the concert. The felt dropped out on the first piece & It took them nearly ¾ hour to get it back so that it would stay. B.F.[2] finally accomplished it by running a pin through it and fastening the head of the pin to side of Diaphragm ring with sealing wax Strange as it may seem it seemed to work all right, though I told him it wouldnt work at all I find the Brass Diaphragms best for Cornet—I suppose the vibrations are too positive for the more delicate copperfoil But for the voice the copper cushioned one you last sent is just Red Hot I cannot do anything on Duet however until I get the other one fixed.

The exhibition Co are in a peck of trouble & have made a strong appeal to the Pride of the Philadelphians for support. They ask for $50,000 & say unless they get it they must close the thing up on 1st Sept. & are even talking about taking it to New York where it would pay.[3] Meanwhile the management are so harrassed they don't know what to do and told m̄e yesterday they couldn't decide at present about a third concert—wanted to know what the prospect was of getting the Talkers[4] &c I told them you were experimenting upon & constantly improving the talking Outfit but that I would write you immediately & see what the prospect was.

Meantime I am going down to Cape May[5] & give a concert on my own a/c. Reed[6] the journalist who does the Blowing for the Ex. Co. & who got all those Puffs inserted in the Philada papers is going with me as Manager advertising agent &c & we divide the profits. Our Expenses—including a W.U. wire from here will be about 100$. We propose to charge 50c admission— Will get the Stockton House Dining Room which holds about 1200— We calculate upon 500 sure—[7] 200 will pay expenses. We may get 1000 as there are 10 000 sojourners there & a thing of this kind is such a great novelty & had been so fully advertized by the Philada Papers that we think by ad. a few days in advance & Reid prevailing on the Cape May papers to reprint the "Press" article[8] on the day of the concert that we will make a hit— In which case I will follow it up, in quick succession from one resort to another— & then I'll divide with Batch & Charlie & perhaps have a nest egg for you. If it takes poorly at Cape May I'll return to the Exhibition at once & not risk a failure Elsewhere—

It will be conducted in a high toned manner throughout— The Stockton is the leading House at Cape May—

Next Wednesday Evening is the day fixed—

What do you think about the 80 miles of wire is it going to reduce me in Volume very much If so advise me quick I can get 400 cells callaud but no carbon—

If I am successful in this tour need I apprehend any annoyance from Caldwell[9] Write me instantly on receipt of this else I shall be off to the shore. The W.U. ask me 30$ for use of the wire but I think when I see Jim Merrihew I can get him to make a reduction— as that is all they charged Gray with his cumbrous apparatus.[10]

Green[11] of the Exhibition Co did the speechifying the other night & told the audience what you proposed to do in the way of recording speech— I am sorry it is not a little later in the season so that Barker & some more of the Scientific Cusses could be here. Truly

E H Johnson

ALS, NjWOE, DF (*TAEM* 14:846).

1. Not found. A report of the second telephone concert, held on 1 August, appeared the following day in the *Public Ledger* (p. 1). Johnson collected a number of newspaper articles about his concerts as "Testimonials" in his *Telephone Handbook* (Johnson 1877; in DF [*TAEM* 162:1042–45]).

2. Benjamin F. Johnson.

3. The U.S. Supreme Court, in a ruling on the Centennial Appropriation by Congress of $1,500,000, deprived the management of $300,000 of their estimated capital. The management of the Permanent Exhibition sought donations from the citizens of Philadelphia to help defray their working expenses of $50,000 (*Public Ledger,* 1 Aug. 1877, p. 2). As of October 1877 the Permanent Exhibition Co. was $84,000 in debt. Among its several efforts to discharge this debt was a grand fête to be held on Thanksgiving (*Public Ledger,* 9 Oct. 1877, p. 2).

4. That is, the "talking" as opposed to the "musical" telephone.

5. Cape May is a resort town at the extreme southern tip of New Jersey, where the Delaware Bay meets the Atlantic Ocean.

6. Unidentified.

7. According to a 13 August 1877 report in the Cape May *Daily Star,* the Stockton "was so crowded with a fashionable audience, desirous of seeing and hearing the far-famed telephone" that the doors into the

dining room had to be thrown open so that more people could enjoy the music. Cat. 1240, item 230, Batchelor (*TAEM* 94:69).

8. It is unclear to which *Philadelphia Press* article Johnson refers.

9. See Doc. 929.

10. Johnson may be referring to the New York–Philadelphia trial of Elisha Gray's harmonic telegraph in the fall of 1876. See Doc. 786 n. 4.

11. C. W. Greene was general manager of the Permanent Exhibition. *Philadelphia Inquirer,* 19 July 1877, p. 8.

**–992–**

*From Henry Draper*

Dobbs Ferry N.Y.[1] Aug 6th 1877

My dear Sir

Your two letters have come safely.[2] I am very glad the speaking telephone has turned out as well as you wished for it obviates the main defect of Bell's viz; having to listen so carefully.

As to "Oxygen in Sun" what you say about the Iron lines is true and I am not yet satisfied why they slant so much When you look at the spark which is in front of the slit[3] the iron appears as a yellow mantle or ruffle round the air spark thus,[4]

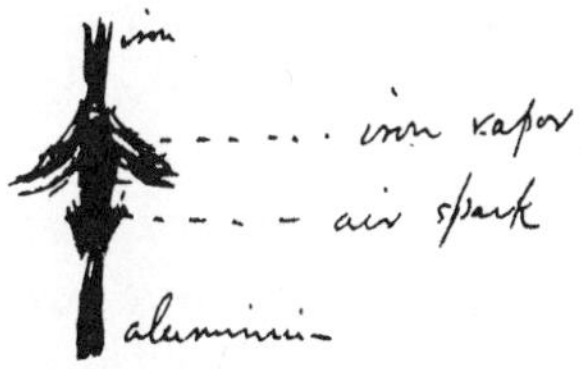

while the air spark is nearly a straight line from pole to pole.[5] As to the delicate lack of coincidence that is due to the fact that it is almost impossible to avoid some flickering of the spark which changes the direction in which the light falls on the slit.[6] Still it is not enough in amount to touch the argument.

As to the value of investigation on the subject of scints the whole matter is so curious that no one can say what may not come out of it.

I hope when we get back to town you will take an early opportunity of coming to 271 Madison Avenue[7] and then you can see and use the fierce currents I get.[8] Yours truly,

Henry Draper

When I am down your way I shall certainly call in. [a]

ALS, NjWOE, DF (*TAEM* 14:115). [a]On two lines spanned by brace.

1. Draper had a country residence in this town, 20 miles (32 km) north of New York City.

2. Docs. 989 and 990.

3. That is, the slit that admits light to the spectroscope.

4. Figure labels are "iron," "iron vapor," "air spark," and "aluminium."

5. See Doc. 989 n. 4.

6. Spectroscopists desired an invariant, electrically produced light source; a large portion of Draper's article had been about dynamos and power sources. He used a Gramme dynamo.

7. Draper's address in New York City. Wilson 1877, 362.

8. There is no evidence that Edison availed himself of Draper's offer to experiment with his dynamo.

## –993–

*From Norman Miller*

New York, Augt. 6th '77

Friend Edison.

I have a package of Telg'f Case testimony, arguments, etc at my office for you. Leave a line here, at Pen office, saying at what hour you are going home if I dont see you.

Hammond[1] has gone up to Illion,[2] will be back this week, and I think will want to make contract with you, and spend a week or two at your shop.

I hope that you will be in [easy?][a] circumstances today.

Mr. O[rton]. said on Thursday that he saw you a moment the day before on way to train but did not get much report on "Speaker" which he is anxious to have.

On Friday Murray promised to bring his[3] in to day—[4] [See?][a] them if he dont—

Mr Prescott has gone to White Mountains—back on Wednesday Yours

N.C.M.

ALS, NjWOE, DF (*TAEM* 14:711). Letterhead of Norman C. Miller. [a]Illegible.

1. James Hammond, a shorthand writer and Civil War correspondent, developed a typewriter design in 1876 that came to the attention of E. Remington & Sons, the Ilion, N.Y., arms manufacturers who were then manufacturing the Sholes typewriter. After working with the Remingtons for over a year, Hammond went on to conduct further experiments elsewhere, eventually bringing out the first Hammond typewriters in 1881. *NCAB* 3:321; Bliven 1954, 83–85.

2. Ilion is a small town in upstate New York, 11 miles (18 km) from Utica.

3. That is, the telephones Joseph Murray was making. See Doc. 982.

4. Miller wrote this on Monday.

**–994–**

*From G. A. Mason*

NEW YORK Aug 7 1877.[a]

Dear Sir:

"The telephone" is something I know very little about and in my present condition I should hardly feel warranted in putting all the money I have or can have, until my father's return from Europe, into anything that I do not know perfectly. I think that if any man could sell it in the Canadas I could and I have little doubt but that I should succeed and yet it is not safe for me to undertake it and leave my family, as I should have to, without money.

On the "horse clipper"[1] I know what I can do and will start out and pay all of my expenses.

I will send you a cut of the machine together with all the information I can get tomorrow[2] Respectfully

G. A. Mason[3]

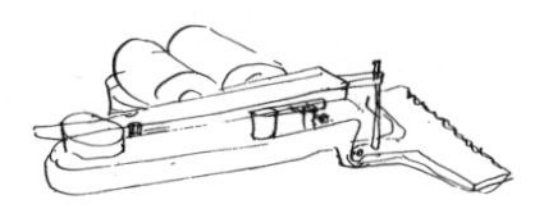

*Edison's sketch of a "Horse Clipper worked by Acoustic Reed."*

ALS, NjWOE, DF (*TAEM* 14:853). Letterhead of R. Henry, Gen. Eastern Agent, Electric Pen and Duplicating Press. [a]"NEW YORK" and "1877." preprinted.

1. Edison first proposed a horse clipping attachment for a portable electric engine in January 1877 (Cat. 997:47, Lab. [*TAEM* 3:375]). On 6 August 1877 he drew a "Horse Clipper worked by Acoustic Reed" (NS-77-001, Lab. [*TAEM* 7:423]). Also see drawings in Vol. 12:89 and NS-Undated-001, both Lab. (*TAEM* 4:36; 8:55). There is no evidence that anything further came of this work.

2. Two days later Mason wrote Edison that "I cannot get a cut of the power horse clippers in the City. They are Manufactured by the American Shear Co. Nashua N.H. and I have sent for circulars, price lists &C &C. Enclosed is the card of a man that makes hand clippers. His price is $3.50." DF (*TAEM* 14:123)

3. Unidentified; possibly Amasa Mason's son.

**–995–**

*From Josiah Reiff*

[New York,] Augt 7/77

Dear T.A.

Do you ever come into town any more?

If so, cannot you drop down occassionally & let me see you.

Have you satisfactorily completed the speaking Telegraph.?

Do you ever give any thought to completing the two (or one) wire Roman letter[1] for a new departure? Has the translator[2] ever been put to work by WU? Do you hear whether it is the impression of WU office that J[ay]. G[ould]. will take control this Fall.?

Could you do anything with use of Electricity in Hydraulic gold mining, where much of the gold [sand?][a] is carried off by the current? Yours tr

J C Reiff

Craig[3] reports that Orton has considered something he has submitted, I presume either belonging to Boyle[4] or Foote & R[5]—that they have been favorably reported on & the committee is now considering what shall be paid for same.

You can ascertain definitely whether that is so. Yours Tr JCR

P.S Did EHJ write you he expected to go to Cape May tomorrow with Telephone?[6]

ALS, NjWOE, DF (*TAEM* 14:713). [a]Illegible.

1. Edison had tried several years earlier to develop an automatic telegraph system that would print incoming messages in roman letters and would use only two (or one) wires for transmission. See, e.g., Docs. 373–74 and 397.

2. Translating embosser.

3. Daniel Craig had been a promoter of Edison's early efforts in automatic telegraphy. See *TAEB* 1 passim; Doc. 700.

4. Robert Boyle, a Brooklyn-based telegraph inventor, was developing an automatic telegraph system. U.S. Pats. 169,513 and 186,104.

5. Theodore Foote and Charles Randall developed an automatic telegraph system that was promoted by Daniel Craig. In April, William Orton had agreed to have the Western Union shop build instruments for tests on Western Union lines. The company declined to adopt the system after testing it in November 1878, and later Craig and others used it on the lines of the American Rapid Telegraph Co. Orton to Craig, 25 Apr., 23 May, and 17 Nov. 1877, LBO 19:205, 301, 20:358, Orton to Reiff, 23 Jan. 1878, LBO 20:456; WU Executive (1875–78): 483, 559, 571, 244, 276–77; Israel 1992, 214 n. 66.

6. See Doc. 991.

[Menlo Park,] Aug 7 1877

*Technical Note: New Force*

Parks[1] Alleged new force

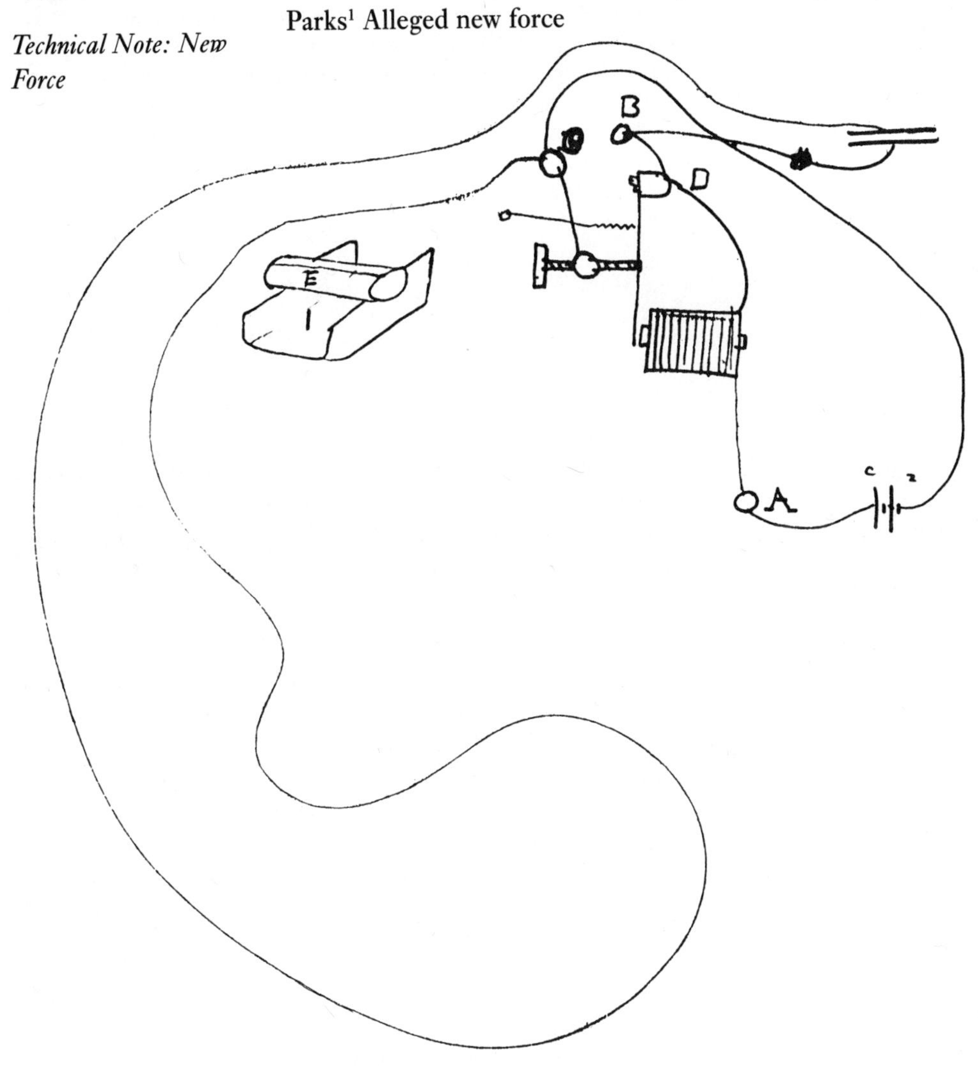


alteration tone vibrator

Only due to effect extra current  Same as I experimented with year or so ago.

X, NjWOE, Lab., Vol. 12:92 (*TAEM* 4:37).

1. Nathaniel Parks, of Deansville in Oneida County, N.Y., first wrote to Edison on 27 July 1877 about his discovery of a "new fourse" like Edison's "Odic fourse," asking Edison if he would "take it in hand" (DF [*TAEM* 14:105]). Edison replied on 29 July, and Parks sent his apparatus to Menlo Park, describing his experimental set-up in some detail in a letter of 31 July (DF [*TAEM* 14:108]). The experiment depicted in this sketch—which largely follows Parks's description—apparently

satisfied Edison that Parks had found nothing new, despite Parks's protestations that this was "the most wonderful machine in the world" (18 Aug. 1877, DF [*TAEM* 14:131]). In the sketch, the apparatus at top center is a vibrator like those Edison used in his etheric force experiments (*TAEB* 2:630 n. 2). The cylinder **E** is a bar magnet whose function is not clear in Parks's description. Parks eventually asked Edison to send the apparatus on to a "tryed friend," a Mr. Herbert of Hoboken, N.J. (11 Sept. 1877, DF [*TAEM* 14:145]).

**–997–**

*Technical Note: Telephony*

[Menlo Park,] Aug 7 1877[1]

Speaking Telegh

I find that with the Sensitive articulater[2] that it is seldom you can put on a metallic diapham without transmitting the ringing Sound of the vibrating metal i.e.[a] harmonics due to inequality in plate, but by using mica or other no ringing Substance this Effect is obviated & this is very important were these ringing sounds to accompany good talking or rather articulation its doubtful if telephone would be a success We tried patent Leather to get rid of them of course this did but all soft substances stretch & give & does not give a proper movement or have stiffness enough to ~~a~~ compress the inferior conducting material hence to make a success such substances as wood—mica—vulcanite etc must I think be used although soft metals like gold & platina might be used—

I find that fine unspun silk if cut up in short lengths Say 1/32 inch & then well plumbagoed will after pressing & refluffing retain the plumbago in spite of the vibration of the diaphragm hence this makes a very good articulater=[3] I find that to get the hissing Sounds of speech good & strong its necessary to use the slots in the mouthpeice even on the new Speakers[b] The hissing sounds striking the sharp edge of the slot somehow are cut into stronger vibrations & these set the diaphragm in vibration

T A Edison

Chas Batchelor
James Adams

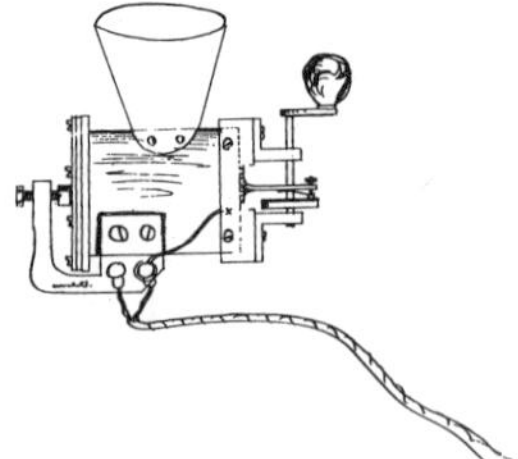

*A portable telephone instrument incorporating an electromotograph.*

X (photographic transcript), NjWOE, TI 2, Edison's Exhibits 93-12–94-12 (*TAEM* 11:415). Document multiply signed and dated. [a]Circled. [b]Obscured overwritten letter.

1. The previous day Edison had drawn several portable telephone receiver/transmitters. Edison's Exhibits 88-12, 90-12–91-12, TI 2 (*TAEM* 11:412–14).

2. A drawing of 3 August is labeled "articulater." Edison's Exhibit 86-12, TI 2 (*TAEM* 11:410).

3. Although Edison testified that this note recorded the first use of a "mass of soft fibre, rubbed with plumbago or other equivalent," which he called "fluff" (TI 1:113 [*TAEM* 11:77]), Batchelor had recorded the use of silk in this manner on 30 July (Doc. 984) and wool the following day (Doc. 986). On 1 August, Edison had stacked a number of "disks of thinnest silk conducterized pbgo—silver—etc" and used them in various transmitter configurations (Edison's Exhibit 84-12, TI 2 [*TAEM* 11:409]). The next day Batchelor had written to a friend in Paterson, N.J. (the center of America's silk industry), asking for some silk cocoons (Batchelor to John Goode, 2 Aug. 1877, Cat. 1238:179, Batchelor [*TAEM* 93:147]); and on 5 August Edison had noted using "fine fibre with plumbago or other semi conductor" in a transmitter (Edison's Exhibit 87-12, TI 2 [*TAEM* 11:411]).

**-998-**

*To Henry Draper*

Menlo Park N.J. Aug 8 1877.

Friend Draper:

After receiving your letter and studying your drawing a little I noticed that the aluminium flame in your sketch[a] darts out sidewise like iron but not so much. on referring to your photo ~~and~~ I[b] notice that the two aluminium lines are also slanting as well as the iron but not nearly so great about in the proportion shewn in your sketch. I have studied your oxygen coincidences, and if anything is certain in Spectroscopy, you have certainly proved the sun guilty of harboring that chemical hyena oxygen There is one thing about this spectral analysis that I cannot get through my head and that is—why does each elementary substance give a number of lines of say different vibrating times. The particle which gives one line Cannot to my mind be like in every particular with the particle that gives a different line. The only way I can figure it out is that there is but one elementary atom, and the aggregate together to form the so called elementary metal or substance[1] for instance iron is composed say of 490 atoms which form a whole, sodium 2 atoms etc. and these great number of lines is due to the break up of a molecule;[2] each taking from the molecule a different vibrating time or harmonic of the fundamental tone of the molecule—[c] When ~~t~~everything that grows or lives can be made of 3 or four substances,[3] as in organics[d] I dont see the wisdom of having so many elementary substances in the inorganics.[4] I will try and call at your place and see how you peek at the almighty through a keyhole Yours

T. A. Edison

P.S. if the Spectra lines of all metals slant and non metals dont it might be used as a distinction E

ALS, NN, HD. [a]"in your sketch" interlined above. [b]Interlined above. [c]"each . . . molecule—" interlined above. [d]"as in organics" interlined above.

1. This general conception, often identified as Prout's hypothesis for the version developed by the British chemist William Prout, was not generally accepted by scientists. Nevertheless, "the idea of a single ultimate unit of matter is met with repeatedly in the physical literature of the second half of the nineteenth century." McGucken 1969, 47.

2. Speculations much like these had recently been published by J. Norman Lockyer, the British astrophysicist who edited *Nature.* He regarded line spectra as evidence for atoms having components and some internal structure. McGucken 1969, 83–101; Meadows 1972, esp. ch. 6.

3. At this time the rapidly growing field of organic chemistry was demonstrating that a very large number of substances were composed of just carbon, hydrogen, oxygen, and nitrogen. Partington 1964, Vol. 4, Part III; Ihde 1964, 161–230, 304–62.

4. For example, among common chemistry books used by Edison, Pepper 1869[?] (16) listed sixty-four elementary substances and Crookes 1871 discussed analytic tests for fifty-six.

**–999–**

*From Henry Draper*

Dobbs Ferry N.Y. Aug 9th 1877

Dear Sir

Your remark about the aluminium flame is true & it does not flare out nearly as much as Iron. I have thought that the slant of those iron lines might give rise to some adverse criticisms on my photograph but you seem to find a pleasure in it. It is proper to say however that if both poles are iron then you can get straight lines easily enough.

As to the relation of position &[a] number of lines and nature of substance quite a good deal has been written and many have felt that there was something in it analogous to fundamental tones & harmonics of sound. But no one has made a clear, precise statement yet.[1] Moreover your idea about compound nature of the so-called elements is just and what is a curious ~~remark~~ fact in that direction is that the nebulae only give about 3 lines and yet they doubtless contain the germs of many of our so-called elements.[2]

I hope you will have time to print something about scints, it is of value theoretically & practically Yours truly

Henry Draper

ALS, NjWOE, DF (*TAEM* 14:121). [a]"position &" interlined above.

1. McGucken 1969 discusses nineteenth-century attempts to understand emission and absorption spectra.

2. For the spectrum of such a nebula, showing just three bright lines, see Roscoe 1870, 282–87, Plate VI.

**-1000-**

*Technical Note: Telephony*

[Menlo Park,] Aug 9 1877

Spkg Telegraph

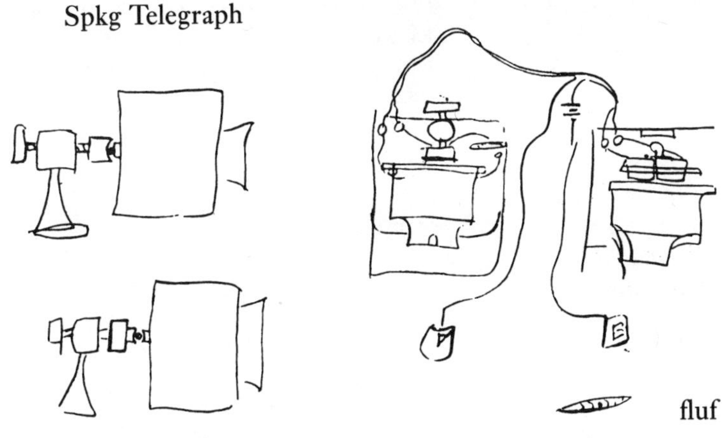


Made tonight

I take unspun silk[1] & pick it out so that fibre will not lay straight & close then take sharp pair scissors & cut it of in peices about ⅛ inch long then plumbago it=& roll it up in cigar shaped rolls & place in the Transmitter It might be wound after rolling with & long single fibre to hold it together well or moulded like a cigar ~~one~~ The magnetic receiver I use wood tube & tin diaphram which I prevent from ringing or ckt vibrating by two deadening rubbers or equivalent constantly pressing diaphram inward or outwards as case may be— tested Magnetic Receiver to NY & back [---][a] tonight drizzled all day had 40 callaud 15 Ohm sounder magnet distance way wire [twas?][b] there & back 69 miles— got ordinary conversation=

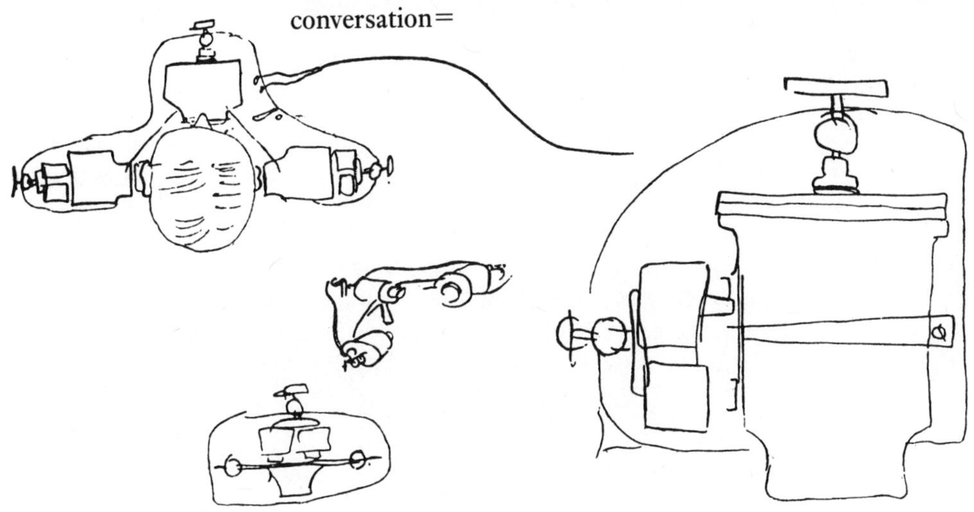

T A Edison

Chas Batchelor
James Adams

X (photographic transcript), NjWOE, TI 2, Edison's Exhibit 98-11 (*TAEM* 11:419). Document multiply signed and dated. [a]Canceled. [b]Illegible.

1. Edison drew a line from "silk" to the drawing of "fluf."

## -1001-

*Technical Note: Telephony*

[Menlo Park,] August 9 1877

Speaking Telegraph

district box[1] with key to call & Telephone in box to tell who it it & what is wanted also call bell in box & the code used to call person where box is from Central Station

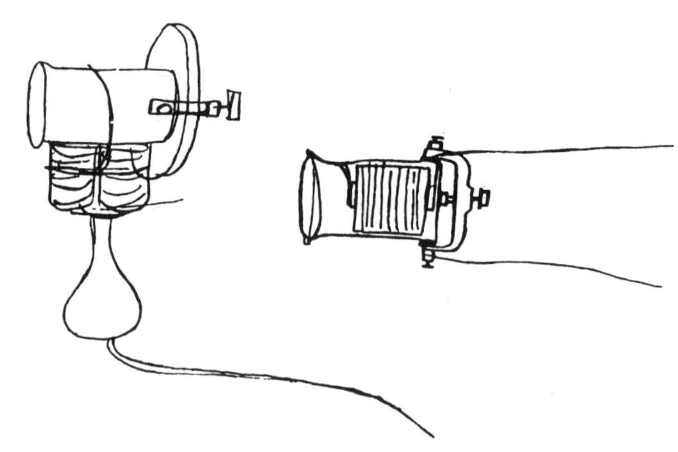

T A Edison

Chas Batchelor
James Adams

X (photographic transcript), NjWOE, TI 2, Edison's Exhibit 96-12 (*TAEM* 11:418). Document multiply signed and dated.

1. Subscribers on district telegraph systems used automatic call boxes to signal a central station for police, fire, medical, messenger, or other services (see *TAEB* 1:411 n. 2). Edison had worked with district systems since early 1872.

**–1002–**

*From Charles Edison*

Port Huron Aug 10 77

Dear Al

Arrived home all OK on Tuesday night and have been spending my time since in looking over the battle field but since the smoke has cleared up I do not see much change for the better in the management of the road as the enemy ~~has~~ are having a cival war among themselves The simple matter is that the whole concern is in a more deplorable condition than it ever was Beard, Cole, Goulden & Atkinson is fighting with Wastell and there is a general eruption in the camp and of course all at the expense of the company. They had a meeting ~~Monday~~ Wednesday and Goulden offered a resolution to pass[1] all the officers & Families connected with the road right in direct opposition to the one that you had passed when you was here and it was unamously passed by every one ecept Wastell and he voted against it and is mad as hell that it was passed I have several things that I want to tell you but I want to get a few more particulars as what I write to you I do not want to misrepresent them in the least will look into that Sarnia matter[2] with Father tomorrow. Love to all

CPE

ALS, NjWOE, DF (*TAEM* 14:583).

1. That is, provide free use of the railway for.
2. See Doc. 1024.

**–1003–**

*Notebook Entry: Telephony*

[Menlo Park,] Aug. 10th 1877.

Speaking Telegraph.

We find that with our transmitting apparatus and a magnet before a diaphragm for a receiver (a plan we frequently tried in the early stage of our experiments) we get splendid talking but low. It is however preferred by Mr David of Pittsburgh who has given it a very thorough test.[1] I made two pair and today we tried it between 197 Bdway[2] and the Clearing house in New York[3] when Mr Camp[4] and several other gentlemen operated on it

Chas Batchelor

X, NjWOE, Batchelor, Cat. 1317:48 (*TAEM* 90:678).

1. In a letter of 21 July discussing the principal virtue (clarity) and shortcoming (inability to work on a grounded circuit) of Bell's telephone, David indicated that he would like to spend 23 July in Menlo Park testing Edison's instruments. He wrote again on 30 July, saying he

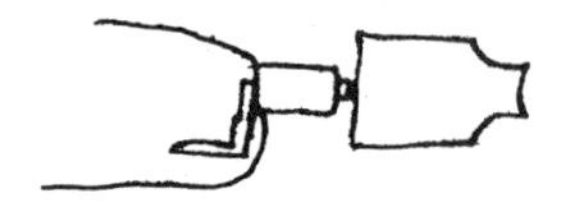

*David's receiver*

hoped to be in Menlo Park on 3 August. In that note he said, "Am very anxious to see what you have done— We have been getting very fair results with Bell's, and if we could get nothing better it would do— I understand its defects pretty well." DF (*TAEM* 14:835, 845).

David first arrived at the laboratory on 4 August. On 9 August, he participated in more tests. Batchelor noted that "He thought it better to abandon the paper and consequently we receive it on a magnet thus [*see left*]. I promised Edison I would make him a full set tomorrow to test and Kreusi & I started at 7 p.m. & worked all night." The next day Batchelor wrote that he had "Worked up till noon finishing 2 sets of speaker telephones and Edison & I went to New York, met David and we got a line. Edison & David stopped at 197 Bdway and I went to Clearing House and we made some good talking. It was some time before we had them all right." Cat. 1233:216, 221–22, Batchelor (*TAEM* 90:161, 163–64).

2. Western Union's main office in New York.

3. The Bank Clearing House was at 14 Pine St., about 1,000 feet (300 meters) from 197 Broadway. Wilson 1877, 14 (City Register).

4. William Camp managed the Bank Clearing House. Wilson 1877, 14 (City Register).

**-1004-**

*Technical Note: Phonograph and Telephony*

[Menlo Park,] Aug 12 1877[1]

Phonograph[2]

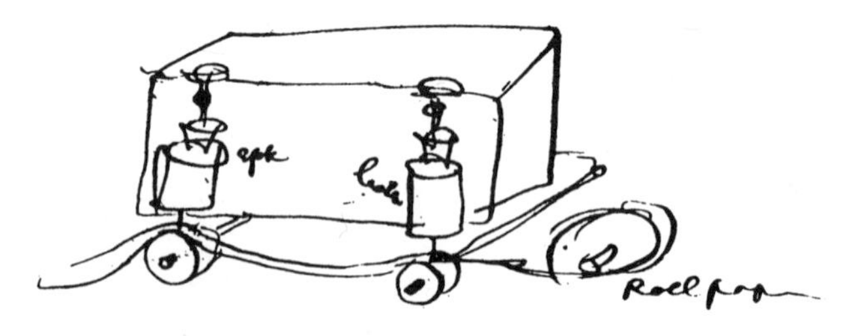

any power to rotate

Imp[ro]v[emen]t in Bells Telephone ~~carry~~ put extension on end of poles & lay clear across diapham.

also

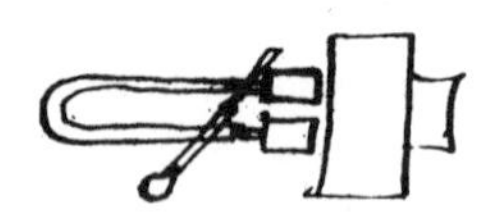

[-----ing?][a] short ckt permanent magnet= use iron filing in cup between diaphram & poles allowing speaking tube be perpendicular—

T A Edison

X, NjWOE, Lab., NS-77-003 (*TAEM* 7:449). This document is photographically reproduced in Charbon 1981, 40. [a]Illegible.

*On the occasion of the fortieth anniversary of the phonograph's invention, Edison signed and (mis)dated this later sketch.*

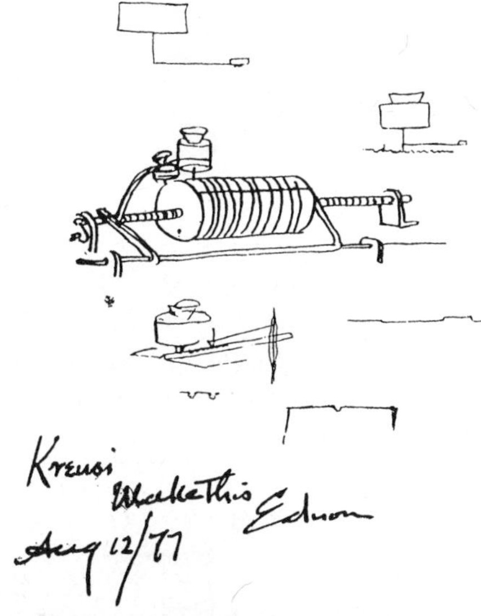


1. Another phonograph sketch incorrectly bearing this date has been widely publicized and reprinted as the first sketch of the phonograph (NS-77-003, Lab. [*TAEM* 7:448]). For the history of that sketch see Koenigsberg 1969, xiii, and Pershey 1989, 110.

2. Figure labels are "spk," "listen," and "Roll paper."

**-1005-**

*Technical Note: Telephony*

[Menlo Park,] Aug 12 1877

Spkg Telegraph

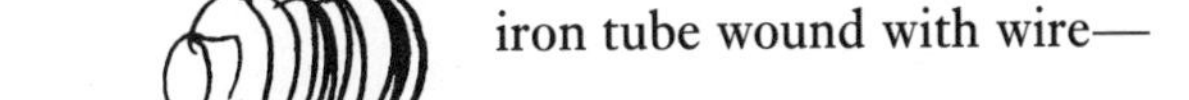

iron tube wound with wire—

Continuous magnet having its core formed of disks of iron vibrations cause these to give a noise=

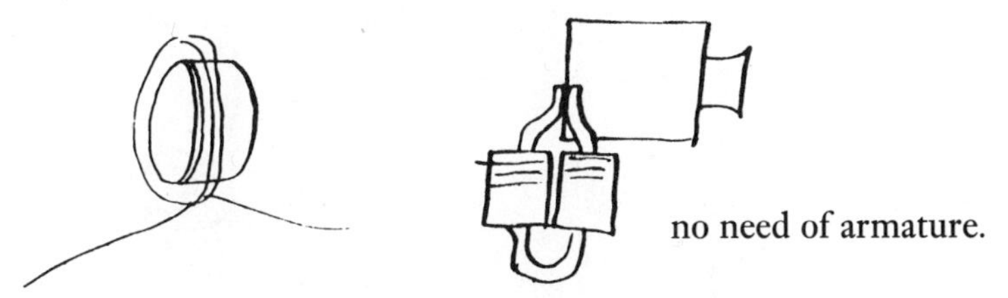

no need of armature.

I will mention here that the conducting fibre can be dispensed with in my telephone & its equivalent substituted ie[a] the clean fibre may have a semiconducting substance included in its folds[b] & that will work=

Even loose plumbago or equivalent will work this

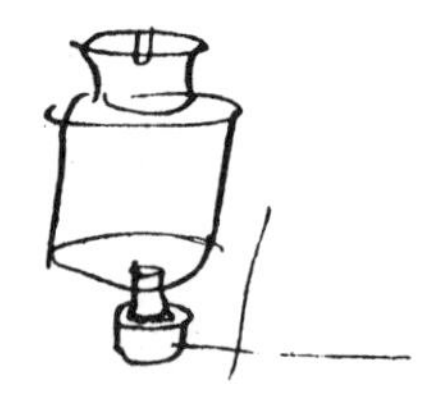

Loose pbgo or Mercury or it may rest on a spring or on top of fibre sponge felt or giving substance or it may be mixed with exceedingly fine iron filings & a permanent magnet placed on diaphram this will attracted it back & forward according to the movement of the diaphragm thus causing variable resistance—

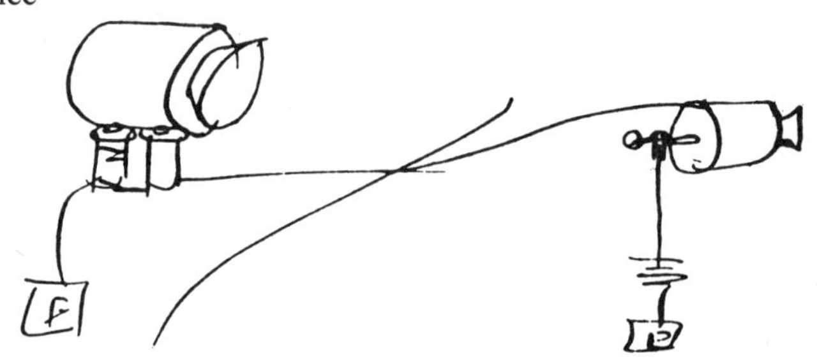


Resonant tube very thin & filled—in with iron filings or tin filings= or lose iron wire.

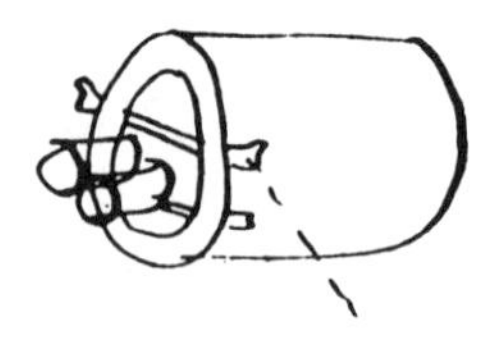

Soft[c] rubber to deaden the diaphram  it wks well, prevents harmonics or excessive vibrations  hard peice of Rubber[b] or equivalent material could be placed on different parts of the diaphragm & held against it by adjusting screws to dampen it=

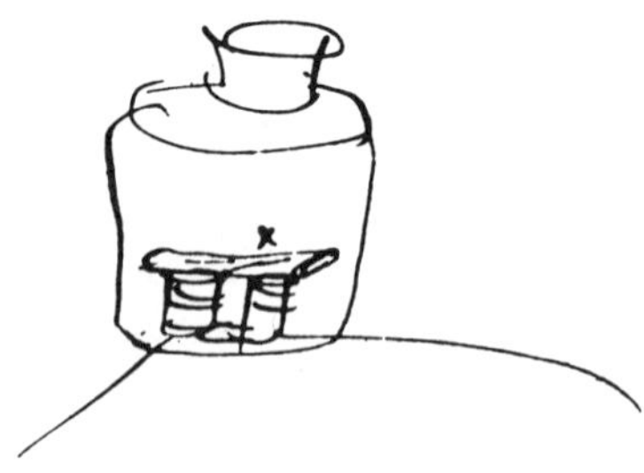

X is peice dry paper with iron filings in its folds  works fairly

Diaphragm with
centre removed

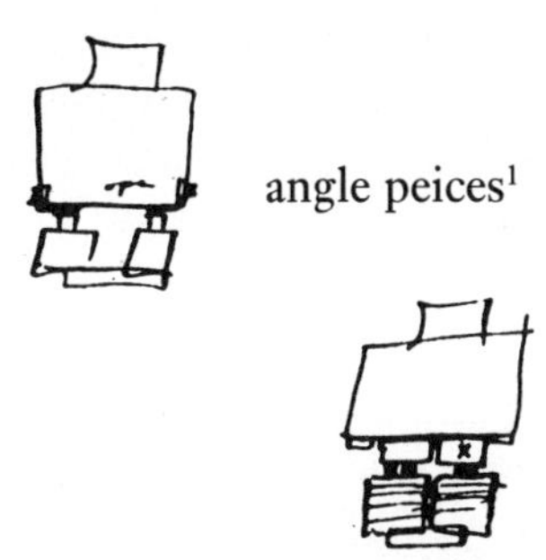


X extra peice going clear across the diaphram

T A Edison

Chas Batchelor
James Adams

X and X (photographic transcript), NjWOE, Lab., Vol. 12:104; TI 2, Edison's Exhibit 105-12 (*TAEM* 4:45; 11:421). Document multiply signed and dated. [a]Circled. [b]Obscured overwritten letters. [c]Text appears at end of dotted line in drawing.

1. Figure label is "open."

**-1006-**

*To Thomas David*

Menlo Park, N.J., Aug. 13, 77.

Friend David:

Just started on a dozen, have beautiful design—[1] have improved articulation so that no harmonics can be detected, and you recognize in an instant who is talking. We use an exceedingly thick Mica diaphram. The dozen will be in N.Y. about Friday.[2] If you wish I can put a complete apparatus in a Domestic Telgh. box, taking out the works and simply adding a call key.[3] Yours,

T. A. Edison.

TL (transcript), NjWAT, Box 1067 (Corp.).

1. Payroll records from this week indicate that John Kruesi was making eight "magnetic" telephones. 77-010, DF (*TAEM* 14:357).

2. That is, 17 August.

3. See Doc. 1001.

**-1007-**

*From Thomas David*

Pittsburgh, Aug 14th 1877[a]

Friend Edison—

Your letter of the 13th is just rec'd, and does me good like a medicine— Your silence after promising to telegraph me I regarded as ominous— Really you have reached the result ~~which~~ by the means[b] I finally tho't you would—a thicker diaphram—

As I have telegraphed,[c] I must have two sets promptly, as I am under obligation to test Bell's,[1] and want to have yours at the same time— Please arrange with Mr. Orton if necessary so that I can have some of the lot now underway— I know he will be agreeable— Besides I have gone thro' the thing so much, that I know pretty well what will do—

As to putting a complete set in a District box—[2] When you can get around to that point I will want to see them, meantime I want to get the other going—

It isn't particularly desirable to have the thing in so small a compass. Good results as you know, in any form, will be the one thing to accomplish—

I wish I could have stayed right with you through these experiments— Yours Truly

T. B. A. David

Regards to Batch, et al—

ALS, NjWOE, DF (*TAEM* 14:858). Letterhead of Central District and Printing Telegraph Co. [a]"Pittsburgh," and "187" preprinted. [b]"by means" interlined above; "the" interlined above that. [c]Obscured overwritten letters.

1. David had been testing Bell's telephones and negotiating with Gardiner Hubbard since the end of June. David to Hubbard, 2, 10, 14, and 20 July 1877, Box 1175, NjWAT.

2. See Doc. 1001.

**-1008-**

*To Thomas David*

[Menlo Park,] Aug 15/77

Friend David

I do not think we shall need a call bell as Hello![1] can be heard 10 to 20 feet away. what you think?

Edison

P.S. first cost of Sender & Receiver to manufacture—is only $7.50—

L (photographic transcript), NjWAT, Box 1067 (Corp.).

1. In his autobiographical reminiscences Edison claimed that he "Invented Hello for telephone" (App. 1.D304). Allen Koenigsberg (1987, 3–9) concludes that Edison was indeed the first to use "Hello" (Alexander Graham Bell preferred "Ahoy").

**–1009–**

*William Orton to Thomas David*

New York 15 Aug 1877.

I have yours of 14th instant, and one of two or three days before.[1]

I have telegraphed Mr Edison to send me the two sets as soon as the present lot is completed.[2] I am greatly in hopes that the new and simplified apparatus will meet your expectations.

I was told today that there were orders for one thousand Bell's Telephone from Boston and vicinity on hand, and it is evidently important that we should commence to occupy the field immediately.

[Wm. Orton][a]

LS (letterpress copy), DSI-NMAH, LBO 20:51. Original unavailable for final proofreading. [a]Signature missing from available copy.

1. Not found.

2. The same day Orton telegraphed Edison that "David wants two setts of the new speakers as soon as ready I think his tests will be valuable and advise sending them." DF (*TAEM* 14:861).

**–1010–**

*To Henry Draper*

Menlo Park N.J. Aug 17. 77

Prof Draper

Is Kirchoffs map in Roscoe the largest published.[1] with my large Eaton Spec[troscope][2] I see lots of lines that are not in Kirchoffs map. for instance at the D lines[3] I see[4]

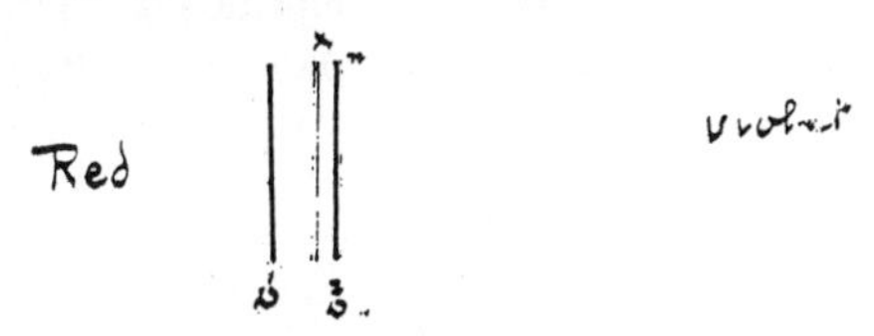


There is a line between D that is double x[a]. it is not central but is nearer the violet end. also see 2 lines between Kirchoffs next line towards the violet. also I think I see, bright lines which are brighter than the general background of the spectrum.[5] I will try and place them on Kirchoffs map.= I am mapping the scints[6] I will have more scints than Kirchoff has lines. Yours

T. A. Edison

ALS, NN, HD. [a]Interlined above.

1. The eminent German physicist Gustav Kirchhoff, who developed fundamental laws of electrical circuits, also pioneered the use of spectroscopy in astronomy and chemistry (*DAB*, s.v. "Kirchhoff, Gustav").

Roscoe 1870 (202–7, Plates III, IV) reprinted Kirchhoff's map of the dark (absorption) lines in much of the solar spectrum.

2. See Doc. 989 n. 3.

3. Early in the nineteenth century the German optician Joseph Fraunhofer diagrammed the relative positions of dark lines seen in the solar spectrum and assigned letter labels to the most conspicuous features of the resultant map. These Fraunhofer lines and their letters became common reference points. "D" is a pair of lines in the orange region. Roscoe 1870, 26–27, 200–201, 224, Plates I, VI.

4. Figure labels are "Red," "x," "n̶," "1D," "2D," and "violet."

5. Bright lines in a spectrum of dark lines were the subject of considerable controversy in spectroscopy. McGucken 1969.

6. See Doc. 988.

**-1011-**

*To William Orton*

Menlo Park NJ Aug 17/77

Wm Orton

The Telephone line spoken of here[1] as running into the Zieutung building[2] runs into Dickenson[3] & Broomans office;[4] is Mr Dickenson attorney for Gardiner G Hubbard & the Telephone?[5] Yours

T A Edison

P.S. I am in interferance in the Patent office with Young Dickenson on a late Quadruplex application=[6] E[dison][a]

ALS, DSI-NMAH, WUTAE. [a]Obscured by attached clipping.

1. A clipping from the *New York Sun*—"The Telephone in New York"—accompanied this letter. It listed the five telephone lines then in operation in the city and reported that one went from the main office of the New York Telephone Co. (the firm then introducing Bell's telephones in New York) to the building "where the attorney of the company has an office."

2. The *New York Staats Zeitung*, on Tryon Row. Wilson 1876, 44 (City Register).

3. Edward Dickerson, Sr., was a technical expert and eminent patent lawyer. He had served as one of the main attorneys for Western Union in the Quadruplex Case, cross-examining Edison for several days three months earlier. *DAB*, s.v. "Dickerson, Edward N."; Dickerson's argument for defendant, Quad. 73.15 (*TAEM* 10:726–72); Quad. 70.7, 71.1 passim (*TAEM* 9:403–762, 800–817; 10:5–202).

4. William Beaman was Edward Dickerson's law partner. Their office was on Tryon Row. Wilson 1877, 81, 334.

5. Gardiner Hubbard, a businessman, patent lawyer, and opponent of Western Union, was Alexander Bell's father-in-law, chief backer, and the organizer of the commercial introduction of Bell's telephone (*TAEB* 1:253 n. 2; Bruce 1973, 83ff). Dickerson was retained by the Bell Telephone Co. in early September 1877 (Charles Cheever to Dickerson and Beaman, 8 Sept. 1877, Box 1198 [Corp.], NjWAT).

6. Edward Dickerson, Jr., was a lawyer and inventor ("E. N. Dickerson," *Elec. Engr.* 9 [1890]: 33). Edison had received a letter dated 13 August announcing an interference involving his Case 139 (which actually was for sextuplex telegraphy) and applications by the younger Dickerson, Franklin Pope, and Henry Nicholson, which began the first of several interferences involving the younger Dickerson (Pat. App. 512,872; Dickerson to TAE, 10 Nov. 1877, DF [*TAEM* 14:504]; on Pope and Nicholson see *TAEB* 1:114, 226; 2:419, 752 n. 2).

**-1012-**

*Charles Batchelor to Vesey Butler*

[Menlo Park,] Aug. 17th [18]77

Dear Sir,

Do you think any thing can be done in the West Indies with the Telephone Mr Edison assisted by myself has just completed a "speaking telegraph" and we are now introducing them. Conversation is carried on with perfect ease over wires from 100 feet to 100 miles in length. The instruments are exceedingly simple and will work on very inferior lines. We are introducing them on the private lines of the Gold and Stock Tel Co who rent out some 400 wires to parties using between their offices, factories, etc.

They are useful for communicating in factories covering large ground, mines, plantations, et[c.][a] We place them all on royalty in this country but probably that could not be done in Cuba. The price per pair of instruments (it requires a pair at each terminal) including royalty is $80 currency. I think you could purchase them from us at this price and charge that amount or a slight advance for placing them on the line and a yearly rental of about 100 to 125 dolls for the use of them

The charges in this country are $25 per p[air?][a] per year. We shall shortly have some circulars[1] on this subject and will then send you some

A number of these instruments rented to responsible parties under a contract would give you a steady income

There is no doubt that the majority of private lines in this country (a great many of which use Edison's printing telegraph)[2] will use the speaking telegraph in place of or in connection with the system they now use. Its low price in comparison with printing instruments[3] will enable parties, who otherwise would not pay the money for printing instruments, or Morse operators, to use a private line.

Please let me know what you think of this project and oblige Yours respectfully

Chas Batchelor

ALS (letterpress copy), NjWOE, Batchelor, Cat. 1238:194 (*TAEM* 93:155). [a]Faint text.

1. Not found.
2. The universal private-line printer (Doc. 262).
3. George Anders's printing telegraph sold for $200 to $250, while Elisha Gray's and Edison's sold for $125 to $150. Israel 1992, 103.

**-1013-**

*Technical Note: Phonograph and Telephony*

[Menlo Park,] Aug 17 1877

Spkg Telegh.

In my apparatus for recording & reproducing the human voice—I propose using a paper coated with a substance which becomes very soft by heat & when cold is extremely hard like sealing wax.[1]

I think a Cork diapham both for receiving and sending is the best thing we have yet struck, on account of an absence of harmonics[a]

Phonograph.

Paper is previously embossed and brought to a knife edge: then the little point ~~is~~on the diaphragm having a knife edge only has to indent this edge which it ought to do very easily.

This edge may be a substance deposited on it ~~the~~ if this embossed edge will work the speech may be retransmitted. over a telegraphic circuit.

Another idea Indent the paper in spiral grooves or on a long strip cover whole of paper with tin foil. The point on the diaphragm wi~~th~~ll then easily indent—

Repeater for talking Telegh

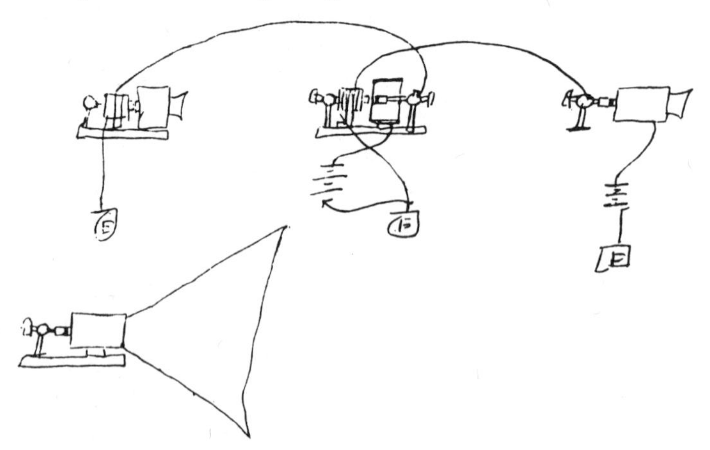

Transmitter by which one can carry on a Communication anywhere in a room

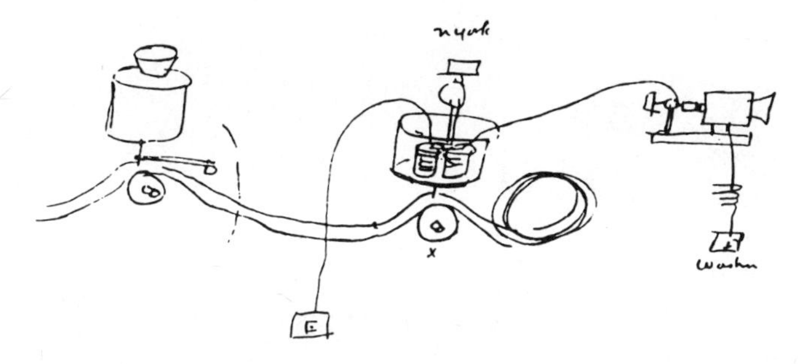


Reproduce in NY[2] the talking recorded on X by slowing up speed so as allow copyist to copy  Record the talking by indenting.[3]

Phonograph[4]

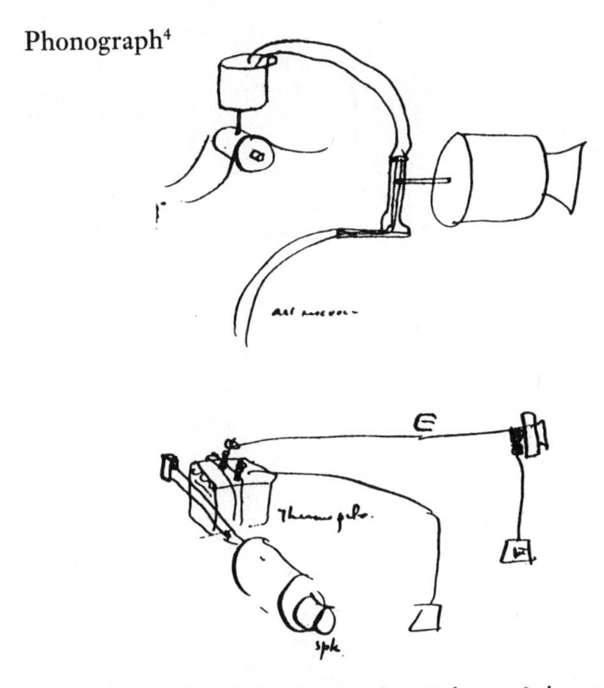


X soft rubber[5] in spkg it stretches & heats & in regaining normal posistn it becomes colder  its only thing I know that loses heat instantly hence its movement by a diaphram would generate heat waves & these acting of Thermopile would generate electric waves; & these acting in magnet & diaphram at other end would be reproduced=[6]

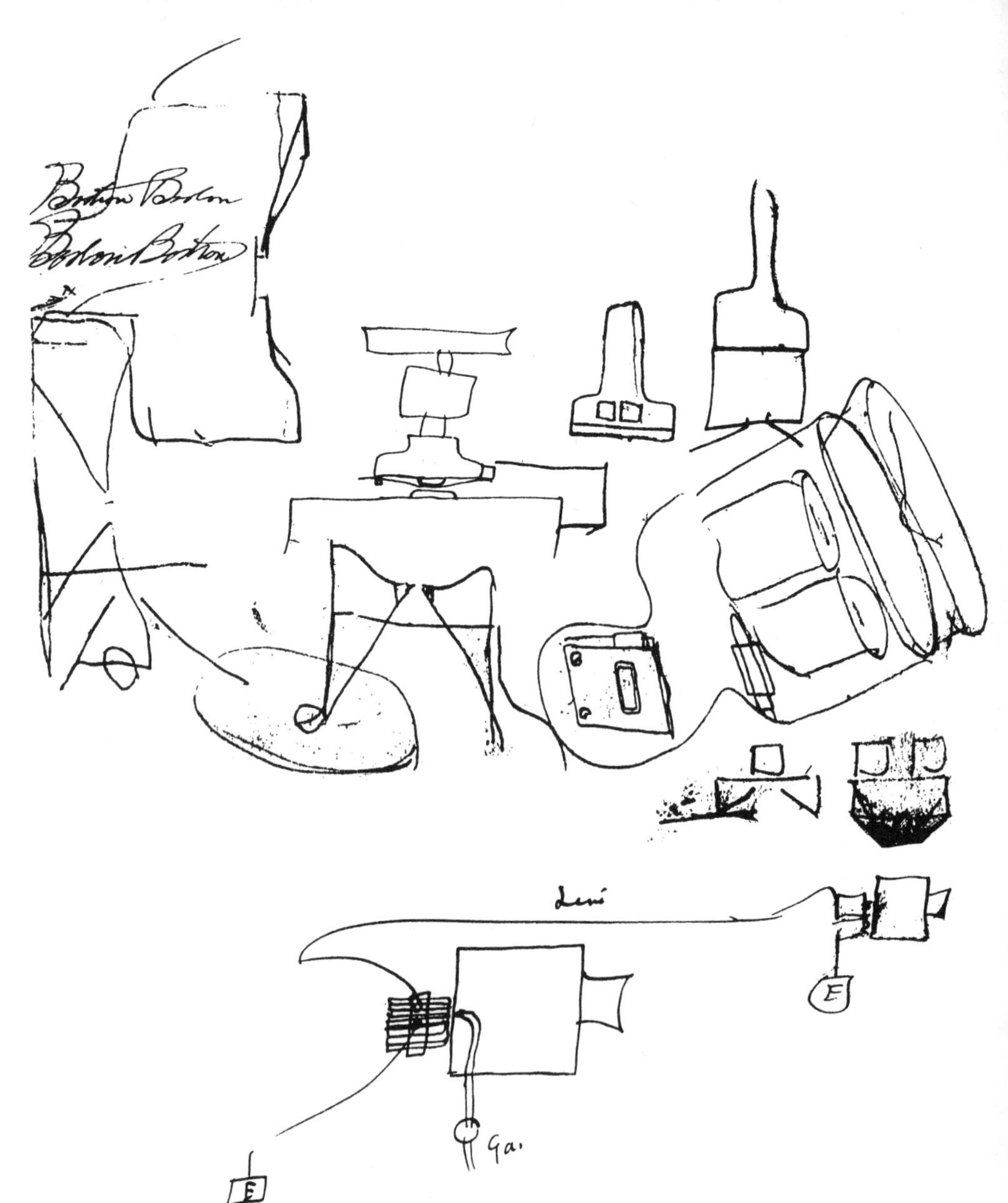


Jet of gas;[7] platina wire kept hot by battery and manny other means keeps the metallic diapham hot  this vibrates in close proximity to an exceedingly delicate thermo pile connected to the line. Perhaps something cool at the other end would be good

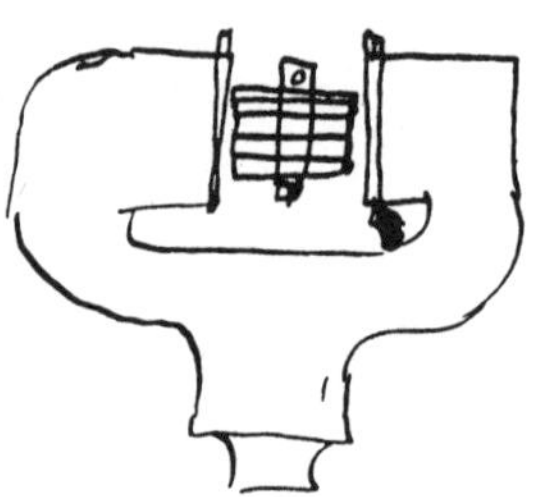

both diaphrams kept hot or one kept hot other cold or can use two diaphrams of[b] which one is ~~na~~always naturally colder than the other=[a]

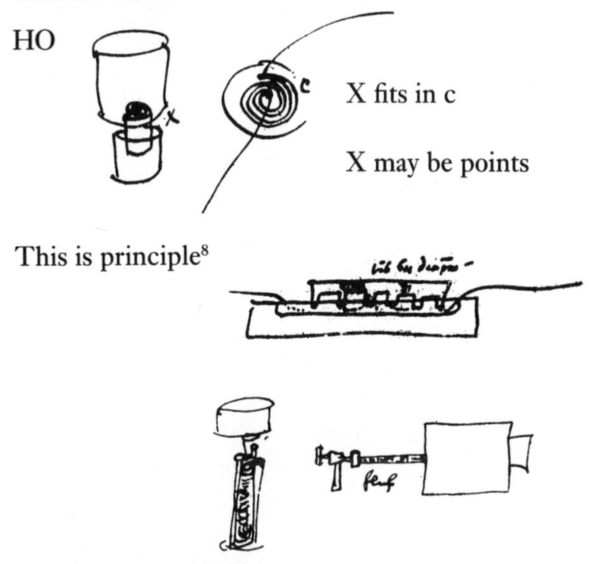


This is principle[8]

fluf[9] coated with a Semi-Condr or wet pla~~t~~ced in or between 3 insulated pins [----ing from a plate?][c] diaphrm has strght pin in it touching fluff & its contracted & expanded by the movement of the diapham

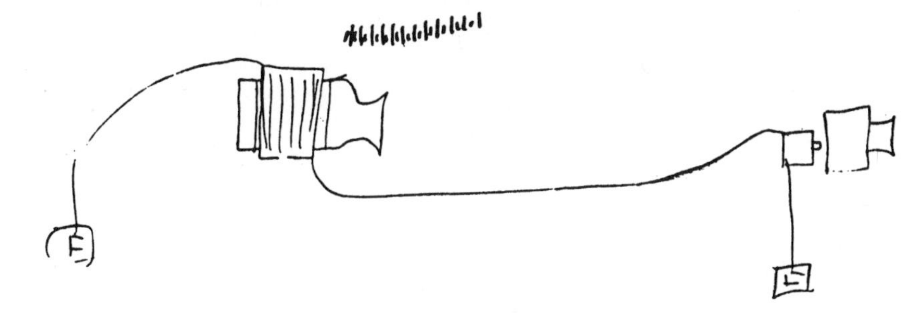

Tin tube closed ends &[b] covered with wire. Talk into it sets whole thing vibration generate magneto currents

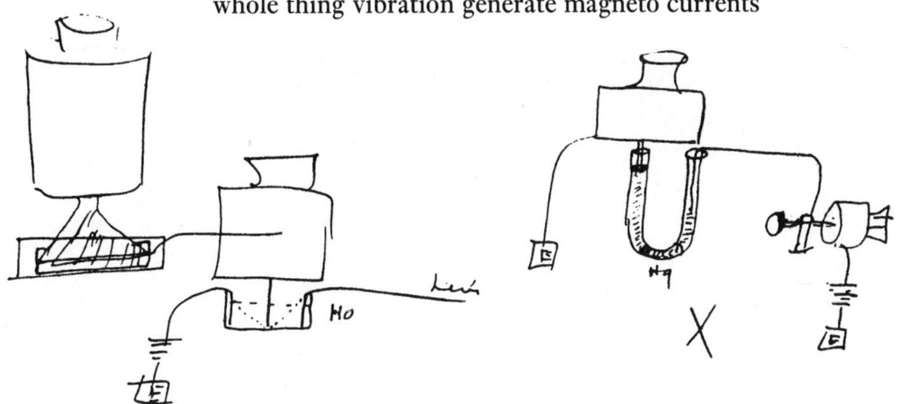

X[10] is an attempt to use the capilliarty movement of mercury to work a diapham[11]

i find that fluf works good if pressed  made thus wks ok[12]

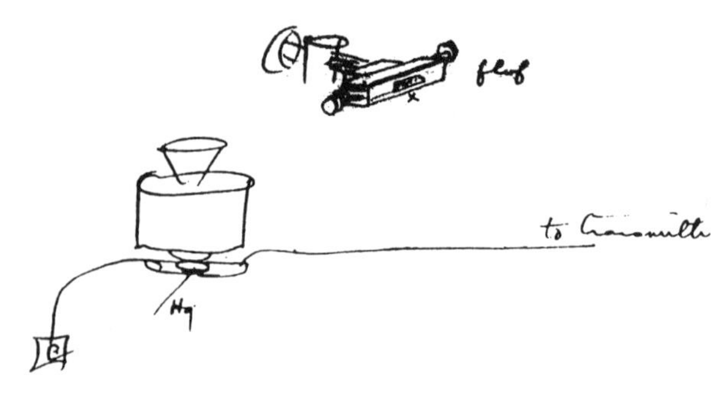


Hg in flat dish with a chem solution.[13] I claim setting a diapharm in motion by a movement drerived from the action of Electrolysis on Mercury=

I use the Expansion of the iron wires to work a diaphram[14]

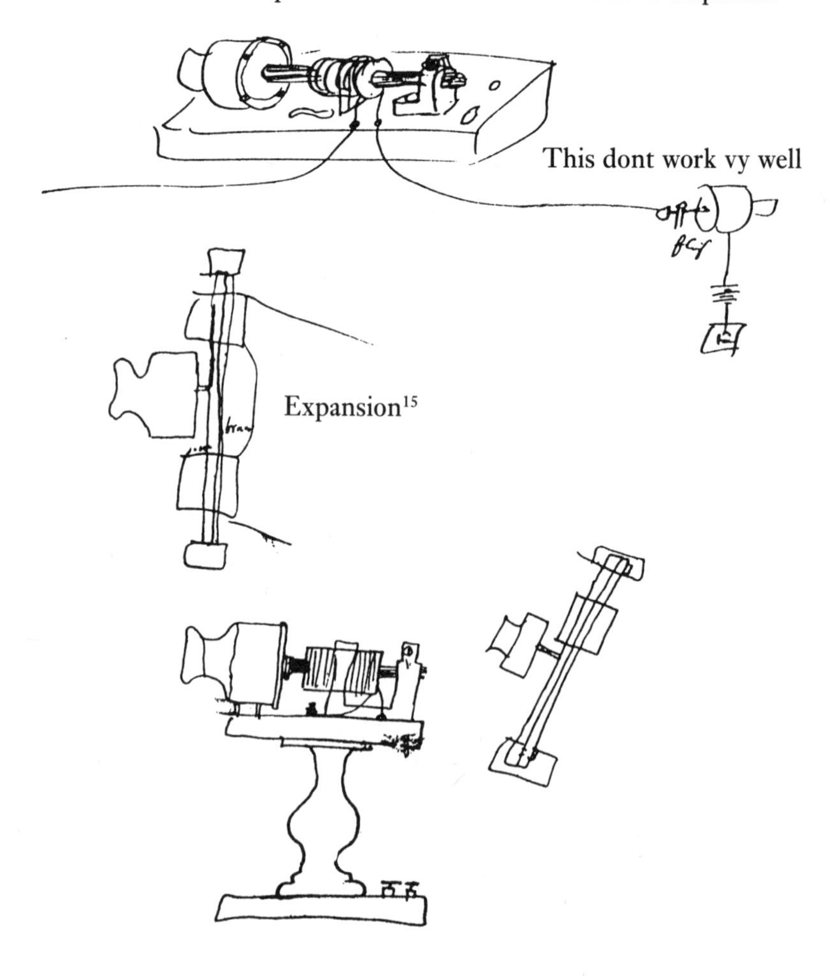

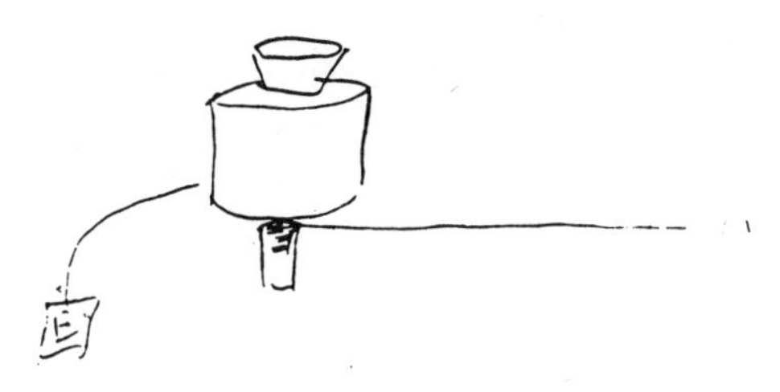

have a very fine diaphm and [insulated?][c] Electrode [except?][c] extremely fine point uninsulated of platina— use a good conducting solution The waves coming over wire will produce gas & this will crackle & give a sound or motion to the diaphram

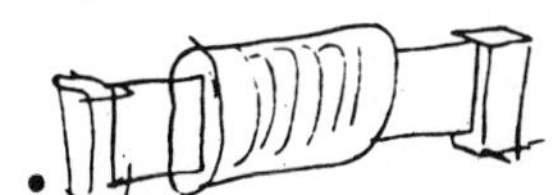

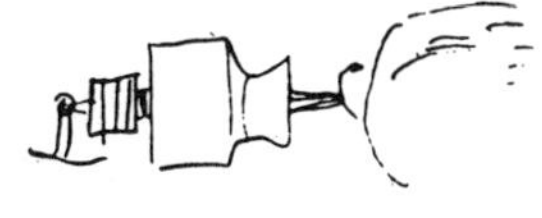

Ear peice to Concentrate Sound[16]

I propose to use Ark[ansas] oil stone & pbgo it & let fluf rest against thus— this will give more margin[17]

pbgo loose talk diphm toss it about & pvt good contact

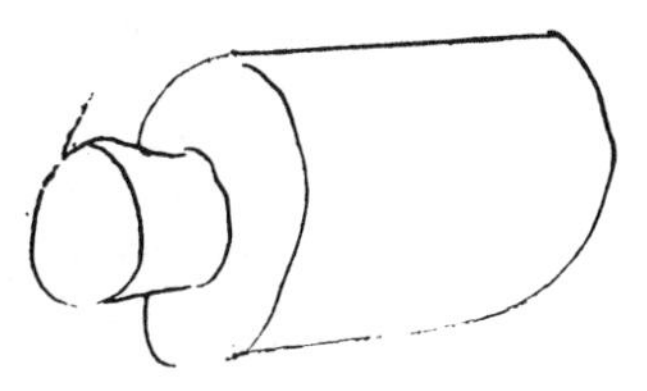

I find that the B. P. F. &. V. give wind rushes different from .C. .G. H J S T X Z

The slot in the mouthpeice is sufficient ~~from~~or the latter but the other rushes go directly in tube hence I propose to put something in tube having sharp edges to cut the wind which rushes in wind not downward like the "sh"

it can be Edge

this this

New Telgh—[18]

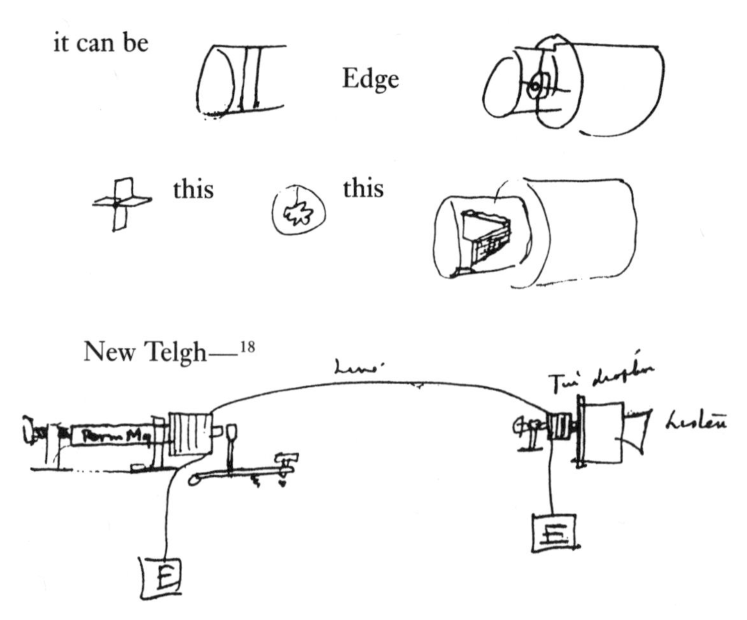

or can dup apparatus at each end
T A Edison

Chas Batchelor
James Adams

X and X (photographic transcript), NjWOE, Lab., Vol. 12:108; TI 2, Edison's Exhibit 115-12; Lab., Vol. 12:109, 112; TI 2, Edison's Exhibit 113-12; Lab., Vol. 12:114, 110; TI 2, Edison's Exhibit 111-12; Lab., Vol. 12:107 (*TAEM* 4:48; 11:425; 4:50, 52; 11:424; 4:53, 51; 11:423; 4:47). Document multiply signed and dated. [a]Followed by centered horizontal line. [b]Interlined above. [c]Illegible.

1. Testifying in 1896, Edison and Batchelor remembered recording sound experimentally on many different materials in the second half of 1877. The exact action of the recording point—embossing, indenting, scraping, or cutting—was the subject of intense legal dispute. Pp. 171–72, 210, 586–92, 599–602, 617–20, 623–25, 644, 647, *American Graphophone v. U.S. Phonograph* (*TAEM* 116:159–60, 179, 367–70, 373–75, 382–87, 397, 398).

2. Figure labels are "N York" and "Washn."

3. Prescott 1879 (549–50) reproduced and described another design for recording telephone messages, a "telephonograph," that Edison drew the same day. The original drawing is lost, but see Vol. 15:6, Lab. (*TAEM* 4:333).

4. Figure label is "air resevoir."

5. Figure labels are "Thermopile" and "Spk." **X** is the thin strip at far left attached to the tube labeled "Spk."

6. Edison had experimented with thermostats using hard rubber and gum copal. Doc. 514; NS-77-004, Lab. (*TAEM* 7:566).

7. Figure labels are "Line" and "Ga[s]."

8. Figure label is "vib by diapm—."

9. Figure label is "fluf."

10. Figure labels are "HO," "Line," and "Hg."

11. See Doc. 846; Prescott 1879, 547; and Jehl 1937–41, 1:144.

12. Figure label is "fluf in at X."
13. Figure labels are "Hg" and "to transmitter."
14. Figure label is "fluf."
15. Figure labels are "iron" and "brass."
16. Figure label is "brass." It seems to indicate a brass wrapper around the oilstone.
17. That is, greater variation of current strength.
18. Figure labels are "Perm Mg," "Line," "Tin diaphm," and "Listen."

**-1014-**

*Charles Batchelor to Charles Edison*

Menlo Park NJ. Aug 18th 1877 saturday night.

Dear Charley,

I have not been able yet to collect anything for ink[1] and Al is perfectly hard up. I cannot get enough to live on. I expect to be better fixed next week. I have asked twice for money but he has none expenses on gas machine[2] & plumbers I guess cleaned him out Yours

C.B.

ALS (letterpress copy), NjWOE, Batchelor, Cat. 1238:201 (*TAEM* 93:159).

1. Duplicating ink.
2. On 22 June, Edison had placed an order with the Combination Gas Machine Co. of Detroit, Mich., for a system to produce illuminating gas from gasoline. After a delay caused by the railroad strike (see Doc. 979), the machine had been delivered to the laboratory and installed during the second week of August; the lights were first lit on 24 August (see Doc. 1020). At this time Edison was still waiting for his insurance companies to grant him permission to use the machine, but he would start using it within a few days. He had already spent $75 on it and owed $300 more. (In 1878 the company would ask Edison to hang a picture of the machine in his office, which he did, promising to call it to the attention of visitors.) *TAEM-G1*, s.vv. "Berry Bros.," "Combination Gas Machine Co."; Cat. 1233:171, 229–30, 236–37, Batchelor (*TAEM* 90:138, 167–68, 171); Humboldt Insurance Co. to TAE, 20 Aug. 1877; American Insurance Co. to TAE, 20 Aug. 1877; Home Insurance Co. to TAE, 21 Aug. 1877; Citizens' Insurance Co. to TAE, 22 Aug. 1877; Germania Insurance Co. to TAE, 22 Aug. 1877; all DF (*TAEM* 14:334–38).

*An engraving from the Combination Gas Machine Co. letterhead.*

**-1015-**

*To George Prescott*

NEW YORK Aug 20 1877[a]

Dear Sir

I have been speaking to Mr. [Gerritt] Smith about the Quadruplex in England.[1] he says that you are inclined to believe that he would do alone over there Now I have had some experience with those people; I know their characteristics from attempt I made myself in trying to sell them something,[2] and I have had conversations with Preece and know his views through his friend Ward of the direct cable; I feel thoroughly convinced that if the Quadruplex is to be sold that two experts with apparatus must be sent from here, and that the apparatus must be placed in actual work with regular business and then your experts must teach 1 or more of their inspectors, any exhibition on a loop, or on bogus business, or in the night will not satisfy them. Those officials are there for life and no one will take the responsibility to reccomend anything without they are absolutely sure that it will work regularly in practice. Preece himself told & Ward attested that people who desire to sell anything to the PO Telgrs must not expect that they were going to great trouble, they parties must prove the worth & practicability of their[b] apparatus and ample facilities would always be given. If you desire this thing to be a success (and it will if it works one week between London & Liverpool) Smith & Hamilton[3] should go over They can do the thing properly and if so the thing is sold; The expense is small if you take into consideration the amount in prospect the responsibility of the parties and the almost certainty that in Smith & Hamiltons hands the thing will be a perfect success— Smith says that everything is ready that he and Hamilton could leave on the 29th of this month August and I think they could be back here in 6 weeks; The simple fares of S & H to London & Return is 530. The living in London is the same as in N.Y. say $15. per week each so you see 700 or 800 will cover all expenses;= I think you ought to say the word go and they will go—

AL, DSI-NMAH, WUTAE. Letterhead of Electrician's Department, Western Union Telegraph Co., George B. Prescott, Electrician. Original unavailable for final proofreading. [a]"NEW YORK" and "187" preprinted. [b]Obscured overwritten letters.

1. Edison, Prescott, and Smith had agreed in May to pool their two British quadruplex patents. Edison received a 35% interest, Prescott 45%, and Smith 20%. Agreement, 31 May 1877, DF (*TAEM* 14:683).

2. For Edison's attempts to sell his automatic telegraph to the British Post Office see *TAEB* 1, chap. 12; 2, chaps. 1–4.

3. George Hamilton (b. 1843) entered telegraphy in 1861 as an operator. In 1867 he was made chief operator and circuit manager of the Pacific and Atlantic Telegraph Co. In 1873 he became an experimental assistant to Moses Farmer, and in 1875 he joined Prescott at Western Union as an assistant electrician. Reid 1886, 674–75; Taltavall 1893, 255.

**–1016–**

*Notebook Entry: Telephony*

[Menlo Park,] Aug 20th 1877.

Speaking Telegraph

I finished to day and shewed with Edison to Mr Orton two pair of our new speaking telegraphs.[1] The transmitting apparatus is improved by coating silk fibre with plumbago (thoroughly) and placing a wad or twist of this material between the spring and the platina adjustable face. It is exceedingly delicate in regard to altering the resistance of circuit by the slightest pressure In order to keep this 'fluff' from falling out it is enclosed by a hard rubber piece as in fig 55 at X.

54 55

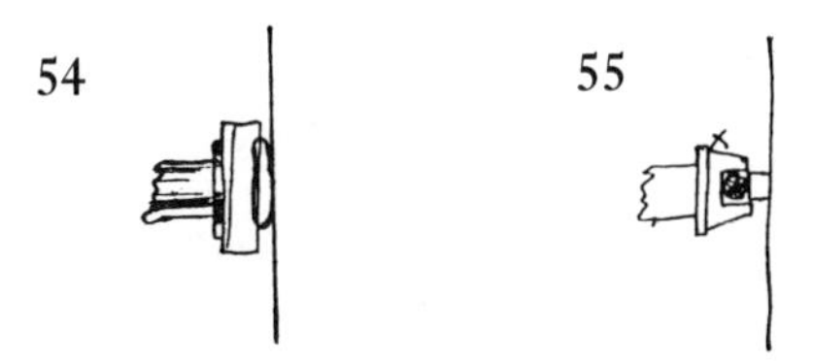

This 'fluff' has considerable resistance and therefore when not talking we cut out this by a switch.

56

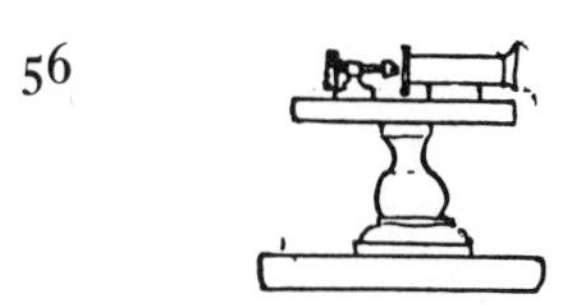

The transmitter is put on a stand as in Fig 56 as also the receiver which is improved by putting a piece of rubber between the diaphragm and frame of magnet which dampens and kills extraneous noises.

Chas Batchelor

X, NjWOE, Batchelor, Cat. 1317:43 (*TAEM* 90:678).

1. In his diary on this day, Batchelor drew the instruments on stands and noted that he had "Finished at 5 AM the first 2 pair of speaking telegraphs and took them to New York and shewed them to Mr Orton Put them up between Prescotts room 39 & Mr Small's room in 197 Bdway. They worked a little weaker than at home." The next day they "tried a little more on same circuit" and then "brought them home and worked all night on trying new fluffs etc." On 22 August they put the instruments on a private line in New York. Batchelor "got [Adams] very good but there was so much noise at his end that it was impossible for him to get all [Batchelor] said." Cat. 1233:232–34, Batchelor (*TAEM* 90:169–70).

**–1017–**

*To William Preece*

[Menlo Park, c. August 20, 1877][1]

Friend Preece=

I forgot to mention that no magnets of any kind must be in the circuit when you are working the Musical Telephone; otherwise the tones will be weakened greatly by the induced Crnt of the magnets; Yours

Edison

P.S. Prescott tells me your people have authorized a test of the Quadruplex= I hope you will insist on Prescott sending over two experts; E

ALS, UkLIEE, WHP.

1. The date "Sept 3" is written on this letter in another hand, possibly indicating when Preece received it. If that were the case then Edison sent the letter about 20 August, which was the day he urged Prescott to send Smith and Hamilton to demonstrate the quadruplex to the British Post Office engineers (see Doc. 1015).

**–1018–**

*From Henry Draper*

Dobbs Ferry N.Y. Aug 22 1877

My dear Sir

The map of Kirchhoff in Roscoe or Schellen[1] is as large as anything published. I think some of the lines you see in D are atmospheric. With my solar spectroscope there are quite a large number toward sunset.—[2] There is no doubt you can see bright lines in the spectrum look for instance near G.[3]

I am glad you are progressing with the scints;[4] it is a curious subject. Yours truly

Henry Draper

ALS, NjWOE, DF (*TAEM* 14:134).

1. Kirchhoff's map of the solar spectrum was also reprinted by the German physicist and educator Heinrich Schellen. Schellen 1872, 164–73, Plates II, III; see Doc. 1010 n. 1.
2. See Roscoe 1870, 224–27.
3. "G" is a Fraunhofer line in the blue region of the solar spectrum.
4. Doc. 988 is the latest record found of such experiments.

**–1019–**

*From George Bliss*

CHICAGO Aug. 23 1877.[a]

Dear Sir:

Beetle[1] wrote from London July 21st as follows.

"Brandon[2] says we are not obliged to work the Pen in Italy until a year from next August."

This does not agree with Mr Serrell's statement.

I hope it is true as it will give more time to perfect an arrangement.

Beetle has gone to Italy to get a start there.

He thinks the London Press & battery[3] will give better satisfaction than ours but I doubt it & shall not think of changing till we consult.

He says the Secy[4] of the English Co. dont amount to much as he is young & is the son of an interested party who has money.

The real head is [Archie Fairlie?][5b] who treated him well and pumped him hard.

They wanted to know where patents had been obtained and applications made. They were about starting in Russia. Beetle shut them off & tried to do so in Australia & other colonies. They sent four Pens to Africa the day Beetle was there. Where wont Edison's goods go? Beetle says they have done more outside of England than in it. Please stop all this if you can. Respfly

Geo. H. Bliss Gen'l Man.

ALS, NjWOE, DF (*TAEM* 14:413). Letterhead of Electric Pen and Duplicating Press. [a]"CHICAGO" and "1877." preprinted. [b]Illegible.

1. Nothing is known of George Beetle apart from his involvement in promoting Edison's electric pen in Europe. See *TAEM-G1*, s.v. "Beetle, George S."; *TAEM-G2*, s.v. "Beetle, George L."

2. David Brandon, an "Englishman educated in France in engineering," was a Paris patent attorney specializing in work for English-speaking inventors. He (or his firm) handled most of Edison's French patents for several decades. "1850–1950: The Centennial of a Firm of Patent Attorneys" (Paris: privately printed for Simonnot, Riuny, Blundell & Pont, 1949).

3. That is, the press and battery that Frederic Ireland and John Breckon had introduced for use with the electric pen. See Doc. 804.

4. Unidentified.

5. Unidentified.

**-1020-**

*Charles Batchelor Diary Entry*

[Menlo Park,] FRIDAY, AUGUST 24, 1877.[a]

Jim took new receiver[1] to New York & went to .E. 25th St & I went to Pearl St.[2] After considerable time I got him well but he could only get me occasionally. Brought all instruments home to alter.[3]

Drew out plan at night for new transmitter & sender so[4]

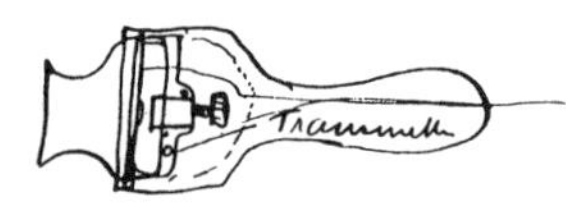

Receiver a magnet inside handle with plate of tin on cores loose & prevented from falling off by small pieces on edge of disc thus These are held loose by staples in edge of wood

~~Receiver~~ Transmitter[b] made so that connections are made in[c] machine proper and flexible cord passes through the handle, if you want to adjust take off shell & slip handle along cord.

Gas started in Laboratory tonight

AD, NjWOE, Batchelor, Cat. 1233:236 (*TAEM* 90:171). [a]"FRIDAY, AUGUST 24, 1877." preprinted. [b]Interlined above. [c]Obscured overwritten letters.

*A "good receiver" designed by Charles Batchelor.*

1. The previous day Batchelor designed the new receiver with a "plain magnet in cylinder with piece of tin laid on top." Cat. 1233:235, Batchelor (*TAEM* 90:170).

2. This is the same private line they used to test instruments on 22 August. See Doc. 1016 n. 1.

3. At some point that day they apparently took the instruments to Western Union headquarters. Cat. 1317:44, Batchelor (*TAEM* 90:679).

4. Batchelor first designed handles for telephones on 12 August (Cat. 1233:224, Batchelor [*TAEM* 90:165]). His drawings of the alterations made on 24 August are Edison's Exhibits 119-12, 121-12, 123-12, 124-12, and 127-12 (TI 2 [*TAEM* 11:427–31]). Batchelor began making instruments on this "new plan" the following day (Cat. 1233:237, Batchelor [*TAEM* 90:171]).

*Charles Batchelor's 24 August 1877 design for a transmitter with a handle.*

**-1021-**

*To Thomas David*

Menlo Park, N.J., Aug. 25, 77.

Friend David:

Yours from Chicago received.[1] Richmond's water dodge[2] was the first thing I worked on, there isn't much good in it. The disengagement of gas from the points due to the decomposition of the water causes a frying pan noise to which Bell's is nothing. I have had an application in pat. office on this long ago.[3] I have made three pair of Telephones thus:[4]

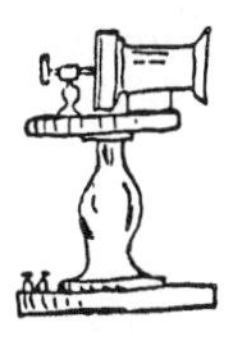

Transmitter Receiver;

They are louder than Bell's and articulation as good, tried them on line in N.Y. 3 miles long, wet day; found no frying pan noise, but the cross fire from the heavy stock circuits was frightful, but strange to say when we commenced talking it went way over these noises so you couldn't hear them when talking, unfortunately one end of the line was in a sugar refinery and the coopering department was next to office; the extraordinary noise of 20 coopers hooping bbls. prevented us from receiving much, but got everything o.k. in Wall street the other. I am told that there is more cross fire on this line than any other in city, also that they couldn't work Morse owing to noise. I now find that these stands although they look very pretty are not convenient. I have withdrawn them and are making some after the following design

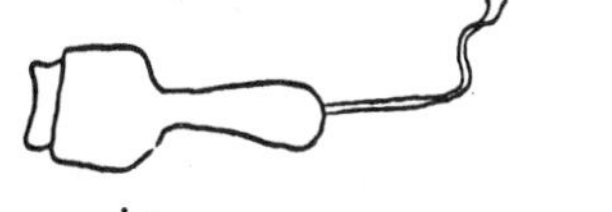

Transmitter Receiver

and this is the reason I have not sent you those promised, the new ones I can make very rapid. The articulation is pronounced perfect much better than Bells. The improvement was made by using no diaphragm. It is also much louder with the new improvement. Will telegraph you when I ship yours. Keep up your faith.

Edison.

Mr. O. has talked on the stand telephone o.k.[5]

TL (transcript), NjWAT, Box 1067 (Corp.). Drawings apparently are tracings.

1. David to TAE, 21 Aug. 1877, DF (*TAEM* 14:879).

2. Both David and George Bliss had written to Edison about George Richmond, of North Lansing, Mich., who claimed to have invented a water telephone for which he filed a patent application on 24 August 1877. He would later be an unsuccessful claimant in seven of the Telephone Interferences. David to TAE, 21 and 25 Aug. 1877; George Bliss

to TAE, 20 and 22 Aug. 1877; all DF (*TAEM* 14:879, 881, 883); TI 1:ii (*TAEM* 11:10); Interferences on Telephones: Brief in Behalf of Thomas A. Edison, TI 5:14 (*TAEM* 11:868).

3. It is not clear to which application Edison refers. His Case 144 described a particular form of water telephone. Although he might have sent it to Lemuel Serrell some time earlier, he did not execute that application until 31 August.

4. A drawing of the stand for these instruments, dated 11 August, is in Vol. 12:101 (Lab. [*TAEM* 4:41]).

5. See Doc. 1016.

**-1022-**

*To Thomas David*

Menlo Park NJ Aug 27/77

Friend David

Just finished our new set of Telephones 2 receivers and two senders they are now as small as Bells and much louder than you have ever heard, and the articulation is so perfect that everybody gets what is said without any effort. The strain on the ears appears to have departed at least in my case; You know I have only Batch & Kreusi that can work so you see that it will be a few days before I can send you ~~my~~ your[a] pair. Yours

Edison

L (photographic transcript), NjWAT, Box 1067 (Corp.). [a]Interlined above.

**-1023-**

*From Thomas David*

Pittsburgh Aug 27 1877[a]

Friend Edison—

Why this delay? No Telephone coming, neither do I hear from you, and I begin to regard the silence ominous— How is it? I am hopeful of your Instrument, and am awaiting impatiently the result—

The Bell Telephone is being brought rapidly into use, and it behooves us to be spry—[1]

I understand there are applications in for patents that cover your principle of "internal resistance"—[2] It is said that Bell in his earlier experiments tried the same thing—[3] I drop these suggestions for your information— I suppose you got my letter from Chicago—[4] Yours Truly

T. B. A. David

ALS, NjWOE, DF (*TAEM* 14:884). Letterhead of Central District and Printing Telegraph Co. [a]"Pittsburgh," and "187" preprinted.

1. David's correspondence with Gardiner Hubbard and Thomas Sanders indicates that his Central District and Printing Telegraph Co. was using Bell telephones by the beginning of September. Box 1175 (Corp.), NjWAT.

2. Bell's telephones used electromagnetic induction to vary the signal current, whereas Edison's successful telephone designs changed the signal current by varying the transmitter's resistance. Edison's first telephone patent application (executed 18 April 1877; U.S. Pat. 474,230) employed variable resistance in this way. George Richmond's water telephone was a variable-resistance design, as was the telephone described in a caveat filed by inventor Emile Berliner on 14 April 1877, which the Bell interests later used against Edison's telephone claims. This principle was the subject of one of the major cases in the Telephone Interferences. TI 1:i–vi (*TAEM* 11:9).

3. Bell's first telephone patent (U.S. Pat. 174,465) mentioned this principle, and he had displayed one device of this type at the Centennial. The first apparatus with which he transmitted any articulate speech was actually of this type, although he had not said so publicly. Bruce 1973, 164, 178–81.

4. David to TAE, 21 Aug. 1877, DF (*TAEM* 14:879).

**–1024–**

*From Charles Edison*

Port Huron Aug 27th 77

T.A.

We went over to Sarnia as I told you we were going and of which Father has written you[1] and when the question was put to symingtons as to how they could seize your stock for the deficiancys if any in Fathers account they admitted that they could not do it  all the differance between the accounts is that they do not intend to allow him any salary or for a large quantity of oak timber that he bought and paid for out of his own money and put into the road  well now you know they have got to do it because they knew just how it stood when they bought.

but you may be easy about your stock for they acknowledged that they could not touch it  they said that probly in the course of a few months they would buy it from you. and as I said before the whole thing was hatched up as a scare in order to get you discouraged and sell your stock cheap just as on this side. I aint going to write you any about this road but will put the whole thing before you when I get back, write me and let me know what you think of the matter.[2]

Charley

ALS, NjWOE, DF (*TAEM* 14:586).

1. In that letter Pitt indicated that the dispute concerned his account with the Sarnia Street Railway Co. and the company stock he and Edison held. 20 Aug. 1877, DF (*TAEM* 14:585).

2. Pitt wrote Edison an account of the meeting with the Symingtons on 30 August. He also said that "Charly is well and home sick to get back." DF (*TAEM* 14:588).

**–1025–**

*To Thomas David*

Menlo[a] Park N.J. Aug 29/77

Friend David

~~Deal~~ Delay cannot be helped= it requires time. Bell had a very easy job compared to what I have had. ~~It was~~ He had nothing to discover is the means of doing it= all he had to ascertain was if well known means would do it. at the Centennial May 10 he exhibited his speaker & there aint much improvement or change been made; over a year; whereas I had to create new things, and many obscure defects in applying my principle. besides I am so deaf that I am debarred from hearing all the finer articulations & have to depend on the judgement of others.

I had scarcely got t~~o~~he principle working before there is pressure in NYork to introduce it immediately  I made 2 or 3 pair but found they were unhandy after they were made; that delayed. I have finished a new pair and they have been working 2 days with no change of adjustment. they are much louder than the loudest you heard here and everyone concurs in saying that the articulation is perfect and they do not see what improvement is required  I have them working on line to Depot with 2500 ohms in ckt thats good 100 miles and everybody gets it, There is no noises whatsoever heard and I have 12 cells of Callaud battery in circuit. I have my man making a model for patent office[1] which is essential I should get in, and I and Batchelor must go to NYork to show it there  so you see I have my hands full ~~every~~ even working 22 hours per day— The Water Resistance man[2] & Gray are behind the age. When Mr Orton sent me translation of Reiuss telephone in 1875 I tried that. I have patent on it ~~usin~~ used in another apparatus issued in 1873;[3] besides its frightfully inconstant owing to decomposition of the water, again, Mr Bells records in his interference case with Gray which gives the whole history of his work in Telephony with dates makes no mention of having employed my principle,[4] & even if he had he loses by ~~not~~ lack of due diligence.

So there is no fear of any one in this Connection; please

have a little patience & I will soon have you a pair= I suppose you could get along with a single receiver & sender for a while. Yours

Edison

L (photographic transcript), NjWAT, Box 1067 (Corp.). This document is photographically reproduced in Simonds 1940, facing p. 161. [a]"“A”" in top margin in unknown hand.

1. Edison executed the application (Case 144) on 31 August and filed it and the patent model on 5 September. The patent issued as U.S. Patent 492,789. Pat. App. 492,789.

2. George Richmond.

3. U.S. Pat. 141,777; see *TAEB* 2:772.

4. That is, varying the resistance in a telephone transmitter to make the signal current vary; see Doc. 1023 n. 2.

**–1026–**

*From Robert Watson, Jr.*

Montreal Augst 29/77

Sir

Since writing you on Monday[1] I have arranged with a gentleman[2] here who has a large Capital from $50.000 to $75.000 who is willing to go in with me and work up a business here for the Telephone and give you any reasonable guarantees that the business will be pushed forward for all interests concerned— I told him your terms viz $5. per year Royalty on each Sett of instruments and $15 for the instruments themselves payable in advance.[3] That you desired all Contracts to be made directly with you that is all parties leasing Telephones through us to sign a contract with you agreeing to pay your royalty as you proposed to me. Should you decide upon coming to an arrangement with us telegraph me to the above address[4] here and we will both go down and close matters at once A large business can be done here even though trade is dull and I am holding some parties desiring private lines until negotiations with you are concluded. If you can possibly send me a Sett of Telephones do so at once. Write me your terms in full giving all particulars so that I may lay them before the party I mention in writing[a] whose standing here I can satisfy you of through Dun Barlow & Co[5] agency N.Y. should you require such guarantee but a personal interview would be more satisfactory so if you will telegraph on receipt of this we will go down at once. Write at all events. We will make the Erection of Telephonic lines our business in case we come to an arrangement with you and push to your satisfaction and as the fall trade is about commencing the sooner we can get to work the better for all. So hoping to hear either by mail or telegraph from you ~~at~~on receipt. I remain Yours &c

Robert Watson Junr[6]

P.S. If you telegraph to come down we will bring everything necessary to close matters in New York at once.

ALS, NjWOE, DF (*TAEM* 14:887). [a]"in writing" interlined above.

1. Watson to TAE, 27 Aug. 1877, DF (*TAEM* 14:885). Watson had been in Menlo Park on 25 August "to see Edison about the introduction of Speaking telegraph in Canada." Cat. 1233:237, Batchelor (*TAEM* 90:171).

2. Unidentified.

3. Edison's draft of the terms, dated 29 August, is in DF (*TAEM* 14:889). It calls for "the actual Cost for the manfr & . . . a yearly royalty in advance of 5 dols gold on each pair of telephones which consists of a trans & receiver that would be ten dolls for 2 terminal stations."

4. The letter's return address is 174 Saint James St., Montreal, the office of the Canadian District Telegraph Co. Pamphlet, Cat. 1240, item 261, Batchelor (*TAEM* 94:78).

5. Dun, Barlow & Co., styling itself the Mercantile Agency, shared an address with "Robert Graham Dun, agent" (335 Broadway). Its exact corporate relationship to R. G. Dun & Co., which also called itself the Mercantile Agency, is unclear. Wilson 1877, 371; on R. G. Dun & Co. see *TAEB* 1:469 n. 1.

6. Watson was the Superintendent of Construction and Apparatus for the Canadian District Telegraph Co. Pamphlet, Cat. 1240, item 261, Batchelor (*TAEM* 94:78).

## -1027-

*From Edward Johnson*

Saratoga N.Y Aug 30 1877[a]

Concert tonight immense success got distinguishable words from New York.[1] Made quite an improvement By shutting[2] receiver with magnet. I announced amid cheers that on next Sunday evening I would have you bring your improved transmitter to the New York end and give us some words.[3] it needs but the merest trifle to make them clear.

Johnson

L (telegram), NjWOE, DF (*TAEM* 14:892). Message form of Western Union Telegraph Co. [a]"187" preprinted.

1. A letter from a Clarence Rathbone to the editor of the *Albany Argus,* 30 August 1877, discusses this concert. Cat. 1240, item 242, Batchelor (*TAEM* 94:72).

2. Shunting.

3. The *Troy Sunday Trojan* printed what appears to be an account of this concert on 2 September. That article, entered into evidence in the Telephone Interferences (TI 2:633 [*TAEM* 11:734]), does not indicate that Edison was involved. Johnson apparently transmitted another con-

cert from New York to Tweedle Hall, Albany, on 5 September (advertisement, TI 2:657 [*TAEM* 11:742]).

**–1028–**

*From Robert Watson, Jr.*

Montreal August 30/77

Dear Sir

I wrote you yesterday in reference to a party who was willing to join with me in procuring your agency ~~for~~in Canada for the Telephone. Since then however I have succeeded in getting two other practical Telegraph men to join me in the following proposal to you. First as to the personnel of the proposed Company F. H Badger the present Superintendent of the Montreal Fire Alarm System who has been connected also with the Fire System of Boston & well known to Chas T Chester of New York.[1] Next E A. Baynes Present Manager of the Canadian District Telegraph Co here and lastly myself. We are all practical Telegraph men & at present in positions here but think there is a good field for an extensive business in the Telephone. ~~Furt~~ We each put in sufficient capital to work the thing in good shape and to show that we mean business will place Two hundred & fifty $250 in your hands at once and take telephones & Royalty against it. That is you deduct your Royalty & the price of Telephones provided us to that amount. We would propose styling ourselves (in case you accept our proposition) The Canadian Telephone & Telegraph Construction Co. and you may see that we intend pushing business as we all throw up permanent positions to undertake your agency for the Dominion. In the meantime in case matters do not come to anything between us please do not mention names as it would only endanger our positions here. Mr Chester would give you all information in reference to Mr Badgers whose reputation stands highest in Canada as an Electrician— Further we are prepared at once to put up the money and close matters at once and would like a Sett of instruments to thoroughly test them as soon as you can let me have them. Address me as below.

Let me hear from you at once on these points as I have to go to New York early[a] next week[2] and if an arrangement is to to be come to would like to go prepared to close the contract under the terms you proposed to me at Menlo Park Yours Very Truly

Robert Watson Junr

address Telephones to Robert Watson Junior Montreal Canada

ALS, NjWOE, DF (*TAEM* 14:890). [a]Interlined above.

1. New York telegraph manufacturer and inventor Charles Chester marketed the fire alarm telegraph he had developed with his brother Stephen. See *Gamewell v. Chester* and *Gamewell et al. v. Chester and Chester.*

2. Watson was in Menlo Park on 1 September and examined Edison's telephone again two days later. Cat. 1233:244, 246, Batchelor (*TAEM* 90:175, 176).

**-1029-**

*Technical Note: Telephony*

[Menlo Park,] Aug 30 1877

Spkg Telephone Domestic Tel[1]

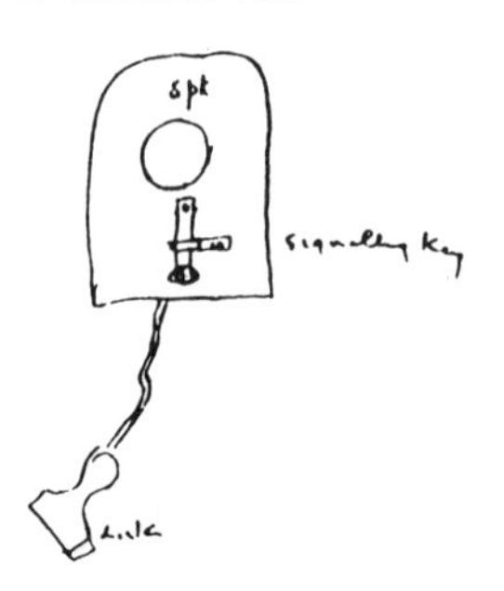


The telephone could be applied to fire Boxes and used as a police telegraph & for summoning more engines etc[2] I propose to include it in the same circuit with the fire signalling apparatus short cktg the fluf & Receiving magnet only when talking

New Spkr.[3]

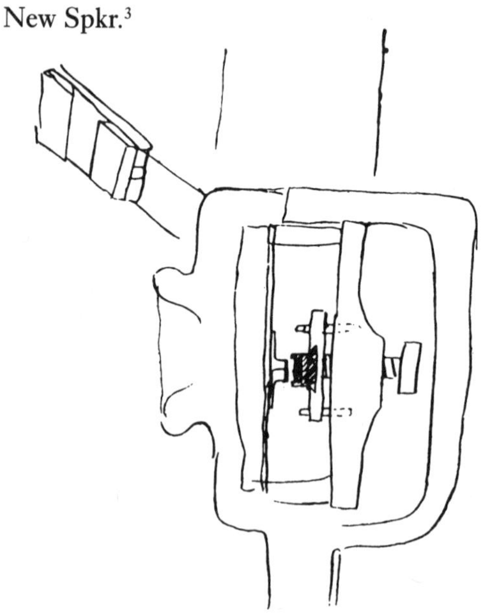

Receiving instruments

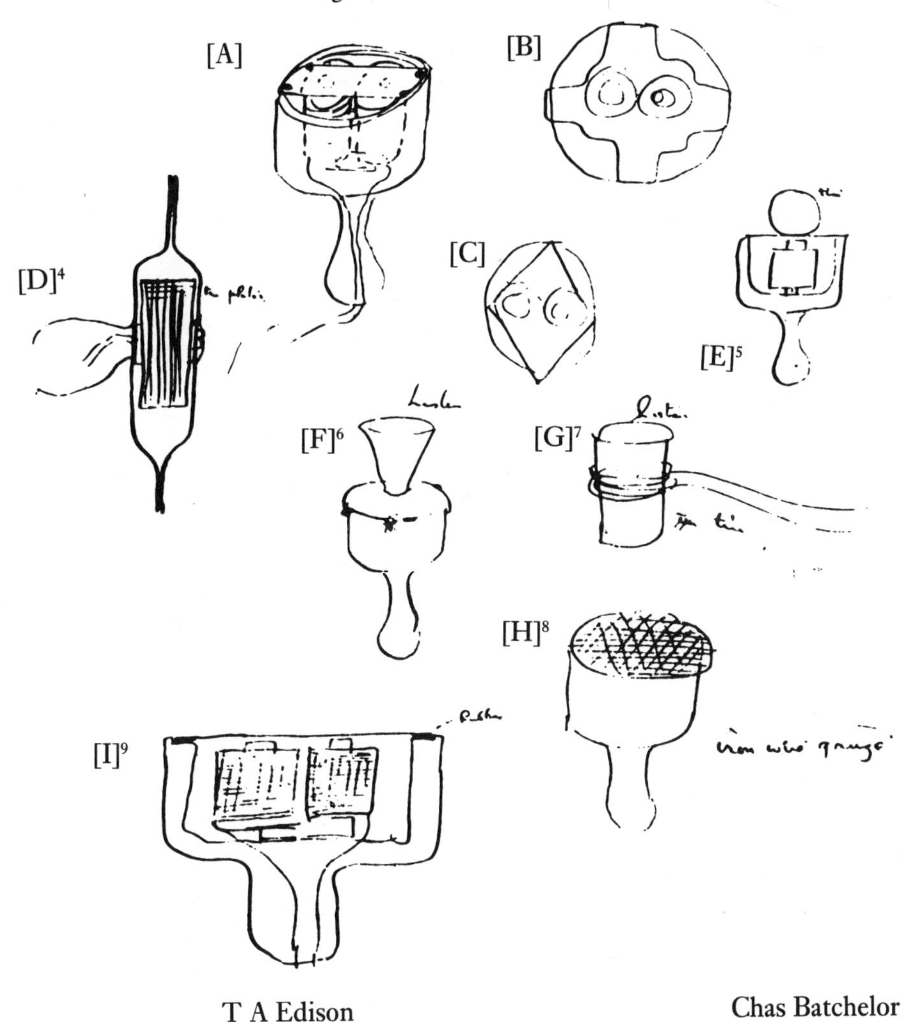


T A Edison

Chas Batchelor
James Adams

X (photographic transcript), NjWOE, TI 2, Edison's Exhibits 135-12–136-12 (*TAEM* 11:436). Document multiply signed and dated.

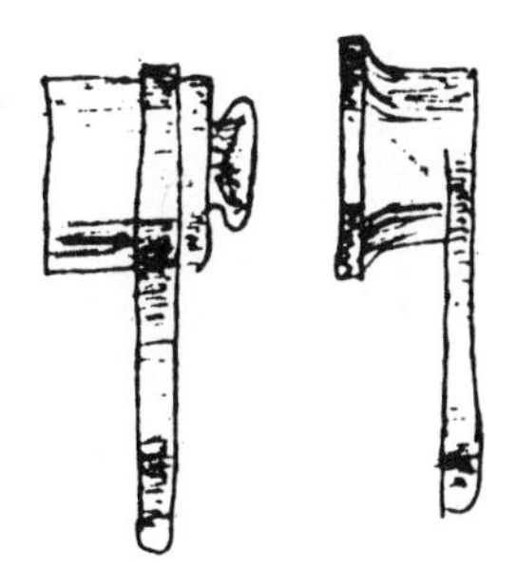

*The new telephones with flat handles.*

1. Figure labels are "spk," "signalling key," and "listen."

2. Police telegraphs using telephones for such purposes became common after John Barrett, superintendent of the Chicago fire-alarm telegraph, designed his system in 1880. Israel 1992, 112.

3. Batchelor began making instruments with this configuration—a flat handle at 90° to the mouthpiece—the next day and finished them on 2 September, noting they were a "success." This transmitter may have included a new fluff holder, designed 28 August, that contained "5 times as much fluff as before." Cat. 1233:240, 243, 245, Batchelor (*TAEM* 90:173–75); see also Edison's Exhibit 132-12, TI 2 (*TAEM* 11:434).

4. Figure label is "tin plates."

5. Figure label is "tin."
6. Figure label is "Listen."
7. Figure labels are "listen" and "[---] tin."
8. Figure label is "iron wire gauze."
9. Figure label is "Rubber."

## –1030–

*To William Orton*

[New York?,][1] August 31, 1877.[a]

Wm Orton.

I shall have the other pair of Telephones[2] over tomorrow I think, that will make the 3 pair. The new pair will be about as perfect in design, convenience and perfection of articulation as I shall probably be able to make them. If you are in no haste for any more those 3 pair can remain on test, while I go ahead with the "embosser" which shall it be= more Telephone, or the embosser!

(signed) Edison

L (transcript), DSI-NMAH, WUTAE. Original on memorandum form of Western Union Telegraph Co. secretary. [a]"(Copy)" at top of page.

1. Edison wrote on a memorandum form from the Western Union secretary's office.

2. According to Charles Batchelor's testimony, these instruments were of the design shown in Edison's British Patent 2,909 (1877). TI 1:354 (*TAEM* 11:154).

*British patent drawings of Edison's hand-held telephone.*

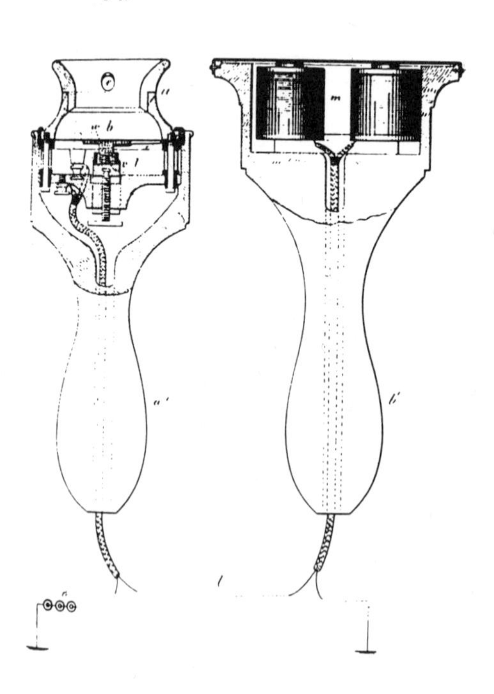

**-1031-**

*From Enos Barton*

Chicago[a] August 31, 1877

Dear Sir:—

Speaking of magneto machines, I find that the Franklin Institute of Philadelphia[1] has recently got from Cleveland one of Brush's machines.[2] They call it a 5000 candle machine.[3] It takes from two to ten horse-power to run it, according to the amount of mechanical resistance that is produced. Their price for the machine is $750.00.

The Washington University, of St. Louis, has just got a 1000 candle machine of the same sort, price $450.00. It takes about three horse-power for 1200 revolutions, to give what they call a 1000 candle light. Weight about 200 pounds.

Stockly,[4] on his return from settting this up, had his eyes all bunged up from exposure to the light.

Brush has invented a new regulator for carbons,[5] and proposes to invent a candle that will not consume.[6] This "candle that will not consume" is the thing required.

Dr. Herz[7] and S. D. Field,[8] of the new California Electric Works, successors to the Elec[trical]. Con[struction]. & M[aintenance]. Co., of San Francisco, are on their way to Europe to get the Jablochkoff candle.

There is a great demand for apparatus for the electric light, and it is one of the best things out to get the best.

For lighting mines &c. it is of great value, and when you get so you can light the streets of a city you have something worth while.

Let us send you one of the Brush machines, go over and look at the one in Philadelphia, and get a machine and invent your candle.

We can get something off from the prices given for the machines. Yours truly,

E. M. Barton, Sec'y

TL, NjWAT, Box 92 02 163 01. Typed in uppercase; letterhead of Western Electric Manufacturing Co. [a]Preprinted.

1. Originally founded as a mechanics' institute, the Franklin Institute was now a nationally prominent center for science and technology. Sinclair 1974.

2. Electrical inventor Charles Brush had recently designed a dynamo as part of his new arc light system, which was manufactured by the Telegraph Supply Co. of Cleveland. Carlson 1991, 82; *DAB*, s.v. "Brush, Charles Francis."

3. Arc lights were rated by candlepower.

4. George Stockly was vice president of the Telegraph Supply Co. Carlson 1991, 82.

5. Brush's lamp used a solenoid to regulate the distance between the carbons. Carlson 1991, 124–25.

6. Although sought after by arc light inventors, such a lamp was not attainable with carbon electrodes.

7. In a letter of introduction to William Preece, Edison referred to Herz as "Dr Cornelius Herz member of the Board of Health of San Francisco Cal." TAE to Preece, 28 Sept. 1877, WHP.

8. Electrical inventor Stephen Field (see *TAEB* 2:418 n. 2) was electrician of the California Electrical Works. On that company see the testimony of Paul Seiler, *Field v. Pope,* and the testimony of George Ladd, *Ladd v. Seiler.*

**-1032-**

*From George Bliss*

CHICAGO Sept 1 1877[a]

Dear Sir:

The W[estern]. E[lectric]. send $175.00 today. Richmond has not turned up and may have struck a snag. Gray thinks it remains to be seen whether he is shut out of the telephone field. Holland is here—He thinks more favorably than I expected about going in on the telephone patents providing all the money does not have to be put up at once. He considers the interest offered too small.[1] It looks as if Beetle could get your patents cheaper on the other side than you are paying. At any rate we can have him[b] put the telephone on the market there at once when you are ready. Expect to leave for the east one week from Monday if not detained. Resptfly

Geo. H. Bliss. Gen'l Man.

ALS, NjWOE, DF (*TAEM* 14:894). Letterhead of Electric Pen and Duplicating Press Co. [a]"CHICAGO" and "1877" preprinted. [b]Interlined above.

1. See Doc. 1056 n. 2.

**-1033-**

*To William Preece*

Menlo Park N.J. Sept 3 1877.

Friend Preece,

I am very sorry you did not receive the Telephone in time. I will send you pair of speakers in a week or two The WU take them from me. The solicitors who took out patent is Brewer & Jensen Chancery Lane London.[1]

Comparative trials between mine and Bells on the lines of the Gold and Stock Telegraph Co[2] proved 1st that the articulation of mine was far better,= 2nd that it was 4 times louder. 3rd that there was no noises, 4th that it worked on every wire tried and but few[a] wires were found that Bells could be worked on owing to cross leakage; on a majority of wires the leakage sounds overpowered the talking= My experiments prove that mine is practical on 60 miles of wire ie[b] Menlo Park to NY & return with 40 cells Callaud battery. The are about the same size as Bells Thus

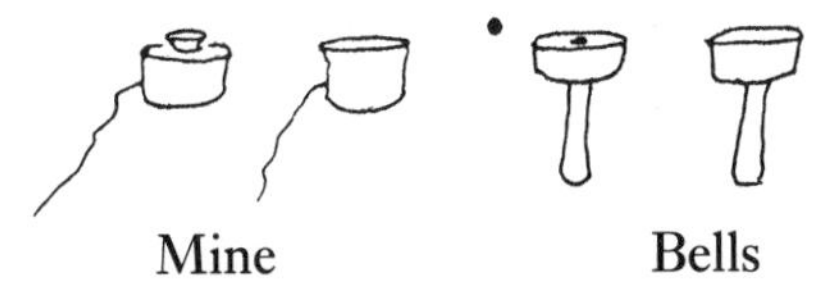


cost about $3. each. I have entered in to contract with parties in Canada who pay me royalty of $5. per year on each terminal station ie[b] transmitter and receiver, that would be my price to your party.[3] I think it reasonable as it has caused me a deal of labor— I have a new thing which I will describe in my next letter I will send the cylinders. Let me know how you get along with the musical Telephone. E H Johnson who is exhibiting it at Saratoga, has had 3 successful exhibitions there from NY distance 180 miles connections thus:

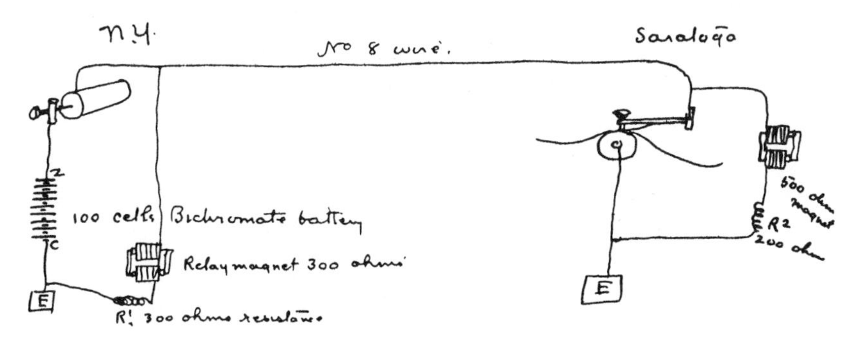


You can vary $R^1$ & $R^2$ to suit— the tones were reproduced so they were distinctly heard in every party of the large dining halls of the hotels. He used almost <u>dry</u> paper. Yours

T. A. Edison

P.S. Smith & Hamilton leave on 5th with Quad= I hope you will give them a wire that don't run on the Bridgewater Canal like Lumsden gave me=[4] give my respects to Lumsden E

ALS, UkLIEE, WHP. [a]Interlined above illegible canceled letters. [b]Circled.

1. Edward Brewer and Peter Jensen were London patent solicitors used by Lemuel Serrell in connection with Edison's British patent applications. See, e.g., Brit. Pats. 4,502 (1878) and 3,880 (1880), and correspondence to and from Brewer and Jensen (*TAEM-G1*, *TAEM-G2*, s.v. "Brewer & Jensen").

2. On 30 August, Batchelor wrote in his diary, "Went to New York

and put our 2 speakers on between 197 Bdway and 180 South Street Worked very well I was at South Street and spoke with Orton, Eckert, and many others With Bell's instruments in we could get nothing." Cat. 1233:242, Batchelor (*TAEM* 90:174).

3. Unidentified.

4. See Doc. 319.

## -1034-

*Technical Note: Telephony and Miscellaneous*

[Menlo Park,] September 4 1877.

Speaking Telegraph.

Experiments with metalization of silk fibre= We find that following the directions given in Rossiliers book on Electro-Metallurgy[1] that it does not work satisfactory= with 1 gram silver in 20 grams water and silk well fluffed and exposing same to a saturated solution of Phosphorus in BiSulphide Carbon that only portions blacken after 2 or three days—and even this shining black is a very poor conductors no where near as good as with plumbago merely rubbed on— We are now trying a stronger silver solution with well picked fibre previously soaked all night in water which at the time of putting it in was boiling hot. after being in[a] the silver solution about ½ hours we took it out and dried & repicked it then took a plate about 7 inches diameter filled it to depth of ¼ inch with Saturated solution phosphorus in Carbon Bisulphide over this a wire gauze about ¼ inch from the surface of the liquid= over the guaze the silk well fluffed— over this the big air pump receiver sealed by beeswax at the bottom so as to exclude all air—

We are trying to reduce the nitrate silver to metallic silver on the silk by passing Hydrogen through a tube Containing the silk thus[2]

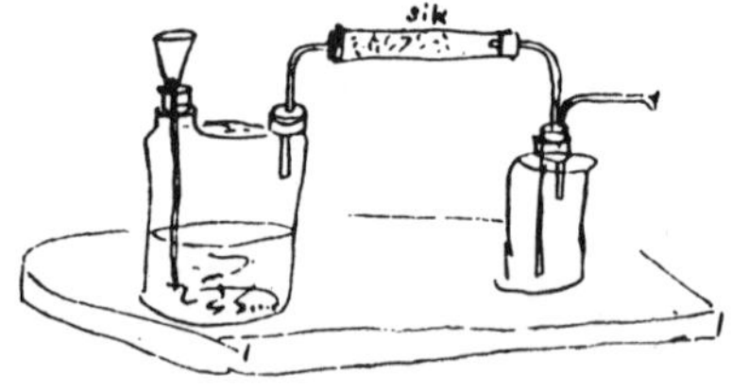


We are also trying to pass Sulphuretted hydrogen through the tube containing the silk to reduce the nitrate to sulphide of silver which is probably a Conductor=

We propose to precipitate conducting salts on the silk such as Iodide of Copper etc=

Soak silk in 10 grammes Sulphate Copper 10 grammes sulphate iron in 1 fluid oz water, then dry= ~~im~~ pass vapors of iodine[b] through silk= to another part immerse in a Saturated Solution Iodide Potassium

Call Bell.[3]

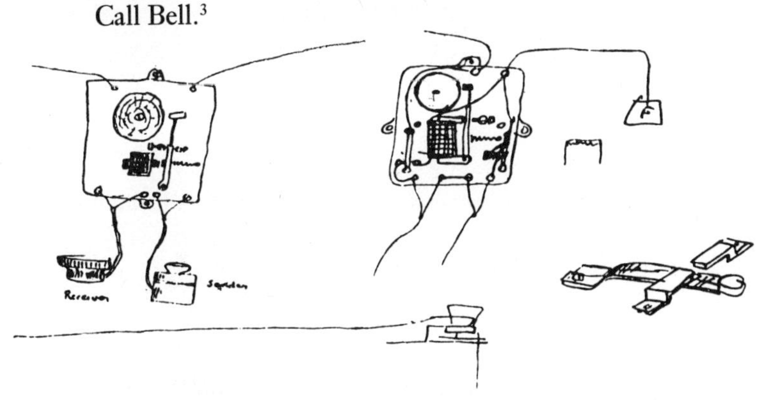

I find that the silk if properly reduced by phosphorus is a fair conductor & with the proper amount of Nitrate Silver & well reduced is as good if not better than plumbago. This will make Speaking Telegraph Constant, never need adjustment except at first few days—[4]

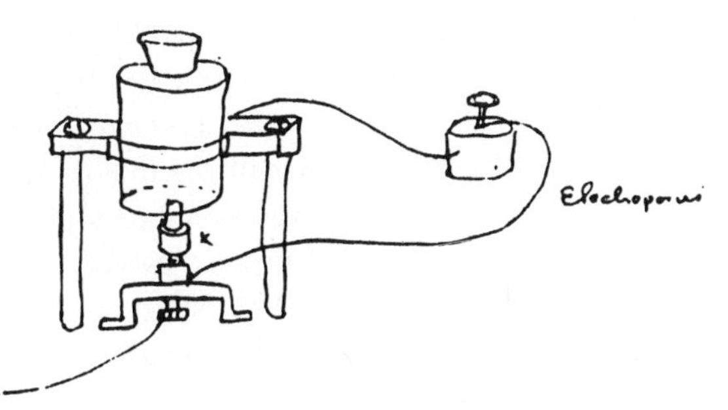


Loose Semi conducting powder in X and plate on diaphragm kept full particles by static electric attraction.

Perhaps plumbagoed mustard seed would work in .X.[5]

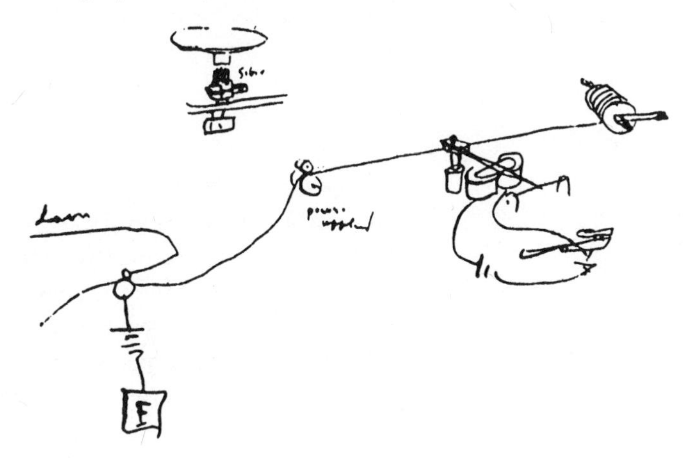

Silk or other thread plumbagoed by an Electro magnet into dots & dashes & used again for transmitting rapidly or receiving & retranslating

T A Edison

Chas Batchelor
James Adams

X (photographic transcript), NjWOE, TI 2, Edison's Exhibits 144-12–145-12, 142-12–143-12 (*TAEM* 11:441–42, 439–40). Document multiply signed and dated. [a]Obscured overwritten letter. [b]"o" interlined above.

1. Unidentified.

2. Figure label is "silk."

3. Figure labels are "Receiver" and "Sender."

4. Figure labels are "x" and "Electrop[h]orus." An electrophorus is a device invented by Alessandro Volta for producing static electricity by induction. Atkinson 1890, 720.

5. Figure labels are "fibre," "Line," and "power applied."

**–1035–**

*Draft Patent Application: Chemistry and Telephony*

[Menlo Park,] September 5th 1877.

Speaking Telegraph.

We find that silk may be easily silvered and made a conductor to be used in my[a] speaking Telegraph for varying the resistance of the line by the movement of the diaphragm. I copy here a specification which I have prepared with the idea of obtaining a patent.[1]

"The object of this invention is to coat various materials especially materials which are non conductors of electricity, with metals.

The invention consists in depositing a salt of the metal upon the material to be coated by immersing it either in a hot or cold solution of a salt of the same in any proper solvent afterwards drying it when it is placed in a tight receiver box or closet or other proper receptable within which is a quantity of moist Phosphide of Calcium or other Phosphide of an element which ~~is~~ slowly evolves phosphuretted hydrogen by moistening it.

The operation is as follows  If say a silk garment is to be silvered, I immerse it in strong hot solution of a salt of silver preferably the nitrate of silver allowing it to remain a few minutes or such a length of time as will in extended experience be found to give the best result

It is taken therefrom and thoroughly dried, afterwards it is taken to the tight closet & suspended over a large flat dish which is covered to the depth of say 1/16 inch with moist phosphide of Calcium= By the action of the water the phosphide of Calcium is decomposed and phosphuretted Hydrogen set free; this acting upon the nitrate of silver reduces the silver upon the silk to a metallic state. The time required to effect perfect reduction is only a few minutes if the phosphuretted Hydrogen is given off in great volume;

I am aware that it is not new to reduce a silver salt upon materials by phosphorus. I am also aware that Phosphorus disolved in its solvents such as Ether, or Bisulphide of Carbon have been placed in evaporating dishes in sealed jars over which was suspended an article containing salts of a metal to be reduced. But the use of phosphorus in solution is quite dangerous: unreliable in its action, inconvenient and very slow and it does not penetrate like the phosphuretted hydrogen to every part of the material. Neither is it so energetic I will mention that liquid phosphide of Hydrogen may be used=

I claim as my invention

Reducing metals to the metallic state[b] from their salts upon any material by exposing the same to the fumes of Phosphide of Hydrogen

2nd Reducing metals to the metallic state from the salts upon material[c] by exposing the same to the fumes which arise from a phospide of an element which is slowly[d] decomposed by moistening it

T A Edison

[Witness:] Chas Batchelor

ADS, NjWOE, TI 2, Edison's Exhibits 146-12–148-12 (*TAEM* 11:443). [a]Obscured overwritten letters. [b]"to the metallic state" interlined above. [c]"upon material" interlined above. [d]Interlined above.

1. A draft of this specification, differing only in a few words, appears in NS-Undated-002, Lab. (*TAEM* 8:90).

**-1036-**

*To William Preece*

Menlo Park NJ. Sept 6/77

Dear Sir

After raining 3 days we find Bells telephone isnt a telephone on over 20 wires none over 3 miles in length. The leakage Currents entirely drowned the speaking, with mine no leakage currents are heard. I see in some of the papers over there that Bells has been at, Sir Wm Thomson's Laboratory.[1] Have you thought over Bells principle. I have figured it out that perfect articulations can never be had= the law of the square interferes, for instance if you double the amplitude of his diaphragm by speaking twice as loud you increase the induction current 4[a] times—owing to the magnetic Square;[2] whereas if the law of the square did not come in play the current ~~tone~~ would be proportional to the amplitude of the diaphragm which is necessary for perfect articulation. You will notice in speaking that the low or weak[b] tones are very weak while the loud tones are exceedingly loud[c] hence the talking is very uneven in strength, and the accenting of a word is unreliable. Yours

Edison

ALS, UkLIEE, WHP. [a]Underlined twice. [b]"or weak" interlined above. [c]"loud" underlined twice.

1. See Bruce 1973, 250–51.
2. See Doc. 1074.

**–1037–**

*From George Bliss*

CHICAGO Sept 6 1877[a]

Dear Sir:

The W[estern]. E[lectric]. will send you some more money the latter part of this week.[1] Please use it to fix your German patent and get that rotary press going. Am glad to hear your telephone is panning out so well. If ready to go on the market it ough to scoop the field. I see Preece did not give your telephone much of a lift in his recent address but then he did not see it perfected.[2] Bell seems to be experimenting with Sir Wm Thompson. Holland has been here and gone. I expect to see you the middle or last of next week. Wheeler is going over to Canada to start up the Pen business. Respectfully—

Geo. H. Bliss

ALS, NjWOE, DF (*TAEM* 14:141). Letterhead of Electric Pen and Duplicating Press. [a]"CHICAGO" and "1877" preprinted.

1. On 14 September, Edison recorded a $200 check from Western Electric in his account book. Cat. 1185:88, Accts. (*TAEM* 22:595).
2. See Preece 1877.

**–1038–**

*From Josiah Reiff*

New York, Sept 6th 1877[a]

My dear Edison—

I want to see you & if you are not coming in I must come out. I find in looking over copy of our contract[1] with Smith, Fleming & Co. signed by You, Harrington & me, that everything is included present & future that may be or can be made[b] applicable to Land line Telgraphs as distinguished from Submarine Cables—the latter being included in another provision— We dont want to get into any farther snarl with these people— So we must consult & if necessary for peace & success we must make some compromise with Prescott. Of course for your sake, I will consent that he shall get what he is not Entitled to, but [-- ---][c] Anderson, Pender, Puleston & S F & Co will certainly prevent anything being done unless some understanding is reached— We simply want to consider what is best to be done under the circumstances. This is only another instance of the necessity of seeking advice legal & having before you, your relations with other people, before entering new combinations— I dont believe you ever intended to deliberate wrong a human soul but we are all full of liability to error. Yr

JCR.

ALS, NjWOE, DF (*TAEM* 14:719). Letterhead of J. C. Reiff. [a]"New York," and "187" preprinted. [b]"or can be made" interlined above. [c]Illegible.

1. Doc. 350.

**-1039-**

*Draft Article: Phonograph*

[Menlo Park,] Sept 7 1877

Edison Phonograph.[a]

An apparatus for recording automatically the human voice and reproducing the same at any future period.

Mr Edison the Electrician has not only succeeded in producing a perfect articulating telephone, which comparative ~~practical~~[b] tests, upon the lines of the Western Union Telegraph Co.[c] have proved to be far superior and much more ingenious than the telephone of Bell. ~~but has actually~~ and has been adopted ~~upon~~ for use upon the 1300 private wires operated by the Gold & Stock Telegraph Company of New York but has gone into a new and entirely unexplored field of acoustics which is nothing less than an attempt to record automatically[b] ~~a~~the speech of a very rapid speaker upon paper; from which ~~by~~ [~~exer~~?][d] he reproduces the same Speech immediately or year's[e] afterwards ~~or at any future~~ preserving the characteristics of the speakers voice so that persons familiar with it would at once recognize it

It would seem that so wonderful result as this would require elaborate mechanery on the Contrary the apparatus although crude as yet is ~~of~~ wonderfully simple I will endeavor to convey the principle ~~of~~ by the use of an illustration which although not exactly the apparatus used by Mr Edison will enable the reader to grasp the idea at once.[1]

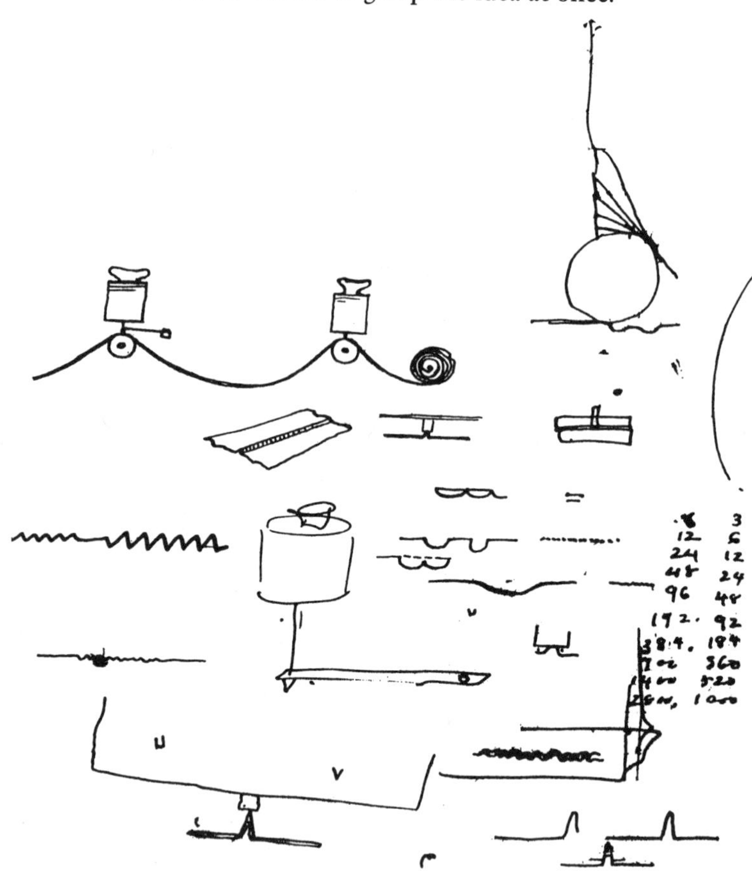

T A Edison

[Witnesses:] James Adams Chas Batchelor J Kruesi
~~G E Carman~~[2] ~~M N. Force~~[3]

ADfS, NjWOE, Lab., Vol. 17:13 (*TAEM* 4:882). [a]Underlined twice. [b]Interlined above. [c]"upon . . . Co." interlined above. [d]Canceled. [e]"or year's" interlined above.

1. The following sketches appear to be rough designs that Edison then drew in more detail in Doc. 1040.

2. According to Jehl 1937–41 (1:318), George Carman did odd jobs around the laboratory. His name appears in the accounts by the end of April 1877 (Cat. 1185:141–42, 247, Accts. [*TAEM* 22:617–18, 66]).

3. Martin Force was a carpenter who helped build the Menlo Park laboratory in winter 1876. In spring 1876 he apparently worked in the electric pen factory packing pens and occasionally running the steam engine. Sometime in late 1877 or early 1878 he became a general handyman around the laboratory. Force's testimony, pp. 98–99, *Sawyer and Man v. Edison* (*TAEM* 46:232); Cat. 1185:141, Accts. (*TAEM* 22:617).

*Technical Note: Phonograph*[1]

Phonograph

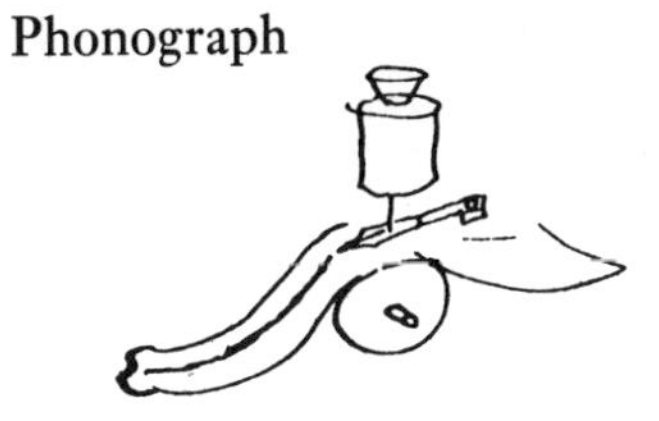

Recorder

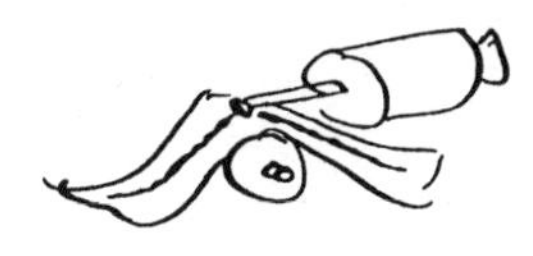

Receiver worked by diference of friction in the ink

plumbago on a spring on diaphragm makes light & heavy marks which have more or less lubrication Receiver worked by difference friction

Rough paper—Knock[a] down fibre or knock down a crystallized salt such as PbCl.[2]

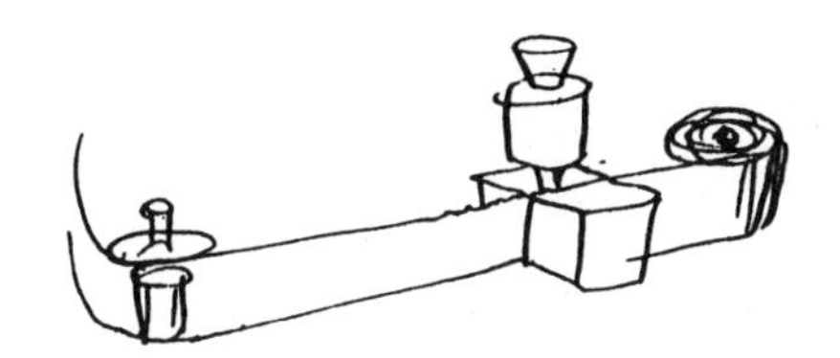

Edge embossed

—Emboss it—Emboss or knock down the previous boss[3]

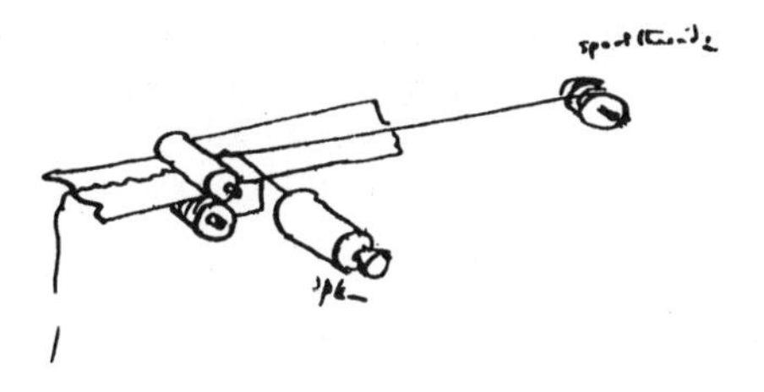

ink= wkg by dif friction of the ink[4]

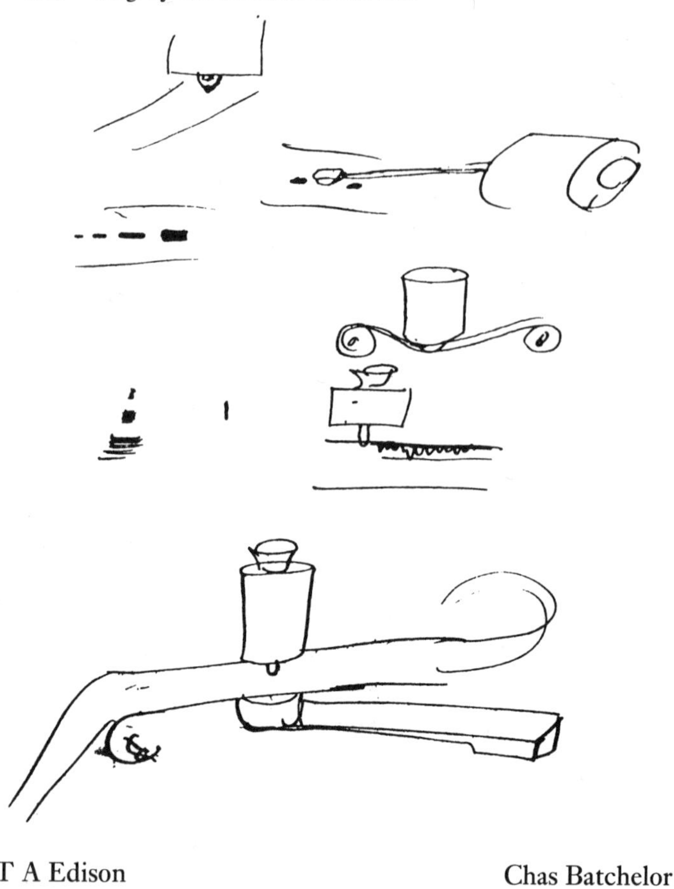

T A Edison
James Adams
~~G E Carman~~

Chas Batchelor
J Kruesi
~~M N Force~~

X, NjWOE, Lab., Vol. 17:15, 14 (*TAEM* 4:886, 885). Document multiply signed and dated. [a]Obscured overwritten letters.

1. A number of the designs sketched in this technical note and in notes from 8 and 9 September (NS-77-003, Lab. [*TAEM* 7:453–54]) were later embodied in Edison's British Patent 2,909 (1877), figures 4–8. See also NS-77-003, Lab. (TAEM 7:450).

2. On 9 September, Edison suggested using paraffin or other soft substances. NS-77-003, Lab. (*TAEM* 7:453).

3. Figure labels are "spk—" and "spool thread." On the use of thread between two rollers to emboss a signal see Doc. 196. The advantage of this method is that the signal does not need to supply the embossing power; it simply moves the thread.

4. Edison tried to record using ink marks on 9 September and noted that it worked "OK." NS-77-003, Lab. (*TAEM* 7:453).

**–1041–** [Menlo Park,] Sept 8 1877

*Technical Note: Telephony*

Speaking Telegraph.

I propose to cover the fibre with metals[1] by the process described in the September number of the American Journal of Arts and Science No 81 Vol XIV. New Haven 1877= page 169 to 178.[2] This will get rid of the effect of the acids used with the phosphorus plan I find that by my small coil it can be done to a certain extent but it will require a larger one to do it well.[3]

This wks ok=[4] Working by induction[5]

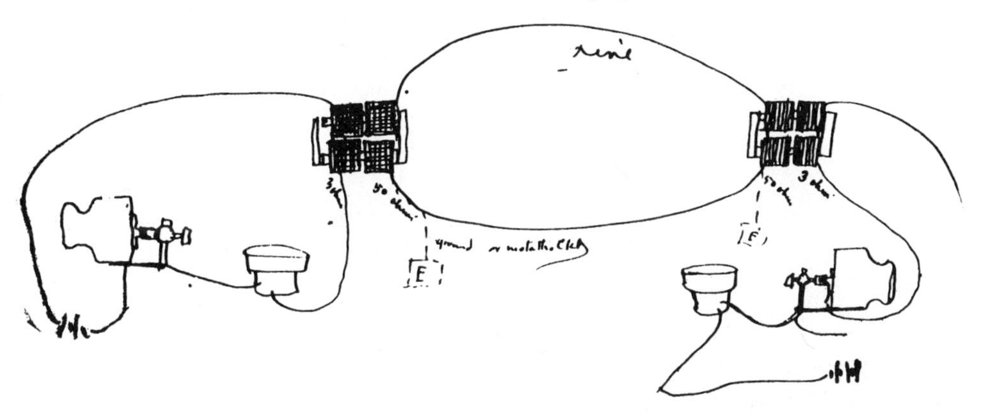

4 Callaud arranged as one Callaud for quantity[6] The resistance of metalized fibre brought down to 10 ohms

T A Edison

Chas Batchelor
James Adams

X, NjWOE, Lab., Vol. 12:149 (*TAEM* 4:70).

1. On 3 September, Batchelor "Went to different Electrotypers etc to see if we could find anybody that would metalize our fibre but could not. Worked at night trying metalization of fibre." He worked on it the next day as well. On 6 September the staff "Made some phosphide of Calcium at night and reduced the silver on silk saturated with nitrate of silver by placing it above the calcium phosphide which it dampened. This reduces almost instantly but it does not work well in Speaking telephone by reason of noises made by the nitric acid (being left in the silk) when affected by Battery current." Cat. 1233:246–47, 249, Batchelor (*TAEM* 90:176, 177).

2. That article was "On a new Process for the Electrical Deposition of Metals, and for constructing Metal-covered Glass Specula," by Arthur Wright, professor of physics at Yale. It described the deposition of metals (particularly platinum) on glass in an evacuated tube.

3. Wright's article called for "an induction-coil capable of giving sparks four or five centimeters in length" (p. 171).

4. Figure labels are "3 ohms," "50 ohms," "ground or metallic ckt," "line," "50 ohms," and "3 ohms."

5. Edison later said of using fluff in the transmitters,

Upon investigation, the difference of resistance produced by the varying pressure upon the semi-conductor was found to be exceedingly small, and it occurred to me that as so small a change in a circuit of large resistance [i.e., the line] was only a small factor, in the primary circuit of an induction coil, where a slight change of resistance would be an important factor, it would thus enable me to obtain decidedly better results at once. The experiment, however, failed, owing to the great resistance of the semi-conductors then used.

After further experimenting in various directions, I was led to believe, if I could by any means reduce the normal resistance of the semi-conductor to a few ohms, and still effect a difference in its resistance by the pressure due to the vibrating diaphragm, that I could use it in the primary circuit of an induction coil. Having arrived at this conclusion, I constructed a transmitter in which a button of some semi-conducting substance was placed between two platinum disks, in a kind of cup or small containing vessel. [Prescott 1879, 224]

On the "button," see Doc. 1091; on Edison's further use of induction coils, see Docs. 1107, 1112, and 1121. It is worth noting that in the circuit shown here the receiver and transmitter are both in the primary circuit of the coil; similar circuits in late November have the receiver in the main line (e.g., Doc. 1121). Edison had drawn a telephone circuit including an induction coil with his electromotograph receiver on 19 May 1877 (Vol. 11:94, Lab. [*TAEM* 3:969]).

6. Arranging the batteries this way, given the small amount of resistance in the local circuit, would result in quite a strong current and a strong magnetic field in the induction coil.

**–1042–**

*Draft to Thomas Eckert*

NY Sept 9 77

T.T.E.

I hereby resign my position as Electrician of ur Co to take effect Sept[a] 11th 1877

T A Edison

ADfS, NjWOE, DF (*TAEM* 14:757). [a]Interlined above.

**–1043–**

*To William Preece*

Menlo Park NJ Sept 9/77

Friend Preece:

I send you a little slip from the Tribune.[1] This is the work of Prof Dolbear of Tufts College;[2] the work has been ~~surp~~ suppressed.[3] I have seen a proof of it[4] and from it I learn that Dolbear was the original inventor of the Bell apparatus which you have and that Bell took the whole thing bodily from him[5] it appears it exhibited it before Bell had it and now produces a paper read before some scientific body describing it I think.[6] He also has a letter from G Hubbard who is somehow connected with Bell's Co in which Hubbard urges him to renounce his claim in the name of science,[7] and also offers to submit the whole question[8] to arbitration selecting Dr Miner as the arbitration said Miner being a friend of Hubbards.[9] however the book will not appear[10] and the whole "biz" appears Strange to me= Please let me know when the next Conversasionye of the Royal Society takes place; I shall probably be able to send you for Exhibition something so novel that it will startle them out of their boots. Yours

T A Edison

ENCLOSURE[a]

[New York, September 1877?]

"The[b] Telephone: an account of the phenomena of electricity, magnetism and sound, as involved in its action: with directions for making a speaking telephone," is the comprehensive title of Prof. ~~K~~Dolbeare's[c] coming book (Lee & Shepard).

ALS, UkLIEE, WHP. [a]Enclosure is a PD; place and date taken from attribution in main document. [b]Entire paragraph embraced by hand-drawn parentheses. [c]Correction done by hand.

1. The enclosed clipping has not been found in issues of the *New York Tribune* available on microfilm from August or early September 1877.

2. Amos Dolbear (1837–1910) worked as a machinist in his youth. He completed his college education at 29, taught at various institutions, and had served as professor of physics and astronomy at Tufts since 1874. Bruce 1973, 263–64; *NCAB* 9:414–15.

3. Dolbear stated in a 1 August letter to Gardiner Hubbard that his book was in press and would probably appear in about ten days; in a letter of 10 August he said the book was printed and awaited binding (both letters transcribed in Dolbear's deposition [originally from *Bell Telephone Co. et al. v. Peter A. Dowd*], pp. 24, 27, *Telephone Interferences*). See note 10.

4. Edison's source for this would have been Western Union. Dolbear had negotiated with the Gold and Stock Telegraph Co. during August and reached an agreement in September that gave him funding and the assistance of Franklin Pope, as patent solicitor, to prepare, file, and argue for a patent application on his claimed invention, and gave Gold and Stock (and thus Western Union) ownership of any resulting patent. Dolbear's deposition, p. 9, *Telephone Interferences;* Bruce 1873, 266–67.

5. Dolbear claimed to have invented a telephone instrument that used permanent magnets, rather than electromagnets, before Bell. The published version skipped much of what Edison reports here. On Dolbear's dispute with Bell, see Prescott 1972, 19, 74–75, 347–48, 460–65; Bruce 1973, 206, 264–67; Dolbear to Gardiner Hubbard, 1 Aug. 1877, transcribed in Dolbear's deposition, p. 24, *Telephone Interferences;* and Dolbear 1877, 110–16.

6. No such paper, lecture, or public demonstration on this by Dolbear prior to Bell's January 1877 patent has been identified. Dolbear did exhibit his instruments in some manner at Tufts early in April 1877. Dolbear, "Researches in Telephony," *Proc. Am. Acad. Arts Sci.* 14 (1879): 77–91.

7. In letters responding to Dolbear on 5, 9, and 14 August 1877, Hubbard asserted Bell's priority, urged that any effort to dispute that priority be made through the patent application process and the courts, and urged that Dolbear "as a professional gentleman" publish no claims that might damage Bell's reputation without prior full investigation to be sure he wouldn't say things that would damage himself. All letters transcribed in Dolbear's deposition, pp. 25–28, *Telephone Interferences.*

8. It was not clear whether this was a question of money, recognition, or both. See Dolbear to Hubbard, 1 and 10 Aug. 1877, transcribed in Dolbear's deposition, *Telephone Interferences;* Prescott 1972, 75, 465; and Bruce 1973, 265–66.

9. Alonzo Miner, social reform advocate, prolific journalist, and pastor of Boston's Columbus Ave. Universalist Church since 1848, had been president of Tufts University from 1862 to 1875. *NCAB* 1:315.

10. Dolbear's *Telephone* was published later in 1877 by Lee and Shepard of Boston, and Charles Dillingham of New York. The preface (p. vi) states that "steps have already been taken" to get a patent; Dolbear's application was filed on 31 October. A disclaimer in the preface also suggests that the delay of publication until after the patent application and the absence of references to prior public presentations of Dolbear's work were deliberate, both serving to avoid any suggestion that Dolbear's invention had been released into the public domain.

## ELECTRIC LIGHTING Docs. 1044, 1048, 1078, 1098, and 1136

Edison had briefly experimented with arc lights in late January and early February 1877.[1] In early September he returned to these experiments, beginning five months' intermittent investigation of electric lighting.[2] Edison's renewed interest in this subject was probably spurred by reports of the recently introduced Jablochkoff candle.[3] Stephen Field and Cornelius Herz, who were going to Europe to obtain Jablochkoff equipment, saw some of Edison's experiments when they stopped in Menlo Park.[4] According to Field, Edison's power source was a bichromate battery, not the magnetoelectric machine (generator) indicated in these documents; there is in fact no evidence that Edison had such a machine before the fall of 1878.[5] It is noteworthy that in Docs. 1044 and 1098 Edison drew arc lights in both parallel and series circuits, because contemporary arc lights were wired in series and Edison's parallel arrangement of lamps would be an important aspect of his later, successful incandescent lighting system.

Edison described some of the experiments that began in September during testimony in *Sawyer and Man v. Edison.*

> Two rods of brass, sliding in bearings forming the two poles of the battery had upon their ends small clamps in which different substances could be clamped. In these clamps strips of carbonized paper were placed about an eighth of an inch wide and two inches long. The paper used for carbonizing was bristol board. The carbon was brought up to incandescence, but quickly oxydized and was destroyed, as it was in the open air. Attempts were made to coat the carbon with powdered glass so it would melt and run over the carbon and thus preserve it. This did not work. Then experiments were tried on pieces of silicon, which the books stated did not oxidize when in incandescence in the open air. Also upon boron; but these did not succeed well, as they were in very small pieces, and we could not make good contact at the electrodes. Afterwards we tried the experiment in vacuo with a common air pump, but the vacuum that we were able to get was so poor that the carbon oxidized almost as rapidly as it did in the air.[6]

Edison said that in these experiments

> we were trying to subdivide the electric light into a small number of burners, where the circuit was closed by solid conductors, and the reason why experiments were conducted with boron and silicon was because they were not subject to oxidation like carbon, which we had previously tried, and which did not last as long at a white incandescence as pieces of graphitoidal silicon. The results of the carbon experiments, and also of the boron and silicon experiments, were not considered sufficiently satisfactory, when looked at in the commercial sense, to continue them at that time, and they were laid aside.[7]

1. Doc. 855; John Kruesi's testimony, pp. 131–33, *Sawyer and Man v. Edison* (*TAEM* 46:241–42).

2. Edison continued to experiment on electric lighting until the end of January 1878 (Vol. 16:1–4, Lab. [*TAEM* 4:479–81]). Docs. 1044, 1098, and 1136 are pages 1–3 of a group of electric lighting notes that had been collected together and numbered by September 1878. At that time, Edison's bookkeeper, William Carman, copied those notes into Experimental Researches Vol. 1 (Cat. 994, Lab. [*TAEM* 3:192]). Except for Doc. 1044 and those documents (including Docs. 1098 and 1136) introduced into evidence in *Sawyer and Man v. Edison,* loose pages of electric lighting notes from 1877 and 1878 were subsequently assembled into Vol. 16, Lab. (*TAEM* 4:478).

3. See Docs. 970 and 1031.

4. See Doc. 1056.

5. Friedel and Israel 1986, 23.

6. P. 3008, *Sawyer and Man v. Edison (U.S.)* (*TAEM* 48:11); see also Field's testimony, pp. 183–84, *Sawyer and Man v. Edison* (*TAEM* 46:244–45).

7. Edison's testimony, pp. 3020–21, *Sawyer and Man v. Edison (U.S.)* (*TAEM* 48:17).

**–1044–**

*Technical Note: Electric Lighting*[1]

[Menlo Park,] Sept 9 1877

Electric Light[2]

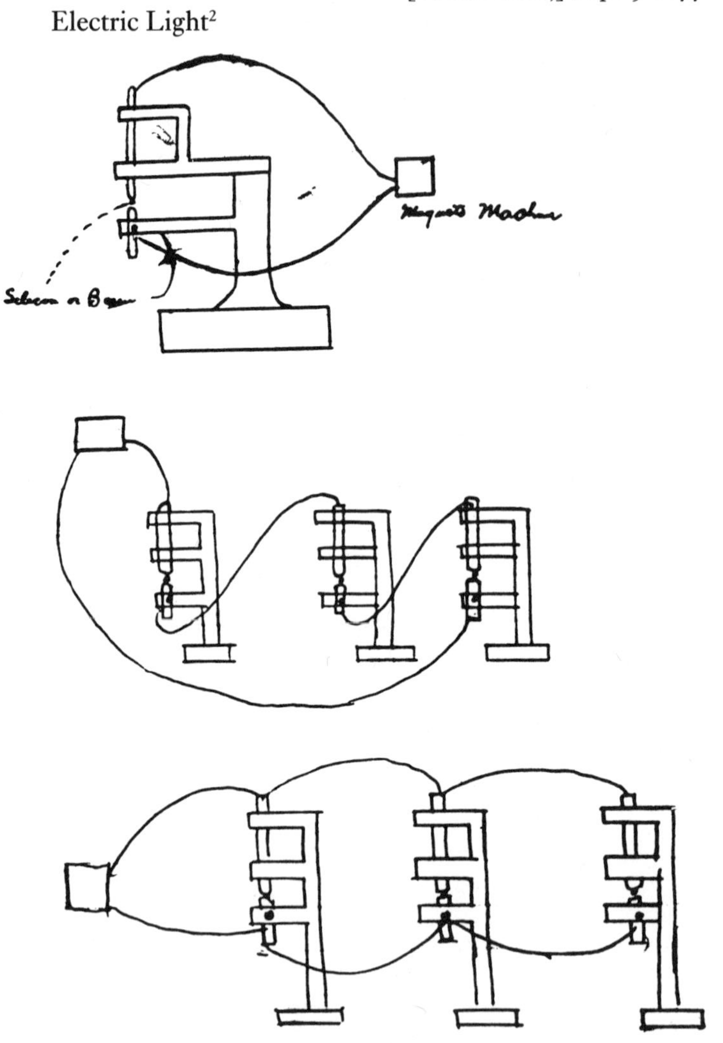

The employment of Silicon or Boron= (Metallic) ~~e~~serves to ~~i~~ensure the continuity ~~&~~by their conductivity and still have resistance enough to allow the arc to form—[3] besides the peice of silicon keeps the carbon points at a proper distance by its presence between them[4]

T A Edison — Chas Batchelor

James Adams — J Kruesi

~~M N Force~~

X, NjWOE, Lab., Vol. 16:9½ (*TAEM* 4:489). "(Co~~pp~~pied in book No 1 page 114 Sept 28 1878 Wm Carman)" written at top of page; that copy is in Cat. 994:114, Lab. (*TAEM* 3:226).

1. See headnote above.

2. Figure labels are "Silicon or Boron" and "magneto machine"; and "magneto machine."

3. The arc lights in use at this time were connected in series (as in the upper right sketch). The arrangement Edison describes here does not need a separate mechanism to insure the continuity of the circuit in the event of the failure of one light in the system.

4. See Doc. 855.

**-1045-**

*From George Barker*

Philadelphia Sept. 10, 1877.

My dear Sir:—

I have just reached home from my vacation in the country and find your letter of ~~a~~Aug. 2d upon my table.[1]

I am delighted at what you tell me of your great success in perfecting the articulation of your telephone. From what you say it must be far better than Bell's. When in Boston the other day the manager of the N. E. Telephone Co.[2] told me they had already 2000 Bell telephones in use. You know the Co. only rents them ($35 a year) and will not sell one to anybody. They were very short with me when I made some inquiries about Bell's patents. I hope I may have your assistance in giving that lecture to my class this fall from my house five squares away. Will you send me by Express as soon as you can, that small receiving apparatus which you said you would loan me? I mean the one which goes by a magnetic engine. I want to try it as it is and then if I can't use it, I will put on a crank myself. As to a transmitter, what can you send me to experiment on? Anything that will do, that you can spare as well as not. By the way, Mr. Richards the mechanical engineer of Colt's Armory[3] Hartford told me the other day that graphite does not diminish in resistance by compression. It is true it appears to if you squeeze in a vise a cylinder of graphite between two strips of copper; but if the circuit through the graphite is made by placing the wires near, not at, the ends thus, there is no increase of conductivity during compression. He hence believes that the effect you get, which is undoubted, is due simply to improved contact at the first and last surfaces.

Can you get for me, or have fitted up for me, one of your automatic instruments? I want simply the metal parts. I will put them on a table. I want to experiment to determine the accuracy of some of Koenig's forks.[4]

The phenomenon you mention of india rubber is well known & is mentioned by Tyndall in "Heat"[5] and others. As it heats by being stretched, unlike other bodies, so unlike them, when it is heated it contracts; the two phenomena being of the same order.

The facts concerning those scintillations are very interesting. When I get so good a condenser I shall try the experiments you mention. When at Mr. Wallace's the other day he heated for us with his new magneto 18 feet of ¼ inch copper wire bright red in daylight! Our friend Draper has struck a good thing has he not? You have heard of course that he has discovered a third satellite of Mars.[6] When you are here come & see me. Cordially yours

Geo F. Barker.

ALS, NjWOE, DF (*TAEM* 14:142).

1. Not found.

2. Barker is probably referring to Frank Gower, who had the general agency for the Bell telephone in New England. There was no formally organized New England Telephone Co. until February 1878. Tosiello 1979, 187–207.

3. Charles Richards, assistant superintendent of the Colt Armory, was a prominent consulting mechanical engineer and later an engineering professor at the Sheffield Scientific School of Yale University. *DAB*, s.v. "Richards, Charles Brinckerhoff."

4. Karl Koenig studied physics under Hermann Helmholtz at the University of Berlin before moving to Paris where he became a scientific instrument maker. He was known for his acoustical instruments, especially precision tuning forks, and had had a large exhibit at the 1876 Centennial Exhibition. *DSB*, s.v. "Koenig, Karl Rudolph"; U.S. Centennial Commission 1880, 488–89.

5. Barker is referring to the 1862 lectures on heat by British physicist John Tyndall, published by D. Appleton and Co. as *Heat Considered as a Mode of Motion* in several American editions beginning in 1863. Tyndall describes these phenomena on pp. 82–84 of the 1873 edition.

6. On 26 and 27 August 1877 Draper and Edward Holden observed what they believed to be a third satellite of Mars. (Holden was a former assistant to Asaph Hall, who earlier in August 1877 had announced the discovery of two Martian satellites.) This proved to be incorrect. Plotkin 1972, 47–51.

**–1046–**

*From Charles Barnard*

New York.[a] [September 10, 1877][b]

Dear Sir,

I am in want of some information in regard to the practical working of your telephone and I write to ask if you can name a day and hour when I may find you at your office Western Union Building. I wish a few facts for publication and if you can inform me when I can see you at convenience I shall be greatly obliged[1] Yours respectfully

Charles Barnard[2]

ALS, NjWOE, DF (*TAEM* 14:247). Letterhead of the Editorial Room of *Scribner's Monthly.* Second page of document overwritten by a sketch of same date signed by Edison. [a]"New York." preprinted. [b]Date from docket.

1. Barnard wrote to Edison on 13 September to arrange an afternoon meeting at the electrician's office on 17 September. Barnard also later planned to meet Edison at Menlo Park on 13 October. Barnard to TAE, 13 Sept. and 4 Oct. 1877, DF (*TAEM* 14:249).

2. Charles Barnard (1838–1920) edited the "World's Work" department of *Scribner's* from 1875 to 1884. Among his writings were articles on scientific and technical subjects. *NCAB* 13:64–65.

**–1047–**

*From Jerome Smith*

New York, Sept. 10th 1877—[a]

Dear Sir—

Last season you honored the Polytechnic Association with an exhibition of the wonders of the telephone;[1] and the reputation since accorded to you as a genius and a progressive philosopher, has created a str~~i~~ong desire to have you again appear before the members with the improvements and suggestions which are heralded in this country and Europe, the results of subsequent investigations—[2b]

Fix upon any Thursday evening in October you may select, which will give us an opportunity of arranging for the particular night you may se[l]e[c]t. If you could give us seasonably the title of the lecture or exhibition of apparatus, it would greatly convenience the public. Very respectfully yours

J. V. C. Smith, Chairman.[3]

ALS, NjWOE, DF (*TAEM* 14:443). Letterhead of Secretary's Office of the Board of Managers, American Institute. [a]"New York," and "187" preprinted. [b]Followed by "(over)" to indicate page turn.

1. Nothing is known of this exhibition.

2. A month previous the general superintendent of the American Institute's 46th Grand National Exhibition had requested that Edison ex-

hibit his telephone. Edison referred this to William Orton who agreed to the exhibit. Charles Hull to TAE, 9 Aug. 1877, DF (*TAEM* 14:441).

3. In *TAEB* 2 and *TAEM-G1* Jerome V. C. Smith's signature was misread as S. V. C. Smith. A medical doctor and professor of anatomy and physiology, Smith (1800–1879) served as Health Officer and later mayor of Boston. He edited a number of medical publications and also authored books on natural history and anatomy. Retiring to New York City in 1870, he became president of the Polytechnic Association of the American Institute of New York in 1875. Obituary, *New York Times*, 22 Aug. 1879, 5.

**–1048–**

[Menlo Park, c. September 10, 1877][2]

*Technical Note: Electric Lighting*[1]

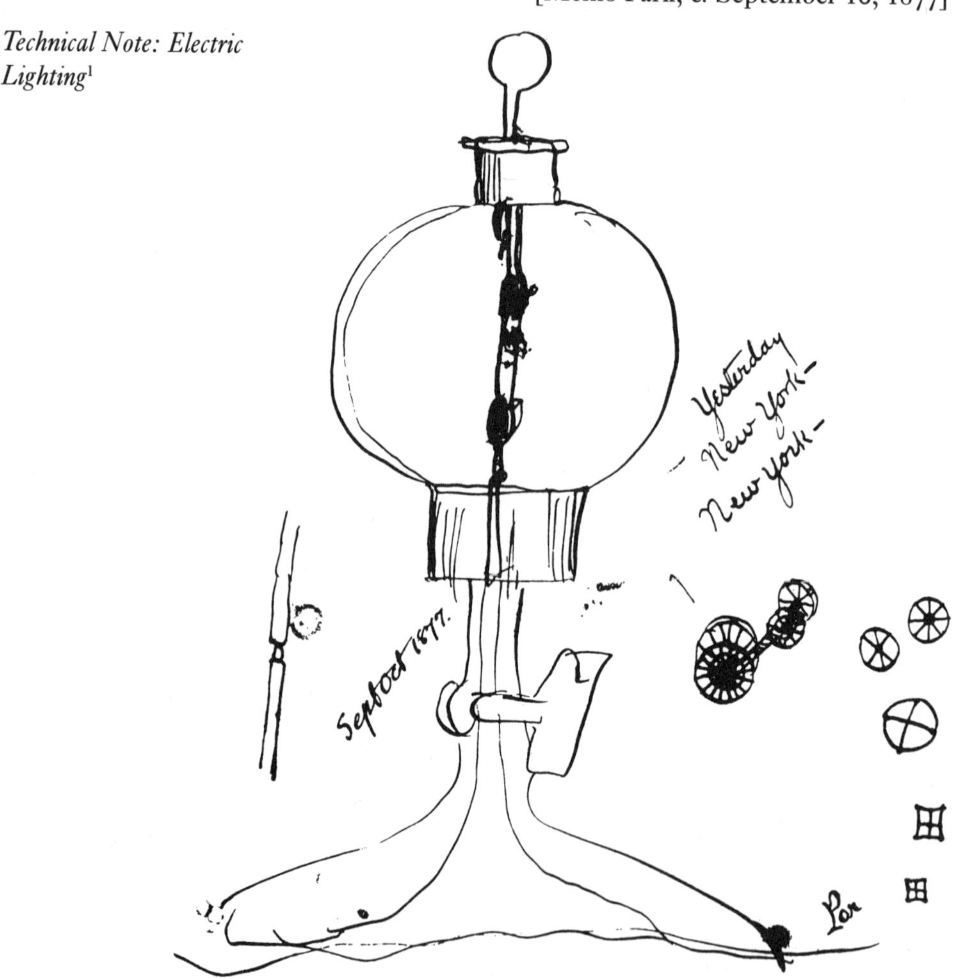


1 carbon paper carbon ⅛ wide—¾ wide thickness blotting pap[3a]

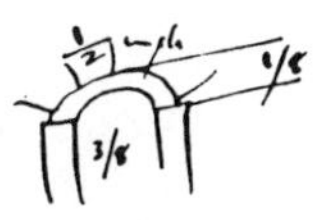

Another Carbon—½ inch[4]

~~1/45~~ from 1/12 to 1/20 wide. Several thicknesses of white[b] blotting paper

X, NjWOE, Lab., Cat. 1146 (*TAEM* 6:627). This document is photographically reproduced in Friedel and Israel 1986, 95. [a]Edge of page cut off. [b]"Several . . . white" interlined below.

1. See headnote, p. 540.

2. The dates "Sept Oct 1877" and "Aug 1878" appear on this page. Both Edison and Charles Batchelor testified about an electric light introduced in *Sawyer and Man v. Edison* as "Edison's Exhibit First Incandescent Lamp," which appears to be the one drawn here. Although Edison dated this lamp as September or October 1877 and Charles Batchelor thought it was summer or fall, Batchelor also remembered that Edison had been experimenting for "about a day on electric lighting previous to his telling me to put this carbon in the globe." If Batchelor's memory is correct then this lamp was likely made soon after Doc. 1044. Both Edison and Batchelor testified that incandescent lamp experiments using carbon predated those using boron and silicon, and Stephen Field saw experiments around 19 September involving silicon crystals brought to incandescence. Field's testimony, pp. 183–84, *Sawyer and Man v. Edison* (*TAEM* 46:244–45); testimony of Edison and Batchelor, pp. 3018, 3020–21, 3148–49, 3151, *Sawyer and Man v. Edison (U.S.)* (*TAEM* 48:16, 17, 81, 82).

In his testimony Batchelor described how the lamp was made from a Gassiot tube that had been in the laboratory since 1875. He explained that

> It was a difficult and tedious process to put the carbon in there, but I did it by unscrewing the ball from the top of the rod and also unscrewing the globe from the holder above the cock; also unscrewing the cock from the base; also unscrewing the packing cap. When these are all apart the top rod can be left in the part having the cock. The carbon was now screwed to the clamp of the bottom rod whilst lying on the table. The other clamp was then screwed to the other end of the carbon, and all three together lifted and turned, so that the part having the cock would be topmost. The lamp was also turned upside down, and the rods and carbons carefully dropped through it. The top rod was then held until the packing and packing cap were put on, when the whole was screwed together again, and the ball replaced. The binding posts were put on the lamp to hold the connections from the battery. . . .
>
> After putting the carbon into the lamp the lamp was placed on the plate of an ordinary air-pump and the bulb exhausted as well as we could do it with that pump. The current of electricity was then applied to the carbon, and heated the same for some short time. [Pp. 3149–50 (*TAEM* 48:81–82)]

*An electric egg.*

Edison described the vacuum lamp experiment as follows:

> The apparatus originally was one for illustrating Geissler tube action in vacuo. The base of the apparatus fitted over the hole in the platen of the air pump. It was then exhausted and the cock turned to preserve the vacuum in the globe of this lamp. We did not succeed in getting a higher vacuum than 2½ millimeters on the mercury gauge, and we could not make the carbons burn more than a few minutes at a time. Some of the carbons were brought up to brilliant incandescence, and probably gave thirty or forty candles of light. The carbons were brought up to various degrees of incandescence. [P. 3018 (*TAEM* 48:16)]

The lamp appears to be what was commonly known as an "electric egg," used to study electric discharge in rarefied gases (Guillemin 1873, 637). Atkinson 1886 (887) shows such lamps and notes that the stratification of the electric light resulting from the discharge of a Ruhmkorff coil through rarefied gases was initially investigated by John Peter Gassiot using "sealed glass tubes first constructed by Geissler, of Bonn, and generally known as *Geissler's tubes.*"

3. Batchelor remembered that Edison originally requested that he use hard carbon, but he found it difficult to make a piece small enough to fit in the tube. Edison then suggested using carbonized paper, which was a common material in the laboratory. Batchelor did not remember what kind of paper he used except that the carbon sheet was "thin compared with others we had, . . . about seven or eight thousandths." He thought the other dimensions were ¾ inch long by about 1/16 inch wide. Batchelor's testimony, p. 3149, *Sawyer and Man v. Edison (U.S.)* (*TAEM* 48:81–82).

Edison recalled that

> The carbons were made of sheet paper, of various widths and thickness. I think they were made of Bristol board. They were from three-sixteenths to a sixteenth wide, and probably from eight to fifteen thousandths thick. I believe they were carbonized in tubes made of gas pipe. I cannot remember whether they were prepared at the time or were on hand; we had an immense collection of carbonized paper and wood on hand, which we used in our telephonic experiments, in 1877. [Edison's testimony, p. 3018, ibid. (*TAEM* 48:16)]

4. Figure labels are (clockwise from top) "½ inch," "⅛," and "⅜."

**–1049–**

*From George Prescott*

New York, Sep 13, 1877.

Friend Edison,

I would be glad to see you at my office next Tuesday at eleven o'clock to meet some gentlemen who are going to Europe and would like to arrange to introduce the quadruplex on the continent.[1] I have made an appointment for you to meet them at that time, and if anything will prevent your coming please let me know. Yours truly

Geo. B. Prescott

ALS, NjWOE, DF (*TAEM* 14:720).

1. On 18 September, Edison and Prescott agreed to assign to Cornelius Herz and Stephen Field their patents for the quadruplex telegraph in Austria, Spain, France, and Belgium. Agreement, DF (*TAEM* 14:722).

**-1050-**

*To Franklin Badger*

Menlo Park, N.J. Sept 17 1877[1]

Dear Sir,

We had a final test to day of the merits of the Edison and the Bell Telephones at which Mr Orton was present[2] He expressed himself decidedly in favor of mine and said there was no comparison between the two. He had no idea that he could get it so loud and good as on the line we worked. Bell's was very indistinct & troubled greatly from extraneous noises. He gave an order immediately after for 150[a] one hundred and fifty sets to be made immediately.[3] I expect to have 6[a] six full setts by Saturday, when I shall send you 2[a] two stations, and perhaps shall be able to send you call bells also.[4]

There are a few little points about them which I forgot to tell you of when here.[5] It is however better perhaps for you to find them out for yourself. I am writing a small book of instructions for them in which every fault likely to occur will be fully shewn and the remedy pointed out.[6]

Hoping to hear from you soon as to their working Respectfully yours

Thos. A. Edison per Batchelor

P.S. As the silks in your machines were made in a hurry they may not articulate so well, and as we have now a little machine for making them you may find the ones I send on Saturday better.[7] T.A.E.

L (letterpress copy), NjWOE, Lbk. 1:285 (*TAEM* 28:161). Written by Batchelor. [a]Circled.

1. Two days earlier Edison had signed an agreement with Badger appointing him sole agent for Edison's telephone inventions in Canada. The agreement was made with Badger alone in an effort to avoid public conflict between Robert Watson, E. A. Baynes, and the Canadian District Telegraph Co. Watson to TAE, 4 and 10 Sept. 1877; Canadian District to TAE, 8 Sept. 1877; TAE to Canadian District, 15 Sept. 1877; TAE agreement with Badger, 15 Sept. 1877 (and drafts); all DF (*TAEM* 14:897, 903, 908, 910, 911–24).

2. They "put a pair on South Street line with call bells etc Stager, Gray, Orton & many others worked on it." Afterwards Gray and George Bliss went to the Menlo Park laboratory. Cat. 1233:260, Batchelor (*TAEM* 90:183).

3. According to Joseph Murray's testimony, he received an order from William Orton on this same day for one hundred instruments. He noted that they were not all made at once, but "sometimes in pairs, and in different numbers as they were called for, Mr. Edison usually took them over. They were tested and slight changes made from time to time to make them more perfect" (TI 1:360–61 [*TAEM* 11:157]). However, there appear to have been two orders; see Docs. 1073 and 1085 n. 1. Charles Batchelor's diary entry of 17 September notes that "Orton gave Murray order for 100 sets at $11—of Edison Telephones" (Cat. 1233:260, Batchelor [*TAEM* 90:183]).

4. There is an incomplete memorandum of agreement between Edison and Badger, dated 15 September 1877, that indicates Edison was to ship instruments for an unspecified number of stations (probably the two indicated in this letter) within thirty days. Edison was to receive a royalty of fifty dollars for these instruments. DF (*TAEM* 14:925).

5. Badger had been at Menlo Park on 5, 14, and 15 September. Cat. 1233:248, 257–58, Batchelor (*TAEM* 90:177, 181–82).

6. There is no evidence that Edison wrote such a book.

7. Edison wrote this on Monday. Doc. 1058 indicates that Edison did not yet have this machine.

**-1051-**

*To William Orton*

Menlo Park N.J. Sept 17/77

Dear Sir

In looking over my old caveats tonight[1] I have found in an Acoustic Caveat, practically the same thing as Dolbear & Bell have, ie[a] a tube, a plate, a permanant magnet, and no battery used.[2] This is on record in the patent office and goes back of both.[3] therefore if Dolbear dont catch him I can.[4] Pope I think has copy of caveat.[5] Yours

T. A. Edison

ALS (letterpress copy), NjWOE, Lbk. 1:288 (*TAEM* 28:164). [a]Circled.

1. Edison's acoustic telegraphy caveats (Docs. 664, 708–9, and 715; and see headnote, *TAEB* 2:709) contained material relevant to several of his current patent applications.

2. See Doc. 1043 n. 5.

3. Edison included something approximating this description in two of his caveats on acoustic telegraphy (Caveat 74, figs. 3 and 4; Caveat 76. fig. 5 [see illustrations for *TAEB* 2:665 n. 3, 723 n. 14; also Doc. 675]). However, Edison's devices were designed only as receivers, whereas Bell's and Dolbear's designs were also to work as transmitters; Edison's plan called for the instrument to respond strongly only to a specific tone; and Edison's circuits did involve a battery, even though the receivers were operated by induction coils in local circuits without any battery. Edison had executed the caveats on 13 January 1876 and Lemuel Serrell had filed them the next day; the Bell patent in question dated from January 1877; and Dolbear dated his invention to the fall of 1876 (Edison Caveats 74 and 76; Doc. 1043 n. 5). In preliminary state-

ments for two of the Telephone Interferences, Edison asserted that he had conceived this arrangement in September 1875 (Edison's preliminary statements for interferences E and L, TI 1:xiii, xviii [*TAEM* 11:16, 19]).

4. On 19 September, Edison sent Western Union a drawing from which George Phelps was to construct a patent model. The associated patent application covered telephones conforming to Edison's acoustic telegraph instrument design. President Orton authorized the expense on 20 September, Phelps produced the model, and an application was drawn up (Case 145), although it was not executed and filed until December 1877 ("Model Speaking Telephone," TAE to Phelps, Box 92 02 163 01, NjWAT; Edison's Case No. 145, TI 2:25 [*TAEM* 11:198]; figure labels are "brass tube," "iron to represent a permanant magnet," and "tube").

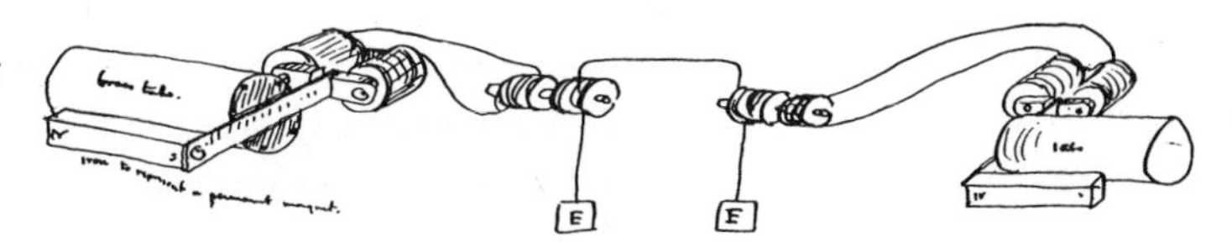

*Edison's sketch of a telephone patent model to be made at the Western Union shop.*

Also on 19 September, Edison had James Adams make a device of this type, very similar to the instrument depicted in Doc. 675 (Vol. 12:155, Lab. [*TAEM* 4:72]).

Edison's Case 145 described a magnetic diaphragm or armature affected by an electromagnet in a circuit with any acoustic transmitting instrument; it also covered operating such a receiver in a local circuit by induction from the line current. The possibility of using the device as a transmitter—the primary dispute between Bell and Dolbear—was mentioned but not claimed. However, Edison's design did compete with various other arrangements Bell, Dolbear, and Elisha Gray each claimed, and the application became part of the Telephone Interferences (Subject Matter, TI 1:iv, vi [*TAEM* 11:11, 12]). Edison never obtained a patent from this application.

Edison's review of his January 1876 caveats may also have led to consideration of a claim for "the first telephone on record," as figure 12 from Caveat 75 (Doc. 708) was redrawn and so labeled (NS-76-003, Lab. [*TAEM* 7:403]). Edison never actually presented such a claim, however. Although he maintained that he preceded Bell in building devices that were essentially magneto telephones, he acknowledged that he did not think to use them as voice transmitters until after Bell did (Prescott 1879, 222).

5. In connection with Dolbear's agreement with Gold and Stock Telegraph Co., Franklin Pope had begun acting on behalf of Western Union as Dolbear's patent attorney. Doc. 1043 n. 4; Bruce 1973, 266–67; Answers 27–33 to Cross-Interrogation, p. 9, Dolbear's deposition (originally from *Bell Telephone Co. et al. v. Peter A. Dowd*), Telephone Interferences.

**-1052-**

*From Franklin Badger*

Montreal Sept. 17th 1877

Dear Sir

I received your note of 16th inst. today— all right.[1] I am working things up— I have tested Telephone on our lines and over a line of 2 miles or so. I astonished the natives. Am to test it tomorrow on some private lines and expect similar results.

I use plenty of battery in testing as it gives much better results I find the contact variable in these I have, necessitating a change of adjustment. I think if you can manage to make the silk pads more compact with ~~pressure~~ greater[a] pressure, that better results would be obtained.

I have been told that Bell's Telephone was tried on one of her Majesty's war ships for purposes of signaling to the men in different parts of the ship and it utterly failed on account of external and internal noises I presume the internal noises were caused by the pulsation of the circulatory apparatus of the barnacles on the ship's bottom.

When you send documents let the proposition you made me for the sale of patent right be embodied in a separate paper. Please send me another sett of apparatus for 2 terminals as soon as possible as I desire to use them in Quebec. Very truly yrs

F. H. Badger

ALS, NjWOE, DF (*TAEM* 14:928). [a]Interlined above.

1. Not found.

**-1053-**

*Technical Note: Telephony*

[Menlo Park,] Sept 17 1877

Speaking Telegraph.

I find that a large number of Silk plumbagoed disks say 8 or 10 hard pressed makes speaking painfully loud—and it may be so adjusted as to be very sensitive. transmitter placed 25 feet from speaker conversation can be carried on and I have no doubt but if a very sensitive Receiving instrument was used that the voice of the speaker in any part of the Senate ~~C~~or House Chamber could always be transmitted to a distance providing the telephone was placed permanently on the speakers desk

I propose to dampen the diaphragm in a Bell telephone by strips[a] of rubber stretched over them= this will [~~we~~?][b] prevent Harmonical Sounds

X, NjWOE, Lab., Vol. 12:153 (*TAEM* 4:71). [a]Obscured overwritten letter. [b]Canceled.

## COMMERCIAL TELEPHONE TRANSMITTER AND RECEIVER Doc. 1054

In mid-September, Edison tested a telephone design for Western Union President William Orton, who placed an order for 150 sets.[1] Joseph Murray began tooling up to manufacture them and delivered the first sets by the end of the month.[2] The receiver was very much like Bell's—a metal diaphragm vibrated by an electromagnet—except that the diaphragm was loosely mounted and the listener damped any excessive vibration by pressing it against an ear. The transmitter, however, embodied Edison's signal contribution to telephone technology—a battery-powered circuit that passed through a material (in this case "carbon fluff," silk coated with plumbago) whose resistance changed as it was variably compressed by sound waves in the mouthpiece.

George Prescott, who entered the transmitter and receiver in Doc. 1054 as evidence in the Telephone Interferences, testified that he experimented with two sets of these instruments in the late summer or early fall of 1877, and that they were "operative instruments."[3] Even as Murray made these instruments for Western Union, Edison continued his transmitter experiments to improve the signal's clarity. By early October Edison had changed the design in several particulars and Murray changed the production instruments accordingly.[4]

1. See Doc. 1050. Edison seems to have arrived at the basic design of these instruments at the end of August (see Doc. 1030).

2. See Doc. 1059. Joseph Murray identified the instruments as having been made for the order of 17 September (TI 1:361 [*TAEM* 11:157]), although he mistakenly also remembered having altered them later in October (they do not exhibit the alterations Murray alluded to). See Doc. 1081.

3. TI 1:351–52 (*TAEM* 11:152–53).

4. See Doc. 1081.

*Production Model: Telephony*[1]

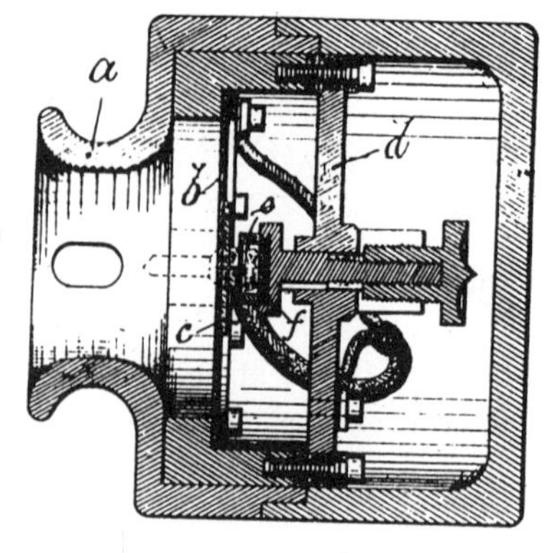


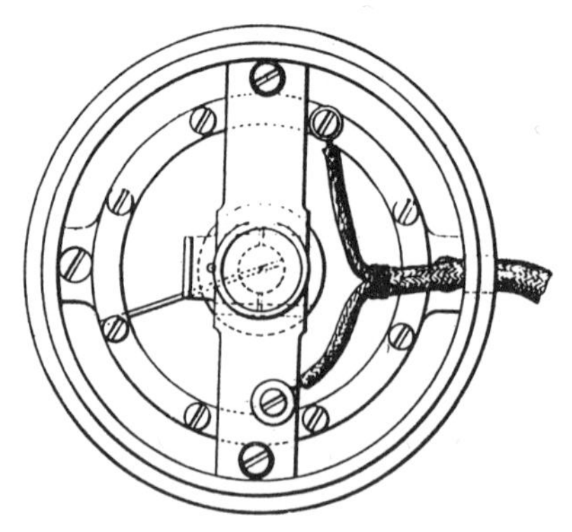

a. wood
b. mica diphragm
c. cork

d. brass
e. hard rubber
f. carbon fluff

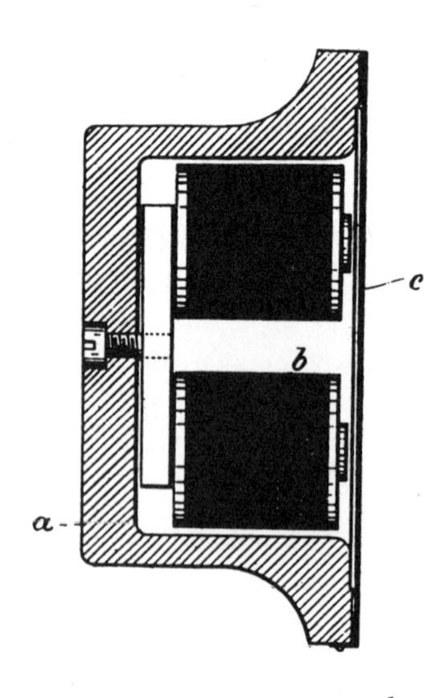


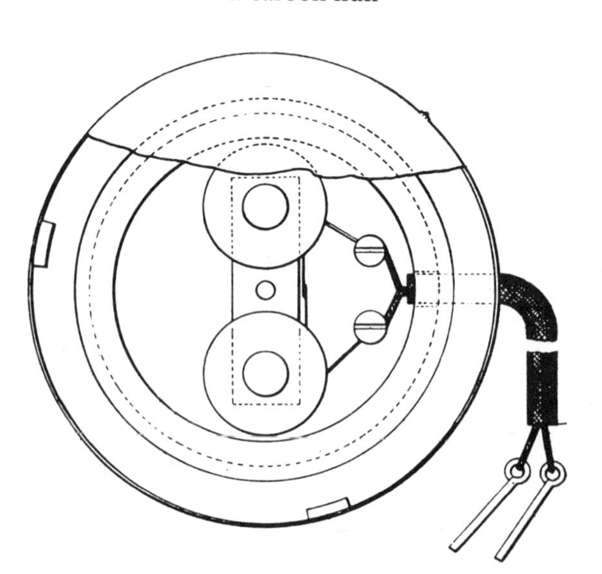

a. wood
b. electro magnet
c. Iron diaphragm

M (historic drawing) (transmitter [top]: 7.6 cm dia. × 7.8 cm; receiver [bottom]: 10.6 cm dia. × 5.2 cm), NjWOE, TI 2:540, 539 (*TAEM* 11:668, 667). Photolithograph of drawing of exhibited instrument, prepared for printed record.

1. See headnote above.

**–1055–**

*To William Preece*

Menlo Park N.J. Sept 19/77

Friend Preece,

I will ship you Sept 26th Speaking Telephones for two stations. I do not know what the effect of the voyage will be upon them but I guess they will be ok. I will try and send you four more stations by 10th of Oct. WU Co. are having a pile made, but I cannot get any of them and am compelled to make them myself.[1] If you would like a pair for lecturing, I can make them so a moderately sized audience can hear all the conversation, and in a small room you can conduct a conversation when you are not within 10 or 15 feet from the transmitter. Yours

T A Edison

P.S. Remember on the Tone-Telephone that the zinc ~~a~~end of the battery goes to the arm; the C end is very weak.[2] E

ALS, UkLIEE, WHP.

1. A John Kruesi time sheet for the dates 14 and 27 September 1877 shows work on six and twelve telephone stations respectively. DF (*TAEM* 14:359).
2. Figure labels are "zinc" and "copper."

**–1056–**

*Draft Agreement with George Bliss, Cornelius Herz, and Stephen Field*

Menlo Park, N.J. September 19, 1877[a]

This memorandum of agreement entered into this nineteenth day of September, eighteen seventy-seven by and between Thos A Edison of Menlo Park, N.J. party of the first part, Geo H Bliss of Chicago Illinois[b] ~~& Chas E Holland, of~~ party of the ~~second~~ third part, and Cornelius Herz & Stephen D. Field of San Francisco, parties of the ~~3rd third~~ second part, witnesseth:[1]

Whereas—the said ~~C~~Party of the first part is the inventor of a Speaking Telephone for which he is desirous of applying for patents in and selling such [-- --][c] patents in[d] the following Countries to wit. Russia, Spain, Italy, Austria, Germany France, & Belgium,[2] and Whereas the said parties of the second & third parts are desirous of introducing & selling[e] the ~~said~~ patent right in the above mentioned countries, Therefore be it agreed that for and in consideration of the sum of one dollar in hand paid by the said party of the second & third parts to the said party of the first part the receipt of which is hereby acknowledged and In consideration[f] of the payment within ten days from date[g] of all fees & expenses incurred in procuring patents in the above mentioned Countries, which sum is to be ~~lodged~~ paid[h] into the hands of L W Serrell patent solicitor 76 Chambers St New York[3] [--------][c] ~~the day~~ by the said parties of the ~~third~~ 2' Second[i] part, and ~~for a further in~~ for a further[j] Consideration ~~of~~ on the part of the[k] the said parties of the second & third parts ~~do~~ that they shall[l] use due diligence, in offering for sale & endeavoring to sell such patents in the said Countries which endeavors shall be[m] at their own expense: ~~The said party of the first part will upon the performance of such~~ The said party of the first part in consideration of these ~~premises heret~~ the performance of the above stipulations[n] grants to the said parties of the 2nd & 3rd parts ~~two th~~ one third of all[h] the proceeds from the sale of such patents in such countries, ~~and shall upon deliver proper assignments and patents papers upon the payment~~ But it is expressly agreed that the patent for France shall not be sold for less than $8000. That the patent for ~~sp~~Germany shall not be sold for less than 15 000. that the patent for Italy shall not be sold for less than 5000, that the patent for Spain shall not be sold for less than ~~for~~ 5000, that the patent for Russia shall not be sold for less than 4000 that the patent for Belgium shall not be sold for less than 4000, ~~from~~ of[h] ~~the~~which sums ~~that sha~~ the said parties of the second & 3rd parts are to receive 1/3.

~~It is further agreed that,~~

~~It is hereby agree~~ It is hereby agreed ~~that in case both of the said parties of the second & 3rd parts fail to dispose of such patent~~ that all the parties to this agreement shall endeavor to sell such patents and if the said party of the first part shall succeed in selling the whole or any one of such patents at a price not less than above mentioned then the said party of the first part is to have 7/9 of the proceeds, and ~~such~~ the part~~y~~ies of the 2nd part 1/9 and the party of the 3rd part 1/9, and in case the parties of the 2nd part succeed in selling the whole or any one of such patents at ~~the~~ a[h] prices not less than above mentioned then the said party of the first part is to receive 2/3 the party of the 2nd part 2/9[o] and the said party of the 3rd part 1/9, and in case the said party of the 3rd part should succeed in selling the whole or any one of ~~the~~ such patents at a price not less than above mentioned then the said party of the first part is to receive 2/3 the said parties of the 2nd part 1/9 and the said party of the 3rd part 2/9.

It is further agreed that the said parties of the 2nd & 3rd parts shall not transfer their interest without the consent of the said party of the first part

If it can be shewn to the satisfact~~ary~~ion of the said party of the first part that the sums herein mentioned for the patents ~~are too great~~[p] in all or any country are too great,[q] or that it would be advantageous, to sell at a lesser sum or that it would be necessary to take a lesser sum to secure a sale; then the the said party of the first part shall name ~~th~~a lesser sum

It is further understood that the portion of the patents which pertain to apparatus for recording & reproducing the Human Voice & other Sounds Locally ~~s~~is not to be sold with the telephonic apparatus but in any Contract assignment or transfer of the patent for telephonic apparatus th~~a~~e part above mentioned is to be retained & reserved[r] and shall belong to the parties to this contract in the proportions & in the countries[s] herein[t] mentioned are[h] and when perfected shall be the subject of another Contract for its sale

ADf, NjWOE, DF (*TAEM* 14:935). [a]Place and date taken from document, form altered. [b]"of Chicago Illinois" interlined above. [c]Canceled. [d]"patents in" interlined above. [e]"& selling" interlined above. [f]"of the sum . . . consideration" overwritten by large flourishes. [g]"within . . . date" interlined above. [h]Interlined above. [i]"2' Second" interlined above. [j]"~~in~~ . . . further" interlined above. [k]"on . . . the" interlined above. [l]"that they shall" interlined above. [m]"which . . . be" interlined above. [n]"in . . . stipulations" interlined above. [o]"the party . . . 2/9" interlined above. [p]"~~are too great~~" interlined above. [q]"are too great" interlined above. [r]"& reserved" interlined above. [s]"& in the countries" interlined above. [t]Obscured overwritten letters.

1. On 16 August, Edison had drafted a proposition to George Bliss and Charles Holland regarding his telephone patents for various European countries. On 18 September, when Bliss was in Menlo Park, Edison drafted an agreement calling for Bliss and Holland to provide Lemuel Serrell with sufficient funds to procure Edison's patents in various European countries in return for which Edison would sign an agreement with them regarding the division of profits from sale of the patents (DF [*TAEM* 14:862, 933–34]). That night, Bliss "stayed at Edison's house Talking over matters relative to pen & Telephone till 2 AM" (Cat. 1233:261, Batchelor [*TAEM* 90:183]). On 26 September, Batchelor recorded, "Dr Herz & Mr Field here from California they pay the European patents for the ~~pen~~ Telephone & get 1/9 out of Bliss's 1/5" (Cat. 1233:269, Batchelor [*TAEM* 90:187]). The 18 September agreement with Bliss was voided by the 17 December contract between Edison, Bliss, and Theodore Puskas (Kellow [*TAEM* 28:1195]; see Doc. 1153 n. 5).

2. The previous day, on a page headed "foreign Telephone," Edison had sketched several telephone and phonograph devices with labels like "mention this in Canada & England" and "mention Cork diaphragm." Edison's Exhibit 154-12, TI 2 (*TAEM* 11:448).

3. On the same day Serrell wrote Edison to inform him of the cost of acquiring and maintaining speaking telegraph patents equivalent to

his Canadian patent application (which Serrell was just finishing) in France, Belgium, Austria, Germany, and Italy. DF (*TAEM* 14:496).

**–1057–**

*Technical Note: Telephony*

[Menlo Park,] Sept. 20 1877

Speaking Telegraph

In place of the rubber diaphm[a] for preserving the regular[a] diaphragm in the same plane & also allowing it to be free & thus counteract expansion  I can use & find it o.k. silk tissue, silk net & cloth with large interstices like sieves made of say celluloid, asbestos silk etc.

The diaphragm is loose, may have its harmonics deadened by gluing cloth Rubber or other dead substance to it=[1]

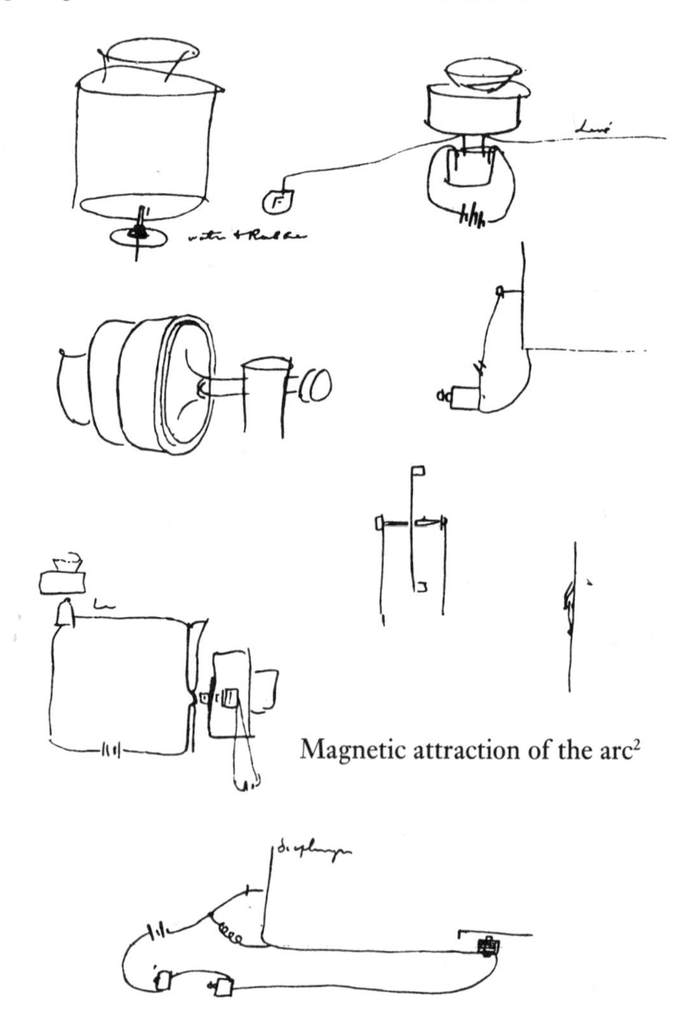

Magnetic attraction of the arc[2]

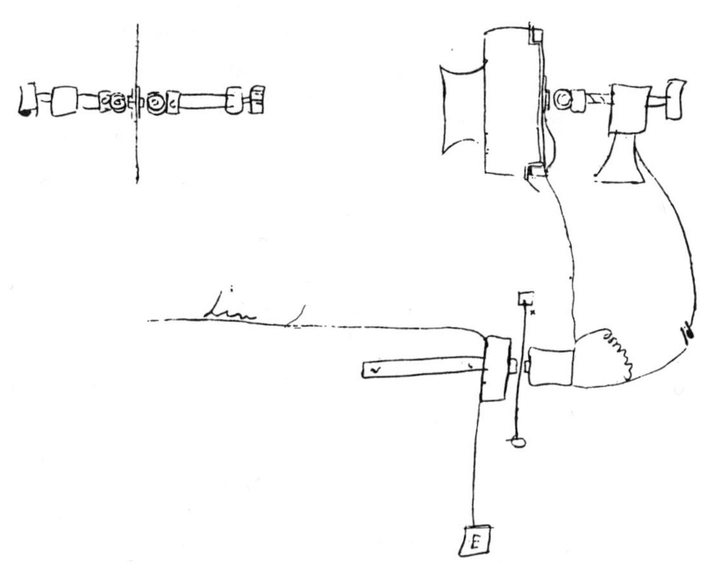

Reuiss principle: New Idea! X perhaps rubber would make it better[3]

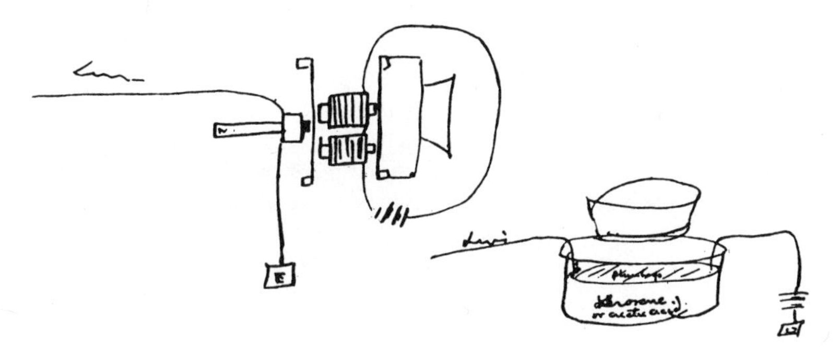

Plumbago surface floating spkg against it varies the resistance=[4]

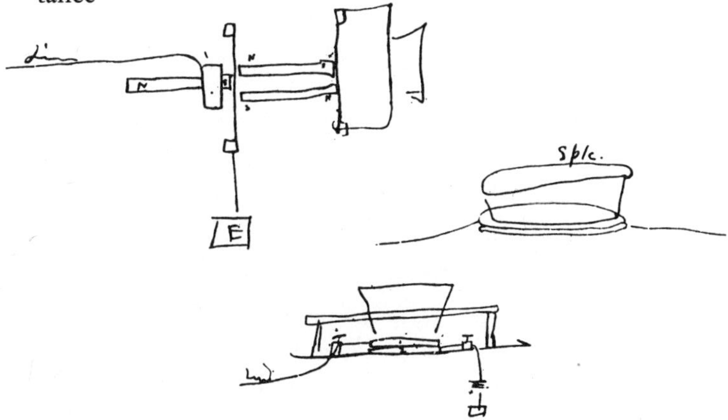


1 & 2 are plumbagoed or conducting cloth spk against it=[5]

We now dispense with the silk fluff or loose fibre with intermixed plumbago and use cloth, perferably silk cloth impregnate it with a syrupy solution of plumbago in Dextrin Water. The impregnation is made more thorough by jabbing it in by needles it is allowed to dry when the cloth exhibits a beautiful surface, washers or rather discs are cut from it and these disks generally have 13 ohms resistance though much lower when the plumbago is well worked in the Dextrin causes it to adhere with great tenacity to the fibre. Instead of Dextrin any sticky substance may be used, or matter which when mixed with the plumbago will cause ~~the~~ it to adhere to the fibre. Velvet may be used the plumbago being forced through the body & the Extending plushy fibre is also coated= of course peroxide lead & other compound conductors or finely divided metals may be used=

T A Edison Chas Batchelor
James Adams

X (photographic transcript), NjWOE, TI 2, Edison's Exhibit 159-12 (*TAEM* 11:449). Document multiply signed and dated. [a]Interlined above.

1. Figure labels are "water & Rubber" and "Line."

2. An electromagnet inside the transmitter (at right) induces a magnetic field in the piece of soft iron attached to the center of the diaphragm. As the diaphragm vibrates it changes the distance of the magnetized iron piece from the electric arc, causing the arc to change its shape and presumably altering the resistance of the line.

3. In figure 25 of his British Patent 2,909 (1877) Edison proposed a similar design that used inductive effects to turn the make-and-break signal of the Reis telephone into a continuous, varying signal. Also see Doc. 1079.

4. Figure labels are "Line," "plumbago," and "Kerosene or acetic acid."

5. See Prescott 1879, 527–28; and Jehl 1937–41, 1:148.

**-1058-**

*To Thomas David*

Menlo Park N.J. [c. September 20, 1877][1]

Friend David,

I think those Telephones must have got banged around as they worked beautifully in N York all day and a similar pair taken by my montreal man are working in ~~m~~Montreal O.K. one pair on Dutton & Townsends[2] line in NY 2½ miles has been working perfectly 13 days to date[3] and they have noted it on their slips, which Mr Orton saw= You must bear with me a little while until I can finish my machine for making those silks properly.[4] I had to make yours with a pair of plyers and two peices of hard rubber. The proper way to adjust them if they are way out is to adjust one rather close and the other far away bringing it gradually closer until you hear the other man say all right  you can then turn around & adjust yours away & come closer until he says all right= but I think adjusting will not be necessary when my silk resistances machine is ok. Yours

T A Edison

ALS (letterpress copy), NjWOE, Lbk. 1:290 (*TAEM* 28:165).

1. This letter is a reply to David's letter of 19 September. DF (*TAEM* 14:940); also see note 4.

2. Dutton and Townsend are listed as stave dealers at 70 Beaver St. Wilson 1878, 389.

3. Charles Batchelor's scrapbook contains the transcript of a telephone conversation between himself and Adams during a test of the Dutton and Townsend private line on 17 September (Cat. 1240, item 250, Batchelor [*TAEM* 94:74–75]). Batchelor had installed the instruments on 5 September (Cat. 1233:248, Batchelor [*TAEM* 90:177]).

4. A technical note of 21 September shows the design of a mandrel for making silk fluffs (Vol. 12:169, Lab. [*TAEM* 4:79]). On 23 September, Batchelor noted "Worked all Day and night on Speaking telephone fluffs  Made winding machine for fluffs  Made stamping press for fluffs" (Cat. 1233:266, Batchelor [*TAEM* 90:186]).

**–1059–**

*To Franklin Badger*

Menlo Park, N.J. Sept. 21, 1877

F. H. Badger,

The Contracts were sent= you must bear with the Telephones for a few days until we can make a machine to properly make the silk flufs then I dont think you will need any adjustment= Ive got a better plan even now to make them and have them in a pair in N.Y. on one mile line 12 cells; Every one pronounces; the talking to be "painfully" loud & they dont hold receiver to ears; I will try and send you 2 stations tuesday and 4 stations saturday following  from these you will soon be able to report defects ~~of~~ in[a] practical use etc and then perhaps an alteration or two when I guess we are on the right track for biz. Mr Murray is making a set of tools[1] so he will be able to make 100 per week. Yours

T A Edison

ALS (letterpress copy), NjWOE, Lbk. 1:292 (*TAEM* 28:167). [a]Interlined above.

1. Manufacturers usually had a special set of tools for each instrument they made, including jigs to guide machines; gauges to check work; cutting blades for milling machines, shapers, and lathes; and fixtures to hold each piece in place as the work was being done. *TAEB* 1:518 n. 16.

**-1060-**

*To James MacKenzie*

Menlo Park N.J. Sept 21/77

Friend Mackinzie,

I will show your card to Mr Orton and put in a good word. I know he intends to appoint agents for the introduction of my speaker, and I think he is going to sell them outright instead of renting— What are Bells terms to agents. Jimmy[1] was only 2½ feet high when I used to plague him You want to ring in on this Telephone biz, it has a great future. Keep writing letters. Write Wm Orton direct, also write Stager at Chicago. Your old friend

Thos. A. Edison

ALS (letterpress copy), NjWOE, Lbk. 1:291 (*TAEM* 28:166).

1. In 1862 Edison had snatched three-year-old Jimmy MacKenzie from the path of a freight car; in gratitude, his father, James, had taught Edison railroad telegraphy (*TAEB* 1:8). According to his father's letter of 18 September, "Jimmy is within a very little of being as tall as his dad now & has been the assistant line man for over a year for [the American District Telegraph] Co." (DF [*TAEM* 14:931–32]).

**-1061-**

*From Franklin Badger*

Montreal, Sept. 21st 1877[a]

Dear Edison

I sent you last evening a local paper[1] containing a paragraph in notice of a trial of the Telephone on a line of the Dominion Tel. Co. to Quebec. We had two trials, one in the forenoon and one at 5 P.M The first trial was more satisfactory than the last, and I will here mention a point I observed.— While talking to Quebec the signals would come at times very distinct so much so that in answer to a question I recognized the voice of the particular person speaking while at other times and without changing the adjustment at either end the signal or articulation would become indistinct Now it seems to me that this is due entirely to the silks being too soft and lacking homogenity, as the resistance of them seems to be changed by the vibration of the mica diaphragm.

The clearness of articulation at times, even at that distance of 200 miles, indicates the possibilities of the apparatus, if this variability of the silk pad under vibration, can be overcome

There were noises enough on the line but these don't seem to interfere with the s vocal sounds materially.

The day before, I tested on a short line of five miles running out into the country, and as the wind was blowing freshly I could hear sounds in the receiver exactly like a continual rushing ~~sho~~ such as the wind makes in rushing through the bare branches of trees, or through a lot of wires, and I feel convinced that it must have been the molecular tremor caused by the wind, which I heard in the ~~transmitter~~ receiver. On the Water line we have here, of twelve miles I heard no such noise, nor on the Quebec line—~~because~~ of 200 miles, and it was quite calm ~~dur~~ during these last tests. In the above instance, where the rushing sound was heard, I could here other disturbances of a purely electrical kind caused by proximity of other lines, but these were emphatic and interrupted by lapping &c. easily distinguished while the other sound was continuous, rising, and falling, just like the wind. It did not interfere with the talking however.

But to come to more practical points; if you can only manage to improve the pads to get rid of the trouble I refer to a grand point will be attained, and I am awaiting anxiously the receipt of another sett, as soon as you can send them.

With this I send the contract papers with my signature thereon.

Let us have that book you referred to as soon as possible.

With my best wishes for your success I remain Yrs. &c

Fn. H. Badger

ALS, NjWOE, DF (*TAEM* 14:945). Letterhead of Fire Alarm Department, City Hall. [a]"Montreal," and "187" preprinted.

1. Not found.

## –1062–

*Technical Note: Telephony and Phonograph*

[Menlo Park,] Sept. 21 1877

Speaking Telegraph[1]

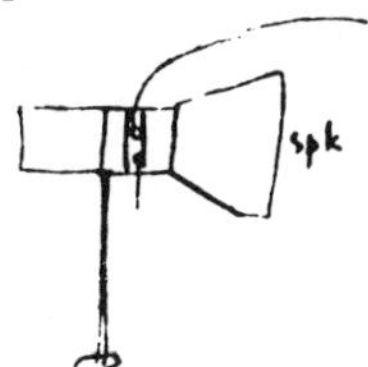

I think if you speak into a thermo pile[2] that the undulations will take heat waves to the pile & this will generate currents that may possibly be received upon an exceedingly delicate diaphragm & magnet.

I propose to use[a] batteries upon the principle of Pulvermacher chains[3] for my regular Telephone=[4]

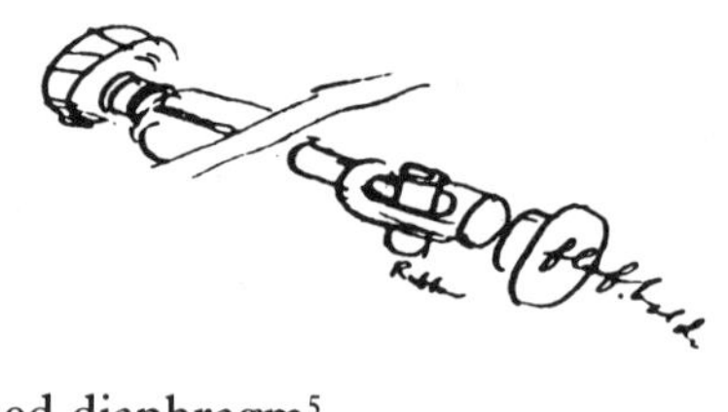

~~In~~ Dampened diaphragm[5]

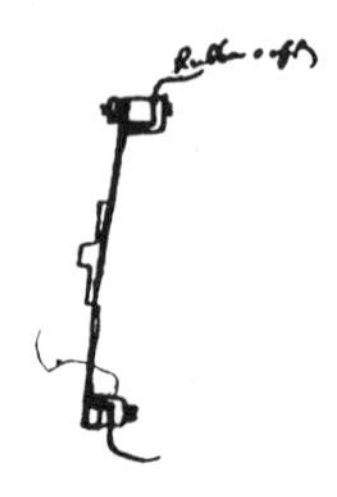

Mica diaphragm loose & laid against a stretched rubber diaphragm this prevents the effect of expansion & kills harmonics both low & high it works beautifully we made a lot of telephones this way and they all showed that it did the biz hunky Dori—[6]

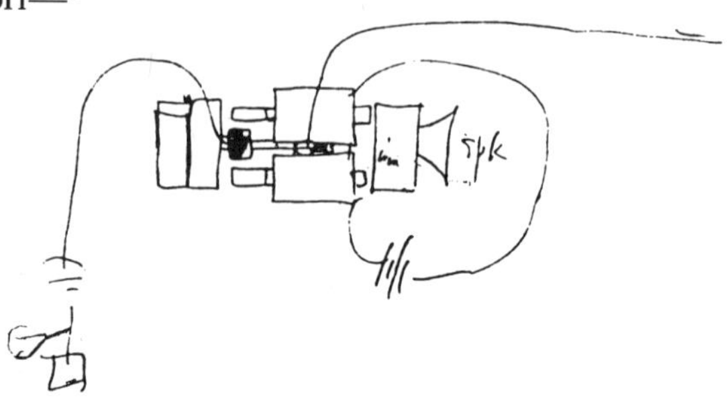


Bell, modification:

Cork diaphragm for ~~Be~~Magneto principle with very thin iron foil on it—[7]

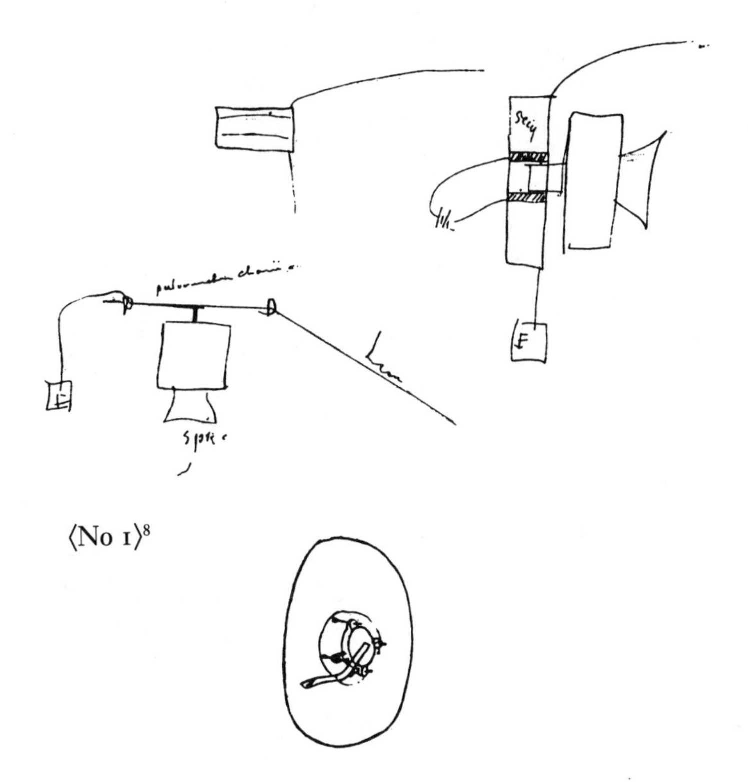

⟨No 1⟩[8]

Working by momentum no adjusting screw

⟨No 2⟩[9] ⟨No 3⟩[10] ⟨No 4⟩[11]

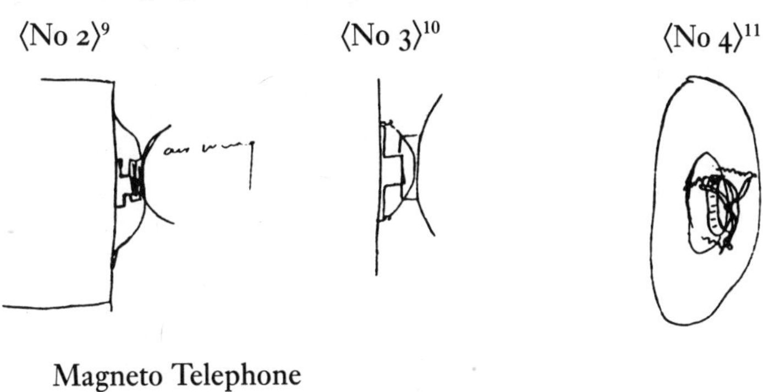

Magneto Telephone

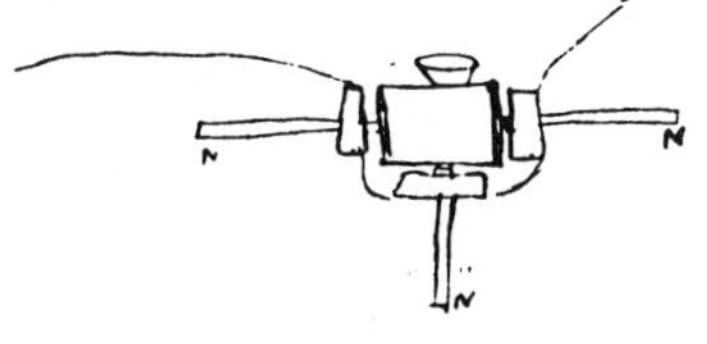

Magneto Telephone

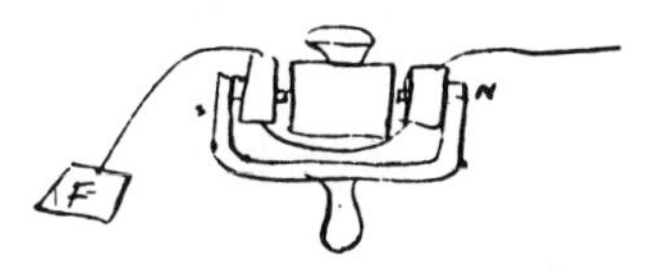

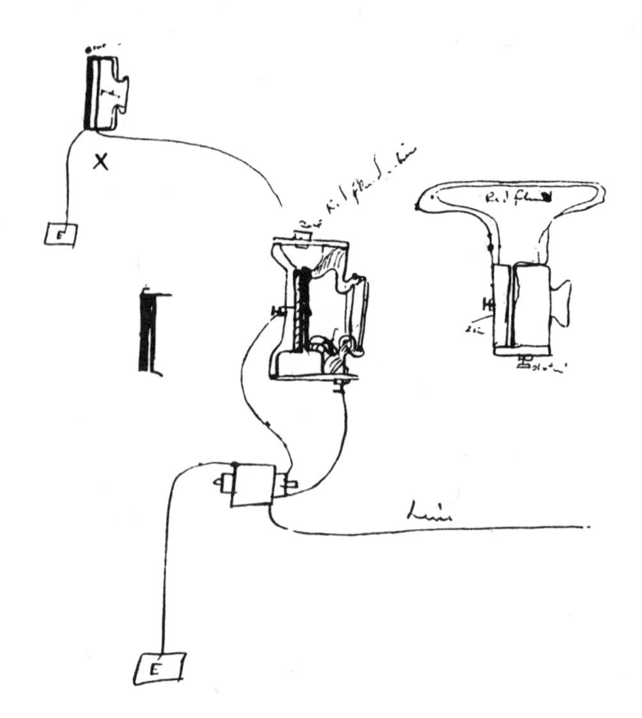


Water[12] ~~may~~ or other fluid may be used in place[a] of Red fluid[13] & platina in place zinc using it as main resistance [varier=?][b] of course a local battery would be necessary

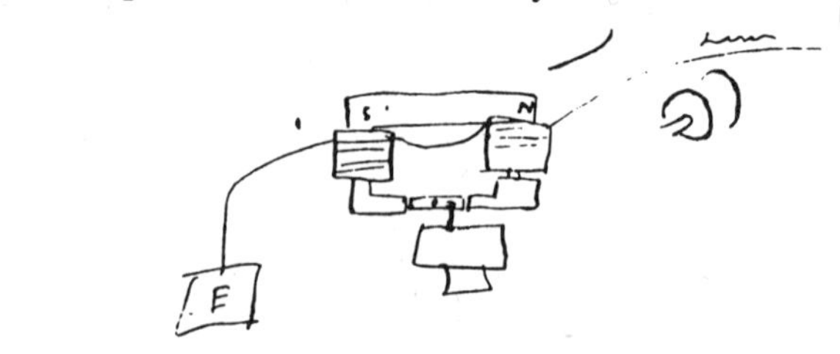


Magneto Telephone

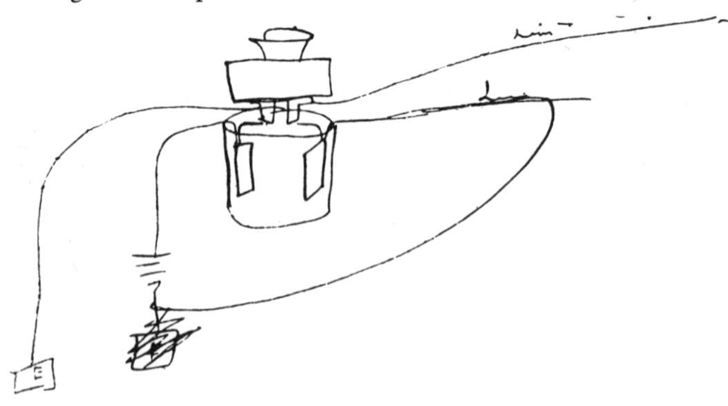

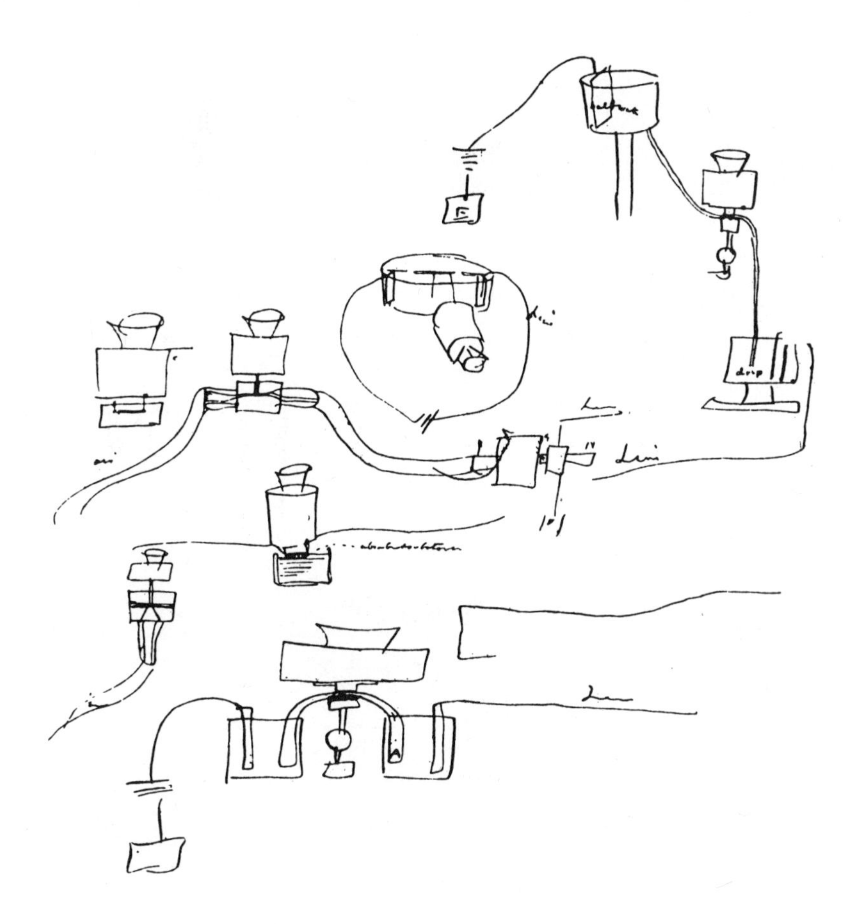

A[14] absorbent spongy material—

Sending by fine german silver wire cut out more or less by a metal spring press [against it?][b] by diaphragm =[15]

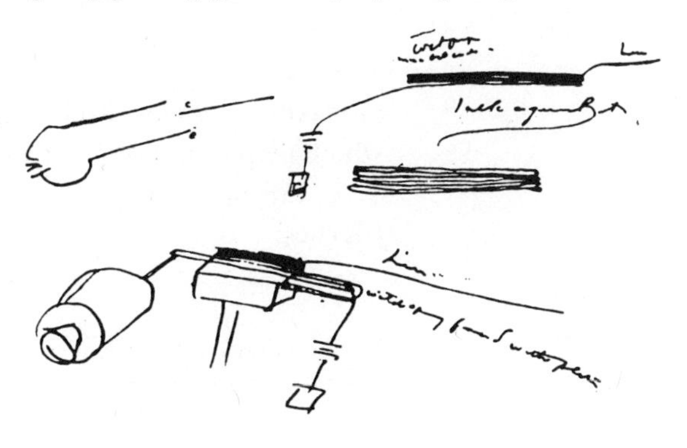

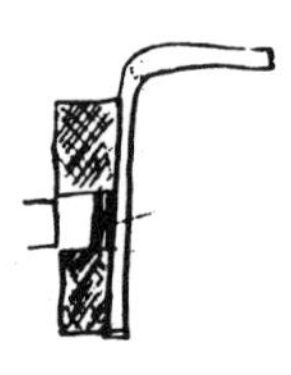

spring to give & disk on spring— 20 or 30 very thin discs of hard pressed plumbago bet[ween] this & the disk on the Cork

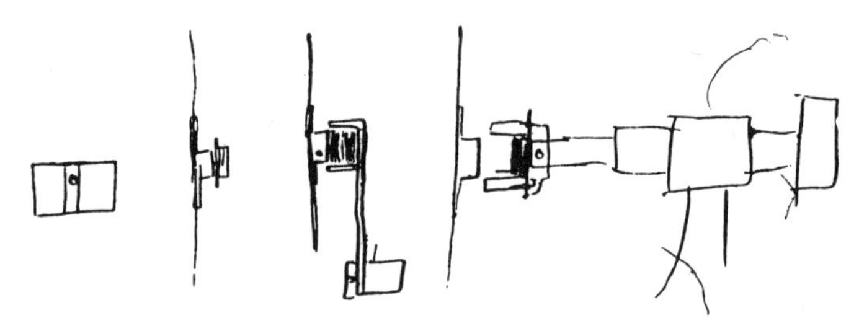

Phonograph[16]

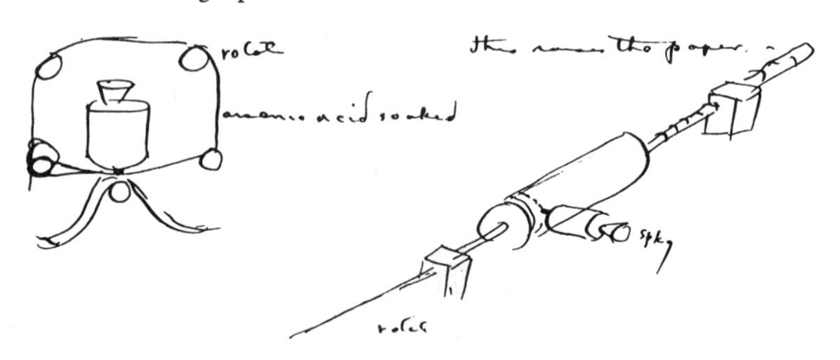

Record by embossing or otherwise  say with string or arsenic acid=

Oh heres an Idea.[17]

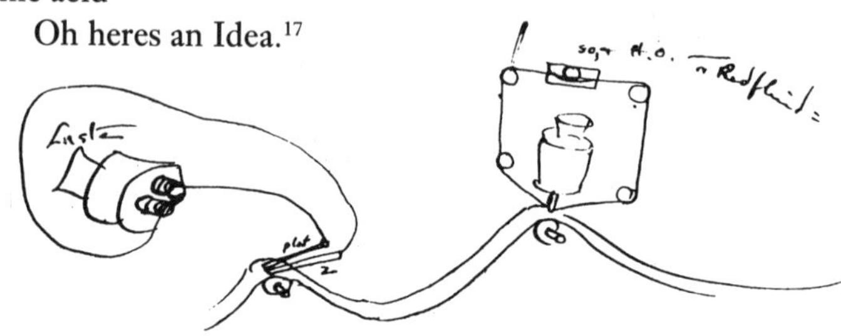

makes a battery which varies according to the pressure which diaphragm puts on string= & squeezes out fluid=[a]

T. A. Edison  Chas Batchelor
J Kruesi  ~~G E Carman~~
~~M N Force~~

X (photographic transcript) and X, NjWOE, TI 2, Edison's Exhibits 164-12, 173-12, 165-12, 167-12, 168-12, 166-12, 171-12; Lab., Vol.

17:17 (*TAEM* 11:454, 160, 455, 457, 458, 456, 259; 4:887). Document multiply signed and dated. [a]Obscured overwritten letters. [b]Illegible.

1. Figure label is "spk."

2. Edison had drawn telephones incorporating thermopiles on 18 July and 17 August (Docs. 972 and 1013). On 5 November he drew another design for a "Thermo Electric Telephone" (Edison's Exhibit 60-13, TI 2 [*TAEM* 11:524]). He patented a design like the one described here in his British Patent 2,909 (1877). There he noted that

> I have discovered that vulcanite or hard rubber produces cold and heat waves on its surface by the slightest movement, and in some cases I employ the same for the diaphragm of the resonant case . . . and place a thermo-electric pile . . . in very close proximity to the rubber, so that the vibration of the diaphragm . . . producing changes of temperature will cause currents of electricity to be set up within the thermo-battery, and these passing over the line . . . and acting upon a very delicate electro-magnet and diaphragm will reproduce the sounds acting upon the rubber diaphragm.

3. In 1853 British inventor Isaac Pulvermacher patented a battery using chains of alternating positive and negative metal elements (U.S. Pat. 9,571). In 1876 he patented modifications, including designs using thread-wrapped wire (U.S. Pat. 177,273).

4. Figure labels are "Rubber" and "fluf holder."

5. Figure label is "Rubber soft."

6. Figure labels are "iron" and "spk." On 29 September, Edison

> went to Gold & Stock & found that the [transmitters] we last took would not work We came home & worked all night & found that the mica diaphragm was very sensitive to heat. Put in brass one & also made one from a tintype twice as thick. Brass one articulation perfect, Iron one not quite so good but more permanent not being affected by heat like the brass. The mica being worse by about 5 times than brass. [Cat. 1233:272, Batchelor (*TAEM* 90:189)]

7. Figure labels are (below left) "pulvermacher chains=," "spk," and "line"; and (below right) "sec[ondar]y."

8. Edison numbered drawings 1–4 during his testimony in the Telephone Interferences (TI 1:55–56 [*TAEM* 11:48–49]). He described No. 1 as showing "a diagram near the center of which there are four thin pins, between these pins is a disk of plumbago, having flanges cut upon it, through these flanges were holes, through which the four pins passed. A spring was secured near the outer periphery of the diaphragm, which served to press the plumbago against the diaphragm."

9. Figure label is "air [ring?]." In testimony Edison described No. 2 as showing

> a diaphragm, in the center of which is fixed a piece of metal; upon this metal there was held a piece of plumbago by means of a thread; upon the outer surface of the plumbago was a concave paper disk, the object of which was to resist the sudden movement of the plumbago when an outward movement of the diaphragm took place, thus modifying the degree of pressure between the plumbago and the diaphragm, both by the inertia of the plumbago

and the resistance offered to the sudden concussion by the paper disk . . . the diaphragm being connected to one pole of the circuit, and the plumbago to the other. [TI 1:55 (*TAEM* 11:48)]

10. In testimony Edison described this drawing as a modification of No. 2 and introduced a diaphragm that he thought had been used in experiments with this design. TI 1:55; Edison's Diaphragm 165-12, TI 2:526 (*TAEM* 11:48, 654).

11. In testimony Edison described this design as similar to one from 5 July 1877 (Edison's Exhibit 2-12, TI 2 [*TAEM* 11:353]), except that here "the plumbago was held against the diaphragm by means of spiral springs" instead of strings attached to pins in the plumbago disk. He could not recall whether this design was ever made (TI 1:55–56 [*TAEM* 11:48–49]). Figure labels in the illustration are "P—," "X strings hold it on lightly—," and "spk."

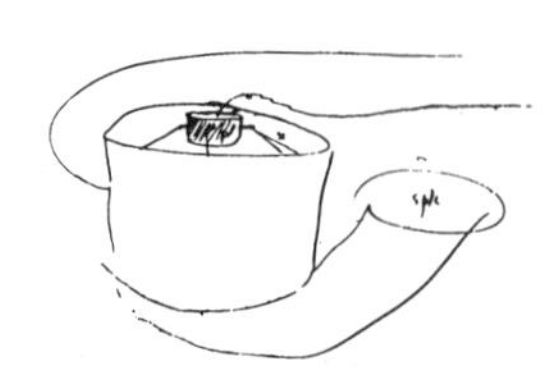

*In this design the plumbago was to be held "lightly" against the diaphragm by strings.*

12. Figure labels are (from left) "zinc," "platina," "Cork," "Red Fluid in here," and "line"; and "Red fluid," "zinc," and "platina."

13. An instruction pamphlet for the electric pen describes "red fluid."

> Procure a common jug of 1½ gallons capacity. Pour into this about one gallon of water (hot if possible) put in one lb pound of powdered Bichromate of Potash, and allow it to dissolve, shaking the liquid frequently. After it is dissolved pour slowly into the jug 3 pounds of Sulphuric Acid, and then allow it to stand 1 hour, shaking occasionally. It is then ready for use. This quantity of fluid will charge the battery about 15 times and if it is not in constant use should last 2 months, costing 10¢ per week. Wholesale Druggists charge 35¢ per lb for bichromate of potash, Retail Druggist 60¢ it is advisable to purchase from the former several pounds at a time.
>
> Common Sulphuric acid can be bought in 3 pound bottles and should not cost more than 8¢ per pound. A common teaspoon should be procured to measure the quantity of acid put in the glass jars The battery should be renewed at the sink and the water kept running [Cat. 1144, Scraps. (*TAEM* 27:310)]

14. Figure labels above are (at top) "Line," "batt water"; (middle) "air," "Line," "Line," "Line," and "drip"; and (below) "absorbent substance," "A," and "Line."

15. Figure labels below are "wet paper [mica bet cards—?]," "talk against it," and "line"; and "line" and "watch spring faced with platina."

16. Figure labels are "rotate," "arsenic acid soaked," and "this raises the paper"; and "spkg" and "rotate."

17. Figure labels are (from left) "Listen," "plat[ina]," "z[i]n[c]," and "$SO_3$ & H.O.—or Red fluid=."

**–1063–**

*From Thomas David*

Pittsburgh, Sept 21st 1877[a]

Friend Edison—

I regret to have to report that we do not get ~~sa~~results that are entirely satisfactory— It is true we can talk where Bell's cannot be heard on a/c of the "frying sound" and the induction, but the articulations are very imperfect—

The sound is not nearly so good as that we obtained when I was at Menlo— I have been trying a hollow wooden tube pressed against the diaphram of the "Listener", and with better results than by simply holding my ear against ~~the~~it— I think an arrangement something like that we had when I visited you will come nearer the mark—

The "Talker" will have to be stationary, so that it can be adjusted with one hand, while the "Listener" is held to the ear with the other—

I see you have not been able to get rid of the harmonics—

Give me some idea of the number of cells require on the average—

I will make a drawing (very rude sketch,) of what I think the "Listener" should be, in order to secure all the sounds—
Yours Truly

T. B. A. David

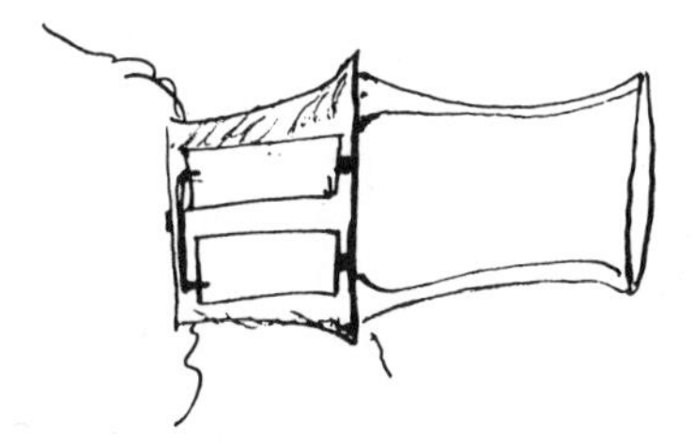

screw[1] it on just here, which will enable one to regulate the pressure—

ALS, NjWOE, DF (*TAEM* 14:943). Letterhead of the Central District and Printing Telegraph Co. [a]"Pittsburgh," and "187" preprinted.

1. This text is connected to the line coming down from the center of the drawing.

**-1064-**

*To Thomas David*

New York Sept 22 1877[a]

Harmonics is new to us must be something radically wrong will send another

Edison

L (telegram), NjWOE, DF (*TAEM* 14:948). Message form of Western Union Telegraph Co. [a]"187" preprinted.

**-1065-**

*From Thomas David*

[Pittsburgh, September 22, 1877?][1]

I shall have the above with joy— Really we have too much trouble with this one— It is tongue tied half the time— There is nothing like as perfect articulation as comes through Bell's— A great deal of the extraneous difficulties experienced with Bell's is avoided tho'— We talked today where it would have been next to impossible to have used Bell's.—

When you have the thing perfected, the outfit should be a bracket with the speaker fitted on it, and at the back a switch, bell[a] & key to call with & then to turn on the Telephone— Yrs

T. B. A. David

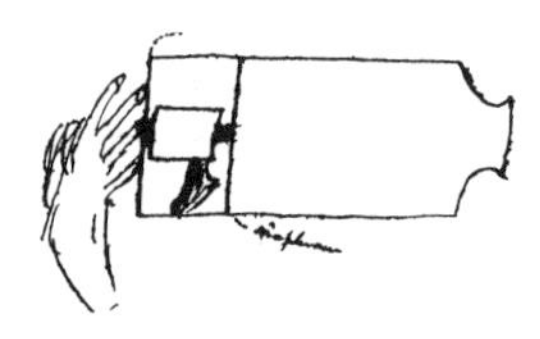

*Thomas David's suggestion for Edison's telephone receiver.*

ALS, NjWOE, DF (*TAEM* 14:948). On bottom and back of Doc. 1064. [a]Obscured overwritten letters.

1. David wrote this on Doc. 1064. On 22 September he also wrote to suggest a receiver design that would "bring us back to where we were when I experimented with you at Menlo, and got such good results" (DF [*TAEM* 14:950]). In that design the listener used a hand to press the magnet against the diaphragm.

**-1066-**

*Technical Note: Telephony*

[Menlo Park,] Sept 22 1877

Spkg Telegraph

We find that the diaphragms change their position so easily by heat of the mouth that it is impossible to use a rigid fluf holder hence we go back to the old spring lever device and we find that the fluf works ok with it although we had bad luck with it on solid plumbago we find articulation ok & moderately loud with resistance in fluf of on 15 ohms

fluf

Single plumbagoed fibre wound on thin stiff disks of plumbagoed paper.— also square disks

Also wound like a magnet with a stiff hog bristle in centre thus[1]

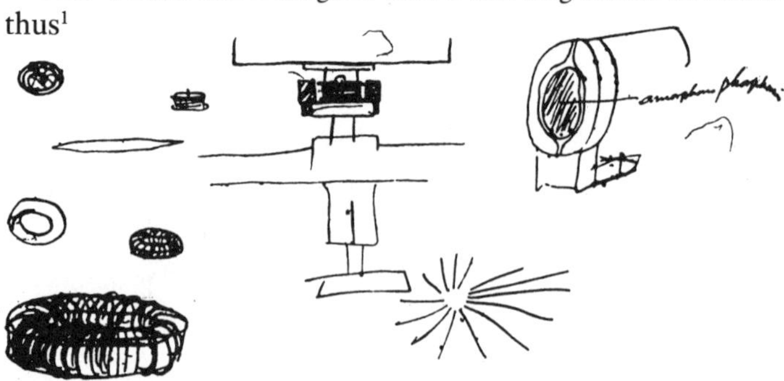

Chas Batchelor

X (photographic transcript), NjWOE, TI 2, Edison's Exhibits 174-12–175-12 (*TAEM* 11:461). Document multiply dated.

1. Figure label is "amorphous phosphorous."

**-1067-**

*Technical Note: Telephony*

[Menlo Park,] Sept 24 1877

Speaking Telegraph.
We have got the fluf biz dead-to rights— thus[1]

X flufs[2]

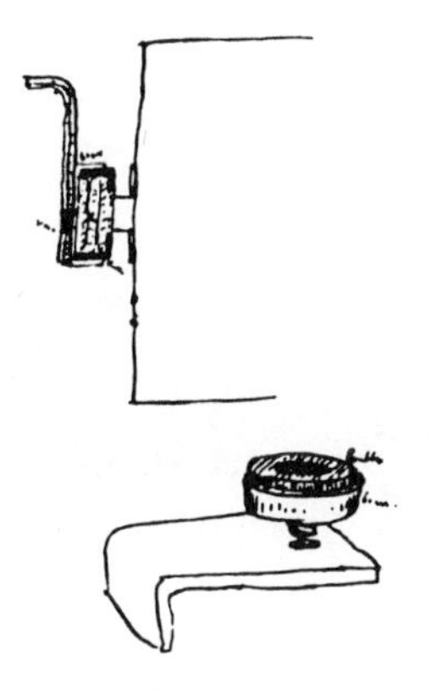

We propose to send several of these with Each machine.—[3]
T A Edison Chas Batchelor

X (photographic transcript), NjWOE, TI 2, Edison's Exhibit 177-12 (*TAEM* 11:464).

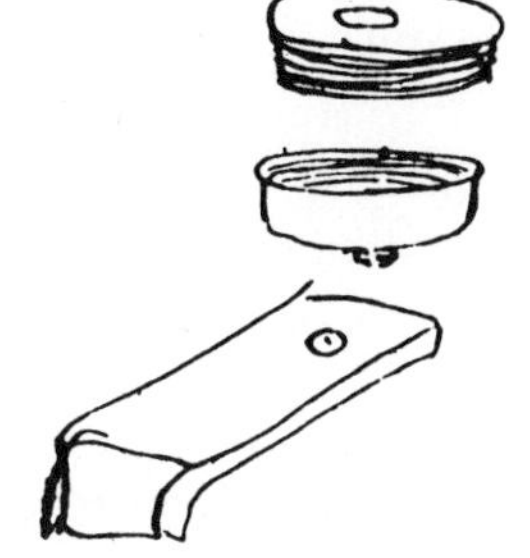

*An exploded view of the redesigned fluff holder.*

1. On 28 September, Edison noted that "The lever with fluf holder works good nearly if not loud enough and articulation better than those we have in NYork." Edison's Exhibit 181-12, TI 2 (*TAEM* 11:467).

2. Figure labels are "[iron?]," "brass," and "rubber." "X" is in the center of the speckled rectangle.

3. Figure labels are "Rubber" and "brass." In the Telephone Interferences, James Murray introduced into evidence a model of the new fluff holder (brass with a vulcanite cover) he received on 8 October along with instructions from Batchelor for other alterations on the telephones he was manufacturing (Murray's testimony, TI 1:361; Murray's Exhibit Edison's Fluff Holder, TI 2:533 [*TAEM* 11:157, 661]; see also Doc. 1081). A drawing of 29 September shows the fluff holder in exploded detail (Edison's Exhibit 186-12, TI 2 [*TAEM* 11:471]).

**-1068-**

*Technical Note: Telephony*

[Menlo Park,] Sept 24 1877

Spkg Telgh

Tonight while testing speakers with ordinary fluffs we had them quite loud when I had receiver away from Batch & he noticed or thought he heard Jims talking in transmitter upon investigation we found that it was really so had jim try it & he heard & conversation was kept up few minutes had 12 cells battery & limit was 1200 ohms; but at 100 could do biz on it on line as it was a little low the extraneous sounds would be relatively low or couldnt be heard at all I propose to conduct a series of experiments on a receiver based on the expansion principle thus

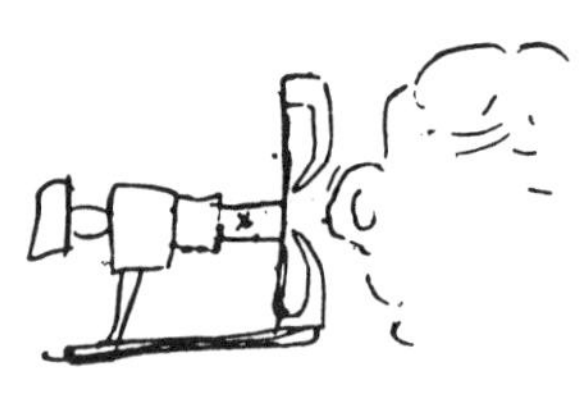

X is different things, say[a]

[~~Bals~~?][b] I find that it is almost impossible to find a substance that will stick plumbago to silk for instance white of an egg— Dextrin Balsam Peru; sugar; water glass;[1] Boil it in Chloride zinc= wont stick it as in picking it comes off but I find that Asphaltum varnish sticks it on beautifully but unfortunately when the flufs are made and put in the transmitter and adjusted the talking soon goes low due to the effect of the heated plumbago on the Asphaltum expanding the compressed fluffs— as this regains its natural possition the idea occurred that certain material such as rubber etc sensitive to heat could be ~~make~~ mixed with plumbago ~~and~~ in a cylinder & if it was headed with a disk of metal & point with a screw point in front it might be used as a relay by passing the current through the cylinder & expanding to close circuit

T A Edison Chas Batchelor

X (photographic transcript), NjWOE, TI 2, Edison's Exhibits 178-12–179-12 (*TAEM* 11:465). [a]Obscured overwritten letters. [b]Canceled.

1. Water glass, or soluble glass, an alkaline silicate, found wide commercial and industrial use in the late nineteenth century. Knight 1876, s.vv. "Soluble glass," "Water-glass."

**-1069-**

*From Franklin Badger*

Montreal, Sept. 25th 1877[a]

Dear Sir

I recd a letter from you yesterday[1] and in reply will say that I am anxious to get 2 setts of apparatus as perfect as possible We are arranging for a grand exhibition here of the Telephone, before a number of our most influential citizens, and

I feel very anxious to produce results that will astonish them, and which I believe we are able to do.

The working of the pair I have is on the whole not very satisfactory as they are so variable in working, and consequently I have deferred any further public exhibition of them, awaiting ~~further~~ another[b] consignment.

Please send as soon as you can and let them be equal to the "painfully loud" ones you described in your last letter, or what you consider good. Very truly yrs.

Fn. H. Badger

I have a party who is going to Paris in a week or two and if you have not made any disposal of France we can push matters there. F.H.B.

ALS, NjWOE, DF (*TAEM* 14:953). Letterhead of Fire Alarm Department, City Hall. [a]"Montreal," and "187" preprinted. [b]Interlined above.

1. Doc. 1059.

**–1070–**

*From Charles Edison*

Port Huron, Mich., Sept 26 1877[a]

Al

I suppose you expected as I said, that I would be back before this but I have had the ague for the last monthe and I am just getting well now and as soon as I can arrange some matters I will start   I dont think it is necessary for me to write you in regard to Rail Rail but will take the whole thing in and post you up when I come to Menlo   business is very good on the road but it aint doing it any good as old beard is squandering it as fast as it can make it   I think I can show you some good specimens of good Management on his part   Beard & Wastell is fighting over the Control of the roads and beard is running it & Wastell is crazy. Beard is rebuilding the whole business and in such a manner that it is not any better than it was before but I will give you full particulars soon[1]

Every body well. Love to all

Charley

ALS, NjWOE, DF (*TAEM* 14:594). Letterhead of Huron House. [a]"Port Huron, Mich.," and "187" preprinted.

1. In a 10 November 1877 letter to Edison, Wastell decried Beard's profligacy and dishonesty. He also reported that Pitt Edison had "started an omnibus and luggage waggon" that cut "into the earnings of the road from 11 to 12 dollars pr day," but that Pitt would "take Buss off the rout when the road is as it should be in proper hands." DF (*TAEM* 14:599–600).

**–1071–**

*From William Preece*

London 28th September 1877.

Dear Edison,

I was disappointed on the arrival of Gerritt Smith and Hamilton to find that they had not brought with them a sample of your talking telephone—

We are very anxious indeed in England to see what you have done, particularly with the transmitting portion, and I shall be extremely obliged if as soon as you are in a position to do so, you will send me at least the transmitting portion if not the whole.—

Bell has done nothing as yet in England. The instrument does not surmount the difficulty of cross fire which with us is very severe indeed in our town work. It is due more to induction than to leakage for our gutta percha wires in dry pipes induce currents[a] in each other to a very large extent—

I have tried mica for a vibrating diaphragm on your musical telephone and it comes out very much better—. I have not the slightest doubt that the next time I exhibit it that it will produce a very powerful effect. I have not liked to do anything with it while Bell is here pushing his own concern, but as soon as Bell retires I shall probably then have some opportunity of bringing the matter forward publicly— Indeed I have already promised to give three or four different lectures—[1]

I am anxiously looking forward for the other new thing that you promised, and shall always be pleased to hear of or from you.

Smith and Hamilton have started the quad[a] between London and Liverpool and it has gone off first rate.[2] Smith is very confident of ultimate success, but he has not yet seen what an English wet day is— We have given him a good wire and he certainly has received a very much better reception and has had more facilities given him than you had when you were over here—[3] In point of fact you were unfortunate enough to tumble into the wrong hands— Yours very truly,

Wm. Preece

ALS, NjWOE, DF (*TAEM* 14:152). [a]Repeated at end of one page and beginning of next.

1. For example, on 19 January 1878 Preece spoke about telephones at the Physical Society of London, where Bell had lectured in December ("Physical Society," *Nature* 17 [1877–78]: 295). He was also to be the Friday evening lecturer on 1 February 1878 at the Royal Institution (see Doc. 1157).

2. The previous day George Prescott had telegraphed Edison,

"Cable from London Says Quadruplex all right between London and Liverpool  Smith confident of Success." DF (*TAEM* 14:725)

3. See *TAEB* 1, chap. 12.

**-1072-**

*From Anson Stager*

Chicago Ill [September][a] 28 1877[b]

when may I expect the Telephones you promised me very important to Have them Here at the earliest moment possible

Anson Stager

L (telegram), NjWOE, DF (*TAEM* 14:955). Message form of Western Union Telegraph Co. [a]From docket. [b]"187" preprinted.

**-1073-**

*From Franklin Badger*

Montreal, Sept. 29th 1877[a]

Dear Edison

Yrs. of the 26th recd.[1] In my last I wrote you that the apparatus I have was found to be exceedingly variable and to come to more tangible results I tested them on a Differential Galvanometer. Of course there was a constant in the Receiver of some 23 ohms but the Transmitters gave these results.

As I had them adjusted with 8 cells gravity[2] at first contact they gave (one) about 260 ohms, on speaking with a continuous tone resistance rose to 380—and with cessation of tone fell to sometimes 260 then again to 240–280, 300, and so on rarely coming back to first condition. of course this I expected but I also found that the slightest increase of temperature by breathing gently into Receiver sent the resistance up 100 to 150 ohms which fell back when and as it cooled.— This effect I attributed to the expansion of the mica disc rendering it concave towards the mouth and thus lessening pressure on the silk. The effects were constantly observed and the disc seemed to be very sensitive, as after ~~the~~ it had remained quiescent for some[b] time the gentlest breath of warm air ~~with~~ from the mouth ~~to~~close to the Receiver the apparatus not being touched, would send the needle of Galv. flying—showing the increased resistance.— It could not be from change of temperature in the silk pad as the effect was to instantaneous for that, and it seems to me referable only to the expansion of the mica as above indicated.

I thought perhaps this point had not been noticed by you—and you can judge if it is worth any attention as affecting the adjustment and consequently the articulation of the Transmitter.

I like the new plan described in your last and hope you will forward 2 terminals ~~Co~~ (2 complete setts) as soon as you can as many are waiting results. Yrs &c

F. H. Badger

I will mention that when the air was blown on the mica from a distance so as to cool it, the effects were reversed, that is the resistance became less.

As the plates of mica are not perfect planes but have more or less a natural curved surface, the tendency of variations of temperature is to lessen or increase the curve it being fixed at the edge— Also in another Transmitter the effects I observed might be reversed by the disc being placed with the slightly concave side towards the mouth.

⟨Say that we used mica because it was least effected by heat and had no harmonics. The metals were frightfully variable by heat= say that it has bothered us so that we have altered the method of holding the silks on all our telephones so that expansion & contraction of the diaphragm to any extent makes no difference those[b] to be sent will be the new arrangement & many thanks ~~for the points~~ for the points. Murray is making 100 for us besides 150 for W.U.= hope send you in 3 or 4 days⟩[3]

ALS, NjWOE, DF (*TAEM* 14:961). Letterhead of the Fire Alarm Department, City Hall. [a]"Montreal," and "187" preprinted. [b]Obscured overwritten letters.

1. Not found.

2. On the gravity battery see *TAEB* 2:79 n. 2.

3. Batchelor sent a letter to Badger based on these notes (TAE to Badger, 2 Oct. 1877, Lbk. 1:296 [*TAEM* 28:170]). Nothing else is known of the one hundred instruments for Edison. See Doc. 1050 for the Western Union order.

**–1074–**

*From George Barker*

PHILADELPHIA.[a] Sept. 30. 1877.

My dear Sir,

I am much obliged for your letter of the 15th[1] and for your promise to fit me out for the lecture.[2] Would it be possible for you to send me the transmitter and receiver some time this week so that I can experiment some with it before using it before my class? I suppose that receiver with the electric engine which you showed me would do perfectly well. And as for the transmitter I would like one in which I could use either a plumbago or a platinum point. You did not tell me whether you could have made for me one of your automatic instruments for tuning fork experiments. I am very glad to know of the progress your speaking telephone is making and am sure it must supersede Bell's for general use. As to the point you raise, while all you say is true enough, you should go one step further back. It is true as you state that the energy of the induced current is proportional to the square of the amplitude of vibration of the plate. To go back a step further, however, the intensity of the sound is also proportional to the square of the amplitude. A sound of 4, for example, will produce a vibration in the plate of amplitude 2, and this vibration of amplitude 2 will in its turn, produce a current of intensity 4, corresponding to the original intensity of the sound. A sound double this in intensity will produce an amplitude of $\sqrt{8}$ and this amplitude of current of 8 as before. Precisely the same result is reached if you reason upon the phenomena taking place at the receiving end, where the order of events is exactly reversed, and, except for the loss by resistance, the phenomena of the sending end are exactly reproduced in the inverse order. If the current were as the square of the amplitude at the sending end, the sound at the receiving end would be as the square root of the current and things would be restored to their normal condition.

By the way, and between ourselves, these telephone concerts are doing you no good. Johnson may do his best & yet that man Read don't understand his business. The concert at the Exhibition failed from a want of arrangement on his part. I hear on my Bell telephone all the music which goes over the wires on same poles.

Your rubber phenomenon is of an entirely different order from that observed in elastic rubber, mentioned in your former letter.[3] While elastic rubber contracts on heating, hard rubber expands; hence both phenomena. When the hard rubber plate is bent, the convex surface is extended, the concave compressed, as is the case with beams &c under similar strain. Now expanding a body which heating expands, cools it; and conversely compression heats it. The phenomenon may be observed with any elastic material which is a poor conductor of heat. My paper knife, which is of ivory, shows the same heating and cooling, though not to the same degree. Your suggestions relative to the production of alternate heat & cold waves is, like all your ideas, extremely original & ingenious.

At noon there is still air and moisture between us & the sun though less than at sunset. There are seven atmospheric (probably moisture) lines between the two D lines, beside the nickel line. Kirchhoff's map has many air lines in it, but there are vastly more lines in the spectrum than there are on Kirchhoff's map.

Come & see me when you can & we will talk these things over. Cordially yours,

George F. Barker.

ALS, NjWOE, DF (*TAEM* 14:155). Letterhead of University of Pennsylvania. [a]Preprinted.

1. Not found.

2. About this time Barker also wrote Elisha Gray to request his telephone instruments. Gray responded on 3 October stating that he would have his assistant William Goodridge "fit up some of the traps and send to you right away." He also informed Barker that Edison had conceded priority to Gray in three acoustic telegraph patent interferences (see Pat. Apps. 186,330, 198,087, and 198,089). Gray to Barker, 3 Oct. 1877, GFB.

3. Not found.

**-1075-**

*Draft to [Jay Gould?]*[1]

[Menlo Park, August–September 1877?][2]

You did

I see you ~~didnt~~ did not choose to wait the other 4 years to make A&P a success; WU ~~is~~ will be[a] the Telegraph Co ~~and always will be~~ for the next 40 years after that ~~th~~we

I see you have awakened out of your Telegraph Sleep: given the boys the vision and ~~you have~~ gone for the reality. Now that you have left, (if you have left [E?][3b])[c] have you any idea that I shall ever obtain any compensation whatever for my Automatic Interest from the A&P people: Yours

T. A. Edison

P.S. Have just Completed my Speaking Telegraph which talks as loud & as plain as you & I do= also nearly completed a machine which records the human voice on paper from which after the lapse of any time the same voice can be reproduced at any speed & with all its fine inflection.

ADfS, NjWOE, DF (*TAEM* 14:767). [a]"will be" interlined above. [b]Illegible. [c]"(if . . . [E?])" interlined above.

1. The references to Atlantic and Pacific Telegraph Co. indicate that this was written to Jay Gould.

2. References to the consolidation of Atlantic and Pacific and Western Union suggest that this was probably written sometime after the pooling agreement was reached between the companies on 20 August 1877, possibly about the time that Edison formally resigned as electrician of Atlantic and Pacific (Doc. 1042). Edison was still using paper as his recording surface for the phonograph at the end of September.

3. Possibly "E" as a reference to Thomas Eckert, whom Gould had made president of Atlantic and Pacific in 1875. See *TAEB* 2:368 n. 7.

# -7- October–December 1877

Edison and his staff worked intensely to improve his commercial telephone in the last quarter of 1877. They spent hundreds of hours in the laboratory, redesigning the instruments they had and creating new ones, and they frequently traveled to lower Manhattan to test their telephones on working telegraph lines. Edison continued his intermittent experiments on electric lighting, and, at the end of November, found time to turn his idea for a sound recorder into a working phonograph, but the overwhelming portion of the fall was devoted to the telephone.

During October the staff modified the design of the instruments Joseph Murray was producing for Western Union. They altered the transmitter mouthpiece and inserted a small rubber tube between the diaphragm and the plumbago "fluff" to squeeze the fluff and damp harmonics. By the end of the month they had abandoned the fluff, using instead solid disks of plumbago mixed with rubber. After just a few days they replaced the plumbago in the disk with lampblack.

Edison also explored alternative transmitters through November, but by the end of the month he had settled on the design developed in October, using a carbon disk and the rubber tube. He found that squeezing the disk to lower its internal resistance and using an induction coil to amplify the line signal gave louder, clearer reception. As he wrote to a friend, "Ive got it now without fail. Constant and good articulation with the regular Carbon but I work it on an entirely new principle."[1] In December he also explored circuits that would reduce interference from nearby wires and tested batteries for several days to find the one best suited for telephones.[2]

There was much public exhibition of Edison's telephones during these months. Edward Johnson continued giving concerts, including one in Washington, D.C., attended by Joseph Henry—who praised Edison—and government figures including the Vice President. George Barker and William Preece both pressed Edison to send them instruments for public lectures, but he seems to have put off doing so until he felt satisfied with the design, which was too late for Barker's presentation. A reporter from *Scribner's Monthly* visited Menlo Park in early October and published a description of Edison's (and Bell's) telephone in December.

On 17 November, Western Union established the American Speaking Telephone Company, which combined its interest in Edison's telephone patents with the Harmonic Telegraph Company's ownership of Elisha Gray's patents. The combination gave the Gold and Stock Telegraph Company, a Western Union subsidiary, the exclusive right to manufacture, sell, and lease telephone instruments protected by these patents. They immediately contracted with the Western Electric Manufacturing Company to make 100 telephone sets. At the very end of the year, Edison made Theodore Puskas, a Hungarian promoter, his agent for the sale of Continental phonograph and telephone patents.

Occasional notes from these months attest to Edison's continuing interest in developing his phonograph. Edward Johnson announced the invention in the pages of *Scientific American* in early November. Edison envisioned many uses for the phonograph, but had no practical instrument until the beginning of December. In the first few days of that month John Kruesi built a cylinder phonograph that recorded on tin foil, a design conceived in early November. On 7 December, Edison, Batchelor, and Johnson demonstrated the new marvel at the *Scientific American* office in New York. Although the cylinder would eventually become the standard design, Edison still considered other forms and by the end of the year he had designed a disk phonograph. He was unsure of his business plans for the phonograph, but in late December he received an offer to establish the "Talking Toy" Company. At the same time Johnson was encouraging Uriah Painter and Bell Telephone Company President Gardiner Hubbard to acquire rights to the invention.

Edison's reputation spread considerably outside the telegraph industry. The editor of *Appleton's Cyclopedia of Applied Mechanics* asked him to write the entry on telegraphy for the new edition. Edison and astronomer Samuel Langley began a correspondence about developing a new heat-measuring device; Langley suggested that Edison try to create something a hundredfold more sensitive than Langley's best thermopile. Edison's name was proposed as a scientific expert to represent the United States at the 1878 Paris Universal Exposition. The phonograph would soon make his name known world-wide.

1. Doc. 1131.
2. For the battery tests, see Vol. 13:184–200, Lab. (*TAEM* 4:129–45).

**–1076–**

*From James Hammond*

Ilion, N.Y. Oct. 1st. [1877][1a]

My Dear Sir:

I expected when I returned to Ilion to make arrangements by which I could avail myself of your assistence in perfecting my type writer. I found some difficulty in doing so however. The firm are very short of funds and not seeming to favor the proposition I abandoned it for the time. I intended to have written to that effect, but you probably know what becomes of an inventors good intentions. I am getting along slowly but quite satisfactorily. I am glad to hear through the papers of the success of your telephonic experiments.

I am very well satisfied here. The people are of the right sort, although their affairs are in rather an uncertain condition which makes me a little uncertain of the future of my machine in connection with this eztablishment.

I am writing this on a Shoals machine[2] from which you will conclude that I have not one of my own to use.

Regards to Mr Miller. Yours truly,

Jas. B. Hammond

TL, NjWOE, DF (*TAEM* 14:158). Letterhead of Remington's Armory. Typed in uppercase. [a]"Ilion, N.Y." preprinted.

1. Year handwritten, probably by an archivist. This letter is also dated by reference to other correspondence (e.g., Doc. 993).

2. Christopher Sholes's typewriter was being manufactured at the Remington Armory. See *TAEB* 1:245 n. 6.

**–1077–**

*From William Preece*

General[a] Post Office [London,] Octr 3. 77.

My dear Edison:

I am very glad indeed to learn from you that you have sent me your speaking telephone. I shall watch for it with great anxiety. Bell has not been very successful at present and I am much inclined to think that if yours does all you say ~~you~~ it does it will be a useful thing. There will be no difficulty in making terms satisfactory to you, but of this I will let you know more when I have seen the apparatus. I should very much like a pair for lecturing and especially if you can make them so that the audience can hear what takes place. I am engaged to give several lectures during the ensuing winter and I want to get together all I can  yrs &c

(signed) W H Preece

L (copy), UkLIEE, WHP. [a]"Copy" appears in top margin.

**-1078-**

*Technical Note: Electric Lighting*[1]

[Menlo Park,] Oct 5 1877[2]

Electric Light

No 1[3]

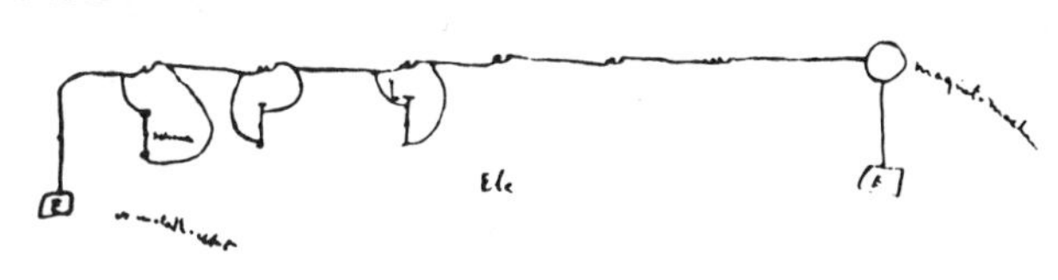

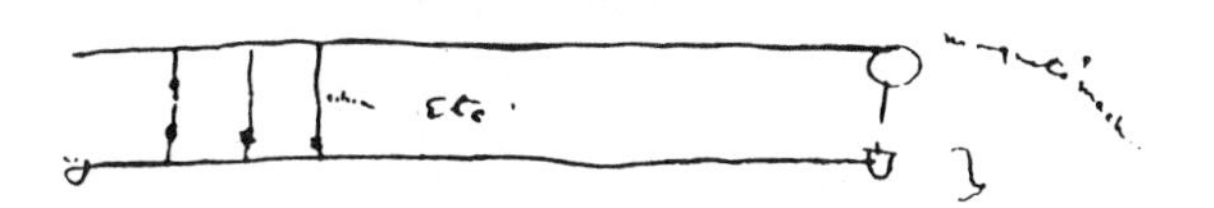

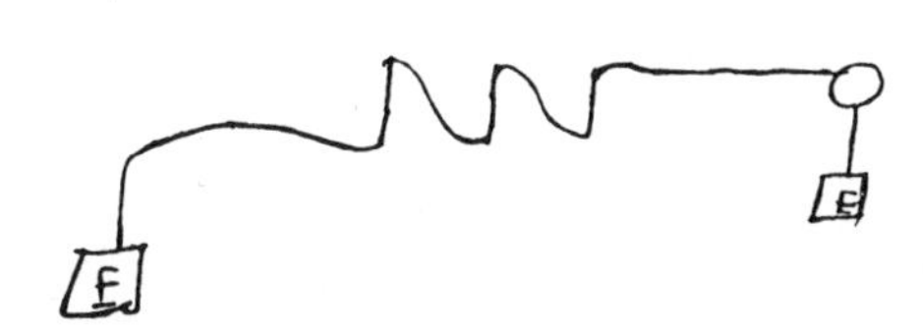

Mix metallic Silicum with Kaoline[4] or Lime which will cause them to glow

T A Edison Chas Batchelor

X (photographic transcript), NjWOE, Lit., *Sawyer and Man v. Edison*, Edison's Exhibit 13 (*TAEM* 46:270).

1. See headnote, p. 540.

2. Both Edison and Stephen Field testified that experiments similar to those depicted here and in Doc. 1098 took place when Field and Cornelius Herz were at the Menlo Park laboratory in mid-September. Field stated that the second and third drawings in this document showed circuit designs like those he had seen in September, except that Edison

used a bichromate battery instead of a magneto machine. Edison's testimony, p. 3008, *Sawyer and Man v. Edison (U.S.)* (*TAEM* 48:11); Field's testimony, p. 183, *Sawyer and Man v. Edison* (*TAEM* 46:244).

3. Figure labels are (top) "silicum," "or metallic ckt," "etc," and "magneto machine"; and (center) "silicum," "etc," and "magneto machine."

4. The Jablochkoff candle used a kaolin-based nonconductor between the carbon points. A 13 September 1877 article in *Nature* discussed the use of kaolin in the manner described here by Edison. Munro 1877, 423; see also 30 June 1877 clipping from the British journal *Iron,* Cat. 1240, item 207, Batchelor (*TAEM* 94:63).

**–1079–**

*Technical Note: Telephony*

[Menlo Park,] Oct 5/1877

Spkg Telegh

We at last adopt the small sharp edged hole to speak through[1] have tried every conceivable kind of hole in thick & thin stuff also various shapes of holes & groups of holes slots etc. round 3/16 hole pbly best

We have adopted a new method of working the Semi-Conductor[2]

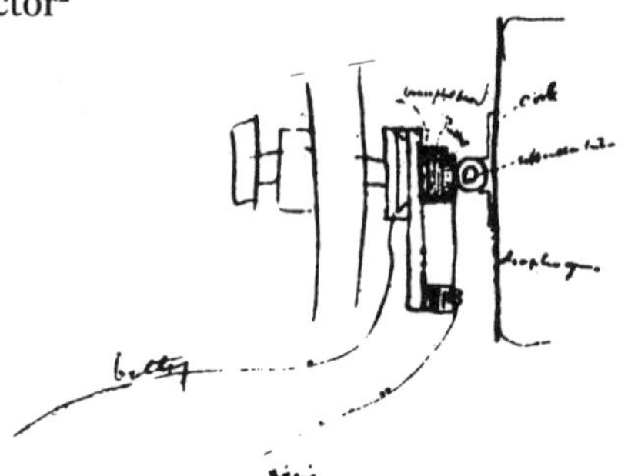

Soft rubber to back=[3]

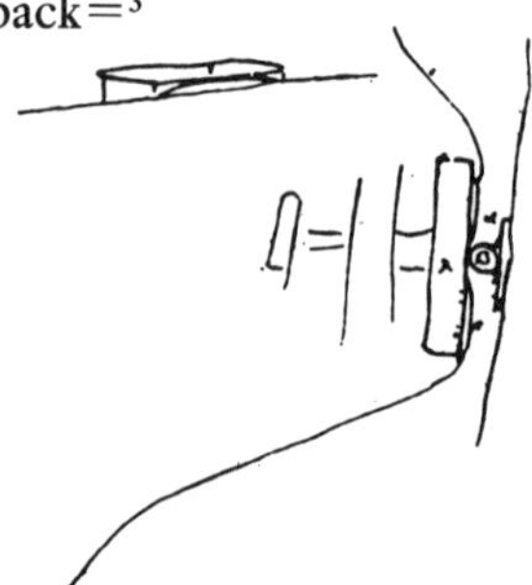

rubber ring [coated?][a] with metal foil X [rubber][b] insulator with gold leaf etc pbgo etc ftin etc surface from c to d.[4]

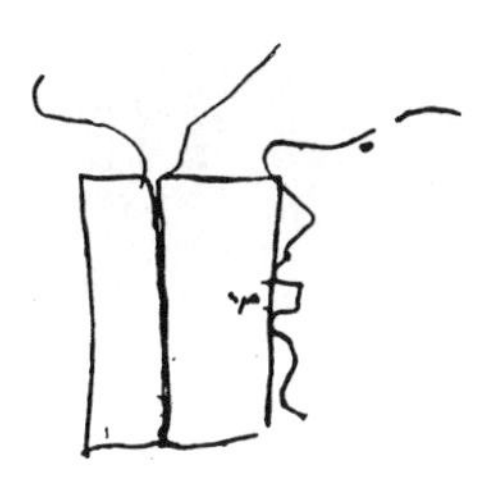

two[5] plumbagoed diaphragm

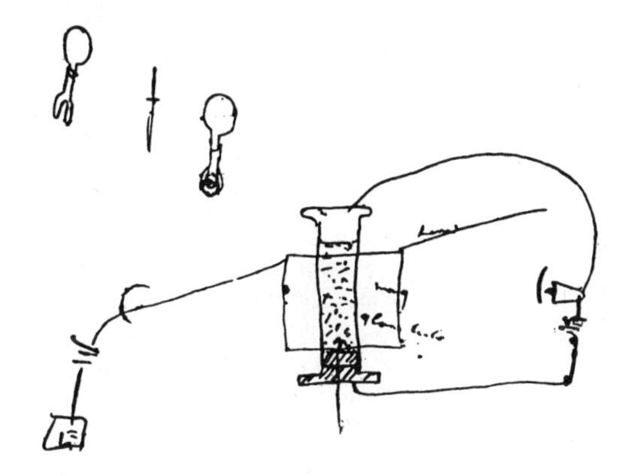

filled[6] with iron dust coated with a semi Condr like plumbago  passage current through coil causes magnetic attractn & particles adhere resistance lessened & local current strengthened

We have just made complete instrument for working on the new principle of turning straight waves from a Reuiss point into undulatory waves[7]  it works red hot[8]

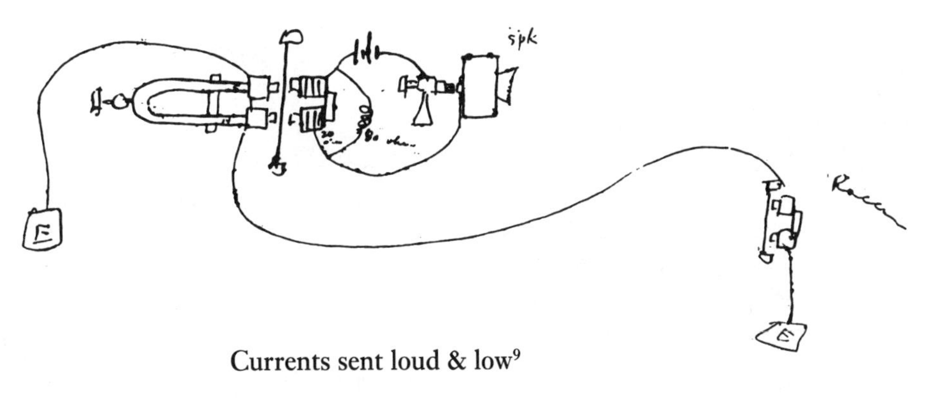


Currents sent loud & low[9]

Same strength
in magnet, shunt [-----][a]
diaphram insertion
Recd

T A Edison  Chas Batchelor

X (photographic transcript), NjWOE, TI 2, Edison's Exhibits 16-13, 10-13, 13-13 (*TAEM* 11:486, 482, 484). Document multiply signed and dated. [a]Illegible. [b]Ditto marks in original.

1. See Doc. 1081. Experiments with Alexander Graham Bell's telephone transmitter undertaken by John Peirce and others at the physical laboratory of Brown University had led to similar design changes in that apparatus in April 1877. That result was publicized prominently around this time. Prescott 1879, 274–75; "The Future of the Telephone," *Frank Leslie's Illustrated Newspaper,* 29 Sept. 1877.

2. Figure labels are (from top) "brass plat[inized] base," [------], "cork," "soft rubber tube," "diaphragm," "battery," and "line." This design first appeared two days earlier (Edison's Exhibits 5-13, 7-13, TI 2 [*TAEM* 11:477, 479]) and a few days later Murray introduced it into the instruments he was manufacturing (Doc. 1081). Edison's patent application of 8 December 1877 (U.S. Pat. 203,013), which contained designs from the spring of 1877, also included this arrangement. On 2 October, Edison had inserted a piece of rubber tubing lengthwise behind the fluff holder to damp the spring that the holder was mounted on (Edison's Exhibit 1-13, TI 2 [*TAEM* 11:474]).

3. That is, a mica diaphragm backed by a diaphragm of stretched soft rubber. See Cat. 1233:278, Batchelor (*TAEM* 90:192); and headnote, p. 590.

4. That is, on the diaphragm from above the rubber ring to below it.

5. Preceding figure label is "Spk."

6. Figure labels are "Line," "mag," and "glass tube"; the square shape ("mag") represents a coil.

7. Cf. Doc. 1057.

8. Figure labels are "20 ohms," "80 ohms," "spk," and "Receiver." The upper arm of the magnet is labeled "S"; the lower arm, "N."

9. The following four lines illustrate the progressive steps in "turning straight waves from a Reuiss point into undulatory waves."

## -1080-

*From Charles Batchelor*

Menlo Park Oct 8 1877

T.A.E.

Just after you had gone[1]—3PM—today a man came up from Scribner's[2] to see you he said you had made an engagement with ~~you~~ him. I showed him the Telephone & he got every word of it and took notes on the principle etc I also stuffed him with the pressure relay & Electromotograph. I offered to write a small article which he could correct himself for the magazine but he said he could do that himself I told him you would be back in a few days and he will see you himself before the article is published—[3a]

Have already started Kruesi but I shall have to get him some more Mica I think. I shall have Murray make his own model for ring as Kruesi does not know what size Bradley[4] has bored out the rings I will attend to that all right—[5b]

6.30 p.m. Johnson has been here & taken speaker we gave it good trial & it ~~sa~~ was very good he was delighted. I fix up the other tonight & take to Murray in morning to rush him on them— The talker had not altered in resistance at all but was just the same as when left on galvanometer. I intend to find out why that fluff gave out so quick I cant believe they go so soon.

Batch

P.S. Gas machine gave out, (want of gasoline,) put in 2 barrels more. We measured it & found about 1 barrel in it that is using 3 barrels in 45 days or about $5.26 per week I think it is because we had so many ~~burners~~ lights[c] without burners[6] we must be more careful in future B

ALS, NjWOE, DF (*TAEM* 14:974). [a]Dash continued to end of line. [b]Dash continued to end of line; followed by centered horizontal line. [c]Interlined above.

1. Edison went "West to bring his wife home," according to Batchelor's diary. Although there are technical notes dated 9 and 10 October and signed by Edison, he did not return to Menlo Park until the night of 11 October. Cat. 1233:281, 284, Batchelor (*TAEM* 90:193, 195); Edison's Exhibits 19-13, 20-13, TI 2 (*TAEM* 11:489–90).

2. Charles Barnard of *Scribner's* had written Edison on 4 October asking for an appointment the following Saturday (6 October). DF (*TAEM* 14:249).

3. The December 1877 issue of *Scribner's* contains a description of the Edison and Bell telephones in a section entitled "The World's Work" (15:285). The nature of that discussion, as well as an accompanying description of A. W. Wright's experiments in the electrical deposition of metals (see Doc. 1041), indicates that the article resulted from this visit.

4. James Bradley (1845–1925) first worked as a machinist for Edison at the Ward St. shop in 1872. Following Edison's removal to Menlo Park, Bradley continued working at Ward St. under Joseph Murray. Bradley then took charge of the Newark factory of electrical manufacturer Edward Weston for about a year before returning to Edison's employ in the Menlo Park machine shop at the end of 1879. He helped set up the Edison lamp factory at Menlo Park in 1880 and later worked at Edison's Harrison, N.J., lamp factory. "James J. Bradley," Pioneers Bio.

5. See headnote, p. 590.

6. This indicates that the gas lights were being lit without jet-piece burners attached to the outlets. For descriptions of common gas burners see Knight 1877, s.v. "Gas-burner."

## IMPROVED TELEPHONE TRANSMITTER Doc. 1081

Edison continued experimenting with his telephone transmitter design during autumn, and as he improved it he ordered several changes in the instruments Joseph Murray was making for Western Union. In late September he devised a new holder for the semiconducting fluff,[1] and at the very end of the month he began mounting the fluff holder against a spring.[2] During the first week of October he mounted a piece of soft rubber tubing between the diaphragm and the fluff.[3] This tube, with the transmitter's adjusting screw, allowed an easier adjustment of the pressure on the fluff and also damped high-frequency vibrations of the diaphragm.[4] (Although the drawing in Doc. 1081 has only a mica diaphragm labeled, other evidence indicates that the mica was laid over a stretched rubber diaphragm.)[5]

On 8 October, Charles Batchelor wrote out instructions for Murray describing design alterations to be incorporated in the transmitters.[6] The instrument shown in the document below reflects those changes. The Gold and Stock Telegraph Company received seven of these transmitters in mid-October (with receivers like those shown in Doc. 1054), including the one from which the drawings were made, and began using them immediately between their main and branch offices. Edward Johnson then took two from Gold and Stock for his exhibitions. According to a company official, "These instruments worked well while in adjustment; they had to be adjusted frequently. . . . In comparison with the magneto telephones that had before been made use of, these carbon telephones were louder, but not as clear."[7] The staff, meanwhile, continued to experiment with and modify the transmitters.[8]

*A telephone transmitter with a piece of rubber tube squeezing the carbon fluff.*

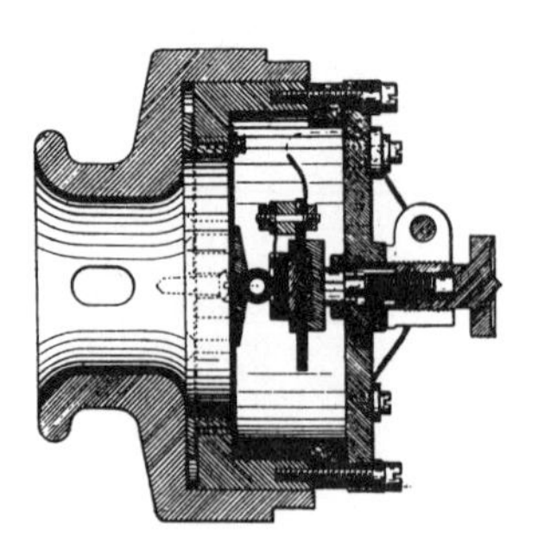

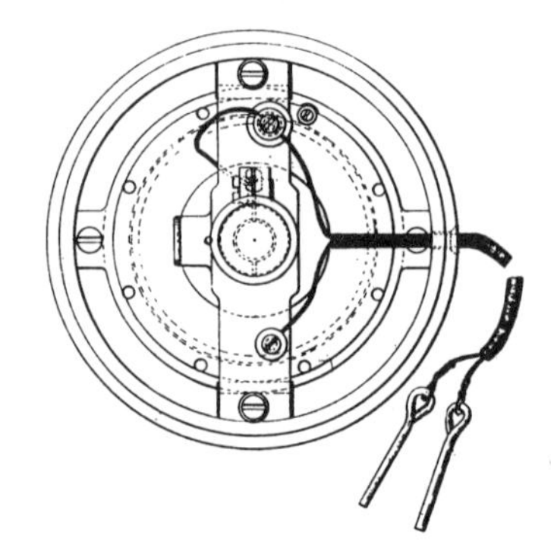

1. See Doc. 1067.

2. Edison's Exhibits 188-12, 189-12, and possibly 185-12, all TI 2 (*TAEM* 472–73, 470); Cat. 1233:273, 276, Batchelor (*TAEM* 90:189, 191). The use of springs was the principal issue of Telephone Interference No. 1, *Edison v. Blake v. Irwin v. Voelkers* (TI 1:vi½ [*TAEM* 11:12]).

3. See Doc. 1079. Edison embodied the rubber-tube configuration in the instrument introduced in the Telephone Interferences as "Edison's Articulating Transmitter" (TI 2:529 [*TAEM* 11:657]). That instrument was almost certainly made in the first week of October, as it does not reflect the alterations shown in Doc. 1081.

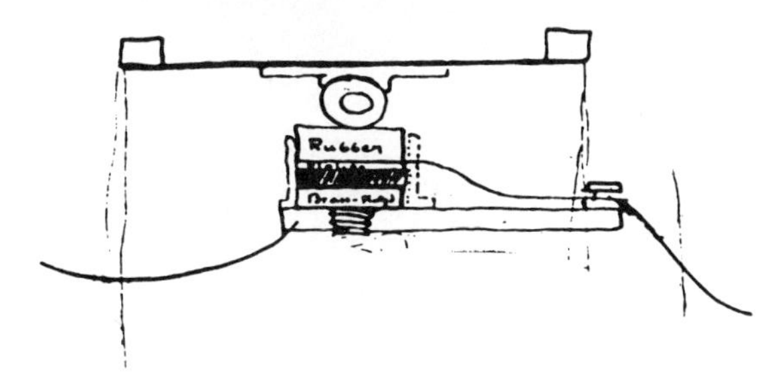


*A detailed drawing of the fluff in its holder.*

4. Edison's testimony, TI 1:70 (*TAEM* 11:56). On 24 October, Edison made a series of sketches for George Wiley, a Gold and Stock Telegraph Co. employee, to explain the principle and structure of the transmitter (Wiley's testimony, TI 1:348–49; Edison's Exhibit, TI 2:607 [*TAEM* 11:151, 727]). One of those drawings details the variable carbon resistor Edison used in these transmitters, showing the diaphragm with attached rubber tubing pressing on a stack of materials in the fluff holder: a piece of hard rubber, a thin piece of platina (one electrode), the carbon fluff, and a platinized brass screw (the other electrode).

5. Charles Batchelor testified (TI 1:241 [*TAEM* 11:97]) that the telephone made in the first week of October (see note 3) embodied the principle shown in Doc. 1062—that is, a "mica diaphragm loose & laid against a stretched rubber diaphragm." Edison had apparently considered a stretched rubber diaphragm as an improvement on Bell's transmitter on 20 September (Vol. 12:157, Lab. [*TAEM* 4:75]). Batchelor's 8 October instructions to Joseph Murray also specify a "ring for rubber diaphragm" (Edison Exhibit, TI 2:609 [*TAEM* 11:729]).

6. Edison's Exhibit, TI 2:609 (*TAEM* 11:729). He took the instructions to Murray the next day. Cat. 1233:282, Batchelor (*TAEM* 90:194).

7. Wiley's testimony, TI 1:347–48 (*TAEM* 11:150–51).

8. For example, on 24 October, Batchelor "took [to New York] 2 telephones with slightly new principle worked tolerably well" (the "principle" was the flat disc of plumbago and rubber described in Doc. 1091). Cat. 1233:297, Batchelor (*TAEM* 90:201).

**–1081–** [Menlo Park, c. October 8, 1877]

*Production Model: Telephony*[1]

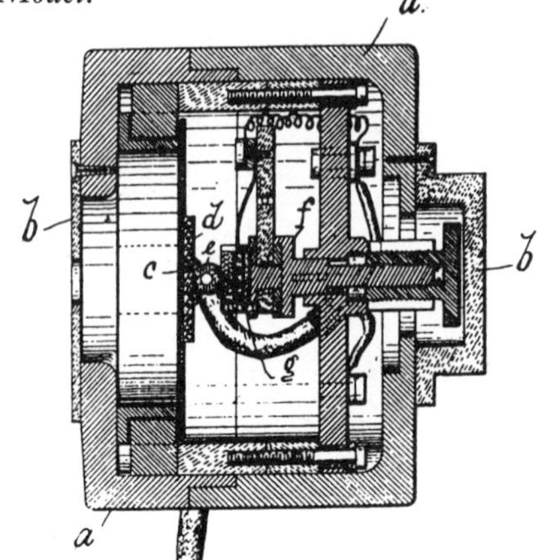


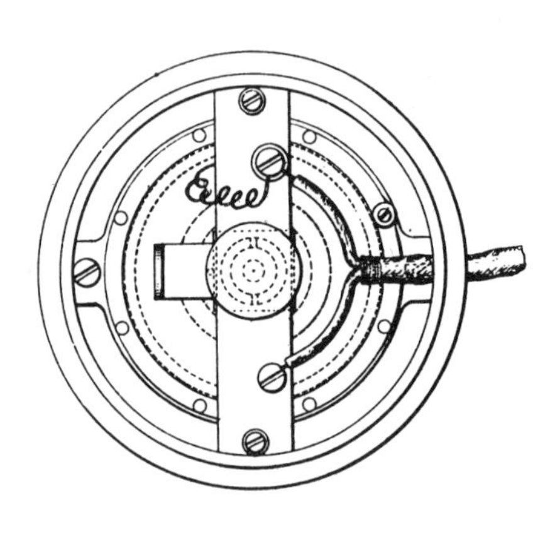

| | |
|---|---|
| a wood | e. soft rubber |
| b hard rubber | f. brass |
| c. mica diphragm | g. carbon fluff |
| d. cork | |

M (historic drawing) (7.6 cm dia. × 7.0 cm), NjWOE, TI 2:538 (*TAEM* 11:666). Photolithograph of drawing of exhibited instrument, prepared for printed record.

1. See headnote above.

**–1082–** Menlo Park N.J. [c. October 10, 1877][1]

*To William Preece*

Friend Preece.

Many thanks for the Polarascope;[2] unfortunately two of the thin glasses were broken in transit but I can probably obtain them from Mr Ladd. Have you made any tests with the Tone Telephone.[3] Johnson is having some very successful Exhibitions here; I have not sent you speakers yet because when I do I want to send you the final result or nearly so, probably in 10 days I shall send you a lot for trial. Bell uses 2 wires ie[a] metallic circuit for all his in N York & isolates these from all others. The extraneous currents do not effect mine. Mr Prescott has referred me the paper about the speed of E[lectricity] over wires received from England wishes me to conduct the experiments.[4] as soon as I get over my Speaking Telephone flurry & can get a second beating clock[5] I will do so[6] Yours

T A Edison

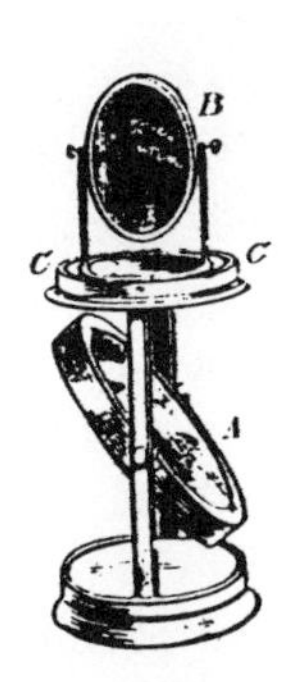


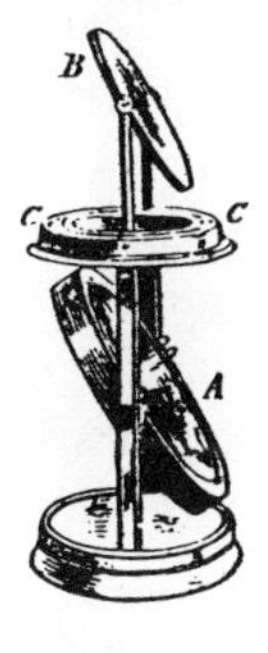


*A polariscope.*

ALS, UkLIEE, WHP. [a]Circled.

1. On 20 September, William Ladd & Co., a supplier of scientific instruments in London, sent Edison a "pocket polarising apparatus, with Mr W. H. Preece's compliments." By 27 October, Ladd & Co. had received an order for (and sent) replacements for the broken glasses mentioned in this document. Edison's question about the "Tone Telephone" indicates he had not yet received Preece's last letter (Doc. 1071). W. Ladd & Co. to TAE, 20 Sept. and 27 Oct. 1877, DF (*TAEM* 14:148, 171).

2. A polariscope, of which there were several kinds, revealed the polarization of light by passing it through particular materials or reflecting it at particular angles (Knight 1877, s.v. "Polariscope"). Preece had brought one with him on his trip to the United States (Baker 1976, 147).

3. This was Preece's preferred term for "instruments employed in the transmission of musical sounds" rather than speech, including Edison's musical telephone, Gray's telephone devices, and the Reis telephone. Preece 1877, 403.

4. The only experiments explicitly on the speed of electricity found from this time period are on transmission of a spark along overhead telegraph wires, performed by Werner Siemens and reported in "Science in Germany" by S.W. (*Nature* 14 [1876]: 358 [originally in the *Monatsberichte* of the Berlin Academy of Sciences for December 1875]). However, closely related subjects were taken up in Robert Sabine, "On a Method of Measuring the Contour of Electric Waves passing through Telegraph Lines," and Oliver Heaviside, "On the Speed of Signalling through Heterogeneous Telegraph Circuits," *Phil. Mag.* (5th ser.) 2 (1876): 321–31; 3 (1877): 211–21. The contents of the papers do not determine just what "the experiments" would have been, but Sabine presented specifics about apparatus and techniques while Heaviside theoretically determined results of potential practical significance for Western Union and its International Ocean Telegraph Co. cables to Cuba. Preece was hostile toward Heaviside's theoretical approach to telegraphy (Baker 1976, 109; Hunt 1983; Nahin 1988, 59–78).

5. Although studies like Sabine's involved measurements of $1/1000$ second or less, for calibration and regulation of the apparatus it was enough to measure seconds accurately. Sabine, for example, relied upon a disk rotating twice per second. One of Charles Batchelor's scrapbooks contains a reprint of R. W. H. P. Higgs, "Electric Motor Pendulum" (*Nature* 15 [1877]: 98), which described a convenient way to make a seconds-beating clock. Cat. 1240, item 55, Batchelor (*TAEM* 94:22).

6. No records have been found of any such experiments by Edison.

**–1083–**

*To Franklin Badger*

Menlo Park NJ Oct 13/77

Dear Sir,

Yours of the 6th received.[1] My Solicitor L. W. Serrell of 76 Chambers Street has the receipt for the patent from the Dominion authorities at Ottowa so there is no doubt about that.= As for Bell claiming a plate vibrated by a magnet, that is absurd as he states in his "interferance case" with Gray[2] of Chicago that he got his idea from a book called "Wonders of Electricity" from the French of J Baille, published by Scribner Armstrong & Co NY 1872. In which is described a speaking Telegh having plates set in vibration by the magnet and plates actuated by sound, also fully described in "DuMoncel On the Applications of Electricity" an extract from DuMoncel you will find in either the Oct or Nov 1877[a] number of Sillimans American Jnl of Art & Science; also in "Nature" within the last 6 weeks;[3] also in the Journal of the Austrian Telegraph Society 1862. [b] Bell has done absolutely nothing new over Reuiss except to turn Reuiss from a Contact Breaking Telephone into a magneto Non Contact breaking Telephone; with permanent magnet & worked the thing up to a success; the records of the patent office will show that Myself Gray & Bell started nearly together on Acoustic Telegraph for Morse working; that Bell & myself dropped this for speaking Acoustic & that I dropped it first & was working on it before Bell. however Bell got ahead of me by striking a principle of easy application[c] whereas I have been plodding along in the correct principle but harder of application. you will find when the thing is sifted in the patent office how little Bell can claim except his magneto principle & even that is questioned by Prof Dolbear of Tufts College & John Cammack of London England,[4] but I guess Bell is all right on that point. I hope he is & will stick to that principle. I have got my Telephone Loud & Constant now as you will see when you arrive; Yours

Edison

ALS, MiDbEI, ER&RI. [a]Interlined above. [b]Underlined twice. [c]"easy application" underlined twice.

1. DF (*TAEM* 14:902).

2. See Bruce 1973, 104–5.

3. The extract published in the 30 August 1877 issue of *Nature* (16:359–60) was taken from volume 3, page 110 of the 1857 edition of Théodose Du Moncel's *Exposé des Applications de l'Electricité*. It was extracted, and presumably translated, by Paget Higgs. The *American Journal of Science and Arts* ([3d ser.] 14 [1877]: 312) reproduced the *Nature* extract.

4. In a letter published in the 16 September 1877 issue of the *English Mechanic,* John Cammack provided an illustrated account of his attempts to transmit sound dating back to 1860. This prompted Edison to publish the following note in the 9 November issue:

> To John Cammack— I notice your communications on the telephone in the English Mechanic. Had you ever previous to 1876 published in any printed publication any account of your telephone

where magneto-inductive currents were tried—*i.e.*, a constant, closed circuit, a battery, two diaphragms, with magnets in front to work inductively? If so, you can undoubtedly make a deal of money here.—

In the 16 November issue Cammack replied to Edison that he had never published a description of his work, although he had exhibited it to fellow students in Manchester in 1860. After hearing of Bell's work he had contacted a patent agent, but was advised not to seek a patent. He also stated that he had given up the idea of working by induced currents because they would be too weak to transmit a signal any distance. Cat. 1240, items 284, 299, 312, Batchelor (*TAEM* 94:86, 91, 95).

**-1084-**

*To Benjamin Butler*

Menlo Park NJ Oct 13, 1877.

Dear Sir,

You probably remember our conversation about printing the human voice at 5th ave hotel.[1] I have not done that but I have succeeded in recording the voice at 150 words per minute on a paper from which I can reproduce the same voice at any future time, so you can listen and recognize the voice of the original speaker, as yet its not perfect in articulation.

If you want the thanks of every scientific man in the country many of which outside of Colleges are very poor please do something towards having the outrageous duty of 40 pc on ~~int~~ instruments for scientific research[2] taken off or reduced. no protection is needed because no one here pretends to make or would even attempt to make such instruments. These instruments are the means of reaching great results of benefit to the country for all time and the men who use them are poor, and ask no reward. Yours Truly

T. A. Edison

ALS (letterpress copy), NjWOE, Lbk 1:306 (*TAEM* 28:179).

1. See Doc. 921.
2. See Doc. 758 n. 2.

**-1085-**

*To William Orton*

[Menlo Park,] Oct 15 1877

Wm Orton

I have today delivered Six Telephones with call bells[1] I deliver two tomorrow. I believe these will prove satisfactory The rubber I use is pure rubber, the same as th~~ose~~at used in the valves of moledoians vavles in Gasolene gas machines. I have a diaphragm streatched on a Helmholz resonator made in ~~A~~September 1875 used in acoustic, it has not altered. I think no trouble will be had with it, as it is neither exposed to the sun or to oils.[2] Murray makes the slight alteration [~~add~~?][a] or rather addition an excuse for his delay whereas he has not today obtained his boxes or double conducting cord I shall hurry him up.

T A Edison

PS Johnson gets perfect music now with the rubber and no more adjustment is required whereas before this was momentarily necessary.[3]

ALS (photographic transcript), NjWOE, TI 2:606 (*TAEM* 11:726). A copy is in WUTAE. [a]Canceled.

1. See headnote, p. 590. Western Union (for the Gold and Stock Telegraph Co.) had placed an order for 150 telephones in September (see Doc. 1050). However, on 16 October, Charles Batchelor answered an inquiry regarding Edison's telephone by noting that "the Gold and Stock telegraph Co are making a hundred sets They have two lines in regular use and they will soon put up more" (Batchelor to Horatio Reed, Cat. 1238:219, Batchelor [*TAEM* 93:164]).

2. According to Edison's testimony in the Telephone Interferences, "One crotchet of Mr. Orton was his dislike of the use of soft rubber in a telephone or in any telegraphic instrument. He said he had been in the rubber business and knew all about it, and that it was unreliable, and that he would not have any instrument with soft rubber in it." TI 1:88 (*TAEM* 11:65).

3. The following day Edison traveled to Jersey City to prepare for Johnson's 18 October telephone concert at the Tabernacle there (Batchelor's testimony, TI 1:236; Edison's Exhibits 22-13–27-13, TI 2 [*TAEM* 11:95, 492–97]). The advertisement for the concert indicates that music was to be transmitted from the Western Union office in Philadelphia (Edison's Exhibit, TI 2:640 [*TAEM* 11:735]). According to Johnson's testimony Edison himself was in Jersey City and spoke and sang over the line, using the carbon transmitter and receiving with magneto receivers (see Doc. 1054) and electromotograph instruments. The music from Philadelphia was received with the electromotograph (TI 1:287–91 [*TAEM* 11:120–22]).

## –1086–

*Charles Batchelor to Joseph Murray*

Menlo Park N.J. Oct 15th 1877

Dear Sir,

In the new improvement on the telephones Jim Bradley proposed to split the rubber in the (fluff holders). This however Edison does not like and consequently he will have to make them push in tight like the model.

Edison tells me that you say I never explained this alteration to you or Jim at your shop but only left a letter there for you.[1] He has evidently mistaken you in that respect for you know I left the letter Tuesday night and saw you there on

Wednesday morning and at that time you told Jim he had better push it through & I then I explained it thoroughly to him I wish you would set me right in this respect for he infers from what you have said to him that I have lied to him. Yours

Batchelor

ALS (letterpress copy), NjWOE, Cat. 1238:218, Batchelor (*TAEM* 93:163).

1. See headnote, p. 590.

## POLARIZED TELEPHONE TRANSMITTER Doc. 1087

The transmitter shown in Doc. 1087 is designed so that the diaphragm is free to vibrate, as in Alexander Bell's transmitter. Edison tried to augment the volume of the signal by using his pressure-sensitive carbon fluff and magnetizing the iron diaphragm as one pole of a permanent magnet.[1] In Edison's words,

> This instrument had a carbon button placed between an adjusting screw and the lever of the diaphragm, and provided with a spiral adjusting spring,[2] so that the initial pressure between the electrodes and the carbon could be regulated.
>
> The diaphragm which was of iron[3] was polarized north by the U shaped permanent magnet,[4] while the iron lever forming one of the electrodes was polarized south by the same magnet, hence there was an attraction between the diaphragm and the iron electrode arm or lever, which in a state of rest was constant.
>
> In the vibration of the diaphragm, which in this case was perfectly free, it never coming in contact with anything,[5] its approach and recession from the iron lever electrode produced more or less attraction upon it, contrary to that of the spiral spring, hence the initial pressure upon the carbon button was varied at each vibration.
>
> This instrument worked but was not loud.[6]

1. Edison made several drawings of this configuration on 16, 17, and 18 October (Edison's Exhibits 21-13–27-13, TI 2 (*TAEM* 11:491–97). He had earlier designed other telephone transmitters that used polarized magnets in different ways (Edison's Exhibit 185-11 [25 June 1877], TI 2 [*TAEM* 11:338]; 19 and 21 Sept. 1877, Vol. 12:155, 172, Lab. [*TAEM* 4:72, 81]), including a modification of Bell's basic design which he sketched again in early November (Edison's Exhibit 173-12 [21 Sept.

1877], TI 2 [*TAEM* 11:460]; 1 Nov. 1877, Vol. 13:52, Lab. [*TAEM* 4:87]).

2. In Doc. 1087 this spring is at lower center pulling on lever **c**.

3. In the document the diaphragm **a** is described as "tinned iron." The plate at the far left (**e**) is a cover for the diaphragm with a hole in the middle to admit sound.

4. The magnet is **b** in the document.

5. The adjusting screw in Doc. 1087 appears to be screwed so far forward that lever **c** touches the diaphragm.

6. Edison's testimony, TI 1:99–100 (*TAEM* 11:70–71).

**–1087–**

[Menlo Park,] Oct 17 1877

*Experimental Model: Telephony*[1]

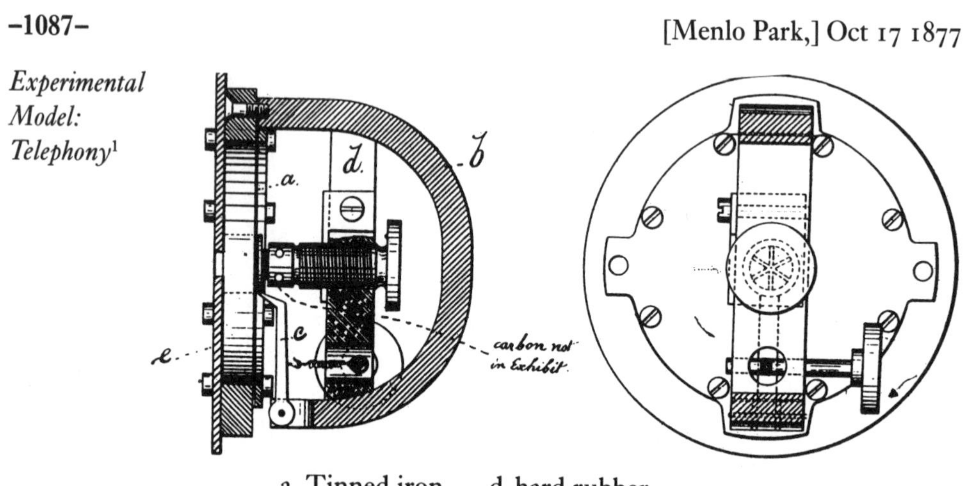


a. Tinned iron  d hard rubber
b. steel  e. Brass
c. "

M (historic drawing) (8.5 cm dia. × 6.0 cm), NjWOE, TI 2:526 (*TAEM* 11:654). Photolithograph of drawing of exhibited instrument, prepared for printed record.

1. See headnote above.

**–1088–**

*From Franklin Badger*

Montreal, Oct. 20th 1877[a]

Dear Edison

I have been waiting patiently for apparatus, expecting its receipt daily for the last two weeks but it don't come worth a cent. Your letter of 14th[1] was satisfactory in regard to patents. you said Serrell had receipt for Canadian Patent: what do you mean by that? If receipt for application fees I understand; otherwise am in obscure. Please explain more fully in yr next. I hope you will be able to send me two Transmitters soon, as I am getting into a delicate position with people here who desirous to see something practical have received promises from me until they half suspect I am a humbug. Please help me out as soon as you can.

I have been told that Bell intends to pounce down on me as soon as I offer your telphs. in market. I dont credit report but shall give him an opportunity very soon after you send apparatus. Please reply and oblige Yours sincerely

Badger

ALS, NjWOE, DF (*TAEM* 14:976). Letterhead of Fire Alarm Department, City Hall. [a]"Montreal," and "187" preprinted.

1. Edison's letter was dated 13 October (Doc. 1083).

**-1089-**

*Charles Batchelor to George Bliss*

[Menlo Park,] Oct 20th [187]7

Dear Sir

Yours of Oct 5th, 10th, & 18th to hand.[1] We have been so very busy on the Telephone that I have not as yet made the ink feeding attachment.[2] The new ink also has not been made Dont you think the bottle that was made had better be given a good trial, and if you conclude it is all right we can make it in one night?

I dont understand what diagram you mean in letter of 18th I dont remember you showing me any diagram please explain.

In regard to letter of 10th please dont forget me this month.

We have worked almost continuously the last two weeks on telephone the difficulty being the immense amount of induction and sympathy on the Gold and Stock wires to be worked over. Respectfully yours

Chas Batchelor

P.S. Please send me ~~directions~~ name of Lawyer to whom I can trust Wheeler's case.[3]

P.S. What is price of Byrne Battery[4] as sold by you? C.B.

ALS (letterpress copy), NjWOE, Cat. 1238:220, Batchelor (*TAEM* 93:165).

1. Not found.
2. See Doc. 1124.
3. See Doc. 951.
4. The Byrne battery was covered by U.S. Patents 182,101 and 183,798.

**-1090-**

*To Samuel Edison*

Menlo Park Oct 21 [1877][1]

Dear Father,

Yours received=[2] As you can always find a home at Menlo Park you need not work so hard. I have had pretty hard Luck with my speaking Telegh ~~not b~~ but I think it is OK now= regarding the road it dont make much difference to me what they do as they have ruined the thing anyway and have as much money in it as I have; I consider my railroad investments in that region as a dead lost $30 000, and I would not come to Port Huron again for the whole of it and I do not think that any living human being will ever see me there again= I am at present very hard up for cash but if my speaking telegraph is OK I shall get an advance of money & send you some on account= Tom & Dot are well & thriving. Tom is a smart one; and no mistake. Take care of yourself & when you want to come to Menlo write    Your Son

T A Edison

ALS, MiDbEI, EP&RI.

1. The reference to Edison's telephone work indicates that this is 1877.

2. Not found.

**-1091-**

*Technical Note: Telephony*

[Menlo Park,] Oct 21 1877

Spkg Telgh

Tests with ~~f~~perfectly flat plumbago[1] mixed with dif substances with flat discs on it held there by diaphragm and peices of rubber tubing.

144:[2a] 15 grammes Plumbago. 6 grammes Rubber in Bi Sul Carbon

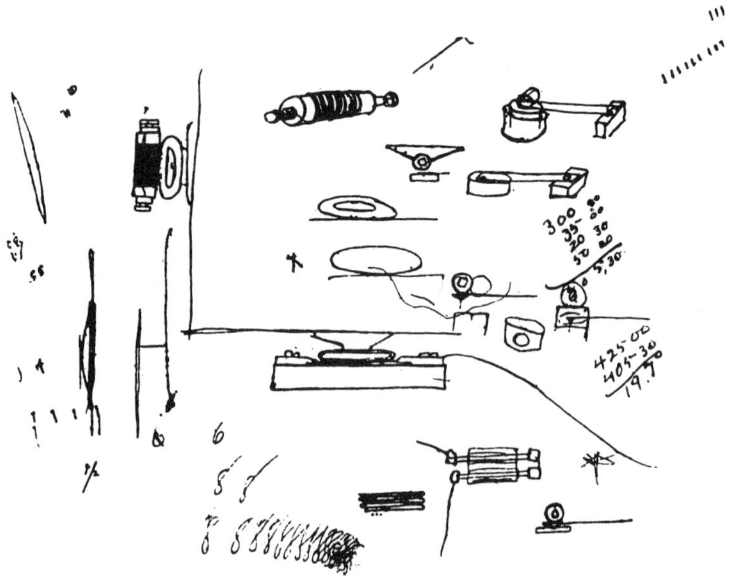

Magneto[3]

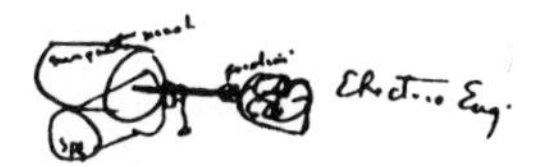

Diaphragm increases & decreases by friction revolution shaft=[4] hence magneto currents varied in strength[5]

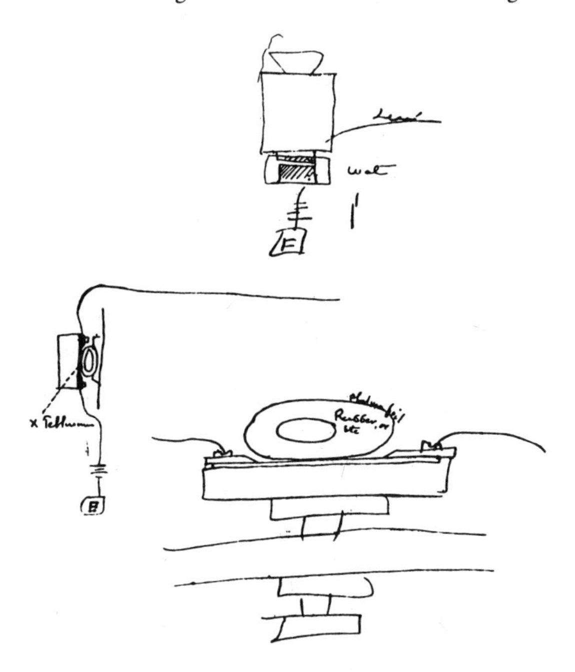

Instead of Tellurium; Red phosphorus[6] may be used
Magneto[7]

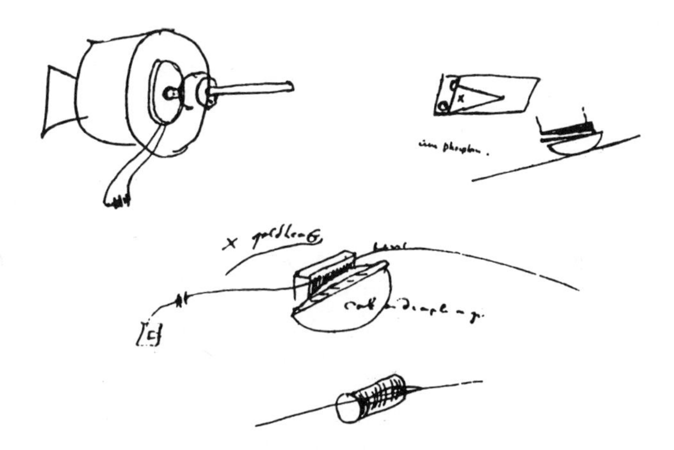

gold leaf on Rubber & put on lath[8] a fine knife edge tool run on finest screw gear feed from end to end so that there[b] will be 50 feet of gold that current must pass through  the spring on diaphragm serves to cut in & out more or less of it[9]

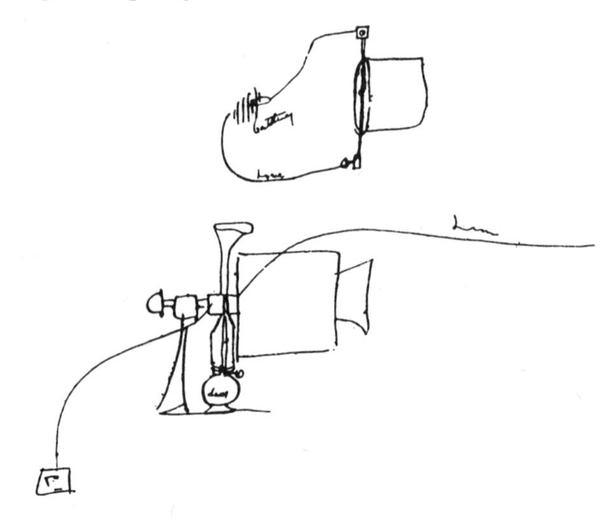

Lampblack formed all time.[10]

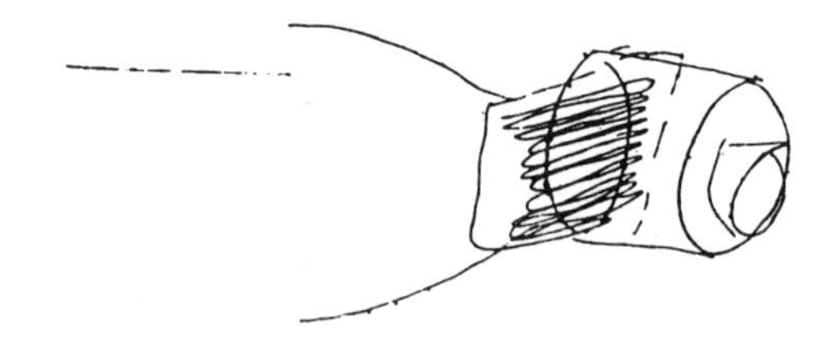

Hard Rubber diaphragm gives heat and Cold waves  these acting on Gold or other foil in a plate scratched so that about 50 feet of it is exposed to Rubber & through which the current passes permanently  the heat & cold waves increases & decreases the distance & thus give waves

T A Edison  Chas Batchelor

X (photographic transcript), NjWOE, TI 2, Edison's Exhibits 32-13, 28-13, 30-13, 33-13 (*TAEM* 11:502, 498, 500, 503). Document multiply signed and dated. [a]Spanned by brace in left margin. [b]Obscured overwritten letter.

1. The next day Batchelor "made a fixture for filing & polishing the discs exactly 1/16 thick." Cat. 1233:295, Batchelor (*TAEM* 90:200).

2. This mixture is listed in Doc. 941.

3. Figure labels are "Magneto mach," "spk," "friction," "Electric eng."

4. This diagram shows the mechanism Edison is discussing. The small cylinder **A**, rotating on a shaft driven by the "Electric Eng[ine]," slows down and speeds up as it is rubbed more or less hard by the vibrating

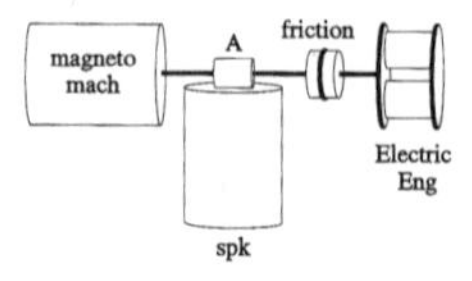

diaphragm (the "friction" is a clutch that slips as the diaphragm presses harder against A). The "Magneto mach[ine]" generates a current proportional to the speed of the shaft, thereby creating a telephonic signal.

Earlier in the year Edison had conceived a telephone transmitter that looked like this one except that there was no magneto at the end of the shaft. Instead, the surfaces of the diaphragm and the cylinder rubbing against it were both platinized. Since "contacts at high speeds offer a resistance in proportion to the pressure . . . the louder you talk the stronger the pressure & you get articulation." Edison's Exhibit 179-11, TI 2 (*TAEM* 11:332).

5. Figure labels are "X Tellurium," "platina foil," "Rubber, or etc," "line," and "water."

6. Red (amorphous) phosphorous is much less volatile than other forms and has a high electrical resistance.

7. Figure labels are "X gold leaf," "line," "[cork on?] diaphragm," and "am[orphous] phosphorous."

8. Probably should be "lathe."

9. Figure labels are "battery" and "line."

10. Figure labels are "Lamp" and "Line." An undated description of a lamp chimney cleaner (possibly from a patent or patent application), which may be related to Edison's interest in obtaining lampblack for the telephone, is in NS-Undated-001, Lab. (*TAEM* 8:61).

**-1092-**

*From Benjamin Butler*

Washington, D.C. Oct 23 1877

My dear Edison:—

Tell me something more about your wonderful invention in recording the human voice. I need not say that you had better keep it perfectly secret. It is so remarkable that I do not understand it at all, and it is an invention or discovery, which, if you have made it you need go no further. The recording of the words on paper would be sufficient, provided they are recorded, without being reproduced articulately.

In regard to the scientific instruments you are very right, and I will do everything I can to have that duty taken off.

Please write me also that you will call upon me at the Fifth Avenue Hotel when I am there and I will have you notified when I am there, so that we can talk the matter over. I am very much interested. Yours truly,—

Benj. F Butler

ALS, NjWOE, DF (*TAEM* 14:166).

**-1093-**

*From Joseph Murray*

N.Y Oct 23rd 77

Sir

I receved letter from <u>Norris</u>[1] he got Judgement for us against the Domestic they did not appear or defend there case I Paid your Cost and told him to push them as fast as possible and he will do so. he put it in sheriff hand to levy on there offices so our time has come for settlement with them they cant put if off—[a]

Everything goes wrong in telephone business but I am near the end and ready for the next with experience that will help me in future

Murray

Edison see cord tell me will it do or shall I take same as you got from N.Y that I returned to Bliss Henry[2] has sample[3]

ALS, NjWOE, DF (*TAEM* 14:168). [a]Followed by centered horizontal line.

1. George Norris was a New York attorney hired by Edison and Murray to sue Domestic Telegraph Co. for money still owed them by the company. He filed a statement for judgment in the case of *Thomas A. Edison v. the Domestic Telegraph Co.* on 29 September 1877. (Edison's name replaced Murray's on the file wrapper; see also Murray to TAE, 5 Sept. 1877, DF [*TAEM* 14:139].) Edison and Murray claimed that the company owed them $1,723.14 plus court costs. As evidence for their claims, Edison and Murray submitted an account, in Charles Batchelor's hand, covering the period 13 March 1874 to 12 January 1876. On 20 October 1877 they received a judgment in their favor. Subsequent correspondence from Norris indicates that they had some difficulty collecting; Edison finally received payment on 11 April 1878. *Thomas A. Edison v. the Domestic Telegraph Co.*, Superior Court, County of New York, #1877-225; *TAEM-G1*, s.v. "Norris, George"; A&P Executive (1878–79): 7–8.

2. Robert Henry was the General Eastern Agent for the Electric Pen and Duplicating Press, having replaced Ezra Gilliland on 1 June. On 20 November he was in turn replaced by William Wheeler. Cat. 1233:152, 324, Batchelor (*TAEM* 90:129, 215); letterhead, G. A. Mason to TAE, 7 Aug. 1877, DF (*TAEM* 14:853).

3. The same day Murray wrote another letter to Edison stating that he had ordered cord (i.e., flexible, insulated wire) from both George Bliss and L. G. Tillotson but that neither was like the sample provided by Edison. He also had a sample from a Providence manufacturer, probably Eugene Phillips, which was single rather than double cord. He asked Edison to decide which cord to use. DF (*TAEM* 14:979); see also Phillips to TAE, 21 Dec. 1877, DF (*TAEM* 14:1008).

On 23 October, Bliss also wrote to Edison about Murray's order for cord. He noted that Murray had talked separately to Henry, who ordered pen cord from Western Electric, which Bliss then told Western Electric to make heavier. Murray apparently then complained to William Orton about the Western Electric cord and showed him a sample of cord which Bliss explained he had never seen. Bliss expressed to Edison his displeasure with Murray for telling Orton that he needed an advance of $150 because of Western Electric's error on the cord order.

Bliss complained that as a result Orton took the "occasion to rap the W.E. on the knuckles," which tarnished his own reputation as well. DF (*TAEM* 14:977).

**-1094-**

[Menlo Park,] Oct 2~~9~~6 1877

*Technical Note: Battery*

Dynamo Voltaic Apparatus

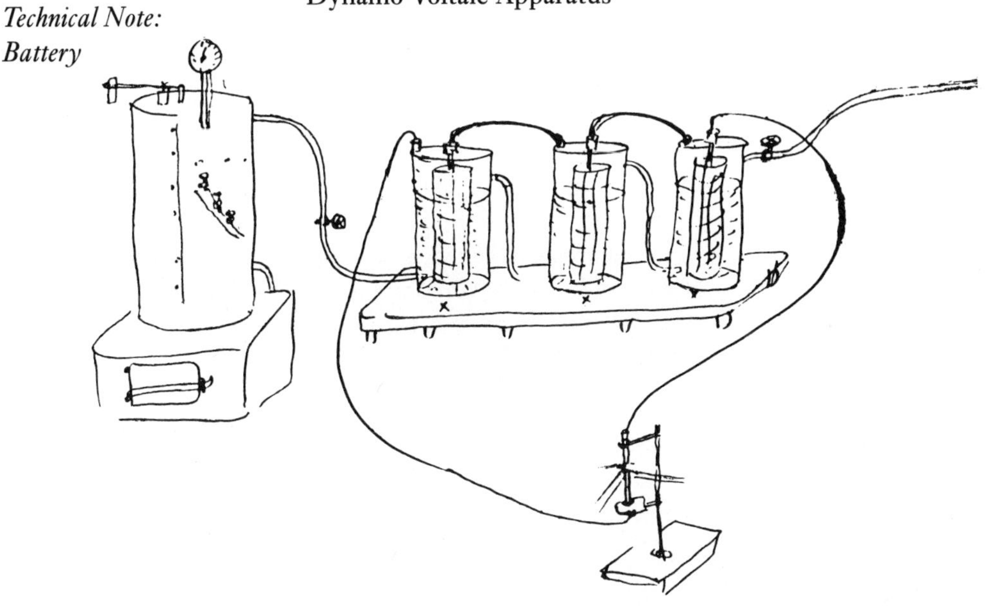


X X X are cast iron boilers with a cylindrical rack in middle filled with pea size gas retort Carbon The iron cylinders are about 7 feet high 3 feet diameter— they are connected together[a] in usual way for batteries. they are filled ⅔ full salt water or water containing any other excitant or conductor such as $SO_3$= Leading from a steam boiler are pipes which passes steam[b] through the liquid and agitate it keeping it depolarized Our experiment tonight was successful with this in a small way= polarization was stopped entirely and constant deflection was had running from 10 on Galvanometer without steam to 50 with steam & agitation of water

T A Edison Chas Batchelor

X, NjWOE, Lab., Cat. 1146 (*TAEM* 6:623). [a]Obscured overwritten letters. [b]Interlined above.

**–1095–**

*Technical Note: Telephony*

[Menlo Park,] October 26 1877

Speaking Telegraf[1]

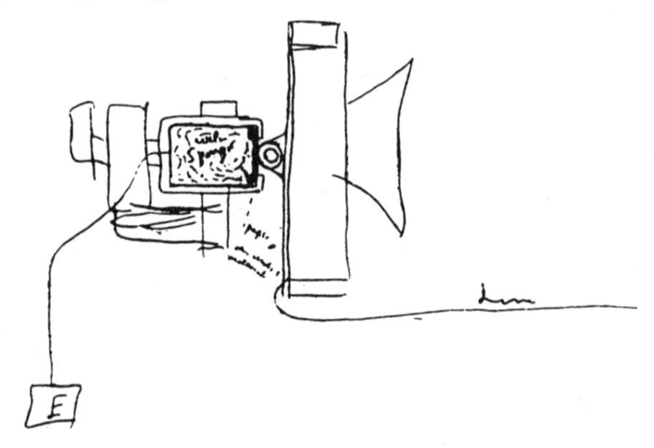


Telephone which generates its own current    Platina faced Tube of Rubber on diaphragm    If this not strong enough and it appears to be from the experiments we tried, then it may be arranged for intensity

tube on diaphragm covered with zinc & platina [two pair?][a] elements each touching against moist paper

We tried Experiment tonight that was very successful in this point

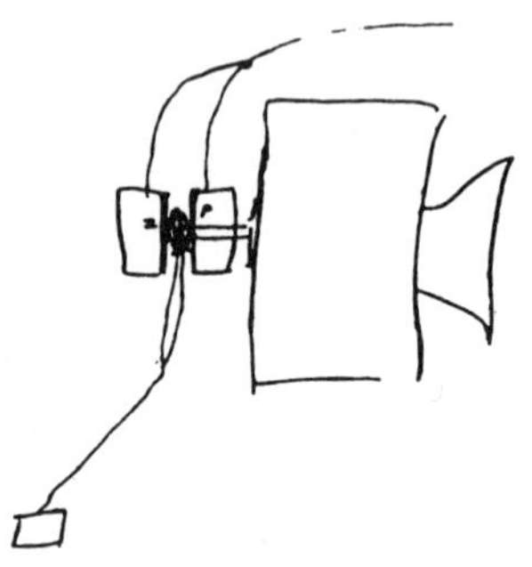

Telephone which generates its own current and also which sends positive & negative currents in the Line=

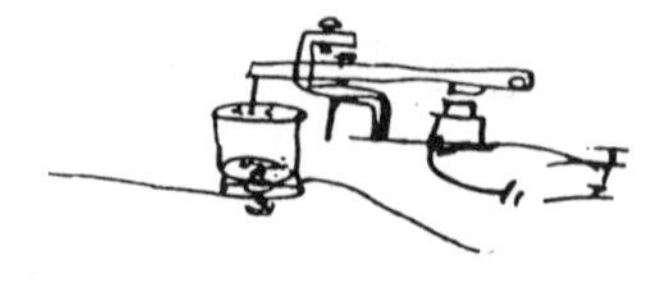

Put several telephones in ckt with extra diaphm connected to sounders— then sevl msags can be sent at one t[b] opers picking out their sounder after little practice

Sextuplex

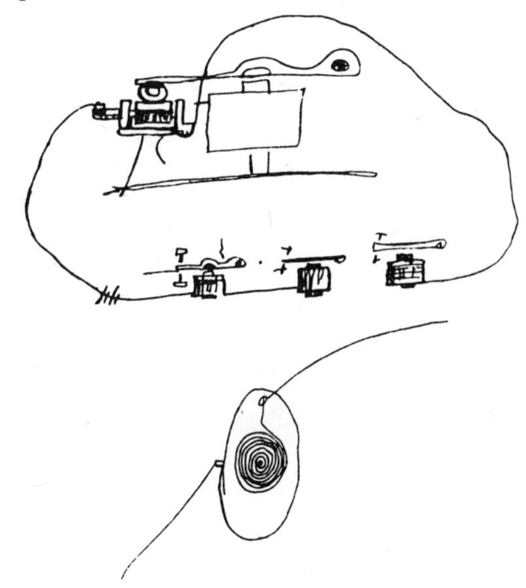

glass or ivory[c] disc coated with gold leaf by Wrights process[2] and then cut in a volute spiral the diaphragm having same with constant current passing through this acts inductively in other, or it may be a steel diaphragm permanently magnetized=[3]

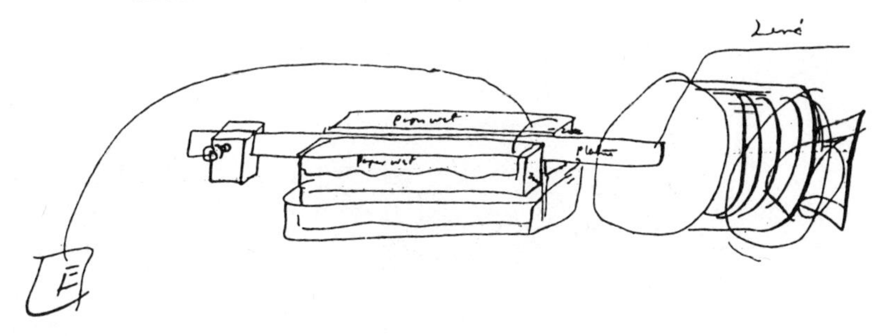

Lamp Capilliary Telephone:

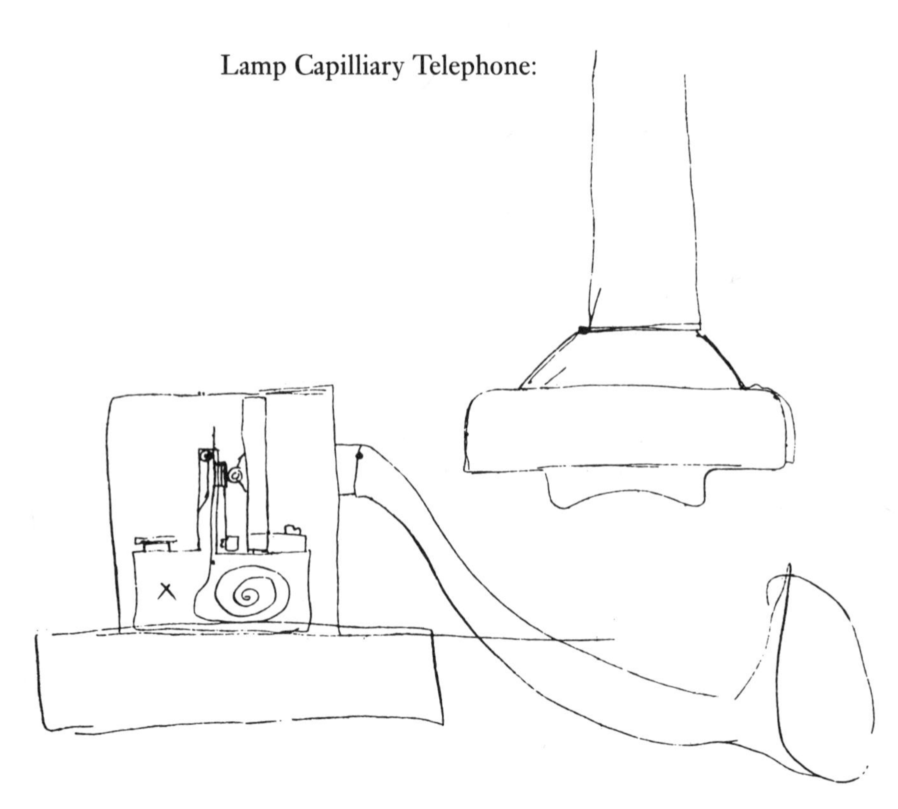


X has supply of the Conducting Liquid  $c^4$ is an absorbant material passing upward, where pressure between 2 Electrodes may be varied by diaphragm  whole placed in box with flex tube

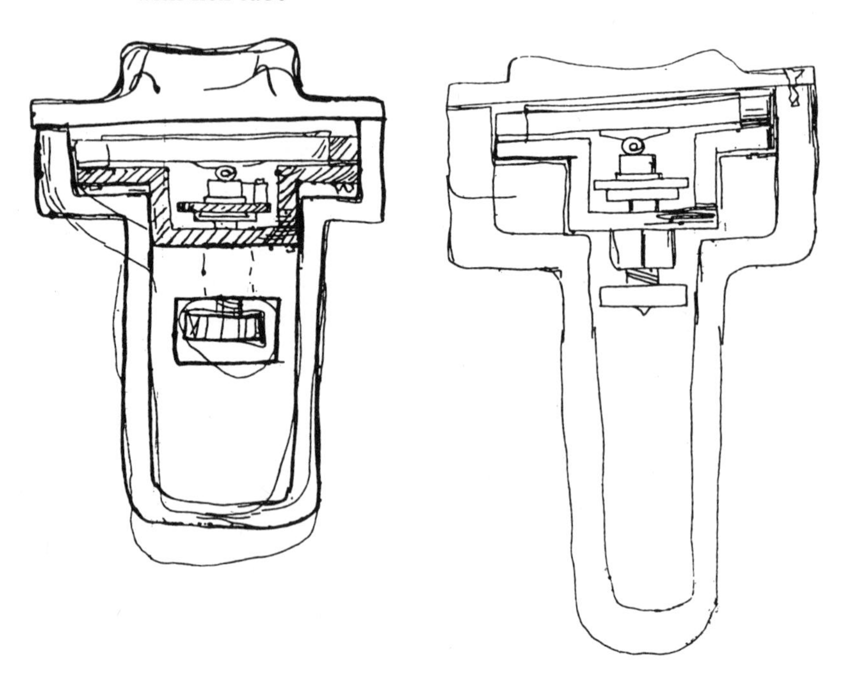

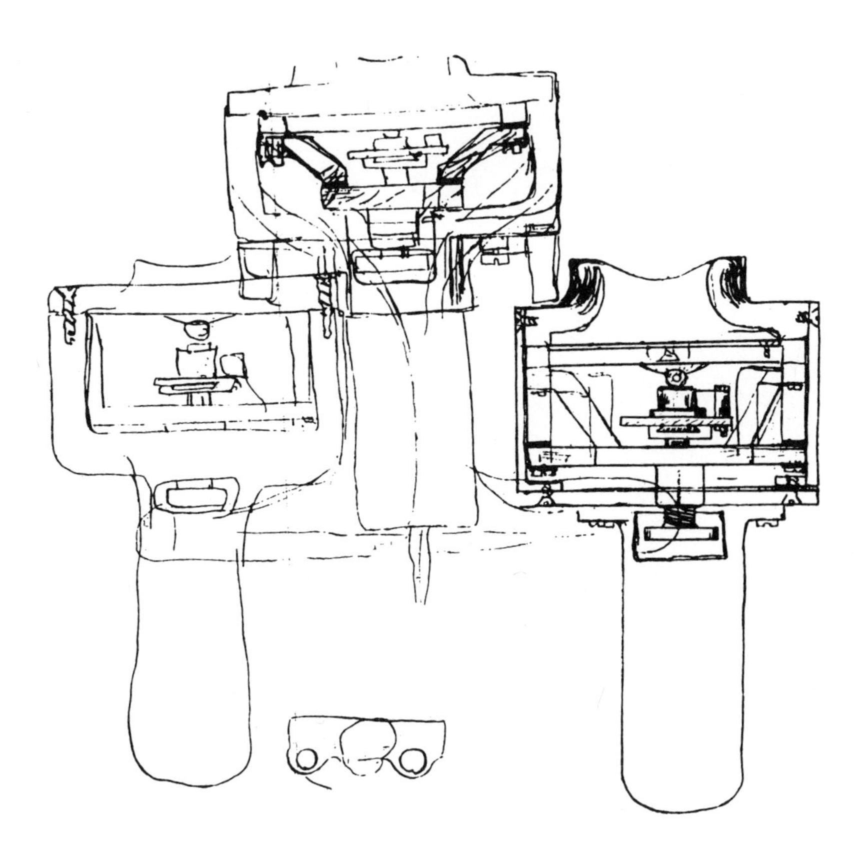

T. A. Edison Chas Batchelor

X, NjWOE, TI 2, Edison's Exhibits 43-13, 42-13, 38-13, 41-13, 39-13, 40-13 (*TAEM* 11:512, 511, 507, 510, 508, 509). Document multiply signed and dated. [a]Illegible. [b]"at one t" interlined above. [c]"or ivory" interlined above.

1. Figure labels are "wet sponge," "paper other absorbent materials," and "line."

2. See Doc. 1041.

3. Figure labels are "paper wet," "paper wet," "zinc," "zinc," "platina," and "line." This is a variation on a design Edison had tried in March; see Doc. 874.

4. **c** is the spiral line near **x**.

**–1095A–**

*Draft to Samuel Langley*[1]

Menlo Park N.J. Oct 27/77

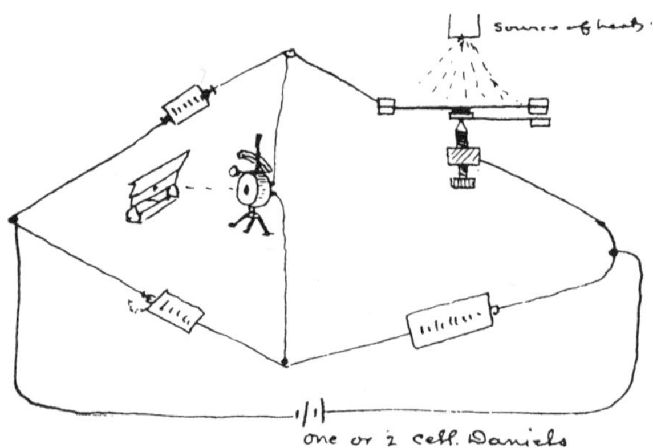

X Gold foil on mica used at contact point or platina foil may be used.[2]

Dear Sir,

I think the thermo apparatus could be arranged in the balance as shewn using one of Thomsons ~~r~~high resistance reflecting galvanometers for indicating.[3] Arranged in this manner I believe it would be very difficult to handle owing to extreme sensibility. I do not see how the source of error due to the varibility of the carbon discs can be compensated for yet it might be possible to obtain the relative heating effects in all parts of the spectrum ~~by rapid~~ by a number of efforts; I suppose that vulcanized rubber would be even more sensitive than mica. I see that Tyndal has measured the relative heating effects of different parts of the spectrum,[4] I suppose through a prism  Is it so much more difficult to do it with the spectrum from lines on the metal you speak of.[5] I should think that if the Red yellow green etc spectrum could be made off the same length, that the heat in each could be more acurately investigated. Yours

T A Edison

ADfS, DF, Supp. III (*TAEM* 162:1018).

1. Samuel P. Langley (1834–1906) was director of the Allegheny Observatory and professor of physics and astronomy at Western University of Pennsylvania in Pittsburgh. At this time his primary research involved the measurement of radiation. He was later secretary of the Smithsonian Institution. *DSB*, s.v. "Langley, Samuel Pierpont."

Edison sent Langley a letter very much like this one on 29 October (see Doc. 1135).

2. Figure labels are "Sources of heat" and "one or 2 cell Daniels." Edison and Batchelor later signed this drawing, which marks the conception of what would become the tasimeter. A Wheatstone bridge ar-

rangement like this one was the common method for precisely measuring variations in resistance. In the early 1870s Charles William Siemens had investigated the use of a platinum wire coil in such an arrangement to measure temperature. Siemens 1874.

The small circular sketch shows a mica diaphragm with conducting foil; that diaphragm is drawn edge-on in the "thermo apparatus" in the upper right branch of the bridge circuit.

3. The galvanometer is the device on the vertical branch of the circuit. To its left is the scale upon which the galvanometer's mirror reflects a beam of light.

4. John Tyndall discussed this issue in some detail in his *Heat Considered as a Mode of Motion;* see, for example, chapters 8 and 13 of the 1873 edition (New York: D. Appleton and Co.).

5. That is, with a spectrum created by a diffraction grating.

**-1096-**

*From Josiah Reiff*

New York, Octo 31 1877[a]

My dear Edison—

Please advise me whether you will come to see me here, or whether I must come to Menlo Park—

It is necessary to take some immediate action to protect the "new application"—[1]

The Omnibus Suit is busted & even Lowrey now[b] admits there is no injunction—[2] I shall arrange to get extension of time in the Gerritt Smith interference case.[3]

I was in Washington yesterday & saw the Comsr of Patents.[4]

I only want fair play from WU. but I dont propose they shall get any advantage of me. Roscoe Conkling may be very powerful under some circumstances but he dont run the present Comsr of Patents or the Hayes administration.[5] Yrs

J C Reiff

ALS, NjWOE, DF (*TAEM* 14:728). Letterhead of J. C. Reiff. [a]"New York," and "187" preprinted. [b]Interlined above.

1. Presumably this refers to the quadruplex telegraph patent application of 30 June 1876 (Case 121), which Reiff had filed as Edison's patent attorney. *TAEB* 2:223 n. 2; Doc. 734; table in Edison's testimony, TI 1:108 (*TAEM* 11:75).

2. This refers to *Western Union v. Harrington & others* (see Chapter 4 introduction). With the trial of *Atlantic & Pacific v. Prescott & others* finished but the decision reserved by the judge, and with the agreement effected pooling patent rights, traffic, and revenues between Western Union and Atlantic and Pacific, most of the issues in the omnibus suit were rendered moot and it never went to trial. See Doc. 985 n. 1; Reiff to TAE, 28 July and 14 Nov. 1877, and n.d., DF (*TAEM* 14:707, 506, 738).

3. That is, extension of the date by which preliminary statements were to be filed at the Patent Office. Reiff was acting as Edison's patent attorney in this interference, which involved relays for multiple telegra-

phy (see Doc. 717 and *TAEB* 2:353 n. 3). An extension was granted into December; later that month Smith conceded priority on the general point at issue to Edison. Reiff to TAE, 14 Nov. and 26 Dec. 1877; Lemuel Serrell to TAE, 27 Dec. 1877; all DF (*TAEM* 14:506, 517).

Edison's sextuplex patent applications (Cases 138–140) led to interference proceedings over similar multiple telegraph relays and circuits with Franklin Pope, Henry Nicholson, and Edward Dickerson, Jr. Pat. Apps. 452,913, 453,601, 512,872.

4. General Ellis Spear, former patent examiner-in-chief and civil service examiner for the Interior Department, was Commissioner from January 1877 until November 1878. *NCAB* 13:364.

5. Conkling was a Republican Senator from New York, an attorney for Western Union, and a political ally of Zachariah Chandler, the Secretary of the Interior who had decided against Harrington and Edison in their appeal of patent right assignments for the quadruplex. President Hayes's Secretary of the Interior, Carl Schurz, reversed Chandler's policies and practices, particularly his treatment of government jobs as political resources. Reference to Conkling's vivid antagonism to Hayes over such civil service reform was common in the public discourse of the time. *TAEB* 2:374, 770 n. 6; "A Reform in the Civil Service," *Scribner's Monthly* 15 (1877–78): 271; Oberholtzer 1926, 3:342, 349–50.

**–1097–**

*To William Orton*

[New York,] ⟨Oct 77⟩[a]

Wm Orton

I think there will be no difficulty with the instrument now being made by Mr Murray. I have made a very great improvement in the way of constancy and hence articulation. Have also made another kind altogether which works well. Murrays machines will I think work well & give satisfaction. If I was sure of what I have I would be more enthusiastic, but experience in articulating telephones have learned me better.

T A Edison

ALS, DSI-NMAH, WUTAE. On memorandum form of Western Union Telegraph Co. Secretary. [a]Dated by Orton.

**–1098–**

*Technical Note: Electric Lighting*[1]

[Menlo Park,] Nov 1 1877

Electric Light[2]

have tried Boron Ruthenium Chromium and the almost infusible metals for separators in my Electric Light device= Boron is very high resistance and would do if arranged thus

⟨Fig 1 TAE⟩[3]

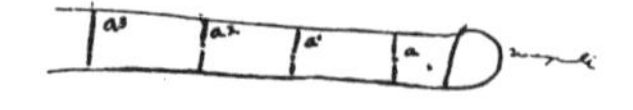

Silicum on the other hand is very low resistance & would have to be arranged thus

⟨Fig 2 TAE⟩[4]

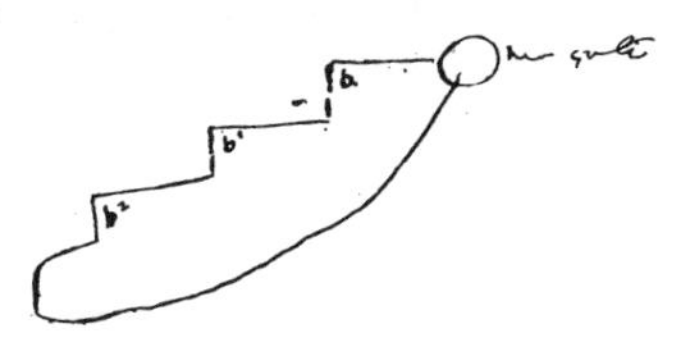

I think powdered silicum mixed with lime or other very infusible non conductor or semi conductor would be good

T A Edison

Chas Batchelor
J Kruesi

X (photographic transcript), NjWOE, Lit., *Sawyer and Man v. Edison (U.S.)*, Edison's Exhibit 4 (*TAEM* 48:759). "Coppied in Vol No 1 Experimental Researches page 115 Sept 28, 1878 Wm Carman" written at bottom of page; that copy is in Cat. 994:115, Lab. (*TAEM* 3:226).

1. See headnote, p. 540.
2. Stephen Field later testified that he had seen circuit arrangements like this in September, with a bichromate battery instead of a magneto machine. P. 183, *Sawyer and Man v. Edison* (*TAEM* 46:244).
3. That is, in parallel. Figure labels are "$a^3$," "$a^2$," "$a^1$," "a" (all apparently added later by Edison), and "magneto."
4. That is, in series. Figure labels are "$b^2$," "$b^1$," and "b" (all apparently added later by Edison), and "magneto."

## –1099–

*Technical Note: Phonograph*

[Menlo Park,] November 1 1877

Phonograph[1]

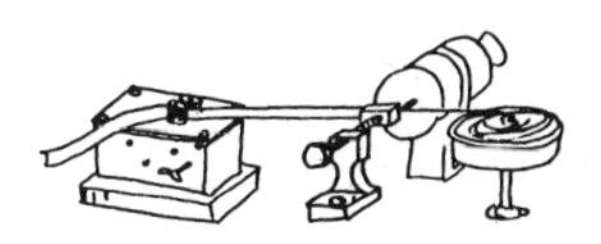

Paper softr and coated heavily with a compound of Beeswax & Parafin, or other soft substance

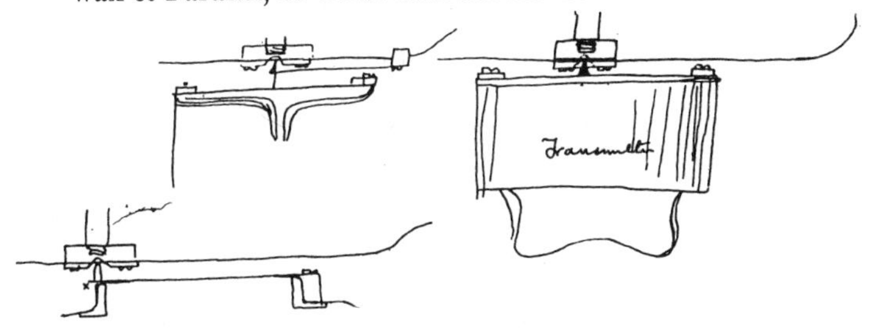

X Reed diaphragm backed with rubber.

T. A. Edison
J Kruesi

Chas Batchelor.
~~G. E. Carman~~
~~M. N. Force~~

X (photographic transcript), NjWOE, NS-77-003 (*TAEM* 7:455).

1. The clockwork at left, which pulls the paper past the phonograph mouthpiece, is similar to those used by Edison for automatic and domestic telegraphy. See, e.g., Docs. 458 and 615.

**-1100-**

*Technical Note: Telephony*

[Menlo Park,] Nov 3 1877

Spkg Telgh

Faraday Page 27 Vol II Researches[1]

Good Conductors= Galena— Sulphuret of iron= Arsenical pyrites= native sulphuret Copper & iron native gray artificial sulphuret of ~~iron~~ copper Sulphuret of Bismuth Iron & Copper globules of oxide of burnt iron, oxide of iron by heat or scale iron= Conduct current single thermopile well= native[a] Peroxide of Manganese & peroxide of lead conduct moderately well.

A solution of sulphuret of Potassium is a remarkably good conductor also, greenish nitrous acid= Sulpt Pot. has as[2] action on the ~~metals such as~~ iron & platinum together

Idea=

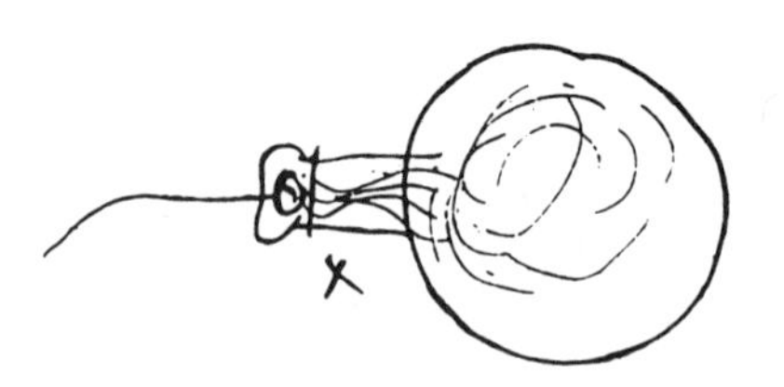


disk of Lampblack[3] thinnest tin foil laid on it talk against it or coat it with thick gold foil continuing foil on to paper .X. the foil will adhere then varish foil & talk against it

T A Edison

X (photographic transcript), NjWOE, TI 2, Edison's Exhibit 53-13 (*TAEM* 11:520). [a]Interlined above.

1. Edison's notes (to ". . . platinum together") are drawn nearly verbatim from Faraday 1965, 2:27–28. These pages are concerned with the conductivity of various electrolytes.

2. This exhibit was made from a tracing. The original probably said "no action" as does Faraday 1965 (28).

3. On 26 October, Batchelor wrote, "Made solid cylinders of Lamp Black and rubber & much more delicate than Plumbago & rubber." On 30 and 31 October he "worked on machine for pressing Lamp black and rubber." Cat. 1233:299, 303–4, Batchelor (*TAEM* 90:202, 204–5).

**–1101–**

*Technical Note: Telephony and Phonograph*

[Menlo Park,] Nov 5 1877

Speaking Telegraph

On phonograph I propose having a cylinder 10 threads or gembossing grooves to the inch cylinder 1 foot long on this ~~th~~ tin foil of proper thickness arranged with cylinder is transmitter with embos'g point running in these grooves or over them when cylinder rotated. I have tried various experiments with wax chalk[a] etc to obtain an easy indenting surface but find that tin foil over a groove is the easiest of all= this cylinder will indent about 200 spoken words & reproduce them from same cylinder=

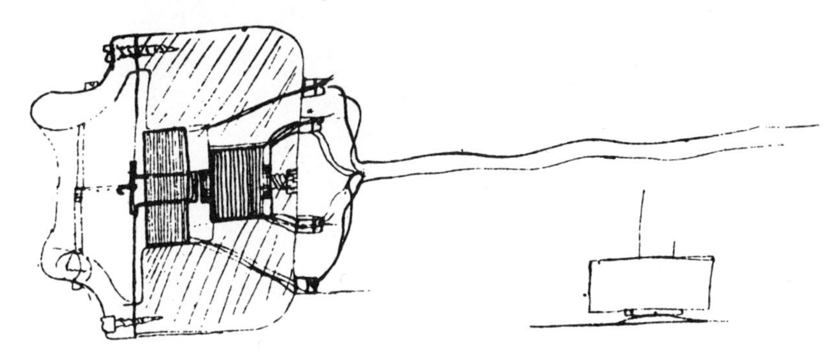

X (photographic transcript), NjWOE, TI 2, Edison's Exhibit 54-13 (*TAEM* 11:521). [a]Interlined above.

**–1102–**

*Edward Johnson to the Editor of the* Scientific American

New York, November [6?], 1877[1]

*To the Editor of the Scientific American:*

In your journal of November 3, page 273,[2] you made the announcement that Dr. Rosapelly and Professor Marey have succeeded in graphically recording the movements of the lips, of the vail[3] of the palate, and the vibrations of the larynx, and you prophesy that this, among other important results, may lead possibly to the application of electricity for the purpose of transferring these records to distant points by wire.

Was this prophecy an intuition? Not only has it been fulfilled to the letter, but still more marvelous results achieved by Mr. Thomas A. Edison, the renowned electrician, of New Jersey, who has kindly permitted me to make public not only the fact, but the *modus operandi.*[4] Mr. Edison in the course of a series of extended experiments in the production of his speaking telephone, lately perfected, conceived the highly bold and original idea of recording the human voice upon a strip of paper, from which at any subsequent time it might be automatically re-delivered with all the vocal characteristics of the original speaker accurately reproduced. A speech delivered into the mouthpiece of this apparatus may fifty years hence—long after the original speaker is dead—be reproduced audibly to an audience with sufficient fidelity to make the voice easily recognizable by those who were familiar with the original. As yet the apparatus is crude, but is characterized by that wonderful simplicity which seems to be a trait of all great invention or discovery. The subjoined illustration, although not the actual design of the apparatus as used by Mr. Edison, will better serve to illustrate and make clear the principle upon which he is operating.

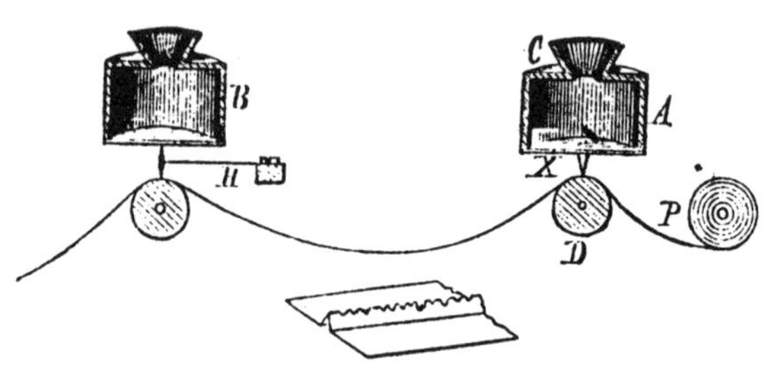


A is a speaking tube provided with a mouthpiece, C—: X is a metallic diaphragm which responds powerfully to the vibrations of the voice. In the center of the diaphragm is secured a small chisel-shaped point. D is a drum revolved by clockwork, and serves to carry forward a continuous fillet of paper, having throughout its length and exactly in the center a raised V-shaped boss, such as would be made by passing a fillet of paper through a Morse register with the lever constantly depressed. The chisel point attached to the diaphragm rests upon the sharp edge of the raised boss. If now the paper be drawn rapidly along, all the movements of the diaphragm will be recorded by the indentation of the chisel point into the delicate boss—it, having no support underneath, is very easily indented; to do this, little or no power is required to operate the chisel. The tones of small amplitude will be recorded by slight indentations, and those of full amplitude by deep ones. This fillet of paper thus receives a record of the vocal vibrations or air waves from the movement of the diaphragm; and if it can be made to contribute the same motion to a second diaphragm, we shall not only see that we have a record of the words, but shall have them re-spoken; and if that second diaphragm be that of the transmitter of a speaking telephone, we shall have the still more marvelous performance of having them re-spoken and *transmitted by wire at the same time to a distant point.*

The reproductor is very similar to the indenting apparatus, except that a more delicate diaphragm is used. The reproductor, B, has attached to its diaphragm a thread which in turn is attached to a hair spring, H, upon the end of which is a V-shaped point resting upon the indentations of the boss. The passage of the indented boss underneath this point causes it to rise and fall with precision, thus contributing to the diaphragm the motion of the original one, and thereby rendering the words again audible. Of course Mr. Edison, at this stage of the invention, finds some difficulty in reproducing the finer articulations, but he quite justified by results obtained, from his first crude efforts, in his prediction that he will have the apparatus in practical operation within a year. He has already applied the principle of his speaking telephone, thereby causing an electro-magnet to operate the indenting diaphragm, and will undoubtedly be able to transmit a speech, made upon the floor of the Senate, from Washington to New York, record the same in New York automatically, and by means of speaking telephones redeliver it in the editorial ear of every newspaper in New York. In view of the practical inventions already contributed by Mr. Edison, is there any one who is prepared to gainsay this prediction? I for one am satisfied it will be fulfilled, and that, too, at an early date.

EDWARD H. JOHNSON, Electrician.

PL, *Sci. Am.* (n.s.) 37 (1877): 304.

1. The date on the issue of the *Scientific American* was 17 November but it was printed and available by 6 November, on which day it was reprinted in a New York newspaper (the *Sun*) as "Echoes from Dead Voices. Wonderful possibilities of Mr. Edison's Latest Invention" (Cat. 1240, item 269, Batchelor [*TAEM* 94:80]). The newspaper report spread widely and apparently more rapidly than the journal, eliciting both skeptical and jocular initial responses. See "A Singular Invention," *New York World,* 7 Nov. 1877; "The Phonograph," *New York Times,* 7 Nov. 1877; "The Inventor of the Age. An Afternoon in the Laboratory of Prof. Thomas A. Edison," *New York Sun,* 29 Apr. 1878 (Cat. 1240, item 561, Batchelor [*TAEM* 94:186]). *Scientific American* prefaced the letter with several laudatory editorial paragraphs, all but one of which were also reprinted in the *Sun.*

2. "Speech Automatically Transmitted in Shorthand by the Telegraph."

3. The velum, or soft palate.

4. Johnson had mentioned Edison's invention in his telephone exhibitions as early as the beginning of August, and the *Philadelphia Record* had reported the invention on 14 August. His pamphlet on the telephone, published about this time, also contained a discussion similar to this letter. Doc. 991; "Edison at Home," *Philadelphia Record*, [6 June 1878?] (Cat. 1240, item 648 [*TAEM* 94:222]); Johnson 1877, 10–12; Johnson, "Address to the 11th National Electric Light Association Meeting," *Elec. W.* 15 (1890): 154.

**-1103-**

*From George Barker*

[Philadelphia, November 8, 1877] Thursday morn.[1]

Nothing yet from you.[2] Can you not loan me a singing transmitter, a talking transmitter and a motorgraph receiver? It will be of no use unless I get it Tuesday. Do not do anything elaborate. I want only to show the principle Yrs

G.F.B.

ALS, NjWOE, DF (*TAEM* 14:177).

1. Postmarked November 9, which that year was a Friday.

2. Barker had been at Menlo Park on the night of 27 October. Cat. 1233:300, Batchelor (*TAEM* 90:203).

**-1104-**

*From Josiah Reiff*

NY Nov 8/77

I saw Norris today— Dont fail to bring in all your papers especially copy of Domestic contract[1] if you have one. I must see you. I want to talk about the Phonograph. Yr

JCR

Where is EHJ.?

ALS, NjWOE, DF (*TAEM* 14:177).

1. Doc. 415.

**-1105-**

*From Josiah Reiff*

New York, Novr 9th 1877[a]

My dear T.A.

I learn from London the Quad works perfectly—[1] they will probably adopt it & pay somebody something, but in their expressive language they remark "we'll take damned good care not to pay the wrong man—

You are in a dilemma[b] for regardless of me, you got certain money from G[ould],[b] & besides he as atty transferred the foreign rights to Mills[2] a/c. to A&P so surely G.B.P. can do nothing toward [fixing?][c] a title.

If Prescott will be half decent, all can be adjusted, a/c simply transfer of title. Yours

JCR

P.S. I go to Washington tonight be back about Tuesday

ALS, NjWOE, DF (*TAEM* 14:730). Letterhead of J. C. Reiff. [a]"New York," and "187" preprinted. [b]Obscured overwritten letters. [c]Illegible.

1. According to the *Journal of the Telegraph,* "The quadruplex apparatus which [Gerritt Smith and George Hamilton] so successfully introduced on the English lines continues to work well under the management of the employés who were instructed by them during their brief stay in England." "Homeward Bound," 1 Dec. 1877, Cat. 1240, item 279 (*TAEM* 94:85).

2. On 6 January 1875 Jay Gould had transferred his assignment of Edison's duplex and quadruplex patents to Samuel Mills, who transferred it to the Atlantic and Pacific Telegraph Co. five days later. Exhibits 10–11, Bill of Complaint, 1:64–70, Box 17B, *Harrington v. A&P.*

**-1106-**

*Charles Batchelor to Vesey Butler*

Menlo Park NJ. Nov 9th 1877

Dear Sir,

When I wrote you about the Telephone we had some in New York working very satisfactorily. Time has however shewn us some defects in it and we are now experimenting with it with a view to make it more perfect. I hope to be able to send you better news in regard to it shortly and probably a pair of telephones Respectfully yours

Chas Batchelor

ALS (letterpress copy), NjWOE, Cat. 1238:230 (*TAEM* 93:167).

**-1107-**

*Notebook Entry: Telephony*[1]

[Menlo Park,] Nov. 9th 1877.

Speaking Telegraph

Oour fluff holder contained silk saturated with Plumbago; but this we find shakes out with time, and the articulation consequently deteriorates; we have substituted silk cloth covered over on both sides with a thick paste of plumbago and dextrin; this silk is cut in discs, and a number of them is put in the holder, and the resistance of the circuit is altered ~~p~~by pressing them together and opening them out again. We thought this an improvement but these also deteriorated not by losing their plumbago but by continual tamping they get perfectly flat and without spring.

Mr Edison also found out that plumbago does not alter its

resistance by pressure as we at first thought, but the increased pressure made better contact and as this was the case we proposed to used hard (fluffs) made of a combination of plumbago and other materials.

The best combination we found is 1 gram lampblack (best) and 250 milgram Rubber dissolved in Bisulphide carbon this is very sensitive and we make it into a small cake perfectly flat 61 [drawing] on both sides as in 61. The construction of the telephone was slightly altered for this surface contact and as we make it now is shown in 62

62.[2]

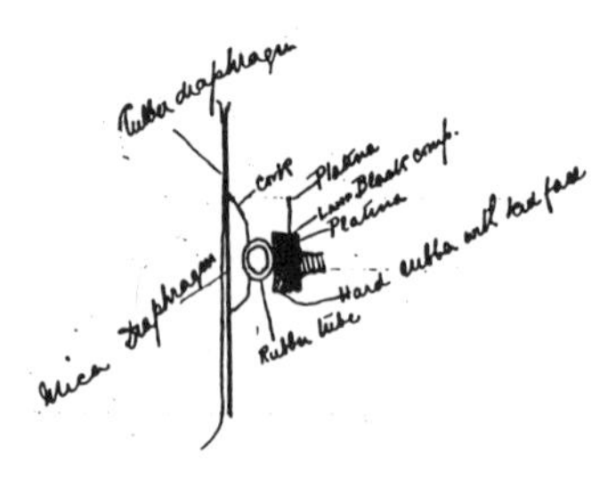


You speak against a soft rubber diaphragm on the other side of which lays a mica diaphragm loosely, only fastened at one point to keep it in place; on the centre of the mica is placed a disc of cork to which is shallac'd a piece of flexible tube which acts as a peculiar spring, this pushes against a small disc of rubber faced with kid, next to the kid to a disc of platinum foil and next to that our composition cake of lampblack and rubber which rests upon a platina surface. The faces of these discs must be exceedingly true in order to get more points of contact. We have now one hundred made after this plan and they work well we have yet to see whether they will stand

Another Speaking telegraph we made as shewn in fig 63.[3]

63[a]

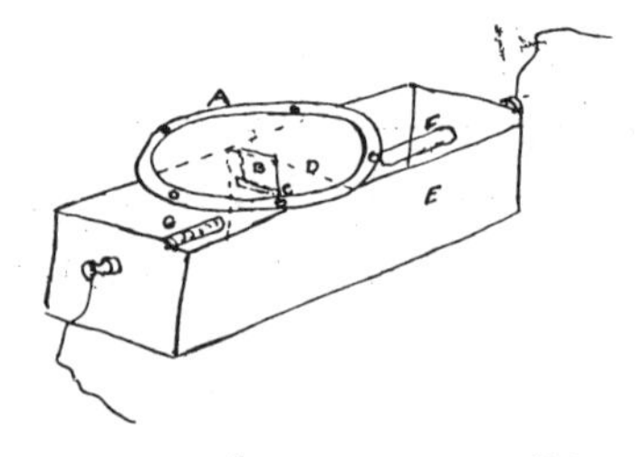


A is a diaphragm mounted over a two cell box E in which is two electrodes F & G in the partition D is a fine slot which is closed or opened by the plate B which is fastened to the diaphragm thus altering the resistance of the circuit on each movement of the diaphragm This works pretty well the articulation being good but low.

We also made another telephone thus:—

64

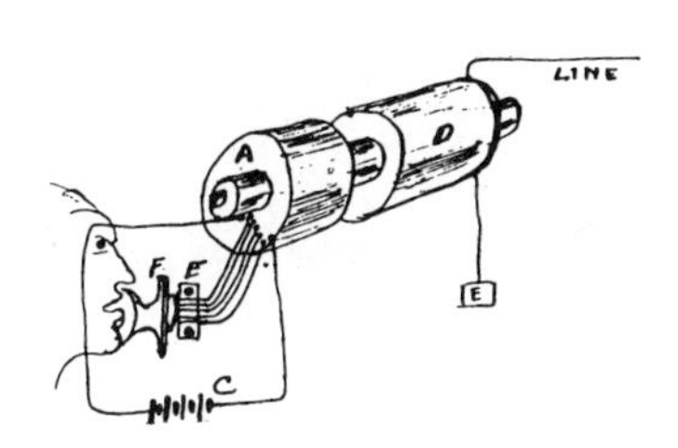


Fig 64 A is a spool round core B which is kept permanently magnetised by battery C; this by induction sends currents over the line through spool D. In spool A each layer is connected by wire to a platina point in an insulated stand E so that you have a row of points very close together each one connected to its respective layer of the magnetising spool A. You now place a diaphragm with an elliptic spring on its face in front of these points so that it shortcircuits the magnet and gives variable magnetism In order to give all the finer sounds the magnet A should have a great number of layers or the wires should be connected to the stand very frequently during the winding of the magnet.

Chas Batchelor

X, NjWOE, Batchelor, Cat. 1317:46 (*TAEM* 90:680). [a]Underlined twice.

1. This entry is continued in Doc. 1112.

2. Figure labels are (clockwise from top) "Rubber diaphragm," "Cork," "Platina," "LAMP Black comp.," "Platina," "Hard rubber with kid face," "Rubber tube," and "Mica Diaphragm."

3. This was made on 7 November. Cat. 1233:311, Batchelor (*TAEM* 90:208).

## -1108-

*From Joseph Murray*

Newark, N.J., Nov 10th 1877[a]

T A Edison

I am very short the men are all out of money and patience if you endorse this bill[1] I will try get some money from WU— Yours

JTM

Return it by bearer shall I ship[2] all my men or not tell me what the future prospect is if possibble

ALS, NjWOE, DF (*TAEM* 14:989). Letterhead of J. T. Murray. [a]"Newark, N.J.," and "187" preprinted.

1. Not found.

2. Dismiss. Farmer and Henley 1970, s.v. "Ship."

## -1109-

*Technical Note: Telephony and Phonograph*

[Menlo Park,] Nov 10 1877

Speaking Telegraph

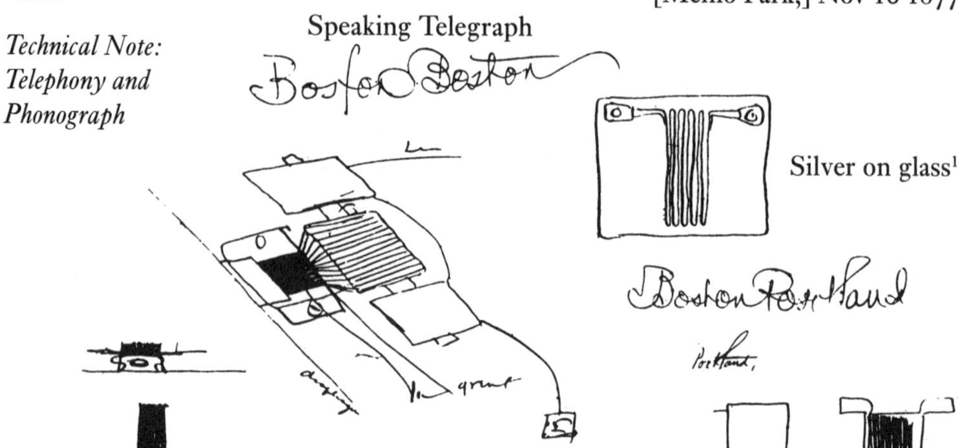

Silver on glass[1]

Phonograph  tissue ppr & tinfoil over it  10 th[rea]ds to inch on Cylinder

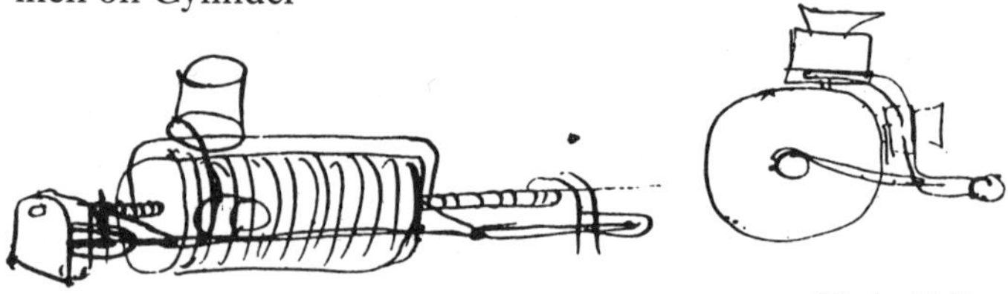

T. A. Edison

X (photographic transcript), NjWOE, TI 2, Edison's Exhibit 86-13 (*TAEM* 11:545).

1. Figure labels are "diaphragm," "[grenet?]," and "Line."

## -1110-

*From Thomas David*

Pittsburgh, Nov 14th 1877[a]

Friend Edison—

I wish you would send me those two sets of Telephones— If you have not been able to make them, please send me the two stationary upright talkers that you made first— I want something to hold the Talker while we adjust it—

Your idea that it will not require re~~quire~~adjusting after being once adjusted[b] wont do, as the jaring from wagons etc will gradually shake them loose—especially about our mills—

Let me hear from you promptly— Yours Truly

T. B. A. David

ALS, NjWOE, DF (*TAEM* 14:990). Letterhead of Central District and Printing Telegraph Co. [a]"Pittsburgh," and "187" preprinted. [b]Obscured overwritten letters.

-1111-

*To William Orton*

Menlo Park N.J. Nov 15 77

Dear Sir,

Allow me to introduce Mr Jas Mackinzie[1] who is manager of the District Telgh Co in Detroit and has been very successful with Bell Telephone; He has been to my place and I exhibited to him several varities of Telephones; one of which I had abandoned owing to fact I could not hear it well. To my sup~~p~~rise he asserts that it is louder and superior in articulation to Bells;[2] It is based on a principle which is entirely different from anything now out and is perfectly Constant. Yours

T. A. Edison

ALS, NjWOE, DF (*TAEM* 97:596).

1. MacKenzie was at the laboratory overnight. He returned the night of 17 November and was there all the following day, and returned once more on 23 November. Cat. 1233:319–22, 327, Batchelor (*TAEM* 90:212–14, 216).

On 29 September, MacKenzie had written to Edison discussing the possibility of his becoming the agent for Edison's telephone. Edison had previously suggested that MacKenzie contact Orton or Anson Stager, but MacKenzie had hesitated because of a potential conflict with G. W. Balch, president of American District telegraph of Detroit, whose Detroit Electrical Works claimed to have local rights to Edison's telephone. DF (*TAEM* 14:957).

2. Batchelor recorded that MacKenzie tested "our water telephone & [said] it was far better than Bell's." When he returned on 17 November he brought two Bell telephones, which he and the staff compared to Edison's the next day. Cat. 1233:320, 322, Batchelor (*TAEM* 90:213, 214).

-1112-

*Notebook Entry: Telephony*[1]

[Menlo Park,] Nov 15th 1877

Speaking Telegraph.

The principle of increasing and decreasing a local polarising circuit as shewn in Fig 64 we have tried as shown in Fig 65.[2]

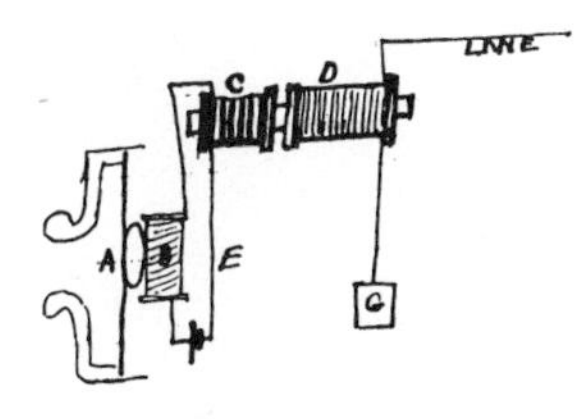


A is a diaphragm on which is fastened an elliptical spring which is pressed on to the spool of fine platina wire B shortcircuiting the coils and continually altering the resistance of local polarising circuit E. D is a magnet on line working by induction from C. We tried this and the spool B measured (35) ohms and with very loud talking it was reduced to 5[a] ohms.

66

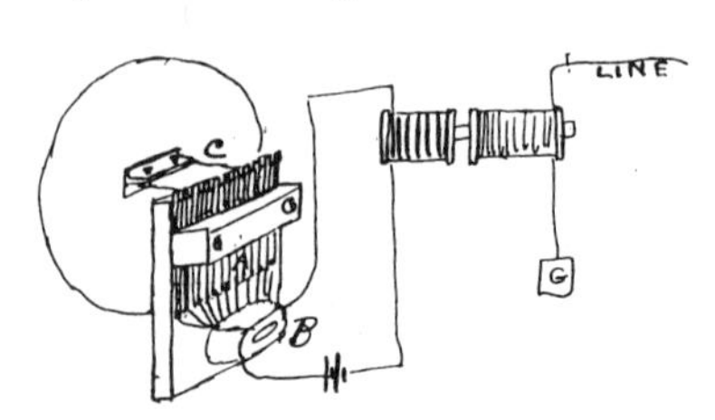


Fig 66 is also another way of doing the same thing but a more reliable one A is a lot of springs each connected with resistance coil B so that when knife edge C on diaphragm touches points of springs it will shortcircuit some thus increasing and decreasing local circuit Plate C must have its edge about 1/200 of an inch out of square so that it will touch those at one end first and then each successive one

67.

From the device in 67 we got very fair talking.[3] Diaphragm vibrating caused the platinum spring to alter resistance and we got talking pretty fair. Spring shakes about a great deal when talking

Chas Batchelor

X, NjWOE, Batchelor, Cat. 1317:48 (*TAEM* 90:681). [a]Circled.

1. This entry is a continuation of Doc. 1107 and is continued in Doc. 1116.

2. Edison had considered this use of an induction coil, with the receiver and the carbon resistance of the transmitter in circuit with the primary coil, in September (Doc. 1041). Although he later recalled using induction coils in this way throughout the fall, this document is the first evidence of their use since September. Prescott 1879, 224–25.

3. Figure labels are "Diaphragm" and "spiral spring of platinum."

**-1113-**

*From Charles Leyenberger*

Newark, Nov 16 1877[a]

Dear Sir:

Jos T. Murray's fire insurance policyies expired at noon today.

Will you see that the premiums are paid if I hold the insurance good?

Please let me know what to do. Yours truly

C Leyenberger.[1]

ADDENDUM[b]

[Menlo Park, November 16, 1877?]

Cant do it just now  Very much busted!

Edison[c]

ADDENDUM[d]

[Newark, November 17, 1877?]

If you will agree to pay next month, I will hold the insurance good[2]

Leyenberger

ALS, NjWOE, DF (*TAEM* 14:341). Letterhead of Provident Savings Life Assurance Society of New York. [a]"Newark," and "187" preprinted. [b]Addendum is an ALS; written below Leyenberger's letter. [c]Followed by "—over—" in Leyenberger's hand to indicate page turn. [d]Addendum is an ALS; written on back.

1. Charles Leyenberger, identified on the letterhead as manager for New Jersey of the Provident Savings Life Assurance Society of New York, acted as Edison's insurance agent. *TAEM-G1*, s.v. "Leyenberger, Charles."

2. This matter was still unresolved at the end of December. Leyenberger to TAE, 27 Dec. 1877, DF (*TAEM* 14:345).

**-1114-**

*Technical Note: Telephony*

[Menlo Park,] Nov 16 1877

Spkg Telegh

Found[a] that in Bells telephone a copper or mica diaphragm could be used in transmitter in place of iron  Wherefore how does it work deponant sayeth not except that the vibration of diaphragm disturbs the molecular quietness of the permanent magnet hence currents= we are now making a number of experiments to devise a telephone based on the molecular disturbance of a permanent mag or electromagnet  We find that words spoken in a Bell telephone is received with ½ the loudness ~~of~~on the copper diaphragm apparatus that they would if there was an iron diaphragm= This is a new advance in the art—hence patentable & gets clear of Bells fearful claws[1]

T A Edison

X, NjWOE, Lab., Vol. 13:111 (*TAEM* 4:104). [a]Obscured overwritten letter.

1. Edison did not patent this.

**–1115–**

*Charles Batchelor to George Bliss*

Menlo Park, N.J. Nov 17, 1877

Dear Sir

Yours of Nov 6th to hand. For the last month I have worked incessantly on Telephone and although a small job I really have not had time to make the ink attachment. Our telephones made by Murray are very good but it wants something much louder than either those or Bell's to work on the Gold and Stock wires. We have made six or eight new principles, three of which Mr Phelps is now making[1] and it is Edison's intention to make 'the' telephone that will work under all conditions

Of course you know that the Gold & Stock wires (owing to the number of wires on the poles and the heavy batteries on them) are entirely different from an isolated wire and when I tell you that between 197 Broadway and 23rd Street it is very difficult indeed to get any thing at all on the Bell instrument you will see the difficulty we have to contend with[2] Do you not have trouble in Chicago on city lines with it?

Thanks for your attention to our Wheeler bill

If it would not inconvenience you I should very much like to have balance of first ink for I can assure you I am very hard up.

Had a card from Col Beetle[3] I see he has got into quarters at last Very respectfully yours

Chas Batchelor

ALS (letterpress copy), NjWOE, Batchelor, Cat. 1238:233 (*TAEM* 93:169).

1. In a letter, which William Orton dated 17 November 1877, Edison informed the Western Union president that he "went to Phelps on Monday. said could do nothing= since then he has seen you says OK I took apparatus there today." WUTAE.

2. The previous day Edison had recorded an idea for overcoming induction between two telephone lines, but subsequent experiments showed it did not work. Vol. 12:103, Lab. (*TAEM* 4:103).

3. Not found.

**–1116–**

*Notebook Entry: Telephony*[1]

[Menlo Park,] Nov 17 1877.

Speaking Telegraph.[2]

I proposed to Edison that we should make a telephone on the principle of Fig 66 but made in a circle, as one great trouble with 66 was that there was too much strain on one side of the diaphragm. This we have done in the following manner

68

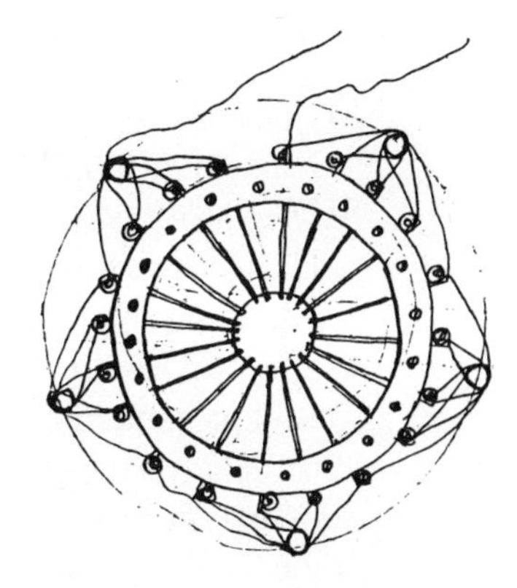

68 is the arrangment of springs which we call the 'sunflower'[3] all having their ends bearing on the edge of a hole perfectly true on the face. The diaphragm is so arranged that in the 1/100 of an inch movement it closes circuit with them all one after another. These springs are hair spring and faced on their points with platina

Ways of fastening connection to hair spring

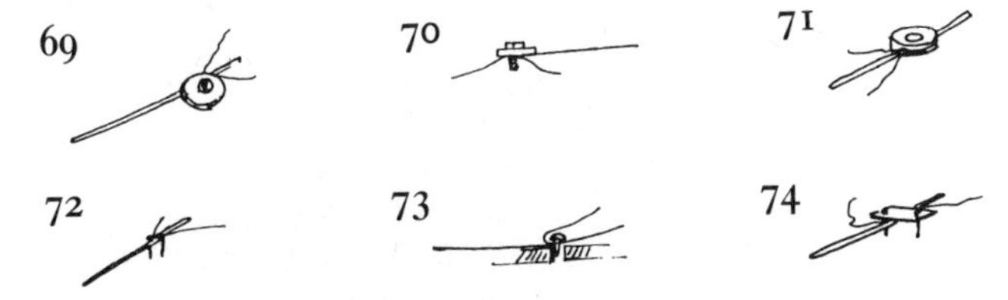


69 screw & washer clamp
70 spring soldered to washer
71 split washer clamp
72 staple
73 spring bent & wedged in hole
74 Cap. held by pins

We however found that the rebound of these springs prevented perfect articulation, and as yet we have not found any means of dampening the rebound sufficiently to get it

X, NjWOE, Batchelor, Cat. 1317:49 (*TAEM* 90:681).

1. This entry is a continuation of Doc. 1112 and is continued in Doc. 1121.

2. What are probably Batchelor's original notes for this entry are Edison's Exhibits 114-13–116-13, TI 2 (*TAEM* 11:564).

3. Other sketches of the sunflower design from 15 and 16 November are Edison's Exhibits 108-13 and 110-13, TI 2 (*TAEM* 11:561–62); Edison described its circuit design and performance in Prescott 1879 (232–33; see also Jehl 1937–41, 1:152). The sunflower design had been used by George Anders in one of his dial telegraphs, which he had first shown to Edison in 1869, and in printing telegraphs during the early 1870s. Anders's testimony, *Anders v. Warner;* on Anders see *TAEB* 1:104.

-1117-

*From Josiah Reiff*

Washington D.C Nov 18/77

My dear T.A.

I learn that Prof Henry paid you a great compliment at the EHJ[ohnson] concert here—[1]

He said you were the most ingenious inventor in this country, & hesitating then said or in any other—

The opportunity to say this was because U.H.P[ainter] took special pains to interest Prof H, as EHJ can explain— UHP.s effort to advertise you is worth a note of thanks—

Dont fail to get preliminary statement ready[2] I will be home Tuesday, please let me see you.[3] Surely you are not so busy that you cannot give me a moment for I am working as hard for you as you are for me. Yrs trl

JCR

ALS, NjWOE, DF (*TAEM* 14:181).

1. In that concert, given on 16 November at a church, vocal and cornet music was transmitted from Philadelphia to Washington. Henry gave "a brief explanatory adress with reference to the great progress in telegraphic inventions, &c." Also in attendance were Vice President William Wheeler, Secretary of the Navy Richard Thompson, Benjamin Butler and other members of Congress, and "many other distinguished people." "Johnson's Washington Exhibition of Telephone (from the *Evening Star*, Washington, Nov. 17, 1877)," TI 2:667 (*TAEM* 11:752).

2. See Doc. 1096 n. 3.

3. Reiff went to Menlo Park on Thursday, 22 November. Cat. 1233:326, Batchelor (*TAEM* 90:216).

-1118-

*From George Barker*

PHILADELPHIA.[a] Nov. 20, 1877.

Dear Sir:—

I am more disappointed than I can tell you at your failure to send me even the least thing which I can use to illustrate your telephone. I saw plenty of things in your shop with the loan of which I should have been quite satisfied. You need not have spent a moment more than the time necessary to order them boxed and expressed. There was a motorgraph receiver there which would have answered my purpose entirely and this I wanted most. Then there were lots of your lampblack transmitters one of which might certainly have been loaned me for a fortnight. As to a singing transmitter, I could get along without that perhaps. But now I have to go to Indiana on Thursday and my lectures can be only advertisements for Bell's telephone. I say I am exceedingly disappointed at this result, because I spoke to you about this matter in July last; and I reason that what the past has not permitted you to do, the future will not either. You are always full of business you know. I would not have gone back on you this way for the world. Yours truly

George F. Barker

ALS, NjWOE, DF (*TAEM* 14:183). Letterhead of University of Pennsylvania. [a]Preprinted.

**-1119-**

*Technical Note: Phonograph*

[Menlo Park,] Nov 23 1877

Phonograph

I propose to apply the phonograph principle to make Dolls speak sing cry & make various sounds also apply it to all kinds of Toys such as Dogs' animals, fowls reptiles ~~m~~human figures: to cause them to make various sounds to steam Toy Engine imitation of exhaust & whistele= to ~~seem~~ reproduce from sheets music both orchestral instrumental & vocal the idea being to use a plate machine with perfect registration & stamp the music out in a press from a die or punch previously prepared by cutting in steel or from an electrotype or cast from the original on tin foil=A family may have one machine & 1000 sheets of this music thus giving endless amusement I also propose to make toy music boxes & toy talking boxes playing several tunes & speaking several sentences also to clocks and watches for calling out the time of day or ~~walkin~~ waking a person up for advertisements rotated continuously by clockwork to call the attention of passersby. I propose to make it loud by building up or enlarginng the original vibrations or indentations by laying them out & cutting on steel or building up by electrolysis, & other ways= The method for preparing a wheel or plate for toys is by stamping or casting the same, if a wheel a band is stamped & put[a] around it

It may replace a man on a telephone ~~and a telephone man~~ at either end by holding the Telephone to the spkg tube= or the receiving magnet may be connected to ~~the~~ work[a] the iron diaphragm. with toys or apparatus in which only reproduction is required the characters may ~~be~~ either be indented or raised—

For taking down testimony where it is impracticable for the witness to speak right in the tube I propose to arrange a large chamber of metal

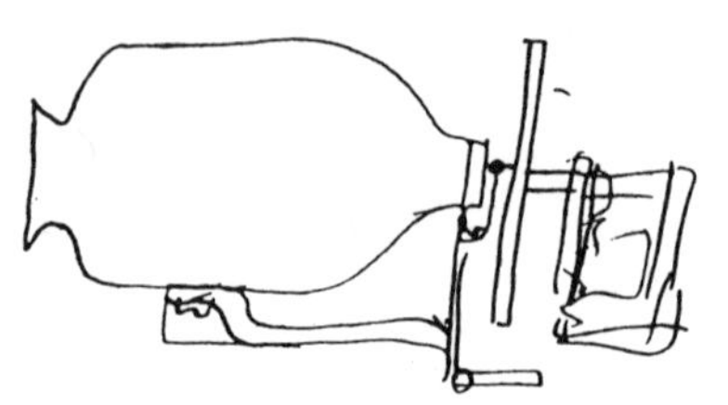

or to shew the principle better[1]

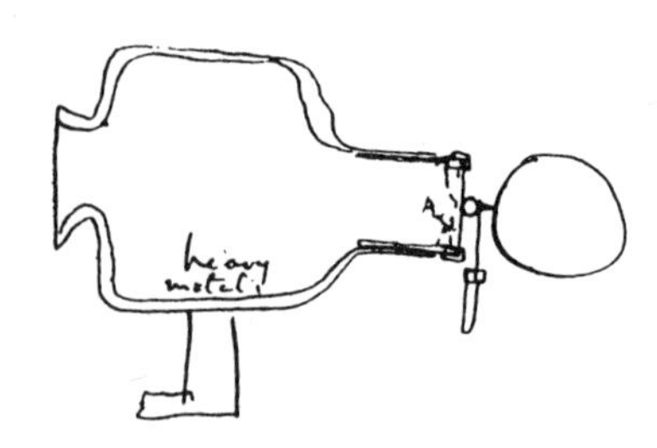


This is  This large chamber will resound more powerfully to a voice at a distance than a small chamber—

| | |
|---|---|
| T A Edison | Chas Batchelor |
| J Kruesi | James Adams |
| ~~G E Carman~~ | ~~M N Force~~ |

X, NjWOE, Lab., Vol. 17:18 (*TAEM* 4:888). Document multiply signed and dated. [a]Obscured overwritten letter.

1. Figure labels are "heavy metal" and "[---]."

**-1120-**

*Technical Note: Telephony*

[Menlo Park,] Nov 23 1877[1]

Speaking Telegraph

[A][2] [B][3]

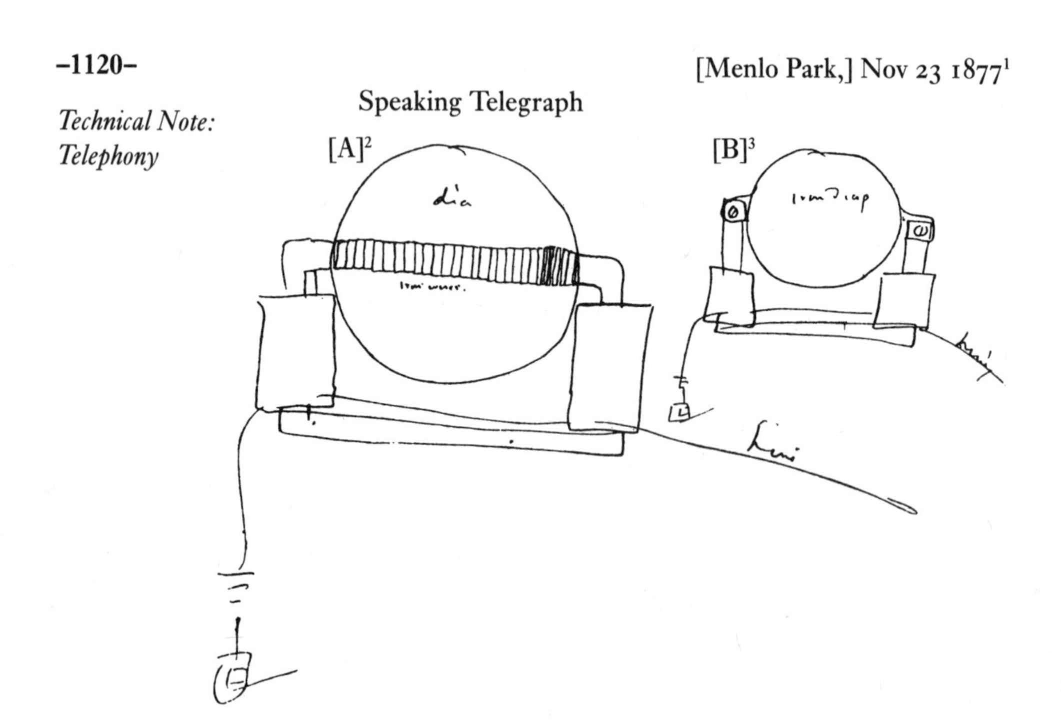

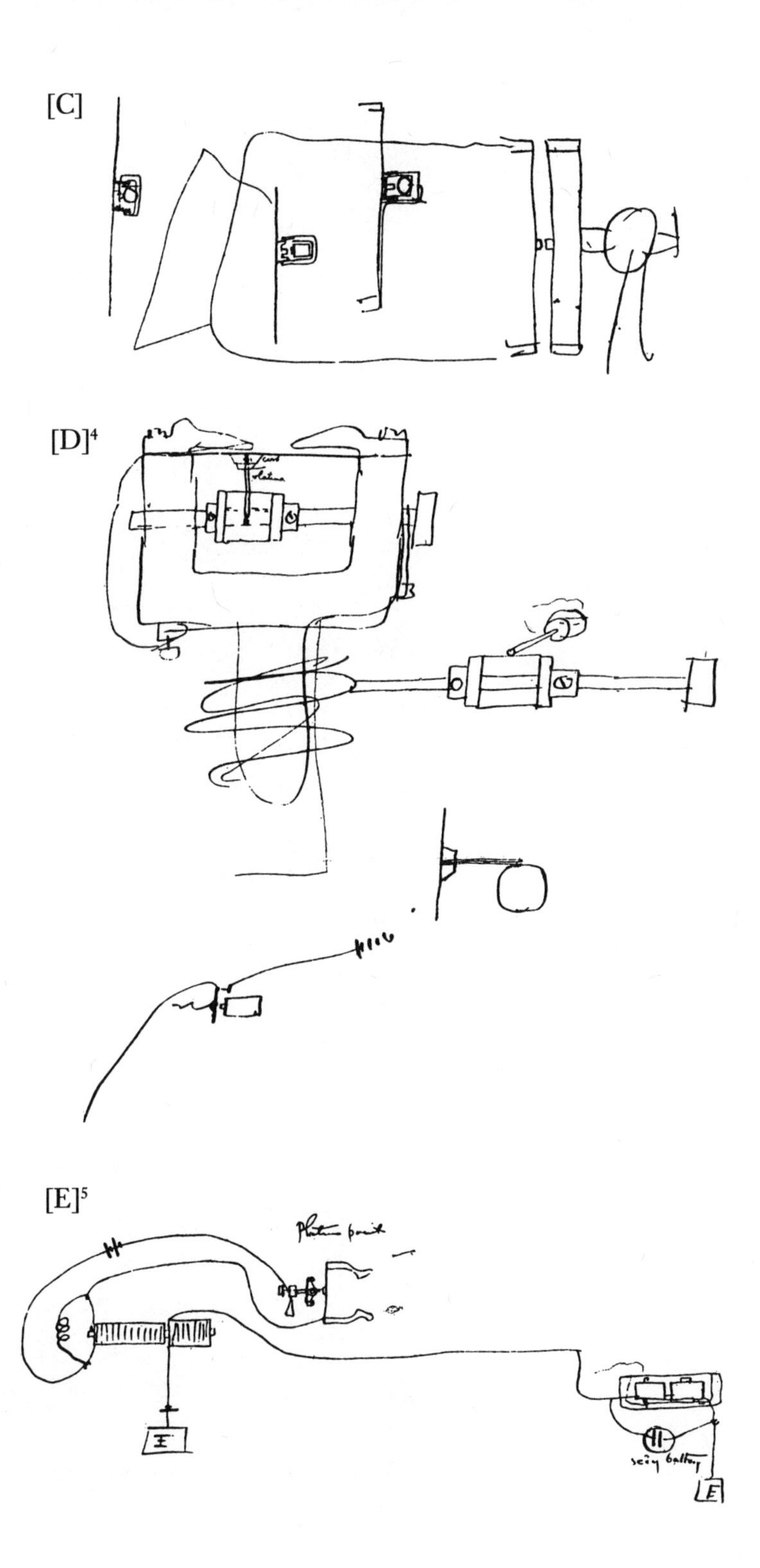
[C]
[D][4]
[E][5]
Platina point

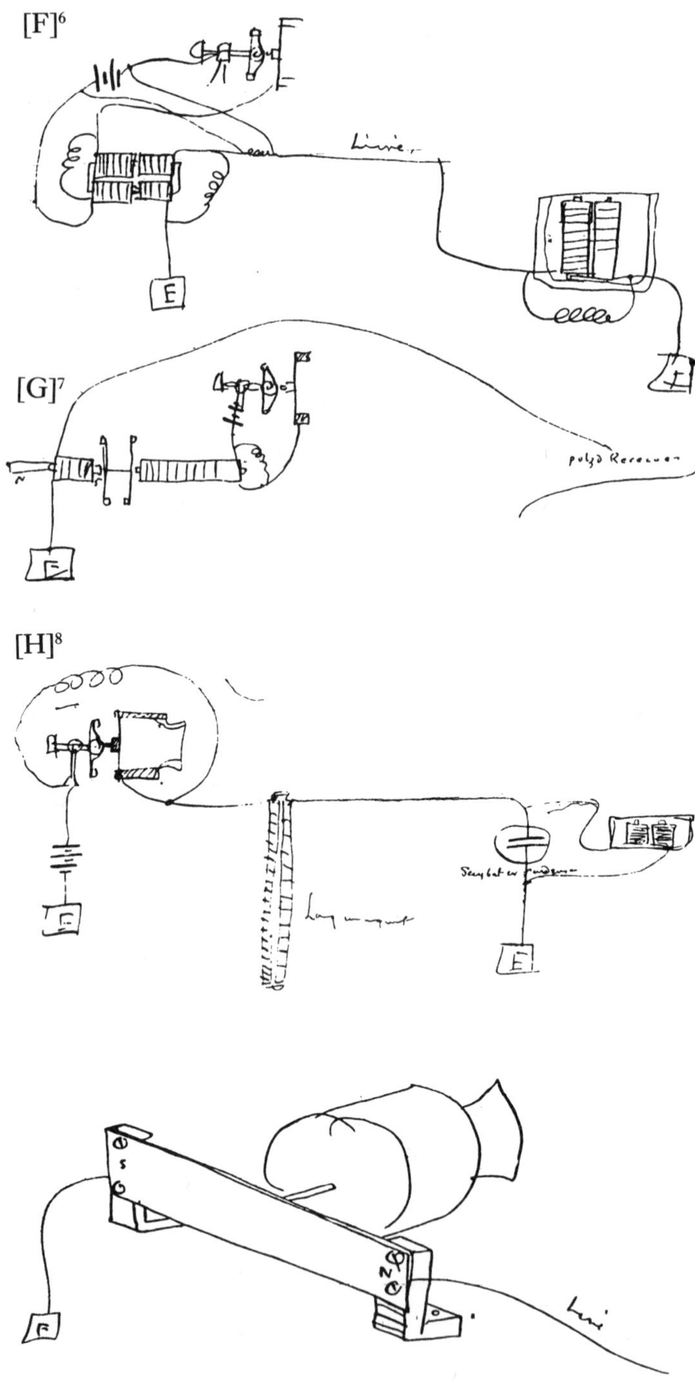


Molecular disturbance[9] generates Currents

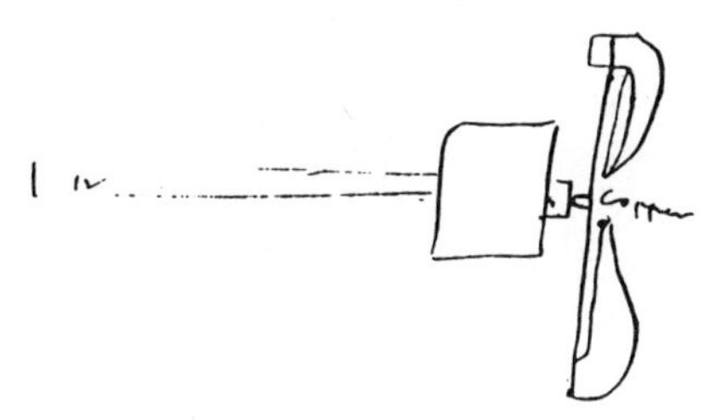

point[10] hits end of permanent magnet & send currents by molecular disturbance

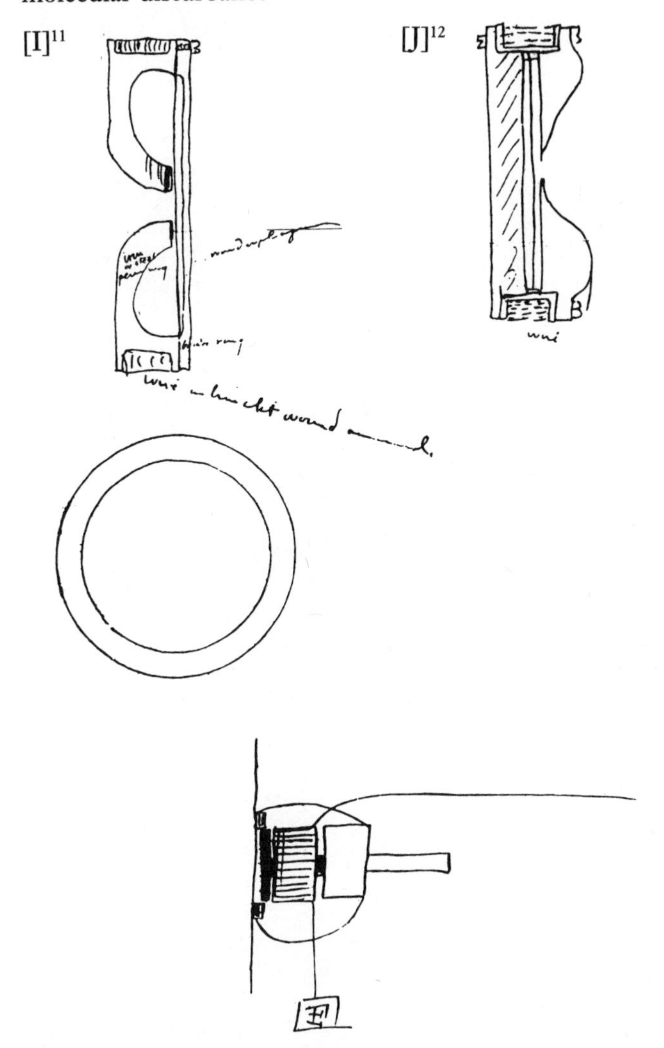


Magneto multiplication of Siemens & Wheatstone applied to telephony[13] for causing the human voice to generate it primary magnetization to allow of induction currents being generated.

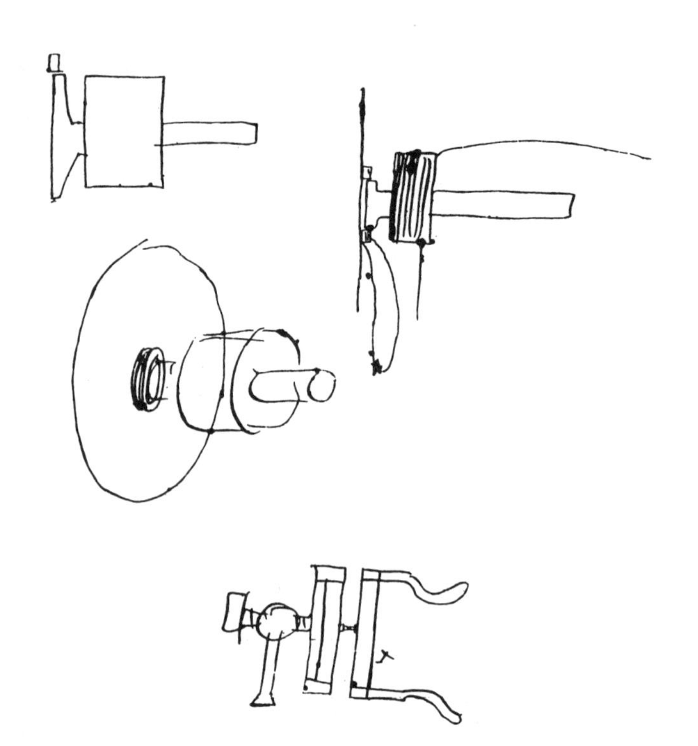

X mica diaphgm to prevent heat effecting iron diaphm in front

[K][14]

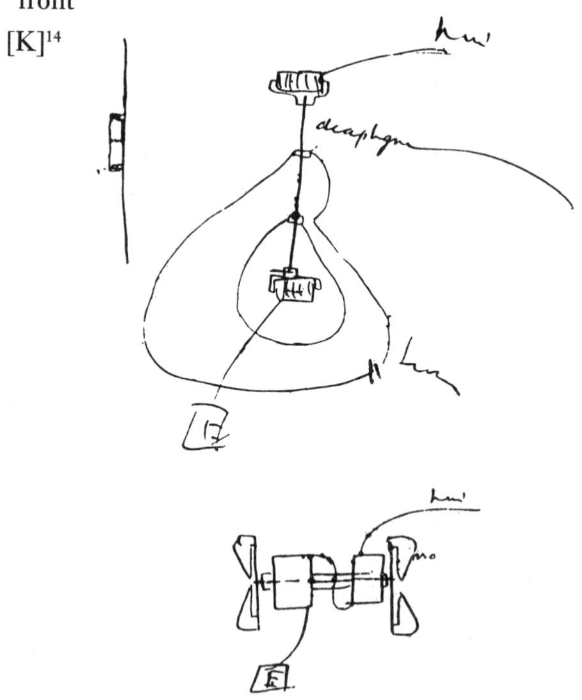

2 diaphgm connected together balanced telephone hence allow to work closer to mag & have double power[15]

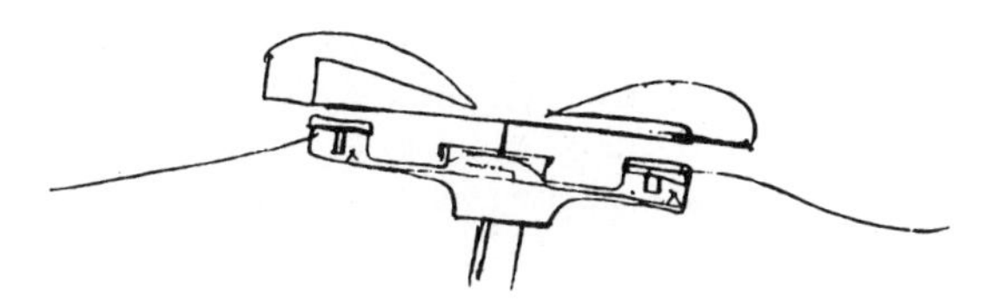

in X[16] a chem sol working a diaphagm by the capilliary Electromotive principle  you can also send as this moving of mercury produces a current=

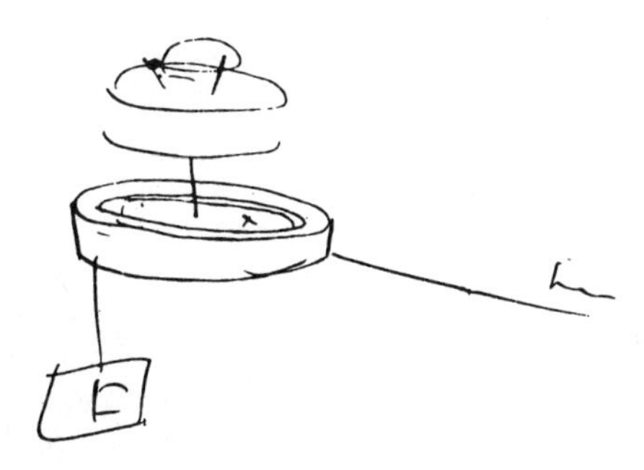

Hg in dish within a coil, point from diaphagm dips in Hg= Hg has lot iron disolved in it by use of sodium [ama~~g~~lgmtn?][a] The waves given this mercury ~~p~~by point from diaphragm set up induction currents.[17]

[L][18]

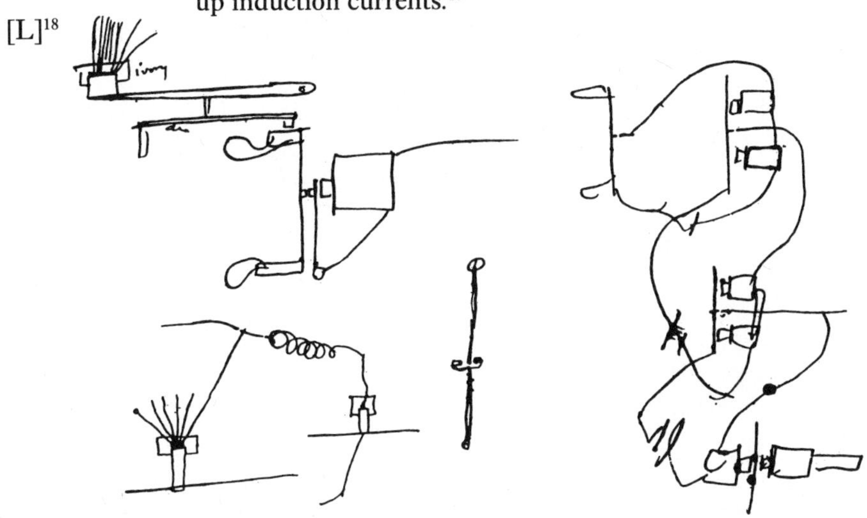


T A Edison

Chas Batchelor
James Adams

X (photographic transcript), NjWOE, TI 2, Edison's Exhibits 131-13–133-13, 136-13–141-13, 143-13 (*TAEM* 11:577). [a]Illegible.

1. This document represents much other work from November on alternative telephone designs.Some of those efforts reworked older de-

signs; others were new ideas (see *TAEM-G1*, p. 129, for a listing of other technical notes from November). The next day Batchelor wrote in his diary, "Took models & specifications to Serrel of first water telephone and springs with resistance in" (Cat. 1233:328, Batchelor [*TAEM* 90:217]). Those specifications were for Cases 146 and 147, for which Edison executed applications on 8 December (U.S. Pats. 203,013 and 203,018).

2. Figure labels are "dia," "iron wires," and "line."

3. Figure labels are "iron diap" and "line."

4. Figure labels are "cork" (the diaphragm) and "platina" (the electromotograph lever). This is described in Prescott 1879 (549–50). Edison returned to this design the next year using moistened chalk for the drum of the electromotograph.

5. Figure labels are "Platina point" and "sec[ondar]y battery."

6. Figure label is "line."

7. Figure label is "polzd Receiver."

8. Figure labels are "long magnet" and "secy bat or condenser"; indistinct word appears at bottom of figure. In the fall of 1873 Edison had used long magnets in connection with his automatic telegraph experiments. See, for example, Docs. 363, 375, and 386.

9. Figure label is "line"; the oblong in front is a bar magnet, with the ends labeled "S" and "N." For Edison's ideas on molecular disturbance see Doc. 1114.

10. Figure label is "copper."

11. Figure labels are "iron or steel perm mag," "iron diaphragm," "brass ring," and "wire in line ckt wound around." **J** and **K** appear to be magneto transmitters in which the line wire forms a helix around the diaphragm.

12. Figure label is "wire."

13. Edison seems to be thinking of an analogy to the dynamo principle, which had been discovered independently by Werner Siemens and Charles Wheatstone. See *TAEB* 2:13 n. 2.

14. Figure labels are "line," "diaphragm," and "line."

15. Figure labels are "line" and "mo[uthpiece?]."

16. Each side chamber is labeled "X"; the label in the center is "mercury."

17. Figure label is "line."

18. Figure labels are "ivory" and "dia."

**-1121-**

*Notebook Entry: Telephony*[1]

[Menlo Park,] Nov 24th 1877

We have found that our telephones as shown in Fig 62[2] works best when they are placed in a polariz~~ed~~ing circuit and the line worked inductively.[3] The margin of adjustment is very great working as well when the little rubber ring is flat with pressure as when in its ordinary state. The full range of adjustment does not seem to alter it more than one ohm and consequently with plenty of pressure it is not so liable to alter and get out.[4] Fig 75 shews it.

Fig 75

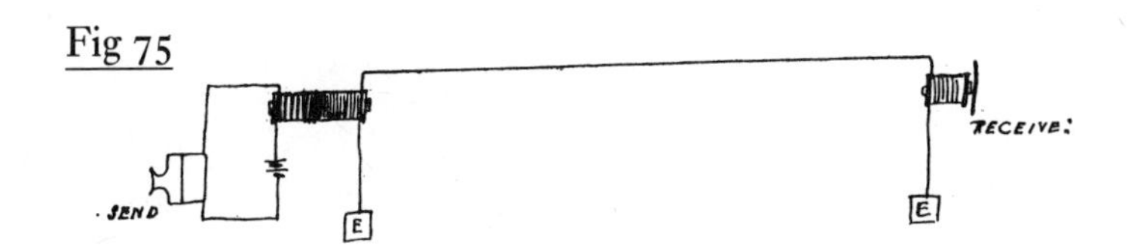


I have also tried to work on this local polarizing principle platina points arranged as in 76 and 77.

76

In 76 A is agate and B is platina and spring on diaphragm strikes no point but rubs on the surface.

77

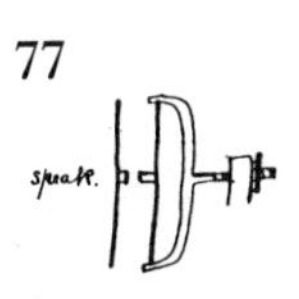

In 77 we have two diaphragms each with a platina point on and one adjustable when you speak against one the other has commenced to move and consequently the points do not strike so sharp. I have also put the whole thing on the diaphragm thus Fig 78 and have tried 79 two diaphagms one with point and the other with a small piece of rubber tube round which was stretched fine platina foil

78 79

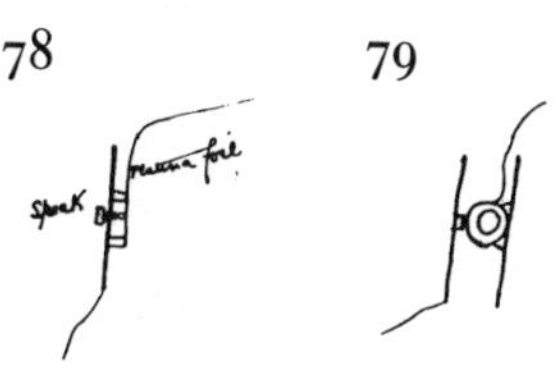

Chas Batchelor

X, NjWOE, Batchelor, Cat. 1317:49 (*TAEM* 90:681).

1. This entry is a continuation of Doc. 1116.

2. See Doc. 1107.

3. Drawings from the following days show telephones with induction coils in several configurations. Within two days Edison had sketched a single board that held a coil, a call bell worked by a polarized relay, and a switch for the receiver and transmitter (Edison's Exhibit 147-13, TI 2 [*TAEM* 11:587]; see also Edison's Exhibit 177-13, TI 2 [*TAEM* 11:610], which shows a similar board from 15 December). A 29 November drawing of an induction coil may be the design that was built and tested on 2 and 3 December (Edison's Exhibit 155-13, TI 2 [*TAEM* 11:594]; Cat. 1233:336–37, Batchelor [*TAEM* 90:221]).

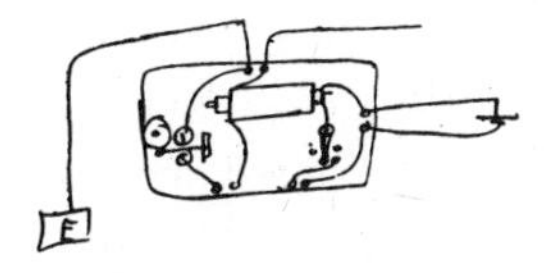


*An induction coil–call bell arrangement for the telephone, with a switch for the transmitter. The terminals at lower right are for the transmitter; those at lower left, for the receiver.*

Around this time Edison drew several other circuits that incorporated induction coils, including one intended as a patent model in which the coil was built into the handle of the receiver (Edison's Exhibits 152-13, 163-13–165-13, TI 2 [*TAEM* 11:592, 601–3]; see Doc. 1154). Although that design was never patented, on 28 February 1878 Edison executed two patent applications that specified the use of induction coils in telephone circuits for the main signal, by which time this had become standard practice on lines employing his telephones (U.S. Pats. 203,016 and 203,017). Thomas Watson's U.S. Patent of 8 January 1878 (No. 199,007, filed on 5 December 1877) also called for induction coils in telephone circuits for the call signal, to allow the telephone to substitute for a bell when not in normal operation.

4. Edison later wrote

> It was soon found upon investigation, that the resistance of the [lampblack] disk could be varied from 300 ohms to the fractional part of a single ohm by pressure alone, and that the best results were obtained when the resistance of the primary coil, in which the carbon disk was included, was 6/10 of an ohm, and the normal resistance of the disk itself was 3 ohms. [Prescott 1879, 225]

**-1122-**

*To Thomas David*

[Menlo Park,] Nov 25, 1877

Friend David,

I think I have struck the Carbon Telephone now that is Constant.[a] few days more will Tell the tale— Your

Edison

ALS, NjWOE, DF (*TAEM* 14:994). [a]Underlined twice.

**-1123-**

*To William Preece*

Menlo Park N.J. Nov 25 1877.

Confidential[a]

Friend Preece.

Yours received.[1] The Telephones brought over by Dr Herz were given him by mistake. Those which should have been given him were left behind. I wrote him immediately to your care[2] about the mistake, promising to send him the others. since then, I have found great difficulty in making them hold their adjustment. when adjusted properly they were very loud and the articulation about the same as Bells, but soon changed adjustment losing both loudness and articulation. Therefore I delayed sending any until I was satisfied with them. I have attained I believe constancy of adjustment, and am still engaged on volume of sound. It is now about 3 times the volume of Bells on his own receiver, but to be practical on ordinary lines it must be at least 6 times louder than Bells:= I have had a very hard time of it,= but it will come out all right. I shall send you a pair the very moment I am satisfied with them= I wish you were here now. I could show a great many new things stumbled on in the last 3 months. Apps[3] or Ladd I forget which, sent me the broken glasses and my polariscope is in good health and gets plenty of exercise.[4] I notice in Telegraphic journal an extract from Muirheads patent given to show that Muirhead had noticed the Electromotograph action, says patent dated 1873. I find by referring to patent that it is dated 1874, Oct I think,[5] whereas it was well known here in 1872–3, and full description was printed in many scientific papers before the date of Muirheads patent.[6] I am going to get E. H. Johnson to write you his experience in working the Tone Telephone: He has got it to perfection. you should have transmitters like his.= I will try and fix you out for the lecture. What do you think of transmitting magnetism over iron wires to distances. I have signs of it=[7] My phonograph is going to beat the Telephone; when I get the scratching noises out of it= I made a pair of Telephones that work with Copper diaphragms its on that revolving copper disc principle of Arago'= I find that a copper diaphragm may replace the iron ~~o~~in Bells= copper must be 1/32 thick. its very very low with copper in both, but if receiver is one of the regular kind and transmitter has ~~an~~ copper diaphragm you can carry on conversation with ease both ways= but with the pair I have made[b] the talking is loud, as I have several dodges on it=[8]

I gave Dr Herz the agency for the continent (he has nothing to do with England) because Prescott said that I had better as he was a big gun and had influence in France etc. I did not expect nor do I now expect he will do anything, as he talked too much, mostly about Dr Herz.

I see that your lecture before the B.A. was the centre of interest.[9] You put it to them excessively plain.[10] if I could only write as clearly and consecitive as you lectured, I believe I would brush up cheek enough to send a paper to the Philosophical Magazine. Yours

T. A. Edison

ALS, UkLIEE, WHP. [a]Added in upper left. [b]Interlined above.

1. Not found.

2. Not found.

3. Alfred Apps, scientific instrument maker, located at 433 Strand, London. Edison had purchased a battery from Apps when he was in London in 1873. Hackmann 1985, 59, 69, 88; Dyer and Martin 1910, 150.

4. The glasses were thin plates in his polariscope (see Doc. 1082). However, no record has been found of observations or experiments with the polariscope.

5. John Muirhead was a prominent British telegraph engineer and inventor. He filed his initial specification for the patent in question (Brit. Pat. 3,663 [1874]) on 23 October 1874. The patent dealt with sev-

eral unrelated points in the field of telegraphy. The extract Edison mentions is in a scrapbook (Cat. 1240, item 265, Batchelor [*TAEM* 94:78]).

6. Edison's earliest record of any experiment or observation of the phenomenon is dated 10 April 1874 (Doc. 419; *TAEB* 2:46, 173). Edison's August 1874 account of his electromotograph (Doc. 476) appeared in the 5 September 1874 *Scientific American* and elsewhere, after he had filed his patent application (U.S. Pat. 158,787; see Doc. 463 n. 2).

7. No record of such observations has been identified.

8. The motion of a copper mass in a magnetic field creates eddy currents of electricity in the copper, which in turn have magnetic effects; this phenomenon was initially explored in a famous experiment in 1824 by the French scientist François Arago. Edison's "dodges" on this have not been identified. Cf. Doc. 1114.

9. Preece spoke about telephones twice at the Plymouth meeting of the British Association for the Advancement of Science, and in a meeting noted for sparse attendance the audience at his 17 August report was "crowded to excess" while his 18 August popular lecture audience "crammed almost to suffocation" its large hall. "The British Association," *Nature* 16 (1877): 341–42.

10. Preece emphasized the practical limitations of Bell's telephones and presented Edison as the one who was in the process of remedying the defects with his alternative designs. He mentioned speaking on a Bell telephone over 32 miles, then reported tests indicating Edison's could cover 1,000 miles. Preece 1877, 403–4.

## -1124-

*From George Bliss*

Chicago, Nov. 26 1877[a]

Dear Sir:

The statement sent you was correct and Barton's letter[1] was wrong. This fall I have been selling off the over stock accumulated at the agencies in the summer. September was a better month than July or August & October better still for actual sales to customers. You have no idea what a tremendous effect the hot weather, strike & change in price had on the business. For a while it looked to me as if the thing was literally going to smash. I pitched in and have sent out thousands of circulars all over the country & am advertising today in a hundred newspapers. My bills of every kind are enormous & sometimes it seems as if the public never would give down. The work done with the Pen is of the most beautiful character and is coming in from all sources from San Francisco to Maine & from Lake Superior to Louisiana. The samples which are coming over from Brussels & Vienna are wonderfully fine considering the short time they have had the Pens. The testimonials coming in are stronger than ever. In spite of all this people seem to have been mad for the Papyrograph. In the midst of all this you have neglected me— You promised that rotary press without fail also the ink attachment for check work and that 100 copy duplicating ink—[b] How on earth is a man to win in this contest unless you will help him? Please do something about these matters right away. I have more faith in the Pen than ever and it is certain to come into universal use. There has got to be a first class pen office in every large city with a good stock of Pens and first class penmen to do printing cheap & lots of it. We must have that rotary press. I have just learned after much experimenting how to advertise with effect. I shall be out this year including Beetle $[2]0,000$^{00}$/00[2c] but am going to have it back next. The scheme for a large company to work the business is progressing favorably & I hope will consumate. Today I had a hundred letters & eight sales. In time I shall run it up a 1000 letters a day. In the Pen you have got a certainty and why not take the time to keep me in the advanced field. It will pay you better than anything you have ever had. We have made an improvement in the Reed pen & for fast penman[d] it is going to be first class. How is it that the W.U. have ordered Gray to make 100 telephones? Richmond of Lansing has worked 230 miles with his telephone. I know you will scoop the field for a down right practical instrument and no man will be more delighted than myself but I beg of you to help me out with that rotary press  I cant come down there now to stay & see it done & must depend on you to not neglect me. All & more than was ever hoped for the Pen can be realized with lots of energy & hard work. Please write  Respectfully

Geo. H. Bliss  G.M.

ALS, NjWOE, DF (*TAEM* 14:422). Letterhead of Electric Pen & Duplicating Press. [a]"Chicago," and "187" preprinted. [b]"ink—" interlined above. [c]Illegible. [d]Obscured overwritten letters.

1. Not found.
2. Possibly $10,000.00.

## -1125-

*Technical Note: Telephony*

[Menlo Park,] Nov. 26th 1877

Speaking telegraph[1]

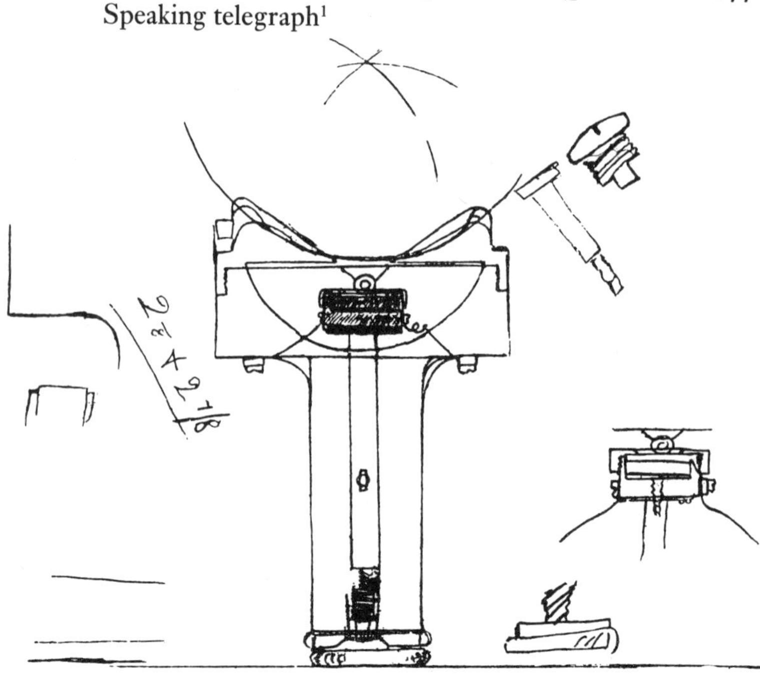

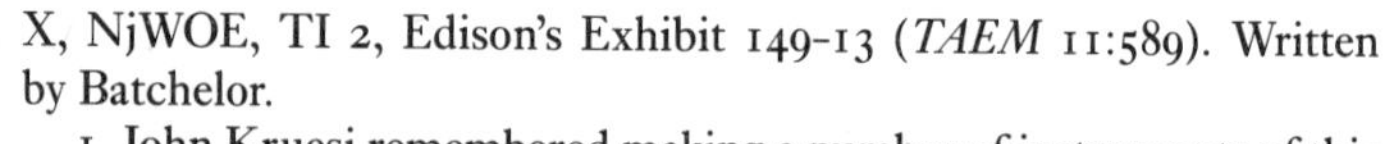


X, NjWOE, TI 2, Edison's Exhibit 149-13 (*TAEM* 11:589). Written by Batchelor.

1. John Kruesi remembered making a number of instruments of this design on or about 26 November. They were apparently made with different mouthpieces, as indicated in the drawing (cf. Edison's Exhibit 149-13 and Bentley's Exhibits Nos. 1 and 2, TI 2:531, 534–35 [*TAEM* 11:659, 662–63]). Batchelor began making 6 transmitters on 11 December and worked on them for at least a week (Cat. 1233:345–50, Batchelor [*TAEM* 90:225–28]); see also Batchelor's drawings of 1 December (Cat. 1317:51, Batchelor [*TAEM* 90:682]). The instruments had hard rubber cases and, according to Batchelor, worked "first rate in every respect, with the exception that a difference of temperature after talking a little while alter[ed] the adjustment slightly by expanding the case. There were, therefore, not a great many of these instruments made, the case shortly being changed to an iron one." In early 1878 Edison gave Henry Bentley, president of the Philadelphia Local Telegraph Co., a number of these transmitters, which were subsequently altered during experiments. Bentley found that they frequently got out of adjustment as "the hard rubber of which they were made expanded very greatly from the warmth of the hand, and contracted greatly after the warmth of the hand was removed from them, and they were laid down and not used for a short period, so that the rapid changes from heat and cold constantly interfered with the proper adjustment of the carbon button." Testimony of Batchelor, Kruesi, and Bentley, TI 1:238, 266, 317–23 (*TAEM* 11:96, 110, 135–38).

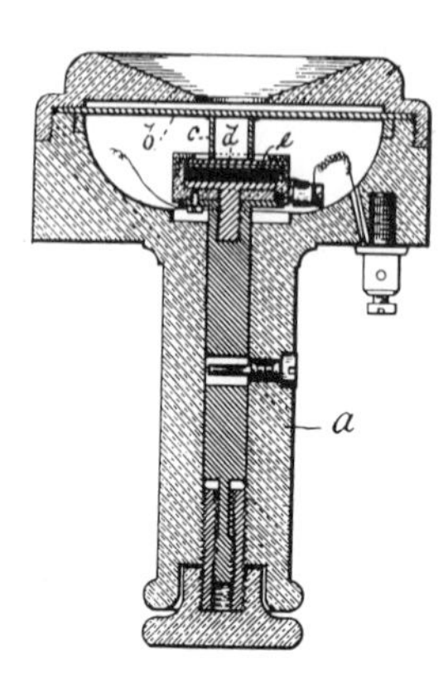


*One of the hard-rubber-case carbon transmitters based on the 26 November 1877 design and later modified by Henry Bentley.*

On 26 December, Batchelor drew up instructions for altering these instruments, which he probably gave to Murray when he took an order for twelve of the new design to Murray's shop in Newark on 2 January 1878. Murray, who delivered the instruments on 11 February, recalled that they were "larger and better proportioned" than the instruments made in November. Within a few days of their delivery Murray participated in successful, practical tests with these telephones over a line between New York and Philadelphia. During these tests Edison, Murray, William Orton, and George Phelps were in New York and Batchelor and Bentley were in Philadelphia. Vol. 13:201, Lab. (*TAEM* 4:146); Murray's testimony, TI 1:364 (*TAEM* 11:159); Cat. 1233:2, Batchelor (*TAEM* 90:54).

**-1126-**

*From James MacKenzie*

Detroit Nov 27th 1877[a]

My dear Al

I arrived at home on Sunday pm about 11 K and found everything OK. Mr Andrews informed me on Saturday that the consolidation of the Harmonic Dolbear and Edison Telephones under the name of the American Telephone Co really the Gold & Stock Co was actually completed and that instructions had then gone forward to the Western Electric at Chi to begin the manufacture of the different kinds now in working condition[1]

I enclose you a clipping[2] from the Detroit Evening news of the 23d inst for your edification. I have not yet put your Telephs in circuit have been too busy at other things but will put them on soon and rept to you the result. I am anxious to know what you have accomplished since I left. I am anxiously & confidently expecting to hear of your having struck a flowing Telephone Oil well and day & feel sure you will notify me of the same when done

I have spoken of the Musical concert business to my prominent friends and am assured of liberal support now Al I want you to take steps at once to have a complete outfit sent on by Mr Johnson or yourself and I will be willing to give any reasonable share of the profits to Mr Johnson or you as the case may be. If you will put this in shape for me immediately Al I will feel exceedingly obliged to you. when you get a choice Telephone just to suit you I hope you will send 1 along as you promised if only for a day or two. I have told some heavy yarns about you & want to give illustrative proof of same— present my regards to Mrs Dot and Tommy also Mr Adams & Mr Batchelor and do not loose sight of the fact that you owe to your family & yourself a duty which is to preserve your bodily & intellectual health to the best of your ability which can best be done by being temperate in all things & in your case paricularly I would ask you to get to work as early as you please quit for meals regularly on the stroke of the clock take sufficient time at your meals to enjoy your food and carry on enjoyable conversation with the members of your family shunning business topics if possible and devote your evenings to domestic pleasure and enjoyment—never work on Sunday! I suppose you think me officious but Al I assure you I do this with the best intention in the world & I feel sure you can and will benefit much if you will follow this advice— give me your visions on the subject when you write & let me hear from you soon   Yours Sincerely

J. U MacKenzie

ALS, NjWOE, DF (*TAEM* 14:997). Letterhead of American District Telegraph Co. of Detroit. [a]"Detroit" and "187" preprinted.

1. Elisha Andrews, former president of the Gold and Stock Telegraph Co., was president of the Harmonic Telegraph Co., which controlled Elisha Gray's acoustic telegraph and telephone inventions (see *TAEB* 1:153 n. 2). On 17 November the company completed an agreement with Gold and Stock to form the American Speaking Telephone Co., which was to control both companies' telephone patents. Gold and Stock was to have the exclusive right to manufacture, sell, and lease telephone instruments protected by these patents. An addendum to that agreement noted that "The Gold and Stock Telegraph Co will give the Harmonic Telegraph Company a private memo agreeing to secure to the American Speaking Teleph Co any patents or inventions of Edison relating to Speaking Telephones which they may acquire through the W. Union Teleg. Co." Gold and Stock also controlled Amos Dolbear's telephone patents (see Doc. 1043 n. 4). A 22 November agreement between Harmonic Telegraph, Gold and Stock, and the newly organized American Speaking Telephone formalized this consolidation of telephone interests. These agreements also included provisions for the Bell Telephone Co. to acquire a one-fourth interest in American Speaking Telephone in exchange for its patents; Western Union and Bell had been negotiating for more than a month. American Speaking Telephone Minutes, 3–5, 17–22; Gardiner Hubbard to Percival Richards, 13 Oct. 1877, General Managers Letterbooks, NjWAT.

2. Not found.

## –1127–

*From Thomas David*

[Pittsburgh, November 27, 1877?][1]

Dear Edison

This was tried in a small way by one of our men here—[2] my advice to you is not to say anything about this new appliance until you have made it successful, and then secure this claim— keep out of the newspapers

I sincerely hope you will accomplish the thing yet—

I am still experimenting with yours— I have it very conveniently mounted on a bracket, and am having little keys arranged so that we can shunt the Transmitters as we receive—

I want you to pull through on this thing— Regards to Batch

T.B.A.D.

Could you let me have a large Electromotograph machine to give a private amateur concert with?

ALS, NjWOE, DF (*TAEM* 14:994). See note 1.

1. Written on bottom and back of Doc. 1122.
2. See Doc. 1128.

**-1128-**

*To Thomas David*

Menlo Park Nov 28/77

Friend David=

Dont understand about one[a] your men trying carbon telephone=[1] What I mean by Carbon telephone is solid disks of either plumbago, Lampblack or Gas retort Carbon same as I have always used= I wish you were here I believe I have got the telephone now that is perfectly satisfactory in every degree; I use one cell of battery and the talking is twice as loud as Bells through 70,000. ohms; Bells being on short circuit= or in other words mine has the same loudness as Bells' when Bells is on 100 ohms & mine is on 140,000 ohms; I using one cell of battery. Yours

Edison

ALS, NjWOE, DF (*TAEM* 14:1002). [a]Interlined above.

1. See Docs. 1122 and 1127. In response to this letter, David wrote at the bottom, "What I mean is that he used carbon where you use plumbago. It was simply an experiment and not [very?] satisfactory nor well tried— . . ." DF (*TAEM* 14:1002).

**-1129-**

*From H. H. Duncklee*

New York, Nov 28/77

Mr Dear Sir—

Our mutual friend Mr McKenzie from Detroit has kindly given me a letter of introduction to you. He says you have a successful motor of sufficient power for running Sewing Machines and other light machinery and that your Battery is simple & easily cared for.[1] I am not an electrician but know from my connection with telegraph matters, that if you have such a motor that it can be made of value by proper handling— I have spent considerable money in that direction already and the drawback has been the question of battery. If you will please let me know on what days of say next week you will be at home and have a little leisure to show me your motor I would like to visit you.[2] Yours truly—

H. H. Duncklee[3] Sec'y Harmonic Telegraph Co.

ALS, NjWOE, DF (*TAEM* 14:187).

1. For the sewing machine motor see Doc. 802. The battery referred to is probably the battery used with the electric pen.

2. Edison wrote Duncklee on 29 November concerning the motor, and Duncklee wrote back on 6 December requesting an opportunity to see the motor when Edison had finished developing it. Nothing seems to have come of this. Duncklee to TAE, 6 Dec. 1877, DF (*TAEM* 14:197).

3. Nothing is known of Duncklee apart from his connection with the Harmonic Telegraph Co.

**-1130-**

*From Henry Law*

New York, Nov 28 1877[a]

Dr Sir

I sent you some time ago a piece of glass foiled in the hope that you would by electricity cut a sharp line in the foil for the purpose of glass ornamenting &c I suppose you have been so much occupied that you have not given it much attention. Could you spare time to call on me I could show you the work as now done by hand & you could get a better idea than in any other way— if in seeing it you thought it feasible I would myself spend a little money to make the preliminary trial for an interest in it. & since I started the idea I have heard of a party in Newark[1] who has carried the idea into the making of brass stencils. of these a great many are made for a variety of purposes. Made very largely but I doubt of his success as I had his address given me & I invited him to call & show us what he had with a view of buying but he does not call & I have heard no further of its success but it seems to me feasible & it would open a wide field which might be done by you with your experience which I doubt he can do. I would like to revive this whole matter with you, if you have time to attend to it, though I dont know how much there is in it.[2] Yours truly

Henry W. Law

It may be for Stencil work he is a head of us but if you can cut thru brass 1/16 in I would soon find out where he was & what he had & if he cant do it apply at once for a patent I may be mistaken but I really think it worth your attention

ALS, NjWOE, DF (*TAEM* 14:189). Letterhead of New York Sand Blast Works. [a]"New York," and "187" preprinted.

1. Unidentified.

2. Edison invited Law to Menlo Park the following week where they discussed the subject of stenciling glass as well as Edison's new phonograph. Law and Edison continued to correspond periodically about the stencil project over the course of the next year but nothing ever came of it. *TAEM-G1*, s.v. "Law, Henry A."; *TAEM-G2*, s.v. "Law, Henry W."

**-1131-**

*To James MacKenzie*

Menlo Park N.J. Nov 29/77

Dear Mac'=

Ive got it now without fail. Constant and good articulation with the regular Carbon but I work it on an entirely new principle. last night worked through 480 000 ohms with one cell of battery 480 000 is equal to sixteen thousand miles. It dont need adjusting and I think it will work on any line without trouble from noises— That Richmond telephone is the one I invented in 1875 as ~~you~~ I explained to you= loop circuits are bad tests= My carbon telephone was wkd from Montreal to Quebec with 50 cells callaud & conversation carried on by several gentlemen= more anon= Yrs

T A Edison

Ill take you advise about health=

ALS, NjWOE, DF (*TAEM* 14:1003).

**-1132-**

*From Franklin Badger*

Montreal Nov 29th 77

Dear Edison

I recd a letter from you yesterday and was overjoyed to find you are still in the flesh. It has been so long since you wrote me last that I began to think you had evolved into electricity yourself and had either gone to "earth" or been dissipated into space and I began to think I had better "test" to see where you was grounded, or by an electroscope to see if I could detect your presence in the atmosphere. However your letter has rendered this unnecessary and I am glad to know your solid flesh has not melted yet. If you have got your telephone "constant" there is a "chiel"[1] up here will be tickled about it, and after you are satisfied thereof just send them along and ~~I~~we'll sett them a buzzing. Don't go to the trouble just now of fitting up a "Singing Telephone" as I can wait your greater convenience. Yrs. "constant"

Fn. H. Badger

ALS, NjWOE, DF (*TAEM* 14:1004).

1. A Scottish word for a fellow, lad, or child.

**-1133-**

*Technical Note: Phonograph*

[Menlo Park,] Nov 29th 1877

Phonograph[a]

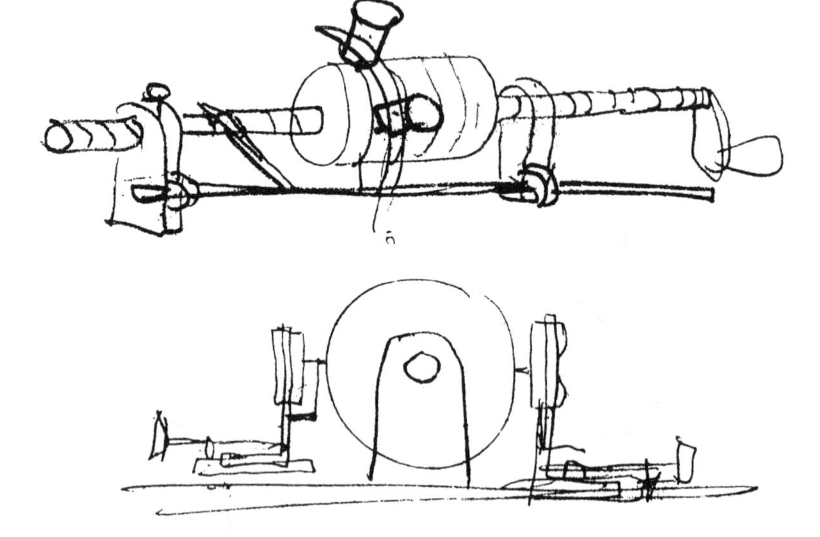

T A Edison

Chas Batchelor
J. Kruesi

X, NjWOE, Lab., NS-77-003 (*TAEM* 7:451). This document is photographically reproduced in *TAEM-G1*, following p. 30; Charbon 1981, 43; Frost 1969, 68; and Koenigsberg 1969, xiv. [a]Heading written by Batchelor.

## TIN-FOIL PHONOGRAPH Doc. 1134

According to payroll time sheets, John Kruesi spent the first six days of December making the first tin-foil, cylinder phonograph.[1] Charles Batchelor described the instrument in detail in his notebook.[2] On 7 December, Edison, Batchelor, and Edward Johnson gave the first public demonstration of the phonograph at the *Scientific American* office in New York.

In 1880, Edison either loaned or gave the first tin-foil phonograph to H. Stuart Wortley of the Patent Office Museum in South Kensington, London (now the Science Museum). The exact character of the Museum's acquisition became a point of contention in 1922 when the Edison Pioneers and the Association of Edison Illuminating Companies sought to establish a permanent collection of Edison's inventions. This led to several years of correspondence and, in 1927, negotiations with the British Ambassador to the United States that resulted in the phonograph's return to Edison at his West Orange Laboratory on 28 October 1928.[3]

1. 77-010, DF (*TAEM* 14:360–61).
2. Doc. 1140.
3. *TAEM-G2*, s.v. "Wortley, H. Stuart"; Nominal File 614, Museum Registry, UkLS; other material concerning the return of the phonograph is in DF for the years 1922–27.

**-1134-** [Menlo Park, c. December 1, 1877]

*Experimental Model: Phonograph*[1]

M (37 cm × 30 cm × 13 cm), NjWOE, Cat. 034.

1. See headnote above.

**-1135-**

*From Samuel Langley*

Allegheny, Penn'a.[a] Dec. 3. 1877

Dear Sir

Your letter[1] of Octr 29 came duly, but since its receipt I have been absent, and unable[b] to give it the attention your valued suggestions demand. The second suggestion, (that of replacing the thermopile by a very small coil of high resistance, whose small fluctuations of tempaerature are made sensible through a Wheatstone Bridge, & Galvanometer, recording the variations in the transmitted current of an independent battery) would, though apparently promising, hardly work for the gold leaf conductor would register the fluctuations in temperature of the surrounding air, ~~of neighbouring~~ etc as well as of the subject under study. This difficulty exists in the Bismuth-Antimony pile in some degree; but is eliminated by taking care that alien sources of heat shall[c] affect all parts of the pile alike,—those under examination, one face only and if possible for a very brief time.

It seems to me that the first idea involving the use of the ~~of the~~ accumulation principle is a very promising one indeed, (though the increased time it demands is a difficulty) and I would rather look for success in the direction of your first suggestion.

I have thought you might take sufficient interest in the subject to warrant my sending you one of my thermopiles; as it is of unusually delicate construction, containing as you see, forty elements each of about one-quarter the common size. The linear thermopile works through a slit about 0.50 in[c] by 0.01 in[c]; and should be a great deal more sensitive than even the one I send to do what I would like with it. If you could make something in the same compass say one hundred times as sensitive ~~and~~equally prompt, and capable of being comparatively uninfluenced by alien radiations you would not perhaps produce anything commercially paying, but you certainly would ~~do~~confer a precious gift on science

You can return the little accompanying[b] pile at anytime, or I will sometime call for it in passing. With thanks for your kind suggestions I am Yours truly

S. P. Langley[d]

P.S. I omitted to say in answer to your questions that 1[e] (in the linear thermopile especially) it is an object, though usually of secondary importance, to have the exposed area small. In my linear pile it is about 0.005 sq. in. [f]

2.[e] The heat from the grating (1st spectrum) is theoretically less than 1/10 that from a diathermanous prism.[2]

3.[e] I do not think we are warranted in saying that the heat in every color is the same, nor even that it is less in a given length of the violet than in the same length of the red using the normal spectrum. But I cannot either say (nor I believe any one at present) authoratatively that the latter is not true. We wish to determine these very points, and our only delicate instrument is the thermopile[3] S.P.L.

ALS, NjWOE, DF (*TAEM* 14:194). Letterhead of Allegheny Observatory. [a]"Allegheny, Penn'a" preprinted. [b]Obscured overwritten letters. [c]Interlined above. [d]Followed by "Over" to indicate page turn. [e]Circled. [f]"sq. in." interlined above.

1. Not found (but see Doc. 1095A).

2. That is, one made of a substance that transmits infrared radiation.

3. Sometime during December or early 1878 "the heat measurer was described and experiments shown to Professor Langley," who suggested "that it ought to be worked up as a very valuable boon to science, especially in the line he was investigating (that is, measuring the heat of the stellar spectra)." Edison then "tried many experiments with it." "Note by Professor Edison," in "Edison's Micro-Tasimeter," *Sci. Am.* 38 (1878): 385; clipping is Cat. 1240, item 681, Batchelor (*TAEM* 94:233).

## -1136-

*Technical Note: Electric Lighting*[1]

[Menlo Park,] Dec 3 1877

Electric Light

further experiments with Ruthenium & Silicon prove their adaptability to subdivision of the Light= It is possible to get good results by powdering the silicon & putting it in a glass tube passing the current through the powder[a]

T A Edison — Chas Batchelor

M N Force — J Kruesi

X (photographic transcript), NjWOE, Lit., *Sawyer and Man v. Edison,* Edison's Exhibit No. 5 (*TAEM* 46:266). "Copy of this on page 116, Vol 1 Exp Researches this 28 day of Sept. 1878 Wm Carman" written at top of page (that copy is in Cat. 994:116, Lab. [*TAEM* 3:227]); "3" written in upper left and right corners. [a]Followed by centered horizontal line.

1. See headnote, p. 540.

**-1137-**

*Technical Note: Phonograph*

[Menlo Park,] Dec 3 1877

Have tried lot of experiments with different thickness of tin foil. its the best material yet for recording.

This is better than direct=

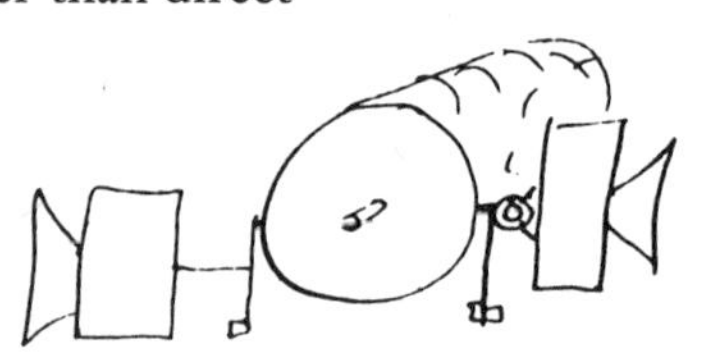

Trans[lating] Imbosser[1]

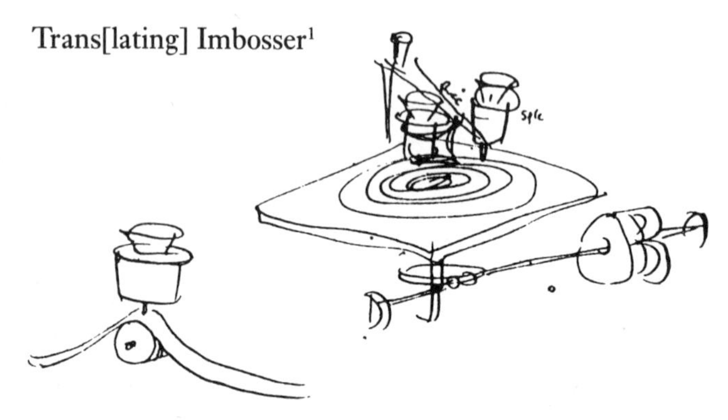


Tin-foil paper ie[a] tissue backed with tin foil=

T A Edison — Chas Batchelor

James Adams — ~~J Kruesi~~

~~G E. Carman~~ — ~~M. N Force~~

X, NjWOE, Lab., Vol. 17:22 (*TAEM* 4:891). This document is photographically reproduced in Charbon 1981, 48. [a]Circled.

1. Figure labels are "Rec" and "spk."

**–1138–**

*From George Bliss*

Chicago, Dec. 4 1877[a]

Dear Sir:

What have you done about the combination mentioned by Breckon in the enclosed letter.[1] If you fix it so stencil can be made by writing on a key board wouldnt we make it lively[b] fr printers? Respfly

Geo H. Bliss G.M.

ALS, NjWOE, DF (*TAEM* 14:424). Letterhead of Electric Pen and Duplicating Press. [a]"Chicago," and "187" preprinted. [b]Obscured overwritten letters.

1. In his letter to Bliss of 17 November 1877, Breckon wrote that Edison's patent combined the electric pen and the typewriter (DF [*TAEM* 14:425]). Edison's British patent for autographic printing stated that "In cases where the letters are made by a type writing machine with keys that are depressed to compose the words, the type are to be composed of numerous needle points arranged in the form of letters, so that the printing therefrom perforates the paper" (Brit. Pat. 3,762 [1875]). Edison did not pursue this idea.

**–1139–**

*Edward Johnson to Uriah Painter*

New York Dec 4/77

My Dr U.H.

I'm sorry to have seen so little of you when in Washn. but you disappeared just as the senate adjourned and I was too sick to essay a search for you in the coriders, or to venture an intrusion on the Ladies at the House— I had an Instrument for you & brought it away in my valise without once thinking of it. Will now send it to you by express.

I wanted to tell you that I have put myself in the hands of The American literary bureau of NYork (a 1st class institution) they to get me engagements on best terms possible none less than $50. & RRd & Hotel fares for self— As I have so many applications from churches &c who have felt that 100 to 125$ was too much, I expect so soon as they get their circulars out I shall have plenty to do especially as I agree to go anywhere—and shall within a week have a perfect working Edison Speaking Telephone to add to my attractions. I saw him at Menlo yesterday & he was just completing a 3 days trial of a new thing with most satisfactory results My predictions have thus been fulfilled I'll explain it to you— You remember the carbon disc arrangement—the trouble with it was that it had no margin of adjustment If it was adjusted up to a point where it would always retain the[a] same adjustment it was too weak to be heard distinctly & if adjusted the slightest degree away from that point to get more sound the adjustment would vary so as to render it impracticable. The problem then was to get sufficient volume of sound when adjusted to the ~~p~~staying point— this is what he had done & this is how he has done it—

Old plan[1]

in this you see the vibrations of the Diaphragm had to create sufficient variation of the resistance of the carbon to effect a variation of the total resistance of the line— Now as the variation of the carbon resistance is within the limit of one ohm—you will readily see that the effect upon the total of the line was ridiculously small & its only a marvel that any effect could be had upon the receiver at all—

This was especially so when the carbon was adjusted close to the Diaphragm—the only position in which it was reliable—

New Plan—[2]

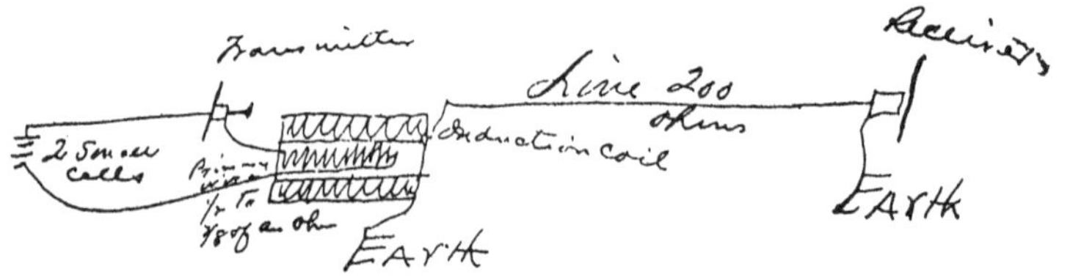

Now you see the transmitter being placed in the primary circuit of an ordinary induction coil which has a total resistance of less than 1 Ohm the variation caused by the slightest possible movement of the Diaphragm causes a variation of the ckt in which it is placed amounting to almost an absolute interruption— hence the inductive effect in the intensity circuit, which constitutes a portion of the line, is very great—& the line—having no battery on it is very readily affected. The result being a sound far greater than Bells because the Inductive currents used have their origin from a battery source (like those which annoy you on the Bell Telephone) and not as in Bells from a magneto source—[3] The use of th~~is~~ese inductive currents does not bring him in conf~~ict~~lict with Bell as he produces them with Battery and entirely outside of the magneto principle.— I am of the opinion too that this arrangement will work my motograph on a short ckt loud enough to be audible to an audience— If so (I will try it in a day or so) then am I truly happy—as such an exhibition would be a monopoly of the world—[4]

Edison is also constructing for me a Phonograph—which he is willing to bet me heavy, "is going to work perfectly satisfactory from the word go." with this in addition to the other things my fortune is made— The only trouble I have is to tide over the interval—& even that would give me none but for an unfortunate loan I effected a short time ago when I essayed to make a tour on my own hook—& which resulted in the series of mishaps of which you are aware— The money ($250.) is returnable on Jany 1st—and I am in the dark as to how it is to be done— The few engagements I may get this month will not net me sufficient over & above my living necessities to realize the sum— Now in view of what is in the near future for me—can you not either yourself or in the person of some heavier capitalist come to my relief by Jany 1st, with an amount that will enable me to meet this obligation If you will I will assign to you one tenth of the net amount I receive on each & every engagement I obtain— that will never be less than $5. which if they occur with ever so moderate a frequency will net a handsome return on the investment & if I do get even one of the additional features Edison promises me—the results would be "Bonanzian"—for I could upon giving one successful Exhibition of either the loud talking Telephone or the Phonograph, command any price I might, chose to ask for my Exhibitions

If you are sufficiently satisfied of the soundness of my Expectations to recommend any one to make the investment I have scarcely a doubt but that Gen Butler—who appreciates Edison's wonderful fertility, & I think is not unkindly disposed toward me,—would venture $500. as a "war risk" for the triple purpose of, an investment—lending a helping hand to me—and aiding me in establishing Edisons claims to Public recognition—

I contribute this "Hours loaf" without much dizziness of head as to Expectations of its early return in the form of Fishes & Baskets of bread—but rather[b] as a straw Yours Ever

E H Johnson

ALS, PHi, UHP. [a]Obscured overwritten letters. [b]Interlined above.

1. Figure labels are "20 cells," "transmitter," "carbon disc," "line say 200 ohms," and "Receiver."

2. Figure labels across the top are "Transmitter," "Line 200 ohms," and "Receiver"; and below are "2 small cells," "Primary wire— ½ to ⅜ of an ohm," "Earth," "induction coil," and "Earth."

3. Batchelor's diary for 7 December shows an induction coil and call bell in the circuit of a telephone tested on a line in New York that "worked much better than before." Cat. 1233:341, Batchelor (*TAEM* 90:223).

*A telephone circuit tested in New York, showing the call bell and induction coil with the receiver in the main line.*

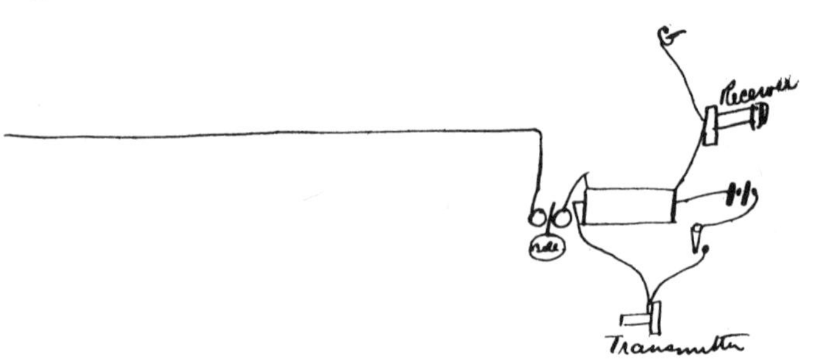


4. On 13 December, Johnson wrote Painter:

> Edisons Speaking Tel. is proving immense success—been on trial in W.U. 3 days—works through 125,000 ohms without apparent loss of sound—(going to import extra barrel ohms to see where it will stop)— Bells ceases to act at 40,000— we been talking backward & forward with Menlo Park (35 miles by W.U. wire—40 on Poles)—right along —Electricians dept compelled to acknowledge its superiority—against the grain too—the're all in favor of Bell as against E— All my prophecies are completely fulfilled in it— [UHP]

## -1140-

*Notebook Entry: Phonograph*

[Menlo Park,] Dec 4th 1877

Phonograph:

This machine we devised for the recording and reproduction of the human voice; it consists in moving a sheet of tinfoil in front of a diaphragm having an indenting point in its centre, which when vibrated by the voice indents the number of vibrations accurately on its surface. This indented sheet is afterwards moved in front of another diaphragm to which is attached a point on a delicate spring. The movement of the spring in passing over the indents on the tinfoil transmits to its diaphragm the rates of vibrations recorded there and the diaphragm gives forth the sounds originally spoken. The machine proper is shown in Fig

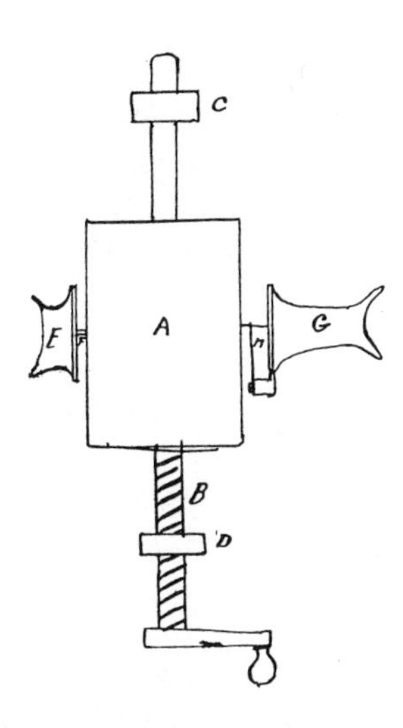


A is a cylinder fast on the shaft B. This shaft has a thread on one end which engages in bearing D the other end slides freely in bearing C

The cylinder A has a groove or thread cut on its face to allow the tinfoil which is put round it to be indented

E is the speaking diaphragm provided with an indenting point F. G is the reproducing diaphragm which receives its vibrations from spring H. This works well and the plain "How do you get that?" comes very plainly

Chas Batchelor.

X, NjWOE, Batchelor, Cat. 1317:52 (*TAEM* 90:683).

**-1141-**

*To Charles Darwin*[1]

Menlo Park N. Jersey U.S.A. Dec 7 1877.

Dear Sir,

Several small green colored insects were caught by me this summer having come into my laboratory windows at night. The peculiarity of these insects are that they give off when bothered an exceedingly strong smell of napthaline. No difference can be detected between the odor from the insect and the crystals of napthaline except that the odor from the insect is much more powerful. I suppose this odor is used as a means of defence like that from the skunk. I thought this would interest you if you were not already aware of such an insect. I could procure some next summer and send by mail If[a] you desire them[2] Yours

Thomas. A. Edison Telegraph Engineer

ALS, UkCU, CRD. [a]Obscured overwritten letters.

1. Charles Darwin, English scientific writer and thinker, was at this point world-famous. Noted for his contributions to evolution and natural selection, geology, and botany, he was most well known for *On the Origin of Species* (1859) and *The Descent of Man* (1871). *DSB,* s.v. "Darwin, Charles Robert."

2. Edison received a reply dated 5 January 1878 from Charles's son Francis. The young Darwin declined Edison's offer, saying that his father was already "at work on different subjects." DF (*TAEM* 15:176).

**-1142-**

*From Norman Miller*

[New York?,] Friday, Decr. 7/'77.

T.A.E.—

You will find with this my letter of Saturday last.[1]

The situation has not changed, and in the continued absence of Mr. O[rton]— you seem to be about the only friend that I have to rely upon.

Johnson tells us good things about the latest developments in the Telephone. Hope you will draw for last week and this today. Yours

N.C.M.—

ALS, NjWOE, DF (*TAEM* 14:199).

1. Not found.

**-1143-**

*Charles Batchelor to George Bliss*

[Menlo Park,] Dec 6[7]th[1] [1877]

Dear Sir,

Your letter of Nov 26th to hand[2] I have not drawn, as I have raised enough to last me a few weeks longer. Thanks all the same. I appreciate the necessity for the getting out of the "Rotary," the Ink attachment, and also the hundred copy ink; and have talked the thing over with Edison, and he has decided to start immediately he gets rid of the model telephone which I think will be very shortly now. We tried a new one of his in New York today and with much better results. You probably remember when you were down here about Edison's idea of recording the human voice and afterwards reproducing it. Well we have done it and have today shown it in New York to the Scientific American people who are now sketching the apparatus for a future issue[3] As a proof that the principle is correct it actually worked the first time we tried it. Wheeler & Mac were very much astonished to hear it speak out, if possible I will get a piece of the record on the tinfoil and send you tomorrow.

I am delighted to hear such good news from Europe  I felt sure from the first time I saw Beetle your judgement was sound in selecting him for such an important position.[4]

In speaking with Wheeler today he said that there were three accounts (for $4.97, $1.35 & $4.50) against me. The last time I saw Mr Henry we settled this matter, as I showed him Gilliland[s][a] receipt for $5.85 and this money I paid him covered two of these bills, and the other $4.97 of which I knew nothing, Henry said was not my bill, but something sent abroad by E.T.G.[illiland] to be charged by right to Bliss, but if you can get a letter to Henry he can settle that immediately. Mr Wheeler also wanted to know if Edison's account must be charged to me. I would rather this were not done as it mixes my matters so, if you will render bill to Edison of his account, I will get him to give you order on W.E. unless he wants me to take it on mine. I left with Mr Henry a short time ago a European Directory for which I paid (duty & freight included) $26 50/100  I now have an offer of $15— for it and if you do not want it at that price I will sell it, have you seen it? & do you want it?

Edison tells me that Herz did not pay those European patents  probably he thought them worthless  Well we will see!!!

By the way (between us) I think Herz is not very well liked over there  The 'Phonograph' is going to be a magnificent success!!!!!! Yours respectfully

Chas Batchelor

ALS (letterpress copy), NjWOE, Batchelor, Cat. 1238:237 (*TAEM* 93:172). [a]Edge of page.

1. According to his diary entry (Cat. 1233:341, Batchelor [*TAEM* 90:223]) and Edward Johnson's letter (Doc. 1147), the phonograph was not taken to the *Scientific American* office until 7 December.

2. Doc. 1124.

3. Charles Batchelor made an entry in his diary for Friday, 7 December 1877, in which he recorded, "Went to New York today . . . Took Phonograph to Scientific American." Cat. 1233:341, Batchelor (*TAEM* 90:223).

4. Batchelor meet Beetle in New York on 13 May when he "Gave him all foreign European letters and my answers to them together with a short account of all sales made by us." Cat. 1233:135, Batchelor (*TAEM* 90:120).

*Charles Batchelor to the Editor of the* English Mechanic

[Menlo Park,] Dec 7th [1877]

Dear Sir

Phonograph:— Mr Thos A Edison of New York a well known Electrician has just devised a method of recording and reproducing the human voice. It has the merit of extreme simplicity and is entirely a mechanical device.

A sheet of tinfoil is made to move in front of a diaphragm provided with an embossing point in its centre, at a uniform speed. When the diaphragm is vibrated by the human voice the ever varying rate of vibration is accurately recorded by indents in the tinfoil; this indented sheet is made to move at the same speed in front of, and in contact with, a delicate spring which is ~~also~~ connected to another diaphragm and to which it transmits the same rates of vibrations under the same conditions and in consequence gives forth the same sounds as those spoken in the first place. It has been exhibited for the last few days in New York and has excited the admiration of many scientific men.

The following sketch will probably explain it better:—

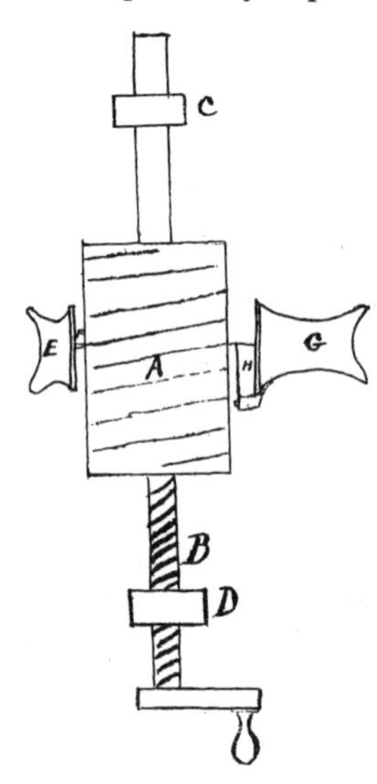


A is a cylinder fast on the shaft B, which has a thread on one end that engages in the bearing D whilst the other end is plain and slides in bearing C. E is a diaphragm with mouthpiece attached which can be adjusted to and from the cylinder, and having an indenting point fast in its centre. G is another diaphragm made similarly adjustable but receiving its vibrations from a delicate spring H which is in contact with the cylinder when ~~receive~~ reproducing. The surface of the cylinder A has a groove or thread cut in in order to give the stylus on E room to indent the tinfoil. The operation of the machine is as follows: The cylinder A is covered with a sheet of tinfoil securely held by a simple device at each end. The diaphragm E is adjusted so that the point indents the paper slightly all the time and as the handle is turned the cylinder revolves and moves forward at the same time, thus keeping a fresh place in front of the indenting point. When the cylinder has received all it will hold, and after the diaphragm E[a] has been adjusted back it is run back, and the diaphragm G with spring H is adjusted so that the spring is moved by every indent on the record. Now when the cylinder is moved forward again the plain "*How do you do*" astonishes everybody

After a sheet of tinfoil has been indented it can be reproduced almost any number of times and the impressions do not get worn down as anybody would naturally suppose.

Some of these sheets of tinfoil after having a sentence recorded on them have been straightened by Mr Edison and plaster casts taken of them In this state the indents made on the foil by the diaphragm form an interesting study.[1]

Asor—[2]

AL (letterpress copy), NjWOE, Batchelor, Cat. 1238:240 (*TAEM* 93:174). [a]Interlined above.

1. This letter was published in the *English Mechanic* 26 (4 Jan. 1878). Cat. 1240, item 324, Batchelor (*TAEM* 94:97).
2. Batchelor's wife's name was Rosa.

**-1145-**

*Edward Johnson to Uriah Painter*

New York Dec 7 1877[a] 1.39 pm

Phonograph Delivered to me today. Complete success Talks plainer than telephone inform Henry & Butler

E H Johnson

L (telegram), PHi, UHP. Message form of Western Union Telegraph Co. [a]"187" preprinted.

**-1146-**

*Technical Note: Telephony*

[Menlo Park,] Dec 7 1877

Speaking Telgh

Idea occurred to me that the plan I thought of for compensating for sympathy on Quad line from Pitts to Chicago[1] several months ago[a] would be just the thing for destroying the Extra noises on a Telephone wire. Tried experiment today in New York on a telephone wire to 23rd st from 198

Brdway placed one magnet on a disturbing circuit facing a similar magnet in the telphone ckt on working disturbing circuit induction due[a] to proximity of wires outside was neutralized by the induction of the magnet when in a particular adjustment Thus[2]

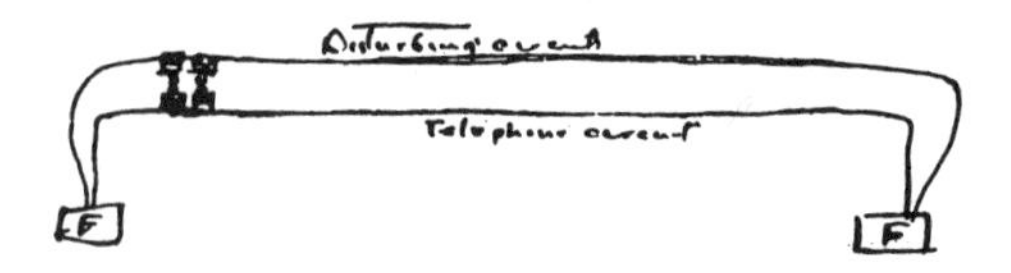

Compensation[3]

I propose to Carry out this principle of Compensation in telegraphs of all Kinds. If 4 lines disturb a particular one I use it thus[4]

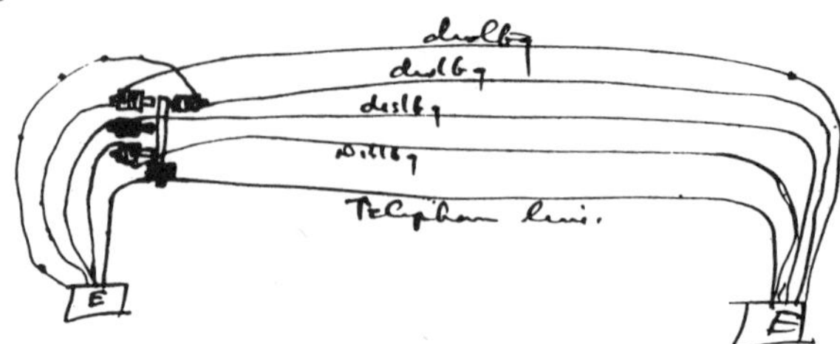

or

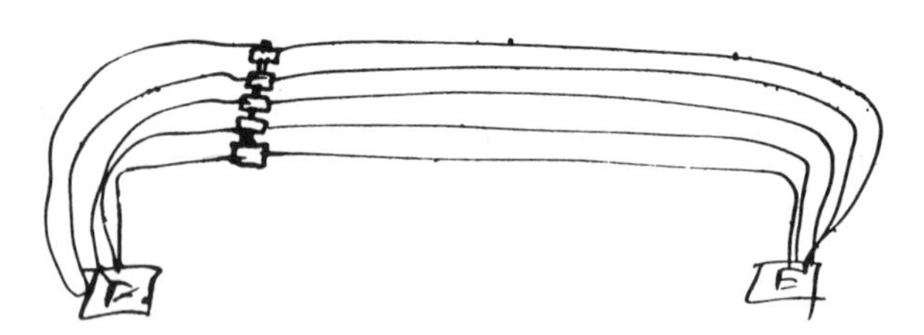

Quad line way stations

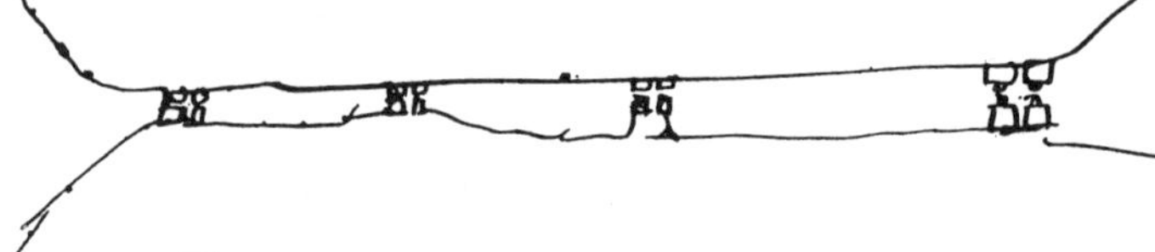

The compensation may even be in the helices of the signalling insts instead of the line of course the principle being Cureing one Evil by another "similia similibus Curantur[5b] it may be Carried out in numerous ways'=[6]

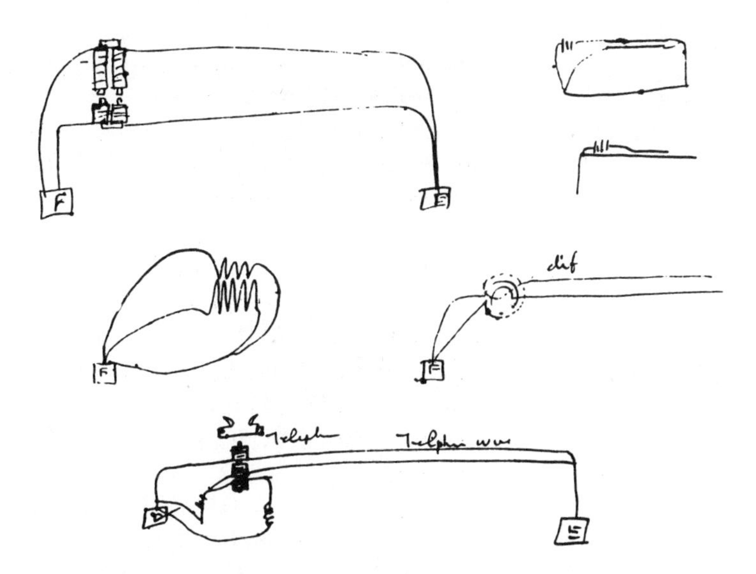


I think coarse wire must be used in magnet placed in circuit to be compensated for if such ck is of low resistance;

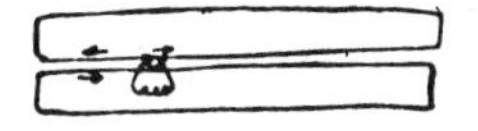

T A Edison

X (photographic transcript) and X, NjWOE, TI 2, Edison's Exhibit 169-13; Lab., Vol. 13:173; TI 2, Edison's Exhibits 168-13, 175-13 (*TAEM* 11:607; 4:123; 11:606, 608). Document multiply signed and dated. [a]Obscured overwritten letters. [b]Multiply underlined.

1. See Docs. 910 and 935.
2. Figure labels are "Disturbing circuit" and "Telephone circuit."
3. Edison drew another compensating circuit design on 10 December (Edison's Exhibit 176-13, TI 2 [*TAEM* 11:609]), and there is an undated sketch of one in Vol. 15:171 (Lab. [*TAEM* 4:474]). Edison also made a drawing for a patent model of compensating circuits on 9 December (Vol. 15:120, Lab. [*TAEM* 4:432]). He executed the application for this patent on 13 February 1878 (U.S. Pat. 203,019).
4. Figure labels are "dist[ur]b[in]g," "distbg," "distbg," "distbg," and "Telephone line."
5. "Like is cured by like," the basic principle of homeopathic medicine.
6. Figure labels are "dif"; and "Telephone" and "Telephone wire."

**-1147-**

*Edward Johnson to Uriah Painter*

New York Dec 8/77.

My Dr U.H.P.

The party from whom I borrowed the money is a timid young man[1] who boards in the same house ~~as~~that I do,[2] and is making things rather unpleasant for my wife who occasionally hears from one source or another that he expresses (evidently born of a fear) the opinion that he will lose it by my inability to pay. hearing of this I took him in hand and gave him to understand that inasmuch as his note was not due until Jany 1st he had better keep silent as to my action Else he might make a sure thing of what he evidently now deemed a probability. In view of this I am very anxious to meet the note promptly and count him out & some reasonable man in just at the time when the investment promises the best return. I saw him this Eve. & made the proposition to him that in as much as a Bird in the hand &— I would now give him $100. & renew my note for the remainder from Jany 1st to March 31st= or would let the thing stand as it is & trust to my Earning the whole amount by Jany 1st & liquidating his claim then. He at once chose the former plan—which is much preferable to me as it at once releases me from all obligations to pay him anything from the proceeds of my concerts as interest—and gives me 3 months more to raise the remaining principal.

Now what I propose to you is this. That instead of asking Butler or any one else to put up $500. you take the interest represented by that sum, & let me have a trifle occasionally as I may urgently need it. For Instance now—a check for $100. ~~to~~a/c of the note—(as your initiating fee)—the check for—15. asked for in todays letter for purchase of some stock for the Phonograph—[3] And such subsequent small individual instalments as the exigencies of my "Pursuit of the Hussy" may require. This will be an easy way for you to acquire an interest and will be of more real value to me than a lump sum as such are apt to fade quickly ~~in~~from my hands without leaving an adequate imprint.

It would be understood that I should call upon you only in cases of real emergencies & never for money for personal use.

The moral[a] result of such an arrangement would be good as between you and Edison—in this way. When I returned from Washn I told him how I came out whole He remarked "What was his object" Oh says I, "none thats his way" then I gave him some West Chester stories—and went on to demonstrate that all this was "Casting bread upon the Waters"—in that you were already reaping[b] fruit from some[c] seed planted in that little borrough.

This he could understand, and when I told him that I could call upon you as occasion might require for a few dollars in the same way he seemed to realize the beauty of the thing.

And to put the climax today he asked me for 50 cts to go get some lunch. I told him I hadn't had a penny this 3 days. After we "financiered" awhile & got the requisite "50" we lunched—& he there remarked that if he had the money to share he would buy some Brass tubing some stub steel[4] and clock work & make me an instrument that would talk to an audience—& one that would be more fit for Exhibition than the patched up & unreliable one he had (the 1st born) I remarked (nonchalantly) "You make me one & I'll find the stubb"— Quoth he why I thought you hadn't a penny"— Neither have I says I—"but I know where to get it"— Quoth he Reiff?— I—(deprecatingly)—No!— He—"Oh! Painter." I nod— All right says he. You get the stock— I'll do the work & give you a nice machine in time for your Exhibition on the 14th Inst. (given by me (at cost)[d] in connection with another concern for purpose of getting the notice from the N.Y. papers desired by my Literary Bureau for adv.)

My Idea is to put this thing in on that occasion create a sensation— then push things heavy & profit by the run— I don't mean to lose my head & go again into the thing on my own hook—but will of course see that my agents keep my price at its full market value— They will naturally do so however, as their pay is in percentage (10)

Thus you see Edison finds me equal to some degree of self help ~~support~~ in a case ~~in~~where I am to receive a benefit—and consequently of assistance to him. (2 Birds)

Its a drop—but its frequency will have the usual effect especially if he fails to see (as I mean he shall)—that other than the Philanthropical—and the "returning bread—causes produce the effect.

Now perhaps you would like to know something about the Phonograph— I'll try to depict it on the next page—

Phonograph

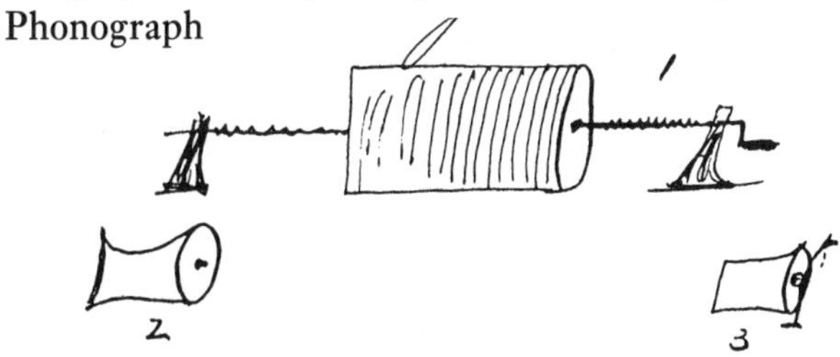


(Handsome?)

1 is a brass cylinder—having a spiral groove cut from end to end—& made to revolve by the hand crank—(must use a clock work for steady motion—else words are snapped out

like a fish womans) The thread on the shaft carries cylinder from one Post to other & back again —No. 2— is a Mouth Piece & vibrating diaphragm— The diaphram has a stylus or chisel point in its centre— No 3, is a similar contrivance Except that the Diaphragm is of paper, very easily vibrated—and the point is on a thin watch spring—the spring being fastened at the base and resting against the Diafram at the centre—

Now place a sheet of tin foil—quite heavy—around the grooved cylinder—& set No 2—so that the stylus rests on the foil with sufficient pressure to make a slight indentation if the cylinder is revolved— Now revolve No 1—& talk in No 2— result an indentation having all the characteristics of the vocal waves— The same No—per second—time the same degree of amplitude shown by depth of indentation—intensity & of[a] the same duration (form)—as shown by a long or short indent.— Quality Now take No 2 away— Put No 3 in its place adjust until point on end of spring is pushed back so that spring stands nearly perpendicular— Now grind— result is—indentations communicate to Paper Diafram all their peculiarities & of necessity produce in it a vibration—the exact ~~double~~ duplicate of that which made them—hence setting in motion the original[e] air waves—excepting only the loss resultant from mechanical defects— These defects extraordinary as it may seem exist in this 1st crude apparatus, in such modified form as to permit this invention to do what no other invention of Edisons ever did viz—Perform its functions practically upon the very first trial— Its talking is sufficiently plain with the hand crank for me to read it at once off hand without even stooping down to ear the Instrument—& Edison assures me that it is not to be compared with the effect when a clock work is used—as then the tone is accurately preserved & the utterances not only more distinct but the voice easily recognizable

I have a plaster cast of some of Edisons speech—from which I can at any time take tin foil impressions by the Dozen— So if you want "a doz of Edison" just send in your order & dont forget to say whether you want it dry or sour—The Times' funny article is realized within a fortnight—are facts stranger than fiction? Can the fools poke fun at inventive genius after such an experience—[5]

I'll come to Washn anytime you say after next friday if you "Pervide the Sinoos"[6]

The Telephone— This superanuated invention is O.K. at last. Edison claims it will work N.Y. to Washn Easily— When

I come down He agrees to try it with me on a WU wire— He is now working on G&S wires to get rid of induction— Not to make his telephone work over it—but to destroy it altogether and has got an idea—which he communicated to me this PM in strict confidence that demme if I don't believe will do it— Its so simple that it wont do to whisper it—till he has secured it— When secured—& tested practically & thus proven effective its application will be universally made. especially in Europe— &it can be done throughout the Telegraph system of the world in a fortnight It applies only to the Induction between wires—not between the earth & the wire thats another thing—so dont go wild over cables—

I've neglected to send your Instrument been so occupied with this other thing—but will do so monday sure—

The Scientific American was all ready to go to Press when I took the machine up there yesty. They stopped it—took a sketch of the machine made an engraving of it last night (engraver boards at our House & sat up all night in his room working on it) & will issue one day later in consequence Get a copy— get also last issue (Dec 15th)[7] & if you haven't got it—the one previous with my original article—[8] keep them.

If you want to know anything else—Say so.

—Yes I heard of it—& thought of you & Philadelphia at the time How's the family—Big little & medium Yours

E H Johnson

ALS, PHi, UHP. [a]Interlined above. [b]"a" interlined above. [c]Obscured overwritten letters. [d]"(at cost)" interlined above. [e]"the original" interlined above.

1. Unidentified; see Doc. 1139 for a discussion of this loan.

2. 101 Waverly Pl., New York. Wilson 1878, 703.

3. Five days later Johnson thanked Painter for his "big enclosure" and indicated that he had "consummated the bargain with it & now have that matter staved off till Apl 1st." Johnson to Painter, 13 Dec. 1877, UHP.

4. Stubbs steel was a fine spring steel made in England. Rufus Wilson, "Edison on Inventions," *Monthly Illustrator* 9 (1895): 342; in Unbound Clippings (*TAEM* 146:657).

5. A month earlier the writer of "The Phonograph" (*New York Times*, 7 Nov. 1877) had reported the announcement of the phonograph (Doc. 1102) and joked that, if true, this would lead to "bottled" speeches and sermons, which could be ordered by the dozen.

6. "Sinews of war," meaning ready money. Farmer and Henley, s.v. "rhino"; *OED*, s.v. "sinew," 4b.

7. An article in that issue of the *Scientific American*, titled "Machines That Hear And Write" (p. 376), included a brief reference to "purely mechanical means of registering sound, to which class belong the Edison and other phonographs."

8. See Doc. 1102.

-1148-

*From William Wastell*

Port Huron Dec 11/1877

Dear Sir

We had a meeting last evening the parties present James— Goulden, James Beard, Wm Stewart, and Wm Wastell, Mr Sanborn did not appear Cole was away in the woods and yourself absent—so that only (4) were present— the meeting was a talk over what Mr Beard had done and how well he had done it a report was read and accepted and Mr Beard talked and acted as if he was the sole owner of the whole road he offered a resolution that Antwine Marontate[1] be hired to take charge of road office act as Secty and run the machine—@ 20.00 pr week— it was discussed and closed by giving Beard authority to hire a manager of the road— I asked if a statement was made for each stockholder Mr Voorhees said it took too much time but he would send one to you— I had paid on receipt of Freight monies about One Hundred dollars pr month— and the earnings of the road as you will see are good— the expenses have been cut down to the lowest figure & I as far as I have had control have done the best I could working all ways to reduce expenses and increase earning Mr Beard caused the 25¢ fare at last train at night to be cut to 10¢ which was was unnecessary— passes are given to any one he sees fit and tickets given away freely he sells what he pleases at just what price he pleases without asking any questions. his road building is the worst botch I ever saw no mechanism or neatness being thought of with few exceptions his work is done over the second time and in some instances the third time. he goes to work without any regard to expense or knowlege of what he is about to do he started a ditch from St Clair river to the corner of Stone St to be put down 5 feet as it passed Wrights Store in Gratiot This was stoped he going East I used [means?][a] to stop the work since which time he has been at swords points with me This road if you and I controlled it will make (4) to (5) thousand dolars a year clear I believe or at very lowest estimate (3) thousand and you can set it down that with Beard running it he will not make a dollar Beard is as ignorant as he is [imprudent?][a] scarcly having a common laborers education— and is the worst specimen of a man I ever saw I am not discouraged I believe we shall yet control this road and make some money You and I control this road as far as stock is concerned and we should elect a Board of Directors to suit ourselves and make money it can be done W.P.E. promises me the Bus shall be put in shape to satisfy you and I if we could have the road run as it should be square, honest, without the appearance of trickery—and in a business like manner;

I believe the calculation is to drive you or me to sell so that the control would be out of our hands  I will never sell to place either your stock or my own in any other place but the control if I loose all— I believe you to be a true square man & had I had your confidence in years gone by you and I today would have made money in this road— I hold myself in position to act with you to use my best endeavors to make the most of our stock, in all cases I have endeavoured to make true plain statements without regard to anything but the truth.— and you will find every man in the road excepting J. P. Sanborn is ready to do anything to annoy you or me  we have the controlling stock and that is what hurts—but without the control your stock and mine would be worthless—

I consider it proper and necessary for me to write you this plain and positive opinion of the affairs of the road and time will prove that I have been at least honest in all transactions connected with it.

Goulden Beard Cole, Atkinson, Stewart [Linabury?][2a] are all mad because they think and know your stock with mine controls.

I send your last letter to W.P.E and he said he would write you and explain Omnibus— hoping you understand me and my position with you in this matter  I remain yours

W Wastell

| | |
|---|---|
| Earning of Cars for 13 weeks from Aug 4th to Nov 3d | 1863.49 |
| Earning for Carring Mail for same time | 1217.50 |
| Total[b] Earnings | 1990.99. |
| Average pr day 78 Days | 25.52 |
| Contra | |
| Wages of men in connection with Cars pr day | 7.50. |
| Feed of Horses in connection with Cars pr day | 2.25 |
| Oil &c | 25 |
| Average Cost pr day | $10.00[c] |
| Freight | |
| Earnings from March 10th to November 12 | $811.67 |
| 34½ weeks pr week | $23.53 |
| Contra | |
| Cost of Feed for 2 Horses pr week | 4.20 |
| One Mans wages pr week | 9.00 |
| Total for Frt Exp | $13.20 |
| Bal in favor of Frt pr week | $10.33 |

I give these figures so that you may see how statement will be when you receive it these are as near as I can judge in many places in statement when the expense of building road is made it should be tearing up and removing old road which I did the most off W

ALS, NjWOE, DF (*TAEM* 14:606). [a]Illegible. [b]Preceded by horizontal line. [c]Followed by '"over"' to indicate page turn.

1. Marontate was a Canadian-born lumber dealer who settled in Port Huron, where he managed the firm of William Sanborn & Co. He later formed a lumber partnership with W. B. and J. Hibbard. *History of St. Clair County, Michigan* 1883, 584.

2. Isaac Linabury owned 20 shares in the Port Huron Railway Co. Jenks n.d., 14.

**-1149-**

*To James MacKenzie*

Menlo Park N.J. Dec 12/77

Dear Mac'=

The Phonograph is a perfect success; and soon as I have cheapened the mechanism down a little, I want you to take charge of the whole bu~~i~~siness for th~~e~~is ~~C~~[---][a] & all other Countries; But you must think it over well and be ready as we must move quick The new telephones are also a big success; ~~I~~ Write me— See Scientific American of Dec 22[1] you will get it. next Saturday or Monday—[2]

Edison

ALS, NjWOE, Supp. II, 1877 (*TAEM* 97:601). This document is photographically reproduced in Frost 1969, 66. [a]Canceled.

1. Doc. 1150.

2. That is, on 15 or 17 December.

**-1150-**

*Article in the* Scientific American

New York, December [14?], 1877[1]

THE TALKING PHONOGRAPH.

Mr. Thomas A. Edison recently came into this office, placed a little machine on our desk, turned a crank, and the machine inquired as to our health, asked how we liked the phonograph, informed us that *it* was very well, and bid us a cordial good night. These remarks were not only perfectly audible to ourselves, but to a dozen or more persons gathered around, and they were produced by the aid of no other mechanism than the simple little contrivance explained and illustrated below.

Fig. 1.

The principle on which the machine operates we recently explained quite fully in announcing the discovery.[2] There is, first, a mouth piece, A, Fig. 1, across the inner orifice of which is a metal diaphragm, and to the center of this diaphragm is attached a point, also of metal. B is a brass cylinder supported on a shaft which is screw-threaded and turns in a nut for a bearing, so that when the cylinder is caused to revolve by the crank, C, it also has a horizontal travel in front of the mouthpiece, A. It will be clear that the point on the metal diaphragm must, therefore, describe a spiral trace over the surface of the cylinder. On the latter is cut a spiral groove of like pitch to that on the shaft, and around the cylinder is attached a strip of tinfoil. When sounds are uttered in the mouthpiece, A, the diaphragm is caused to vibrate and the point thereon is caused to make contacts with the tinfoil at the portion where the latter crosses the spiral groove. Hence, the foil, not being there backed by the solid metal of the cylinder, becomes indented, and these indentations are necessarily an exact record of the sounds which produced them.

It might be said that at this point the machine has already become a complete phonograph or sound writer, but it yet remains to translate the remarks made. It should be remembered that the Marey and Rosapelly,[3] the Scott, or the Barlow apparatus, which we recently described, proceed no further than this. Each has its own system of calligraphy, and after it has inscribed its peculiar sinuous lines it is still necessary to decipher them. Perhaps the best device of this kind ever contrived was the preparation of the human ear made by Dr. Clarence J. Blake, of Boston, for Professor Bell, the inventor of the telephone.[4] This was simply the ear from an actual subject, suitably mounted and having attached to its drum a straw, which made traces on a blackened rotating cylinder. The difference in the traces of the sounds uttered in the ear was very clearly shown. Now there is no doubt that by practice, and the aid of a magnifier, it would be possible to read phonetically Mr. Edison's record of dots and dashes, but he saves us that trouble by literally making it read itself. The distinction is the same as if, instead of perusing a book ourselves, we drop it into a machine, set the latter in motion, and behold! the voice of the author is heard repeating his own composition.

The reading mechanism is nothing but another diaphragm held in the tube, D, on the opposite side of the machine, and a point of metal which is held against the tinfoil on the cylinder by a delicate spring. It makes no difference as to the vibrations produced, whether a nail moves over a file or a file moves over a nail, and in the present instance it is the file or indented foil strip which moves, and the metal point is caused to vibrate as it is affected by the passage of the indentations. The vibrations, however, of this point must be precisely the same as those of the other point which made the indentations, and these vibrations, transmitted to a second membrane, must cause the latter to vibrate similar to the first membrane, and the result is a synthesis of the sounds which, in the beginning, we saw, as it were, analyzed.

Fig. 2

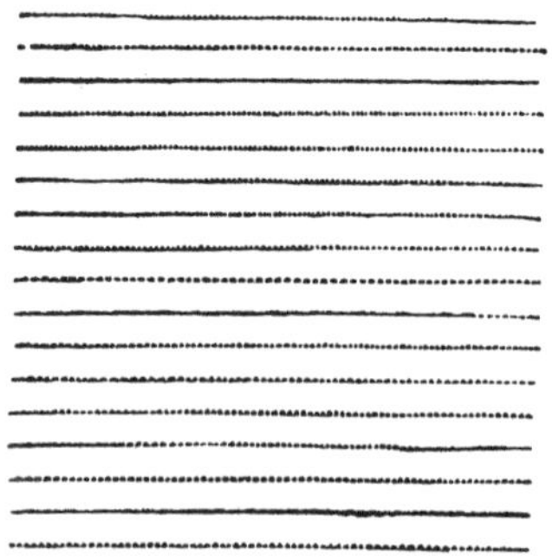

In order to exhibit to the reader the writing of the machine which is thus automatically read, we have had a cast of a portion of the indented foil made, and from this the dots and lines in Fig. 2 are printed in of course absolute facsimile, excepting that they are level instead of being raised above or sunk beneath the surface. This is a part of the sentences, "How do you do?" and "How do you like the phonograph?" It is a little curious that the machine pronounces its own name with especial clearness. The crank handle shown in our perspective illustration of the device does not rightly belong to it, and was attached by Mr. Edison in order to facilitate its exhibition to us.

In order that the machine may be able exactly to reproduce given sounds, it is necessary, first, that these sounds should be analyzed into vibrations, and these registered accurately in the manner described; and second, that their reproduction should be accomplished in the same period of time in which they were made, for evidently this element of time is an important factor in the quality and nature of the tones. A sound which is composed of a certain number of vibrations per second is an octave above a sound which registers only half that number of vibrations in the same period. Consequently if the cylinder be rotated at a given speed while registering certain tones, it is necessary that it should be turned at precisely that same speed while reproducing them, else the tones will be expressed in entirely different notes of the scale, higher or lower than the normal note as the cylinder is turned faster or slower. To attain this result there must be a way of driving the cylinder, while delivering the sound or speaking, at exactly the same rate as it ran while the sounds were being recorded, and this is perhaps best done by well regulated clockwork. It should be understood that the machine illustrated is but an experimental form, and combines in itself two separate devices—the phonograph or recording apparatus, which produces the indented slip, and the receiving or talking contrivance which reads it. Thus in use the first machine would produce a slip, and this would for example be sent by mail elsewhere, together in all cases with information of the velocity of rotation of the cylinder. The recipient would then set the cylinder of his reading apparatus to rotate at precisely the same speed, and in this way he would hear the tones as they were uttered. Differences in velocity of rotation within moderate limits would by no means render the machine's talking indistinguishable, but it would have the curious effect of possibly converting the high voice of a child into the deep bass of a man, or *vice versa.*

No matter how familiar a person may be with modern machinery and its wonderful performances, or how clear in his mind the principle underlying this strange device may be, it is impossible to listen to the mechanical speech without his experiencing the idea that his senses are deceiving him. We have heard other talking machines. The Faber apparatus for example is a large affair as big as a parlor organ. It has a key board, rubber larynx and lips, and an immense amount of ingenious mechanism which combines to produce something like articulation in a single monotonous organ note. But here is a little affair of a few pieces of metal, set up roughly on an iron stand about a foot square, that talks in such a way, that, even if in its present imperfect form many words are not clearly distinguishable, there can be no doubt but that the inflections are those of nothing else than the human voice.

We have already pointed out the startling possibility of the voices of the dead being reheard through this device, and there is no doubt but that its capabilities are fully equal to other results just as astonishing. When it becomes possible as it doubtless will, to magnify the sound, the voices of such singers as Parepa and Titiens[5] will not die with them, but will remain as long as the metal in which they may be embodied will last. The witness in court will find his own testimony repeated by machine confronting him on cross-examination—the testator will repeat his last will and testament into the machine so that it will be reproduced in a way that will leave no question as to his devising capacity or sanity. It is already possible by ingenious optical contrivances to throw stereoscopic photographs of people on screens in full view of an audience. Add the talking phonograph to counterfeit their voices, and it will be difficult to carry the illusion of real presence much further.

PD, *Sci. Am.* (n.s.) 37 (1877): 384.

1. The date on this issue of the *Scientific American* was 22 December, but issues were "for" the weeks ending with their listed dates and were apparently available before the start of the week (see Doc. 1102 n. 1). This issue had been ready for printing on 7 December but was delayed by a day to prepare this article; Edison may well have had a copy by 12 December (see Docs. 1147 and 1149).

2. See Doc. 1102. The publication of these two accounts led to considerable legal difficulties for Edison when he tried to obtain patents in some other countries. See, e.g., Lemuel Serrell to TAE, 16 Dec. 1878, DF (*TAEM* 18:845).

3. The apparatus was described in an article titled "Speech Automatically Transmitted in Shorthand by the Telegraph." *Sci. Am.* (n.s.) 37 (1877): 273.

4. Bruce 1973, 121.

5. Euphrosyne Parepa-Rosa and Teresa Caroline Johanna Titiens were both operatic sopranos. *DNB,* s.vv. "Parepa-Rosa, Euphrosyne"; "Titiens, Teresa Caroline Johanna."

**-1151-**

*From Park Benjamin*

N.Y. Dec 16th 1877

Dear Sir,

I send herewith pages of old edition of Appleton's Dictionary on Telegraphy. It is desirable to use just as much of this as you possibly can.

There is no need of retaining portions of less than half a page.

Larger pieces (the matter being stereotyped) can be reused at considerable saving of time.

I send pages in duplicate so that you can easily incorporate them in your manuscript.[1]

Will you please return me the Electricity article at your earliest possible moment?

My printer is now waiting for copy and I can give him none until that article is ready. Address me in future as below[2] Yours &c

Park Benjamin[3] per W.E.B.[4]

L, NjWOE, DF (*TAEM* 14:250).

1. Edison was writing the section on telegraphy for *Appleton's Cyclopedia of Applied Mechanics,* which was edited by Park Benjamin but not published until 1880. A draft in Edison's hand, docketed "Copy of Article for Park Benjamin 1878," is in DF (*TAEM* 17:33). On 26 March 1878 Benjamin wrote Edison to inform him that he needed the article by 6 April and Edison replied that he would try to send it by then (Benjamin to TAE, DF [*TAEM* 17:32]). The article as published follows Edison's manuscript very closely. However, sections on Cowper's writing telegraph and on telegraph construction, as well as a bibliography, were apparently added by Benjamin. As a result the article is signed "T.A.E. (in part)." Edison contributed a description of his inventions for the article on telephony written by Benjamin, and he may have contributed to the sections on electricity and the phonograph, also written by Benjamin (Benjamin to TAE, 4 Jan. 1878, DF [*TAEM* 17:2]).

2. 107 East Thirty-Fifth St., N.Y.

3. Park Benjamin, associate editor of *Scientific American,* was an author and patent attorney who wrote extensively on technical and scientific subjects. *DAB,* s.v. "Benjamin, Park."

4. Unidentified.

**-1152-**

*From Thomas David*

[Pittsburgh?,] Dec' 17" 77—[1]

Dear Edison—

I find that "twist"[2] will not avail me anything— I shall have to content myself in aiding you in getting your Telephone to do what we want— Push Batch up—

Go for that "compensater"! There is as much in that as in the Telephone almost— I will try it anytime you are ready, and I have a favorable place Yours Truly,

T. B. A. David

ALS, NjWOE, DF (*TAEM* 14:1007).

1. David had been at Menlo Park on the afternoon of 14 December. On 17 December, Batchelor finished making two telephone transmitters for him, possibly two of the new hard-rubber instruments. Cat. 1233:348, 351, Batchelor (*TAEM* 90:227–28).

2. Unidentified.

**-1153-**

*Edward Johnson to Uriah Painter*

New York Dec 17th 1877.

My Dear U.H.

Page 3—[1] was returned to me this am by J B Knight Secy Franklin Institute Pha in whose letter I accidentally enclosed it— I enclose it here— I dont remember suggesting to you to put singer in connection with Induction Coil—though since you suggest it, I think it would prove very effective If you use a Bell Telephone for receiver— I'll get you one in a day or so from Edison he is having quite a No of them made for his new speakers— The speaker continues to hold its own & now that he has got the principle almost any crudely constructed Instrument works 1st rate. He is putting all the small Telegraph instrument makers about town to work on his Telephone & will this week have quite a number out.[2]

The Phonograph grows slowly, man at work on it. The old one delights visitors to the Laboratory. The concert at Chick. Hall was very successful,[3] but papers although fully represented make no mention of it—excepting the Herald—& it only in a small way— Guess they have all been seen by the Bell People—

Edison has had a Hungarian in tow for several days, who is apparently completely captured. He is either a Count—or a representative of the Countess Touscas—[4] (thats the pronunciation) Took ~~to~~E—— to the U.S. Rolling Stock Co to ascertain his standing— Prest. that Co. told E. didn't personally know the man—but that he was introduced to him—the

Prest—by one of their heaviest[a] foreign stockholders—& that in the two years of their business relations he had invested for him over $200,000—that he either had an immense amt. of[b] his own—or some one else's money at command. In course of negotiation with E. it was necessary to send a Telegram to Europe which Edison was trying to boil down—expecting to pay for it himself—but the party wouldn't have it so— made a long one of it—& paid it, displaying a handfull of $1000. bills in so doing. at same time sent a message to the Countess Touscas signing it "Theodore"— Thats all E. knows about him. He's somewhat of a scientist—is very quiet & unassuming—& is quite carried away with Edisons Laboratory— Drove out there from NYork yesterday with a magnificent Tandem Cart to show a friend what a "~~n~~nest" he had found, and stayed all day following E. like a child through every nook & crannie of his establishment & enjoying every experiment apparently with the true scientific fervor— Talked Hungarian to his friend on the Phonograph & went as wild as boys on hearing it repeat it. Now with Edison in town today to pony up the money to bind the negotiation which are these[5]

He— to take out for E Patents on all E's inventions in all Foreign Countries (where the cost of patents you know is much greater than in our own)—paying all the costs therefor—in consideration of which

E—— gives him 1/20 interest in the product of the same—

A special agreement[b] is also being made with E—— by him—for use of E's Speaking Telephone in Belgium— It seems this party has a con~~ec~~cession from the Belg. Govt. on which he is organizing a sort of general Telegraph Burglar & Fire alarm system in Brussels—like this—

a central & several sub stations are located—with wires radiating to Houses, Stores, Offices, Courts etc etc—throughout city— a House—say 195—wants another—say 900— 195 calls a central or sub station & says want 900— He is at once connected by switch board with 900—& thus has a private wire for the time being to that particular point— Its a splendid Idea—as it gives a subscriber a private wire from his ofs or House to that of Any other subscriber, at will.—[6] His 1st Idea was to work Printing Instruments on it, but he saw the Bell Telephone & wavered—but owing to certain Inductive defects was left still in doubt— He has now seen the Edison—& is finally decided— He wants some one to go over with him & put it in operation (now your getting at the kernel)—& Edison has suggested me at $200. Gold per month—

If its all OK, it will be closed up within 3 days & I may start in a week or 10 Days— What do you think— I am starving here— The W.U. is antagonistic to E. in the Dept. I would want to go in.[7] There is nothing else for me to look to.

Some talk is again on foot about Automatic being revived in London Also some talk about my going to Paris[8] in somebodys interest with Electrical apparatus— I think automatic represented there would capture Europe—with a simple Perforator & Transmitter & Receiver— All in all my prospects for immediate usefulness to some one—seem to lie in direction of the rising sun— hence if I can make a bargain & go with this man I'll be on the ground to do anything that may be required of me

How!

E—— thinks his clock work will work all right— dont want the special one.[9] In order to make sure of getting this Instrument however I want to pay everything except the Labor If you can therefor send me $10. without inconvenience I want to repay E—— that amount borrowed from him to buy some parts which my money failed to cover— I want to go to Philada on Saturday next & would be glad to see you between that & Jany 1st— Expect to have the phonograph with me— All this of course providing I dont "go to Europe" Yours Truly

E. H. Johnson

ALS, PHi, UHP. [a]Inserted in margin. [b]Obscured overwritten letters.

1. See Doc. 1147.

2. No records exist of any such orders. However, about ten days later Batchelor sent Joseph Murray a model for an order for twelve carbon transmitters (the order was apparently formalized 2 January 1878). Vol. 13:201, Lab. (*TAEM* 4:146); Murray's testimony, TI 1:364 (*TAEM* 11:159).

3. The *New York Herald* of 15 December reported on the telephone concert, which was held on the previous evening at Chickering Hall. Cat. 1240, item 295, Batchelor (*TAEM* 94:91); also see Johnson to Painter, 13 Dec. 1877, UHP.

4. The Hungarian telecommunications promoter and inventor Theodore Puskas (1844–1893) was a business associate of Edison's for many years. He first came to the laboratory on 13 December and returned three days later. Cat. 1233:347, 350, Batchelor (*TAEM* 90:226, 228); Kenyeres 1969, s.v. "Puskás Tivadar."

5. This same day Edison signed two agreements with Puskas and George Bliss making Puskas his agent for the sale of patent rights to his telephone and phonograph in Russia, Spain, Italy, Austria, Germany, France, and Belgium (Kellow [*TAEM* 28:1195, 1204]). After seeing Edison on 29 December, Uriah Painter wrote Gardiner Hubbard that he

had sent Johnson to see Puskas to determine if they could purchase any European rights from him. Noting that Johnson was also to talk to Edison about his British telephone patents, Painter told Hubbard that "I believe a monopoly of Edison & Bell here, & in Europe would easily nett million dollars in next 5 year— It is the big thing to do, competition & legal contests only eat it all up for both, & I think I made a very strong impression on [Edison], to that effect. He will have time to trade on U.S. but must move promptly on Europe— Send me a plan for basis of monopoly in Europe" (Painter to Hubbard, 30 Dec. 1877, UHP).

6. Small, local telegraph exchanges serving groups of banks or other businesses existed in the 1870s. Alexander Graham Bell had already recognized the potential of a central switching station for his instrument, and rudimentary telephone exchanges had been established at a very few, scattered locations in 1877 by Bell licensees, but "the world's first commercial telephone exchange" would be established in New Haven, Conn., in January 1878. Tosiello 1979, 119–25; cf. App. 1.G18.

7. This refers to the Western Union electricians' antagonism toward Edison and his telephone. See Johnson to Painter, 13 Dec. 1877, UHP, quoted in Doc. 1139 n. 4.

8. For the Universal Exposition to be held there in 1878.

9. Edison was making Johnson a phonograph with clockwork for exhibitions. On 13 December, Johnson had written Painter that this phonograph was "going to cost me more than I thought as I have to buy a clock work movement (special style) which will cost me $25.— Edison will do all he can at the shop— He has it about half done now & is trying to make a clock work he has do but thinks it wont work—in which case I'll have to buy one already selected." UHP.

**-1154-**

*To Thomas David*

Menlo Park NJ Dec 22/77

Friend David[1]

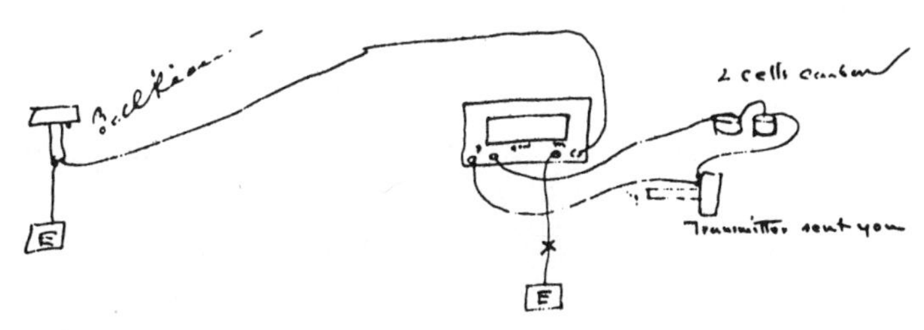


X at this point you might insert a Bell to receive on

Since I sent you the rubber transmitter we have improved the Lampblack by getting it purer and it has made it louder. Also have improved coil. Have also a Rubber receiver that is elegant and Monday or Wednesday will have another receiver with Є Induction Coil in the receiver=[2]

When you wish to take off cup to see diaphagm adjust screw way out, that will carry the cup away from diaphragm= (The new phonograph is finished except the clockwork & I have made a vast improvement=)[3] I speak & receive on one receiver & it is so loud that it can be heard the length of the laboratory I received you telegram tonight saying look for letter at 20 new church.[4] as luck will have it I suppose those things were broken or something is wrong so you cant make it work I hope not therefore I am anxious to see your letter Yours

T A Edison

If anything wrong dont make a report but write trouble & I will remedy it immediately=

L (photographic transcript), NjWAT, Box 1059 (Corp.).

1. Figure labels are "Bell Receiver—," "2 cells carbon," and "Transmitter sent you." The "X" is on the wire to earth at lower center. The rectangle within a rectangle represents an induction coil on a wooden base, as shown in Edison's Exhibits 155-13 and 177-13, TI 2 (*TAEM* 11:594, 610).

2. Edison sketched this idea on 4 December. On 26 December, Batchelor noted that a "new receiver with switch" had been finished. Edison's Exhibits 164-13–165-13, TI 2 (*TAEM* 11:602–3); Cat. 1233:360, Batchelor (*TAEM* 90:233).

3. This is probably the long-cylinder, single-diaphragm phonograph that Edison experimented with and displayed through the first half of 1878. Batchelor's testimony, pp. 590–91, *American Graphophone v. U.S. Phonograph* (*TAEM* 116:369).

4. Not found.

## -1155-

*From Park Benjamin*

New York, Dec 24 1877[a]

My dear Sir:

No one could be more disgusted than myself at the way Appletons have rushed me for this article—but their printers took a sudden spurt and finished all the matter on hand and of course demanded more copy—

You will have abundant opportunity however to add in new things, to the proof— Go on and write whatever you want to put in and send me the copy I will weave it into the proof which first comes back and then send you the final proof of the whole for examination—

There is no hurry at all about Telegraphy—only as soon as you decide upon having a cut send me the sketch or whatever you want it made from and I will send you back a proof of the engraving to paste in your MSS— In that way you can have about 40 days time—and the article when it comes from you can go at once to the printer without delay—

Any time within the next fortnight will do to send me the additions to the "electricity" article—

Your phonograph is a marvel— Perhaps you will gives us an opportunity to hear it in its improved form, at this office someday—Yours very truly

Park Benjamin

ALS, NjWOE, DF (*TAEM* 14:252). Letterhead of *Scientific American* Patent Office Department. [a]"New York," and "187" preprinted.

**-1156-**

*From Henry Morton*

Hoboken. N.J. December 24th/77

My dear Sir

I send you inclosed a translation of part of a letter which I have just received from my friend Dr. Schellen Director of the Technical School at Cologne.

He is the author of numerous work of high reputation, such as his great work on Spectrum Analysis; and is a very nice man in all respects. Anything you can do[a] for him will be esteemed a personal favor by  Yours &c.

Henry Morton.

P.S. I should like very much to visit your establishment myself if I knew how to get there and when I should be likely to find you[1]

ENCLOSURE[b]

Cologne Nov. 19, 1877.

Dear Sir—

I see in several of our papers that a Mr. Edison of New York has constructed a telephone differing somewhat in its arrangement from those of Messrs. Elisha Gray and Graham Bell.

Intending to treat very fully of the Telephone in my forthcoming public lectures on Sound, I am at present gathering all attainable information on the subject.[2] I am very desirous of becoming acquainted with Mr. Edison's form of the telephone and would like to obtain if possible such an apparatus for our Physical Museum.

I would be under many obligations to you if you will be kind enough to let me know at your earliest convenience whether Mr. Edison manufactures and sells telephones suitable for lecture purposes and what his address is. In order to lose as little time as possible it might be well, if agreeable, to inform Mr. Edison of my wishes so that he can write me directly.[3]

I am moderately well acquainted with English, so that Mr. Edison need not put himself to the trouble of writing in German.

H. Schellen

Director of the Realschule Cologne. Germany.

ALS, NjWOE, DF (*TAEM* 14:1010). [a]Repeated at end of one page and beginning of next. [b]Enclosure is a D (abstract), not in Morton's hand except for transcribed signature.

1. Edison evidently replied favorably on 29 December, suggesting that Morton bring Prof. Alfred Mayer of Stevens Institute along with him, which Morton agreed to do. Morton saw the phonograph demonstrated at the Western Union headquarters in New York in early 1878. Morton to TAE, 2 Jan. 1878, DF (*TAEM* 15:168); Cat. 1233:365, Batchelor (*TAEM* 90:235); Morton's testimony, p. 672, *American Graphophone v. U.S. Phonograph* (*TAEM* 116:410).

2. In early October, Elisha Gray had a set of his telephone instruments made for Schellen. Gray to George Barker, 3 Oct. 1877, GFB.

3. No direct response has been found, but Edison did agree to provide apparatus for Schellen. Morton to TAE, 11 Feb. 1878, DF (*TAEM* 15:233).

**-1157-**

*From William Preece*

[London?,] Dec 24, 77

Dear Edison

I have only time to say that I lecture before the Royal Institution on Feby 1st— Whatever you have to send me pray send by then.

I will make a great sensation over the talking phonograph and should like to bring it specially before the Royal Society.

I will see you put straight about the Quadruplex when my turn comes.

With all the compliments of the season I am Yrs very truly

Wm Preece

ALS, NjWOE, DF (*TAEM* 14:218).

**-1158-**

*Edward Johnson to Uriah Painter*

1 PM train 12/24/77

My Dr U.H.

My summary is "we can do it—but it will take active work"—

Bliss was out at Menlo all day yesterday[1] & He & E went in with me this am—[2] nothing had been concluded—but E. was very sanguine of what B. could do—& very skeptical of us.

The tender of 500 or even 1000 to prove "business" had no effect Bliss had heard a boast (I think this a legend)—that a Director of the Bell Co was reported to have made to effect that they owned the Phonograph—& had gone out to see E. at once. He has evidently got a strong hold on E. & has been working on his fears of effect dealings with Hubbard would have upon his W.U. contract & now brings up as his greatest obstacle the fact that if it was sold to Hubbard & should prove a success—W.U. folks would be mad because he hadnt offered them a chance— My reply to this was ~~to~~ "go right up there now & tell them you are offered a good price & ask them what they will give— (They wont give him 5000) This[a] he agrees to do— Then again I offered to make up a Co—outside of the Bell people—or at least with their names not immediately prominent—or again—to use a medium to transfer it from him to them & thus avoid the direct dealing by him with Hubbard= He called ~~b~~Bliss to the conference told him of our offer & sought his opinion Bliss took it sore, but apparently molified by E's assurance that he should be cared for— Edison took Bliss entirely into his bosom & gave him to understand that no scheme would be considered that did not count him in—

Even told me in his presence that he preferd to deal with Bliss—would in fact rather have the $30,000 Co & 15 per cent Royalty from Bliss than $100,000 cash from Hubbard—& yet he wants now an offer from us to purchase outright— says on any royalty basis he would have no guarantee that he would get his royalty. Wants a plump offer of $100,000 from us for outright purchase or $10,000 down and royalty of 15% & a share of the Stock of the Company— ~~or~~ He gave me the enclosed paper as a basis for you to work on & get him a definite proposition Bliss[a] read it—& says "if you can get a better offer than I have made you I wont stand in your way"

He is however no doubt now urging Edison hard not to antagonize Orton by dealing with his Enemies— You can see his game by the enclosed memo[3] handed E in the Pen ofs in N.Y.—& which E. gave me— I pointed out to E the value to him of forming a new money centre for future operations & in fact every argument I could use but no avail as against Bliss spectre of Ortons wrath— His response to my personal appeal was that he would get Bliss to do something for me If he dealt with him. (which he wont do)— Of course that is not what I want—& I stuck to him till I got him to promise to commit himself to no one till he got a square proposition from us—then if he could do better, we would withdraw— this I promised him he should have by Wednesday or Thursday—[4] you must come on & get it— I'll be in N.Y. Wednesday morning—

This thing must be ours in 10 Days or it will be beyond the reach of even 100,000— He holds now a proposition to purchase the right for Toys—from a party who propose to organize the Talking Toy Co[5]—with a capital of $100,000— Get that Item in the papers once & up goes the value of the invention to a quarter of million—

Now all this looks gloomy—but I know Edison—& I tell you if you will come and get things in shape & hand him a check for $10,000 the thing can be secured & I dont believe otherwise  I will have the machine in N.Y. to show—

Dont stop to ask if its worth $100,000— its worth <u>10 times</u> that amount

E—— has no faith in ~~me~~us—has known ~~me~~us too long without seeing any of our money— Devote ourselves to one successful operation which shall net him some real stuff & henceforth our task will be an~~y~~ easy one—& our fortunes made—but we have got to work hard to get the first prize

E—— proposes to use Bliss to transfer the thing to the Bell people—that is make simultaneous contracts with Bliss & us—to cover his tracks from the W.U.— (He is deadly afraid of that ghost)— Bliss says his personal interest in his own Co would be 1/5—or 20%— hinted that he would act in a transfer for 1/10— I guess he would do it for much less— If we cant do anything else we can run up the price & make E divide— thats worth your time— telegh me at 314 North 20th tomorrow when to meet you in NYork— I only alluded to telephone to show that this deal would secure that  Yours

Johnson

ENCLOSURE[b]

[Menlo Park,] ⟨12/24-77⟩[c]

Mr Bliss ~~who~~ has ~~sp~~ been interested for some time in the Phonograph and who is desirous of introducing it through the medium of a Company which he is about to from at Chicago= I said to Mr Bliss that if he could form a solid Co with not less than $30 000 in bank to open offices in all the principale cities and Carry a stock of Machines that I would ~~b~~feel disposed to let him work the Phonograph on a royalty of 15 to 21 per cent  selling price of machine $100. costing say $25 to make:[6] Mr Bliss has had my Electric pen for over one year and has established a permanent business in the same bringing me a large and sure royalty, and this in my eyes is a good guarantee of his ability to put the phonograph on the market in a legitimate manner which would if managed like the Electric pen prove the source of a large income to me; Mr Painter has had some conversation with me on the subject of forming a Company in New York to work the phonograph in the U.S. on a much larger scale than that contemplated by Mr Bliss and I have said that if such a Company could be organized and with the object of introducing the article instead of Stock manipulation, and would put up sufficient money that if Mr Bliss was properly recognized that I would be willing to enter into a contract w[-----][d] in[e] two or three ways.

1st to have $10 000 to conduct experiments a portion of the stock and a royalty of 15 per cent—

2nd to sell outright for a sum in cash.

This is the whole status=

I have a contract with the WU Telegraph which covers inventions useful upon the land lines & cables of the W.U. Co. & no other which contract will be shewn.

ALS, PHi, UHP. [a]Preceded by illegible canceled letter. [b]Enclosure is an AD written by Edison. [c]Dated by Uriah Painter. [d]Document damaged. [e]Obscured overwritten letters.

1. Bliss was also at the laboratory on the evenings of 16, 17, 27 December and the morning of the 28th. Cat. 1233:350–51, 361–62, Batchelor (*TAEM* 90:228, 233–34).

2. Johnson was also at Menlo Park on 22 December with Hilborne Roosevelt, a New York organ manufacturer and co-manager of the New York Telephone Co., which was introducing Bell's telephones. Cat. 1233:356, Batchelor (*TAEM* 90:231); *DAB*, s.v. "Roosevelt, Hilborne."

3. This second enclosure has not been found.

4. On Thursday, 27 December, Johnson returned to Menlo Park with Gardiner Hubbard and Uriah Painter. According to Batchelor, they "wanted Edison to bust his contract with W. U. Tel. Co." (Cat. 1233:361, Batchelor [*TAEM* 90:233]). In a 30 December letter to Hubbard, Painter said

> I went to see E. Saturday [29 Dec.] & find he has not seen O[rton]. yet. & will probably not before Wed. or Thursday— So he had no figures to make on Phonogph. but I saw some new changes he has made by which you speak close up to the diaphragm & the embossing being made more distinct, you now get the words far more clearly and boldly— He is still at work on it! [UHP]

Johnson had accompanied Painter on that visit also. On 31 December, Batchelor noted in his diary that "Kruesi took over phonograph & showed to Orton" (Cat. 1233:363, 365, Batchelor [*TAEM* 90:234, 235]).

5. Edison signed an agreement with Oliver Russell on 7 January 1878 which gave Russell the "exclusive license and right to manufacture" toys incorporating his phonograph. Kellow (*TAEM* 28:1064); see also Doc. 1119.

6. Edison made this proposition in a 19 December letter to Bliss in which he stated that "I should much rather place it in your hands than any of the eastern merchandising Companies working on royalties." DF, Supp. III (*TAEM* 162:1020).

**–1159–**

*Notebook Entry: Phonograph and Telephony*

[Menlo Park,] Dec 26th 1877

We tried a very interesting experiment today  Sent man to New Brunswick[1] to work one end of telephone during experiments we spoke a sentence to the Phonograph and afterwards held the transmitter to it as it reproduced them  the sentences were received perfectly at the other end by several people.[2]

Chas Batchelor

X, NjWOE, Batchelor, Cat. 1317:53 (*TAEM* 90:683). Written by Batchelor.

1. New Brunswick, N.J., is about eight miles (13 km) from Menlo Park.

2. Batchelor described this experiment in a 26 December letter to the *English Mechanic*, which he signed "Asor" (a reversal of his wife's name, Rosa). In his diary he noted that Theodore Puskas was at Menlo Park that day. Cat. 1238:247; Cat. 1233:360; both Batchelor (*TAEM* 93:179; 90:233).

**–1160–**

*Technical Note: Phonograph*

[Menlo Park,] Dec 28 1877

Phonograph

I propose to apply the principle of the Phonograph to various purposes such as cast wheels or wheels containing electrotyped embossed indented & other metallic bands containing a speech or sentence to be applied to Dolls & various toys to be turned by hand or by other motor  also I propose to apply these wheels to Clocks Watches to call the hour etc. for advertisements  for calling out directions automatically  Delivering Lectures, explaining the way  ~~deliver~~ as a musical instrument where sheets of indented or[a] embossed music may be sold ~~&~~cheaply by producing them by stamping from a d~~u~~ie or mould etc.

Gutta Percha sheet is 1st class for indenting

T A Edison

Chas Batchelor

J Kruesi

X (photographic transcript), NjWOE, Lab., NS-77-003 (*TAEM* 7:456).
[a]Interlined above.

**-1161-**

*To Frank Foell*[1]

Menlo Park N.J. Dec' 29/77

Dear Sir=

Your letter and Sketch received[2] am much obliged for the suggestion and it would be valuable but owing to the inconvenence of placing the paper on the Cylinder I had to adopt a revolving plate with a Volute spiral and have one now working Satisfactorily.[3] The reproduction of the voice is now absolutely perfect. Yours

Thos. A. Edison

ALS, MiDbEI, EP&RI.

1. Nothing is known of Frank Foell apart from his address on this letter's envelope (510 Second St., Philadelphia).

2. In a letter of 21 December (DF [*TAEM* 14:213]), Foell suggested ways to amplify the sound produced by the phonograph, which he had read about in *Scientific American* (Doc. 1150). Foell wrote to Edison again on 16 April 1878 (Unbound Documents, Batchelor [*TAEM* 92:312]).

3. Although few details about Edison's initial plate or disk phonograph are certain, he clearly based it on his translating embosser design and was thinking about it at least as early as November (Docs. 857 and 1119). Aspects of the design and construction of this machine may be shown in Doc. 1137, in undated material, and in drawings of 7 and 23 December 1877 (Vol. 17:7, 11, 23, 27, Lab. [*TAEM* 4:879, 881, 892, 897]). However, the undated items may relate to another design, as on 28 December Edison wrote of a "new plate machine" in which "I propose to detach the plate while rotating & without stopping clockwork or electric eng and provide the motor with work to do equal to that which is required to rotate the plate" (Vol. 17:30, Lab. [*TAEM* 4:899]; cf. Vol 17:38, Lab. [*TAEM* 4:907]).

**-1162-**

*From Thomas Jenkins*

Washington Dec 29th 77

My dear Sir—

I desire to apologise for leaving you so abruptly the other day, but as I had to return to N.Y by the 3 o'clock train to keep an engagement at New York, and then one at Baltimore the next day and one in this city immediately thereafter, I was compelled to catch the 3 o'clock train at Menlo Park. I must thank you for your attentions while at your laboratory and wish you a long life of success and honor—

Since coming here and having an interview with one of the attachés of the Government, I mentioned your invention of the Phonograph Telephone, Quadruplex &c and suggested that you be appointed one of the Scientific Experts to go to the Paris Exposition next year and report on the progress of Electrical Science— The suggestion was very well received indeed and I was asked to propose your name— I said I would write you about it—

The position of paid Commissioner for N.J. I think is provided for but one or two perhaps honorary commissionerships for that state are yet open— Let me hear from you by return mail—[1]

If you have not patented the Phonograph in France and other European countries, I would advise you to do it as soon as you think it is in a practical form and condition, for I think it is destined to play a very important part in the future— I would ask if I may send[a] a short report of it to a scientific society in Paris of which I am one of the Foreign Correspondents? or to M. Koenig the famous acoustician who is an old acquaintance and friend— I am yours truly—

Thos. E. Jenkins[2]

ALS, NjWOE, DF (*TAEM* 14:222). [a]Interlined above.

1. Edison's initial response was favorable. However, after Jenkins told him he should "at once" get testimonial letters from Frederick Barnard, George Barker, and Western Union official Norvin Green; from New Jersey's governor, senators, and congressional representatives; and from as many additional appropriate people "from New York, Phila, Boston and elsewhere" as possible, Edison declined. He discussed the possibility in a 7 January letter to Uriah Painter, and decided, "Lord If I have got to get all these letters to do it I dont want to be an honarary Comr or any kind of Comr." TAE to Painter (Jenkins to TAE, 4 Jan. 1878, enclosed), UHP; see also Jenkins to TAE, 5, 7, and 8 Jan. 1878, DF (*TAEM* 18:386–89).

2. Edison referred to Jenkins in a letter as a Lousiville chemist "who called at my laboratory some days ago with letter of introduction from Norvin Green= he is nice man all science." TAE to Painter, 7 Jan. 1878, UHP.

**–1163–**

*From Enos Barton*

Rochester N.Y. Dec 31st 1877

Dear Sir:

I expect to arrive in New York Wednesday morning.[1] I saw Bliss Saturday Evening on his way home as he passed through here

I should like to see your Telephone as early as practicable after I get to New York. If you will send word to me at 20 New Church St stating when & where I can see it and the phonograph I will make that my first business on arrival Yours,

Enos M Barton.

ALS, NjWOE, DF (*TAEM* 14:1016).

1. Barton wrote this on Monday. On Wednesday, 2 January, he was in New York, and on 4 January he was at Edison's laboratory all day. Cat. 1233:2, 4, Batchelor (*TAEM* 90:54, 55).

# Appendix 1

## *Edison's Autobiographical Notes*

In 1908 and 1909 Edison wrote a series of autobiographical notes whose origin and nature are described in the first appendices of Volumes One and Two.[1] Three of the documents discussed there contain sections related to events of the period of Volume Three; those sections are published here.[2] Notes datable only as pertaining to sometime during his years at Menlo Park are also included. Edison sometimes referred in the same paragraph to the periods covered by more than a single volume; such notes will be reprinted as appropriate. Each document has been designated by a letter and each paragraph sequentially numbered.

1. A general discussion of these is in *TAEB* 1:627–28 and specific information on six of the autobiographical documents, designated A–F, follows there. A seventh autobiographical document, G, is described in *TAEB* 2:781. An eighth autobiographical document, presumably created in the same period, has not been found, although it survived long enough to be used in Josephson's 1959 biography of Edison (Josephson 1992; it is cited as "Edison's Notes for Meadowcroft, Book I"). It has been identified only through quotations from it that differ significantly in subject or wording from the entries in any other document; however, several of its sections closely overlap portions of document A.

2. The autobiographical documents designated A, C, E, and F do not refer to the period of this volume.

### B. FIRST BATCH

The following is from a typescript that Edison revised ("Pencil indicates Mr. Edison's revision" is written on the front); see Volume One, p. 646. Two of its 81 paragraphs ostensibly deal with the period covered by this volume; however, there is no evidence to corroborate the date of section 51, and in section 59 Edison misdated his acoustic transfer telegraph experiments and the visit of William Thomson to 1879.

[51][a] When I started at Menlo Park, I had an electric furnace for melting rare metals I did not know about clearly. I was in the dark room, when I had a lot of chloride of ~~silver~~sulphur,[b] a very corrosive liquid. I did not know that it would decompose by water. I poured a beaker full of water, and the whole thing exploded, and threw a lot of it into my eyes. I ran to the hydrant and leaned over backwards, opened my eyes and ran the hydrant water right into them. But it was two weeks before I could see.

REMINISCENCE OF KELVIN.

[59] The first time I saw Lord Kelvin he came to my laboratory at Menlo Park in 1879. I was then experimenting with sending eight messages simultaneously over a wire by means of synchronizing tuning forks. I would take a wire with similar apparatus at both ends, and would throw it over on one set of instruments, take it away and get it back so quickly that you could not miss it; thereby taking advantage of the rapidity of electricity to perform operations. On my local wire I got it to work very nicely. When Sir William Thomson came in the room he was introduced to me and had a number of friends with him. He said "What have you here?" I told him briefly what it was. He then turned around to my great surprise, and explained the whole thing to his friends. Quite a different exhibition was given two weeks later to another well known Englishman also an electrician,[c] who came in with his friends, and I was trying for two hours to explain to him and failed.

TD (transcript), NjWOE, Meadowcroft. [a]This is the second paragraph following the heading "Four Instances of Personal Danger." [b]Correction in pencil. [c]"also an electrician," interlined above in pencil.

## D. BOOK NO. 2

This undated notebook contains a mix of narratives, questions, and notes in Edison's hand; see Volume One, p. 648. Ten items pertain to the period covered by this volume; section 225 also refers to work with an elevated railway years later.

[167] Trouble with Bear

[225] When TCM 1st met u fall 1877 u wr rather deaf

whats true story of beginning of this deafness its attributed to injury to ear drum by angry conductor or train official in 1877 it didn't interfere in any way with work on telephone or phonogh nor with tests for elevated RR stoping noise.[a]

[248] Kelvin says referg to his visit Phila Centennial 1876— "saw Es auto telgh delivering 1015 words in 57 sec Can you tell us anything abt this exhibit it was never referred to before—

[249] Have you any personal reminiscences of taking up telephone—

[253] ~~K~~ Killing potato bugs Bisulphate[b]

[292] 1st Exhibit in Scientific Ams ofs[b]

[304] Invented Hello for telephn[c]

[328] One week blinded kby putting water in Chl Sulpher[c]

[340] 1876— Exhibted 8 message at once over wire tuning forks Foreign electns couldnt grasp Kelvin did at once & commenced Explain it to his friends—[b]

[341] [---] Bear got loose Menlo—[b]

AD, NjWOE, Lab., N-09-06-27. [a]Followed by centered horizontal lines. [b]Entry overwritten with a large "X". [c]Entry overwritten with a large check mark.

## G. MR. EDISON'S NOTES

The following is a transcription of relevant portions of a typescript titled "Mr. Edison's notes in Book No. 2," probably prepared by William Meadowcroft in 1908 or 1909 (see Volume Two, p. 781). Six of its sections either involve the period of Volume Three to some extent or refer to Menlo Park but are of otherwise indefinite date. Another section, 19, which confuses together accurate and highly inaccurate points relating to widely disparate years, is included here for its reference to Jay Gould's conflict with Western Union Telegraph Co.

[17] In 1876, I started again to experiment for the W.U. and Mr. Orton, this time it was the telephone. Bell invented the first telephone, which consisted of the present receiver, which was used both as a transmitter and a receiver. It was attempted to introduce it commercially, but failed on account of its faintness and the extraneous sounds which came in on the wires from various sources. Mr. Orton wanted me to take hold of it and make it commercial. As I had been also working on a

telegraph system employing tuning forks simultaneously with both Bell and Gray, I was pretty familiar with the subject. I started in and soon produced the carbon transmitter now universally used (see final litigation on telephone—Berliner—Judge's opinion).

[18] Tests were made between N.Y. and Phila. Also between N.Y. and Washington, using regular W.U. wires. The noises were so great that not a word could be heard with the Bell receiver when used as a transmitter between N.Y. and Newark. Mr. Orton and W. A. Vanderbilt, and the Board of Directors witnessed and took part in the tests. The W.U. then started in to put them on private lines. Mr. Theodore Puskas of Budapest, Hungary was the first man to suggest a telephone exchange and soon after exchanges were established. The telephone department was put in the hands of Hamilton McKay Twombly, Vanderbilt's ablest son-in-law, who made a success of it. The Bell Company in Boston, also started an exchange, and the fight was on the W.U. pirating the Bell receiver and the Boston Co. pirating the W.U. Transmitter. About this time, I wanted to be taken care of. I threw out hints of this desire. Then Mr. Orton sent for me. He had learned that inventors didn't do business by the regular process and concluded he would close it right up. He asked me how much I wanted. I had made up my mind that it certainly was worth $25,000, if it ever amounted to anything for central station work, so that was the sum I had made up my mind to stick to and get obstinate; still it had been an easy job and only required a few months and I felt a little shaky and uncertain. So I asked him to make me an offer. He promptly said he would give me $100,000. All right I said, it yours on one condition and that is that you do not pay it all at once, but pay it to me at the rate of $6,000. per year for 17 years—the life of the patent. He seemed only too pleased to do this and it was closed. My ambition was about four sizes too large for my business capacity and I knew that I would soon spend this money experimenting if I got it all at once, so I fixed it so I could'nt. I saved 17 years of worry by this stroke.

[19] Soon after, the Page patent, which had been in the patent office for years was finally issued. It covered the use of a magnet contact point and sub-magnet. There was no known way, whereby this patent could be evaded and its possessor would eventually control the use of what is known as the relay and sounder and this was vital to telegraphy. Gould was pounding the W.U. on the exchange, disturbing its railroad

contracts and being advised by his lawyers that this patent was of great value, bought it. The moment Mr. Orton heard this, he sent for me and explained the situation and wanted me to go to work immediately to see if I could'nt evade or discover some other means that could be used in case Gould sustained the patent. It seemed a pretty hard job, because there was no known means of moving a lever at the other end of a telegraph wire except the use of a magnet. I said I would go at it that night. In experimenting some years previously, I discovered a very peculiar phenomenon, and that was that when a piece of metal connected to a battery was rubbed over a moistened piece of chalk, resting on a metal connected to the other pole, that when the current passed the friction was greatly diminished, and when the current was reversed the friction was greatly increased over what it was when no current was passing. Remembering this, I substituted a piece of chalk rotated by a small electric motor for the magnet and connecting a sounder the combination claim of Page was made worthless, a hitherto unknown means was introduced in the electric art. Two or three of the devices were made and tested by the Company's expert. Mr. Orton, after he had me sign the patent and got it in the Patent Office, wanted to settle for it at once. He asked my price. Again I said—make me an offer; again he named $100,000. I accepted providing he would pay it at a rate of $6,000 a year for 17 years. This was done, and this with the telegraph money received $12,000. yearly for the period from the W.U. Tel. Co.

[24] In 1877, I invented the phonograph. The invention was brought about in this way. I was experimenting on an automatic method of recording telegraph messages on a disk of paper laid on a revolving platten, exactly the same as the disk talking machine of today. The platten had a volute spiral groove on its surface, like the disk. Over this was placed a circular disk of paper, an electromagnet with an embossing point connected to an arm travelled over the disk and any signals given the magnets was embossed on the disk of paper. If this disk was removed from the machine, and put on another similar machine provided with a contact point, the embossed record would cause the signals to be repeated into another wire. The ordinary speed of telegraphic signals is 35 to 40 words a minute, but with this machine several hundred words were possible. From my experiments on the telephone I knew of the power of a diaphragm to take up sound vibrations, as I had made a little toy which when you recited loudly in the funnel

would work a pawl connected to the diaphragm and this engaging in a ratchet wheel served to give continuous rotation to a pulley. This pulley was connected by a cord to a little paper toy representing a man sawing wood. Hence, if one shouted Mary had a little lamb, etc., the paper man would start sawing wood I reached the conclusion that if I could record the movements of the diaphragm properly I could cause such record to reproduce the original movements imparted to the diaphragm by the voice and thus succeed in recording and reproducing the human voice.

[25] Instead of using a disk, I designed a little machine using a cylinder provided with grooves around the surface. Over this was to be placed tin-foil, which easily received and recorded the movements of the diaphragm. A sketch was made and the piece work price $18. was marked on the sketch. I was in the habit of marking the price I would pay on each sketch. If the workman lost, I would pay his regular wages; if he made more than the wages he kept it. The workman who got the sketch was John Kreuzi, who in after years became Chief Engineer of the General Electric Company. I did'nt have much faith that it would work, expecting that I might possibly hear a word or so that would give hope of a future for the idea. Kreuzi, when he had nearly finished it, asked what it was for. I told him that I was going to record talking, and then have the machine talk back. He thought it absurd. However, it was finished, the foil put on; I then shouted Mary had a little lamb, etc. I adjusted the reproducer and the machine reproduced it perfectly. I never was so taken back in my life. Everybody was astonished. I was always afraid of things that worked the first time. Long experiments proved that there was great drawbacks generally found before they could be got commercial, but here was something that there was no doubt of.

[26] I worked at it all night and we fixed it up to get the best results. That morning I took it over to N.Y. and walked into the office of the Scientific American, walked up to Mr. Beech's desk and said I had something new to show him. He asked what it was. I told him I had a machine that would record and reproduce the human voice. I opened the package set up the machine and recited Mary, etc., then I reproduced it so it could be heard all over the room. They kept me at it until the crowds got so great that Mr. Beech was afraid the floor would collapse and we were compelled to stop. The papers next morning contained columns. None of the writers seemed to understand how it was done. I tried to explain it was so very

very simple, but the results were so surprising that they probably made up their mind beforehand that they could never understand it, and they did'nt.

[37] At Menlo Park one day a farmer came in and asked if I knew any way to kill potato bugs; he had 20 acres of potatoes and the vines were being destroyed. I sent men out and culled two quarts of bugs and tried every chemical I had to destroy them. Bisulphide of Carbon was found to do it, instantly. I got a drum and went over to the potato farm and sprinkled it on the vines with a sprinkling pot; every bug dropped dead. The next morning the farmer came in very excited and reported that the stuff had killed the vines as well. I had to pay $300. for not experimenting properly.

TD (transcript), NjWOE, Meadowcroft. Because this transcription of Edison's manuscript is presented only as a reference text, typographical errors have not been reproduced or noted.

# Appendix 2

## *Charles Batchelor's Recollections of Edison*

The following are two of five reminiscences written by Charles Batchelor concerning his work with Edison. He entered them in one of his daily journals (Cat. 1339) between October 1906 and February 1908. Although they were published in Welch 1972 (app. 1), we have decided to present new transcriptions of them in the appropriate volumes of this work.

The recollections presented below discuss Edison's invention of the phonograph (in which Batchelor conflates events from July through December 1877) and his attitude regarding an 1876 award for his electric pen.

### #1 THE INVENTION OF THE PHONOGRAPH

This occurred at Menlo Park N.J. in the Edison Laboratory, about the middle of the month of November 1877. I was Mr. Mr Edison's chief assistant at that time and had been so for some years— We had been at work off & on for years previous to this time and had developed a system of automatic telegraphy, one of the instruments for which consisted of a rapidly running small wheel carrying forward a strip of paper, with a stylus resting on it to record chemically the dots & dashes that came over the line— Some of these instruments we had in the laboratory & much of the paper— We had also for a long time been developing the 'Edison Carbon telephone,' an instrument in which a diaphragm was made to put a varying pressure upon a button of pressed carbon by the vibrations produced by the human voice— Many of these instruments were in the laboratory at the time and we used them daily— Some years previous to this date we had designed and made some machines for coating paper with parafin (similar to the paper now used to wrap candy in) for making condensers for Electrical work and a large lot of variable thicknesses of this paper coated and uncoated was stocked away in the cupboards—

When making different sized telephone diaphragms it was a very common usage to mount them in a frame with a mouthpiece, hold them up, and talk to them in a loud or low voice; at the same time putting a finger close to the centre to feel how much vibration was communicated to them

One night, after supper (which was prepared for us at midnight) and at which all the principal workers sat down together; Mr Edison who had been trying different diaphragms in this manner suddenly remarked "Do you know Batch I believe if we put a point on the centre of that diaphragm and talked to it whilst we pulled some of that waxed paper under it so that it could indent it, It would give us back talking when we pulled the paper through the second time'— The brilliancy of the suggestion did not at first strike any of us— It was so obvious that it would do so that everyone said 'Why of course it must!!'

I said We'll try it mighty quick! and we went to work— Mr Kruesi the Chief Mechanician took the diaphragm to solder on to it at the middle a needle point about ¼″ long; he also took one of the automatic telegraph wheels and stands to fasten the diaphragm to so that we could draw the paper through easily—

I cut and got ready some strips of paper of different thicknesses of parafin coating— It was a matter of an hour or so when we all got together again to make a trial— We fixed the instrument on to a table and I put in a strip of paper and adjusted the needle point down until it just pressed lightly on the paper— Mr Edison sat down and putting his mouth to the mouthpiece delivered one of our favorite stereotyped sentences used in experimenting on the telephone "Mary had a little lamb" whilst I pulled the paper through—

We looked at the strip and noticed the irregular marks, then we put it in again and I pulled it through as nearly at the same speed as I had pulled it in the first place and we got "ary ad elll am" something that was not fine talking, but the shape of it was there, and so like the talking that we all let out a yell of satisfaction and a "Golly it's there"!! and shook hands all round— We tried it many times and in many different ways continually improving the apparatus during the early morning— During the time that some of these changes were being made Edison & I would talk about the possibilities of such an invention and it was then that we fully realized the brilliancy of the suggestion and the magnitude of its possible applications— Before breakfast the next morning we had reproduced almost perfect articulation from a strip of the waxed paper which I had embossed as it were with a ridge in the middle running the whole length, the needle point in this case was ground chisel shaped.

Before the next night we had reproduced speech from a strip of tinfoil using again a rounded point needle, this was so remarkable that we decided to design a machine to experiment with In a few days about the beginning of Dec 1877 we had this instrument finished. It consisted of a cylinder of brass turned by hand that was provided on its surface with a spiral groove running the whole length and being about ⅛″ apart; the shaft also was cut the same pitch so that when the handle was turned the cylinder moved forward uniformly

A talking diaphragm was mounted on one side of the cylinder to record the speech, and a much more delicate diaphragm was mounted on the other side to reproduce the same— Each diaphragm could be moved away from the cylinder at will so that only one was in operation at a time.

The nut that the screw thread on the shaft engaged with, could also be disengaged so that the cylinder could be set back quick.

The cylinder was covered with a sheet of tinfoil and a suitable device was provided to hold it— This sheet could be put on and reproduced many times— The needle most generally used was a rounded point— Many thousands of experiments were made with this machine, and similar ones made immediately after; some of which were exhibited in different parts of the country & Europe whilst great crowds of people came almost every day to Menlo Park to hear with astonishment the reproduction of their own

The original instrument here described is now in the South Kensington Museum London, Edison having presented it to that Institution

AD, NjWOE, Batchelor, Cat. 1339:57 (*TAEM* 90:580).

## #3 THE CENTENNIAL MEDAL FOR THE ELECTRIC PEN

In 1876 Mr Edison made an exhibit of his 'Electric Pen and Duplicating Press at the Centennial Exhibition— This was a novel device for making any number of copies from a single writing in a short time— We had built a factory for this at Menlo Park N.J. and I had an office for the sale of the same at 41 Dey Street New York City— I attended to the factory and the office in the daytime and then joined Mr E. at the Laboratory for the best part of the remaining 24 hours. of [1?] It was a very useful novelty and took well with the public so much so that I had customers in almost all the civilized parts of the world.

The Exhibition at Philadelphia thought well of it and bestowed on us a bronze medal— One day this was presented to me at the office by some government officer— On that day Mr. E. happened to be in the office and was waiting for me to go out on my regular train— As was usual in such cases all the way to the Ferry we talked over the different inventions of his that we were developing at the Laboratory and so interested and bound up in the subject were we that I got up from the boat and left the medal in its morocco case on the seat in the boat— We had almost got to the train at Jersey City before I found my loss, when I realized that I had left it I mentioned it to him & started back to get it; but he called me back saying 'Don't bother some one will surely have picked it up'— and we talked the inventions all the way out to Menlo Park. I doubt if he has ever given it a thought since—

AD, NjWOE, Batchelor, Cat. 1339:76 (*TAEM* 90:585).

# Appendix 3

## *Edison's U.S. Patents, April 1876–December 1877*

The following list contains all patents for which Edison executed an application in the period covered by Volume Three. It is arranged in chronological order by execution date, which is the date on which Edison signed the application and the date in the patenting process that comes closest to the time of actual inventive activity. The application date is the date on which the Patent Office received and recorded the application. The case numbering system, which Edison used throughout his career, seems to have originated in Lemuel Serrell's office as a means of ordering Edison's applications as they arrived. The full Patent Office files containing each original application, any amendments, and related correspondence are at MdSuFR. Edison's American patents from this period are in *TAEM* 1–2; British patents are in Cat. 1321, Batchelor (*TAEM* 92:54–77).

| | *Exec. Date* | *Appl. Date* | *Issue Date* | *Pat. No.* | *Case No.* | *Title* |
|---|---|---|---|---|---|---|
| 112. | 04/03/76 | 04/06/76 | 12/11/77 | 198,088 | 115 | Telephonic Telegraphs |
| 113. | 04/03/76 | 04/06/76 | 12/11/77 | 198,089 | 116 | Telephonic or Electro-Harmonic Telegraphs |
| 114. | 05/09/76 | 05/16/76 | 10/10/76 | 182,996 | 119 | Acoustic Telegraphs |
| 115. | 05/09/76 | 05/16/76 | 01/16/77 | 186,330 | 117 | Acoustic Electric Telegraphs |
| 116. | 05/09/76 | 05/16/76 | 12/11/77 | 198,087 | 118 | Telephonic Telegraphs |
| 117. | 05/09/76 | 05/18/76 | 01/23/77 | 186,548 | 120 | Telegraphic Alarm and Signal Apparatus |
| 118. | 08/16/76 | 08/31/76 | 12/19/76 | 185,507 | 122 | Electro-Harmonic Multiplex Telegraphs |
| 119. | 08/26/76 | 09/18/76 | 03/05/78 | 200,993 | 124 | Acoustic Telegraphs |
| 120. | 08/26/76 | 09/30/76 | 12/07/80 | 235,142 | 125 | Acoustic Telegraph |
| 121. | 10/30/76 | 11/01/76 | 02/05/78 | 200,032 | 126 | Synchronous Movements for Electric Telegraphs |

| | | | | | | |
|---|---|---|---|---|---|---|
| 122. | 10/30/76 | 11/11/76 | 03/05/78 | 200,994 | 127 | Automatic-Telegraph Perforator and Transmitter |
| 123. | 02/03/77 | 03/26/77 | 03/25/79 | 213,554 | 128 | Automatic Telegraphs |
| 124. | 02/03/77 | 03/26/77 | 06/25/78 | 205,370 | 129 | Pneumatic Stencil-Pens |
| 125. | 04/18/77 | 04/23/77 | 11/06/77 | 196,747 | 131 | Stencil-Pens |
| 126. | 04/18/77 | 04/23/77 | 05/07/78 | 203,329 | 134 | Perforating Pens |
| 127. | 04/18/77 | 04/27/77 | 05/03/92 | 474,230 | 130 | Speaking-Telegraph |
| 128. | 05/08/77 | 05/10/77 | 08/03/80 | 230,621 | 137 | Addressing-Machine |
| 129. | 05/08/77 | 05/14/77 | 07/22/79 | 217,781 | 133 | Sextuplex Telegraphs |
| 130. | 05/08/77 | 05/14/77 | 02/07/88 | 377,374 | 132 | Telegraphy |
| 131. | 05/31/77 | 06/02/77 | 05/26/91 | 452,913 | 138 | Sextuplex Telegraph |
| 132. | 05/31/77 | 06/02/77 | 06/02/91 | 453,601 | 140 | Sextuplex Telegraph |
| 133. | 05/31/77 | 06/02/77 | 01/16/94 | 512,872 | 139 | Sextuplex Telegraph |
| 134. | 07/09/77 | 07/20/77 | 05/03/92 | 474,231 | 141 | Speaking-Telegraph |
| 135. | 07/16/77 | 07/20/77 | 04/30/78 | 203,014 | 136 | Speaking-Telegraphs |
| 136. | 07/16/77 | 07/20/77 | 09/24/78 | 208,299 | 135 | Speaking-Telephones |
| 137. | 08/16/77 | 08/22/77 | 02/04/90 | 420,594 | 142 | Quadruplex Telegraph |
| 138. | 08/16/77 | 08/28/77 | 04/30/78 | 203,015 | 143 | Speaking-Telegraphs |
| 139. | 08/31/77 | 09/05/77 | 03/07/93 | 492,789 | 144 | Speaking-Telegraph |
| 140. | 12/08/77 | 12/13/77 | 04/30/78 | 203,013 | 146 | Speaking-Telegraphs |
| 141. | 12/08/77 | 12/13/77 | 04/30/78 | 203,018 | 147 | Telephones or Speaking-Telegraphs |
| 142. | 12/15/77 | 12/24/77 | 02/19/78 | 200,521 | 149 | Phonograph or Speaking Machines |

# Bibliography

Atkinson, E., trans. and ed. 1886. *Elementary Treatise on Physics, Experimental and Applied.* From *Ganot's Eléments de Physique.* 12th ed. New York: William Wood and Co.

———. 1890. *Elementary Treatise on Physics, Experimental and Applied.* From *Ganot's Eléments de Physique.* 13th ed. New York: William Wood and Co.

Avery, Elroy M. 1885. *Elements of Natural Philosophy.* New York and Chicago: Sheldon and Co.

Baker, Edward C. 1976. *Sir William Preece, F.R.S.: Victorian Engineer Extraordinary.* London: Hutchinson and Co.

Bergen, Teunis G. 1878. *Genealogy of the Lefferts Family, 1650–1878.* Albany, N.Y.: Joel Munsell.

Bliven, Bruce, Jr. 1954. *The Wonderful Writing Machine.* New York: Random House.

*Boston Directory.* (Printed annually.) Boston: Sampson, Davenport and Co.

Bright, Charles. 1974 [1898]. *Submarine Telegraphs: Their History, Construction, and Working.* New York: Arno Press.

Bruce, Robert V. 1959. *1877: Year of Violence.* Indianapolis: Bobbs-Merrill.

———. 1973. *Bell: Alexander Graham Bell and the Conquest of Solitude.* Boston: Little, Brown and Co.

Brush, Stephen G. 1976. *The Kind of Motion We Call Heat: A History of the Kinetic Theory of Gases in the 19th Century.* Vol. 6 of *Studies in Statistical Mechanics.* New York: North-Holland Publishing Co.

Butrica, Andrew J. 1986. "From *Inspecteur* to *Ingénieur:* Telegraphy and the Genesis of Electrical Engineering in France, 1845–1881." Ph.D. diss., Iowa State University.

Cardwell, D. S. L. 1972. *Technology, Science, and History.* London: Heinemann.

Carlson, W. Bernard. 1991. *Innovation as a Social Process: Elihu Thomson and the Rise of General Electric, 1870–1900.* Cambridge: Cambridge University Press.

Carlson, W. Bernard, and Michael E. Gorman. 1992. "A Cognitive Framework to Understand Technological Creativity: Bell, Edison, and the Telephone." In *Inventive Minds: Creativity in Technology,* ed. Robert J. Weber and David N. Perkins. Oxford: Oxford University Press.

Carosso, Vincent P. 1987. *The Morgans: Private International Bankers, 1854–1913.* Cambridge, Mass.: Harvard University Press.

Cassino, Samuel, ed. 1883. *The International Scientists' Directory.* Boston: S. E. Cassino and Co.

Cohen, Morton N. 1976. "The Electric Pen." *Illustrated London News,* Christmas number, p. 33.

Cooper, Grace Rogers. 1976. *The Sewing Machine: Its Invention and Development.* Washington, D.C.: Smithsonian Institution Press.

Crookes, William. 1871. *Select Methods in Chemical Analysis (Chiefly Inorganic).* London: Longmans, Green, and Co.

Currie, A. W. 1957. *The Grand Trunk Railway of Canada.* Toronto: University of Toronto Press.

Dolbear, Amos E. 1877. *The Telephone.* New York: Lee and Shepard.

Draper, Henry. 1877. "Discovery of Oxygen in the Sun by Photography, and a New Theory of the Solar Spectrum." *American Journal of Science and Arts,* 3d ser., 14:89–96.

Du Moncel, Theodose. 1974 [1879]. *The Telephone, the Microphone, and the Phonograph.* New York: Arno Press.

Dyer, Frank, and T. C. Martin. 1910. *Edison: His Life and Inventions.* 2 vols. New York: Harper and Bros.

Endlich, Helen. 1981. *A Story of Port Huron.* Port Huron, Mich.: Privately published.

Faraday, Michael. 1965 [1844]. *Experimental Researches in Electricity.* New York: Dover Publications.

Farmer, John S., and W. E. Henley, comps. and eds. 1970 [1890–1904]. *Slang and Its Analogues.* New York: Arno Press.

Favre-Perret, Edouard. 1877. *Exposition de Philadelphie, 1876.* Winterthur, Switzerland: J. Westfehling.

Finn, Bernard S. 1989. "Working at Menlo Park." In *Working at Inventing: Thomas A. Edison and the Menlo Park Experience,* ed. William S. Pretzer. Dearborn, Mich.: Henry Ford Museum and Greenfield Village.

Fischer, Henry C., and William H. Preece. 1877. "Joint Report upon the American Telegraph System." Archives, Post Office, London, UK.

Fleming, John Ambrose. 1977 [1921]. [Telephones] (Extract from *Fifty Years of Electricity: The Memories of an Electrical Engineer*). In *The Electric Telegraph: An Historical Anthology,* ed. George Shiers. New York: Arno Press.

Friedel, Robert. 1983. *Pioneer Plastic: The Making and Selling of Celluloid.* Madison: University of Wisconsin Press.

Friedel, Robert, and Paul Israel. 1986. *Edison's Electric Light: Biography of an Invention.* New Brunswick, N.J.: Rutgers University Press.

Gopsill, James, ed. (Printed annually.) *Gopsill's Philadelphia City Directory.* Philadelphia: James Gopsill.

Gray, Elisha. 1977 [1878]. "Experimental Researches in Electro-Harmonic Telegraphy and Telephony." In *The Telephone: An Historical Anthology,* ed. George Shiers. New York: Arno Press.

Greer, William. 1979. *A History of Alarm Security.* Washington, D.C.: National Burglar and Fire Alarm Association.

Guillemin, Amédée. 1873. *The Forces of Nature: A Popular Introduction to the Study of Physical Phenomena,* ed. J. Norman Lockyer, trans. Mrs. Norman Lockyer. New York: Scribner, Welford, and Armstrong.

———. 1977 [1891]. "Telegraphic Apparatus for Rapid Transmission." In *The Electric Telegraph: An Historical Anthology,* ed. George Shiers. New York: Arno Press.

Hackmann, W. D. 1985. "The Nineteenth-Century Trade in Natural Philosophy Instruments in Britain." In *Nineteenth-Century Scientific Instruments and Their Makers,* ed. P. R. de Clerq. Amsterdam: Rodopi.

Heaviside, Oliver. 1873. "On Duplex Telegraphy." *Phil. Mag.,* 4th ser., 45:426–32.

*History of St. Clair County, Michigan.* 1883. Chicago: A. T. Andreas and Co.

Hounshell, David. 1975. "Elisha Gray and the Telephone: On the Disadvantages of Being an Expert." *Technology and Culture* 16:133–61.

Hunt, Bruce. 1983. " 'Practice vs. Theory': The British Electrical Debate, 1888–1891." *Isis* 74:341–55.

———. 1992. "The Ohm Is Where the Art Is: British Telegraph Engineers and the Development of Electrical Standards." Paper delivered at History of Science Society Meeting, 27–30 December 1992, Washington, D.C.

Ihde, Aaron. 1964. *The Development of Modern Chemistry.* New York: Harper and Row.

Israel, Paul B. 1992. *From Machine Shop to Industrial Laboratory: Telegraphy and the Changing Context of American Invention, 1830–1920.* Baltimore: Johns Hopkins University Press.

Jehl, Francis. 1937–41. *Menlo Park Reminiscences.* 3 vols. Dearborn, Mich.: Edison Institute.

Jenks, William Lee. 1912. *St. Clair County, Michigan: Its History and Its People.* 2 vols. Chicago: Lewis Publishing Co.

———. n.d. "History of Port Huron Street Railways." William Lee Jenks Collection, Burton Historical Collection, Detroit Public Library, Detroit, Mich.

Johnson, Edward H. 1877. *Telephone Hand Book.* New York: Russell Brothers, Printers.

Josephson, Matthew. 1992 [1959]. *Edison: A Biography.* New York: John Wiley and Sons.

Kenyeres, Ágnes, ed. 1969. *Magyar Életrajzi Lexicon.* 2 vols. Budapest: Akadémiai Kiadó.

King, W. James. 1962a. "The Development of Electrical Technology in the Nineteenth Century: 1. The Electrochemical Cell and the Electromagnet." *United States Museum Bulletin 228.* Washington, D.C.: Smithsonian Institution.

———. 1962b. "The Development of Electrical Technology in the Nineteenth Century: 2. The Telegraph and the Telephone." *United States Museum Bulletin 228.* Washington, D.C.: Smithsonian Institution.

———. 1962c. "The Development of Electrical Technology in the Nineteenth Century: 3. The Early Arc Light and Generator." *United States Museum Bulletin 228.* Washington, D.C.: Smithsonian Institution.

Knight, Edward H. 1876–77. *Knight's American Mechanical Dictionary.* 3 vols. New York: Hurd and Houghton.

Koenigsberg, Allen. 1969. *Edison Cylinder Records, 1889–1912.* New York: Stellar Productions.

———. 1987. "The First 'Hello!' Thomas Edison, the Phonograph and the Telephone." *Antique Phonograph Monthly* 8:3–9.

Landes, David. 1983. *Revolution in Time.* Cambridge, Mass.: Harvard University Press.

Lathrop, George Parsons. 1890. "Talks with Edison." *Harper's Monthly Magazine* 80:425–35.

Lines, Robert B. 1876. *Report on Telegraphs and on the Telegraphic Administration.* Washington, D.C.: GPO.

McGucken, William. 1969. *Nineteenth-Century Spectroscopy: Development of the Understanding of Spectra, 1802–1897.* Baltimore: Johns Hopkins Press.

McHugh, Jeanne. 1980. *Alexander Holley and the Makers of Steel.* Baltimore: Johns Hopkins University Press.

Maver, William, Jr. 1892. *American Telegraphy: Systems, Apparatus, Operation.* New York: J. H. Bunnell and Co.

Maxwell, James Clerk. 1873. "Molecules." *Nature* 8:437–41.

Meadows, A. J. 1972. *Science and Controversy: A Biography of Sir Norman Lockyer.* London: Macmillan and Co.

Munro, James. 1877. "New Electric Lights." *Nature* 16:422–23.

Nahin, Paul J. 1988. *Oliver Heaviside: Sage in Solitude.* New York: IEEE Press.

Niaudet, Alfred. 1884. *Elementary Treatise on Electric Batteries,* trans. L. M. Fishback. New York: John Wiley and Sons.

Nuf Ced [Edward O. Chase]. 1876. "The Display of Telegraphic Apparatus at the Centennial." *Operator,* 1 July 1876, 4–5.

Oberholtzer, Ellis Paxson. 1926. *A History of the United States since the Civil War.* 5 vols. New York: Macmillan Co.

Oppenheim, Janet. 1985. *The Other World: Spiritualism and Psychical Research in England, 1850–1914.* Cambridge: Cambridge University Press.

Partington, J. R. 1964. *A History of Chemistry.* London: Macmillan and Co.

Peck, William G. 1866. *Introductory Course of Natural Philosophy, Edited from Ganot's Popular Physics.* New York: A. S. Barnes and Co.

Pepper, John H. [1869?]. *Chemistry: Embracing the Metals and Elements Which Are Not Metallic.* London: Frederick Warne and Co.

Pershey, Edward Jay. 1989. "Drawing as a Means to Inventing: Edison and the Invention of the Phonograph." In *Working at Inventing: Thomas A. Edison and the Menlo Park Experience,* ed. Wil-

liam S. Pretzer. Dearborn, Mich.: Henry Ford Museum and Greenfield Village.
Plotkin, Howard. 1972. "Henry Draper: A Scientific Biography." Ph.D. diss., Johns Hopkins University.
Plum, William R. 1882. *The Military Telegraph during the Civil War in the United States.* 2 vols. Chicago: Jansen, McClurg and Co.
Podmore, Frank. 1897. *Studies in Psychical Research.* London: Kegan Paul, Trench, Trubner and Co.
Pole, William. 1888. *The Life of Sir William Siemens, F.R.S., D.C.L., LL.D.* London: John Murray.
Pope, Franklin. 1872. *Modern Practice of the Electric Telegraph: A Handbook for Electricians and Operators.* 7th ed. New York: D. Van Nostrand.
Preece, William H. 1877. "The Telephone." *Nature* 16:403–4.
Preece, William H., and J. Sivewright. 1891. *Telegraphy.* 9th ed. London: Longmans, Green, and Co.
Prescott, George B. 1877. *Electricity and the Electric Telegraph.* New York: D. Appleton and Co.
———. 1879. *The Speaking Telephone, Electric Light, and Other Recent Electrical Inventions.* New York: D. Appleton and Co.
———. 1885. *Electricity and the Electric Telegraph.* 6th ed. New York: D. Appleton and Co.
———. 1972 [1884]. *Bell's Electric Speaking Telephone: Its Invention, Construction, Application, Modification, and History.* New York: Arno Press.
Pretzer, William S., ed. 1989. *Working at Inventing: Thomas A. Edison and the Menlo Park Experience.* Dearborn, Mich.: Henry Ford Museum and Greenfield Village.
Reid, James D. 1879. *The Telegraph in America.* New York: Derby Bros.
———. 1886. *The Telegraph in America.* Rev. ed. New York: John Polhemus.
Roscoe, Henry E. 1870. *Spectrum Analysis: Six Lectures.* 2d ed. London: Macmillan and Co.
Scharf, J. Thomas, and Thompson Westcott. 1884. *History of Philadelphia.* 3 vols. Philadelphia: L. H. Everts and Co.
Schellen, Heinrich. 1872. *Spectrum Analysis,* trans. [William] Lassell. 2d ed. New York: D. Appleton and Co.
Shaffner, Taliaferro P. 1859. *The Telegraph Manual.* New York: Pudney and Russell.
Shaw, G. M. 1878. "Sketch of Edison." *Popular Science Monthly* 13:487–91.
Siemens, Charles William. 1874. "On the Dependence of Electrical Resistance on Temperature." *J. Soc. Teieg. Eng.* 3:297–338.
Siemens, Werner von. 1968. *Inventor and Entrepreneur: Recollections of Werner von Siemens,* trans. W. C. Coupland. New York: Augustus M. Kelley.
Sinclair, Bruce. 1974. *Philadelphia's Philosopher Mechanics: A History of the Franklin Institute, 1824–65.* Baltimore: Johns Hopkins University Press.

Smith, Crosbie, and M. Norton Wise. 1989. *Energy and Empire: A Biographical Study of Lord Kelvin.* Cambridge: Cambridge University Press.

Sprague, John T. 1875. *Electricity: Its Theory, Sources, and Applications.* London: E. & F. N. Spon.

Stock, John T., and Mary Virginia Orna, eds. 1989. *Electrochemistry, Past and Present.* ACS Symposium Series 390. Washington, D.C.: American Chemical Society.

Taltavall, John B. 1893. *Telegraphers of Today.* New York: John B. Taltavall.

Thompson, Silvanus P. 1876. "On Some Phenomena of Induced Electric Sparks." *Phil. Mag.*, 5th ser., 2:191–98.

Thomson, William. 1869–70. "The Size of Atoms." *Nature* 1:551–53.

Tosiello, Rosario J. 1979. *The Birth and Early Years of the Bell Telephone System, 1876–1880.* New York: Arno Press.

Tunzelmann, G. W. de. 1900. *Electricity in Modern Life.* New York: P. F. Collier and Son.

Tyler, David B. 1972. *Steam Conquers the Atlantic.* New York: Arno Press.

U.S. Bureau of the Census. 1970. *Population Schedules of the Tenth Census of the United States, 1880.* National Archives Microfilm Publications Microcopy no. T9. Washington, D.C.: National Archives.

U.S. Centennial Commission. 1880. *International Exhibition, 1876: Reports and Awards Vol. VII, Groups XXI–XXVII,* ed. Francis A. Walker. Washington, D.C.: GPO.

Weiher, Sigfrid von, and Herbert Goetzeler. 1977. *The Siemens Company: Its Historical Role in the Progress of Electrical Engineering,* trans. G. N. J. Beck. Berlin: Siemens Aktiengesselleschaft.

Western Union Telegraph Co. (Printed annually.) *Annual Reports.* New York: Western Union Telegraph Co.

Wheeler, Everett Pepperrell. 1927. *Reminiscences of a Lawyer: A Few Pages from the Record of a Busy Life.* Poughkeepsie, N.Y.: A. V. Haight Co.

Wilson, H., comp. (Printed annually.) *Trow's Business Directory of New York City.* New York: John F. Trow.

Wilson, Joseph M. 1876–78. *The Masterpieces of the Centennial Exhibition.* Philadelphia: Gebbie and Barrie.

# Credits

Reproduced with permission of the AT&T Corporate Archive: Docs. 877, 908, 1006, 1008, 1021, 1022, 1025, 1031, 1154; illustration on p. 551. Reproduced with permission of the Syndics of Cambridge University: Doc. 1141. Reproduced with permission of the City Archives of Philadelphia, Pennsylvania: Doc. 757. Reproduced with permission of the Royce Miscellaneous Papers, Manuscript Department, The Filson Club, Louisville, Ky.: Docs. 751, 753. From the collections of the Henry Ford Museum and Greenfield Village: Docs. 849, 959, 1083, 1090, 1161; illustrations on pp. 154 (neg. B111600), 188 (neg. 0-17017), 469 (neg. B103571). Reproduced with permission of the Historical Society of Pennsylvania: Docs. 1139, 1145, 1147, 1153, 1158. Reproduced with permission of the Institution of Electrical Engineers, London: Docs. 987, 1017, 1033, 1036, 1043, 1055, 1077, 1082, 1123. Courtesy of the New-York Historical Society: Docs. 741, 746, 748, 752. Reproduced with permission of the Henry Draper and Anna Palmer Draper Papers, Rare Books and Manuscripts Division, The New York Public Library, Astor, Lenox and Tilden Foundations: Docs. 989, 990, 998, 1010.

Courtesy of Edison National Historic Site (all designations are to *TAEM* reel:frame unless otherwise indicated): illustrations on pp. 116 (8:488), 130 (3:363), 254 (11:191), 509 (14:113), 524 (92:63).

# Index

*Boldface page numbers signify primary references or identifications; italic numbers indicate illustrations. Page numbers refer to headnote or document text unless the reference appears only in a footnote.*